PREFACE

Introductory Algebra: A Text/Workbook, Fourth Edition, has been written and designed to address the particular needs of students taking their first college course in algebra. The only prerequisite is a knowledge of arithmetic. For students whose background is not solid or who have not studied the prerequisites recently, the first two sections review the key ideas of fractions, decimals, and percents that students will need in algebra.

The book's clear explanations, precise learning objectives, more than 650 detailed examples, carefully graded exercise sets referenced to examples, and open, accessible design make this an ideal text for either a traditional lecture class or individualized instruction.

NEW CONTENT FEATURES

Changes in the fourth edition include the following.

- New examples and exercises have been added, and the applications have been extensively rewritten and updated. The integrity of the applications in previous editions has been retained, but we have provided a freshness for instructors who have taught from those editions.
- At appropriate points in the text we discuss estimation and determining whether an answer is reasonable.
- New intuitive introductions to some topics have been included to help students make connections with the real world. For example, illustrations and real examples of slope have been added in Chapter 7 to make the idea more concrete.
- In many sections, summaries of procedures have been moved earlier in the section, so that students can refer to the summary while working through the examples.
- We have increased the emphasis on relating words and symbols in Section 1.3.
- In Chapter 2 we have added examples and new exercises on translating words to symbols. We also have given an introduction to the properties of real numbers that appeals more to intuition than was the case in the third edition.
- Chapter 3 has been rewritten to present word problems in a more effective way, grouping them by type and keeping them simpler. More examples have been included, and the exercises are carefully referenced to the examples. We have included new examples of linear equations with no solutions or with infinitely many solutions, as well as corresponding exercises.
- The first two sections in Chapter 4 have been rearranged to present the product rule and the three power rules in Section 4.1. Then in Section 4.2, negative and zero exponents are defined, and the quotient rule is introduced in just one form.
- Divisibility rules are now given in Chapter 5 to make it easier for students to factor large numbers.
- Section 10.4 has been extensively rewritten to show how to graph parabolas by finding intercepts, finding the vertex with a formula, finding additional points, and then plotting and connecting the points. Tables of values have been included with each figure.

NEW FEATURES

All the successful features of the workbook format in the previous edition are carried over in the new edition: learning objectives for each section, careful exposition and fully developed examples, sample problems in the margins for immediate feedback, carefully graded exercises with work space, and a clear, functional design. Screened boxes that set off important definitions, formulas, rules, and procedures further aid in learning and reviewing the course material.

In addition, the following new features have been developed to enhance the pedagogy and usefulness of the text.

Example Titles

Each example now has a title to help students see the purpose of the example. The titles also facilitate working the exercises and studying for examinations.

Cautionary Remarks

Common student errors and misconceptions, or difficulties students typically encounter, are highlighted graphically and identified with the headings "Caution" or "Note."

Skill Sharpeners

These short sets of review exercises at the end of each exercise set beginning with Chapter 2 provide students with a quick review of skills that will be needed in the next section.

Chapter Summaries

This new study aid at the end of each chapter provides a glossary of key terms with section references, a list of new symbols introduced in the chapter, and a "Quick Review" featuring section-referenced capsule summaries of key ideas accompanied by worked-out examples.

Chapter Review Exercises

A thorough set of chapter review exercises follows each chapter summary. Most of these exercises are organized by section, but a concluding section, titled "Mixed Review Exercises," includes randomly mixed exercises from the entire chapter. This helps students practice identifying problems by type. As in the previous edition, a sample test concludes each chapter.

Historical Reflections

These brief items, designed to emphasize the human dimension of mathematics, introduce eminent mathematicians and the important ideas they advanced. They are included as enrichment, to interest and motivate students.

Success in Algebra

This foreword to the students provides additional support by offering suggestions for successful study of the course material.

Section References in Answer Section

As another study aid, section references are given in the answers to the chapter tests and final examination at the back of the textbook. Students then can readily go back to study any section that gave them difficulty when they took the sample chapter test or final examination.

Glossary
A comprehensive glossary is placed at the end of the book. Each term in the glossary is defined and then referenced to the appropriate section in the text, where students may find more detailed explanations of the term.

SUPPLEMENTS

Our extensive supplemental package includes an annotated instructor's edition, testing materials, solutions, software, videotapes, and audiotapes.

Annotated Instructor's Edition
This edition provides instructors with immediate access to the answers to every exercise in the text; each answer is printed in color next to the corresponding text exercise.

Instructor's Testing Manual
The Instructor's Testing Manual includes suggestions for using the textbook in a mathematics laboratory; short-answer and multiple-choice versions of a diagnostic placement test; six forms of chapter tests for each chapter, including four open-response and two multiple-choice forms; two forms of a final examination; and an extensive set of additional exercises, providing 10 to 20 exercises for each textbook objective, which can be used as an additional source of questions for tests, quizzes, or student review of difficult topics.

Student's Solutions Manual
This book contains solutions to half of the odd-numbered section exercises (those not included at the back of the textbook) as well as solutions to all chapter review exercises, chapter tests, and cumulative review exercises.

Instructor's Solutions Manual
Available at no charge to instructors, this book includes solutions to all the margin problems in the textbook and solutions to the even-numbered section exercises. The two solutions manuals plus the solutions given at the back of the textbook provide detailed, worked-out solutions to each exercise and margin problem in the book.

HarperCollins Test Generator for Mathematics
Available in Apple, IBM, and Macintosh versions, the test generator enables instructors to select questions by objective, section, or chapter, or to use a ready-made test for each chapter. Instructors may generate tests in multiple-choice or open-response formats, scramble the order of questions while printing, and produce multiple versions of each test (up to 9 with Apple, up to 25 with IBM and Macintosh). The system features printed graphics and accurate mathematics symbols. It also features a preview option that allows instructors to view questions before printing, to regenerate variables, and to replace or skip questions if desired. The IBM version includes an editor that allows instructors to add their own problems to existing data disks.

Interactive Tutorial Software
This innovative package is also available in Apple, IBM, and Macintosh versions. It offers interactive modular units, specifically linked to the text, for reinforcement of selected topics. The tutorial is self-paced and provides unlimited opportunities to review lessons and to practice problem solving. When students give a wrong answer, they can

request to see the problem worked out. The program is menu-driven for ease of use, and on-screen help can be obtained at any time with a single keystroke. Students' scores are automatically recorded and can be printed for a permanent record.

Audiotapes

A set of audiotapes, one per chapter, is available free to departments upon adoption of the text. The tapes guide students through each topic, allowing individualized study and additional practice for troublesome areas. They are especially helpful for visually impaired students.

Videotapes

A new videotape series, *Algebra Connection: The Introductory Algebra Course,* has been developed to accompany *Introductory Algebra,* Fourth Edition. Produced by an Emmy Award–winning team in consultation with a task force of academicians from both two-year and four-year colleges, the tapes cover all objectives, topics, and problem-solving techniques within the text. In addition, each lesson is preceded by motivational "launchers" that connect classroom activity to real-world applications.

ACKNOWLEDGMENTS

We wish to thank the many users of the third edition for their insightful suggestions on improvements for this book.

We also wish to thank our reviewers for their contributions: Connie A. Bish, Las Positas College; Warren J. Burch, Brevard Community College; Jorge F. Cossio, Miami-Dade Community College–North; Robert L. Davidson, East Tennessee State University; John A. Dersch, Jr., Grand Rapids Junior College; William Edgar, American River College; Frank D. Farmer, Arizona State University; Mike Farrell, Carl Sandburg College; Dauhrice Gibson, Gulf Coast Community College; Steve E. Green, Tyler Junior College; Julie A. Guelich, Normandale Community College; Robert Kaiden, Lorain County Community College; J. Paul Ketron, Cleveland State Community College; Linda Kodama, Kapiolani Community College; Nenette Loftsgaarden, University of Montana; F. Arnold Lowry III, East Tennessee State University; Timothy F. Magnavita, Bucks County Community College; Mary C. Metzner, University of Louisville; Kathy Monaghan, American River College; Henry J. Pronovost, Mattatuck Community College; Sharon Taylor Riley, Wayne County Community College; John Samoylo, Community College of Delaware County; Winona S. Sathre, Valencia Community College–East; Lauren A. Stodden, Olympic College; Lillian M. Thiel, William Rainey Harper College; Tommy Thompson, Brookhaven College; Paul Van Erden, American River College; David Zerangue, Nicholls State University.

We wish to thank Janet Krantz of American River College, who did an excellent job of checking all the answers.

Paul J. Eldersveld, College of DuPage, deserves heartfelt thanks for undertaking the enormous job of coordinating all of the print ancillaries for us.

We also want to thank Tommy Thompson, Seminole Community College, for his suggestions about the essay to the students on studying algebra at the beginning of the book.

Our special thanks go to our editors, Bill Poole, Linda Youngman, and Janet Tilden. As always, they have done a wonderful job under difficult circumstances.

Margaret L. Lial
E. John Hornsby, Jr.

CONTENTS

TO THE STUDENT: SUCCESS IN ALGEBRA

The main reason students have difficulty with mathematics is that they don't know how to study it. Studying mathematics *is* different from studying subjects like English or history. The key to success is regular practice.

This should not be surprising. After all, can you learn to play the piano or to ski well without a lot of regular practice? The same thing is true for learning mathematics. Working problems nearly every day is the key to becoming successful. Here is a list of things you can do to help you succeed in studying algebra.

1. Pay attention in class to what your teacher says and does, and make careful notes. In particular, note the problems the teacher works on the board and copy the complete solutions. Keep these notes separate from your homework to avoid confusion when you read them over later.
2. Don't hesitate to ask questions in class. It is not a sign of weakness, but of strength. There are always other students with the same question who are too shy to ask.
3. Before you start on your homework assignment, rework the problems the teacher worked in class. This will reinforce what you have learned. Many students say, "I understand it perfectly when you do it, but I get stuck when I try to work the problem myself."
4. *Read your text carefully*. Many students read only enough to get by, usually only the examples. Reading the complete section will help you to be successful with the homework problems. As a bonus you will be able to do the problems more quickly if you have read the text first. As you read the text, work the sample problems given in the margin and check the answers to see if you have worked them correctly. This will test your understanding of what you have read.
5. Do your homework assignment only *after* reading the text and reviewing your notes from class. Check your work with the answers in the back of the book. If you get a problem wrong and are unable to see why, mark that problem and ask your teacher about it.
6. Work as neatly as you can. Write your symbols clearly, and make sure the problems are clearly separated from each other. Working neatly will help you to think clearly and also make it easier to review the homework before a test.
7. After you have completed a homework assignment, look over the text again. Try to decide what the main ideas are in the lesson. Often they are clearly highlighted or boxed in the text.
8. Keep any quizzes and tests that are returned to you and use them when you study for future tests and the final exam. These quizzes and tests indicate what your teacher considers most important. Be sure to correct any problems on these tests that you missed, so you will have the corrected work to study.
9. Don't worry if you do not understand a new topic right away. As you read more about it and work through the problems, you will gain understanding. Each time you look back at a topic you will understand it a little better. No one understands each topic completely right from the start.
10. What a great feeling you will experience when, after a lot of time and hard work, you can say, "I really do understand this now."

name date hour

DIAGNOSTIC PRETEST

Ask your instructor whether or not you should work this pretest. It is designed to tell what material in the course may already be familiar to you. The actual course begins on page 1.

1. Write $\frac{96}{144}$ in lowest terms.

2. Add: $\frac{3}{4} + \frac{11}{12} + \frac{7}{8}$.

3. Divide: $\frac{15}{16} \div \frac{25}{24}$.

Simplify each of the following.

4. $2(5 + 6) + 7 \cdot 3 - 4^2$

5. $\frac{4(5 + 3) + 3}{2(3) - 1}$

6. $|-5|$

7. $-|-14|$

Add.

8. $-2 + (-9)$

9. $-8 + 15 + 13 + (-4 + 9)$

Subtract.

10. $-7 - (-15)$

11. $8 - [-7 - (-4)]$

Solve each equation.

12. $4r + 5r - 3 + 8 - 3r = 5r + 12 + 8$

13. $4(k - 3) - k = k - 6$

14. The perimeter of a rectangle is 68 meters. The length of the rectangle is 7 meters more than the width. Find the width of the rectangle.

1. ______
2. ______
3. ______
4. ______
5. ______
6. ______
7. ______
8. ______
9. ______
10. ______
11. ______
12. ______
13. ______
14. ______

15. ______

16. ______

17. ______

18. ______

19. ______

20. ______

21. ______

22. ______

23. ______

24. ______

25. ______

26. ______

27. ______

28. ______

29. ______

30. ______

15. Solve: $3z + 2 - 5 > -z + 7 + 2z$.

16. Solve: $-\frac{2}{3}y \leq 6$.

17. Evaluate 5^4. Name the base and the exponent.

Simplify and write each expression in Exercises 18 and 19 without negative exponents.

18. $(2x^7)(-4x^8)$

19. $\left(\frac{3^2y^{-2}}{4^{-1}y^3}\right)^{-3}$

20. Subtract $6x^3 - 4x^2 + 2$ from $11x^3 + 2x^2 - 8$.

21. Multiply $3p - 5$ and $2p + 6$.

22. Divide $2x + 3$ into $8x^3 - 4x^2 - 14x + 15$.

Write each of the following in prime factored form.

23. 50

24. 320

Factor each trinomial in Exercises 25–27.

25. $m^2 + 9m + 14$

26. $12a^2 - ab - 20b^2$

27. $121z^2 - 44z + 4$

28. Solve the quadratic equation $x^2 - 5x = -6$.

29. The product of two consecutive odd integers is 1 less than five times their sum. Find the integers.

30. Find the product: $\frac{x^2 + 3x}{x^2 - 3x - 4} \cdot \frac{x^2 - 5x + 4}{x^2 + 2x - 3}$.

name date hour

31. Add: $\frac{x}{x^2 - 1} + \frac{x}{x + 1}$.

31. ______________

32. Solve: $\frac{2}{r^2 - r} = \frac{1}{r^2 - 1}$.

32. ______________

Graph each of the following.

33. $4x - 5y = 20$

33.

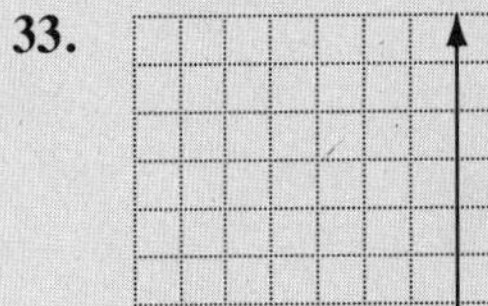

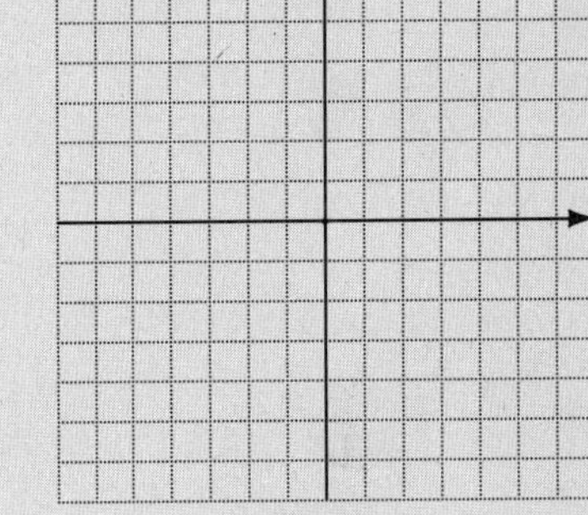

34. $x = 3$

34.

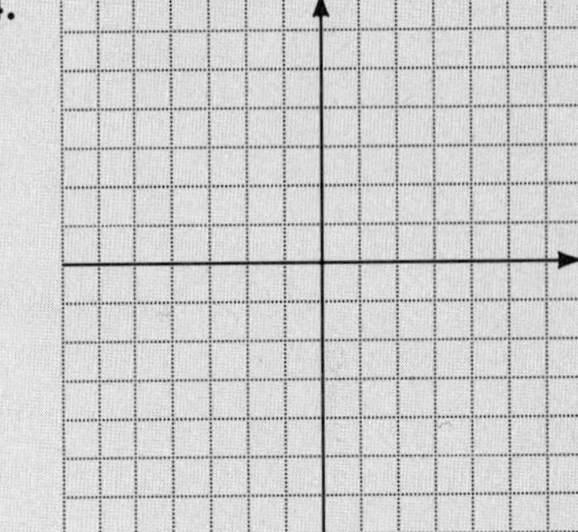

35. $2x - 5y \geq 10$

35.

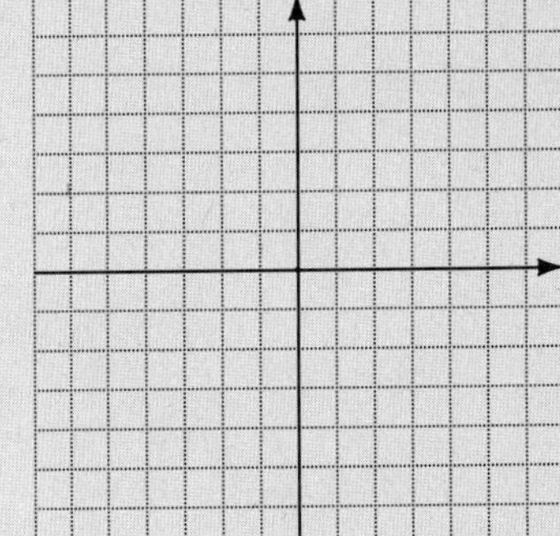

36. Find the slope of the line through the points $(8, -3)$ and $(-4, 7)$.

36. ______________

37. ____________

38. ____________

39. ____________

40. ____________

41. ____________

42. ____________

43. ____________

44. ____________

45. ____________

46. ____________

47. ____________

48. ____________

49. ____________

50.

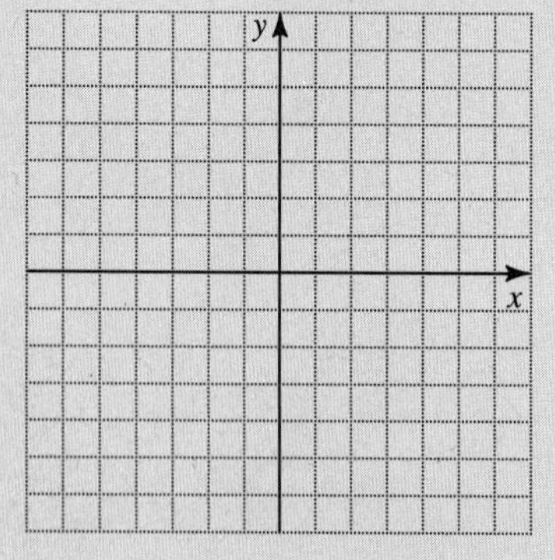

Solve each of the following systems of equations.

37. $2x + 3y = -15$
$5x + 2y = 1$

38. $3x - y = 4$
$-9x + 3y = -12$

Find each of the following roots.

39. $-\sqrt{1024}$

40. $\sqrt[3]{216}$

Simplify each of the following.

41. $\sqrt{72}$

42. $\sqrt[3]{40}$

43. $2\sqrt{12} + 3\sqrt{75}$

44. $\sqrt{18} \cdot \sqrt{75}$

Rationalize each denominator.

45. $\dfrac{14}{\sqrt{7}}$

46. $\dfrac{-2}{3 - \sqrt{2}}$

Solve each equation in Exercises 47–49.

47. $3\sqrt{y} = \sqrt{y + 8}$

48. $z^2 = 2z + 1$

49. $\dfrac{1}{10}t^2 = \dfrac{2}{5} - \dfrac{1}{2}t$

50. Graph $y = (x - 2)^2$.

1

NUMBER SYSTEMS

1.1 FRACTIONS

As preparation for the study of algebra, this section begins with a brief review of arithmetic.

OBJECTIVES

1. Learn the terminology of fractions.
2. Learn the definition of *factor.*
3. Write fractions in lowest terms.
4. Multiply and divide fractions.
5. Write a fraction as an equivalent fraction with a given denominator.
6. Add and subtract fractions.

1 In everyday life, the numbers seen most often are the **whole numbers,**

$$0, 1, 2, 3, 4, 5, \ldots$$

and the **fractions,** such as

$$\frac{1}{3}, \frac{2}{3}, \text{ and } \frac{11}{12}.$$

Of course, whole numbers may be written as fractions: 5 can be written as $\frac{5}{1}$, 6 can be written as $\frac{18}{3}$, and so on.

In a fraction, the number on top is called the **numerator** and the number on the bottom is the **denominator.** A fraction is a **proper fraction** if the numerator is smaller than the denominator; otherwise it is an **improper fraction.**

EXAMPLE 1 Identifying Proper and Improper Fractions

The numbers $\frac{3}{4}$, $\frac{7}{8}$, $\frac{9}{10}$, and $\frac{125}{126}$ are proper fractions, and $\frac{5}{4}$, $\frac{17}{15}$, and $\frac{28}{9}$ are improper fractions. ■*

An improper fraction can be written as a **mixed number,** the sum of a whole number and a proper fraction. For instance, the improper fraction $\frac{4}{3}$ can be written as $1\frac{1}{3}$, since $1\frac{1}{3}$ equals $\frac{3}{3} + \frac{1}{3} = \frac{4}{3}$. In algebra, the improper fraction form is usually preferred.

WORK PROBLEM 1 AT THE SIDE.

1. Which of these fractions are proper fractions?

$$\frac{1}{2}, \frac{5}{3}, \frac{7}{4}, \frac{10}{9}, \frac{8}{15}, \frac{13}{13}$$

2 In the statement $2 \times 9 = 18$, the numbers 2 and 9 are called **factors** of 18. Other factors of 18 include 1, 3, 6, and 18. The result of the multiplication, 18, is called the **product.**

To factor a number means to write it as the product of two or more numbers. For example, 18 can be factored in several ways, as $6 \cdot 3$, or $18 \cdot 1$, or $9 \cdot 2$, or $3 \cdot 3 \cdot 2$. In algebra, raised dots are usually used instead of the $\times$ symbol to indicate multiplication.

*This symbol, ■, indicates the end of an example.

ANSWER

1. $\frac{1}{2}$ and $\frac{8}{15}$ are proper since the numerator is smaller than the denominator in each; the others are improper.

2. Factor each number as the product of prime factors, using exponents as needed.

(a) 90

(b) 48

A whole number (except 1) is **prime** if it has only itself and 1 as factors. (By agreement, the number 1 is not a prime number.) The first dozen primes are listed here.

$$2,\quad 3,\quad 5,\quad 7,\quad 11,\quad 13,\quad 17,\quad 19,\quad 23,\quad 29,\quad 31,\quad 37$$

In algebra, it is often useful to find all the **prime factors** of a number—those factors that are prime numbers. For example, the only prime factors of 18 are 2 and 3, and 18 can be written as the following product of its prime factors: $18 = 2 \cdot 3 \cdot 3$.

A product such as $2 \cdot 3 \cdot 3$ can be written in a shorter form by using *exponents*. For example, $3 \cdot 3$ is written as 3^2, where the 2 shows that there are two factors of 3. In 3^2, the number 2 is called the **exponent** and 3 is called the **base.**

EXAMPLE 2 Factoring Numbers

Factor each number as the product of prime factors, using exponents as needed.

(a) 35

Factor 35 as the product of the prime factors 5 and 7, or as

$$35 = 5 \cdot 7.$$

(b) 24

Divide by the smallest prime, 2, to get

$$24 = 2 \cdot 12.$$

Now divide 12 by 2 to find factors of 12.

$$24 = 2 \cdot 2 \cdot 6$$

Since 6 can be factored as $2 \cdot 3$,

$$24 = 2 \cdot 2 \cdot 2 \cdot 3 = 2^3 \cdot 3$$

where all factors are prime. ■

WORK PROBLEM 2 AT THE SIDE.

3 Prime factors are used to write fractions in **lowest terms.** A fraction is in lowest terms when the numerator and the denominator have no factors in common (other than 1). Use the following steps to write a fraction in lowest terms.

WRITING A FRACTION IN LOWEST TERMS

Step 1 Factor the numerator and the denominator as the product of prime factors.

Step 2 Replace each pair of common factors in the numerator and denominator with 1.

Step 3 Multiply the remaining factors in the numerator and in the denominator.

ANSWERS

2. (a) $2 \cdot 3^2 \cdot 5$ (b) $2^4 \cdot 3$

EXAMPLE 3 Writing Fractions in Lowest Terms

Write each fraction in lowest terms.

(a) $\frac{10}{15} = \frac{2 \cdot 5}{3 \cdot 5} = \frac{2}{3} \cdot \frac{5}{5} = \frac{2}{3} \cdot 1 = \frac{2}{3}$

Since 5 is the only common factor of 10 and 15, dividing both numerator and denominator by 5 gives the fraction in lowest terms.

(b) $\frac{15}{45} = \frac{3 \cdot 5}{3 \cdot 3 \cdot 5} = \frac{1 \cdot 3 \cdot 5}{3 \cdot 3 \cdot 5} = \frac{1}{3} \cdot \frac{3}{3} \cdot \frac{5}{5} = \frac{1}{3} \cdot 1 \cdot 1 = \frac{1}{3}$

Multiplying by 1 in the numerator does not change the value of the numerator and provides a numerator to go with the first 3 in the denominator.

(c) $\frac{150}{200}$

It is not always necessary to factor into *prime* factors in Step 1. Here, if you see that 50 is a common factor of the numerator and the denominator, factor as follows:

$$\frac{150}{200} = \frac{3 \cdot 50}{4 \cdot 50} = \frac{3}{4} \cdot 1 = \frac{3}{4}. \quad \blacksquare$$

NOTE When you are factoring to write a fraction in lowest terms, look for the largest common factor in the numerator and denominator. If none is obvious, factor the numerator and the denominator into prime factors.

WORK PROBLEM 3 AT THE SIDE.

4 Multiplication of fractions is defined next.

MULTIPLYING FRACTIONS

To multiply two fractions, multiply the numerators and multiply the denominators.

EXAMPLE 4 Multiplying Fractions

Find the product of $\frac{3}{8}$ and $\frac{4}{9}$, and write it in lowest terms.

$$\frac{3}{8} \cdot \frac{4}{9} = \frac{3 \cdot 4}{8 \cdot 9}$$ Multiply numerators. Multiply denominators.

$$= \frac{3 \cdot 4}{2 \cdot 4 \cdot 3 \cdot 3}$$ Factor.

$$= \frac{1}{2 \cdot 3} = \frac{1}{6}$$ Write in lowest terms. ■

WORK PROBLEM 4 AT THE SIDE.

3. Write each fraction in lowest terms.

(a) $\frac{8}{14}$

(b) $\frac{35}{42}$

(c) $\frac{9}{18}$

(d) $\frac{12}{20}$

4. Find each product, and write it in lowest terms.

(a) $\frac{5}{8} \cdot \frac{2}{10}$

(b) $\frac{3}{4} \cdot \frac{2}{3}$

(c) $\frac{1}{10} \cdot \frac{12}{5}$

(d) $\frac{7}{9} \cdot \frac{12}{14}$

ANSWERS

3. (a) $\frac{4}{7}$ (b) $\frac{5}{6}$ (c) $\frac{1}{2}$ (d) $\frac{3}{5}$

4. (a) $\frac{1}{8}$ (b) $\frac{1}{2}$ (c) $\frac{6}{25}$ (d) $\frac{2}{3}$

5. Find each quotient, and write it in lowest terms.

(a) $\frac{9}{10} \div \frac{3}{5}$

(b) $\frac{3}{4} \div \frac{9}{16}$

(c) $\frac{1}{2} \div \frac{1}{8}$

(d) $\frac{2}{3} \div 6$

ANSWERS

5. (a) $\frac{3}{2}$ (b) $\frac{4}{3}$ (c) 4 (d) $\frac{1}{9}$

Two fractions are **reciprocals** of each other if their product is 1. For example, $\frac{3}{4}$ and $\frac{4}{3}$ are reciprocals because

$$\frac{3}{4} \cdot \frac{4}{3} = 1.$$

The numbers $\frac{7}{11}$ and $\frac{11}{7}$ are reciprocals also. Reciprocals are used in dividing fractions.

DIVIDING FRACTIONS

To divide two fractions, multiply the first fraction and the reciprocal of the second.

The reason this method works will be explained later. The answer to a division problem is called the **quotient.** For example, the quotient of 20 and 10 is 2, since $20 \div 10 = 2$.

EXAMPLE 5 Dividing Fractions

Find the following quotients, and write them in lowest terms.

(a) $\frac{3}{4} \div \frac{8}{5} = \frac{3}{4} \cdot \frac{5}{8} = \frac{3 \cdot 5}{4 \cdot 8} = \frac{15}{32}$

Multiply by reciprocal of second fraction.

(b) $\frac{3}{4} \div \frac{5}{8} = \frac{3}{4} \cdot \frac{8}{5} = \frac{3 \cdot 8}{4 \cdot 5} = \frac{24}{20} = \frac{6}{5}$

(c) $\frac{2}{30} \div \frac{6}{15} = \frac{2}{30} \cdot \frac{15}{6} = \frac{2 \cdot 15}{30 \cdot 6} = \frac{30}{180} = \frac{1}{6}$

(d) $\frac{5}{8} \div 10 = \frac{5}{8} \div \frac{10}{1} = \frac{5}{8} \cdot \frac{1}{10} = \frac{1}{16}$ ■

Write 10 as $\frac{10}{1}$.

WORK PROBLEM 5 AT THE SIDE.

5 All the fractions in an addition or subtraction problem must be written with the same denominator. If some of the fractions have different denominators, rewrite all the fractions with a new common denominator. For example, to rewrite $\frac{3}{4}$ as a fraction with a denominator of 32, find the number that can be multiplied by 4 to give 32. Since $4 \cdot 8 = 32$, use the number 8. Keep the value of the original fraction, $\frac{3}{4}$, the same by multiplying $\frac{3}{4}$ by the fraction $\frac{8}{8}$, which equals 1.

$$\frac{3}{4} = \frac{3}{4} \cdot \frac{8}{8} = \frac{3 \cdot 8}{4 \cdot 8} = \frac{24}{32}$$

Multiplying by 1 does not change the value.

EXAMPLE 6 Changing the Denominator

Write each of the following as a fraction with the given denominator.

(a) $\frac{5}{8} = \frac{}{72}$

Since 8 must be multiplied by 9 to get 72, multiply $\frac{5}{8}$ by $\frac{9}{9}$.

$$\frac{5}{8} = \frac{5}{8} \cdot \frac{9}{9} = \frac{5 \cdot 9}{8 \cdot 9} = \frac{45}{72}$$

(b) $\frac{2}{3} = \frac{}{18}$

Since $3 \times 6 = 18$, multiply by $\frac{6}{6}$.

$$\frac{2}{3} = \frac{2}{3} \cdot \frac{6}{6} = \frac{2 \cdot 6}{3 \cdot 6} = \frac{12}{18}$$ ■

WORK PROBLEM 6 AT THE SIDE.

6 The result of adding two numbers is called the *sum* of the numbers. For example, since $2 + 3 = 5$, the sum of 2 and 3 is 5. The sum of two fractions is found as follows.

ADDING FRACTIONS

To find the **sum** of two fractions with the same denominator, we add their numerators, keeping the same denominator.

EXAMPLE 7 Adding Fractions with the Same Denominator
Add. Write sums in lowest terms.

(a) $\frac{3}{7} + \frac{2}{7} = \frac{3 + 2}{7} = \frac{5}{7}$ Denominator does not change.

(b) $\frac{2}{10} + \frac{3}{10} = \frac{2 + 3}{10} = \frac{5}{10} = \frac{1}{2}$ Write in lowest terms. ■

WORK PROBLEM 7 AT THE SIDE.

If the two fractions to be added do not have the same denominator, first rewrite them with a common denominator and then use the rule above.

EXAMPLE 8 Adding Fractions with Different Denominators
Add.

(a) $\frac{1}{2} + \frac{1}{3}$

To add these fractions, first find a **least common denominator,** the smallest number that both denominators divide into without remainder. Here the least common denominator is $2 \cdot 3 = 6$. Write both $\frac{1}{2}$ and $\frac{1}{3}$ as fractions with a denominator of 6.

$$\frac{1}{2} = \frac{1}{2} \cdot \frac{3}{3} = \frac{1 \cdot 3}{2 \cdot 3} = \frac{3}{6} \quad \text{and} \quad \frac{1}{3} = \frac{1}{3} \cdot \frac{2}{2} = \frac{1 \cdot 2}{3 \cdot 2} = \frac{2}{6}$$

Now add.

$$\frac{1}{2} + \frac{1}{3} = \frac{3}{6} + \frac{2}{6} = \frac{3 + 2}{6} = \frac{5}{6}$$

6. Write as fractions with the given denominators.

(a) $\frac{9}{10} = \frac{}{40}$

(b) $\frac{4}{5} = \frac{}{60}$

(c) $\frac{1}{7} = \frac{}{49}$

(d) $\frac{3}{2} = \frac{}{8}$

7. Add. Write sums in lowest terms.

(a) $\frac{1}{3} + \frac{1}{3}$

(b) $\frac{4}{7} + \frac{1}{7}$

(c) $\frac{1}{9} + \frac{5}{9}$

(d) $\frac{1}{4} + \frac{1}{4}$

ANSWERS
6. (a) $\frac{36}{40}$ (b) $\frac{48}{60}$ (c) $\frac{7}{49}$ (d) $\frac{12}{8}$
7. (a) $\frac{2}{3}$ (b) $\frac{5}{7}$ (c) $\frac{2}{3}$ (d) $\frac{1}{2}$

8. Add.

(a) $\frac{2}{3} + \frac{1}{12}$

(b) $\frac{3}{4} + \frac{1}{6}$

(c) $\frac{7}{30} + \frac{2}{45}$

(d) $2\frac{1}{8} + 1\frac{2}{3}$

ANSWERS

8. (a) $\frac{3}{4}$ (b) $\frac{11}{12}$ (c) $\frac{5}{18}$ (d) $\frac{91}{24}$ or $3\frac{19}{24}$

(b) $\frac{3}{10} + \frac{5}{12}$

The least common denominator often can be found by inspection. If not, first write each denominator as the product of prime factors.

$$10 = 2 \cdot 5 \quad \text{and} \quad 12 = 2 \cdot 2 \cdot 3$$

The least common denominator must have as factors all factors of both numbers. Therefore, the least common denominator must have factors of 2 and 5 (from 10) as well as another 2 and a 3 (from 12). The least common denominator is

$$2 \cdot 2 \cdot 3 \cdot 5 = 60.$$

Now write each fraction with a denominator of 60.

$$\frac{3}{10} = \frac{3 \cdot 6}{10 \cdot 6} = \frac{18}{60} \quad \text{and} \quad \frac{5}{12} = \frac{5 \cdot 5}{12 \cdot 5} = \frac{25}{60}$$

Finally,

$$\frac{3}{10} + \frac{5}{12} = \frac{18}{60} + \frac{25}{60} = \frac{18 + 25}{60} = \frac{43}{60}.$$

(c) $3\frac{1}{2} + 2\frac{3}{4}$

Change both mixed numbers to improper fractions, as follows.

$$3\frac{1}{2} = 3 + \frac{1}{2} = \frac{3}{1} + \frac{1}{2} = \frac{6}{2} + \frac{1}{2} = \frac{6 + 1}{2} = \frac{7}{2}$$

Also,

$$2\frac{3}{4} = 2 + \frac{3}{4} = \frac{8}{4} + \frac{3}{4} = \frac{8 + 3}{4} = \frac{11}{4}.$$

Now add. By inspection, the least common denominator is 4.

$$3\frac{1}{2} + 2\frac{3}{4} = \frac{7}{2} + \frac{11}{4} = \frac{14}{4} + \frac{11}{4} = \frac{25}{4} \quad \text{or} \quad 6\frac{1}{4} \quad ■$$

WORK PROBLEM 8 AT THE SIDE.

The *difference* between two numbers is found by subtracting the numbers. For example, $9 - 5 = 4$, so the difference between 9 and 5 is 4. Find the difference between two fractions as follows.

SUBTRACTING FRACTIONS

To find the **difference** between two fractions with the same denominator, we subtract their numerators, keeping the same denominator.

EXAMPLE 9 Subtracting Fractions

Subtract. Write differences in lowest terms.

(a) $\frac{5}{8} - \frac{3}{8} = \frac{5 - 3}{8} = \frac{2}{8} = \frac{1}{4}$ Lowest terms

(b) $\frac{3}{4} - \frac{1}{3}$

First write the fractions with a common denominator. The least common denominator is 12.

$$\frac{3}{4} - \frac{1}{3} = \frac{9}{12} - \frac{4}{12} = \frac{9 - 4}{12} = \frac{5}{12}$$

(c) $\frac{7}{9} - \frac{1}{6} = \frac{14}{18} - \frac{3}{18} = \frac{14 - 3}{18} = \frac{11}{18}$

Write each fraction with the least common denominator, 18.

(d) $2\frac{1}{2} - 1\frac{3}{4}$

First, change the mixed numbers $2\frac{1}{2}$ and $1\frac{3}{4}$ into improper fractions.

$$2\frac{1}{2} = 2 + \frac{1}{2} = \frac{4}{2} + \frac{1}{2} = \frac{5}{2}$$

$$1\frac{3}{4} = 1 + \frac{3}{4} = \frac{4}{4} + \frac{3}{4} = \frac{7}{4}$$

Now subtract.

$$2\frac{1}{2} - 1\frac{3}{4} = \frac{5}{2} - \frac{7}{4}$$

$$= \frac{10}{4} - \frac{7}{4} \quad \text{Get a common denominator.}$$

$$= \frac{3}{4}$$ ■

WORK PROBLEM 9 AT THE SIDE.

9. Subtract.

(a) $\frac{9}{11} - \frac{3}{11}$

(b) $\frac{2}{3} - \frac{1}{2}$

(c) $\frac{3}{10} - \frac{1}{4}$

(d) $2\frac{3}{8} - 1\frac{1}{2}$

EXAMPLE 10 Solving a Word Problem with Fractions

Wing's favorite recipe for barbecue sauce calls for $2\frac{1}{3}$ cups of tomato sauce. The recipe makes enough barbecue sauce to serve 7 people.

(a) How much tomato sauce is needed for 1 serving?

To find the amount of tomato sauce needed to serve 1 person, divide $2\frac{1}{3}$ by 7, since $2\frac{1}{3}$ is the total amount.

$$2\frac{1}{3} \div 7 = \frac{7}{3} \div 7 = \frac{7}{3} \cdot \frac{1}{7} = \frac{7 \cdot 1}{3 \cdot 7} = \frac{1}{3}$$

(b) How much tomato sauce should Wing use to make enough barbecue sauce to serve 18 people at a party?

Wing needs $\frac{1}{3}$ cup of tomato sauce for each person. To serve 18 people, he needs

$$\frac{1}{3} \cdot 18 = \frac{1}{3} \cdot \frac{18}{1} = \frac{1 \cdot 18}{3 \cdot 1} = \frac{18}{3} = 6$$

cups of tomato sauce. ■

WORK PROBLEM 10 AT THE SIDE.

10. If an upholsterer needs $2\frac{1}{4}$ yards of fabric to re-cover a chair, how many chairs can be re-covered with $23\frac{2}{3}$ yards of fabric?

ANSWERS

9. (a) $\frac{6}{11}$ (b) $\frac{1}{6}$ (c) $\frac{1}{20}$ (d) $\frac{7}{8}$

10. 10 chairs (Some fabric will be left over.)

Historical Reflections

Historical Reflections

WOMEN IN MATHEMATICS: Mary Somerville (1780–1872)

Mary Somerville, a self-educated Scottish mathematician, had a deep understanding of science and was able to communicate its concepts to the general public. One of her major accomplishments was her explanation of Pierre-Simon Laplace's (1749–1827) *Celestial Mechanics*.

In her childhood Somerville was free to observe nature. She studied Euclid thoroughly and perfected her Latin so that she could read Newton's *Principia Mathematica*. In about 1816 she went to London and soon became part of its literary and scientific circles.

Somerville's book on Laplace's theories came out in 1831 with great acclaim. Her next published work was a panoramic book entitled *Connection of the Physical Sciences* (1834). A statement in one of its editions suggested that the irregularities in the orbit of Uranus might indicate that a more remote planet, not yet seen, existed. This statement caught the eye of the scientist who worked out the calculations for Neptune's orbit.

Art: David Eugene Smith Collection/Columbia University

name date hour

1.1 EXERCISES

Write each fraction in lowest terms. See Example 3.

1. $\frac{7}{14}$ **2.** $\frac{3}{9}$ **3.** $\frac{10}{12}$ **4.** $\frac{8}{10}$

5. $\frac{16}{18}$ **6.** $\frac{14}{20}$ **7.** $\frac{50}{75}$ **8.** $\frac{32}{48}$

9. $\frac{72}{108}$ **10.** $\frac{96}{120}$ **11.** $\frac{120}{144}$ **12.** $\frac{77}{132}$

Find the products or quotients. Write answers in lowest terms. See Examples 4 and 5.

13. $\frac{3}{4} \cdot \frac{3}{5}$ **14.** $\frac{3}{8} \cdot \frac{5}{7}$ **15.** $\frac{1}{10} \cdot \frac{6}{5}$ **16.** $\frac{6}{7} \cdot \frac{1}{3}$

17. $\frac{9}{4} \cdot \frac{8}{15}$ **18.** $\frac{3}{5} \cdot \frac{20}{15}$ **19.** $\frac{3}{8} \div \frac{5}{4}$ **20.** $\frac{9}{16} \div \frac{3}{8}$

21. $\frac{5}{12} \div \frac{15}{4}$ **22.** $\frac{21}{16} \div \frac{7}{8}$ **23.** $\frac{28}{3} \div \frac{7}{6}$ **24.** $\frac{121}{9} \div \frac{11}{18}$

25. $3\frac{15}{16} \div \frac{15}{4}$ **26.** $1\frac{15}{32} \div 2\frac{5}{8}$ **27.** $7\frac{4}{5} \div 2\frac{3}{10}$

28. $1\frac{5}{9} \cdot 3\frac{7}{10}$ **29.** $6\frac{7}{10} \cdot 4\frac{5}{7}$ **30.** $2\frac{3}{10} \cdot 5\frac{5}{3}$

Add or subtract. Write answers in lowest terms. See Examples 6–9. (Work from left to right in Exercises 49–54.)

31. $\frac{1}{12} + \frac{5}{12}$ **32.** $\frac{2}{5} + \frac{1}{5}$ **33.** $\frac{1}{10} + \frac{7}{10}$ **34.** $\frac{3}{8} + \frac{1}{8}$

35. $\frac{4}{9} + \frac{2}{3}$ **36.** $\frac{3}{5} + \frac{2}{15}$ **37.** $\frac{8}{11} + \frac{3}{22}$ **38.** $\frac{9}{10} + \frac{3}{5}$

39. $\frac{2}{3} - \frac{3}{5}$ **40.** $\frac{8}{12} - \frac{5}{9}$ **41.** $\frac{5}{6} - \frac{3}{10}$ **42.** $\frac{11}{4} - \frac{5}{8}$

43. $3\frac{1}{4} + \frac{1}{8}$ **44.** $5\frac{2}{3} + \frac{3}{4}$ **45.** $4\frac{1}{2} + 3\frac{2}{3}$ **46.** $7\frac{5}{8} + 3\frac{3}{4}$

47. $6\frac{1}{3} - 5\frac{1}{4}$

48. $8\frac{4}{9} - 7\frac{4}{5}$

49. $\frac{2}{5} + \frac{1}{3} + \frac{9}{10}$

50. $\frac{3}{8} + \frac{5}{6} + \frac{2}{3}$

51. $\frac{5}{7} + 3\frac{1}{4} - \frac{1}{2}$

52. $\frac{2}{3} + \frac{1}{6} - \frac{1}{2}$

53. $\frac{3}{4} + \frac{1}{8} - \frac{2}{3}$

54. $\frac{7}{10} + \frac{3}{5} - \frac{5}{8}$

Work each of the following word problems. See Example 10.

55. John Sanchez paid $\frac{1}{8}$ of a debt in January, $\frac{1}{3}$ in February, and $\frac{1}{4}$ in March. What portion of the debt was paid in these three months?

56. The Eastside Wholesale Market sold $3\frac{1}{4}$ tons of broccoli last month, $2\frac{3}{8}$ tons of spinach, $7\frac{1}{2}$ tons of corn, and $1\frac{5}{16}$ tons of turnips. Find the total number of tons of vegetables sold by the firm during the month.

57. A motel owner has decided to expand his business by buying a piece of property next to the motel. The property has an irregular shape, with five sides as shown in the figure. Find the total distance around the piece of property.

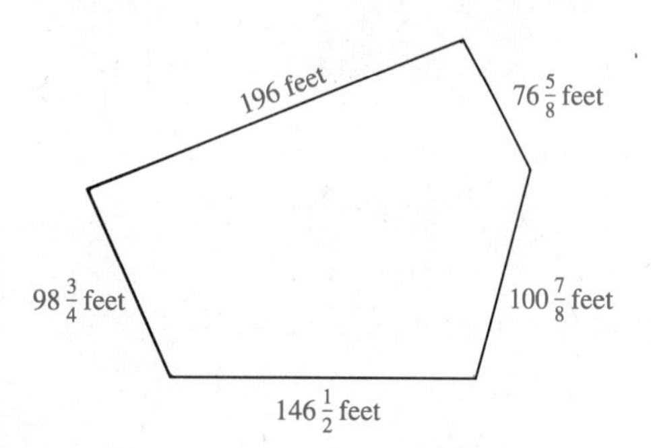

58. A rectangle is $\frac{5}{16}$ yard on each of two sides, and $\frac{7}{12}$ yard on each of the other two sides. Find the total distance around the rectangle.

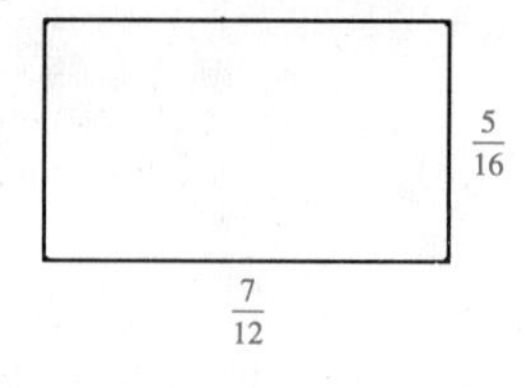

59. Joann Kaufmann worked 40 hours during a certain week. She worked $8\frac{1}{4}$ hours on Monday, $6\frac{3}{8}$ hours on Tuesday, $7\frac{2}{3}$ hours on Wednesday, and $8\frac{3}{4}$ hours on Thursday. How many hours did she work on Friday?

60. A truck is loaded with $9\frac{7}{8}$ cubic yards of topsoil. The driver unloads $1\frac{1}{2}$ cubic yards at the first stop and $2\frac{3}{4}$ cubic yards at the second stop. At the third stop, $3\frac{5}{12}$ cubic yards are unloaded. How much topsoil is left in the truck?

61. A cake recipe calls for $1\frac{3}{4}$ cups of sugar. A caterer has $15\frac{1}{2}$ cups of sugar on hand. How many cakes can he make?

62. Lindsay Branson allows $1\frac{3}{5}$ bottles of beverage for each guest at a party. If she expects 35 guests, how many bottles of beverage will she need?

1.2 DECIMALS AND PERCENTS

Fractions are one way to represent parts of a whole. Another way is with a **decimal fraction** or **decimal,** a number written with a decimal point, such as 9.4. Each digit in a decimal number has a place value, as shown below.

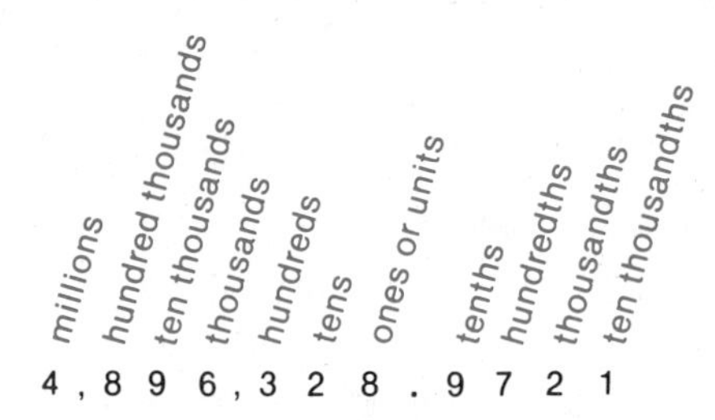

Each successive place value is ten times larger than the place value to its right and is one-tenth as large as the place value to its left.

Prices are often written as decimals. The price $14.75 means 14 dollars and 75 cents, or 14 dollars and $\frac{75}{100}$ of a dollar.

1 We can find a digit's place value in a number by writing the number in **expanded form.** For example, the expanded form for 1729 is

$$1729 = 1000 + 700 + 20 + 9.$$

This expanded form shows that 1 is in the thousands position, 7 is in the hundreds position, 2 is in the tens position, and 9 is in the ones or units position.

EXAMPLE 1 Writing a Number in Expanded Form

Write the following numbers in expanded form.

(a) $98.42 = 90 + 8 + .4 + .02$

(b) $.618 = .6 + .01 + .008$ ■

WORK PROBLEM 1 AT THE SIDE.

2 Place value is used to write a decimal number as a fraction. For example, since the last digit of .67 is in the *hundredths* place,

$$.67 = \frac{67}{\mathbf{100}}.$$

EXAMPLE 2 Writing a Decimal as a Fraction

Write each decimal as a fraction in lowest terms.

(a) .9

Use place value. This decimal is the same as $\frac{9}{10}$.

(b) $.6 = \frac{6}{10} = \frac{3}{5}$

(c) $.25 = \frac{25}{100} = \frac{1}{4}$

(d) $.295 = \frac{295}{1000} = \frac{59}{200}$ ■

WORK PROBLEM 2 AT THE SIDE.

OBJECTIVES

1. Write a number in expanded form.
2. Write decimals as fractions.
3. Add and subtract decimals.
4. Multiply and divide decimals.
5. Write fractions as decimals.
6. Convert percents to decimals and decimals to percents.
7. Find percentages and percents.

1. Write in expanded form.

(a) 478

(b) 15,813

(c) 12.46

(d) .438

2. Write each decimal as a fraction in lowest terms.

(a) .5

(b) .80

(c) .15

(d) .305

ANSWERS

1. (a) $400 + 70 + 8$
   (b) $10{,}000 + 5000 + 800 + 10 + 3$
   (c) $10 + 2 + .4 + .06$
   (d) $.4 + .03 + .008$
2. (a) $\frac{1}{2}$ (b) $\frac{4}{5}$ (c) $\frac{3}{20}$ (d) $\frac{61}{200}$

3. Add or subtract as indicated.

(a)
$$\begin{array}{r} 68.9 \\ 42.72 \\ +\ 8.973 \\ \hline \end{array}$$

(b)
$$\begin{array}{r} 32.5 \\ -21.72 \\ \hline \end{array}$$

(c) 42.83 + 71.629 + 3.074

(d) 351.8 − 2.706

ANSWERS
3. (a) 120.593 (b) 10.78
(c) 117.533 (d) 349.094

3 The operations of addition and subtraction of decimals are explained in the next examples.

EXAMPLE 3 Adding and Subtracting Decimals
Add or subtract as indicated.

(a) 6.92 + 14.8 + 3.217

Place the numbers in a column, with decimal points lined up so that tenths are in one column, hundredths in another column, and so on.

$$\begin{array}{r} 6.92 \\ 14.8 \\ +\ 3.217 \\ \hline 24.937 \end{array}$$

Decimal points lined up

A good way to avoid errors is to attach zeros to make all the numbers the same length. For example,

$$\begin{array}{r} 6.92 \\ 14.8 \\ +\ 3.217 \\ \hline \end{array} \quad \text{becomes} \quad \begin{array}{r} 6.920 \\ 14.800 \\ +\ 3.217 \\ \hline 24.937. \end{array}$$

Attach zeros.

(b) 47.6 − 32.509

Write the numbers in a column, attaching zeros to 47.6.

$$\begin{array}{r} 47.6 \\ -32.509 \\ \hline \end{array} \quad \text{becomes} \quad \begin{array}{r} 47.600 \\ -32.509 \\ \hline 15.091 \end{array}$$

(c) 3 − .253

A whole number is assumed to have the decimal at the right of the number. Write 3 as 3.000; then subtract.

$$\begin{array}{r} 3.000 \\ -\ \ .253 \\ \hline 2.747 \end{array}$$ ■

WORK PROBLEM 3 AT THE SIDE.

4 Multiplication with decimals is done as follows.

MULTIPLYING DECIMALS

Ignore the decimal points and multiply as if the numbers were whole numbers. Then add together the number of **decimal places** (digits after the decimal point) in each number being multiplied. Locate the decimal point in the answer that many digits from the right.

EXAMPLE 4 Multiplying Decimals
Multiply.

(a) 29.3 × 4.52

Multiply as if the numbers were whole numbers.

```
    29.3    1 decimal place in top number
   ×4.52    2 decimal places in second number
     586    |  1 + 2 = 3
    1465    |
   1172     ↓
 132.436    3 decimal places in answer
```

(b) 7.003 × 55.8

```
    7.003    3 decimal places
  ×  55.8    1 decimal place
    56024    |  3 + 1 = 4
   35015     |
  35015      ↓
 390.7674    4 decimal places
```

(c) 31.42 × 65

```
   31.42    2 decimal places
  ×   65    0 decimal places
   15710    |  2 + 0 = 2
  18852     ↓
 2042.30    2 decimal places ■
```

WORK PROBLEM 4 AT THE SIDE.

Division of decimal numbers uses a slightly different process.

DIVIDING DECIMALS

Change the **divisor** (the number being divided by) into a whole number by moving the decimal point as many places as necessary to the right. Move the decimal point in the **dividend** (the number being divided) to the right by the same number of places. Finally, bring the decimal point straight up and divide as with whole numbers.

EXAMPLE 5 Dividing Decimals
Divide.

(a) 279.45 ÷ 24.3

Step 1 Write the problem as follows.

$$24.3\overline{)279.45}$$

4. Multiply.

(a) 2.13 × .05

(b) 69.32 × 1.04

(c) 397.12 × .06

(d) 42,980 × .012

ANSWERS
4. (a) .1065 (b) 72.0928 (c) 23.8272 (d) 515.76

5. Divide.

(a) $32.3\overline{)481.27}$

(b) $.37\overline{)5.476}$

(c) $375.125 \div 3.001$

ANSWERS
5. **(a)** 14.9 **(b)** 14.8 **(c)** 125

Step 2 To change 24.3 into a whole number, move the decimal point one place to the right. Move the decimal point in 279.45 the same number of places to the right to get 2794.5.

$24.3.\overline{)279.4.5}$ Move one decimal place to the right.

To see why this works, write the division in fraction form and multiply the numerator and denominator by $\frac{10}{10}$ or 1.

$$\frac{279.45}{24.3} \cdot \frac{10}{10} = \frac{2794.5}{243}$$

The result is the same as when we moved the decimal point one place to the right in the divisor and the dividend.

Step 3 Bring the decimal point straight up and divide as with whole numbers.

$$\begin{array}{r} 11.5 \\ 243\overline{)2794.5} \\ \underline{243} \\ 364 \\ \underline{243} \\ 1215 \\ \underline{1215} \\ 0 \end{array}$$

Move decimal point straight up.

(b) $73.82\overline{)1852.882}$

Move the decimal point two places to the right in 73.82, to get 7382. Do the same thing with 1852.882, to get 185288.2.

$73.82.\overline{)1852.88.2}$

Bring the decimal point straight up and divide as with whole numbers.

$$\begin{array}{r} 25.1 \\ 7382\overline{)185288.2} \\ \underline{14764} \\ 37648 \\ \underline{36910} \\ 7382 \\ \underline{7382} \\ 0 \end{array}$$

■

WORK PROBLEM 5 AT THE SIDE.

CAUTION In working with decimals, it is helpful to check the work by estimating the answer. For example, in Example 4(a) the answer should be about $30 \times 5 = 150$. The answer, 132.436, is the correct size. Estimate the answers to Example 5 by rounding both numbers to whole numbers with only the first digit nonzero. (*Hint:* In Example 5(a) use $300 \div 20$.)

5 A fraction is written in decimal form as follows.

Write a fraction as a decimal by dividing the denominator into the numerator.

EXAMPLE 6 Writing a Fraction as a Decimal
Write each fraction as a decimal.

(a) $\frac{1}{2}$

Divide the denominator, 2, into the numerator, 1. Attach zeros after the decimal point of the numerator as needed.

Decimal point comes straight up.

$$\begin{array}{r} .5 \\ 2\overline{)1.\mathbf{0}} \\ \underline{10} \\ 0 \end{array}$$

Attach zero.

$$\frac{1}{2} = .5$$

(b) $\frac{3}{8}$

$$\begin{array}{r} .375 \\ 8\overline{)3.\mathbf{000}} \\ \underline{24} \\ 60 \\ \underline{56} \\ 40 \\ \underline{40} \\ 0 \end{array}$$

$$\frac{3}{8} = .375$$

(c) $\frac{2}{3}$

$$\begin{array}{r} .6666 \\ 3\overline{)2.000...} \\ \underline{18} \\ 20 \\ \underline{18} \\ 20 \\ \underline{18} \\ 20 \end{array}$$

The remainder in this division is never 0. This quotient is a **repeating decimal.** Round repeating decimal quotients to as many places as needed. For example, rounding to the nearest thousandth,

$$\frac{2}{3} = .667. \quad ■$$

6. Convert to decimals. Round to the nearest thousandth if necessary.

(a) $\frac{2}{9}$

(b) $\frac{17}{20}$

(c) $\frac{5}{8}$

7. Convert as indicated.

(a) 23% to a decimal

(b) 310% to a decimal

(c) 0.71 to a percent

(d) 1.32 to a percent

ANSWERS
6. (a) .222 (b) .85 (c) .625
7. (a) .23 (b) 3.10 (c) 71%
(d) 132%

The last part of this example used the rule for rounding: if the first digit to be dropped is 5 or more, round the last digit to be kept to the next highest number. If the digit to be dropped is 4 or less, do not round up. For example, to the nearest thousandth,

$$.5555 = .556 \quad \text{and} \quad .3333 = .333.$$

(5 or more; 4 or less)

WORK PROBLEM 6 AT THE SIDE.

6 An important application of decimals is in work with percents. The word **percent** means "per one hundred." Percent is written with the sign %. One percent means "one per one hundred" or "one one-hundredth."

$$1\% = .01 \quad \text{or} \quad 1\% = \frac{1}{100}$$

EXAMPLE 7 Converting Percents and Decimals

(a) Write 73% as a decimal.

Since 1% = .01,

$$73\% = 73 \times 1\% = 73 \times .01 = .73.$$

Also, 73% can be written as a decimal using the fraction form $1\% = \frac{1}{100}$.

$$73\% = 73 \times 1\% = 73 \times \left(\frac{1}{100}\right) = \frac{73}{100} = .73$$

(b) Write 125% as a decimal.

$$125\% = 125 \times 1\% = 125 \times .01 = 1.25$$

(c) Write .32 as a percent.

Since .32 means 32 hundredths, write .32 as 32 × .01. Finally, replace .01 with 1%.

$$.32 = 32 \times .01 = 32 \times 1\% = 32\%$$

(d) Write 2.63 as a percent.

$$2.63 = 263 \times .01 = 263 \times 1\% = 263\% \quad \blacksquare$$

NOTE To change a decimal to a percent or a percent to a decimal, we move the decimal point two places. Move to the left to convert a percent to a decimal. Move to the right to convert a decimal to a percent. Keeping in mind the meaning of percent (per hundred) will help you remember which way to move the decimal point.

WORK PROBLEM 7 AT THE SIDE.

7 A **percentage** is part of a whole. For example, since 50% represents $\frac{50}{100} = \frac{1}{2}$ of a whole, 50% of 800 is half of 800, or 400. Percentages are found by multiplication, as in the next example.

EXAMPLE 8 Finding Percentages
Find the following percentages.

(a) 15% of 600

In examples like this, the word *of* indicates multiplication, so 15% of 600 is found by multiplying .15 and 600.

$$\begin{aligned} 15\% \times 600 &= .15 \times 600 \\ &= 90 \end{aligned}$$

(b) 125% of 80

$$\begin{aligned} 125\% \times 80 &= 1.25 \times 80 \\ &= 100 \end{aligned}$$

(c) A camera with a regular price of $18 is on sale this week at 22% off. Find the amount of the discount and the sale price of the camera.

Find 22% of $18 by multiplying.

$$\begin{aligned} 22\% \times \$18 &= .22 \times \$18 \\ &= \$3.96 \end{aligned}$$

The discount is $3.96. The camera is on sale for $18 − $3.96 = $14.04. ■

WORK PROBLEM 8 AT THE SIDE.

To find the percentage of an amount, we multiplied by the percent. When the percentage and the amount are known, find the percent by *dividing* the percentage by the amount.

EXAMPLE 9 Finding Percent Given the Percentage and the Amount
30 is what percent of 50?

The amount is 50 and the percentage is 30. To find the percent, we must find what part 30 is of 50, so form the fraction $\frac{30}{50}$. Then divide.

$$\frac{\text{Percentage} \rightarrow 30}{\text{Amount} \rightarrow 50} = .60$$

Since .60 = 60%, 30 is 60% of 50. ■

8. Find the percentages.

(a) 20% of 70

(b) 36% of 500

(c) Find the amount of discount on a television set with a regular price of $270 if the set is on sale at 25% off. Find the sale price of the set.

ANSWERS
8. (a) 14 (b) 180
(c) $67.50; $202.50

9. Find each of the following.

(a) 90 is what percent of 270?

(b) What percent of 70 is 14?

10. The interest in one year on deposits of $11,000 was $682. What percent interest was paid?

ANSWERS

9. (a) $33\frac{1}{3}\%$ (b) 20%

10. 6.2%

EXAMPLE 10 Finding Percent Given the Percentage and the Amount

A newspaper ad offered a mountain bike at a sale price of $258. The regular price was $300. What percent of the regular price was the savings?

The savings amounted to $300 − $258 = $42. We find the percent of savings by forming a fraction with the percentage, $42, written above the whole amount, $300.

$$\frac{42}{300} = .14 = 14\%$$

The sale price represented a 14% savings. ■

WORK PROBLEMS 9 AND 10 AT THE SIDE.

The problems involving percent that we have discussed require finding either the *percentage* or the *percent*. These two kinds of problems are summarized below.

Multiply to find percentage.

$$\textbf{Percentage} = \textbf{Percent} \times \textbf{Amount}$$

Divide to find percent.

$$\textbf{Percent} = \frac{\textbf{Percentage}}{\textbf{Amount}}$$

name date hour

1.2 EXERCISES

Write in expanded form. See Example 1.

1. 86 **2.** 15 **3.** 694

4. 856 **5.** 5237 **6.** 4761

7. 36.81 **8.** 78.92 **9.** .567

Write each decimal as a fraction in lowest terms. See Example 2.

10. .2 **11.** .8 **12.** .36 **13.** .72

14. .336 **15.** .215 **16.** .805 **17.** .625

Perform the indicated operations. See Examples 3–5.

18. 14.23 + 9.81 + 74.63 + 18.715 **19.** 89.416 + 21.32 + 478.91 + 298.213

20. 19.74 − 6.53 **21.** 27.96 − 8.39 **22.** 219.4 − 68 **23.** 283 − 12.42

24. $\begin{array}{r} 48.96 \\ 37.421 \\ +\ 9.72 \\ \hline \end{array}$ **25.** $\begin{array}{r} 9.71 \\ 4.8 \\ 3.6 \\ 5.2 \\ +\ 8.17 \\ \hline \end{array}$ **26.** $\begin{array}{r} 8.6 \\ -3.751 \\ \hline \end{array}$ **27.** $\begin{array}{r} 27.8 \\ -13.582 \\ \hline \end{array}$

28. 39.6 × 4.2 **29.** 18.7 × 2.3 **30.** 42.1 × 3.9 **31.** 19.63 × 4.08

32. .042 × 32 **33.** 571 × 2.9 **34.** 24.84 ÷ 6 **35.** 32.84 ÷ 4

36. 7.6266 ÷ 3.42 **37.** 14.9202 ÷ 2.43 **38.** 2496 ÷ .52 **39.** .56984 ÷ .034

Write the following fractions as decimals. Round to the nearest thousandth. See Example 6.

40. $\frac{5}{8}$ **41.** $\frac{3}{8}$ **42.** $\frac{3}{4}$ **43.** $\frac{7}{16}$

44. $\frac{9}{16}$ **45.** $\frac{15}{16}$ **46.** $\frac{2}{3}$ **47.** $\frac{5}{6}$

* **48.** $\frac{7}{13}$ **49.** $\frac{11}{15}$ **50.** $\frac{15}{17}$ **51.** $\frac{12}{19}$

Convert the following decimals to percents. See Examples 7(c) and 7(d).

52. .80 **53.** .75 **54.** .007 **55.** 1.4

56. .67 **57.** .003 **58.** .125 **59.** .983

Convert the following percents to decimals. See Examples 7(a) and 7(b).

60. 53% **61.** 38% **62.** 129% **63.** 174%

64. 96% **65.** 11% **66.** .9% **67.** .1%

Work each of the following. Round your answer to the nearest hundredth. See Examples 8 and 9.

68. What is 14% of 780?

69. Find 12% of 350.

70. Find 22% of 1086

71. What is 20% of 1500?

72. 4 is what percent of 80?

73. 1300 is what percent of 2000?

*Color is used to indicate exercises that are designed for calculator use.

74. What percent of 5820 is 6402?

75. What percent of 75 is 90?

76. 121 is what percent of 484?

77. What percent of 3200 is 64?

78. Find 118% of 125.8.

79. Find 3% of 128.

80. What is 91.72% of 8546.95?

81. Find 12.741% of 58.902.

82. What percent of 198.72 is 14.68?

83. 586.3 is what percent of 765.4?

Solve the word problems. See Examples 8 and 10.

84. A retailer has $23,000 invested in her business. She finds that she is earning 12% per year on this investment. How much money is she earning per year?

85. A family of four with a monthly income of $2000 spends 90% of its earnings and saves the rest. Find the *annual* savings of this family.

86. Jasspreet Kaur recently bought a duplex for $144,000. She expects to earn $23,040 per year on the purchase price. What percent of the purchase price will she earn?

87. A tourist calculated that she had traveled 805 miles by bus on a trip that covered 2300 miles. What percent of the trip was by bus?

88. Capitol Savings Bank pays 8.9% interest per year. What is the annual interest on an account of $3000?

89. Beth's Bargain Basement is having a sale this week. A purchase of $250 was discounted by $37.50. What percent was the discount?

90. Tidwell's Wallpaper installed a total of 4150 square feet of wallpaper in May. Tidwell figured that 249 square feet of this total was wasted. What percent was wasted?

91. Vinh must pay 6.5% sales tax on a new car. The cost of the car is $8600. Find the amount of the tax.

92. An ad for steel-belted radial tires promises 15% better mileage when using them. Alexandria's Escort now goes 420 miles on a tank of gas. If she switched to the new tires, how many extra miles could she drive on a tank of gas?

93. A home worth $77,000 at the beginning of the year has increased in value by $4620 over the last year. What percent of the value at the beginning of the year did the increase represent?

94. A small business takes in $274,600 per year and spends $30,755.20 for advertising. What percent of the income is spent on advertising?

95. A piece of property contains 126,000 square feet. Before the county will give a building permit, the owner of the property must donate 9.4% of the land for a park. How much land must be donated?

96. Self-employed people now must pay a Social Security tax of 15.3%. Find the tax due on earnings of $1756.

97. The Social Security tax on people who work for others is 7.65%. Find the tax on earnings of $2109.

1.3 SYMBOLS

So far only the symbols of arithmetic, such as $+$, $-$, $\times$ (or $\cdot$), and $\div$, have been used. Another common symbol is the one for equality, $=$, which says that two numbers are equal. This symbol with a slash through it, $\neq$, means "is *not* equal to." For example,

$$7 \neq 8$$

indicates that 7 is not equal to 8.

1 If two numbers are not equal, then one of the numbers must be less than the other. The symbol $<$ represents "is less than," so "7 is less than 8" is written

$$7 < 8.$$

Also, we write "6 is less than 9" as $6 < 9$.

The symbol $>$ means "is greater than." We write "8 is greater than 2" as

$$8 > 2.$$

The statement "17 is greater than 11" becomes $17 > 11$.

Keep the symbols $<$ and $>$ straight by remembering that the symbol always points to the smaller number. For example, write "8 is less than 15" by pointing the symbol toward the 8: $8 < 15$.

WORK PROBLEM 1 AT THE SIDE.

2 Word phrases or statements often must be converted to symbols in algebra. The next example shows how to do this.

EXAMPLE 1 Converting Words to Symbols

Write each word statement in symbols.

(a) Twelve equals ten plus two. $12 = 10 + 2$

(b) Nine is less than ten. $9 < 10$

(c) Fifteen is not equal to eighteen. $15 \neq 18$

(d) Seven is greater than four. $7 > 4$ ■

WORK PROBLEM 2 AT THE SIDE.

3 Two other symbols, $\leq$ and $\geq$, also represent the idea of inequality. The symbol $\leq$ means "is less than or equal to," so

$$5 \leq 9$$

means "5 is less than or equal to 9." This statement is true, since $5 < 9$ is true. If either the $<$ part or the $=$ part is true, then the inequality $\leq$ is true.

The symbol $\geq$ means "is greater than or equal to";

$$9 \geq 5$$

is true because $9 > 5$ is true. Also, $8 \leq 8$ is true since $8 = 8$ is true. But $13 \leq 9$ is not true because neither $13 < 9$ nor $13 = 9$ is true.

OBJECTIVES

1. Know the meaning of $<$ and $>$.
2. Translate word phrases to symbols.
3. Know the meaning of $\leq$ and $\geq$.
4. Write statements that change the direction of inequality symbols.

1. Write each statement in words, then decide whether it is true or false.

(a) $7 < 5$

(b) $12 > 6$

(c) $4 \neq 10$

(d) $28 \neq 4 \cdot 7$

2. Write in symbols.

(a) Nine equals eleven minus two.

(b) Seventeen is less than thirty.

(c) Eight is not equal to ten.

(d) Fourteen is greater than twelve.

ANSWERS

1. (a) Seven is less than five. False
   (b) Twelve is greater than six. True
   (c) Four is not equal to ten. True
   (d) Twenty-eight is not equal to four times seven. False
2. (a) $9 = 11 - 2$ (b) $17 < 30$ (c) $8 \neq 10$ (d) $14 > 12$

3. Tell whether each statement is true or false.

(a) $30 \leq 40$

(b) $25 \geq 10$

(c) $40 \leq 10$

(d) $21 \leq 21$

(e) $3 \geq 3$

4. Write each statement with the inequality symbol reversed.

(a) $8 < 10$

(b) $3 > 1$

(c) $9 \leq 15$

(d) $6 \geq 2$

ANSWERS
3. (a) true (b) true (c) false (d) true (e) true
4. (a) $10 > 8$ (b) $1 < 3$ (c) $15 \geq 9$ (d) $2 \leq 6$

EXAMPLE 2 Using the Symbols $\leq$ and $\geq$
Tell whether each statement is true or false.

(a) $15 \leq 20$ — The statement $15 \leq 20$ is true since $15 < 20$.

(b) $25 \geq 30$ — Both $25 > 30$ and $25 = 30$ are false. Because of this, $25 \geq 30$ is false.

(c) $12 \geq 12$ — Since $12 = 12$, this statement is true. ■

WORK PROBLEM 3 AT THE SIDE.

4 Any statement with < can be converted to one with >, and any statement with > can be converted to one with <. We do this by reversing both the order of the numbers and the direction of the symbol. For example, the statement $6 < 10$ can be written as $10 > 6$.

$$6 < 10 \quad \text{becomes} \quad 10 > 6$$

Exchange numbers.

Reverse symbol.

EXAMPLE 3 Converting Between < and >
The following list shows the same statements written in two equally correct ways.

(a) $9 < 16$ $\quad 16 > 9$ **(b)** $5 > 2$ $\quad 2 < 5$

(c) $3 \leq 8$ $\quad 8 \geq 3$ **(d)** $12 \geq 5$ $\quad 5 \leq 12$ ■

WORK PROBLEM 4 AT THE SIDE.

Here is a summary of the symbols discussed in this section.

SYMBOLS OF EQUALITY AND INEQUALITY

Symbol	*Meaning*
$=$	is equal to
$\neq$	is not equal to
$<$	is less than
$>$	is greater than
$\leq$	is less than or equal to
$\geq$	is greater than or equal to

CAUTION The symbols introduced in this section are used to write mathematical *sentences.* They differ from the symbols for operations (+, −, ·, and ÷), discussed earlier, which are used to write mathematical *expressions* that represent a number. For example, compare the sentence $4 < 10$ with the expression $4 + 10$, which equals the number 14.

name date hour

1.3 EXERCISES

Insert < or > to make the following statements true.

1. 6 9 **2.** 5 3 **3.** 12 15 **4.** 8 10

5. 25 12 **6.** 17 9 **7.** 32 50 **8.** 41 72

9. $\frac{3}{4}$ 1 **10.** $\frac{2}{3}$ 0 **11.** $1\frac{5}{8}$ 1 **12.** $3\frac{7}{9}$ 2

Insert ≤ or ≥ to make the following statements true.

13. 12 17 **14.** 28 42 **15.** 16 14 **16.** 39 17

17. 8 28 **18.** 10 15 **19.** 35 42 **20.** 51 62

Which of the symbols <, >, ≤, and ≥ make the following statements true? Give all possible correct answers.

21. 6 9 **22.** 18 12 **23.** 51 50 **24.** 0 12

25. 5 5 **26.** 10 10 **27.** 48 0 **28.** 100 1000

29. 16 10 **30.** 5 3 **31.** $\frac{1}{4}$ $\frac{2}{5}$ **32.** $\frac{2}{3}$ $\frac{5}{8}$

33. .609 .61 **34.** .5 .499 **35.** $3\frac{1}{2}$ 4 **36.** $5\frac{7}{8}$ 6

Write the following word statements in symbols. See Example 1.

37. Seven equals five plus two.

38. Nine is greater than the product of four and two.

39. Three is less than the quotient of fifty and five.

40. Five equals ten minus five.

41. Twelve is not equal to five.

42. Fifteen does not equal sixteen.

43. Zero is greater than or equal to zero.

44. Six is less than or equal to six.

Tell whether each statement is true or false. See Example 2.

45. $8 + 2 = 10$

46. $8 \neq 9 - 1$

47. $12 \geq 10$

48. $45 < 45$

49. $0 < 15$

50. $16 \geq 10$

51. $9 + 12 = 21$

52. $9 < 12$

53. $25 \geq 19$

54. $18 < 5$

55. $9 < 0$

56. $15 \leq 32$

57. $6 \neq 5 + 1$

58. $15 < 21$

59. $11 < 11$

60. $29 \geq 30$

61. $8 \leq 0$

62. $26 \geq 50$

Rewrite the following statements so the inequality symbol points in the opposite direction. See Example 3.

63. $6 < 14$

64. $8 \leq 9$

65. $15 \geq 3$

66. $29 > 4$

67. $9 > 8$

68. $12 < 17$

69. $0 \leq 6$

70. $7 \leq 12$

71. $18 \geq 15$

72. $25 \geq 1$

73. $.481 \geq .439$

74. $.762 < .763$

1.4 EXPONENTS AND ORDER OF OPERATIONS

OBJECTIVES

1. Use exponents.
2. Use order of operations.
3. Use brackets.
4. Insert parentheses to make a statement true.
5. Write a word problem using symbols.

1 As we mentioned earlier, it is common for a number to have the same factor appearing several times. For example, in the product

$$3 \cdot 3 \cdot 3 \cdot 3 = 81$$

the factor 3 appears four times. To save space, repeated factors are written with an *exponent*. For example, in $3 \cdot 3 \cdot 3 \cdot 3$, the number 3 appears as a factor four times, so the product is written as 3^4.

$$3 \cdot 3 \cdot 3 \cdot 3 = 3^4$$

The number 4 is the exponent and 3 is the base in the **exponential expression** 3^4.

EXAMPLE 1 Finding the Value of an Exponential Expression
Find the values of the following.

(a) 5^2

$$\underbrace{5 \cdot 5}_{} = 25$$

5 is used as a factor 2 times.

Read 5^2 as "5 squared."

(b) 6^3

$$\underbrace{6 \cdot 6 \cdot 6}_{} = 216$$

6 is used as a factor 3 times.

Read 6^3 as "6 cubed."

(c) 2^5

$$2 \cdot 2 \cdot 2 \cdot 2 \cdot 2 = 32$$ 2 is used as a factor 5 times.

Read 2^5 as "2 to the fifth power."

(d) 7^4

$$7 \cdot 7 \cdot 7 \cdot 7 = 2401$$ 7 is used as a factor 4 times.

Read 7^4 as "7 to the fourth power."

(e) $\left(\frac{2}{3}\right)^3$

$$\frac{2}{3} \cdot \frac{2}{3} \cdot \frac{2}{3} = \frac{8}{27}$$ $\frac{2}{3}$ is used as a factor 3 times. ■

WORK PROBLEM 1 AT THE SIDE.

1. Find the value of each exponential expression.

(a) 6^2

(b) 3^5

(c) $\left(\frac{3}{4}\right)^2$

(d) $\left(\frac{1}{2}\right)^4$

(e) $(.4)^3$

2 Many problems involve more than one symbol of arithmetic. For example, in finding the value of

$$5 + 2 \cdot 3$$

which should be done first—multiplication or addition? The following **order of operations** has been agreed on as the most reasonable. (This is the order used by most calculators and computers.)

ANSWERS

1. (a) 36 (b) 243 (c) $\frac{9}{16}$ (d) $\frac{1}{16}$ (e) .064

2. Find the value of each expression.

(a) $3 \cdot 8 + 7$

(b) $12 \cdot 6 + 9$

(c) $6 \cdot 11 + 4$

ANSWERS
2. (a) 31 (b) 81 (c) 70

ORDER OF OPERATIONS

If parentheses or fraction bars are present, simplify within parentheses, innermost first, and above and below fraction bars separately, in the following order.

Step 1 Apply all exponents.

Step 2 Do any multiplications or divisions in the order in which they occur, working from left to right.

Step 3 Do any additions or subtractions in the order in which they occur, working from left to right.

If no parentheses or fraction bars are present, start with Step 1.

EXAMPLE 2 Using the Order of Operations

Find the value of $5 + 2 \cdot 3$.

Using the order of operations given above, first multiply 2 and 3, and then add 5.

$$5 + 2 \cdot 3 = 5 + 6 \quad \text{Multiply.}$$
$$= 11 \quad \text{Add.}$$

Therefore, $5 + 2 \cdot 3 = 11$. If the addition had been performed first, the result would have been $7 \cdot 3 = 21$, instead of 11. By the order of operations, only the first result, 11, is correct. ■

WORK PROBLEM 2 AT THE SIDE.

A dot has been used to show multiplication; another way to show multiplication is with parentheses. For example, 3(7) means $3 \cdot 7$ or 21. The next example shows the use of parentheses for multiplication and parentheses and fraction bars for grouping.

EXAMPLE 3 Using the Order of Operations

Find the value of each of the following.

(a) $9(6 + 11)$

Work first inside the parentheses.

$$9(6 + 11) = 9(17) \quad \text{Add inside parentheses.}$$
$$= 153 \quad \text{Multiply.}$$

(b) $$2(5 + 6) + 7 \cdot 3 = 2(11) + 7 \cdot 3 \quad \text{Add inside parentheses.}$$
$$= 22 + 21 \quad \text{Multiply.}$$
$$= 43 \quad \text{Add.}$$

(c) $\dfrac{4(5 + 3) + 3}{2(3) - 1}$

Simplify the numerator and denominator separately.

$$\frac{4(5 + 3) + 3}{2(3) - 1} = \frac{4(8) + 3}{2(3) - 1} \quad \text{Add inside parentheses.}$$

$$= \frac{32 + 3}{6 - 1} \quad \text{Multiply.}$$

$$= \frac{35}{5} \quad \text{Add and subtract.}$$

$$= 7 \quad \text{Divide.}$$

(d) $9 + 2^3 - 5$

According to the order of operations, we calculate 2^3 first.

$$9 + \mathbf{2^3} - 5 = 9 + \mathbf{8} - 5 \quad \text{Use the exponent.}$$

$$= 12 \quad \text{Add, then subtract.}$$ ■

WORK PROBLEM 3 AT THE SIDE.

NOTE Parentheses and fraction bars are used as grouping symbols to indicate an expression that represents a single number. That is why we must simplify first within parentheses and above and below fraction bars.

3 An expression with double parentheses, such as $2(8 + 3(6 + 5))$, can be confusing. We avoid confusion by using square brackets, [ ], in place of one pair of parentheses.

EXAMPLE 4 Using Brackets

Simplify $2[8 + 3(6 + 5)]$.

Work first within the parentheses, until a single number is found inside the brackets.

$$2[8 + 3(\mathbf{6 + 5})] = 2[8 + 3(\mathbf{11})] \quad \text{Add.}$$

$$= 2[8 + \mathbf{33}] \quad \text{Multiply.}$$

$$= 2[\mathbf{41}] \quad \text{Add.}$$

$$= 82 \quad \text{Multiply.}$$ ■

WORK PROBLEM 4 AT THE SIDE.

4 In writing a mathematical statement, it sometimes is necessary to include parentheses to make the statement true.

3. Find the value of each expression.

(a) $2 \cdot 9 + 7 \cdot 3$

(b) $7 \cdot 6 - 3(8 + 1)$

(c) $\dfrac{2(7 + 8) + 2}{3 \cdot 5 + 1}$

(d) $2 + 3^2 - 5$

4. Find the value of each expression.

(a) $4[7 + 3(6 + 1)]$

(b) $9[(4 + 8) - 3]$

ANSWERS
3. (a) 39 (b) 15 (c) 2 (d) 6
4. (a) 112 (b) 81

EXAMPLE 5 Inserting Parentheses to Make a Statement True

Insert parentheses so the following statements are true.

(a) $9 - 3 - 2 = 8$

Use trial and error to see that this statement would be true if parentheses were inserted around $3 - 2$, since

$$9 - (3 - 2) = 9 - 1 = 8.$$

It is not true that $(9 - 3) - 2 = 8$, since $6 - 2 \neq 8$.

(b) $9 \cdot 2 - 4 \cdot 3 = 6$

Since $9 \cdot 2 - 4 \cdot 3 = 18 - 12 = 6$, no parentheses are needed here. If desired, though, parentheses may be placed as follows.

$$(9 \cdot 2) - (4 \cdot 3) = 6 \quad \blacksquare$$

WORK PROBLEM 5 AT THE SIDE

5 The first step in using mathematics to solve a problem stated in words is to write the information given in the problem using symbols.

EXAMPLE 6 Translating Information from Words to Symbols

Maria Gonsalves had quiz scores as follows: 3 eights, 2 nines, and 1 seven. The teacher threw out a quiz on which Maria had a nine. Write an expression for her total quiz scores.

This information can be written using symbols as

$$3 \cdot 8 + 2 \cdot 9 + 1 \cdot 7 - 9. \quad \blacksquare$$

WORK PROBLEM 6 AT THE SIDE.

5. Insert parentheses as necessary to make each statement true.

(a) $14 - 3 - 1 = 12$

(b) $2 \cdot 5 + 3 \cdot 2 = 26$

(c) $3 + 4^2 \cdot 3 = 57$

6. Use symbols to write the information given in the statement below.

Joe Sun bought two sweaters for \$25 each and a pair of slacks for \$31.

ANSWERS

5. (a) $14 - (3 - 1) = 12$
   (b) $(2 \cdot 5 + 3) \cdot 2 = 26$
   (c) $(3 + 4^2) \cdot 3 = 57$
6. $2 \cdot 25 + 31$

name date hour

1.4 EXERCISES

Find the value of each exponential expression. In Exercises 21–24, round to the nearest thousandth. See Example 1.

1. 6^2 **2.** 9^2 **3.** 8^2 **4.** 10^2

5. 17^2 **6.** 22^2 **7.** 5^3 **8.** 7^3

9. 6^4 **10.** 3^4 **11.** 2^5 **12.** 4^5

13. 3^6 **14.** 2^6 **15.** $\left(\frac{1}{2}\right)^2$ **16.** $\left(\frac{3}{4}\right)^2$

17. $\left(\frac{2}{5}\right)^3$ **18.** $\left(\frac{3}{7}\right)^3$ **19.** $\left(\frac{4}{5}\right)^3$ **20.** $\left(\frac{2}{3}\right)^5$

21. $(.83)^4$ **22.** $(.712)^2$ **23.** $(1.46)^3$ **24.** $(2.85)^4$

Find the values of the following expressions. See Examples 2–4.

25. $9 \cdot 3 - 11$ **26.** $6 \cdot 5 - 12$ **27.** $\frac{2}{5} \cdot \frac{11}{3} + \frac{2}{3} \cdot \frac{1}{4}$

28. $\frac{9}{4} \cdot \frac{2}{3} + \frac{4}{5} \cdot \frac{5}{3}$ **29.** $13 \cdot 2 - 15 \cdot 1$ **30.** $(6.1)(5.7) + (3.4)(12)$

31. $(5.4)(8.1) + (10.9)(4)$ **32.** $4[2 + 3(2^2)]$ **33.** $5[2^3 + (2 + 3)]$

34. $3^2[(14 + 5) - 10]$ **35.** $\frac{5(2^2 - 1) + 3}{2 \cdot 4 + 1}$ **36.** $\frac{7(3 + 1) - 2}{5 \cdot 2 + 3}$

37. $\frac{2(5 + 1) - 3(1 + 1)}{5(8 - 6) - 2^3}$ **38.** $\frac{\frac{3}{4} + \frac{7}{8}}{\frac{11}{16} - \frac{1}{2}}$ **39.** $\frac{\frac{5}{6}}{3 - 1\frac{1}{3}}$

Tell whether each statement is true or false. (Hint: First simplify each statement.) See Examples 2–4.

40. $3 \cdot 8 - 4 \cdot 6 \leq 0$

41. $2 \cdot 20 - 8 \cdot 5 \geq 0$

42. $9 \cdot 2 - 6 \cdot 3 \geq 2$

43. $8 \cdot 3 - 4 \cdot 6 < 1$

44. $12 \cdot 3 - 6 \cdot 6 \leq 0$

45. $3[5(2) - 3] > 20$

46. $2[2 + 3(2 + 5)] \leq 45$

47. $3[4 + 3(4 + 1)] \leq 55$

48. $\frac{2(5 + 3) + 2 \cdot 2}{2(4 - 1)} > 4$

49. $\frac{9(7 - 1) - 8 \cdot 2}{4(6 - 1)} > 2$

50. $\frac{3(8 - 3) + 2(4 - 1)}{9(6 - 2) - 8(5 - 2)} \geq 7$

51. $9 \leq 4^2 - 8$

52. $6^2 - 3^2 > 25$

53. $21.92 \leq 7.43^2 - 5.77^2$

54. $.479 > (.841)^3 - (.58)^4$

Insert parentheses in each expression so that the resulting statement is true. Some problems require no parentheses. See Example 5.

55. $10 - 7 - 3 = 6$

56. $16 - 4 - 3 = 15$

57. $3 \cdot 5 + 7 = 22$

58. $3 \cdot 5 + 7 = 36$

59. $3 \cdot 5 - 4 = 3$

60. $3 \cdot 5 - 4 = 11$

61. $3 \cdot 5 - 2 \cdot 4 = 36$

62. $100 \div 20 \div 5 = 1$

63. $360 \div 18 \div 4 = 5$

64. $2^2 + 4 \cdot 2 = 16$

65. $6 + 5 \cdot 3^2 = 99$

66. $3^3 - 2 \cdot 4 = 100$

67. $8 - 2^2 \cdot 2 = 8$

68. $3 \cdot \frac{2}{3} - \frac{1}{4} \cdot \frac{4}{5} = 1$

69. $\frac{1}{2} + \frac{5}{3} \cdot \frac{9}{7} - \frac{3}{2} = \frac{8}{7}$

Write an expression for the final amounts in the following problems using numerical symbols and parentheses. See Example 6.

70. Marjorie Jensen invested \$600. After one year, her investment had tripled. She then took \$150 and paid a car payment.

71. John Wilson had 5 decks of cards, each containing 52 cards. He removed all 4 aces from one deck.

72. A bus has 63 passengers. At one stop, 23 people get off and 17 new people get on.

73. An elevator has 5 passengers. At the first stop 6 get on and 1 gets off. At the next stop 3 get on and 2 get off.

1.5 VARIABLES AND EQUATIONS

OBJECTIVES

1. Define *variable.*
2. Find the value of algebraic expressions, given values for the variables.
3. Convert statements from words to algebraic expressions.
4. Identify solutions of equations.
5. Define and use the domain of a set of numbers.

1 A **variable** is a symbol, usually a letter, such as x, y, or z, used to represent any unknown number. An **algebraic expression** is a collection of numbers, variables, symbols for operations, and symbols for grouping, such as parentheses, square brackets, or division bars. For example,

$$x + 5, \qquad 2m - 9, \qquad \text{and} \qquad 8p^2 + 6(p - 2)$$

are all algebraic expressions. In the algebraic expression $2m - 9$, the expression $2m$ means $2 \cdot m$, the product of 2 and m, and $8p^2$ shows the product of 8 and p^2. Also, $6(p - 2)$ means the product of 6 and $p - 2$.

2 An algebraic expression takes on different numerical values as the variables take on different values.

EXAMPLE 1 Finding the Value of an Expression Given a Value of the Variable

Find the values of the following algebraic expressions if $m = 5$ and if $m = 9$.

(a) $8m$

Replace m with 5, to get

$$8m = 8 \cdot 5 \qquad \text{Let } m = 5.$$
$$= 40. \qquad \text{Multiply.}$$

If $m = 9$,

$$8m = 8 \cdot 9 \qquad \text{Let } m = 9.$$
$$= 72. \qquad \text{Multiply.}$$

(b) $3m^2$

If $m = 5$,

$$3m^2 = 3 \cdot 5^2 \qquad \text{Let } m = 5.$$
$$= 3 \cdot 25 \qquad \text{Square.}$$
$$= 75. \qquad \text{Multiply.}$$

If $m = 9$,

$$3m^2 = 3 \cdot 9^2$$
$$= 3 \cdot 81 = 243. \quad ■$$

CAUTION In Example 1(b), it is important to notice that $3m^2$ means $3 \cdot m^2$; it *does not* mean $3m \cdot 3m$.

WORK PROBLEM 1 AT THE SIDE.

1. Find the value of each expression if $p = 3$.

(a) $6p$

(b) $p + 12$

(c) $5p^2$

(d) $2p^3$

(e) p^4

EXAMPLE 2 Finding the Value of an Expression with More Than One Variable

Find the value of each expression if $x = 5$ and $y = 3$.

(a) $2x + 5y$

Replace x with 5 and y with 3. Multiply first, and then add.

$$2x + 5y = 2 \cdot 5 + 5 \cdot 3 \qquad \text{Let } x = 5 \text{ and } y = 3.$$
$$= 10 + 15 \qquad \text{Multiply.}$$
$$= 25 \qquad \text{Add.}$$

ANSWERS
1. (a) 18 (b) 15 (c) 45 (d) 54 (e) 81

2. Find the value of each expression if $x = 6$ and $y = 9$.

(a) xy

(b) $4x + 7y$

(c) $\dfrac{4x - 2y}{x + 1}$

(d) $x^2 + y^2$

(e) x^3y

ANSWERS

2. (a) 54 (b) 87 (c) $\frac{6}{7}$ (d) 117 (e) 1944

(b) $\dfrac{9x - 8y}{2x - y}$

Replace x with 5 and y with 3.

$$\frac{9x - 8y}{2x - y} = \frac{9 \cdot 5 - 8 \cdot 3}{2 \cdot 5 - 3} \quad \text{Let } x = 5 \text{ and } y = 3.$$

$$= \frac{45 - 24}{10 - 3} \quad \text{Multiply.}$$

$$= \frac{21}{7} \quad \text{Subtract.}$$

$$= 3 \quad \text{Divide.}$$

(c) $x^2 - 2y^2 = 5^2 - 2 \cdot 3^2$ Let $x = 5$ and $y = 3$.

$= 25 - 2 \cdot 9$ Use the exponents.

$= 25 - 18$ Multiply.

$= 7$ Subtract. ■

WORK PROBLEM 2 AT THE SIDE.

3 Variables are used in changing word phrases into algebraic expressions. The next example shows how to do this.

EXAMPLE 3 Using Variables to Change Word Phrases into Algebraic Expressions

Change the following word phrases to algebraic expressions. Use x as the variable to represent the number.

(a) The sum of a number and 9

"Sum" is the answer to an addition problem. This phrase translates as

$$x + 9 \quad \text{or} \quad 9 + x.$$

(b) 7 minus a number

"Minus" indicates subtraction, so the answer is

$$7 - x.$$

Note that $x - 7$ would *not* be correct because we cannot do a subtraction in either order and get the same results.

(c) A number subtracted from 12

Since a number is subtracted *from* 12, write this as

$$12 - x.$$

Compare this result with "12 is subtracted from 25," which is $25 - 12$.

(d) The product of 11 and a number

$$11 \cdot x \quad \text{or} \quad 11x$$

(e) 5 divided by a number

$$\frac{5}{x}$$

(f) The product of 2, and the difference between a number and 8

$$2(x - 8) \quad ■$$

CAUTION Notice that in translating the words "the difference between a number and 8" the order is kept the same: $x - 8$. "The difference between 8 and a number" would be written $8 - x$.

WORK PROBLEM 3 AT THE SIDE.

4 An **equation** states that two expressions are equal. Examples of equations are

$$x + 4 = 11, \quad 2y = 16, \quad \text{and} \quad 4p + 1 = 25 - p.$$

To **solve** an equation, find all values of the variable that make the equation true. The values of the variable that make the equation true are called the **solutions** of the equation.

EXAMPLE 4 Deciding Whether a Number Is a Solution of an Equation

Decide whether the given number is a solution of the equation.

(a) $5p + 1 = 36$; 7

Replace p with 7.

$$5p + 1 = 36$$
$$5 \cdot 7 + 1 = 36 \qquad \text{Let } p = 7.$$
$$35 + 1 = 36 \qquad \text{Multiply.}$$
$$36 = 36. \qquad \text{True}$$

The number 7 is a solution of the equation.

(b) $9m - 6 = 32$; 4

$$9m - 6 = 32$$
$$9 \cdot 4 - 6 = 32 \qquad \text{Let } m = 4.$$
$$36 - 6 = 32 \qquad \text{Multiply.}$$
$$30 = 32 \qquad \text{False}$$

The number 4 is not a solution of the equation. ■

WORK PROBLEM 4 AT THE SIDE.

5 Sometimes the solutions of an equation must come from a certain list of numbers. This list of numbers is often written as a **set,** a collection of objects. For example, the set containing the numbers 1, 2, 3, 4, and 5 is written with **set braces,** { }, as

$$\{1, 2, 3, 4, 5\}.$$

The set of numbers from which the solutions of an equation must be chosen is called the **domain** of the equation.

EXAMPLE 5 Finding the Solution of an Equation from the Domain

Change each word statement to an equation. Use x as the variable. Then find all solutions for the equation from the domain

$$\{0, 2, 4, 6, 8, 10\}.$$

3. Write as an algebraic expression. Use x as the variable.

(a) The sum of 5 and a number

(b) A number minus 4

(c) A number subtracted from 48

(d) The product of 6 and a number

(e) 9 multiplied by the sum of a number and 5

4. Decide whether the given number is a solution of the equation.

(a) $2x = 10$; 5

(b) $p - 1 = 3$; 2

(c) $2k + 3 = 15$; 7

(d) $8p - 11 = 5$; 2

(e) $9k - 8 = 19$; 4

ANSWERS
3. (a) $5 + x$ (b) $x - 4$ (c) $48 - x$ (d) $6x$ (e) $9(x + 5)$
4. (a) yes (b) no (c) no (d) yes (e) no

(a) The sum of a number and four is six.

The word "is" suggests "equals." Let x represent the unknown number and translate as follows.

The sum of a number and four	is	six.
↓	↓	↓
$x + 4$	$=$	6

Try each number from the given domain, {0, 2, 4, 6, 8, 10}, in turn.

$x + 4 = 6$	Given equation
$0 + 4 = 6$	False
$2 + 4 = 6$	True
$4 + 4 = 6$	False
$6 + 4 = 6$	False
$8 + 4 = 6$	False
$10 + 4 = 6$	False

The only solution of $x + 4 = 6$ is 2.

(b) Nine more than five times a number is 49.

"Nine more than" means "nine is added to." Use x to represent the unknown number.

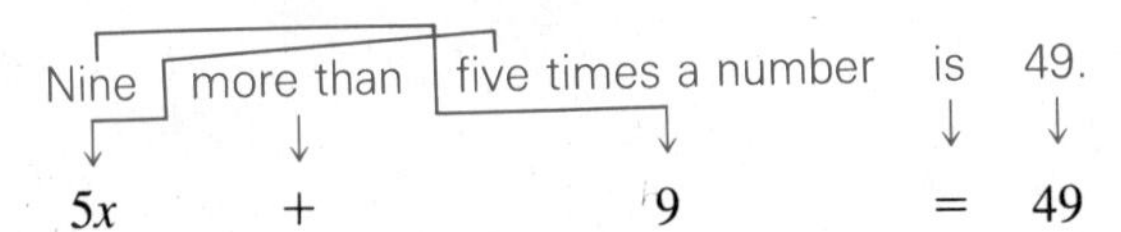

Try each number from the given domain, {0, 2, 4, 6, 8, 10}. The solution is 8, since $5 \cdot 8 + 9 = 49$. ■

WORK PROBLEM 5 AT THE SIDE.

Students often have trouble distinguishing between equations and expressions. Remember that an equation is a sentence; an expression is a phrase.

EXAMPLE 6 Distinguishing Between Equations and Expressions

Decide whether each of the following is an equation or an expression.

(a) $2x - 5y$

There is no equals sign, so this is an expression.

(b) $2x = 5y$

Because of the equals sign, this is an equation. ■

WORK PROBLEM 6 AT THE SIDE.

5. Change each statement to an equation. Find all solutions from the domain {0, 2, 4, 6, 8, 10}.

(a) The sum of a number and 13 is 19.

(b) Twice a number is added to 4, giving 20.

(c) Three times a number is subtracted from 21, giving 15.

(d) The quotient of a number and 2 is added to 5, giving 9.

6. Decide whether each of the following is an equation or an expression.

(a) $2x + 5y - 7$

(b) $\dfrac{3x - 1}{5}$

(c) $2x + 5 = 7$

(d) $\dfrac{x}{y - 3} = 4x$

ANSWERS

5. (a) $x + 13 = 19$; 6
(b) $2x + 4 = 20$; 8
(c) $21 - 3x = 15$; 2
(d) $(x \div 2) + 5 = 9$ or $\frac{x}{2} + 5 = 9$; 8

6. (a) expression (b) expression
(c) equation (d) equation

name date hour

1.5 EXERCISES

Find the numerical values of the following if **(a)** $x = 3$ *and* **(b)** $x = 15$. *See Example 1.*

1. $x + 9$

(a) **(b)**

2. $x - 1$

(a) **(b)**

3. $5x$

(a) **(b)**

4. $7x$

(a) **(b)**

5. $\frac{2}{3}x + \frac{1}{3}$

(a) **(b)**

6. $\frac{9}{4}x - \frac{5}{3}$

(a) **(b)**

7. $\frac{x + 1}{3}$

(a) **(b)**

8. $\frac{x - 2}{5}$

(a) **(b)**

9. $\frac{3x - 5}{2x}$

(a) **(b)**

10. $\frac{x + 2}{x - 1}$

(a) **(b)**

11. $3x^2 + x$

(a) **(b)**

12. $2x + x^2$

(a) **(b)**

13. $6.459x$

(a) **(b)**

14. $.74x^2$

(a) **(b)**

15. $.0745(x^2 + 2)$

(a) **(b)**

16. $.204(3 + x)$

(a) **(b)**

Find the numerical values of the following if **(a)** $x = 4$ *and* $y = 2$ *and* **(b)** $x = 1$ *and* $y = 5$. *See Example 2. Round to the nearest thousandth in Exercises 37 and 38.*

17. $8x + 3y + 5$

(a) **(b)**

18. $4x + 2y + 7$

(a) **(b)**

19. $3(x + 2y)$

(a) **(b)**

20. $2(2x + y)$

(a) **(b)**

21. $x + \frac{4}{y}$

(a) **(b)**

22. $y + \frac{8}{x}$

(a) **(b)**

23. $\frac{y}{3} + \frac{5}{y}$

(a) **(b)**

24. $\frac{x}{5} + \frac{y}{4}$

(a) **(b)**

25. $5\left(\frac{4}{3}x + \frac{7}{2}y\right)$

(a) **(b)**

26. $8\left(\frac{5}{2}x + \frac{9}{5}y\right)$

(a) **(b)**

27. $\frac{2x + 3y}{x + y + 1}$

(a) **(b)**

28. $\frac{5x + 3y + 1}{2x}$

(a) **(b)**

29. $\frac{2x + 4y - 6}{5y + 2}$

(a) **(b)**

30. $\frac{4x + 3y - 1}{x}$

(a) **(b)**

31. $2y^2 + 5x$

(a) **(b)**

32. $6x^2 + 4y$

(a) **(b)**

name date hour

33. $\dfrac{x^2 + y^2}{x + y}$

(a) (b)

34. $\dfrac{9x^2 + 4y^2}{3x^2 + 2y}$

(a) (b)

35. $\dfrac{3x + y^2}{2x + 3y}$

(a) (b)

36. $\dfrac{x^2 + 1}{4x + 5y}$

(a) (b)

37. $.841x^2 + .32y^2$

(a) (b)

38. $\dfrac{3.4x + 2.59y}{.8x + .3y^2}$

(a) (b)

Change the word phrases to algebraic expressions. Use x to represent the variable. See Example 3.

39. Eight times a number

40. Fifteen times a number

41. Five times a number

42. Six added to a number

43. Four added to a number

44. A number subtracted from eight

45. Nine subtracted from a number

46. Eight subtracted from three times a number

47. Six added to two-thirds of a number

48. Three-fourths of a number, added to fifty-two

Decide whether the given number is a solution of the equation. See Example 4.

49. $p - 5 = 12$; 17

50. $x + 6 = 15$; 9

51. $5m + 2 = 7$; 2

52. $3r + 5 = 8$; 2

53. $2y + 3(y - 2) = 14$; 4

54. $6a + 2(a + 3) = 14$; 1

55. $6p + 4p - 9 = 11$; 2

56. $2x + 3x + 8 = 38$; 6

57. $3r^2 - 2 = 46$; 4

58. $2x^2 + 1 = 19;\ 3$

59. $\dfrac{z + 4}{2 - z} = \dfrac{13}{5};\ \dfrac{1}{3}$

60. $\dfrac{x + 6}{x - 2} = \dfrac{37}{5};\ 3\dfrac{1}{4}$

61. $9.54x + 3.811 = .4273x + 16.57718;\ 1.4$

62. $.935(y + 6.1) + .0142 = 7.83y + .2017;\ .8$

Change the word statements to equations. Use x as the variable. Find the solutions from the domain {0, 2, 4, 6, 8, 10}. *See Examples 3–5.*

63. The sum of a number and 8 is 12.

64. A number minus three equals seven.

65. Sixteen minus three-fourths of a number is ten.

66. The sum of six-fifths of a number and 6 is 18.

67. Five more than twice a number is 13.

68. The product of a number and 3 is 24.

69. Three times a number is equal to two more than twice the number.

70. Twelve divided by a number equals three times that number.

71. Twenty divided by five times a number is 2.

72. A number divided by 2 is 0.

Historical Reflections

Historical Reflections

From Tablet to Tally

The clay tablet shown here, despite damage, bears witness to the durability of Babylonian mud. Thousands of years after the tablets were made, Babylonian algebra problems can be worked out from the original writings. Here is an example of a Babylonian problem taken from an actual old text (using modern numerals):

> I have added the area and two-thirds of the side of my square and it is 7/12. What is the side of my square?

Tally sticks like the the one depicted here were used by the English in about 1400 to keep track of financial transactions. Each notch stands for one pound sterling.

Art: Courtesy of the Trustees of the British Museum.

CHAPTER 1 SUMMARY

KEY TERMS

1.1	**numerator**	The numerator of a fraction is the number above the fraction bar.
	denominator	The denominator of a fraction is the number below the fraction bar.
	factored	A number is factored if it is written as a product.
	factor	A factor is a number that divides another number without leaving a remainder.
	prime number	A prime number has only itself and 1 as factors. By agreement, 1 is not a prime number.
	exponent	An exponent is a number that indicates how many times a factor is repeated.
	base	The base is the number that is a repeated factor when written with an exponent.
	lowest terms	A fraction is written in lowest terms when the numerator and denominator have no common factor (except 1).
	reciprocals	Two numbers whose product is 1 are reciprocals.
	least common denominator	The least common denominator is the smallest number that all denominators in a problem divide into without remainder.
1.2	**decimal fraction (decimal)**	A decimal fraction is a number written with a decimal point.
	decimal places	The digits after the decimal point are called decimal places.
	divisor	A divisor is the number by which another number is divided.
	dividend	A dividend is a number that is divided by another number.
	percent	A percent indicates the number of one-hundredths in a quantity.
	percentage	A percentage is part of a total amount.
1.4	**exponential expression**	A number written with an exponent is an exponential expression.
1.5	**variable**	A variable is a symbol, usually a letter, used to represent an unknown number.
	algebraic expression	An algebraic expression is a collection of numbers, variables, symbols for operations, and symbols for grouping.
	equation	An equation is a statement that says two expressions are equal.
	solution	A solution of an equation is any replacement for the variable that makes the equation true.
	domain	The domain of an equation is the set of numbers from which the solution is chosen.

NEW SYMBOLS

a^n	n factors of a
%	percent
=	is equal to
≠	is not equal to
<	is less than
≤	is less than or equal to
>	is greater than
≥	is greater than or equal to
$a(b)$, $(a)(b)$, $a \cdot b$, or ab	a times b
$\frac{a}{b}$ or a/b	a divided by b

QUICK REVIEW

Section	Concepts	Examples
1.1 Fractions	**Operations with Fractions**	
	Addition: Add numerators; the denominator is the same.	$\frac{2}{5} + \frac{7}{5} = \frac{9}{5}$
	Subtraction: Subtract numerators; the denominator is the same.	$\frac{5}{4} - \frac{1}{4} = \frac{4}{4} = 1$
	Multiplication: Multiply numerators and multiply denominators.	$\frac{4}{3} \cdot \frac{5}{6} = \frac{20}{18} = \frac{10}{9}$
	Division: Multiply the first fraction by the reciprocal of the second fraction.	$\frac{6}{5} \div \frac{1}{4} = \frac{6}{5} \cdot \frac{4}{1} = \frac{24}{5}$
1.2 Decimals and Percents	**Operations with Decimals**	
	Addition: Line up the decimal points and add. Bring down the decimal point to the answer.	11.23 + 5.941 17.171
	Subtraction: Line up the decimal points and subtract. Bring down the decimal point to the answer.	10.500 − 3.691 6.809
	Multiplication: Ignore decimal points and multiply. Add the number of decimal places and locate the decimal in the answer that many digits from the right.	4.2 ×1.58 3 36 2 1 0 4 2 6.636

Section	Concepts	Examples
1.2 Decimals and Percents *(continued)*	*Division:* Move the decimal point in both numbers so that the divisor becomes a whole number. Divide as usual. Bring the decimal point straight up to the answer.	$2.6\overline{)157.82}$ = 60.7 156 182 182 0
	Operations with Percents To convert a percent to a decimal, move the decimal point two places to the left.	8.5% = .085 140% = 1.40
	To convert a decimal to a percent, move the decimal point two places to the right.	.165 = 16.5% 2.8 = 280%
	Percentage = Percent × Amount	What is 45% of 60? Percentage = .45 × 60 = 27
	$\text{Percent} = \dfrac{\text{Percentage}}{\text{Amount}}$	What percent **of 75** is 15? $\text{Percent} = \dfrac{15}{75} = .2 = 20\%$
1.4 Exponents and Order of Operations	**Order of Operations** **1.** Apply any exponents. **2.** Simplify within parentheses or fraction bars. **3.** Multiply or divide from left to right. **4.** Add or subtract from left to right.	$36 - 4(2^2 + 3) = 36 - 4(4 + 3)$ $= 36 - 4(7)$ $= 36 - 28$ $= 8$
1.5 Variables and Equations	Evaluate an expression with a variable by substituting a given number for the variable.	Evaluate $2x + y^2$ if $x = 3$ and $y = -4$. $2x + y^2 = 2(3) + (-4)^2$ $= 6 + 16$ $= 22$
	Values of a variable that make an equation true are solutions of the equation.	Is 2 a solution of $5x + 3 = 18$? $5(2) + 3 = 18$ $13 = 18$ False 2 is not a solution.

name date hour

CHAPTER 1 REVIEW EXERCISES

[1.1] *Write each fraction in lowest terms.**

1. $\frac{3}{6}$
2. $\frac{5}{15}$
3. $\frac{20}{36}$
4. $\frac{27}{45}$
5. $\frac{18}{54}$
6. $\frac{60}{72}$
7. $\frac{114}{133}$
8. $\frac{204}{255}$

Find the products or quotients. Write the answers in lowest terms.

9. $\frac{5}{8} \cdot \frac{1}{3}$
10. $\frac{7}{10} \cdot \frac{1}{5}$
11. $\frac{2}{9} \cdot \frac{6}{5}$
12. $\frac{3}{8} \div \frac{1}{4}$
13. $\frac{8}{5} \div \frac{32}{15}$
14. $\frac{19}{115} \cdot \frac{46}{38}$

Add or subtract. Write the answers in lowest terms.

15. $\frac{2}{3} + \frac{5}{3}$
16. $\frac{5}{7} + \frac{3}{7}$
17. $\frac{3}{8} + \frac{1}{12}$
18. $\frac{7}{10} - \frac{1}{4}$
19. $5\frac{3}{4} + 7\frac{2}{3}$
20. $273\frac{3}{4} + 198\frac{5}{12}$

Work the following word problems.

21. John painted $\frac{1}{4}$ of a room on Monday and $\frac{1}{3}$ of the room on Tuesday. What portion was then painted?

22. A contractor installs toolsheds. Each requires $1\frac{1}{4}$ cubic yards of concrete. How much concrete would be needed for 25 sheds?

23. To buy floor tiles Wei needs to know the area of a square floor that is $11\frac{3}{4}$ feet along one side. Find the area. (*Hint:* The area of a square of side s is s^2.)

24. A circular flower bed has a radius of $3\frac{2}{3}$ meters. To buy fertilizer it is necessary to find the area of the bed. What is the area? (*Hint:* The area of a circle with radius r is πr^2. Use 22/7 for π.)

[1.2] *Write the decimals as fractions in lowest terms.*

25. .4
26. .54
27. .85
28. .505
29. 2.345
30. 4.3758

*For help with any of these exercises, look in the section given in brackets.

Write the fractions as decimals. Round to the nearest thousandth, if necessary.

31. $\frac{7}{8}$

32. $\frac{1}{4}$

33. $\frac{11}{16}$

34. $\frac{1}{6}$

35. $\frac{81}{95}$

36. $\frac{152}{233}$

Perform each operation.

37. $18.9 + 5.024 + 7.13 + 256.9$

38. $8.55 + 7.36 + 9.92 + 5.47$

39. $3.4 - 1.725$

40. $(124.9)(8.02)$

41. $4.2282 \div .81$

42. $2.0488 \div .26$

Convert to percents.

43. .96

44. 1.42

45. .5

46. 5.136

Find the following. Round to the nearest hundredth, if necessary.

47. 40% of 9000

48. 124% of 176

49. What percent of 92 is 14?

Solve each word problem.

50. A fireman has $17,800 in the credit union. One year the credit union gives a dividend of 8.4%. Find the dividend received by the fireman.

51. The total cost of a house is $79,500. The salesperson selling the house earns 3% of this amount. Find the fee earned by the salesperson.

[1.3] *Which of the symbols $<$, $\leq$, $>$, and $\geq$ make the following statements true? Give all possible correct answers.*

52. 3 4

53. 9 9

54. 22 15

55. $\frac{5}{8}$ $\frac{1}{2}$

56. .87 .865

57. .94 .904

58. $\frac{2}{3}$ .7

59. $\frac{3}{4}$ .8

name date hour

Rewrite the following so that the inequality symbol points in the opposite direction.

60. $4 \le 9$ **61.** $16 \ge 3$ **62.** $9 < 10$ **63.** $23 \le 25$

[1.4] *Find the following.*

64. 4^3 **65.** 5^2 **66.** 2^5

67. $\left(\frac{5}{8}\right)^2$ **68.** $\left(4\frac{2}{3}\right)^2$ **69.** $\left(2\frac{1}{3}\right)^3$

Decide whether each statement is true or false.

70. $5 \cdot 8 + 4 \ge 45$ **71.** $9 \cdot 5 - 8 \cdot 4 < 15$ **72.** $5 \cdot (4 + 8) + 2 \cdot 3 \le 68$

73. $5[6 - 2(4 - 1)] \le 0$ **74.** $\frac{3(7 + 4)}{2(1 + 5)} > 2$ **75.** $(9 - 3)^2 - 4^2 \ge 5$

[1.5] *Find the numerical value of the given expression if $x = 2$ and $y = 5$.*

76. $3x$ **77.** $x + 3y$ **78.** $\frac{3x - y}{2x}$

79. $7x - y$ **80.** $3(5x + 2y)$ **81.** $(x + 1)^2 + (y - 2)^2$

Write each word phrase or sentence using algebraic expressions. Use x as the variable.

82. 12 times a number

83. The sum of a number and nine

84. The difference between 4 and a number

85. Six subtracted from three times a number

86. Twelve more than a number

87. A number decreased by 3

88. The product of a number and one more than the number

89. The sum of a number and 4 is divided by 8.

90. Three-tenths of a number is subtracted from the sum of the number and 28.

91. The sum of a number and 16 is subtracted from ten-thirds of the number.

MIXED REVIEW EXERCISES*

Perform the indicated operations.

92. $9 \cdot (6 + 2) - 4$

93. $\frac{8}{9} - \frac{1}{6}$

94. $\frac{56}{165} \div \frac{42}{33}$

95. At a ceremony, each scout will get $\frac{3}{8}$ yard of ribbon. How many awards can be made from 15 yards of ribbon?

96. A store is giving 20% off on all purchases. Find the amount paid for a chair with a regular price of $650.

97. 6^3

98. $\begin{array}{r} 8. \\ -\ .0546 \\ \hline \end{array}$

99. $\frac{5}{12} \div \frac{7}{6}$

100. $162\frac{1}{4} - 58\frac{5}{9}$

101. $5^2 + (15 - 11)^2$

102. $(59.4)(3.7)$

103. $\frac{2(1 + 3)}{3(2 + 1)}$

104. $\frac{2x}{y + 1}$ if $x = 2$ and $y = 5$

105. $\left(\frac{1}{2}\right)^4$

106. A savings account of $2300 earned $138 interest in one year. Find the percent of interest earned.

107. A triangle has sides of length $2\frac{1}{4}$ feet, $1\frac{1}{2}$ feet, and $3\frac{1}{6}$ feet. Find the total distance around the triangle.

108. 25% of 892

109. $\frac{3}{7} \cdot \frac{5}{6}$

110. $\frac{2x}{y + 1}$ if $x = 3$ and $y = 4$

111. 100.44 is what percent of 81?

112. $\frac{3x^2 + 5y^2}{7x^2 - y^2}$ if $x = 2$ and $y = 3$

113. $\frac{1}{10} + \frac{2}{5}$

*The order of exercises in this final group does not correspond to the order in which topics occur in the chapter. This random ordering should help you prepare for the chapter test in yet another way.

name date hour

CHAPTER 1 TEST

Write each fraction in lowest terms.

1. $\frac{15}{40}$ **1.** ________

2. $\frac{84}{132}$ **2.** ________

Perform each operation.

3. $\frac{5}{8} + \frac{9}{10}$ **3.** ________

4. $\frac{3}{8} + \frac{7}{12} + \frac{11}{15}$ **4.** ________

5. $21\frac{1}{4} - 7\frac{3}{8}$ **5.** ________

6. $\frac{3}{2} \cdot \frac{4}{9}$ **6.** ________

7. $\frac{6}{5} \div \frac{19}{15}$ **7.** ________

8. Johnson Forest Products needed to raise some money quickly. To do so, the company sold $46\frac{1}{3}$ acres out of a $104\frac{7}{9}$-acre piece of land that it owned. How many acres did the company have left? **8.** ________

9. Write .625 as a fraction in lowest terms. **9.** ________

10. Write $\frac{9}{16}$ as a decimal. **10.** ________

Perform the indicated operations.

11. $9.6 + 8.42 + 3.75$ **11.** ________

12. $123.4 - 98.7$ **12.** ________

13. $21.98 \cdot (.72)$ **13.** ________

14. $252.008 \div 21.8$ **14.** ________

15. ____________________ 15. Convert .19 to a percent.

16. ____________________ 16. Convert 76.2% to a decimal.

Find each of the following.

17. ____________________ 17. 8% of 170

18. ____________________ 18. What percent is 653.2 of 4600?

19. ____________________ 19. A clock radio with a regular price of $48 is on sale this week at 20% off. Find the amount of the discount.

Find the value of each expression.

20. ____________________ 20. $4[5(1) - 3]$

21. ____________________ 21. $\frac{9(4) + 3(2)}{5 \cdot 4 + 1}$

22. ____________________ 22. $6^2 \cdot 2 - 24 - 5^2$

Find the numerical value of the given expression when $x = 3$ and $y = 7$.

23. ____________________ 23. $5x + 2y$

24. ____________________ 24. $x^2 + 3y$

25. ____________________ 25. $\frac{7x - y}{x + 4}$

Write each word phrase as an algebraic expression. Use x as the variable.

26. ____________________ 26. Four times a number

27. ____________________ 27. The sum of twice a number and 11

28. ____________________ 28. The quotient of 9 and the difference between a number and 8

Decide whether the given number is a solution for the equation.

29. ____________________ 29. $6m + m + 2 = 37$; 5

30. ____________________ 30. $8(y - 3) + 2y = 18$; 4

2

OPERATIONS WITH REAL NUMBERS

2.1 REAL NUMBERS AND THE NUMBER LINE

OBJECTIVES

1. Set up number lines.
2. Identify integers, rational numbers, irrational numbers, and real numbers.
3. Tell which of two real numbers is smaller.
4. Find additive inverses.
5. Find absolute values of real numbers.

1 Graphs can be a helpful way to picture sets of numbers. Numbers are graphed on **number lines** like the one shown in Figure 1.

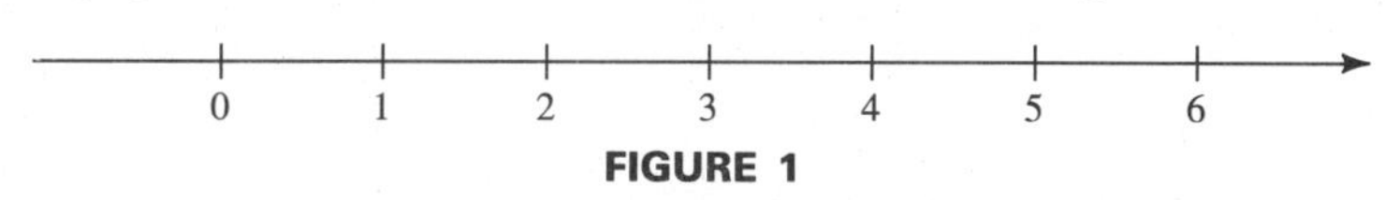

FIGURE 1

Draw a number line by locating any point on the line and calling it 0. Choose any point to the right of 0 and call it 1. The distance between 0 and 1 gives a unit of measure used to locate other points, as shown in Figure 1. The points labeled in Figure 1 correspond to the first few **whole numbers.**

Whole numbers $\{0, 1, 2, 3, 4, 5, \cdots\}$

The three dots show that the list of numbers continues in the same way indefinitely. Think of the graph as a picture of the set of numbers.

All the whole numbers, starting with 1, are located to the right of 0 on the number line. But numbers also may be placed to the left of 0. These numbers, written -1, -2, -3, and so on, are shown in Figure 2. (The minus sign is used to show that the numbers are located to the *left* of 0.)

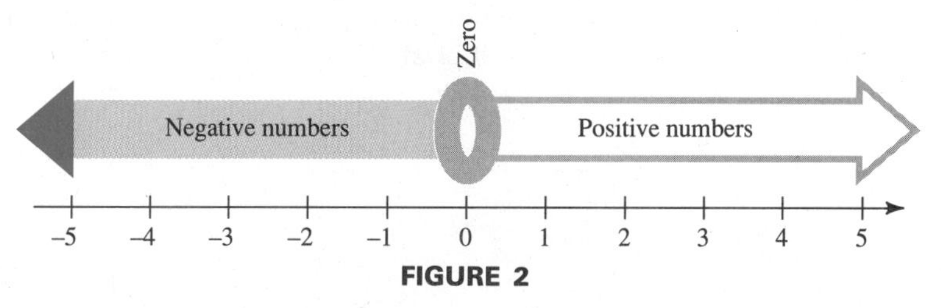

FIGURE 2

The numbers to the *left* of 0 are **negative numbers.** The numbers to the *right* of 0 are **positive numbers.** The number 0 itself is neither positive nor negative. Positive numbers and negative numbers are called **signed numbers.**

There are many practical applications of negative numbers. For example, a temperature on a cold January day might be $-10°$, or 10 degrees below zero. A business that spends more than it takes in has a negative "profit."

1. Identify each number as positive, negative, or neither.

(a) 15

(b) -15

(c) -6

(d) $\frac{3}{4}$

(e) $-\frac{5}{8}$

(f) 0

WORK PROBLEM 1 AT THE SIDE.

2 The set of numbers indicated by the number line in Figure 2 (including positive numbers, negative numbers, and zero) is called the set of **integers.**

Integers $\{\cdots, -3, -2, -1, 0, 1, 2, 3, \cdots\}$

Not all numbers are integers. For example, $\frac{1}{2}$ is not; it is a number halfway between the integers 0 and 1. Also, $3\frac{1}{4}$ is not an integer.

To **graph** a number, we place a dot on the number line at the point that corresponds to the number. Several numbers that are not integers are graphed in Figure 3.

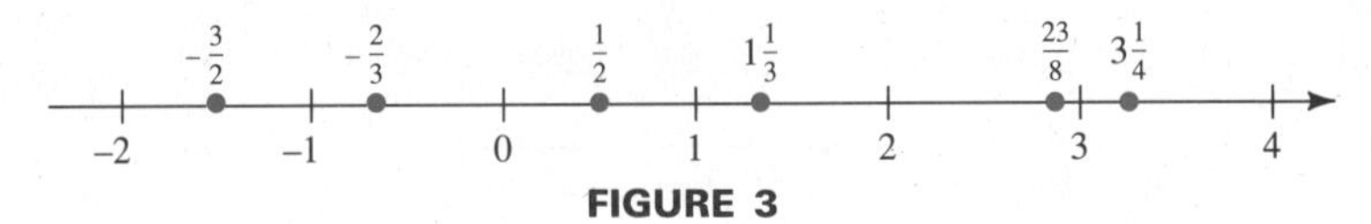

FIGURE 3

All the numbers in Figure 3 can be written as quotients of integers. These numbers are examples of **rational numbers.**

Rational numbers {numbers that can be written as quotients of two integers, with denominator not 0}

Since any integer can be written as the quotient of itself and 1, all integers are also rational numbers. A decimal number that comes to an end (terminates), such as .23, is also a rational number: $.23 = 23/100$. Decimal numbers that repeat in a fixed block of digits, such as .3333. . . and .454545. . ., are also rational numbers. For example, $.3333... = \frac{1}{3}$.

Although a great many numbers are rational, there are also numbers that are not. For example, a floor tile 1 foot on a side has a diagonal whose length is the square root of 2 (written $\sqrt{2}$). See Figure 4. It can be shown that $\sqrt{2}$ cannot be written as a quotient of integers, so it is an example of a number that is not rational. It is **irrational.** The decimal form of an irrational number never terminates and never repeats.

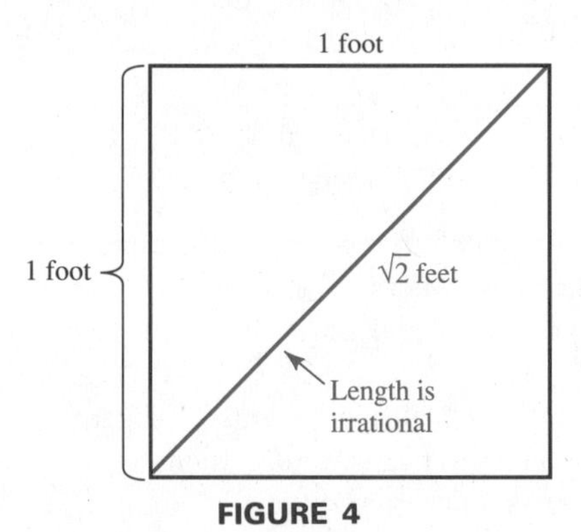

FIGURE 4

ANSWERS

1. (a) positive (b) negative (c) negative (d) positive (e) negative (f) neither

Irrational numbers {non-rational numbers represented by points on the number line}

Examples of irrational numbers include $\sqrt{3}$, $\sqrt{7}$, .10110111011110 . . ., $-\sqrt{10}$, and π, which is the ratio of the distance around a circle to the distance across it.

Finally, *all* numbers that can be represented by points on the number line are called **real numbers.**

Real numbers {all numbers that can be represented by points on the number line}

All the numbers mentioned above are real numbers. The relationships between the various types of numbers are shown in two different ways in Figure 5.

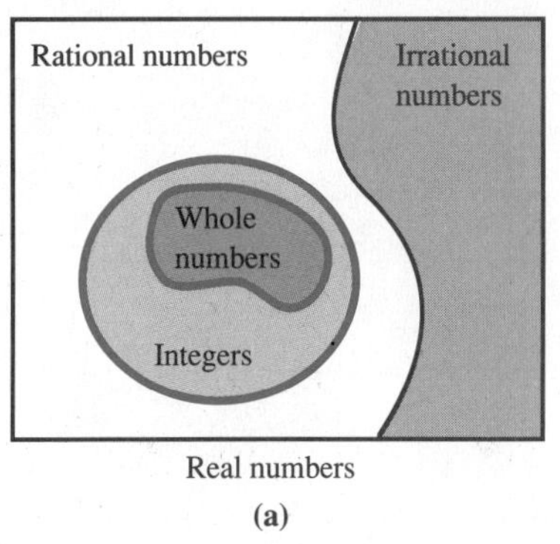

Real numbers

(a)

Irrational numbers

Real numbers

Rational numbers

Integers

Positive integers

Zero

Negative integers

Non-integer rational numbers

(b)

FIGURE 5

EXAMPLE 1 Determining Whether a Number Belongs to a Set

List the numbers in the set

$$\left\{-5, -\frac{2}{3}, 0, \sqrt{13}, 5\frac{1}{4}, 5, 5.8\right\}$$

that belong to each of the following sets of numbers.

(a) Integers

The integers in the set above are -5, 0, and 5.

(b) Rational numbers

The rational numbers are -5, $-\frac{2}{3}$, 0, $5\frac{1}{4}$, 5, and 5.8, since each of these numbers *can* be written as the quotient of two integers. For example, $5\frac{1}{4} = \frac{21}{4}$.

(c) Real numbers

All the numbers in the set are real numbers. ■

WORK PROBLEM 2 AT THE SIDE.

2. (a) Identify each real number in the set below as rational or irrational.

$$\left\{\frac{5}{8}, -7, -1\frac{3}{5}, 0, \sqrt{11}, -\pi\right\}$$

(b) Which of the numbers listed above are *integers?*

ANSWERS

2. **(a)** $\frac{5}{8}$, $-7\left(\text{or } \frac{-7}{1}\right)$, $-1\frac{3}{5}\left(\text{or } \frac{-8}{5}\right)$, and $0\left(\text{or } \frac{0}{1}\right)$ are rational; $\sqrt{11}$ and $-\pi$ are irrational. **(b)** The only integers are -7 and 0.

3. Tell whether each statement is true or false.

(a) $-2 < 4$

(b) $6 > -3$

(c) $-9 < -12$

(d) $-4 \geq -1$

(e) $-6 \leq 0$

ANSWERS
3. (a) true (b) true (c) false (d) false (e) true

3 Given any two whole numbers, we can tell which number is smaller. But what about two negative numbers, as in the set of integers? Moving from zero to the right along a number line, the positive numbers corresponding to the points on the number line *increase*. For example, $8 < 12$, and 8 is to the left of 12 on a number line. This ordering is extended to all real numbers by definition.

THE ORDERING OF REAL NUMBERS

The smaller of any two different real numbers is the one corresponding to the point that is to the left of the other on the number line.

This means that any negative number is smaller than 0, and any negative number is smaller than any positive number. Also, 0 is smaller than any positive number.

EXAMPLE 2 Determining the Order of Real Numbers

Is it true that $-3 < -1$?

To find out, locate -3 and -1 on a number line, as shown in Figure 6. Since -3 is to the left of -1 on the number line, -3 is smaller than -1. The statement $-3 < -1$ is true. ■

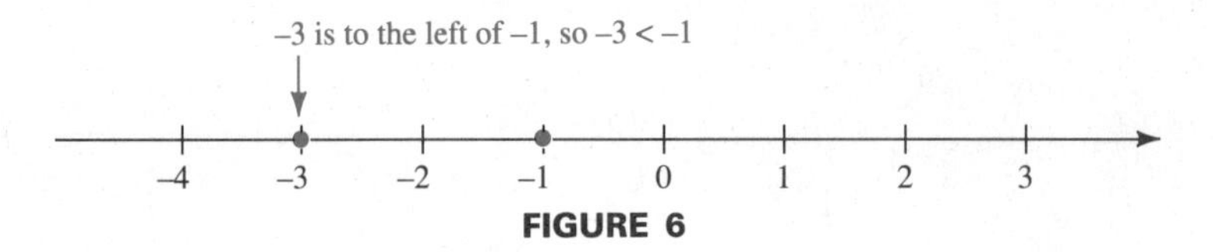

FIGURE 6

WORK PROBLEM 3 AT THE SIDE.

4 By a property of the real numbers, for any real number x (except 0), there is exactly one number on the number line the same distance from 0 as x but on the opposite side of 0.

For example, Figure 7 shows that the numbers 3 and -3 are each the same distance from 0 but are on opposite sides of 0. The numbers 3 and -3 are called **additive inverses, negatives,** or **opposites** of each other. These terms can be used interchangeably.

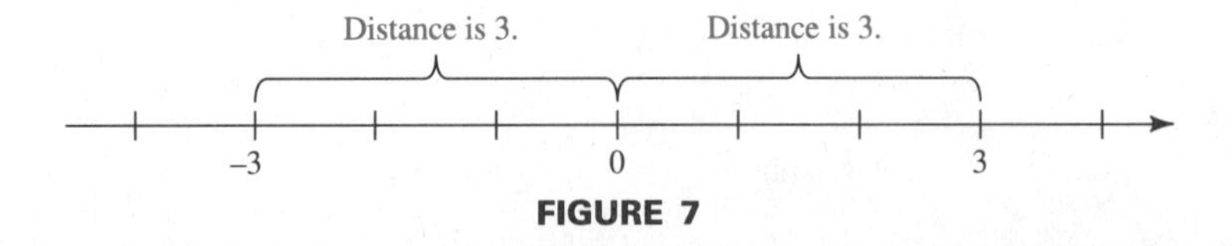

FIGURE 7

ADDITIVE INVERSE

The **additive inverse** of a number *a* is the number that is the same distance from 0 on the number line as *a,* but on the opposite side of 0.

The additive inverse of the number 0 is 0 itself. This makes 0 the only real number that is its own additive inverse. Other additive inverses occur in pairs. For example, 4 and -4, and 5 and -5, are additive inverses of each other. Several pairs of additive inverses are shown in Figure 8.

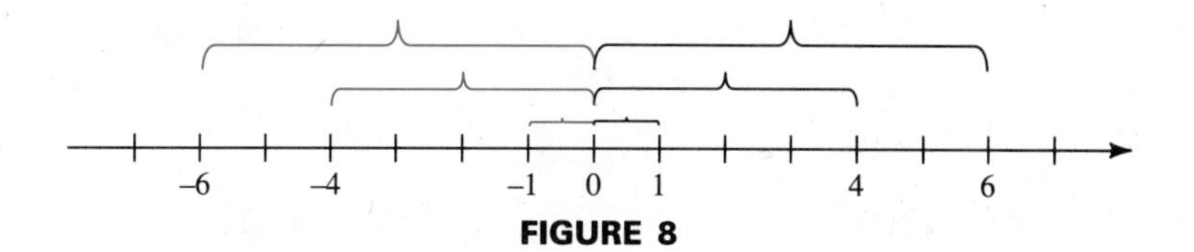

FIGURE 8

The additive inverse of a number can be indicated by writing the symbol $-$ in front of the number. With this symbol, the additive inverse of 7 is written -7 (read "negative 7"). The additive inverse of -4 can be written $-(-4)$. It was suggested in Figure 8 that 4 is an additive inverse of -4. Since a number can have only one additive inverse, the symbols 4 and $-(-4)$ must represent the same number, which means that

$$-(-4) = 4.$$

A generalization of this idea is given below.

DOUBLE NEGATIVE RULE

For any real number a,

$$-(-a) = a.$$

EXAMPLE 3 Finding the Additive Inverse of a Number

The following chart shows several numbers and their additive inverses.

Number	*Additive Inverse*
-3	3
-4	$-(-4)$, or 4
0	0
5	-5
19	-19

■

Example 3 suggests the following rule.

The additive inverse of a number is found by changing the sign of the number.

WORK PROBLEM 4 AT THE SIDE.

4. Find the additive inverse of each number.

(a) 6

(b) 15

(c) -9

(d) -12

(e) 0

ANSWERS
4. (a) -6 (b) -15 (c) 9 (d) 12 (e) 0

5. Simplify by removing absolute value symbols.

(a) $|-6|$

(b) $|9|$

(c) $|-36|$

(d) $-|15|$

(e) $-|-9|$

(f) $-|32 - 2|$

ANSWERS
5. (a) 6 (b) 9 (c) 36 (d) -15 (e) -9 (f) -30

5 As mentioned above, additive inverses are numbers the same distance from 0 on the number line but on opposite sides of 0. This same idea can be expressed by saying that two numbers that are additive inverses have the same absolute value. The **absolute value** of a number is defined as the distance between 0 and the number on the number line. The symbol for the absolute value of the number a is $|a|$, read "the absolute value of a." For example, the distance between 2 and 0 on the number line is 2 units, so

$$|2| = 2.$$

Also, the distance between -2 and 0 on the number line is 2, so

$$|-2| = 2.$$

Since distance is a physical measurement, which is never negative, we can make the following statement.

The absolute value of a number can never be negative.

For example,

$$|12| = 12 \qquad \text{and} \qquad |-12| = 12$$

since both 12 and -12 lie at a distance of 12 units from 0 on the number line. Since 0 is a distance 0 units from 0, we have

$$|0| = 0.$$

EXAMPLE 4 Evaluating Absolute Value

Simplify by removing absolute value symbols.

(a) $|5| = 5$ **(b)** $|-5| = 5$

In parts (c) and (d), find the absolute value first, and then find the additive inverse of the result.

(c) $-|5| = -(5) = -5$ **(d)** $-|-13| = -(13) = -13$

(e) $|8 - 5|$

Simplify within the absolute value bars first.

$$|8 - 5| = |3| = 3 \quad ■$$

Part (e) in Example 4 suggests that absolute value bars are also grouping symbols.

WORK PROBLEM 5 AT THE SIDE.

name date hour

2.1 EXERCISES

Find the additive inverse of each number. For the exercises with absolute value, simplify before deciding on the additive inverse. See Examples 3 and 4.

1. 12 **2.** 9 **3.** -4 **4.** -10

5. $|6|$ **6.** $|0|$ **7.** $-|-18|$ **8.** $-|-3|$

Select the smaller number in each pair. See Example 2.

9. $-10, -3$ **10.** $-8, -15$ **11.** $-7, -2$ **12.** $-13, -14$

13. $5, -6$ **14.** $6, -3$ **15.** $-6, -7$ **16.** $-10, -11$

Write true *or* false *for each of the following statements. See Example 2.*

17. $7 < -4$ **18.** $-5 < -2$ **19.** $-8 < -7$ **20.** $-4 \geq -9$

21. $-7 \geq -10$ **22.** $-12 \leq -18$ **23.** $-9 \leq -(-15)$ **24.** $-7 \leq -(-5)$

For Exercises 25 and 26, see Example 1.

25. List all numbers from the set

$\{-9, -\sqrt{7}, -1\frac{1}{4}, -\frac{3}{5}, 0, \sqrt{5}, 3, 5.9, 7\}$

that belong to each of the following sets of numbers.

(a) Whole numbers

(b) Integers

(c) Rational numbers

(d) Irrational numbers

(e) Real numbers

26. List all numbers from the set

$\{-5.3, -5, -\sqrt{3}, -1, -\frac{1}{9}, 0, 1.2, 1.8, 3, \sqrt{11}\}$

that belong to each of the following sets of numbers.

(a) Whole numbers

(b) Integers

(c) Rational numbers

(d) Irrational numbers

(e) Real numbers

Write true *or* false *for each of the following statements.*

27. All whole numbers are integers.

28. All integers are rational numbers.

29. There is no number that is both rational and irrational.

30. Zero is neither positive nor negative.

Graph each group of numbers on the indicated number line. Simplify the expressions having absolute value symbols before graphing them.

31. $0, 3, -5, -6$

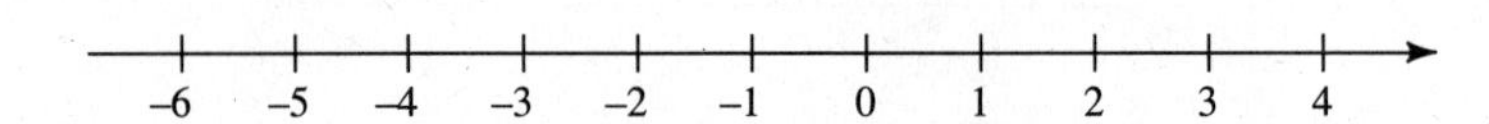

32. $2, 6, -2, -1$

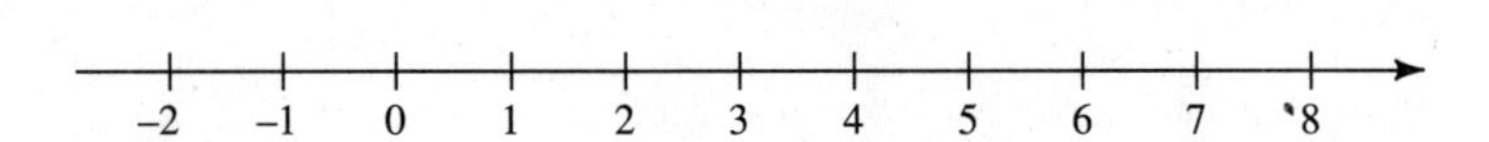

33. $-2, -6, |-4|, 3, -|4|$

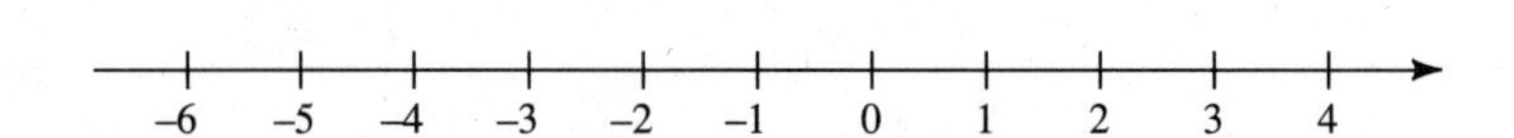

34. $-5, -3, -|-2|, 0, |-4|$

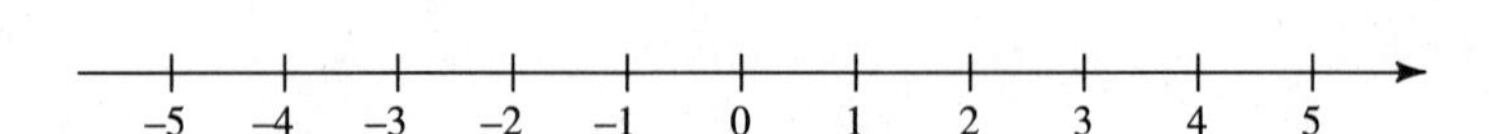

35. $|2|, -|4|, -|-3|, -|-5|$

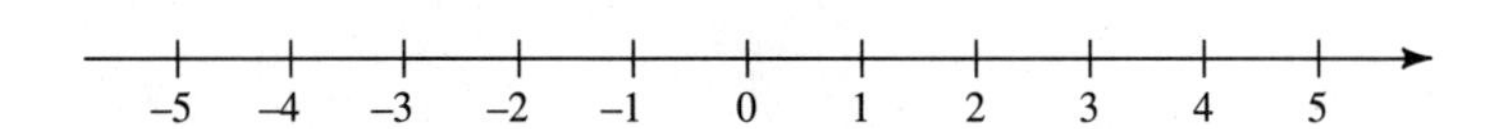

36. $5, -3, -|-4|, -|-1|$

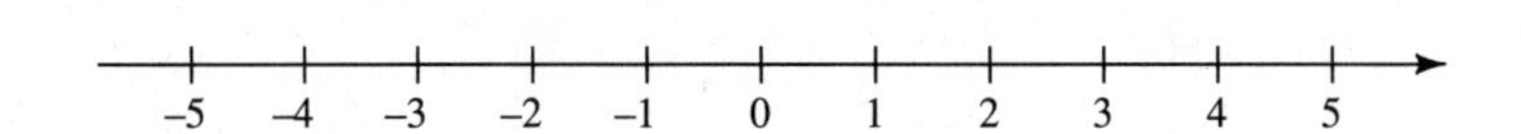

37. $\frac{1}{4}, 2\frac{1}{2}, -3\frac{4}{5}, -4, -1\frac{5}{8}$

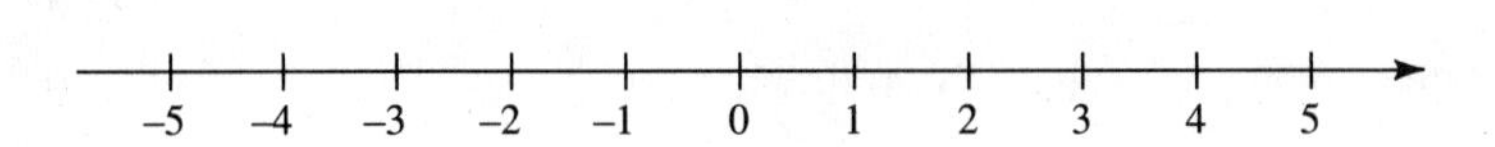

38. $5\frac{1}{4}, 4\frac{5}{9}, -2\frac{1}{3}, 0, -3\frac{2}{5}$

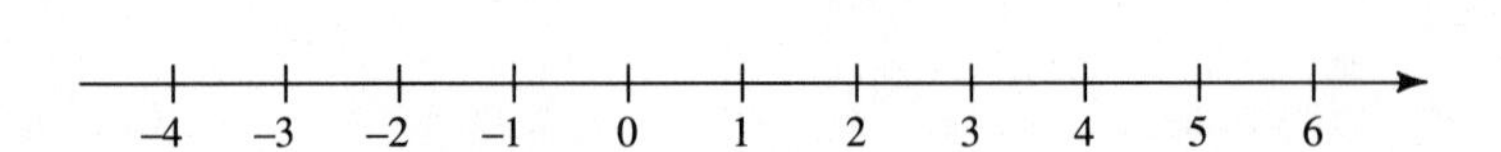

SKILL SHARPENERS

Most of the exercise sets in the rest of the book end with brief sets of "Skill Sharpeners." These exercises are designed to help you review ideas needed for the next few sections in the chapter. If you need help with these, look in the indicated sections.

Add or subtract. Write answers in lowest terms. See Section 1.1.

39. $\frac{2}{3} + \frac{5}{7}$

40. $\frac{4}{9} + \frac{1}{5}$

41. $\frac{3}{4} - \frac{2}{7}$

42. $\frac{8}{9} - \frac{1}{4}$

43. $\frac{2}{3} + \frac{1}{5} - \frac{1}{2}$

44. $\frac{7}{10} + \frac{4}{5} - \frac{5}{9}$

2.2 ADDITION OF REAL NUMBERS

1 The number line can be used to illustrate the addition of real numbers, as in the following examples.

EXAMPLE 1 Adding with the Number Line
Use the number line to find the sum $2 + 3$.

Add the positive numbers 2 and 3 by starting at 0 and drawing an arrow two units to the *right,* as shown in Figure 9. This arrow represents the number 2 in the sum $2 + 3$. Next, from the right end of this arrow draw another arrow three units to the right. The number below the end of this second arrow is 5, so $2 + 3 = 5$. ■

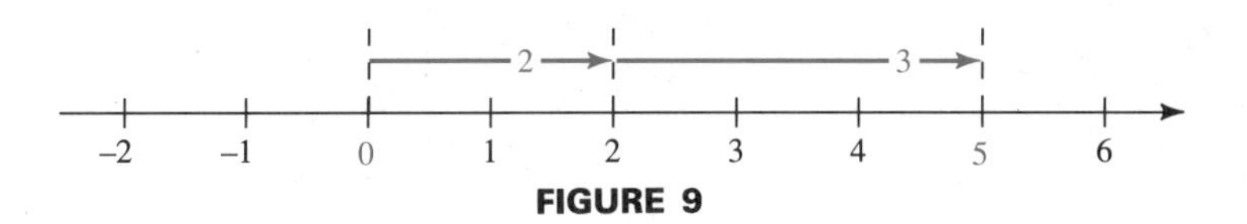

FIGURE 9

EXAMPLE 2 Adding with the Number Line
Use the number line to find the sum $-2 + (-4)$. (Parentheses are placed around the -4 to avoid the confusing use of + and − next to each other.)

Add the negative numbers -2 and -4 on the number line by starting at 0 and drawing an arrow two units to the *left,* as shown in Figure 10. Draw the arrow to the left to represent the addition of a *negative* number. From the left end of this first arrow, draw a second arrow four units to the left. The number below the end of this second arrow is -6, so $-2 + (-4) = -6$.

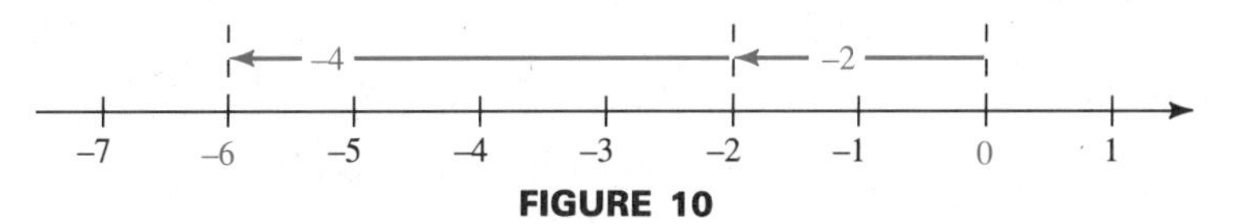

FIGURE 10

WORK PROBLEM 1 AT THE SIDE.

In Example 2, the sum of the two negative numbers -2 and -4 is a negative number whose distance from 0 is the sum of the distance of -2 from 0 and the distance of -4 from 0. That is, *the sum of two negative numbers is the negative of the sum of their absolute values.*

$$-2 + (-4) = -(|-2| + |-4|) = -(2 + 4) = -6$$

Add two numbers having the same signs by adding the absolute values of the numbers. Give the result the same sign as the numbers being added.

EXAMPLE 3 Adding Two Negative Numbers
Find the sums.

(a) $-2 + (-9) = -11$ The sum of two negative numbers is negative.

OBJECTIVES

1. Add two numbers with the same sign on a number line.
2. Add positive and negative numbers.
3. Add mentally.
4. Use the order of operations with real numbers.
5. Interpret gains and losses as positive and negative numbers.
6. Interpret words and phrases that indicate addition.

1. Use the number lines to find the sums.

(a) $1 + 4$

(b) $-2 + (-5)$

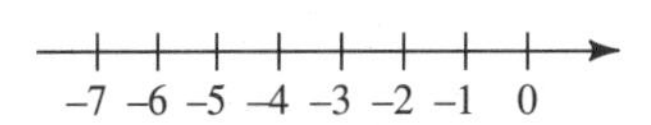

(c) $-3 + (-1)$

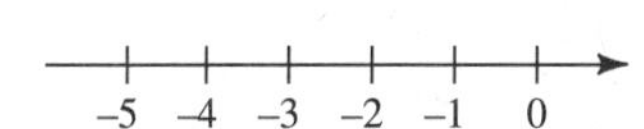

ANSWERS
1. (a) $1 + 4 = 5$

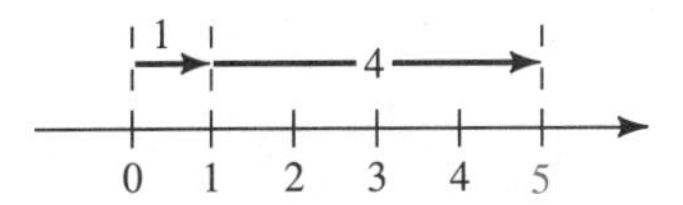

(b) $-2 + (-5) = -7$

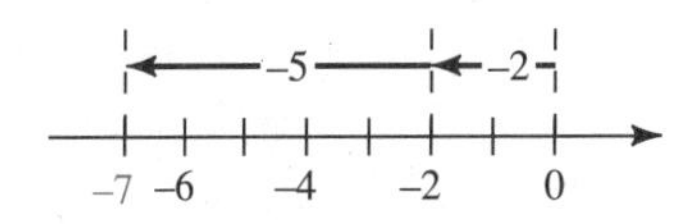

(c) $-3 + (-1) = -4$

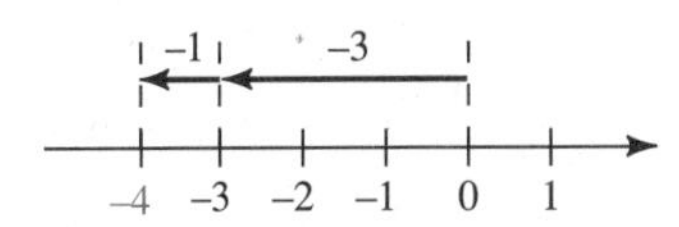

2. Find the sums.

(a) $-7 + (-3)$

(b) $-12 + (-18)$

(c) $-15 + (-4)$

3. Use the number lines to find the sums.

(a) $6 + (-3)$

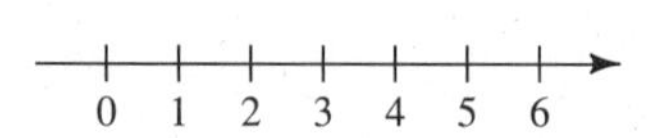

(b) $-5 + 1$

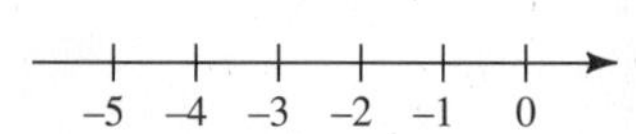

(c) $-7 + 2$

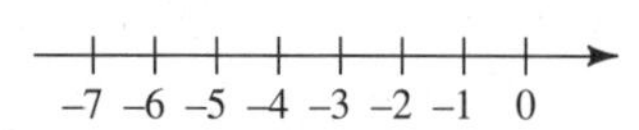

ANSWERS
2. (a) -10 (b) -30 (c) -19
3. (a) $6 + (-3) = 3$

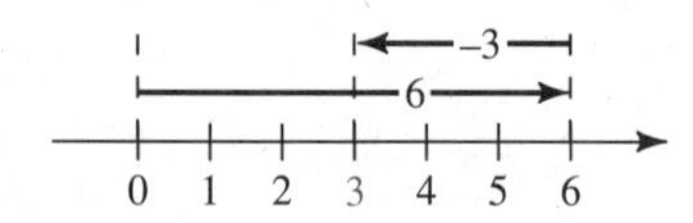

(b) $-5 + 1 = -4$

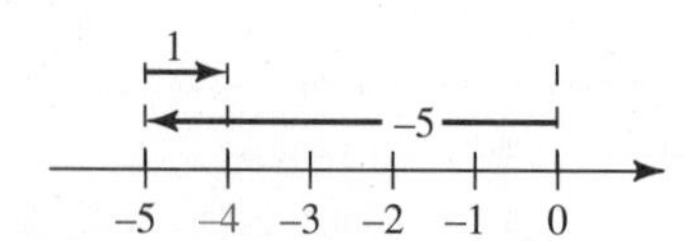

(c) $-7 + 2 = -5$

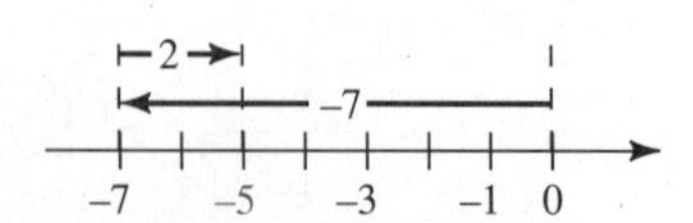

(b) $-8 + (-12) = -20$

(c) $-15 + (-3) = -18$ ■

WORK PROBLEM 2 AT THE SIDE.

2 We can use the number line again to illustrate the sum of a positive number and a negative number.

EXAMPLE 4 Adding with the Number Line

Use the number line to find the sum $-2 + 5$.

Find the sum $-2 + 5$ on the number line by starting at 0 and drawing an arrow two units to the left. From the left end of this arrow, draw a second arrow five units to the right, as shown in Figure 11. The number below the end of this second arrow is 3, so $-2 + 5 = 3$. ■

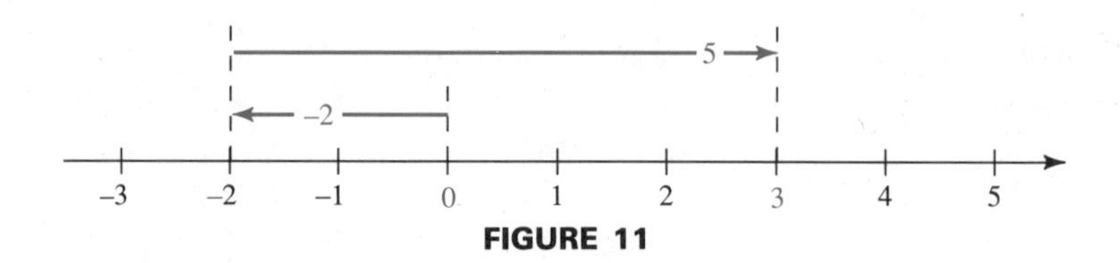

FIGURE 11

WORK PROBLEM 3 AT THE SIDE.

Addition of numbers with different signs also can be defined using absolute value.

Add numbers with different signs by first finding the difference between the absolute values of the numbers. Give the answer the same sign as the number with the larger absolute value.

For example, to add -12 and 5, find their absolute values: $|-12| = 12$ and $|5| = 5$; then find the difference between these absolute values: $12 - 5 = 7$. Since $|-12| > |5|$, the sum will be negative, so the final answer is $-12 + 5 = -7$.

3 While a number line is useful in showing the rules for addition, it is important to be able to find sums mentally.

EXAMPLE 5 Adding a Positive Number and a Negative Number

Check each answer, trying to work the addition mentally. If you get stuck, use a number line.

(a) $7 + (-4) = 3$

(b) $-8 + 12 = 4$

(c) $-\frac{1}{2} + \frac{1}{8} = -\frac{4}{8} + \frac{1}{8} = -\frac{3}{8}$

Remember to find a common denominator first.

(d) $\frac{5}{6} + \left(-\frac{4}{3}\right) = -\frac{1}{2}$

(e) $-4.6 + 8.1 = 3.5$ ■

WORK PROBLEM 4 AT THE SIDE.

The rules for adding signed numbers are summarized below.

ADDING SIGNED NUMBERS

Like signs Add the absolute values of the numbers. Give the sum the same sign as the numbers being added.

Unlike signs Find the difference between the larger absolute value and the smaller. Give the answer the sign of the number having the larger absolute value.

4 Sometimes a problem involves square brackets, [ ]. As mentioned earlier, brackets are treated just like parentheses. Do the calculations inside the brackets until a single number is obtained. Remember to use the order of operations given in Section 1.4 for adding more than two numbers.

EXAMPLE 6 Adding with Brackets

Find the sums.

(a) $-3 + [4 + (-8)]$

First work inside the brackets. Follow the rules for the order of operations given in Section 1.4.

$$-3 + [4 + (-8)] = -3 + (-4) = -7$$

(b) $8 + [(-2 + 6) + (-3)] = 8 + [4 + (-3)] = 8 + 1 = 9$ ■

WORK PROBLEM 5 AT THE SIDE.

5 Gains (or increases) and losses (or decreases) sometimes appear in stated problems. When they do, the gains may be interpreted as positive numbers and the losses as negative numbers.

EXAMPLE 7 Interpreting Gains and Losses

A football team gained 3 yards on the first play from scrimmage, lost 12 yards on the second play, and then gained 13 yards on the third play. How many yards did the team gain or lose altogether?

The gains are represented by positive numbers and the loss by a negative number.

$$3 + (-12) + 13$$

Add from left to right.

$$3 + (-12) + 13 = [3 + (-12)] + 13 = (-9) + 13 = 4$$

The team gained 4 yards altogether. ■

4. Check each answer, trying to work the addition in your head. If you get stuck, use a number line.

(a) $-8 + 2 = -6$

(b) $-15 + 4 = -11$

(c) $17 + (-10) = 7$

(d) $\frac{3}{4} + \left(-\frac{11}{8}\right) = -\frac{5}{8}$

(e) $-9.5 + 3.8 = -5.7$

5. Find the sums.

(a) $2 + [7 + (-3)]$

(b) $6 + [(-2 + 5) + 7]$

(c) $-9 + [-4 + (-8 + 6)]$

ANSWERS

4. All are correct.

5. (a) 6 (b) 16 (c) −15

6. A football team lost 8 yards on the first play from scrimmage, lost 5 yards on the second play, and then gained 7 yards on the third play. How many yards did the team gain or lose altogether?

7. Write a numerical expression for each phrase, and simplify the expression.

(a) 4 more than -12

(b) The sum of 6 and -7

(c) -12 added to -31

(d) 7 increased by the sum of 8 and -3

ANSWERS
6. The team lost 6 yards.
7. (a) $-12 + 4 = -8$
(b) $6 + (-7) = -1$
(c) $-31 + (-12) = -43$
(d) $7 + [8 + (-3)] = 12$

WORK PROBLEM 6 AT THE SIDE.

6 The word *sum* indicates addition. There are other key words and phrases that also indicate addition. Some of these are given in the chart below.

Word or Phrase	*Example*	*Numerical Expression and Simplification*
Sum	The *sum* of -3 and 4	$-3 + 4 = 1$
Added to	5 *added to* -8	$-8 + 5 = -3$
More than	12 *more than* -5	$-5 + 12 = 7$
Increased by	-6 *increased by* 13	$-6 + 13 = 7$

EXAMPLE 8 Interpreting Words and Phrases

Write a numerical expression for each phrase, and simplify the expression.

(a) The **sum of** -8 and 4 and 6

$$-8 + 4 + 6 = [-8 + 4] + 6 = -4 + 6 = 2$$

Notice that brackets were placed around $-8 + 4$ and this addition was done first, using the order of operations given earlier. The same result would be obtained if the brackets were placed around $4 + 6$. (This idea will be discussed further in Section 2.6.)

$$-8 + 4 + 6 = -8 + [4 + 6] = -8 + 10 = 2$$

(b) 3 **more than** -5, **increased by** 12

$$-5 + 3 + 12 = [-5 + 3] + 12 = -2 + 12 = 10$$ ■

WORK PROBLEM 7 AT THE SIDE.

name date hour

2.2 EXERCISES

Find the sums. See Examples 1-6.

1. $5 + (-3)$

2. $11 + (-8)$

3. $6 + (-8)$

4. $3 + (-7)$

5. $-6 + (-2)$

6. $-8 + (-3)$

7. $-9 + (-2)$

8. $-15 + (-6)$

9. $-3 + (-9)$

10. $-11.3 + (-5.8)$

11. $12.6 + (-8.42)$

12. $-10 + 2.531$

13. $4 + [13 + (-5)]$

14. $6 + [2 + (-13)]$

15. $8 + [-2 + (-1)]$

16. $12 + [-3 + (-4)]$

17. $-2 + [5 + (-1)]$

18. $-8 + [9 + (-2)]$

19. $-6 + [6 + (-9)]$

20. $-3 + [4 + (-8)]$

21. $[9 + (-2)] + 6$

22. $[8 + (-14)] + 10$

23. $[(-9) + (-14)] + 12$

24. $[(-8) + (-6)] + 10$

25. $-\frac{1}{6} + \frac{2}{3}$

26. $\frac{9}{10} + \left(-\frac{3}{5}\right)$

27. $\frac{5}{8} + \left(-\frac{17}{12}\right)$

28. $-\frac{6}{25} + \frac{19}{20}$

29. $2\frac{1}{2} + \left(-3\frac{1}{4}\right)$

30. $-4\frac{3}{8} + 6\frac{1}{2}$

31. $7.9 + (-8.4)$

32. $11.6 + (-15.4)$

33. $-6.1 + [3.2 + (-4.8)]$

34. $-9.4 + [-5.8 + (-1.4)]$

35. $[-3 + (-4)] + [5 + (-6)]$

36. $[-8 + (-3)] + [-7 + (-6)]$

37. $[-4 + (-3)] + [8 + (-1)]$

38. $[-5 + (-9)] + [16 + (-21)]$

39. $[-4 + (-6)] + [(-3) + (-8)] + [12 + (-11)]$

40. $[-2 + (-11)] + [12 + (-2)] + [18 + (-6)]$

Write true *or* false *for each statement.*

41. $-4 + 0 = -4$

42. $-6 + 5 = -1$

43. $-8 + 12 = 8 + (-12)$

44. $15 + (-8) = 8 + (-15)$

45. $-9 + 5 + 6 = -2$

46. $-6 + (8 - 5) = -3$

47. $-\frac{3}{2} + \frac{5}{8} = \frac{5}{8} + \left(-\frac{3}{2}\right)$

48. $\frac{11}{5} + \left(-\frac{6}{11}\right) = -\frac{6}{11} + \frac{11}{5}$

49. $\left|-\frac{8}{13} + \frac{3}{4}\right| = \frac{8}{13} + \frac{3}{4}$

50. $|-4 + 2| = 4 + 2$

51. $|12 - 3| = 12 - 3$

52. $|-6 + 10| = 6 + 10$

53. $[4 + (-6)] + 6 = 4 + (-6 + 6)$

54. $[(-2) + (-3)] + (-6) = 12 + (-1)$

55. $-7 + [-5 + (-3)] = [(-7) + (-5)] + 3$

56. $6 + [-2 + (-5)] = [(-4) + (-2)] + 5$

name date hour

57. $-5 + (-|-5|) = -10$

58. $|-3| + (-5) = -2$

Write a numerical expression for each phrase, and simplify the expression. See Example 8.

59. The sum of -5 and 12 and 4

60. The sum of -3 and 5 and -7

61. 14 added to the sum of -19 and -3

62. -2 added to the sum of -18 and 10

63. The sum of -4 and -10, increased by -3

64. The sum of -7 and -13, increased by -5

65. 4 more than the sum of 8 and -23

66. 12 more than the sum of -4 and -8

Solve the following word problems. See Example 7.

67. Debbie has $18. She then spends $12. How much does she have left?

68. An airplane is flying at an altitude of 8000 feet. It then descends 3000 feet. What is its new altitude?

69. Tri Huu Pham has $14 in his checking account. He then writes a check for $18. What is his balance? (Write the answer with a negative number.)

70. Malcolm Mwendo is standing 18 feet below ground level in a small canyon. He then descends another 25 feet. Find his final altitude.

71. The temperature at noon on a July day in New Orleans was 92°. After a thunderstorm, it dropped 5°. What was the new temperature?

72. One number of Albert's blood pressure was 118, but then it changed by -6. Find the new number of his pressure.

73. On three successive running plays, Bo Jackson gained 12 yards, lost 3 yards, and gained 4 yards. What was his gain or loss?

74. Jennifer owes $129 to a credit card company. She makes a $14 purchase with the card, and then pays $60 on the account. What amount does she still owe?

75. Kim owes $870.00 on her Master Card account. She returns two items costing $35.90 and $150.00 and receives credits for these on the account. Next, she makes a purchase of $82.50, and then two more purchases of $10.00 each. She finally makes a payment of $500.00. How much does she still owe?

76. A welder working with stainless steel must use precise measurements. Suppose a welder attaches two pieces of steel that are each 3.60 inches in length, and then attaches an additional three pieces that are each 9.10 inches long. She finally cuts off a piece that is 7.60 inches long. Find the length of the welded piece of steel.

Find the solution of each equation from the domain $\{-3, -2, -1, 0, 1, 2, 3\}$ *by guessing or using trial and error.*

77. $x + 2 = 0$

78. $x + 3 = 0$

79. $k + 1 = -2$

80. $k + 2 = -1$

81. $14 + y = 12$

82. $y + 8 = 7$

83. $r + (-4) = -6$

84. $t + (-2) = -5$

85. $-8 + w = -6$

SKILL SHARPENERS

Simplify. See Section 1.4.

86. $10[(8 + 1) - 4]$

87. $6[4 + (3 - 2)]$

88. $5 + 2(3 - 1)$

89. $12 + 3(8 - 6)$

90. $(8 - 4) + (12 - 5)$

91. $(3 - 2) + (4 - 3)$

2.3 SUBTRACTION OF REAL NUMBERS

OBJECTIVES

1. Find a difference on the number line.
2. Use the definition of subtraction.
3. Work subtraction problems that involve brackets.
4. Interpret words and phrases that indicate subtraction.

1 As we mentioned earlier, the answer to a subtraction problem is called a **difference.** Differences between signed numbers can be found by using a number line. Since *addition* of a positive number on the number line is shown by drawing an arrow to the *right, subtraction* of a positive number is shown by drawing an arrow to the *left*.

EXAMPLE 1 Subtracting with the Number Line

Use the number line to find the difference $7 - 4$.

To find the difference $7 - 4$ on the number line, begin at 0 and draw an arrow 7 units to the right. From the right end of this arrow, draw an arrow 4 units to the left, as shown in Figure 12. The number at the end of the second arrow shows that $7 - 4 = 3$. ■

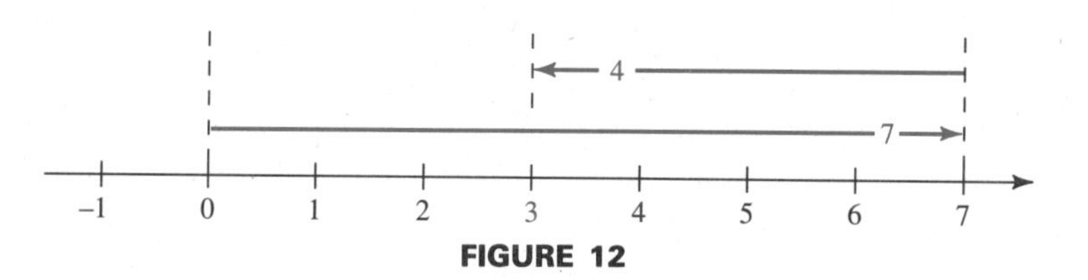

FIGURE 12

WORK PROBLEM 1 AT THE SIDE.

1. Use the number line to find the differences.

(a) $5 - 1$

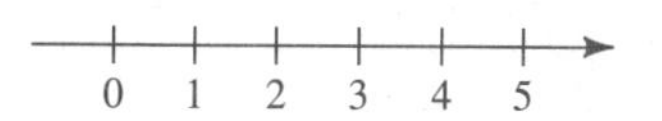

(b) $6 - 2$

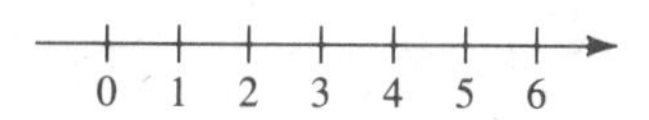

2 The procedure used in Example 1 to find $7 - 4$ is exactly the same procedure that would be used to find $7 + (-4)$, so that

$$7 - 4 = 7 + (-4).$$

This shows that *subtraction* of a positive number from a larger positive number is the same as *adding* the additive inverse of the smaller number to the larger. This result is extended as the definition of subtraction for all real numbers.

DEFINITION OF SUBTRACTION

For any real numbers a and b,

$$a - b = a + (-b).$$

That is, to **subtract** b from a, *add the additive inverse* (or *negative*) of b to a. This definition leads to the following procedure for subtracting signed numbers.

SUBTRACTING SIGNED NUMBERS

Step 1 Change the subtraction symbol to addition.

Step 2 Change the sign of the number being subtracted.

Step 3 Add, as in the previous section.

ANSWERS

1. (a) $5 - 1 = 4$

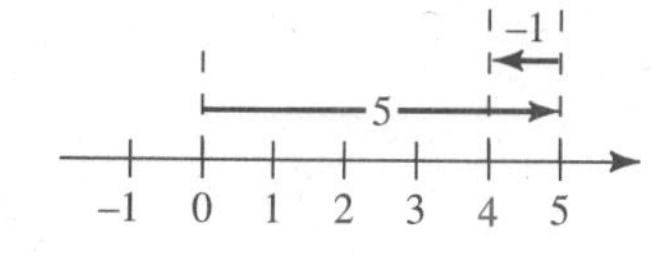

(b) $6 - 2 = 4$

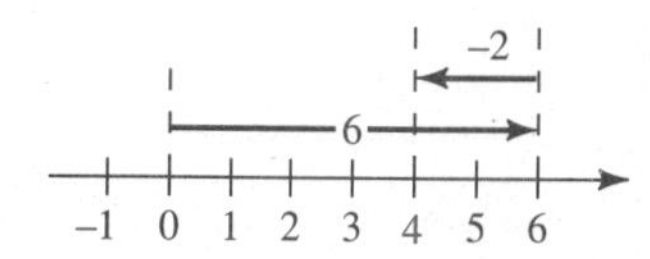

2. Subtract.

(a) $6 - 10$

(b) $-2 - 4$

(c) $3 - (-5)$

(d) $-8 - (-12)$

ANSWERS
2. (a) -4 (b) -6 (c) 8 (d) 4

EXAMPLE 2 Using the Definition of Subtraction

Subtract.

No change (12); Change $-$ to $+$; Additive inverse of 3

(a) $12 - 3 = 12 + (-3) = 9$

(b) $5 - 7 = 5 + (-7) = -2$

(c) $8 - 15 = 8 + (-15) = -7$

No change (-3); Change $-$ to $+$; Additive inverse of -5

(d) $-3 - (-5) = -3 + (5) = 2$

(e) $-6 - (-9) = -6 + (9) = 3$

(f) $8 - (-5) = 8 + (5) = 13$ ■

WORK PROBLEM 2 AT THE SIDE.

Subtraction can be used to reverse the result of an addition problem. For example, if 4 is added to a number and then subtracted from the sum, the original number is the result.

The symbol $-$ has now been used for three purposes:

1. to represent subtraction, as in $9 - 5 = 4$;
2. to represent negative numbers, such as -10, -2, and -3;
3. to represent the additive inverse of a number, as in "the additive inverse of 8 is -8."

More than one use may appear in the same problem, such as $-6 - (-9)$, where -9 is subtracted from -6. The meaning of the symbol depends on its position in the algebraic expression.

3 As before, with problems that have both parentheses and brackets, first do any operations inside the parentheses and brackets. Work from the inside out. Because subtraction is defined in terms of addition, the order of operations rules from Section 1.4 can be used.

EXAMPLE 3 Subtracting with Brackets

Work each problem.

(a)
$$\begin{aligned}-6 - [2 - (8 + 3)] &= -6 - [2 - 11] \\ &= -6 - [2 + (-11)] \\ &= -6 - (-9) \\ &= -6 + (9) = 3\end{aligned}$$

(b)
$$\begin{aligned}5 - [(-3 - 2) - (4 - 1)] &= 5 - [(-3 + (-2)) - 3] \\ &= 5 - [(-5) - 3] \\ &= 5 - [(-5) + (-3)] \\ &= 5 - (-8) \\ &= 5 + 8 = 13\end{aligned}$$ ■

WORK PROBLEM 3 AT THE SIDE.

4 There are several key words and phrases that indicate subtraction. *Difference* is one of them. Some of these are given in the chart below.

Word or Phrase	*Example*	*Numerical Expression and Simplification*
Difference	The *difference* between -3 and -8	$-3 - (-8) = -3 + 8 = 5$
Subtracted from	12 *subtracted from* 18	$18 - 12 = 6$
Less than	6 *less than* 5	$5 - 6 = 5 + (-6) = -1$
Decreased by	9 *decreased by* -4	$9 - (-4) = 9 + 4 = 13$

CAUTION When you are subtracting two numbers, it is important that you write them in the correct order, because, in general, $a - b \neq b - a$. For example, $5 - 3 \neq 3 - 5$. For this reason, it is important to *think carefully before interpreting an expression involving subtraction!* (This problem did not arise for addition.)

EXAMPLE 4 Interpreting Words and Phrases

Write a numerical expression for each phrase, and simplify the expression.

(a) The **difference between** -8 and 5

It is conventional to write the numbers in the order they are given when "difference between" is used.

$$-8 - 5 = -8 + (-5) = -13$$

(b) 4 **subtracted from** the sum of 8 and -3

Here the operation of addition is also used, as indicated by the word *sum*. First, add 8 and -3. Next, subtract 4 from this sum.

$$[8 + (-3)] - 4 = 5 - 4 = 1$$

(c) 4 **less than** -6

Be careful with order here. 4 must be taken *from* -6.

$$-6 - 4 = -6 + (-4) = -10$$

Notice that "4 less than -6" differs from "4 *is less than* -6." The second of these is symbolized as $4 < -6$ (which is a false statement).

(d) 8, **decreased by** 5 **less than** 12

First, write "5 less than 12" as $12 - 5$. Next, subtract $12 - 5$ from 8.

$$8 - (12 - 5) = 8 - 7 = 1 \quad ■$$

WORK PROBLEM 4 AT THE SIDE.

3. Work each problem.

(a) $2 - [(-3) - (4 + 6)]$

(b) $[(5 - 7) + 3] - 8$

(c) $6 - [(-1 - 4) - 2]$

(d) $(-8 - 1) - [(-3 + 2) - (-4 + 1)]$

4. Write a numerical expression for each phrase, and simplify the expression.

(a) The difference between -5 and -12

(b) -2 subtracted from the sum of 4 and -4

(c) 7 less than -2

(d) 9, decreased by 10 less than 7

ANSWERS
3. (a) 15 (b) -7 (c) 13 (d) -11
4. (a) $-5 - (-12) = 7$
(b) $[4 + (-4)] - (-2) = 2$
(c) $-2 - 7 = -9$
(d) $9 - (7 - 10) = 12$

Historical Reflections

Historical Reflections

The Set Theory of Georg Cantor (1845–1918)

The set of counting numbers is infinite; that is, its elements go on and on without coming to an end. Similarly, the real numbers form an infinite set. There is a difference between the two infinities, however—they cannot be matched in a one-to-one correspondence. There are "more" real numbers than counting numbers. This is one of the basic ideas studied in the field of *set theory*.

The symbol used to represent the infinity of the counting numbers is $\aleph_0$, (read "aleph-null"). The infinity of the real numbers is symbolized by the letter c.

A pioneer in the field of set theory was Georg Cantor. This German mathematician was a contemporary of Leopold Kronecker (1823–1891), who was quoted as saying "God made the integers and all else is the work of man." Differences between the beliefs of the two men led to heated debates.

Art: Baveria-Verlag

name date hour

2.3 EXERCISES

Find the differences. See Examples 1-3.

1. $3 - 6$

2. $7 - 12$

3. $5 - 9$

4. $8 - 13$

5. $-6 - 2$

6. $-11 - 4$

7. $-9 - 5$

8. $-12 - 15$

9. $6 - (-3)$

10. $8 - (-5)$

11. $5 - (-12)$

12. $12 - (-2)$

13. $-6 - (-2)$

14. $-7 - (-5)$

15. $2 - (3 - 5)$

16. $5 - (6 - 13)$

17. $-2 - (5 - 8)$

18. $-3 - (4 - 11)$

19. $\frac{1}{2} - \left(-\frac{1}{4}\right)$

20. $\frac{1}{3} - \left(-\frac{4}{3}\right)$

21. $-\frac{3}{4} - \frac{5}{8}$

22. $-\frac{5}{6} - \frac{1}{2}$

23. $\frac{5}{8} - \left(-\frac{1}{2} - \frac{3}{4}\right)$

24. $\frac{9}{10} - \left(\frac{1}{8} - \frac{3}{10}\right)$

25. $3.4 - (-8.2)$

26. $5.7 - (-11.6)$

27. $-6.4 - 3.5$

28. $-4.4 - 8.6$

29. $-4.2 - (7.4 - 9.8)$

30. $(-1.8 - 3.8) - (-9.6)$

31. $[(-2.4) - (-5.9)] - (-6.6)$

32. $[(-2.1 - 1.6) - 7.7] - [-3.2 - (-1.0)]$

Work each problem. See Example 3.

33. $(4 - 6) + 12$

34. $(3 - 7) + 4$

35. $(8 - 1) - 12$

36. $(9 - 3) - 15$

37. $6 - (-8 + 3)$

38. $8 - (-9 + 5)$

39. $2 + (-4 - 8)$

40. $6 + (-9 - 2)$

41. $(-5 - 6) - (9 - 2)$

42. $(-4 - 8) - (6 - 1)$

43. $\left(-\frac{3}{8} - \frac{2}{3}\right) - \left(-\frac{9}{8} - 3\right)$

44. $\left(-\frac{3}{4} - \frac{5}{2}\right) - \left(-\frac{1}{8} - 1\right)$

45. $-9 - [(3 - 2) - (-4 - 2)]$

46. $-8 - [(-4 - 1) - (9 - 2)]$

47. $-3 + [(-5 - 8) - (-6 + 2)]$

48. $-4 + [(-12 + 1) - (-1 - 9)]$

49. $-9.2 + [(-4.9 - 4.1) + 11.3]$

50. $-7.6 - [(-3.9 + 1.4) - 2.4]$

51. $[-12.25 - (8.34 + 3.57)] - 17.88$

52. $[-34.99 + (6.59 - 12.25)] - 8.33$

name date hour

Write a numerical expression for each phrase and simplify. See Example 4.

53. The difference between 4 and -12

54. The difference between 7 and -7

55. 8 less than -5

56. 9 less than -12

57. The sum of 9 and -4, decreased by 5

58. The sum of 12 and -7, decreased by 13

59. 12 less than the difference between 8 and -7

60. 19 less than the difference between 9 and -2

Work the following word problems.

61. Jack owed his brother \$19. He repaid \$12 and later borrowed \$10. What positive or negative number represents his present financial status?

62. The temperature dropped 19° below the previous temperature of $-4°$. Find the new temperature.

63. Tickets to the senior prom cost \$20, and Le Nguyen is \$14 in debt. How much must he earn before he can afford to buy a ticket?

64. The bottom of Death Valley is 282 feet below sea level. The top of Mt. Whitney has an altitude of 14,494 feet above sea level. Find the difference between these two elevations.

65. In 1988, Sonny's Building Specialties showed a "profit" of $-\$13{,}500$. In 1989, the profit decreased by \$3500. Find the profit in 1989.

66. A chemist is running an experiment at a temperature of $-156.3°$. She then raises the temperature by $9.5°$. Find the new temperature.

67. A first reading of a dial was 8.435. A second reading was -4.321. By how much had the reading gone down?

68. One company made a profit of \$35,600, while another company lost \$12,200. Find the difference between these "profits."

SKILL SHARPENERS

Find the value of each expression below when $x = 4$ and $y = 2$. *See Section 1.5.*

69. $3x - 2y$

70. $8x + 3y$

71. $3(2x + 3y)$

72. $5(3x - 3y)$

73. $(x + y)(2x - y)$

74. $(3x - y)(2x + 4y)$

2.4 MULTIPLICATION OF REAL NUMBERS

The rule for multiplying positive numbers is stated below.

The product of two positive numbers is positive.

But what about multiplying other real numbers? Any rules for multiplication of real numbers should be consistent with the rules from arithmetic for multiplication. For example, the product of 0 and any real number (positive or negative) should be 0.

For any number a,

$$a \cdot 0 = 0.$$

OBJECTIVES

1. Find the product of a positive number and a negative number.
2. Find the product of two negative numbers.
3. Identify factors of integers.
4. Use the order of operations.
5. Evaluate expressions that use variables.
6. Interpret words and phrases that indicate multiplication.

1 In order to define the product of a positive and a negative number so that the result is consistent with the multiplication of two positive numbers, look at the following pattern.

$$3 \cdot 5 = 15$$
$$3 \cdot 4 = 12$$
$$3 \cdot 3 = 9$$
$$3 \cdot 2 = 6$$
$$3 \cdot 1 = 3$$
$$3 \cdot 0 = 0$$
$$3 \cdot (-1) = ?$$

The numbers decrease by 3.

What should $3(-1)$ equal? The product $3(-1)$ represents the sum

$$-1 + (-1) + (-1) = -3$$

so the product should be -3. Also,

$$3(-2) = -2 + (-2) + (-2) = -6.$$

WORK PROBLEM 1 AT THE SIDE.

The results from Problem 1 maintain the pattern in the list above, which suggests the following rule.

The product of a positive number and a negative number is negative.

EXAMPLE 1 Multiplying a Positive Number and a Negative Number

Find the following products using the multiplication rule given above.

(a) $8(-5) = -(8 \cdot 5) = -40$

(b) $5(-4) = -(5 \cdot 4) = -20$

(c) $(-7)(2) = -(7 \cdot 2) = -14$

1. Find each product by finding the sum of three numbers.

(a) $3(-3)$

(b) $3(-4)$

(c) $3(-5)$

ANSWERS
1. (a) -9 (b) -12 (c) -15

2. Find the products

(a) $2(-6)$

(b) $7(-8)$

(c) $12(-15)$

(d) $(-9)(2)$

(e) $(-10)(3)$

(f) $(-16)(19)$

3. Find the products.

(a) $(-5)(-6)$

(b) $(-7)(-3)$

(c) $(-8)(-5)$

(d) $(-11)(-2)$

(e) $(-17)(-21)$

(f) $(-82)(-13)$

ANSWERS
2. (a) -12 (b) -56 (c) -180 (d) -18 (e) -30 (f) -304
3. (a) 30 (b) 21 (c) 40 (d) 22 (e) 357 (f) 1066

(d) $(-9)\left(\frac{1}{3}\right) = -\left(9 \cdot \frac{1}{3}\right) = -3$

(e) $(-6.2)(4.1) = -25.42$ ■

WORK PROBLEM 2 AT THE SIDE.

2 The product of two positive numbers is positive, and the product of a positive number and a negative number is negative. What about the product of two negative numbers? Look at another pattern.

$$\begin{aligned}(-5)(4) &= -20\\ (-5)(3) &= -15\\ (-5)(2) &= -10\\ (-5)(1) &= -5\\ (-5)(0) &= 0\\ (-5)(-1) &= ?\end{aligned}$$

The numbers increase by 5.

The numbers on the left of the equals sign (shaded numbers) decrease by 1 for each step down the list. The products on the right increase by 5 for each step down the list. To maintain this pattern, $(-5)(-1)$ should be 5 more than $(-5)(0)$, or 5 more than 0, so

$$(-5)(-1) = 5.$$

The pattern continues with

$$\begin{aligned}(-5)(-2) &= 10\\ (-5)(-3) &= 15\\ (-5)(-4) &= 20\\ (-5)(-5) &= 25\end{aligned}$$

and so on. This pattern suggests that we should multiply two negative numbers as follows.

The product of two negative numbers is positive.

EXAMPLE 2 Multiplying Two Negative Numbers

Find the products using the multiplication rule given above.

(a) $(-9)(-2) = 9 \cdot 2 = 18$

(b) $(-6)(-12) = 6 \cdot 12 = 72$

(c) $(-8)(-1) = 8 \cdot 1 = 8$

(d) $(-15)(-2) = 15 \cdot 2 = 30$ ■

WORK PROBLEM 3 AT THE SIDE.

The products of positive and negative numbers are described below.

MULTIPLYING SIGNED NUMBERS

The product of two numbers having the *same* signs is *positive,* and the product of two numbers having *different* signs is *negative.*

3 In Section 1.1 the definition of a *factor* was given for whole numbers. (For example, since $9 \cdot 5 = 45$, both 9 and 5 are factors of 45.) The definition can now be extended to integers.

If the product of two integers is a third integer, then each of the two integers is a **factor** of the third. For example, $(-3)(-4) = 12$, so -3 and -4 are both factors of 12. The factors of 12 are the numbers $-12, -6, -4, -3, -2, -1, 1, 2, 3, 4, 6$, and 12.

EXAMPLE 3 Identifying Factors of an Integer

The following chart shows several integers and the factors of those integers.

Integer	*Factors*
18	$-18, -9, -6, -3, -2, -1, 1, 2, 3, 6, 9, 18$
20	$-20, -10, -5, -4, -2, -1, 1, 2, 4, 5, 10, 20$
15	$-15, -5, -3, -1, 1, 3, 5, 15$
7	$-7, -1, 1, 7$
1	$-1, 1$

■

WORK PROBLEM 4 AT THE SIDE.

4 The next example shows the order of operations discussed in Chapter 1 used with the multiplication of positive and negative numbers.

EXAMPLE 4 Using the Order of Operations

Simplify.

(a) $(-9)(2) - (-3)(2)$

First find all products, working from left to right.

$$(-9)(2) - (-3)(2) = -18 - (-6)$$

Now perform the subtraction. $\quad -18 - (-6) = -18 + 6 = -12$

(b) $(-6)(-2) - (3)(-4) = 12 - (-12) = 12 + 12 = 24$

(c) $-5(-2 - 3) = -5(-5) = 25$ ■

WORK PROBLEM 5 AT THE SIDE.

5 The next two examples show how numbers may be substituted for variables.

EXAMPLE 5 Evaluating an Expression for Numerical Values

Evaluate the expression $(3x + 4y)(-2m)$ for each set of values.

(a) $x = -1, \quad y = -2, \quad m = -3$

First substitute the given values for the variables. Then find the value of the expression. Put parentheses around the number for each variable.

$$(3x + 4y)(-2m)$$
$$= [3(-1) + 4(-2)][-2(-3)]$$
$$= [-3 + (-8)][6] \qquad \text{Find the products.}$$
$$= (-11)(6) \qquad \text{Use order of operations.}$$
$$= -66 \qquad \text{Multiply.}$$

4. Find all factors of each integer.

(a) 24

(b) 30

(c) 19

(d) 37

5. Perform the indicated operations.

(a) $(-3)(4) - (2)(6)$

(b) $(-5)(-6) - (8)(-3)$

(c) $-7(-2 - 5)$

(d) $-4(-7 - 9)$

(e) $-8[-1 - (-4)(-5)]$

ANSWERS

4. (a) $-24, -12, -8, -6, -4, -3, -2, -1, 1, 2, 3, 4, 6, 8, 12, 24$
(b) $-30, -15, -10, -6, -5, -3, -2, -1, 1, 2, 3, 5, 6, 10, 15, 30$
(c) $-19, -1, 1, 19$
(d) $-37, -1, 1, 37$

5. (a) -24 (b) 54 (c) 49 (d) 64 (e) 168

6. Evaluate the following expressions.

(a) $2x - 7(y + 1)$
if $x = -4$ and $y = 3$

(b) $(-3x)(4x - 2y)$
if $x = 2$ and $y = -1$

(c) $2x^2 - 4y^2$
if $x = -2$ and $y = -3$

7. Write a numerical expression for each phrase and simplify.

(a) The product of 6 and the sum of -5 and -4

(b) Twice the difference between 8 and -4

(c) Three-fifths of the sum of 2 and -7

(d) 20% of the sum of 9 and -4

ANSWERS
6. (a) -36 (b) -60 (c) -28
7. (a) $6[(-5) + (-4)] = -54$
(b) $2[8 - (-4)] = 24$
(c) $\frac{3}{5}[2 + (-7)] = -3$
(d) $.20[9 + (-4)] = 1$

(b) $x = 7, \quad y = -9, \quad m = 5$

Substitute. Put parentheses around -9.

$$(3x + 4y)(-2m)$$
$$= [3 \cdot 7 + 4(-9)](-2 \cdot 5)$$
$$= [21 + (-36)](-10) \quad \text{Find the products.}$$
$$= (-15)(-10) \quad \text{Add.}$$
$$= 150 \quad \text{Multiply.}$$
■

EXAMPLE 6 Evaluating an Expression for Numerical Values

Evaluate $2x^2 - 3y^2$ if $x = -3$ and $y = -4$.

Use parentheses as shown.

$$2(-3)^2 - 3(-4)^2 = 2(9) - 3(16) \quad \text{Square } -3 \text{ and } -4.$$
$$= 18 - 48 \quad \text{Multiply.}$$
$$= -30 \quad \text{Subtract.}$$
■

WORK PROBLEM 6 AT THE SIDE.

6 The word *product* refers to multiplication. The following chart gives some of the key words and phrases that indicate multiplication.

Word or Phrase	*Example*	*Numerical Expression and Simplification*
Product	The *product* of -5 and -2	$(-5)(-2) = 10$
Times	13 *times* -4	$13(-4) = -52$
Twice (meaning "2 times")	*Twice* 6	$2(6) = 12$
Of (used with fractions)	$\frac{1}{2}$ *of* 10	$\frac{1}{2}(10) = 5$
Percent of	12% of -16	$.12(-16) = -1.92$

EXAMPLE 7 Interpreting Words and Phrases

Write a numerical expression for each phrase and simplify. Use the order of operations.

(a) The **product** of 12 and the sum of 3 and -6

Here 12 is multiplied by "the sum of 3 and -6."

$$12[3 + (-6)] = 12(-3) = -36$$

(b) **Twice** the difference between 8 and -4

$$2[8 - (-4)] = 2[8 + 4] = 2(12) = 24$$

(c) Two-thirds **of** the sum of -5 and -3

$$\frac{2}{3}[-5 + (-3)] = \frac{2}{3}[-8] = -\frac{16}{3}$$

(d) 15**% of** the difference between 14 and -2

Remember that $15\% = .15$.

$$.15[14 - (-2)] = .15(14 + 2) = .15(16) = 2.4$$
■

WORK PROBLEM 7 AT THE SIDE.

name date hour

2.4 EXERCISES

Find the products. See Examples 1 and 2.

1. $(-3)(-4)$ **2.** $(-3)(4)$ **3.** $3(-4)$ **4.** $-2(-8)$

5. $(-1)(-5)$ **6.** $(-9)(-5)$ **7.** $(-4)(-11)$ **8.** $(-5)(7)$

9. $(-10)(-12)$ **10.** $9(-5)$ **11.** $(8)(-6)$ **12.** $(13)(-2)$

13. $(-6)(5)$ **14.** $(-9)(0)$ **15.** $0(-11)$ **16.** $(15)(-11)$

17. $\left(-\frac{7}{3}\right)\left(\frac{8}{21}\right)$ **18.** $\left(-\frac{3}{8}\right)\left(-\frac{10}{9}\right)$ **19.** $\left(-\frac{5}{4}\right)\left(\frac{6}{15}\right)$

20. $(-5.1)(.02)$ **21.** $(-3.7)(-2.1)$ **22.** $(-4.6)(-2.4)$

Find all integer factors of each number. See Example 3.

23. 36 **24.** 32

25. 25 **26.** 14

27. 40 **28.** 50

29. 17 **30.** 13 **31.** 29

Perform the indicated operations. See Example 4.

32. $6 - 4 \cdot 5$

33. $3 - 2 \cdot 9$

34. $-9 - (-2) \cdot 3$

35. $-11 - (-7) \cdot 4$

36. $9(6 - 10)$

37. $5(12 - 15)$

38. $-6(2 - 4)$

39. $-9(5 - 8)$

40. $(4 - 9)(2 - 3)$

41. $(6 - 11)(3 - 6)$

42. $(2 - 5)(3 - 7)$

43. $(5 - 12)(2 - 6)$

44. $(-4 - 3)(-2) + 4$

45. $(-5 - 2)(-3) + 6$

46. $3(-4) - (-2)$

47. $5(-2) - (-9)$

48. $(-8 - 2)(-4) - (-5)$

49. $(-9 - 1)(-2) - (-6)$

Evaluate the following expressions if $x = -2$, $y = 3$, and $a = -4$. See Examples 5 and 6.

50. $5x - 2y + 3a$

51. $6x - 5y + 4a$

52. $(2x + y)(3a)$

53. $(5x - 2y)(-2a)$

54. $\left(\frac{1}{3}x - \frac{4}{5}y\right)\left(-\frac{1}{5}a\right)$

55. $\left(\frac{5}{6}x + \frac{3}{2}y\right)\left(-\frac{1}{3}a\right)$

56. $(-5 + x)(-3 + y)(2 - a)$

57. $(6 - x)(5 + y)(3 + a)$

58. $-2y^2 + 3a$

59. $5x - 4a^2$

60. $3a^2 - x^2$

61. $4y^2 - 2x^2$

Write a numerical expression for each phrase and simplify. See Example 7.

62. The product of -9 and 2, added to 6

63. The product of 4 and -7, added to -9

64. The product of -1 and 6, subtracted from -9

65. Twice the product of -8 and 2, subtracted from -4

66. Nine subtracted from the product of 7 and -6

67. Three subtracted from the product of -2 and 3

68. The product of 12 and the difference between 9 and -4

69. The product of -3 and the sum of 4 and -7

70. Four-fifths of the sum of -8 and -12

71. Three-tenths of the difference between -2 and -22

Find the solution for each equation from the domain $\{-3, -2, -1, 0, 1, 2, 3\}$ *by guessing or by using trial and error.*

72. $2x = -4$

73. $3k = -6$

74. $-4m = 0$

75. $-9y = 0$

76. $-8p = 16$

77. $-9r = 27$

78. $2x + 1 = -3$

79. $3w + 3 = -3$

80. $-4a + 2 = 10$

SKILL SHARPENERS

Find the value of each expression when $x = 1$ *and* $y = 5$. *See Section 1.5.*

81. $\dfrac{3x + 5y}{2}$

82. $\dfrac{10x + 4y}{15}$

83. $\dfrac{2(x + 4y)}{2x + y}$

84. $\dfrac{5(8x - y)}{5y - 2}$

85. $\dfrac{x^2 + y^2}{3y - 2}$

86. $\dfrac{3x^2 + 2y^2}{10y + 3}$

2.5 DIVISION OF REAL NUMBERS

1 The difference between two numbers is found by adding the additive inverse of the second number to the first. Division is related to multiplication in a similar way. The *quotient* of two numbers is found by *multiplying* by the *multiplicative inverse*. By definition, since

$$8 \cdot \frac{1}{8} = \frac{8}{8} = 1 \quad \text{and} \quad \frac{5}{4} \cdot \frac{4}{5} = \frac{20}{20} = 1$$

the multiplicative inverse of 8 is $\frac{1}{8}$, and that of $\frac{5}{4}$ is $\frac{4}{5}$.

Pairs of numbers whose product is 1 are called **multiplicative inverses,** or **reciprocals,** of each other.

EXAMPLE 1 Finding the Multiplicative Inverse

The following chart shows several numbers and the multiplicative inverse (if it exists) of each number.

Number	*Multiplicative Inverse (Reciprocal)*
4	$\frac{1}{4}$
-5	$\frac{1}{-5}$ or $-\frac{1}{5}$
$\frac{3}{4}$	$\frac{4}{3}$
$-\frac{5}{8}$	$-\frac{8}{5}$
0	None ■

Why is there no multiplicative inverse for the number 0? Suppose that k is to be the multiplicative inverse of 0. Then $k \cdot 0$ should equal 1. But $k \cdot 0 = 0$ for any number k. Since there is no value of k that is a solution of the equation $k \cdot 0 = 1$, we make the following statement.

0 has no multiplicative inverse.

WORK PROBLEM 1 AT THE SIDE.

2 In a way similar to that used for subtraction, the *quotient* of a and b is defined to be the product of a and the multiplicative inverse of b.

OBJECTIVES

1. Find the reciprocal, or multiplicative inverse, of a number.
2. Divide with signed numbers.
3. Simplify numerical expressions.
4. Interpret words and phrases that indicate division.
5. Translate simple sentences into equations.

1. Complete the chart.

Number	*Multiplicative inverse*
(a) 6	
(b) -2	
(c) $\frac{2}{3}$	
(d) $-\frac{1}{4}$	
(e) 0	

ANSWERS

1. (a) $\frac{1}{6}$ (b) $-\frac{1}{2}$ or $\frac{1}{-2}$ (c) $\frac{3}{2}$
(d) -4
(e) none

2. Find the quotients.

(a) $\frac{42}{7}$

(b) $\frac{-36}{6}$

(c) $\frac{-12}{-4}$

(d) $\frac{18}{-9}$

(e) $\frac{-3}{0}$

ANSWERS
2. (a) 6 (b) −6 (c) 3 (d) −2
(e) undefined

DEFINITION OF DIVISION

For any real numbers *a* and *b*, with $b \neq 0$,

$$\frac{a}{b} = a \cdot \frac{1}{b}.$$

The definition above indicates that *b*, the number to divide by, cannot be 0. The reason is that 0 has no multiplicative inverse, so $\frac{1}{0}$ is not a number.

Division by 0 is undefined and is never permitted.

If a division problem turns out to involve division by 0, write "undefined."

Since division is defined in terms of multiplication, all the rules for multiplication of signed numbers also apply to division.

EXAMPLE 2 Using the Definition of Division

Write each quotient as a product and evaluate.

(a) $\frac{12}{3} = 12 \cdot \frac{1}{3}$
$= 4$

(b) $\frac{-10}{2} = -10 \cdot \frac{1}{2}$
$= -5$

(c) $\frac{8}{-4} = 8 \cdot \left(\frac{1}{-4}\right)$
$= -2$

(d) $\frac{-14}{-7} = -14\left(\frac{1}{-7}\right)$
$= 2$

(e) $\frac{-100}{-20} = -100\left(\frac{1}{-20}\right)$
$= 5$

(f) $\frac{-10}{0}$ Undefined ■

WORK PROBLEM 2 AT THE SIDE.

Using the definition of division directly is awkward. The following rule for division with signed numbers follows from the definition of division and the rules for multiplication with signed numbers.

DIVIDING SIGNED NUMBERS

The quotient of two numbers having the *same* sign is *positive;* the quotient of two numbers having *different* signs is *negative.*

EXAMPLE 3 Dividing Signed Numbers
Find the quotients.

(a) $\frac{8}{-2} = -4$

(b) $\frac{-45}{-9} = 5$

(c) $-\frac{1}{8} \div \left(-\frac{3}{4}\right) = -\frac{1}{8} \cdot \left(-\frac{4}{3}\right) = \frac{1}{6}$ ■

WORK PROBLEM 3 AT THE SIDE.

From the definitions of multiplication and division of real numbers,

$$\frac{-40}{8} = -40 \cdot \frac{1}{8}$$
$$= -5$$

and

$$\frac{40}{-8} = 40\left(\frac{1}{-8}\right)$$
$$= -5$$

so

$$\frac{-40}{8} = \frac{40}{-8}.$$

Based on this example, the quotient of a positive and a negative number can be expressed in any of the following three forms.

For any positive real numbers a and b, with $b \neq 0$,

$$\frac{-a}{b} = \frac{a}{-b} = -\frac{a}{b}.$$

The form $\frac{a}{-b}$ is seldom used.

Similarly, the quotient of two negative numbers can be expressed as the quotient of two positive numbers.

For any positive real numbers a and b, with $b \neq 0$,

$$\frac{-a}{-b} = \frac{a}{b}.$$

3. Find the quotients.

(a) $\frac{-8}{-2}$

(b) $\frac{-16}{2}$

(c) $\frac{1}{4} \div \left(-\frac{2}{3}\right)$

ANSWERS
3. (a) 4 (b) −8 (c) $-\frac{3}{8}$

4. Perform the indicated operations.

(a) $\dfrac{5(-4)}{-2-8}$

(b) $\dfrac{6(-4)-2(5)}{3(2-7)}$

(c) $\dfrac{-6(-8)+(-3)9}{(-2)[4-(-3)]}$

(d) $\dfrac{5^2+3^2}{3(-4)-5}$

ANSWERS

4. (a) 2 (b) $\frac{34}{15}$ (c) $-\frac{3}{2}$ (d) -2

3 The next example shows how to simplify numerical expressions involving quotients.

EXAMPLE 4 Simplifying Expressions Involving Division

Simplify each expression.

(a) $\dfrac{5(-2)-(3)(4)}{2(1-6)}$

Follow the order of operations. Simplify the numerator and denominator separately. Then divide or write in lowest terms.

$$\frac{5(-2)-(3)(4)}{2(1-6)}=\frac{-10-12}{2(-5)}$$ Multiply in numerator. Subtract in denominator.

$$=\frac{-22}{-10}$$ Subtract in numerator. Multiply in denominator.

$$=\frac{11}{5}$$ Express in lowest terms.

(b) $\dfrac{4^2-6^2}{5(-3+2)}$

$$\frac{4^2-6^2}{5(-3+2)}=\frac{16-36}{5(-1)}$$ Square 4 and 6. Add −3 and 2.

$$=\frac{-20}{-5}$$ Subtract in numerator. Multiply in denominator.

$$=4$$ Divide. ■

WORK PROBLEM 4 AT THE SIDE.

The rules for operations with signed numbers are summarized here.

OPERATIONS WITH SIGNED NUMBERS

Addition

Like signs Add the absolute values of the numbers. The result is given the same sign as the numbers.

Unlike signs Subtract the smaller absolute value from the larger absolute value. Give the result the sign of the number having the larger absolute value.

Subtraction

Add the additive inverse, or opposite, of the second number.

Multiplication and Division

Like signs The product or quotient of two numbers with like signs is positive.

Unlike signs The product or quotient of two numbers with unlike signs is negative.

Division by 0 is undefined.

4 The word *quotient* refers to the result obtained in a division problem. In algebra, quotients are usually represented with a fraction bar. The symbol ÷ is seldom used.

The following chart gives some key words and phrases associated with division.

Word or Phrase	*Example*	*Numerical Expression and Simplification*
Quotient	The *quotient* of -24 and 3	$\frac{-24}{3} = -8$
Divided by	-16 *divided by* -4	$\frac{-16}{-4} = 4$

It is customary to write the first number named as the numerator and the second as the denominator when interpreting a phrase involving division. This is shown in the next example.

EXAMPLE 5 Interpreting Words and Phrases

Write a numerical expression for each phrase, and simplify the expression.

(a) The **quotient** of 14 and the sum of -9 and 2

"Quotient" indicates division. The number 14 is the numerator and "the sum of -9 and 2" is the denominator.

$$\frac{14}{-9 + 2} = \frac{14}{-7} = -2$$

(b) The product of 5 and -6, **divided by** the difference between -7 and 8

The numerator of the fraction representing the division is obtained by multiplying 5 and -6. The denominator is found by subtracting -7 and 8.

$$\frac{5(-6)}{-7 - 8} = \frac{-30}{-15} = 2 \quad \blacksquare$$

WORK PROBLEM 5 AT THE SIDE.

5 In this section and the preceding three sections, important words and phrases involving the four operations of arithmetic have been introduced. We can use these words and phrases to interpret sentences that translate into equations. The ability to do this will help us to solve the types of word problems found in Section 3.4.

EXAMPLE 6 Translating Words into an Equation

Write the following in symbols, using x as the variable, and use guessing or trial and error to find the solution. All solutions come from the list of integers between -12 and 12, inclusive.

(a) Three times a number is -18.

The word *times* indicates multiplication, and the word *is* translates as the equals sign (=).

$$3x = -18$$

Since the integer between -12 and 12, inclusive, that makes this statement true is **-6**, the solution of the equation is -6.

5. Write a numerical expression for each phrase, and simplify the expression.

(a) The quotient of 20 and the sum of 8 and -3

(b) The product of -9 and 2, divided by the difference between 5 and -1

ANSWERS

5. (a) $\frac{20}{8 + (-3)} = 4$
(b) $\frac{(-9)(2)}{5 - (-1)} = -3$

6. Write the following in symbols, using x as the variable, and find the solution by guessing or by using trial and error. All solutions come from the list of integers between -12 and 12, inclusive.

(a) Twice a number is -6.

(b) The difference between -8 and a number is -11.

(c) The sum of 5 and a number is 8.

(d) The quotient of a number and -2 is 6.

(b) The sum of a number and 9 is 12.

$$x + 9 = 12$$

Since **3** + 9 = 12, the solution of this equation is 3.

(c) The difference between a number and 5 is 0.

$$x - 5 = 0$$

Since **5** − 5 = 0, the solution of this equation is 5.

(d) The quotient of 24 and a number is -2.

$$\frac{24}{x} = -2$$

Here, x must be a negative number, since the numerator is positive and the quotient is negative. Since $\frac{24}{\mathbf{-12}} = -2$, the solution is -12. ■

WORK PROBLEM 6 AT THE SIDE.

CAUTION It is important to recognize the distinction between the types of problems found in Example 5 and Example 6. In Example 5, the phrases translate as *expressions,* while in Example 6, the sentences translate as *equations*. Remember that an equation is a sentence, while an expression is a phrase.

ANSWERS

6. (a) $2x = -6$; -3
(b) $-8 - x = -11$; 3
(c) $5 + x = 8$; 3
(d) $\frac{x}{-2} = 6$; -12

name date hour

2.5 EXERCISES

Find the multiplicative inverse (if one exists) for each number. See Example 1.

1. 9

2. 8

3. -4

4. -10

5. $\frac{2}{3}$

6. $\frac{3}{4}$

7. $\frac{-9}{10}$

8. $\frac{-4}{5}$

9. 0

10. $\frac{0}{5}$

11. $.2$

12. $.25$

Find the quotients. See Examples 2 and 3.

13. $\frac{-10}{5}$

14. $\frac{-12}{3}$

15. $\frac{18}{-3}$

16. $\frac{24}{-6}$

17. $\frac{-150}{-10}$

18. $\frac{-280}{-20}$

19. $\frac{0}{-2}$

20. $\frac{0}{12}$

21. $-\frac{1}{2} \div \left(-\frac{3}{4}\right)$

22. $-\frac{5}{8} \div \left(-\frac{3}{16}\right)$

23. $(-4.2) \div (-2)$

24. $(-9.8) \div (-7)$

25. $\frac{12}{2 - 5}$

26. $\frac{15}{3 - 8}$

27. $\frac{-30}{2 - 8}$

28. $\frac{-50}{6 - 11}$

29. $\dfrac{-40}{8 - (-2)}$

30. $\dfrac{-72}{6 - (-2)}$

31. $\dfrac{-15 - 3}{3}$

32. $\dfrac{16 - (-2)}{-6}$

33. $\dfrac{-3.8 - (-2.2)}{-.2}$

34. $\dfrac{11.3 - (-2.7)}{.2 - .9}$

Simplify the numerators and denominators separately. Then find the quotients. See Example 4.

35. $\dfrac{-8(-2)}{3 - (-1)}$

36. $\dfrac{-12(-3)}{-15 - (-3)}$

37. $\dfrac{-15(2)}{-7 - 3}$

38. $\dfrac{-20(6)}{-5 - 1}$

39. $\dfrac{-5(2) + 3(-2)}{-3 - (-1)}$

40. $\dfrac{4(-1) + 3(-2)}{-2 - 3}$

41. $\dfrac{-9(-2) - (-4)(-2)}{-2(3) - 2(2)}$

42. $\dfrac{5(-2) - 3(4)}{-2[3 - (-2)] - 1}$

43. $\dfrac{6^2 + 4^2}{5(2 + 13)}$

44. $\dfrac{3^2 + 5^2}{4^2 + 1^2}$

45. $\dfrac{10^2 - 5^2}{8^2 + 3^2 + 2}$

46. $\dfrac{(.6)^2 + (.8)^2}{(1.2)^2 - (-.56)}$

47. $\dfrac{(.3)^3 - .007}{-.03 - (-.01)}$

Find the solution of each equation from the domain $\{-8, -6, -4, -2, 0, 2, 4, 6, 8\}$ *by guessing or by using trial and error.*

48. $\frac{x}{4} = -2$

49. $\frac{x}{2} = -1$

50. $\frac{n}{-2} = 3$

51. $\frac{t}{-2} = -2$

52. $\frac{m}{-2} = -4$

53. $\frac{y}{-1} = 2$

Write a numerical expression for each phrase, and simplify the expression. See Example 5.

54. The quotient of -36 and -6

55. The quotient of -48 and -8

56. The quotient of -12 and the sum of 4 and 2

57. The quotient of -20 and the sum of -12 and 2

58. The sum of 15 and -3, divided by the product of 6 and -2

59. The sum of -18 and -6, divided by the product of 3 and -1

60. The product of 34 and 7 less than 12

61. The product of -25 and 4 less than 9

Write the following in symbols, using x as the variable, and find the solution by guessing or by using trial and error. All solutions come from the list of integers between -12 *and* 12, *inclusive. See Example 6.*

62. Six times a number is -54.

63. Four times a number is -44.

64. The quotient of a number and 3 is -2.

65. The quotient of a number and 4 is -3.

66. 6 less than a number is 3.

67. 7 less than a number is 1.

68. When 5 is added to a number, the result is -3.

69. When 6 is added to a number, the result is -5.

70. When a number is divided by 2, the result is -3.

71. When a number is divided by 4, the result is -2.

SKILL SHARPENERS

Perform the indicated operations. See Sections 2.2 and 2.4.

72. $12 + (-12)$

73. $3 + (-3)$

74. $-\frac{1}{2}(-2)$

75. $-\frac{1}{4}(-4)$

76. $5 + [6 + (-14)]$

77. $(5 + 6) + (-14)$

2.6 PROPERTIES OF ADDITION AND MULTIPLICATION

If you are asked to find the sum

$$3 + 89 + 97$$

it is likely that you would mentally add $3 + 97$ to get 100, and then add $100 + 89$ to get 189. While the rule for order of operations says to add from left to right, it is a fact that we may change the order of the terms and group them in any way we choose without affecting the sum. These are examples of shortcuts that we use in everyday mathematics. These shortcuts are justified by the basic properties of addition and multiplication, which are discussed in this section. In the following statements, a, b, and c represent real numbers.

OBJECTIVES

1. Identify the use of the commutative properties.
2. Identify the use of the associative properties.
3. Identify the use of the identity properties.
4. Identify the use of the inverse properties.
5. Identify the use of the distributive property.

1 **Commutative properties** The word *commute* means to go back and forth. Many people commute to work or to school. If you travel from home to work and follow the same route from work to home, you travel the same distance each time. The commutative properties say that if two numbers are added or multiplied in any order, they give the same result.

$$a + b = b + a$$
$$ab = ba$$

EXAMPLE 1 Using the Commutative Properties

Use a commutative property to complete each statement.

(a) $-8 + 5 = 5 + \_\_\_$

By the commutative property for addition, the missing number is -8, since $-8 + 5 = 5 + (-8)$.

(b) $(-2)(7) = \_\_\_(-2)$

By the commutative property for multiplication, the missing number is 7, since $(-2)(7) = (7)(-2)$. ■

WORK PROBLEM 1 AT THE SIDE.

1. Complete each statement. Use a commutative property.

(a) $x + 9 = 9 + \_\_\_$

(b) $(-12)(4) = \_\_\_(-12)$

(c) $9(-11) = (-11)\_\_\_$

(d) $5x = x \cdot \_\_\_$

2 **Associative properties** When we *associate* one object with another, we tend to think of those objects as being grouped together. The associative properties say that when we add or multiply three numbers, we can group the first two together or the last two together and get the same answer.

$$(a + b) + c = a + (b + c)$$
$$(ab)c = a(bc)$$

EXAMPLE 2 Using the Associative Properties

Use an associative property to complete each statement.

(a) $8 + (-1 + 4) = (8 + \_\_\_) + 4$

The missing number is -1.

(b) $[2 \cdot (-7)] \cdot 6 = 2 \cdot \_\_\_$

The completed expression on the right should be $2 \cdot [(-7) \cdot 6]$. ■

WORK PROBLEM 2 AT THE SIDE.

2. Complete each statement. Use an associative property.

(a) $(9 + 10) + (-3)$
$= 9 + [\_\_\_ + (-3)]$

(b) $-5 + (2 + 8)$
$= (\_\_\_) + 8$

(c) $10 \cdot [(-8) \cdot (-3)]$
$= \_\_\_$

ANSWERS
1. (a) x (b) 4 (c) 9 (d) 5
2. (a) 10 (b) $-5 + 2$
(c) $[10 \cdot (-8)] \cdot (-3)$

By the associative property of addition, the sum of three numbers will be the same no matter which way the numbers are "associated" in groups. For this reason, parentheses can be left out in many addition problems. For example, both

$$(-1 + 2) + 3 \quad \text{and} \quad -1 + (2 + 3)$$

can be written as

$$-1 + 2 + 3.$$

In the same way, parentheses also can be left out of many multiplication problems.

EXAMPLE 3 Distinguishing Between the Associative and Commutative Properties

(a) Is $(2 + 4) + 5 = 2 + (4 + 5)$ an example of the associative property?

The order of the three numbers is the same on both sides of the equals sign. The only change is in the grouping, or association, of the numbers. Therefore, this is an example of the associative property.

(b) Is $6(3 \cdot 10) = 6(10 \cdot 3)$ an example of the associative property or the commutative property?

The same numbers, 3 and 10, are grouped on each side. On the left, however, the 3 appears first in $(3 \cdot 10)$. On the right, the 10 appears first. Since the only change involves the order of the numbers, this statement is an example of the commutative property.

(c) Is $(8 + 1) + 7 = 8 + (7 + 1)$ an example of the associative property or the commutative property?

In the statement, both the order and the grouping are changed. On the left the order of the three numbers is 8, 1, and 7. On the right it is 8, 7, and 1. On the left the 8 and 1 are grouped, and on the right the 7 and 1 are grouped. Therefore, both the associative and the commutative properties are used. ■

WORK PROBLEM 3 AT THE SIDE.

3. Decide whether each statement is an example of a commutative property, an associative property, or both.

(a) $2(4 \cdot 6) = (2 \cdot 4)6$

(b) $(2 \cdot 4)6 = (4 \cdot 2)6$

(c) $(2 + 4) + 6 = 4 + (2 + 6)$

3 **Identity properties** If a child wears a sheet to masquerade as a ghost on Halloween, the child's appearance is changed, but his or her *identity* is unchanged. The identity of a real number is left unchanged when identity properties are applied. The identity properties say that the sum of 0 and any number equals that number, and the product of 1 and any number equals that number.

$$a + 0 = a \quad \text{and} \quad 0 + a = a$$
$$a \cdot 1 = a \quad \text{and} \quad 1 \cdot a = a$$

The number 0 leaves the identity, or value, of any real number unchanged by addition. For this reason, 0 is called the **identity element for addition.** Since multiplication by 1 leaves any real number unchanged, 1 is the **identity element for multiplication.**

EXAMPLE 4 Using the Identity Properties

These statements are examples of the identity properties.

(a) $-3 + 0 = -3$

ANSWERS
4. (a) associative (b) commutative (c) both

(b) $0 + \frac{1}{2} = \frac{1}{2}$

(c) $-\frac{3}{4} \cdot 1 = -\frac{3}{4} \cdot \frac{8}{8} = -\frac{24}{32}$ $\left(\frac{8}{8} = 1\right)$

We use the identity property of multiplication to change a fraction to an equivalent one with a larger denominator. This is often done in adding or subtracting fractions.

(d) $1 \cdot 25 = 25$ ■

WORK PROBLEM 4 AT THE SIDE.

4 Inverse properties Each day before you go to work or school, you probably put on your shoes before you leave. Before you go to sleep at night, you probably take them off, and this leads to the same situation that existed before you put them on. These operations from everyday life are examples of inverse operations. The inverse properties of addition and multiplication lead to the additive and multiplicative identities, respectively. Recall that $-a$ is the **additive inverse** of a and $\frac{1}{a}$ is the **multiplicative inverse** of the nonzero number a. The sum of the numbers a and $-a$ is 0, and the product of the nonzero numbers a and $\frac{1}{a}$ is 1.

$$a + (-a) = 0 \quad \text{and} \quad -a + a = 0$$
$$a \cdot \frac{1}{a} = 1 \quad \text{and} \quad \frac{1}{a} \cdot a = 1 \quad (a \neq 0)$$

EXAMPLE 5 Using the Inverse Properties

The following statements are examples of the inverse properties.

(a) $\frac{2}{3} \cdot \frac{3}{2} = 1$

(b) $(-5)\left(-\frac{1}{5}\right) = 1$

(c) $-\frac{1}{2} + \frac{1}{2} = 0$

(d) $4 + (-4) = 0$ ■

WORK PROBLEM 5 AT THE SIDE.

5 Distributive property The everyday meaning of the word *distribute* is "to give out from one to several." An important property of real number operations involves this idea.

Look at the following statements.

$$2(5 + 8) = 2(13) = 26$$
$$2(5) + 2(8) = 10 + 16 = 26$$

Since both expressions equal 26,

$$2(5 + 8) = 2(5) + 2(8).$$

This result is an example of the *distributive property,* the only property involving *both* addition and multiplication. With this property, a product can be changed to a sum or difference.

4. Use an identity property to complete each statement.

(a) $9 + 0 =$ ____

(b) ____ $+ (-7) = -7$

(c) $\frac{1}{4} \cdot$ ____ $= \frac{3}{12}$

(d) ____ $\cdot 1 = 5$

5. Complete the statements so that they are examples of either an identity property or an inverse property. Tell which property is used.

(a) $-6 +$ ____ $= 0$

(b) $\frac{4}{3} \cdot$ ____ $= 1$

(c) $-\frac{1}{9} \cdot$ ____ $= 1$

(d) $275 +$ ____ $= 275$

ANSWERS

4. (a) 9 (b) 0 (c) $\frac{3}{3}$ (d) 5

5. (a) 6; inverse (b) $\frac{3}{4}$; inverse (c) −9; inverse (d) 0; identity

6. Use the distributive property to rewrite each expression.

(a) $2(p + 5)$

(b) $9(x + 2)$

(c) $-4(y + 7)$

(d) $5(m - 4)$

(e) $9 \cdot k + 9 \cdot 5$

(f) $3a - 3b$

(g) $7(2y + 7k - 9m)$

ANSWERS

6. (a) $2p + 10$ (b) $9x + 18$ (c) $-4y - 28$ (d) $5m - 20$ (e) $9(k + 5)$ (f) $3(a - b)$ (g) $14y + 49k - 63m$

The distributive property says that multiplying a number a by a sum of numbers $b + c$ gives the same result as multiplying a by b and a by c and then adding the two products.

$$a(b + c) = ab + ac \quad \text{and} \quad (b + c)a = ba + ca$$

As the arrows show, the a outside the parentheses is "distributed" over the b and c inside. Another form of the distributive property is valid for subtraction.

$$a(b - c) = ab - ac \quad \text{and} \quad (b - c)a = ba - ca$$

The distributive property also can be extended to more than two numbers.

$$a(b + c + d) = ab + ac + ad$$

EXAMPLE 6 Using the Distributive Property

Use the distributive property to rewrite each expression.

(a) $5(9 + 6) = 5 \cdot 9 + 5 \cdot 6$ — Multiply both terms by 5.
$= 45 + 30$ — Multiply.
$= 75$ — Add.

(b) $4(x + 5 + y) = 4x + 4 \cdot 5 + 4y$ — Distributive property
$= 4x + 20 + 4y$ — Multiply.

(c) $-2(x + 3) = -2x + (-2)(3)$ — Distributive property
$= -2x - 6$ — Multiply.

(d) $3(k - 9) = 3k - 3 \cdot 9$ — Distributive property
$= 3k - 27$ — Multiply.

(e) $6 \cdot 8 + 6 \cdot 2 = 6(8 + 2)$ — Distributive property
$= 6(10) = 60$ — Add, then multiply.

(f) $4x - 4m = 4(x - m)$ — Distributive property

(g) $8(3r + 11t + 5z) = 8(3r) + 8(11t) + 8(5z)$ — Distributive property
$= (8 \cdot 3)r + (8 \cdot 11)t + (8 \cdot 5)z$ — Associative property
$= 24r + 88t + 40z$ ■

WORK PROBLEM 6 AT THE SIDE.

The distributive property is used to remove the parentheses from expressions such as $-(2y + 3)$. We do this by first writing $-(2y + 3)$ as $-1 \cdot (2y + 3)$.

$-(2y + 3) = -1 \cdot (2y + 3)$
$= -1 \cdot (2y) + (-1) \cdot (3)$ — Distributive property
$= -2y - 3$ — Multiply.

EXAMPLE 7 Using the Distributive Property to Remove Parentheses

Write without parentheses.

(a) $-(7r - 8) = -1(7r) + (-1)(-8)$ Distributive property

$= -7r + 8$ Multiply.

(b) $-(-9w + 2) = 9w - 2$ ■

WORK PROBLEM 7 AT THE SIDE.

The properties of addition and multiplication of real numbers are summarized below.

PROPERTIES OF ADDITION AND MULTIPLICATION

For any real numbers a, b, and c, the following properties hold.

Commutative properties $a + b = b + a$ $ab = ba$

Associative properties $(a + b) + c = a + (b + c)$

$(ab)c = a(bc)$

Identity properties There is a real number 0 such that

$a + 0 = a$ and $0 + a = a.$

There is a real number 1 such that

$a \cdot 1 = a$ and $1 \cdot a = a.$

Inverse properties For each real number a, there is a single real number $-a$ such that

$a + (-a) = 0$ and $(-a) + a = 0.$

For each nonzero real number a, there is a single real number $\frac{1}{a}$ such that

$a \cdot \frac{1}{a} = 1$ and $\frac{1}{a} \cdot a = 1.$

Distributive property $a(b + c) = ab + ac$

$(b + c)a = ba + ca$

7. Write without parentheses.

(a) $-(3k - 5)$

(b) $-(2 - r)$

(c) $-(-5y + 8)$

(d) $-(-z + 4)$

ANSWERS

7. (a) $-3k + 5$ (b) $-2 + r$ (c) $5y - 8$ (d) $z - 4$

Historical Reflections

Historical Reflections

WOMEN IN MATHEMATICS: Hypatia, the First Woman Mathematician

The first woman to be identified in the history of mathematics was Hypatia, the daughter of Theon of Alexandria. She lived during the fourth and fifth centuries A.D., and her death at the hands of a mob of fanatical religious zealots in 415 ended the glorious period of Greek mathematics.

Hypatia was proficient in philosophy and medicine as well as mathematics. She wrote commentaries on the works of two important Greek mathematicians, Diophantus and Apollonius. She also assisted her father in writing a revision of Euclid's *Elements*.

Hypatia was a neo-Platonist, and as such her teachings were considered heretical by some Christian leaders. In Howard Eves' *An Introduction to the History of Mathematics (Sixth Edition),* the author writes of Hypatia's death at the hands of a group of followers of Cyril of Alexandria:

> . . . and one day, as Hypatia was driving home, he (Cyril) had her dragged from her chariot, her hair pulled out, her flesh scraped from her bones with oyster shells, and the remnants of her body consigned to flame.

Hypatia's life and tragic death are the subject of a novel written in 1907 by Charles Kingsley, titled *Hypatia, or New Foes with an Old Face*.

Art: Art Resource, NY

name date hour

2.6 EXERCISES

Label each statement as an example of the commutative, associative, identity, inverse, or distributive property. See Examples 1-6.

1. $6 + 15 = 15 + 6$

2. $9 + (11 + 4) = (9 + 11) + 4$

3. $5(15 \cdot 8) = (5 \cdot 15)8$

4. $(23)(9) = (9)(23)$

5. $12(-8 \cdot 3) = [12(-8)] \cdot 3$

6. $(-9)[6(-2)] = [-9(6)](-2)$

7. $2 + (p + r) = (p + r) + 2$

8. $(m + n) + 4 = 4 + (m + n)$

9. $-\frac{6}{5} + \frac{5}{12} = \frac{5}{12} + \left(-\frac{6}{5}\right)$

10. $\left(-\frac{9}{5}\right)\left(-\frac{3}{11}\right) = \left(-\frac{3}{11}\right)\left(-\frac{9}{5}\right)$

11. $6 + (-6) = 0$

12. $-8 + 8 = 0$

13. $-4 + 0 = -4$

14. $0 + (-9) = -9$

15. $3\left(\frac{1}{3}\right) = 1$

16. $-7\left(-\frac{1}{7}\right) = 1$

17. $\frac{2}{3} = \frac{2}{3} \cdot \frac{6}{6} = \frac{12}{18}$

18. $-\frac{9}{4} = -\frac{9}{4} \cdot \frac{3}{3} = -\frac{27}{12}$

19. $6(5 - 2x) = 6 \cdot 5 - 6(2x)$

20. $5(2m) + 5(7n) = 5(2m + 7n)$

Use the indicated property to write a new expression that is equal to the given expression. Simplify the new expression if possible. See Examples 1, 2, 4, 5, and 6.

21. $9 + k$; commutative

22. $z + 5$; commutative

23. $m + 0$; identity

24. $(-9) + 0$; identity

25. $3(r + m)$; distributive

26. $11(k + z)$; distributive

27. $8 \cdot \frac{1}{8}$; inverse

28. $\frac{1}{6} \cdot 6$; inverse

29. $12 + (-12)$; inverse

30. $-8 + 8$; inverse

31. $5 + (-5)$; commutative

32. $-9 + 9$; commutative

33. $-3(r + 2)$; distributive

34. $4(k - 5)$; distributive

35. $9 \cdot 1$; identity

36. $1(-4)$; identity

37. $(k + 5) + (-6)$; associative

38. $(m + 4) + (-2)$; associative

39. $(4z + 2r) + 3k$; associative

40. $(6m + 2n) + 5r$; associative

name date hour

2.6 EXERCISES

Use the distributive property to rewrite each expression. Simplify if possible. See Examples 6 and 7.

41. $5(m + 2)$

42. $6(k + 5)$

43. $-4(r + 2)$

44. $-3(m + 5)$

45. $-8(k - 2)$

46. $-4(z - 5)$

47. $-\frac{2}{3}(a + 9)$

48. $-\frac{3}{7}(p + 14)$

49. $\left(r + \frac{8}{3}\right)\frac{3}{4}$

50. $\left(m + \frac{12}{7}\right)\frac{3}{4}$

51. $(8 - k)(-2)$

52. $(9 - r)(-3)$

53. $2(5r + 6m)$

54. $5(2a + 4b)$

55. $-4(3x - 4y)$

56. $-9(5k - 12m)$

57. $5 \cdot 8 + 5 \cdot 9$

58. $(4.6)(3.54) + (4.6)(8.46)$

59. $(7.12)(2.3) + (7.12)(7.7)$

60. $6x + 6m$

61. $9p + 9q$

62. $8(2x) + 8(3y)$

63. $5(7z) + 5(8w)$

64. $11(2r) + 11(3s)$

Use the distributive property to write each of the following without parentheses. See Example 7.

65. $-(3k + 5)$

66. $-(2z + 12)$

67. $-(4y - 8)$

68. $-(3r - 15)$

69. $-(-4 + p)$

70. $-(-12 + 3a)$

71. $-(-1 - 15r)$

72. $-(-14 - 6y)$

Tell whether or not the events in Exercises 73-76 are commutative.

73. Getting out of bed and taking a shower

74. Putting on your right shoe or your left shoe first

75. Taking English or taking history

76. Putting on your shoe or putting on your sock

77. Evaluate $25 - (6 - 2)$ and $(25 - 6) - 2$. Do you think subtraction is associative?

78. Evaluate $180 \div (15 \div 3)$ and $(180 \div 15) \div 3$. Do you think division is associative?

SKILL SHARPENERS

Perform the indicated operations. See Sections 2.2 and 2.3.

79. $(-8) + 14 + 13 + (-2)$

80. $(-4) + 19 + 7 + (-3)$

81. $14 - [(-6) - (4 - 9)]$

82. $10 - [(-12) - (1 - 7)]$

2.7 SIMPLIFYING EXPRESSIONS

1 The properties of addition and multiplication introduced in the previous section are useful for simplifying algebraic expressions.

EXAMPLE 1 Simplifying Expressions

Simplify the following expressions.

(a) $4x + 8 + 9$

Since $8 + 9 = 17$, $4x + \mathbf{8 + 9} = 4x + \mathbf{17}$.

(b) $4(3m - 2n)$

Use the distributive property.

$$\mathbf{4}(3m - 2n) = \mathbf{4}(3m) - \mathbf{4}(2n)$$
$$= 12m - 8n$$

(c) $6 + 3(4k + 5) = 6 + 3(4k) + 3(5)$ Distributive property

$= 6 + 12k + 15$ Multiply.

$= 21 + 12k$ Add.

(d) $5 - (2y - 8) = 5 - 1 \cdot (2y - 8)$ Write $(2y - 8)$ as $1(2y - 8)$.

$= 5 - 2y + 8$ Distributive property

$= 13 - 2y$ Add. ■

WORK PROBLEM 1 AT THE SIDE.

2 A **term** is a single number, or a product of a number and one or more variables raised to powers. Examples of terms include

$$-9x^2, \quad 15y, \quad -3, \quad 8m^2n, \quad \text{and} \quad k.$$

The **numerical coefficient** of the term $9m$ is 9; the numerical coefficient of $-15x^3y^2$ is -15; the numerical coefficient of x is 1; and the numerical coefficient of 8 is 8.

CAUTION It is important to be able to distinguish between *terms* and *factors.* For example, in the expression $8x^3 + 12x^2$, there are two terms. They are $8x^3$ and $12x^2$. On the other hand, in the expression $(8x^3)(12x^2)$, $8x^3$ and $12x^2$ are *factors.*

OBJECTIVES

1. Simplify expressions.
2. Identify terms and numerical coefficients.
3. Identify like terms.
4. Combine like terms.
5. Simplify expressions from word problems.

1. Simplify each expression.

(a) $9k + 12 - 5$

(b) $7(3p + 2q)$

(c) $2 + 5(3z - 1)$

(d) $-3 - (2 + 5y)$

(e) $-(7 - 6k) + 9$

ANSWERS

1. (a) $9k + 7$ (b) $21p + 14q$ (c) $15z - 3$ (d) $-5 - 5y$ (e) $2 + 6k$

2. Give the numerical coefficient of each term.

(a) $15q$

(b) $-2m^3$

(c) $-18m^7q^4$

(d) $-r$

3. Identify each pair of terms as *like* or *unlike*.

(a) $9x$, $4x$

(b) $-8y^3$, $12y^2$

(c) $7x^2y^4$, $-7x^2y^4$

(d) $13kt$, $4tk$

ANSWERS
2. (a) 15 (b) −2 (c) −18 (d) −1
3. (a) like (b) unlike
(c) like (d) like

EXAMPLE 2 Identifying the Numerical Coefficient of a Term
Give the numerical coefficient of each term.

Term	*Numerical Coefficient*
$-7y$	-7
$34r^3$	34
$-26x^5yz^4$	-26
$-k$	-1
r	1 ■

WORK PROBLEM 2 AT THE SIDE.

3 Terms with exactly the same variables (including the same exponents) are called **like terms.** For example, $9m$ and $4m$ have the same variables and are like terms. Also, $6x^3$ and $-5x^3$ are like terms. The terms $-4y^3$ and $4y^2$ have different exponents and are **unlike terms.**

Here are some examples of like terms.

$$5x \text{ and } -12x \qquad 3x^2y \text{ and } 5x^2y$$

Here are some examples of unlike terms.

$$4xy^2 \text{ and } 5xy \qquad -7w^3z^3 \text{ and } 2xz^3$$

WORK PROBLEM 3 AT THE SIDE.

4 The sum or difference of like terms may be expressed as one term by using the distributive property. For example.

$$3x + 5x = (3 + 5)x = 8x.$$

This process is called **combining terms.**

CAUTION Remember that *only like terms* may be combined.

EXAMPLE 3 Combining Like Terms
Combine terms in the following expressions.

(a) $6r + 3r + 2r$

Use the distributive property to combine like terms.

$$6r + 3r + 2r = (6 + 3 + 2)r = 11r$$

(b) $4x + \mathbf{x} = 4x + \mathbf{1x} = 5x$ Note: $x = 1x$.

(c) $16y - 9y = (16 - 9)y = 7y$

(d) $32y + 10y^2$ cannot be simplified because $32y$ and $10y^2$ are unlike terms. ■

WORK PROBLEM 4 AT THE SIDE.

EXAMPLE 4 Simplifying Expressions Involving Like Terms
Simplify the following expressions.

(a) $14y + 2(6 + 3y) = 14y + 2(6) + 2(3y)$ Distributive property
$= 14y + 12 + 6y$ Multiply.
$= 20y + 12$ Combine like terms.

(b) $9k - 6 - 3(2 - 5k) = 9k - 6 - 3(2) - 3(-5k)$ Distributive property
$= 9k - 6 - 6 + 15k$ Multiply.
$= 24k - 12$ Combine like terms.

(c) $-(2 - r) + 10r = -1(2 - r) + 10r$ $-(2 - r) = -1(2 - r)$
$= -1(2) - 1(-r) + 10r$ Distributive property
$= -2 + r + 10r$ Multiply.
$= -2 + 11r$ Combine like terms.

(d) $5(2a - 6) - 3(4a - 9) = 10a - 30 - 12a + 27$ Distributive property
$= -2a - 3$ ■ Combine like terms.

WORK PROBLEM 5 AT THE SIDE.

4. Combine terms.

(a) $4k + 7k$

(b) $4r - r$

(c) $5z + 9z - 4z$

(d) $8p + 8p^2$

5. Simplify.

(a) $10p + 3(5 + 2p)$

(b) $7z - 2 - 4(1 + z)$

(c) $-(3 + 5k) + 7k$

ANSWERS
4. (a) $11k$ (b) $3r$ (c) $10z$ (d) cannot be combined
5. (a) $16p + 15$ (b) $3z - 6$ (c) $2k - 3$

6. Write the following phrase as a mathematical expression, and simplify by combining terms.

Three times a number, subtracted from the sum of the number and 8

5 The next example shows how to simplify the result of converting a phrase in words to a mathematical expression.

EXAMPLE 5 Converting Words to a Mathematical Expression

Write the following phrase as a mathematical expression and simplify: four times a number, subtracted from the sum of twice the number and 4.

Let x represent the number.

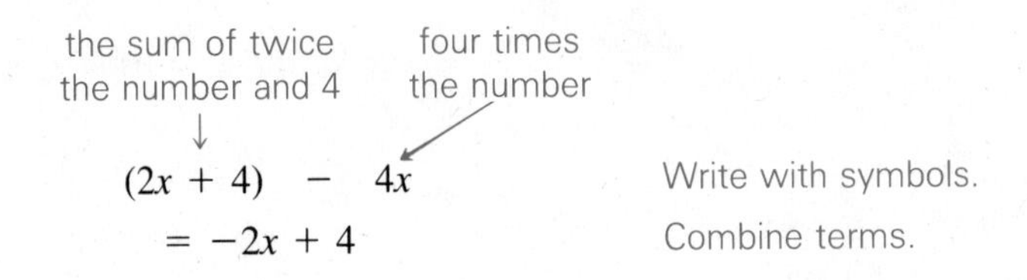

■

WORK PROBLEM 6 AT THE SIDE.

ANSWERS

6. $(x + 8) - 3x$; $-2x + 8$

name date hour

2.7 EXERCISES

Give the numerical coefficient of each of the following terms. See Example 2.

1. $15y$ **2.** $7z$ **3.** $-22m^4$ **4.** $-2k^7$

5. $35a^4b^2$ **6.** $12m^5n^4$ **7.** -9 **8.** 21

9. y^2 **10.** x^4 **11.** $-r$ **12.** $-z$

Write like *or* unlike *for the following groups of terms.*

13. $6m, \quad -14m$ **14.** $-2a, \quad 5a$ **15.** $7z^3, \quad 7z^2$

16. $10m^5, \quad 10m^6$ **17.** $25y, \quad -14y, \quad 8y$ **18.** $-11x, \quad 5x, \quad 7x$

19. $2, \quad 5, \quad -2$ **20.** $-8, \quad 3, \quad 9$ **21.** $p, \quad -5p, \quad 12p$

Simplify the following expressions by combining like terms. See Examples 1 and 3.

22. $9y + 8y$ **23.** $15m + 12m$ **24.** $-4a - 2a$

25. $2k + 9 + 5k + 6$ **26.** $2 + 17z + 1 + 2z$ **27.** $m + 1 - m + 2 + m - 4$

28. $r - 6 - 12r - 4 + 6r$ **29.** $16 - 5m - 4m - 2 + 2m$ **30.** $6 - 3z - 2z - 5 + z - 3z$

31. $-\frac{10}{3} + x + \frac{1}{4}x - 6 - \frac{5}{2}x$ **32.** $-p + \frac{1}{5}p - \frac{3}{5}p - 4 - \frac{1}{3}p$ **33.** $-4.3r + 3.9 - r + .6 + 2.2r$

34. $1.9 + 7.3 + 12x - 1.8 + 1.4x$ **35.** $2p^2 + 3p^2 - 8p^3 - 6p^3$ **36.** $9y^3 + 8y^3 - 7y^2 - 7y^2$

37. $-7.9q^2 + 2.8q - 14.4 + 5.1q^2 - 9.5q - 6.9$ **38.** $12.1m^3 - 3.8m^2 + 5.0 - 3.7m^3 + 4.2m^2 - 1.9$

Use the distributive property and combine like terms to simplify the following expressions. See Example 4.

39. $6(5t + 11)$

40. $2(3x + 4)$

41. $-3(n + 5)$

42. $-4(y - 8)$

43. $5(-2 + t) + 4t$

44. $6t - (3t + 2)$

45. $-3(2r - 3) + 2(5r + 3)$

46. $-4(5y - 7) + 3(2y - 5)$

47. $8(2k - 1) - (4k - 3)$

48. $6(3p - 2) - (5p + 1)$

49. $-2(-3k + 2) - (5k - 6) - 3k - 5$

50. $-2(3r - 4) - (6 - r) + 2r - 5$

51. $-8.8(3y - 4) - 2.6(8 - 4y)$

52. $4.8(7q - 7) + 1.3(4 - 6q)$

Convert the following phrases into mathematical expressions. Use x as the variable. Combine terms when possible. See Example 5.

53. Four times a number, added to the sum of the number and -12

54. Five times a number, subtracted from the sum of the number and 6

55. A number multiplied by 4, subtracted from the sum of 12 and eight times the number

56. A number subtracted from twice the number, subtracted from 7 times the number

57. Six times a number added to -8, subtracted from nine times the sum of three times the number and 5

58. Twelve is multiplied by the sum of four times a number and 3, with the result subtracted from the difference between ten and triple the number. (*Hint: Triple* means "three times.")

SKILL SHARPENERS

Find the additive inverse of each number. See Section 2.1.

59. 6

60. 15

61. -4

62. -8

Add a number to each expression so that the result is just x.

63. $x - 7$

64. $x + 10$

65. $x + 6$

66. $x - 3$

Historical Reflections

Historical Reflections

Mathematics of the Egyptians

At least five thousand years ago, the Egyptians used a method of writing numerals much different from our own. Our Hindu-Arabic system involves place value based on powers of ten (hence the name *decimal system*). The Egyptians used a system based on simple grouping, using the symbols shown below for writing numbers through 9,999,999.

Number	Symbol	Description
1	|	Stroke
10	∩	Heel bone
100	𓍢	Scroll
1000	𓆼	Lotus flower
10,000	𓂭	Pointing finger
100,000	𓆐	Burbot fish
1,000,000	𓁨	Astonished person

A typical Egyptian numeral is shown below:

𓆐𓆐𓆼𓆼𓆼𓆼𓆼𓍢𓍢𓍢𓍢∩∩∩∩∩∩∩∩∩|||||||

There are 2 symbols for 100,000, 5 symbols for 1000, 4 symbols for 100, 9 symbols for 10, and 7 symbols for 1. Thus, the Egyptian numeral above represents

$$2(100{,}000) + 5(1000) + 4(100) + 9(10) + 7(1) = 205{,}497$$

in our system of numeration.

Much of our knowledge of Egyptian mathematics comes from the Rhind papyrus, which dates back some 3800 years. A small portion of the papyrus is shown here.

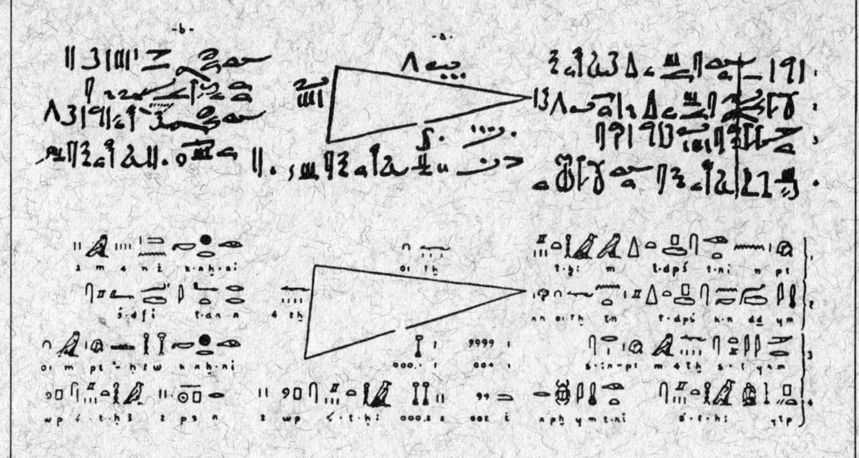

Art: Courtesy of the Trustees of the British Museum

CHAPTER 2 SUMMARY

KEY TERMS

2.1	**whole numbers**	The set of whole numbers is $\{0, 1, 2, 3, 4, 5, \ldots\}$.
	negative number	A negative number is located to the *left* of 0 on the number line.
	positive number	A positive number is located to the *right* of 0 on the number line.
	signed numbers	Signed numbers are either positive or negative.
	integers	The set of integers is $\{\ldots, -3, -2, -1, 0, 1, 2, 3, \ldots\}$.
	rational numbers	Rational numbers can be written as quotients of two integers, with denominator not 0.
	irrational numbers	Irrational numbers are non-rational numbers represented by points on the number line.
	real numbers	Real numbers include all numbers that can be represented by points on the number line.
	additive inverse	The additive inverse of a number a is the number that is the same distance from 0 on the number line as a, but on the opposite side of 0. This number is also known as the **opposite** of a.
	absolute value	The absolute value of a number is the distance between 0 and the number on the number line.
2.2	**sum**	The answer to an addition problem is called the sum.
2.3	**difference**	The answer to a subtraction problem is called the difference.
2.4	**product**	The answer to a multiplication problem is called the product.
2.5	**multiplicative inverse**	Pairs of numbers whose product is 1 are called multiplicative inverses of each other. Multiplicative inverses are also known as **reciprocals.**
	quotient	The answer to a division problem is called the quotient.
2.6	**identity element for addition**	When the identity element for addition, which is 0, is added to a number, the number is unchanged.
	identity element for multiplication	When a number is multiplied by the identity element for multiplication, which is 1, the number is unchanged.
2.7	**term**	A term is a single number, or a product of a number and one or more variables raised to powers.
	numerical coefficient	The numerical factor in a term is its numerical coefficient.
	like terms	Terms with exactly the same variables (including the same exponents) are called like terms.

NEW SYMBOLS

$|x|$ absolute value of x

QUICK REVIEW

Section	Concepts	Examples
2.1 Real Numbers and the Number Line	**Ordering Real Numbers** The smaller of two numbers is the one corresponding to the point that is to the left of the other on the number line.	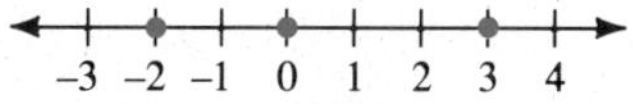 $-2 < 3$ $3 > 0$ $0 < 3$
	The additive inverse of a is $-a$.	$-(+5) = -5$ $-(-7) = 7$ $-0 = 0$
	The absolute value of a is symbolized by $\|a\|$. It is the distance between a and 0 on the number line.	$\|13\| = 13$ $\|0\| = 0$ $\|-5\| = 5$
2.2 Addition of Real Numbers	We add two numbers having the same sign by adding the absolute values of the numbers. The result has the same sign as the numbers being added.	$9 + 4 = 13$ $-8 + (-5) = -13$
	We add two numbers having different signs by first finding the difference between the absolute values of the numbers. Give the answer the same sign as the number with the larger absolute value.	$7 + (-12) = -5$ $-5 + 13 = 8$
2.3 Subtraction of Real Numbers	**Definition of Subtraction** $a - b = a + (-b)$ Subtracting signed numbers: **1.** Change the subtraction symbol to addition. **2.** Change the sign of the number being subtracted. **3.** Add, as in the previous section.	$5 \mathbf{- (-2)} = 5 \mathbf{+ 2} = \mathbf{7}$ $-3 - 4 = -3 + (-4) = \mathbf{-7}$ $-2 \mathbf{- (-6)} = -2 \mathbf{+ 6}$ $= 4$ $13 - (-8) = 13 + 8$ $= 21$

Section	Concepts	Examples
2.4, 2.5 Multiplication and Division of Real Numbers	**Definition of Division** $\frac{a}{b} = a \cdot \frac{1}{b} \quad (b \neq 0)$	$\frac{10}{2} = 10 \cdot \frac{1}{2} = 5$
	Multiplying and Dividing Signed Numbers The product (or quotient) of two numbers having the *same sign* is *positive;* the product (or quotient) of two numbers having *different signs* is *negative*.	$6 \cdot 5 = 30 \qquad (-7)(-8) = 56$ $\frac{10}{2} = 5 \qquad \frac{-24}{-6} = 4$ $(-6)(5) = -30 \qquad (6)(-5) = -30$ $\frac{-18}{9} = -2 \qquad \frac{49}{-7} = -7$
	Division *by* 0 is undefined.	$\frac{0}{5} = 0 \qquad \frac{5}{0}$ is undefined
2.6 Properties of Addition and Multiplication	**Commutative** $a + b = b + a$ $ab = ba$	$7 + (-1) = -1 + 7$ $5(-3) = (-3)5$
	Associative $(a + b) + c = a + (b + c)$ $(ab)c = a(bc)$	$3 + (4 + 8) = (3 + 4) + 8$ $[(-2)(6)](4) = (-2)[(6)(4)]$
	Identity $a + 0 = a \qquad 0 + a = a$ $a \cdot 1 = a \qquad 1 \cdot a = a$	$-7 + 0 = -7 \qquad 0 + (-7) = -7$ $9 \cdot 1 = 9 \qquad 1 \cdot 9 = 9$
	Inverse $a + (-a) = 0$ $-a + a = 0$ $a \cdot \frac{1}{a} = 1 \qquad \frac{1}{a} \cdot a = 1 \quad (a \neq 0)$	$7 + (-7) = 0 \qquad -7 + 7 = 0$ $-2\left(-\frac{1}{2}\right) = 1 \qquad -\frac{1}{2}(-2) = 1$
	Distributive $a(b + c) = ab + ac$ $(b + c)a = ba + ca$	$5(4 + 2) = 5(4) + 5(2)$ $(4 + 2) \cdot 5 = 4(5) + 2(5)$
2.7 Simplifying Expressions	Only like terms may be combined.	$-3y^2 + 6y^2 + 14y^2 = 17y^2$ $4(3 + 2x) - 6(5 - x)$ $= 12 + 8x - 30 + 6x$ $= 14x - 18$

name date hour

CHAPTER 2 REVIEW EXERCISES

[2.1] *Circle the smaller number in each pair.**

1. $-9, \quad 4$ **2.** $3. \quad -5$ **3.** $-8, \quad -7$ **4.** $-\frac{3}{4}, \quad -\frac{7}{8}$

5. $\frac{5}{3}, \quad \left|-\frac{3}{2}\right|$ **6.** $9, \quad |-7|$ **7.** $-|-2|, -|-9|$ **8.** $-|-7|, -|-4|$

Write true *or* false *for each statement.*

9. $-9 < 9$ **10.** $7 < -7$ **11.** $0 \leq -2$ **12.** $-5 \geq -5$

13. $3 \leq -(-5)$ **14.** $9 \geq -(-10)$ **15.** $-3.25 > -2.25$ **16.** $-5.493 < -4.875$

Graph each group of numbers on the indicated number line.

17. $5, \ -4, \ 3, \ -2, \ 0$

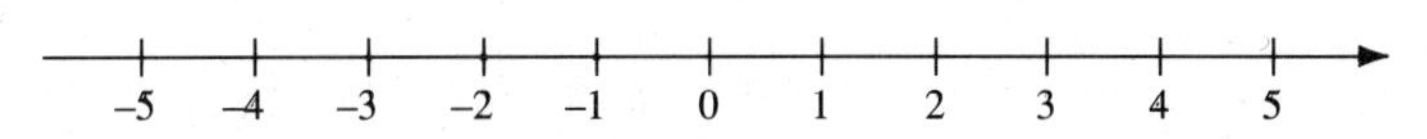

18. $-1, \ -3, \ |-4|, \ |-1|$

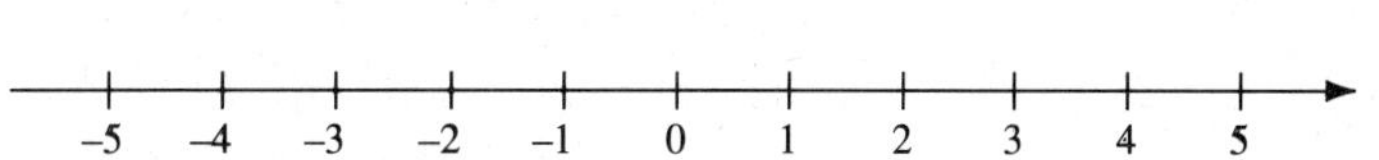

19. $3\frac{1}{4}, \ -2\frac{4}{5}, \ -1\frac{1}{8}, \ \frac{2}{3}$

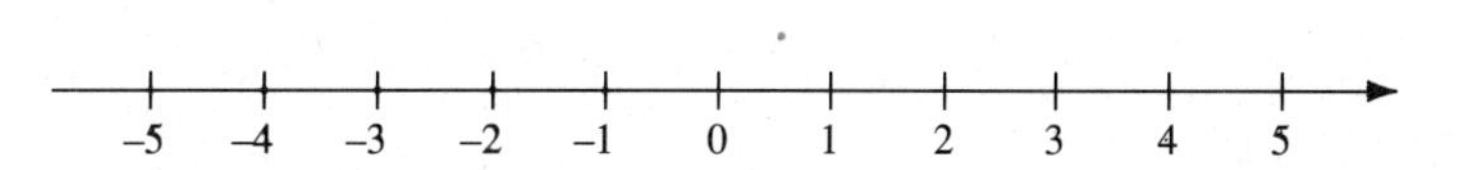

20. $|-2|, \ -|-5|, \ -|3|, \ -|0|$

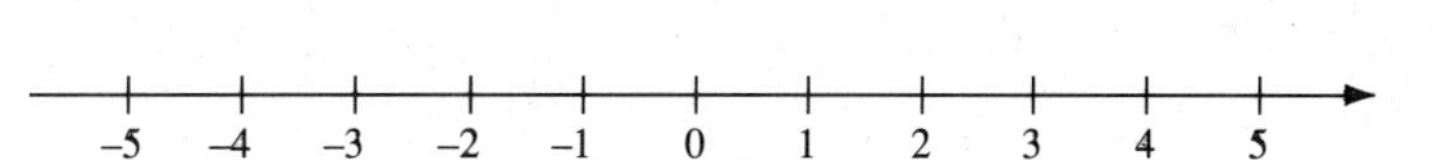

[2.2] *Find the sums in Exercises 21–37.*

21. $-9 + 3$ **22.** $12 + (-15)$ **23.** $-7 + (-8)$

24. $\frac{7}{8} + \left(-\frac{3}{10}\right)$ **25.** $\frac{7}{12} + \left(-\frac{2}{9}\right)$ **26.** $-11.3 + (-2.9)$

27. $1.64 + (-2.97)$ **28.** $-7 + (-2 + 8)$ **29.** $(-9 + 6) + (-10)$

30. $[-3 + (-5)] + (-8)$ **31.** $[(-2) + (-11)] + [7 + (-12)]$

32. $[(-4) + 6 + (-9)] + [-3 + (-5)]$ **33.** $[(-6) + (-7) + 8] + [8 + (-15)]$

*For help with any of these exercises, look in the section given in brackets.

34. Tayari has \$13. She spends \$18 for a new compact disc. What is her new balance?

35. The temperature is 4°. In one hour it drops 12°. What is the new temperature?

36. On a cold day, the temperature was −8°. It then increased by 11°. Find the new temperature.

37. One year, Jeff's Rental Properties, Inc., spent \$2800 on advertising. The next year Jeff changed the amount spent on advertising by −\$1350. How much was spent on advertising during the second year?

[2.3] *Find the differences.*

38. $-6 - (-4)$

39. $-2 - (-11)$

40. $6 - (-10)$

41. $15 - (-3)$

42. $-8 - 9$

43. $-12 - 27$

44. $\frac{3}{4} - \left(-\frac{2}{3}\right)$

45. $-\frac{1}{5} - \left(-\frac{7}{10}\right)$

46. $-12.8 - (-15.4)$

47. $-46.9 - (-21.8)$

48. $(-9 + 6) - (-3)$

49. $(-15 - 7) - (-9)$

50. $-[3 - (-2)] - 9$

51. $-[14 - (-3)] - 10$

[2.4] *Simplify.*

52. $(-11)(-3)$

53. $17(-5)$

54. $(-9)(12)$

55. $-\frac{4}{5}\left(-\frac{10}{7}\right)$

56. $-\frac{3}{8}\left(-\frac{16}{15}\right)$

57. $(-11.3)(2.5)$

58. $(-9.4)(-2.8)$

59. $4(3 - 7)$

60. $(6 - 4)(9 - 11)$

61. $(5 - 1)(3 - 8)$

62. $8(-9) - (6)(-2)$

63. $3(-7) - (-9)$

name date hour

Evaluate the following expressions, given $x = -5$, $y = 4$, and $z = -3$.

64. $5x - 4z$

65. $2y + 7x$

66. $5z + 11y - x$

67. $y^2 - 2z^2$

68. $(2x - 8y)(z^2)$

69. $3z^2 - 4x^2$

[2.5] *Find the quotients.*

70. $\frac{-25}{-5}$

71. $\frac{280}{-7}$

72. $-\frac{2}{3} \div \frac{1}{2}$

73. $44.8 \div (-4)$

74. $\frac{36}{9 + (-3)}$

75. $\frac{-50}{-4 - 1}$

76. $\frac{8 - 4(-2)}{-5(3) - 1}$

77. $\frac{5(-3) - 8(3)}{(-5)(-4) + (-7)}$

78. $\frac{(-1 - 4) - (-2)}{-1 + (-2)}$

79. $\frac{(-12 + 15) - (-4 - 5)}{-3 - (2 - 1)}$

Write a numerical expression for each phrase, and simplify the expression.

80. The quotient of the sum of 13 and -5, and the product of 5 and 8

81. The product of -7 and 4, divided by the difference between -2 and 2

[2.6] *Label each statement as an example of the commutative, associative, identity, inverse, or distributive property.*

82. $8.974 \cdot 1 = 8.974$

83. $-\frac{2}{3} + \frac{2}{3} = 0$

84. $7 + 4m = 4m + 7$

85. $8(4 \cdot 3) = (8 \cdot 4) \cdot 3$

86. $\frac{5}{8} \cdot \frac{8}{5} = 1$

87. $9p + 0 = 9p$

88. $5(2x + 3y) = 5(2x) + 5(3y)$

[2.7] *Combine terms whenever possible.*

89. $15p^2 - 7p^2 + 8p^2$

90. $5p^2 - 4p + 6p + 11p^2$

91. $-2(3k - 5) + 2(k + 1)$

92. $7(2m + 3) - 2(8m - 4)$

MIXED REVIEW EXERCISES*

Perform the indicated operations.

93. $[(-3) + 8 - (-6)] + [-5 - (-9)]$

94. $\left(\frac{6}{7}\right)^2$

95. $-|(-5)(-7)| - (-7)$

96. $(-4)^3$

97. $\frac{6(-4) + 4(-3)}{5(-3) + (-3)}$

98. $\frac{5}{12} - \frac{3}{8}$

99. $\frac{12^2 + 1^2 - 5}{10^2 - (-3)(-20)}$

100. $\frac{8^2 + 6^2}{3^2 + 4^2}$

101. $-12(-3) - 7(-4)$

102. $4\frac{1}{3} - 2\frac{5}{6}$

103. $-2 + [(-3 + 12) - (-3 - 6)]$

104. $\frac{9}{7} \div \left(-\frac{12}{5}\right)$

105. $(-6 - 3) - [(-4 - 10) - (-6)]$

106. $[(-8) + (-2) + 3] + [8 - (-13)]$

107. $\frac{4}{5} \cdot \frac{15}{2}$

108. $|4(-5)| - |-6|$

Write a numerical expression and simplify it, if possible. Use x as the variable, if one is needed.

109. In 1988, a company spent $12,500 on advertising. In 1989, the amount spent on advertising was reduced by $1700. How much was spent on advertising in 1989?

110. The quotient of a number and five more than the number

*The order of exercises in this final group does not correspond to the order in which topics occur in the chapter. This random ordering should help you prepare for the chapter test in yet another way.

name date hour

CHAPTER 2 TEST

Graph each set of numbers on the indicated number line.

1. $-4, 4, -3, 0, 2\frac{1}{2}, -1\frac{7}{8}$

1. 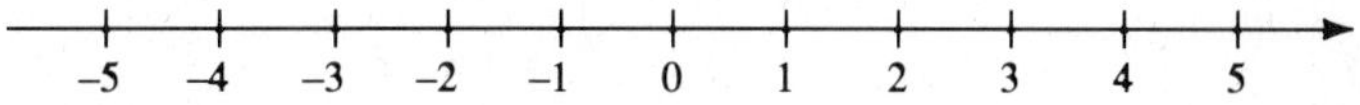

2. $|-2|, -|3|, -2\frac{3}{8}, -|-1|$

2. 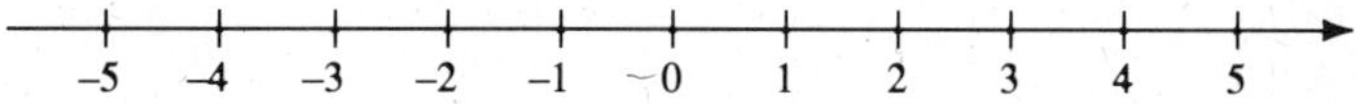

Select the smaller number from each pair.

3. $-.742, \quad -.705$ **3.** ______________

4. $6, \quad -|-8|$ **4.** ______________

Write a numerical expression for each phrase and simplify.

5. Twice 17 subtracted from 11 **5.** ______________

6. The quotient of 9 and the difference between 6 and 8 **6.** ______________

Perform the indicated operations whenever possible.

7. $-9 - (4 - 11) + (-5)$ **7.** ______________

8. $-2\frac{1}{5} + 5\frac{1}{4}$ **8.** ______________

9. $-6 - [-5 + (8 - 9)]$ **9.** ______________

10. $3^2 + (-7) - (2^3 - 5)$ **10.** ______________

11. $(-4)(-12) + 5(-3) + (-5)^2$ **11.** ______________

12. $\dfrac{-7 - (-5 + 1)}{-4 - (-3)}$ **12.** ______________

13. $\dfrac{-6[5 - (-1 + 4)]}{-9[2 - (-1)] - 6(-4)}$ **13.** ______________

14. $\dfrac{15(-4 - 2)}{16(-2) + (-7 - 1)(-3 - 1)}$ **14.** ______________

Find the solution for each equation. Choose solutions from the domain $\{-9, -5, -4, -2, -1, 3, 9\}$.

15. ______ 15. $\frac{t}{-3} = 3$

16. ______ 16. $2x + 1 = -7$

17. ______ 17. $-4x - 2 = 6$

Evaluate the following expressions, given $m = -2$ and $p = 6$.

18. ______ 18. $4m - 3p^2$

19. ______ 19. $\frac{6m + 5p}{p - 3}$

Solve the following problem.

20. ______ 20. Chris owed his brother $12. He repaid $8 and then borrowed $9. What positive or negative number represents his resulting financial status?

Match the property in the first column with the example of it from the second column.

21. ______ 21. Associative **(a)** $8(-3) = (-3)8$

22. ______ 22. Commutative **(b)** $5 + (-4 + x) = [5 + (-4)] + x$

23. ______ 23. Inverse **(c)** $2(9) + 2(5) = 2(9 + 5)$

24. ______ 24. Identity **(d)** $5\left(\frac{1}{5}\right) = 1$

25. ______ 25. Distributive **(e)** $-25 + 0 = -25$

Simplify by combining like terms.

26. ______ 26. $4(2m + 1) - (m + 5)$

3

SOLVING EQUATIONS AND INEQUALITIES

3.1 THE ADDITION PROPERTY OF EQUALITY

OBJECTIVES

1. Identify linear equations.
2. Use the addition property of equality.
3. Simplify equations, and then use the addition property of equality.
4. Solve equations that have no solution or infinitely many solutions.

The word *algebra* is derived from the title of a treatise written in the ninth century by the Persian mathematician al-Khowarizmi. Today, algebra involves much more than equation solving, but during the Middle Ages the word came to mean just that: the science of equations. In order to solve many applied problems, we must be able to solve equations. The simplest type of equation is the *linear equation*. Methods for solving linear equations will be introduced in this section.

1 Before trying to solve a linear equation, we must be able to recognize one.

A **linear equation** can be written in the form

$$ax + b = c$$

for real numbers a, b, and c, with $a \neq 0$.

Linear equations are solved by using a series of steps to produce a simpler equation of the form

$$x = \text{a number.}$$

2 According to the equation $x - 5 = 2$, both $x - 5$ and 2 represent the same number, since this is the meaning of the equals sign. To solve the equation, we change the left side from $x - 5$ to just x. This is done by adding 5 to $x - 5$. To keep the two sides equal, we must also add 5 to the right side.

$x - 5 = 2$	Given equation
$x - 5 + 5 = 2 + 5$	Add 5 to each side.

Here 5 was added to each side of the equation. Next, we simplify each side separately to get $x = 7$. The solution of the given equation is 7. Check by replacing x with 7 in the given equation.

$x - 5 = 2$	Given equation
$7 - 5 = 2$	Let $x = 7$.
$2 = 2$	True

1. Complete each step in solving the following equations.

(a) $r + 11 = 20$

$r + 11 + \_ = 20 + \_$

$r = \_$

(b) $p + 2 = 8$

$p + 2 + \_ = 8 + \_$

$p = \_$

(c) $z + 9 = 3$

$z + 9 + \_ = 3 + \_$

$z = \_$

2. Solve by adding or subtracting the same number from each side of the equation.

(a) $m - 2 = 6$

(b) $y + \frac{3}{4} = \frac{19}{4}$

(c) $a + 2 = -3$

(d) $p + 6 = 2$

ANSWERS

1. (a) −11, −11, 9 (b) −2, −2, 6 (c) −9, −9, −6
2. (a) 8 (b) 4 (c) −5 (d) −4

Since the final statement is true, 7 checks as the solution.

To solve the equation above we added the same number to each side, as justified by the following **addition property of equality.**

ADDITION PROPERTY OF EQUALITY

If *A, B,* and *C* are real numbers, then the equations

$$A = B \quad \text{and} \quad A + C = B + C$$

have exactly the same solution. In other words, we can add the same number to each side of an equation without changing the solution.

The addition property of equality applies to any equation, not just linear equations.

WORK PROBLEM 1 AT THE SIDE.

EXAMPLE 1 Using the Addition Property of Equality

Solve the equation $x - 16 = 7$.

If the left side of this equation were just x, the solution could be found. Get x alone by using the addition property of equality and adding 16 to each side.

$x - 16 = 7$

$x - 16 + 16 = 7 + 16$ Add 16 on each side.

$x = 23$ Combine terms.

Check by substituting 23 for x in the original equation.

$x - 16 = 7$ Given equation

$23 - 16 = 7$ Let $x = 23$.

$7 = 7$ True

Since the check results in a true statement, 23 is the solution. ■

In this example, why was 16 added to each side of the equation $x - 16 = 7$? The equation would be solved if it could be rewritten so that one side contained only the variable and the other side contained only a number. Since $x - 16 + 16 = x + 0 = x$, adding 16 to the left side simplified that side to just x, the variable, as desired.

The addition property of equality says that the same number may be *added* to each side of an equation. As was shown in Chapter 2, subtraction is defined in terms of addition. For this reason, the following rule also applies when solving an equation.

The same number may be subtracted from each side of an equation without changing the solution.

WORK PROBLEM 2 AT THE SIDE.

EXAMPLE 2 Subtracting a Variable Expression

Solve the equation $3k + 17 = 4k$.

As a first step, get all terms that contain variables on the same side of the equation. One way to do this is to subtract $3k$ from each side.

$$3k + 17 = 4k$$

$$3k + 17 - 3k = 4k - 3k \quad \text{Subtract } 3k.$$

$$17 = k \quad \text{Combine terms.}$$

The solution is 17. Check the solution by replacing k with 17 in the original equation. ■

Another way to solve the equation $3k + 17 = 4k$ is to first subtract $4k$ from each side, as follows.

$$3k + 17 = 4k$$

$$3k + 17 - 4k = 4k - 4k \quad \text{Subtract } 4k.$$

$$17 - k = 0 \quad \text{Combine terms.}$$

Now subtract 17 from both sides.

$$17 - k - 17 = 0 - 17 \quad \text{Subtract 17.}$$

$$-k = -17 \quad \text{Combine terms.}$$

This result gives the value of $-k$, but not of k itself. However, this result does say that the additive inverse of k is -17, which means that k must be 17, the same result we obtained in Example 2.

$$-k = -17$$

$$k = 17$$

This situation may be generalized as shown below.

If a is a number and $-x = a$, then $x = -a$.

WORK PROBLEM 3 AT THE SIDE.

3 Sometimes an equation must be simplified as a first step in its solution.

EXAMPLE 3 Simplifying Terms in an Equation

Solve the equation $4r + 5r - 3 + 8 - 3r - 5r = 12 + 8$.

$$4r + 5r - 3r - 5r - 3 + 8 = 12 + 8 \quad \text{Collect like terms.}$$

$$r + 5 = 20 \quad \text{Combine terms.}$$

$$r + 5 - 5 = 20 - 5 \quad \text{Subtract 5.}$$

$$r = 15 \quad \text{Combine terms.}$$

The solution of the given equation is 15. (Check this.) ■

WORK PROBLEM 4 AT THE SIDE.

3. Solve each equation.

(a) $5m = 4m + 6$

(b) $3y = 2y - 9$

(c) $2k - 8 = 3k$

(d) $\frac{7}{2}m + 1 = \frac{9}{2}m$

4. Solve each equation.

(a) $7p + 2p - 8p + 5 = 9 + 1$

(b) $11k - 6 - 4 - 10k = -5 + 5$

(c) $-4 + 3 - 2m + 3m = 10 - 5 - 7$

ANSWERS

3. (a) 6 (b) -9 (c) -8 (d) 1

4. (a) 5 (b) 10 (c) -1

5. Solve each equation.

(a) $2(a + 4) - (3 + a) = 8$

(b) $-(5 - 3r) + 4(-r + 1) = 1$

(c) $-3(m - 4) + 2(5 + 2m) = 29$

6. Solve each equation.

(a) $2(x - 6) = 2x - 12$

(b) $3x + 6(x + 1) = 9x - 4$

(c) $-3(5 - x) = 3x - 3$

(d) $5(2x - 3) - 2(5x + 1) = -17$

ANSWERS
5. (a) 3 (b) −2 (c) 7
6. (a) all real numbers
(b) no solution
(c) no solution
(d) all real numbers

EXAMPLE 4 Using the Distributive Property to Simplify an Equation

Solve the equation $3(2 + 5x) - (1 + 14x) = 6$.

$3(2 + 5x) - (1 + 14x) = 6$	
$3(2) + 3(5x) - 1(1) - 1(14x) = 6$	Distributive property
$6 + 15x - 1 - 14x = 6$	Multiply.
$x + 5 = 6$	Combine terms.
$x = 1$	Subtract 5 from each side.

Check by substituting 1 for x in the original equation. ■

WORK PROBLEM 5 AT THE SIDE.

4 The equations solved so far each have had exactly one solution. Sometimes this is not the case, as shown in the next examples.

EXAMPLE 5 Solving an Equation That Has Infinitely Many Solutions

Solve $5x - 15 = 5(x - 3)$.

$5x - 15 = 5(x - 3)$	
$5x - 15 = 5x - 15$	Distributive property
$5x - 15 + 15 = 5x - 15 + 15$	Add 15 to each side.
$5x = 5x$	Combine terms.
$5x - 5x = 5x - 5x$	Subtract $5x$ from each side.
$0 = 0$	

The variable has "disappeared." When this happens, look at the resulting statement ($0 = 0$). Since the statement is a *true* one, *any* real number is a solution. Indicate the solution as "all real numbers." ■

CAUTION When you are solving an equation like the one in Example 5, do not write "0" as the solution. While 0 is a solution, there are infinitely many other solutions.

EXAMPLE 6 Solving an Equation That Has No Solution

Solve $2x + 3(x + 1) = 5x + 4$.

$2x + 3(x + 1) = 5x + 4$	
$2x + 3x + 3 = 5x + 4$	Distributive property
$5x + 3 = 5x + 4$	Combine terms.
$5x + 3 - 5x = 5x + 4 - 5x$	Subtract $5x$ from each side.
$3 = 4$	Combine terms.

Again, the variable has disappeared, but this time a *false* statement ($3 = 4$) results. When this happens, the equation has no solution. Indicate this by writing "no solution." ■

WORK PROBLEM 6 AT THE SIDE.

name date hour

3.1 EXERCISES

Solve each equation by using the addition property of equality. Check each solution. See Examples 1, 2, 5, and 6.

1. $x - 3 = 7$

2. $x + 5 = 13$

3. $7 + k = 5$

4. $9 + m = 4$

5. $3r = 2r + 10$

6. $2p = p + 3$

7. $7z = -8 + 6z$

8. $4y = 3y - 5$

9. $2p + 6 = 10 + p$

10. $5r + 2 = -1 + 4r$

11. $2k + 2 = -3 + k$

12. $6 + 7x = 6x + 3$

13. $x - 5 = 2x + 6$

14. $-3r + 7 = -4r - 19$

15. $6z + 3 = 5z - 3$

16. $6t + 5 = 5t + 7$

17. $2p = p + \frac{1}{2}$

18. $5m = 4m + \frac{2}{3}$

19. $\frac{4}{3}z = \frac{1}{3}z - 5$

20. $\frac{9}{5}m = \frac{4}{5}m + 6$

21. $6x = 6x + 5$

22. $-3y = -3y + 4$

23. $5x + 1 = 5x + 1$

24. $7 - 2x = 7 - 2x$

Solve the following equations. First simplify each side of the equation as much as possible. Check each solution. See Examples 3, 5, and 6.

25. $4x + 3 + 2x - 5x = 2 + 8$

26. $3x + 2x - 6 + x - 5x = 9 + 4$

27. $9r + 4r + 6 - 8 = 10r + 6 + 2r$

28. $-3t + 5t - 6t + 4 - 3 = -3t + 2$

29. $11z + 2 + 4z - 3z = 5z - 8 + 6z$

30. $2k + 8k + 6k - 4k - 8 + 2 = 3k + 2 + 10k$

31. $5k + 4 - 12k + 4k - 8 = -3k + 2$

32. $6x - 7x + 3x - 4 = 12 + 2x - 5$

Solve the following equations. Check each solution. See Examples 4, 5, and 6.

33. $(5y + 6) - (3 + 4y) = 9$

34. $(8p - 3) - (7p + 1) = -2$

35. $2(r + 5) - (9 + r) = -1$

36. $4(y - 6) - (3y + 2) = 8$

37. $-6(2a + 1) + (13a - 7) = 4$

38. $-5(3k - 3) + (1 + 16k) = 2$

39. $4(7x - 1) + 3(2 - 5x) = 4(3x + 5)$

40. $9(2m - 3) - 4(5 + 3m) = 5(4 + m)$

41. $-2(8p + 7) - 3(4 - 7p) = 2(3 + 2p) - 6$

42. $-5(8 - 2z) + 4(7 - z) = 7(8 + z) - 3$

43. $5(-2x + 1) = -7(x + 2) + 19 - 3x$

44. $4(2 - 3x) = 10(1 - x) - 2x - 2$

45. $2(3x + 4) - 3(7 + x) = 3(x - 4) + 7$

46. $9(4 - 2x) + 3(5 - x) = -7(3x + 2) - 1$

SKILL SHARPENERS

Use the associative, inverse and identity properties to simplify. See Section 2.6.

47. $3\left(\frac{1}{3}m\right)$

48. $5\left(\frac{1}{5}q\right)$

49. $\frac{1}{7}(7y)$

50. $\frac{1}{12}(12z)$

51. $-\frac{1}{9}(-9x)$

52. $-\frac{1}{4}(-4t)$

53. $-\frac{5}{8}\left(-\frac{8}{5}r\right)$

54. $-\frac{9}{10}\left(-\frac{10}{9}p\right)$

3.2 THE MULTIPLICATION PROPERTY OF EQUALITY

The addition property of equality by itself is not enough to solve some equations, such as $3x + 2 = 17$.

$$3x + 2 = 17$$

$$3x + 2 - 2 = 17 - 2 \qquad \text{Subtract 2 from each side.}$$

$$3x = 15 \qquad \text{Simplify.}$$

Instead of just x on the left side, the equation has $3x$. Another property is needed to change $3x = 15$ to $x =$ a number.

OBJECTIVES

1. Use the multiplication property of equality.
2. Simplify equations, and then use the multiplication property of equality.
3. Solve equations such as $-r = 4$.
4. Use the multiplication property of equality to solve equations with decimals.

1 If $3x = 15$, then $3x$ and 15 both represent the same number. Multiplying both $3x$ and 15 by the same number will also result in an equality. The **multiplication property of equality** states that we can multiply each side of an equation by the same number without changing the solution.

MULTIPLICATION PROPERTY OF EQUALITY

If A, B, and C ($C \neq 0$) represent real numbers, the equations

$$A = B \quad \text{and} \quad AC = BC$$

have exactly the same solution.

In other words, we can multiply each side of an equation by the same nonzero number without changing the solution.

This property can be used to solve $3x = 15$. The $3x$ on the left must be changed to $1x$, or x, instead of $3x$. Get x by multiplying each side of the equation by $\frac{1}{3}$. We use $\frac{1}{3}$ because $\frac{1}{3} \cdot 3 = \frac{3}{3} = 1$, since $\frac{1}{3}$ is the reciprocal of 3.

$$3x = 15$$

$$\frac{1}{3}(3x) = \frac{1}{3} \cdot 15 \qquad \text{Multiply each side by } \frac{1}{3}.$$

$$\left(\frac{1}{3} \cdot 3\right)x = \frac{1}{3} \cdot 15 \qquad \text{Associative property}$$

$$1x = 5 \qquad \text{Multiplicative inverse property}$$

$$x = 5 \qquad \text{Multiplicative identity property}$$

The solution of the equation is 5. We can check this result in the original equation.

WORK PROBLEM 1 AT THE SIDE.

Just as the addition property of equality permits *subtracting* the same number from each side of an equation, the multiplication property of equality permits *dividing* each side of an equation by the same nonzero number.

1. Check that 5 is the solution of $3x = 15$.

ANSWER

1. Since $3(5) = 15$, the solution of $3x = 15$ is 5.

2. Solve each equation.

(a) $7m = 56$

(b) $3r = -12$

(c) $8y = 108$

(d) $-2m = 16$

(e) $-6p = -14$

For example, the equation $3x = 15$, which we just solved by multiplication, also could be solved by dividing each side by 3, as follows.

$$3x = 15$$

$$\frac{3x}{3} = \frac{15}{3} \qquad \text{Divide by 3.}$$

$$x = 5$$

We can divide each side of an equation by the same nonzero number without changing the solution.

NOTE In practice, it is usually easier to multiply on each side if the coefficient of the variable is a fraction, and divide on each side if the coefficient is an integer. For example, to solve

$$-\frac{3}{4}x = 12$$

it is easier to multiply by $-\frac{4}{3}$ than to divide by $-\frac{3}{4}$. On the other hand, to solve

$$-5x = -20$$

it is easier to divide by -5 than to multiply by $-\frac{1}{5}$.

EXAMPLE 1 Dividing Each Side of an Equation by a Nonzero Number

Solve the equation $25p = 30$.

Get p (instead of $25p$) on the left by using the multiplication property of equality to divide each side of the equation by 25, the coefficient of p.

$$25p = 30$$

$$\frac{25p}{25} = \frac{30}{25} \qquad \text{Divide by 25.}$$

$$p = \frac{30}{25} = \frac{6}{5} \qquad \text{Reduce to lowest terms.}$$

To check, substitute $\frac{6}{5}$ for p in the given equation.

$$25p = 30$$

$$\frac{25}{1}\left(\frac{6}{5}\right) = 30 \qquad \text{Let } p = \frac{6}{5}.$$

$$30 = 30 \qquad \text{True}$$

The solution is $\frac{6}{5}$. ■

WORK PROBLEM 2 AT THE SIDE.

ANSWERS

2. (a) 8 (b) -4 (c) $\frac{27}{2}$ (d) -8 (e) $\frac{7}{3}$

In the next two examples, multiplication produces the solution more quickly than division would.

EXAMPLE 2 Using the Multiplication Property of Equality

Solve the equation $\frac{a}{4} = 3$.

Replace $\frac{a}{4}$ by $\frac{1}{4}a$, since division by 4 is the same as multiplication by $\frac{1}{4}$. To get a alone on the left, multiply each side by 4, the reciprocal of the coefficient of a.

$$\frac{a}{4} = 3$$

$$\frac{1}{4}a = 3 \qquad \text{Change } \frac{a}{4} \text{ to } \frac{1}{4}a.$$

$$4 \cdot \frac{1}{4}a = 4 \cdot 3 \qquad \text{Multiply by 4.}$$

$$1a = 12 \qquad \text{Multiplicative inverse property}$$

$$a = 12 \qquad \text{Multiplicative identity property}$$

Check the answer.

$$\frac{a}{4} = 3 \qquad \text{Given equation}$$

$$\frac{12}{4} = 3 \qquad \text{Let } a = 12.$$

$$3 = 3 \qquad \text{True}$$

12 is the correct solution. ■

WORK PROBLEM 3 AT THE SIDE.

EXAMPLE 3 Using the Multiplication Property of Equality

Solve the equation $\frac{3}{4}h = 6$.

Get h alone on the left by multiplying each side of the equation by $\frac{4}{3}$. Use $\frac{4}{3}$ because $\frac{4}{3} \cdot \frac{3}{4}h = 1 \cdot h = h$.

$$\frac{3}{4}h = 6$$

$$\frac{4}{3}\left(\frac{3}{4}h\right) = \frac{4}{3} \cdot 6 \qquad \text{Multiply by } \frac{4}{3}.$$

$$1 \cdot h = \frac{4}{3} \cdot \frac{6}{1} \qquad \text{Multiplicative inverse property}$$

$$h = 8 \qquad \text{Multiplicative identity property}$$

The solution is 8. Check the answer by substitution in the given equation. ■

WORK PROBLEM 4 AT THE SIDE.

3. Solve each equation.

(a) $\frac{m}{2} = 6$

(b) $\frac{y}{5} = 5$

(c) $\frac{p}{4} = -6$

(d) $\frac{a}{-2} = 8$

4. Solve each equation.

(a) $\frac{2}{3}m = 8$

(b) $\frac{7}{8}m = 28$

(c) $\frac{3}{4}k = -21$

(d) $-\frac{5}{6}t = -15$

ANSWERS
3. (a) 12 (b) 25 (c) −24 (d) −16
4. (a) 12 (b) 32 (c) −28 (d) 18

2 In the next example, it is necessary to simplify the equation before using the multiplication property of equality.

EXAMPLE 4 Simplifying Terms in an Equation

Solve the equation $5m + 6m = 33$.

$$5m + 6m = 33$$

$$11m = 33 \quad \text{Combine terms.}$$

$$\frac{11m}{11} = \frac{33}{11} \quad \text{Divide by 11.}$$

$$1m = 3 \quad \text{Divide.}$$

$$m = 3 \quad \text{Multiplicative identity property}$$

The solution is 3. Check this solution. ■

WORK PROBLEM 5 AT THE SIDE.

3 The following example shows how to use the multiplication property of equality to solve equations such as $-r = 4$.

EXAMPLE 5 Using -1 with the Multiplication Property of Equality

Solve the equation $-r = 4$.

On the left side, change $-r$ to r by first writing $-r$ as $-1 \cdot r$.

$$-r = 4$$

$$-1 \cdot r = 4 \quad -r = -1 \cdot r$$

$$-1(-1 \cdot r) = -1 \cdot 4 \quad \text{Multiply by } -1\text{, since } -1 \cdot -1 = 1.$$

$$(-1)(-1) \cdot r = -4 \quad \text{Associative property}$$

$$1 \cdot r = -4 \quad \text{Multiplicative inverse property}$$

$$r = -4 \quad \text{Multiplicative identity property}$$

Check this solution.

$$-r = 4 \quad \text{Given equation}$$

$$-(-4) = 4 \quad \text{Let } r = -4.$$

$$4 = 4 \quad \text{True}$$

The solution, -4, checks. ■

WORK PROBLEM 6 AT THE SIDE.

4 The final example shows how to solve equations with decimals.

EXAMPLE 6 Solving an Equation with Decimals

Solve the equation $2.1x = 6.09$.

Divide both sides by 2.1.

$$\frac{2.1x}{2.1} = \frac{6.09}{2.1}$$

$$1x = 2.9 \quad \text{Divide.}$$

$$x = 2.9 \quad \text{Multiplicative identity property}$$

Check that the solution is 2.9. ■

WORK PROBLEM 7 AT THE SIDE.

5. Solve each equation.

(a) $5p + 2p = 28$

(b) $9k - k = -56$

(c) $7m - 5m = -12$

(d) $4r - 9r = 20$

6. Solve each equation.

(a) $-m = 2$

(b) $-p = -7$

7. Solve each equation.

(a) $-1.5p = 4.5$

(b) $12.5k = -63.75$

(c) $-.7m = -5.04$

ANSWERS

5. (a) 4 (b) −7 (c) −6 (d) −4
6. (a) −2 (b) 7
7. (a) −3 (b) −5.1 (c) 7.2

name date hour

3.2 EXERCISES

Solve each equation and check your solution. See Examples 1–6.

1. $5x = 25$ **2.** $7x = 28$ **3.** $2m = 50$ **4.** $6y = 72$ **5.** $3a = -24$

6. $5k = -60$ **7.** $8s = -56$ **8.** $10t = -36$ **9.** $-6x = 16$ **10.** $-6x = 24$

11. $-18z = 108$ **12.** $-11p = 77$ **13.** $5r = 0$ **14.** $2x = 0$ **15.** $-y = 6$

16. $-m = 2$ **17.** $-n = -4$ **18.** $-p = -8$ **19.** $2x + 3x = 20$ **20.** $3k + 4k = 14$

21. $5m + 6m - 2m = 72$ **22.** $11r - 5r + 6r = 84$ **23.** $k + k + 2k = 80$

24. $4z + z + 2z = 28$ **25.** $3r - 5r = 6$ **26.** $9p - 13p = 12$

27. $7r - 13r = -24$ **28.** $12a - 18a = -36$ **29.** $-7y + 8y - 9y = -56$

30. $-11b + 7b + 2b = -100$ **31.** $5x + 3x - 9x = -3$ **32.** $-6y + 8y - 3y = -8$

33. $\frac{x}{7} = 7$ **34.** $\frac{k}{8} = 2$ **35.** $\frac{2}{3}t = 6$ **36.** $\frac{4}{3}m = 18$

37. $\frac{15}{2}z = 20$

38. $\frac{12}{5}r = 18$

39. $\frac{3}{4}p = -60$

40. $\frac{5}{8}z = -40$

41. $\frac{2}{3}k = 5$

42. $\frac{5}{3}m = 6$

43. $-\frac{2}{7}p = -7$

44. $-\frac{3}{11}y = -2$

45. $-\frac{4}{7}r = 2$

46. $-\frac{5}{8}p = 4$

47. $-\frac{7}{9}x = \frac{3}{4}$

48. $-\frac{5}{6}t = \frac{2}{3}$

49. $1.7p = 5.1$

50. $2.3k = 11.04$

51. $-4.2m = 25.62$

52. $-3.9a = -15.6$

SKILL SHARPENERS

Simplify each expression. See Section 2.7.

53. $9(2q + 7)$

54. $4(3m - 5)$

55. $-4(5p - 1) + 6$

56. $-3(2y + 7) - 9$

57. $-(2 - 5r) + 6r$

58. $-(12 - 3y) - 2$

59. $6 - 3(4a + 3)$

60. $9 - 5(7 - 8p)$

3.3 SOLVING LINEAR EQUATIONS

1 In this section we will use *both* the addition and multiplication properties of equality to solve more complicated equations.

SOLVING LINEAR EQUATIONS

Step 1 Combine like terms to simplify each side as much as possible. Use the commutative, associative, and distributive properties as needed.

Step 2 If necessary, use the addition property of equality to simplify further. Transform the equation so that the variable term is on one side of the equals sign and the number is on the other.

Step 3 If necessary, use the multiplication property of equality to get the coefficient of x to be 1. This gives an equation of the form x = a number.

Step 4 Check your result in Step 3 by substituting into the *original* equation. If a true statement results, your solution is correct. (Do *not* substitute into an intermediate step.)

The check is used only to catch errors in carrying out the steps of the solution.

EXAMPLE 1 Using the Four Steps to Solve an Equation

Solve the equation $2x + 3x + 3 = 38$.

Follow the four steps summarized above.

Step 1 Combine like terms.

$$2x + 3x + 3 = 38$$

$$5x + 3 = 38 \quad \text{Combine like terms.}$$

Step 2 Use the addition property of equality. Subtract 3 from each side.

$$5x + 3 - 3 = 38 - 3 \quad \text{Subtract 3.}$$

$$5x = 35 \quad \text{Combine terms.}$$

Step 3 Use the multiplication property of equality. Divide each side by 5.

$$\frac{5x}{5} = \frac{35}{5} \quad \text{Divide by 5.}$$

$$x = 7 \quad \frac{5}{5} = 1;\ 1x = x$$

Step 4 Check the solution. Substitute 7 for x in the given equation.

$$2x + 3x + 3 = 38$$

$$2(7) + 3(7) + 3 = 38 \quad \text{Let } x = 7.$$

$$14 + 21 + 3 = 38 \quad \text{Multiply.}$$

$$38 = 38 \quad \text{True}$$

Since the final statement is true, 7 is the solution. ■

WORK PROBLEM 1 AT THE SIDE.

OBJECTIVES

1. Learn the four steps in solving a linear equation and how to use them.
2. Write word phrases as mathematical phrases.

1. Solve each equation.

(a) $3k + 2k + 7 = 17$

(b) $7m + 9m + 5 = 43$

(c) $9p - 5p + p - 8 = -18$

ANSWERS

1. (a) 2 (b) $\frac{19}{8}$ (c) -2

2. Solve each equation.

(a) $7 + 4p - 3p + 8 = 9p + 7$

(b) $5y - 7y + 6y - 9 = 3 + 2y$

(c) $-3k - 5k - 6 + 11 = 2k - 5$

(d) $2y + 5y - 6 + 8 = 9y - 1$

ANSWERS

2. (a) 1 (b) 6 (c) 1 (d) $\frac{3}{2}$

EXAMPLE 2 Using the Four Steps to Solve an Equation

Solve the equation $3r + 4 - 2r - 7 = 4r + 3$.

Use the four steps again.

Step 1

$$3r + 4 - 2r - 7 = 4r + 3$$

$$r - 3 = 4r + 3 \qquad \text{Combine like terms.}$$

Step 2

$$r - 3 + 3 = 4r + 3 + 3 \qquad \text{Add 3.}$$

$$r = 4r + 6$$

$$r - 4r = 4r + 6 - 4r \qquad \text{Subtract } 4r.$$

$$-3r = 6 \qquad \text{Combine terms.}$$

Step 3

$$\frac{-3r}{-3} = \frac{6}{-3} \qquad \text{Divide by } -3.$$

$$r = -2 \qquad \text{Reduce.}$$

Step 4 Substitute -2 for r in the original equation.

$$3r + 4 - 2r - 7 = 4r + 3$$

$$3(-2) + 4 - 2(-2) - 7 = 4(-2) + 3 \qquad \text{Let } r = -2.$$

$$-6 + 4 + 4 - 7 = -8 + 3 \qquad \text{Multiply.}$$

$$-5 = -5 \qquad \text{True}$$

The solution of the given equation is -2. ■

In Step 2 of Example 2, we added and subtracted the terms in such a way that the variable term ended up on the left side of the equation. Choosing differently would have put the variable term on the right side of the equation. Usually there is no real advantage either way.

WORK PROBLEM 2 AT THE SIDE.

EXAMPLE 3 Using the Four Steps to Solve an Equation

Solve the equation $4(k - 3) - k = k - 6$.

Step 1

$$4(k - 3) - k = k - 6$$

$$4k - 12 - k = k - 6 \qquad \text{Distributive property}$$

$$3k - 12 = k - 6 \qquad \text{Combine terms.}$$

Step 2

$$3k - 12 + 12 = k - 6 + 12 \qquad \text{Add 12.}$$

$$3k = k + 6 \qquad \text{Combine terms.}$$

$$3k - k = k + 6 - k \qquad \text{Subtract } k.$$

$$2k = 6 \qquad \text{Combine terms.}$$

Step 3

$$\frac{2k}{2} = \frac{6}{2} \qquad \text{Divide by 2.}$$

$$k = 3 \qquad \text{Reduce.}$$

Step 4 Check this answer by substituting 3 for k in the given equation. Remember to do all the work inside the parentheses first.

$$4(k - 3) - k = k - 6$$

$$4(3 - 3) - 3 = 3 - 6 \qquad \text{Let } k = 3.$$

$$4(0) - 3 = 3 - 6 \qquad 3 - 3 = 0$$

$$0 - 3 = 3 - 6 \qquad 4(0) = 0$$

$$-3 = -3 \qquad \text{True}$$

The solution of the equation is 3. ■

WORK PROBLEM 3 AT THE SIDE.

EXAMPLE 4 Using the Four Steps to Solve an Equation

Solve the equation $8a - (3 + 2a) = 3a + 1$.

Step 1 Simplify.

$$8a - (3 + 2a) = 3a + 1$$

$$8a - 1 \cdot (3 + 2a) = 3a + 1 \qquad \text{Multiplicative identity property}$$

$$8a - 3 - 2a = 3a + 1 \qquad \text{Distributive property}$$

$$6a - 3 = 3a + 1 \qquad \text{Combine terms.}$$

Step 2 First, add 3 to each side; then subtract $3a$.

$$6a - 3 + 3 = 3a + 1 + 3 \qquad \text{Add 3.}$$

$$6a = 3a + 4 \qquad \text{Combine terms.}$$

$$6a - 3a = 3a + 4 - 3a \qquad \text{Subtract } 3a.$$

$$3a = 4 \qquad \text{Combine terms.}$$

Step 3

$$\frac{3a}{3} = \frac{4}{3} \qquad \text{Divide by 3.}$$

$$a = \frac{4}{3} \qquad \frac{3}{3} = 1;\ 1a = a$$

Step 4 Check that the solution is $\frac{4}{3}$. ■

WORK PROBLEM 4 AT THE SIDE.

EXAMPLE 5 Using the Four Steps to Solve an Equation

Solve the equation $4(8 - 3t) = 32 - 8(t + 2)$.

Step 1

$$4(8 - 3t) = 32 - 8(t + 2)$$

$$32 - 12t = 32 - 8t - 16 \qquad \text{Distributive property}$$

$$32 - 12t = 16 - 8t \qquad \text{Combine terms.}$$

Step 2

$$32 - 12t - 32 = 16 - 8t - 32 \qquad \text{Subtract 32.}$$

$$-12t = -16 - 8t \qquad \text{Combine terms.}$$

$$-12t + 8t = -16 - 8t + 8t \qquad \text{Add } 8t.$$

$$-4t = -16 \qquad \text{Combine terms.}$$

3. Solve each equation.

(a) $7(p - 2) + p = 2p + 4$

(b) $11 + 3(a + 1) = 5a + 16$

(c) $3(m + 5) - 1 + 2m = 5(m + 2)$

4. Solve each equation.

(a) $4y - (y + 7) = 9$

(b) $7m - (2m - 9) = 39$

(c) $4x + 2(3 - 2x) = 6$

ANSWERS

3. (a) 3 (b) −1 (c) no solution

4. (a) $\frac{16}{3}$ (b) 6 (c) all real numbers

5. Solve each equation.

(a) $2(4 + 3r) = 3(r + 1) + 11$

(b) $-3(m + 2) = 4(2m + 1) + 1$

(c) $2 - 3(2 + 6z) = 4(z + 1) + 18$

ANSWERS
5. (a) 2 (b) −1 (c) $-\frac{13}{11}$

Step 3 $\frac{-4t}{-4} = \frac{-16}{-4}$ Divide by − 4.

$t = 4$ Reduce.

Step 4 Check this solution in the given equation.

$4(8 - 3t) = 32 - 8(t + 2)$

$4(8 - 3 \cdot 4) = 32 - 8(4 + 2)$ Let $t = 4$.

$4(8 - 12) = 32 - 8(6)$ Combine terms.

$4(-4) = 32 - 48$ Combine terms.

$-16 = -16$ True

The solution, 4, checks. ■

WORK PROBLEM 5 AT THE SIDE.

2 The next section includes a detailed discussion of the methods of solving word problems. One of the main steps in solving a word problem is converting the phrases in the word problem into mathematical expressions. Once the *expressions* are determined, we must then put them into an *equation* to be solved.

Some key words and phrases indicating the four arithmetic operations were introduced in Chapter 2. They are summarized here.

Addition: *sum, added to, more than, increased by*

Subtraction: *difference, subtracted from, less than, decreased by*

Multiplication: *product, times, twice* (meaning "2 times"), *of* (used with fractions), *percent of*

Division: *quotient, divided by*

CAUTION In translating from phrases to mathematical expressions, the *order* in which addition or multiplication is performed is not important, because these are *commutative* operations. (See Section 2.6.) However, in subtracting or dividing, the order *is* important, because these operations are not commutative. For example, $5 - 3 \neq 3 - 5$ and $2 \div 4 \neq 4 \div 2$.

EXAMPLE 6 Interpreting Phrases as Mathematical Expressions

Write the following phrases as mathematical expressions. Use x to represent the unknown quantity. (Other letters could be used to represent the unknown.)

(a) The sum of a number and 12 $x + 12$ or $12 + x$

(b) 7 more than a number $7 + x$ or $x + 7$

Notice the difference between the wording of part (b) and "7 *is* more than a number," which translates as $7 > x$.

(c) 3 less than a number $\quad x - 3$

Writing $3 - x$ would not be correct here. If you have trouble with this, substitute a number for x. Ask, for example, "What is 3 less than 10?" replacing the unknown with a specific number (10). It is easy to answer this: $10 - 3 = 7$. So $x - 3$ is the correct translation.

(d) A number decreased by 14 $\quad x - 14$ (*not* $14 - x$)

(e) The product of a number and 3 $\quad 3x$ or $x \cdot 3$ ($3x$ is preferred)

(f) -7 times a number $\quad -7x$

(g) Five-ninths of a number $\quad \frac{5}{9}x$

(h) 6% of a number $\quad .06x$ (6% = .06) ■

WORK PROBLEM 6 AT THE SIDE.

CAUTION When a fraction is multiplied by a variable, the variable is written on the same line as the fraction bar, as shown in part (g) of Example 6. It also is acceptable to write the variable in the numerator of the fraction. For example, $\frac{5}{9}x$ may also be written $\frac{5x}{9}$, since $\frac{5}{9}x = \frac{5}{9} \cdot \frac{x}{1} = \frac{5 \cdot x}{9 \cdot 1} = \frac{5x}{9}$. It would be *incorrect* to write $\frac{5}{9}x$ as $\frac{5}{9x}$.

EXAMPLE 7 Interpreting Phrases as Mathematical Expressions

Write the following phrases as mathematical expressions. Use x as the variable.

(a) The quotient of a number and 2 $\quad \frac{x}{2}$

Note that $\frac{2}{x}$ would not be correct here. It is understood that when division is involved, the first number is the numerator of the fraction, and the second number is the denominator.

(b) The reciprocal of a nonzero number $\quad \frac{1}{x}$

(c) Seven less than 4 times a number $\quad 4x - 7$

(d) A nonzero number plus its reciprocal $\quad x + \frac{1}{x}$

(e) Five times the sum of a number and 2 $\quad 5(x + 2)$

(f) A number divided by the sum of 4 and the number $\quad \frac{x}{4 + x}$ ■

WORK PROBLEM 7 AT THE SIDE.

6. Write each phrase as a mathematical expression. Use x as the variable.

(a) 8 more than a number

(b) The sum of a number and 9

(c) 2 less than a number

(d) The product of a number and 5

(e) -3 times a number

(f) Five-eighths of a number

(g) 18% of a number

7. Write each phrase as a mathematical expression. Use x as the variable.

(a) The quotient of a number and 10

(b) 10 added to twice a number

(c) The product of 5 and 2 less than a number

(d) The quotient of 8 plus a nonzero number and 3 times the number

(e) Twice a number, added to the reciprocal of 5

ANSWERS

6. (a) $8 + x$ or $x + 8$ (b) $9 + x$ or $x + 9$ (c) $x - 2$ (d) $5x$ (e) $-3x$ (f) $\frac{5}{8}x$ (g) $.18x$

7. (a) $\frac{x}{10}$ (b) $10 + 2x$ or $2x + 10$ (c) $5(x - 2)$ (d) $\frac{8 + x}{3x}$ (e) $2x + \frac{1}{5}$

Historical Reflections

Historical Reflections

Etymology: Origins of the Words *Algorithm* and *Algebra*

The *Carmen de Algorismo* (opening verses shown here) by Alexander de Villa Dei in the thirteenth century, popularized the art of "algorismus":

> ". . . from these twice five figures
> 0 9 8 7 6 5 4 3 2 1
> of the Indians we benefit . . ."

The *Carmen* related that Algor, an Indian king, invented the art. Actually, the word *algorithm* comes in a roundabout way from the name Muhammad ibn Musa al-Khowârizmi, an Arabian mathematician of the ninth century.

One of the books of al-Khowârizmi, titled *Hisab al-jabr w'al-muqâbalah* is the source of our modern word *algebra*. In the title of the book, written in the ninth century, *jabr* ("restoration") refers to transposing negative quantities across the equals sign in solving equations. From Latin versions of the text, "al-jabr" became the broad term covering the art of equation solving.

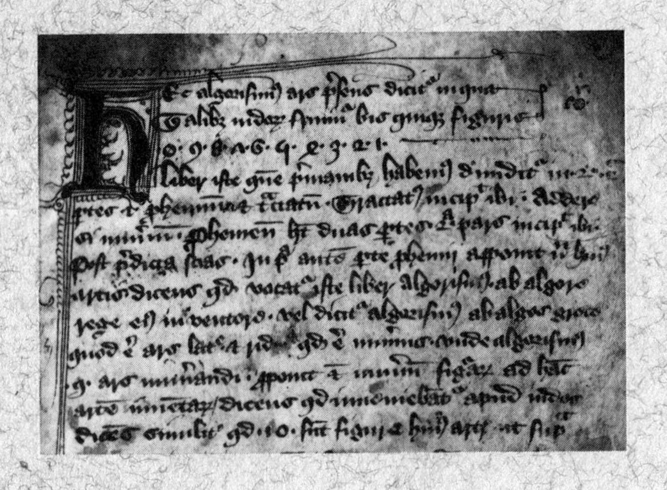

Art: By permission of the Master and Fellows of Trinity College, Cambridge

name date hour

3.3 EXERCISES

Solve each equation and check your solution. See Examples 1–5.

1. $4h + 8 = 16$

2. $3x - 15 = 9$

3. $6k + 12 = -12 + 7k$

4. $2m - 6 = 6 + 3m$

5. $12p + 18 = 14p$

6. $10m - 15 = 7m$

7. $3x + 9 = -3(2x + 2)$

8. $4z + 2 = -2(z + 3)$

9. $2(2r - 1) = -3(r + 3)$

10. $3(3k + 5) = 2(5k + 5)$

11. $\frac{3}{2}\left(\frac{1}{3}x + 4\right) = 6\left(\frac{1}{4} + x\right)$

12. $\frac{4}{3}p + \frac{4}{3} = \frac{2}{3}p - \frac{1}{3}$

13. $3(5 + 1.4x) = 3x$

14. $2(-3 + 2.1x) = 2x + x$

15. $5(4x - 1) = 2(10x - 3)$

16. $3(8x + 1) = 6(4x + 2)$

17. $9(2x - 4) = 18(x - 2)$

18. $5(6 - 3x) = 15(2 - x)$

Combine terms as necessary; then solve the equations. Check your answers. See Examples 3–5.

19. $-5 - 3(2x + 1) = 12$

20. $10 - 2(3x - 4) = 2x$

21. $-5k - 8 = 2(k + 6) + 1$

22. $4a - 7 = 3(2a + 5) - 2$

23. $5(2m - 1) = 4(2m + 1) + 7$

24. $3(3k - 5) = 4(3k - 1) - 17$

25. $5(4t + 3) = 6(3t + 2) - 1$

26. $7(2y + 6) = 9(y + 3) + 5$

27. $5(x - 3) + 2 = 5(2x - 8) - 3$

28. $6(2v - 1) - 5 = 7(3v - 2) - 24$

29. $-2(3s + 9) - 6 = -3(4s + 11) - 6$

30. $-3(5z + 24) + 2 = 2(3 - 2z) - 10$

31. $6(2p - 8) + 24 = 3(5p - 6) - 6$

32. $2(5x + 3) - 3 = 6(2x - 3) + 15$

33. $-(4m + 2) - (-3m - 5) = 3$

34. $-(6k - 5) - (-5k + 8) = -4$

35. $2(4x - 1) - 3(-2x + 5) = 3$

36. $4(3x - 3) - 3(-x - 4) = 20$

37. $3(4x + 2) - 2(5x - 1) = 0$

38. $5(3 - x) + 3(2x - 2) = 2$

39. $5(y - 2) + 6y = 4(y + 3) - 1$

40. $2(3x + 1) - x = 4 - (2x + 3) - 6$

41. $4(x + 8) = 2(2x + 5) + 22$

42. $8(x + 3) = 4(2x + 8) - 8$

43. $9(x + 1) - 3x = 2(3x + 1) - 4$

44. $8(p - 3) + 4p = 6(2p + 1) - 3$

Write each of the following as a mathematical expression. Use x as the variable. See Examples 6 and 7.

45. -1 added to a number

46. A number added to -6

47. A quantity increased by -18

48. The sum of a number and 12

49. A number decreased by 6

50. 5 less than a number

51. 16 fewer than a number

52. Subtract 9 from a number.

53. Double a number

54. The product of a number and 9

55. Three-fifths of a number

56. Triple a number

57. The quotient of -9 and a nonzero number

58. The quotient of a number and 6

59. 7 divided by a nonzero number

60. A number divided by -4

61. A nonzero number subtracted from its reciprocal

62. The product of 8 and the sum of a number and 3

63. Eight times the difference between a number and 8

64. Three times the quotient of a number and 2

SKILL SHARPENERS

Write the following in symbols, using x as the variable. Then solve the resulting equation. See Sections 2.5, 3.1, and 3.2.

65. Seven times a number is -84.

66. A number divided by 3 is -18.

67. 13 more than a number is 9.

68. 3 less than a number is -19.

69. When a number is multiplied by $\frac{1}{2}$, the result is 4.

70. When a number is divided by 12, the result is -2.

3.4 AN INTRODUCTION TO WORD PROBLEMS

As we mentioned in the last section, one of the main skills involved in solving word problems is translating words and phrases into mathematical expressions. To find the solution to a word problem, first read the problem carefully and determine what facts are given and what must be found. Next, use the six steps given below.

OBJECTIVE

1 Translate sentences of a word problem into an equation, and then solve the problem.

SOLVING WORD PROBLEMS

Step 1 Choose a variable to represent the numerical value that you are asked to find—the unknown number. *Write down* what the variable represents.

Step 2 *Write down* a mathematical expression using the variable for any other unknown quantities. Draw figures or diagrams if they apply.

Step 3 Translate the problem into an equation.

Step 4 Solve the equation.

Step 5 Answer the question asked in the problem.

Step 6 Check your solution by using the words of the original problem. Be sure that the answer is appropriate and makes sense.

1 The third step in solving a word problem is often the hardest. Begin to translate the problem into an equation by writing the given phrases as mathematical expressions. Since equal mathematical expressions are names for the same number, translate any words that mean *equal* or *same* as $=$. Forms of the verb "to be" such as *is, are, was,* and *were,* translate this way. The $=$ sign leads to an equation to be solved.

Notice how the six steps for solving word problems are used in the next examples.

EXAMPLE 1 Finding an Unknown Number

If three times the sum of a number and 4 is decreased by twice the number, the result is -6. Find the number.

Let x represent the unknown number. Then

$$3(x + 4) = \text{three times the sum of the number and 4}$$
$$2x = \text{twice the number.}$$

Now write an equation using the information given in the problem.

Three times the sum of a number and 4	decreased by	twice the number	is	-6.
↓	↓	↓	↓	↓
$3(x + 4)$	$-$	$2x$	$=$	-6

1. Write equations for each of the following and then solve.

(a) When 3 times a number is added to 9, the result is 27. Find the number.

(b) If 5 is added to the product of 9 and a number, the result is 19 less than the number. Find the number.

ANSWERS
1. (a) $3x + 9 = 27$; 6
(b) $5 + 9x = x - 19$; -3

Solve the equation.

$$3(x + 4) - 2x = -6$$
$$3x + 12 - 2x = -6 \quad \text{Distributive property.}$$
$$x + 12 = -6 \quad \text{Combine terms.}$$
$$x = -18 \quad \text{Subtract 12.}$$

Check that -18 is the correct answer by substituting this result into the words of the original problem. Three times the sum of -18 and 4 is $3(-18 + 4) = 3(-14) = -42$. Twice -18 is -36; subtract -36 from -42 to get $-42 - (-36) = -6$, as required. ■

WORK PROBLEM 1 AT THE SIDE.

EXAMPLE 2 Finding the Number of Men at a Concert

The audience at a concert included 25 more women than men. The total number of people at the concert was 139. Find the number of men.

Let x = the number of men

$x + 25$ = the number of women.

Now write the equation.

The total	is	the number of men	plus	the number of women.
↓	↓	↓	↓	↓
139	$=$	x	$+$	$(x + 25)$

Solve the equation.

$$139 = 2x + 25 \quad \text{Combine terms.}$$
$$139 - 25 = 2x + 25 - 25 \quad \text{Subtract 25.}$$
$$114 = 2x$$
$$57 = x \quad \text{Divide by 2.}$$

Check the solution in the words of the problem. If the number of men was 57, then the number of women was $25 + 57 = 82$. The total number was $57 + 82 = 139$, as required. There were 57 men at the concert. ■

EXAMPLE 3 Finding the Number of Coffee Orders

The owner of a small cafe found one day that the number of orders for tea was $\frac{1}{3}$ the number of orders for coffee. If the total number of orders for the two drinks was 76, how many orders were placed for coffee?

Let x = the number of orders for coffee

$\frac{1}{3}x$ = the number of orders for tea.

To set up an equation, use the fact that the total number of orders was 76.

The total	is	orders for coffee	plus	orders for tea.
↓	↓	↓	↓	↓
76	$=$	x	$+$	$\frac{1}{3}x$

Now solve the equation.

$$76 = \frac{4}{3}x \qquad \text{Combine terms.}$$

$$\frac{3}{4}(76) = \frac{3}{4}\left(\frac{4}{3}x\right) \qquad \text{Multiply by } \frac{3}{4}.$$

$$57 = x \qquad \text{Combine terms.}$$

There were 57 orders for coffee and $\left(\frac{1}{3}\right)(57) = 19$ orders for tea, giving a total of $57 + 19 = 76$ orders, as required. ■

WORK PROBLEM 2 AT THE SIDE.

Sometimes it is necessary to find three unknown quantities in a word problem. Frequently the three unknowns are compared in *pairs*. When this happens, it is usually easiest to let the variable represent the unknown found in both pairs. The next example illustrates this idea.

EXAMPLE 4 Dividing a Board into Pieces

Maria Gonzales has a piece of board 70 inches long. She cuts it into three pieces. The longest piece is twice the length of the middle-sized piece, and the shortest piece is 10 inches shorter than the middle-sized piece. How long are the three pieces?

Since the middle-sized piece appears in both pairs of comparisons, let x represent the length of the middle-sized piece. We have

x = the length of the middle-sized piece
$2x$ = the length of the longest piece
$x - 10$ = the length of the shortest piece.

A sketch is helpful here.

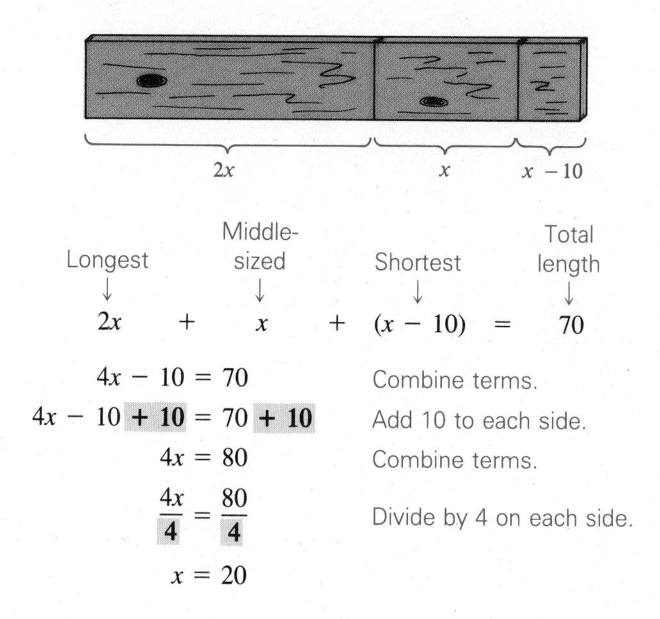

Longest		Middle-sized		Shortest		Total length
↓		↓		↓		↓
$2x$	+	x	+	$(x - 10)$	=	70

$$4x - 10 = 70 \qquad \text{Combine terms.}$$

$$4x - 10 + 10 = 70 + 10 \qquad \text{Add 10 to each side.}$$

$$4x = 80 \qquad \text{Combine terms.}$$

$$\frac{4x}{4} = \frac{80}{4} \qquad \text{Divide by 4 on each side.}$$

$$x = 20$$

2. Solve each word problem.

(a) A farmer has 14 more sheep than goats, with 52 animals in all. How many goats does he have?

(b) On one day of their vacation, Annie drove three times as far as Jim. Altogether they drove 84 miles that day. Find the number of miles driven by each.

ANSWERS
2. (a) 19 goats
(b) 21 miles for Jim; 63 miles for Annie

The length of the middle-sized piece is 20 inches, the length of the longest piece is $2(20) = 40$ inches, and the length of the shortest piece is $20 - 10 = 10$ inches. Check to see that the sum of the three lengths is 70 inches. ■

WORK PROBLEM 3 AT THE SIDE.

EXAMPLE 5 Analyzing a Gasoline/Oil Mixture

A Weedeater (gas-operated trimmer) uses a mixture of gasoline and oil. For each ounce of oil the mixture contains 16 ounces of gasoline. If the tank holds 68 ounces of the mixture, how many ounces of oil and how many ounces of gasoline does it require when it is full?

Let x = the number of ounces of oil required when full

$16x$ = the number of ounces of gasoline required when full.

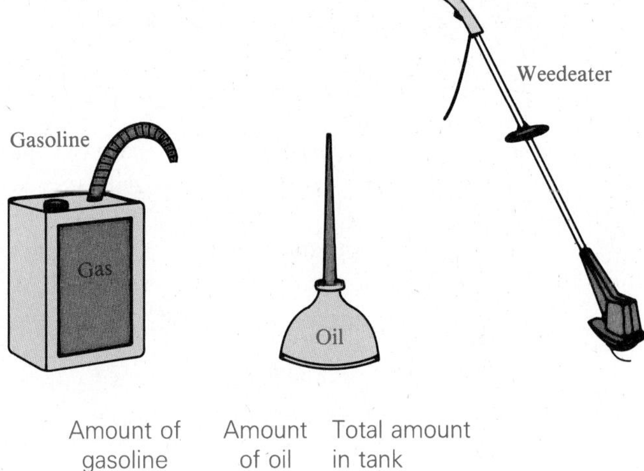

Amount of gasoline ↓ $16x$ + Amount of oil ↓ x = Total amount in tank ↓ 68

$$16x + x = 68$$

$$17x = 68 \quad \text{Combine terms.}$$

$$\frac{17x}{17} = \frac{68}{17} \quad \text{Divide by 17.}$$

$$x = 4 \quad \text{Reduce.}$$

When the tank is full, it holds 4 ounces of oil and $16(4) = 64$ ounces of gasoline. This checks, since $4 + 64 = 68$. ■

WORK PROBLEM 4 AT THE SIDE.

3. Solve each word problem.

(a) A piece of pipe is 50 inches long. It is cut into three pieces. The longest piece is 10 inches more than the middle-sized piece, and the shortest piece measures 5 inches less than the middle-sized piece. Find the lengths of the three pieces.

(b) In one day, Gwen Boyle received 13 packages. Federal Express delivered three times as many as U.P.S., while the Postal Service delivered three more than U.P.S. How many packages did each service deliver to Gwen?

4. Solve the following word problem.

At a meeting of the local coin club, each member brought two non-members. If a total of 27 people attended, how many were members and how many were non-members?

ANSWERS

3. (a) 25 inches, 15 inches, 10 inches
(b) 6 from Federal Express; 2 from U.P.S.; 5 from the Postal Service
4. Nine were members and 18 were non-members.

name date hour

3.4 EXERCISES

When you are solving the problems in this exercise set, follow the six-step method described in this section.

Step 1 Choose a variable to represent the unknown quantity. *Write down* what the variable represents.

Step 2 *Write down* a mathematical expression using the variable for any other unknown quantities. Draw figures or diagrams if they apply.

Step 3 Translate the problem into an equation.

Step 4 Solve the equation.

Step 5 Answer the question asked in the problem.

Step 6 Check your solution by using the words of the original problem. Be sure that the answer is appropriate and makes sense.

Solve the following problems. See Example 1.

1. If 4 is added to a number and this sum is doubled, the result is 1 less than 3 times the number. Find the number.

2. If 8 is subtracted from a number and this difference is tripled, the result is -3 times the number. Find the number.

3. If the sum of a number and 5 is multiplied by 3, the result is the same as if 16 were added to twice the number. Find the number.

4. If 5 is multiplied by the difference between a number and 4, the result is 20 less than the number. Find the number.

Solve the following problems. See Examples 2, 3, and 4.

5. Sprague and Hartman were opposing candidates in a mayoral election. Sprague received 80 more votes than Hartman, with 346 total votes cast. How many votes did Hartman receive?

6. Nagaraj Nanjappa has a board 74 inches long. He wants to cut it into two pieces so that one piece will be 8 inches longer than the other. How long should the shorter piece be?

7. In their daily workout, Katherine did 28 more sit-ups than Brian. The total number of sit-ups for both of them was 80. Find the number of sit-ups that each did.

8. On a geometry test, the highest grade was 66 points higher than the lowest grade. The sum of the two grades was 140. Find the highest and the lowest grades.

9. Captain Tom Tupper gives deep-sea fishing trips. One day he noticed that the boat contained (not counting himself) 2 fewer men than women. If he had 20 customers on the boat, how many men and how many women were there?

10. A farmer has 28 more hens than roosters, with 170 chickens in all. Find the number of hens and the number of roosters on the farm.

11. In a given amount of time, Larry drove twice as far as Jean. Altogether they drove 90 miles. Find the number of miles driven by each.

12. Mark is two years older than Linda. The sum of their ages is 70 years. Find the age of each.

13. A piece of string is 40 centimeters long. It is cut into three pieces. The longest piece is 3 times as long as the middle-sized piece, and the shortest piece is 23 centimeters shorter than the longest piece. Find the lengths of the three pieces.

14. A strip of paper is 56 inches long. It is cut into three pieces. The longest piece is 12 inches longer than the middle-sized piece, and the shortest piece is 16 inches shorter than the middle-sized piece. Find the lengths of the three pieces.

15. During the 1987 baseball season, Wade Boggs had 10 more at-bats than his teammate Dwight Evans. Evans had 17 fewer at-bats than another teammate, Ellis Burks. The three players had a total of 1650 at-bats. How many times did each player come to bat?

16. During the 1987 baseball season, Bret Saberhagen pitched 9 fewer innings than Jack Morris. Mark Langston pitched 6 more innings than Morris. How many innings did each player pitch, if their total number of innings pitched was 795?

17. The sum of the measures of the angles of any triangle is 180 degrees. In triangle ABC, angles A and B have the same measure, while the measure of angle C is 24 degrees larger than each of A and B. What are the measures of the three angles?

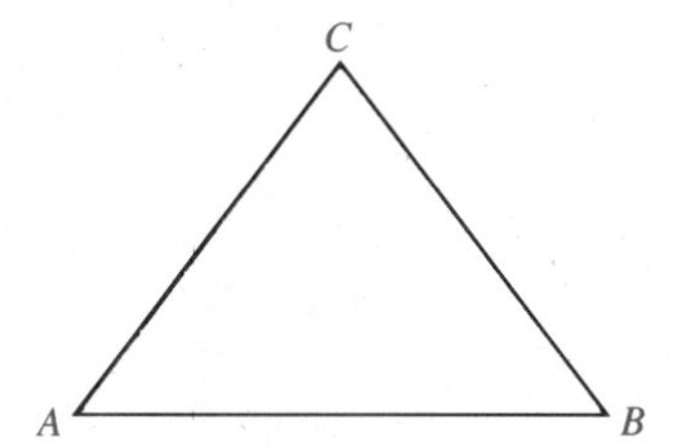

18. (See Exercise 17.) In triangle ABC, the measure of angle A is 30 degrees more than the measure of angle B. The measure of angle B is the same as the measure of angle C. Find the measure of each angle.

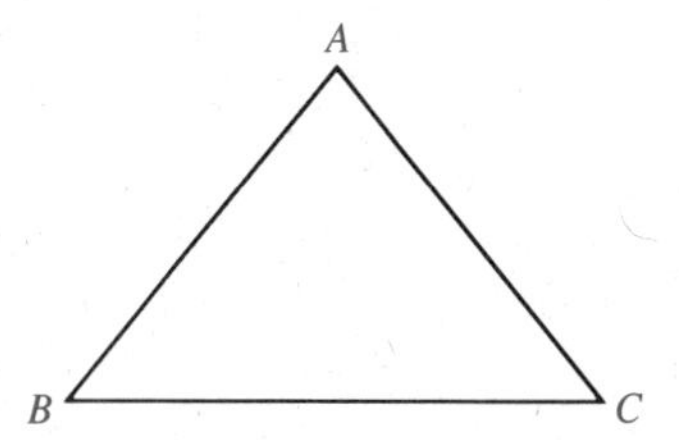

Solve the following problems. See Example 5.

19. A pharmacist found that at the end of the day she had $\frac{4}{3}$ as many prescriptions for antibiotics as she did for tranquilizers. She had 42 prescriptions altogether for these two types of drugs. How many did she have for tranquilizers?

20. In a mixture of concrete, there are 3 pounds of cement mix for every 1 pound of gravel. If the mixture contains a total of 140 pounds of these two ingredients, how many pounds of gravel are there?

21. A mixture of nuts contains only peanuts and cashews. For every ounce of cashews there are 5 ounces of peanuts. If the mixture contains a total of 27 ounces, how many ounces of each type of nut does the mixture contain?

22. An insecticide contains 95 centigrams of inert ingredient for every 1 centigram of active ingredient. If a quantity of the insecticide weighs 336 centigrams, how much of each type of ingredient does it contain?

The following problems are a bit different from those described in the examples of this section. Solve these using the general method described. (In Exercises 23-26, consecutive integers *and* consecutive even integers *are mentioned. Some examples of consecutive integers are* 5, 6, 7, *and* 8; *some examples of consecutive even integers are* 4, 6, 8, *and* 10.)

23. If x represents an integer, then $x + 1$ represents the next larger consecutive integer. If the sum of two consecutive integers is 137, find the integers.

24. (See Exercise 23.) If the sum of two consecutive integers is -57, find the integers.

25. If x represents an even integer, then $x + 2$ represents the next larger even integer. Find two consecutive even integers such that the smaller added to three times the larger equals 46.

26. (See Exercise 25.) Find two consecutive even integers such that six times the smaller added to the larger gives a sum of 86.

27. If x represents a quantity, then $.06x$ represents 6% of the quantity. Louise has 6% of her salary deducted for her daughter's educational fund. After the deduction, one week her check was for $423. How much did she earn before the deduction?

28. (See Exercise 27.) At the end of a day, the owner of a gift shop had $2394 more in the cash register than she had when the shop opened that day. This included sales tax of 5% on all sales. Find the amount of her sales.

29. A store has 39 quarts of milk, some in pint cartons and some in quart cartons. There are six times as many quart cartons as pint cartons. How many quart cartons are there? (*Hint:* 1 quart = 2 pints.)

30. Marin is three times as old as Kasey. Three years ago the sum of their ages was 22 years. How old is each now? (*Hint:* First write an expression for the age of each now, then for the age of each three years ago.)

SKILL SHARPENERS

Use the given values to evaluate each expression. See Section 1.5.

31. LW; $L = 6, W = 7$

32. rt; $r = 25, t = 2.5$

33. prt; $p = 4000, r = .08, t = 2$

34. $\frac{1}{2}Bh$; $B = 27, h = 4$

35. $2L + 2W$; $L = 18, W = 12$

36. $\frac{1}{2}(B + b)h$; $B = 10, b = 7, h = 5$

3.5 FORMULAS

Many word problems can be solved with a formula. Formulas exist for geometric figures such as squares and circles, for distance, for money earned on bank savings, and for converting English measurements to metric measurements, for example. A list of the formulas used in this book is given in an appendix in the back of the book.

OBJECTIVES

1. Solve a formula for one variable, given the values of the other variables.
2. Use a formula to solve a word problem.
3. Solve a formula for a specified variable.

1 Given the values of all but one of the variables in a formula, we can find the value of the remaining variable by using the methods introduced in this chapter.

EXAMPLE 1 Using a Formula to Evaluate a Variable
Find the value of the remaining variable in each of the following.

(a) $A = LW$; $A = 64, L = 10$

As shown in Figure 1, this formula gives the area of a rectangle with length L and width W. Substitute the given values into the formula and then solve for W.

$$A = LW$$

$$64 = 10W \qquad \text{Let } A = 64 \text{ and } L = 10.$$

$$6.4 = W \qquad \text{Divide by 10.}$$

Check that the width of the rectangle is 6.4.

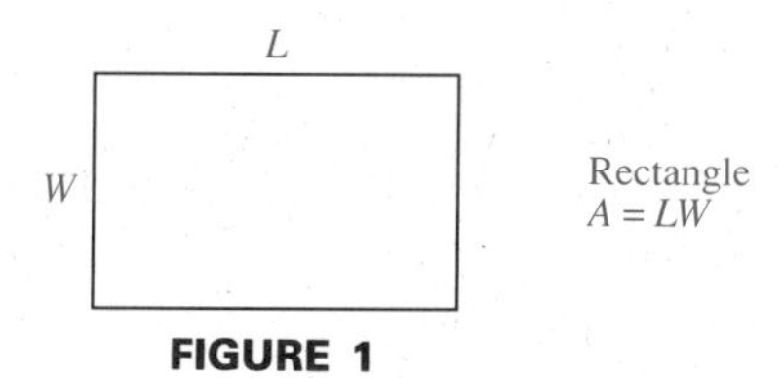

FIGURE 1

(b) $A = \frac{1}{2}(b + B)h$; $A = 210, B = 27, h = 10$

This formula gives the area of a trapezoid with parallel sides of length b and B and distance h between the parallel sides. See Figure 2.

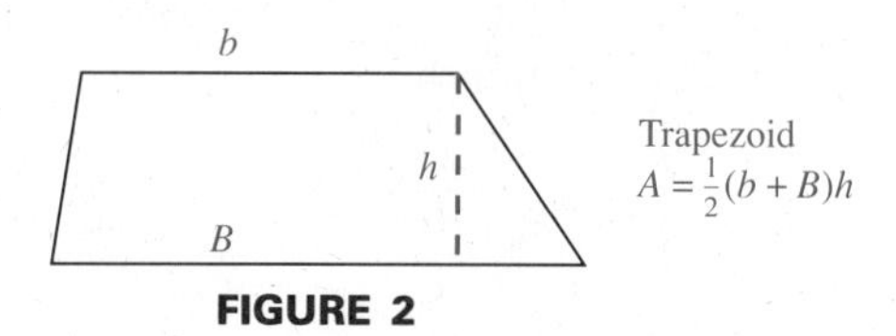

FIGURE 2

Again, begin by substituting the given values into the formula.

$$A = \frac{1}{2}(b + B)h$$

$$210 = \frac{1}{2}(b + 27)(10) \qquad A = 210, \quad B = 27, \quad h = 10$$

Now solve for b.

$$210 = \frac{1}{2}(10)(b + 27) \quad \text{Commutative property}$$
$$210 = 5(b + 27) \quad \text{Multiply.}$$
$$210 = 5b + 135 \quad \text{Distributive property}$$
$$210 - 135 = 5b + 135 - 135 \quad \text{Subtract 135.}$$
$$75 = 5b \quad \text{Combine terms.}$$
$$\frac{75}{5} = \frac{5b}{5} \quad \text{Divide by 5.}$$
$$15 = b \quad \text{Reduce.}$$

Check that the length of the shorter parallel side, b, is 15. ■

WORK PROBLEM 1 AT THE SIDE.

2 As the next examples show, formulas are often used to solve word problems. *It is a good idea to draw a sketch when a geometric figure is involved.*

EXAMPLE 2 Finding the Length of a Side of a Square

The perimeter of a square is 96 inches. Find the length of a side.

Begin by drawing a square. Let s represent the length of each side. See Figure 3.

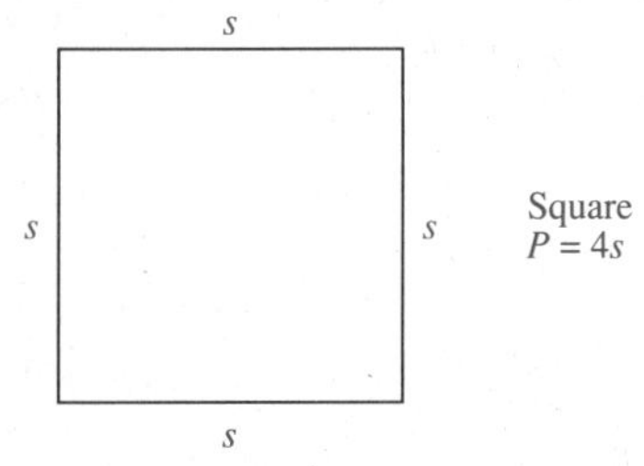

FIGURE 3

The list of formulas in the appendix gives $P = 4s$ for the **perimeter,** or the distance around a square, where s is the length of a side of the square. Here, the perimeter is given as 96 inches, so that $P = 96$. Substitute 96 for P in the formula.

$$P = 4s$$
$$96 = 4s \quad P = 96$$
$$24 = s \quad \text{Divide by 4.}$$

Check this solution to see that each side of the square is 24 inches long. ■

WORK PROBLEM 2 AT THE SIDE.

EXAMPLE 3 Finding the Width of a Rectangle

The perimeter of a rectangle is 80 meters, and the length is 25 meters. Find the width of the rectangle.

Draw a rectangle, as shown in Figure 4. Label the two longer sides L and the two shorter sides W.

1. Find the value of the remaining variable in each of the following.

(a) $I = prt$; $I = \$246$, $r = .06$, $t = 2$

(b) $P = 2L + 2W$; $P = 126$, $W = 25$

2. The perimeter of a triangle is 48 centimeters. One of the sides is 10 centimeters long, and the other two sides are equal in length. Find the length of each equal side.

ANSWERS
1. (a) \$2050 (b) 38
2. 19 centimeters

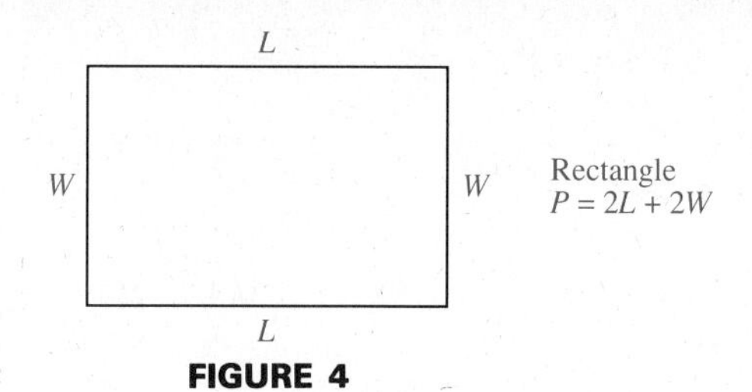

FIGURE 4

To find the perimeter of the rectangle, we add the lengths of the four sides.

$$P = L + W + L + W$$
$$P = 2L + 2W$$

Find the width by substituting 80 for P and 25 for L in the formula $P = 2L + 2W$.

$$P = 2L + 2W$$
$$80 = 2(25) + 2W \qquad P = 80, \quad L = 25$$
$$80 = 50 + 2W \qquad \text{Multiply.}$$
$$30 = 2W \qquad \text{Subtract 50.}$$
$$15 = W \qquad \text{Divide by 2.}$$

Check this result. If W is 15 and L is 25, the perimeter will be

$$2(15) + 2(25) = 30 + 50 = 80$$

as required. The width of the rectangle is 15 meters. ■

WORK PROBLEM 3 AT THE SIDE.

EXAMPLE 4 Finding the Height of a Triangle

The area of a triangle is 126 square meters. The base of the triangle is 21 meters. Find the height.

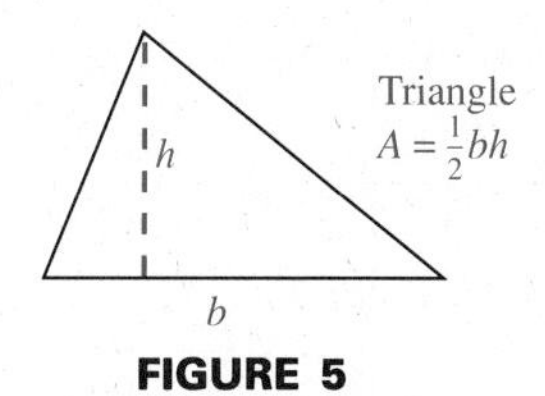

FIGURE 5

The formula for the area of a triangle is $A = \frac{1}{2}bh$, where A is area, b is the base, and h is the height. See Figure 5. Substitute 126 for A and 21 for b in the formula.

$$A = \frac{1}{2}bh$$
$$126 = \frac{1}{2}(21)h \qquad A = 126, \quad b = 21$$

Simplify the problem by eliminating the fraction $\frac{1}{2}$. Multiply both sides of the equation by 2.

3. Solve the following word problems.

(a) The width of a swimming pool is 40 feet and the length is 50 feet. Find the perimeter of the pool.

(b) A farmer has 800 meters of fencing material to enclose a rectangular field. The width of the field is 175 meters. Find the length of the field.

ANSWERS
3. (a) 180 feet
(b) 225 meters

$$2(126) = 2\left(\frac{1}{2}\right)(21)h \quad \text{Multiply by 2.}$$

$$252 = 21h \quad \text{Multiply.}$$

$$12 = h \quad \text{Divide by 21.}$$

Check in the words of the given problem that the height of the triangle is 12 meters. ■

WORK PROBLEM 4 AT THE SIDE.

3 Sometimes it is necessary to solve a large number of problems that use the same formula. For example, a surveying class might need to solve several problems that involve the formula for the area of a rectangle, $A = LW$. Suppose that in each problem the area (A) and the length (L) of a rectangle are given and the width (W) must be found. Rather than solving for W each time the formula is used, it would be simpler to rewrite the *formula* so that it is solved for W. This process is called **solving for a specified variable.** As the following examples will show, solving a formula for a specified variable requires the same steps used earlier to solve equations with just one variable.

In solving a formula for a specified variable, remember the following: treat the specified variable as if it were the *only* variable in the equation, and treat the other variables as if they were numbers.

EXAMPLE 5 Solving for a Specified Variable

Solve $A = LW$ for W.

Think of undoing what has been done to W. Since W is multiplied by L, undo the multiplication by dividing both sides of $A = LW$ by L.

$$A = LW$$

$$\frac{A}{L} = \frac{LW}{L} \quad \text{Divide by } L.$$

$$\frac{A}{L} = W \quad \frac{L}{L} = 1;\ 1W = W$$

The formula is now solved for W. ■

WORK PROBLEM 5 AT THE SIDE.

EXAMPLE 6 Solving for a Specified Variable

Solve $P = 2L + 2W$ for L.

Begin by subtracting $2W$ on both sides.

$$P = 2L + 2W$$

$$P - 2W = 2L + 2W - 2W \quad \text{Subtract } 2W.$$

$$P - 2W = 2L \quad \text{Combine terms.}$$

$$\frac{P - 2W}{2} = \frac{2L}{2} \quad \text{Divide by 2.}$$

$$\frac{P - 2W}{2} = L \quad \frac{2}{2} = 1;\ 1L = L$$

The last step gives the formula solved for L, as required. ■

WORK PROBLEM 6 AT THE SIDE.

4. Solve the following word problems.

(a) The base of a triangle is 45 centimeters. The height is 50 centimeters. Find the area of the triangle.

(b) The area of a triangle is 120 square meters. The height is 24 meters. Find the length of the base of the triangle.

5. Solve each formula for the specified variable.

(a) $d = rt$ for t

(b) $I = prt$ for t

(c) $P = a + b + c$ for a

6. Solve each formula for the specified variable.

(a) $A = p + prt$ for t

(b) $y = mx + b$ for x

ANSWERS

4. (a) 1125 square centimeters (b) 10 meters

5. (a) $t = \frac{d}{r}$ (b) $t = \frac{I}{pr}$ (c) $a = P - b - c$

6. (a) $t = \frac{A - p}{pr}$ (b) $x = \frac{y - b}{m}$

3.5 EXERCISES

In the following exercises a formula is given, along with the values of all but one of the variables in the formula. Find the value of the variable that is not given. See Example 1.

1. $P = 2L + 2W$; $L = 5$, $W = 3$

2. $P = 4s$; $s = 32$

3. $A = \frac{1}{2}bh$; $b = 9$, $h = 24$

4. $A = \frac{1}{2}bh$; $b = 6$, $h = 12$

5. $d = rt$; $d = 100$, $t = 5$

6. $d = rt$; $d = 8$, $r = 2$

7. $A = \frac{1}{2}bh$; $A = 30$, $b = 6$

8. $A = \frac{1}{2}bh$; $A = 20$, $b = 5$

9. $P = 2L + 2W$; $P = 180$, $L = 50$

10. $P = 2L + 2W$; $P = 40$, $W = 6$

11. $V = \frac{1}{3}Bh$; $V = 52$, $h = 13$

12. $V = \frac{1}{3}Bh$; $V = 80$, $B = 24$

13. $C = 2\pi r$; $C = 25.12$, $\pi = 3.14$*

14. $C = 2\pi r$; $C = 9.42$, $\pi = 3.14$

15. $A = \pi r^2$; $r = 15$, $\pi = 3.14$

16. $A = \pi r^2$; $r = 9$, $\pi = 3.14$

17. $I = prt$; $I = 60$, $p = 150$, $r = .08$

18. $I = prt$; $I = 100$, $p = 500$, $r = .10$

19. $V = LWH$; $V = 800$, $L = 40$, $W = 10$

20. $V = LWH$; $V = 150$, $L = 10$, $W = 5$

21. $A = \frac{1}{2}(b + B)h$; $A = 70$, $b = 15$, $B = 20$

22. $A = \frac{1}{2}(b + B)h$; $A = 42$, $b = 5$, $B = 7$

*Actually, π is approximately equal to 3.14, not *exactly* equal to 3.14.

Use a formula to write an equation for each word problem; then solve the equation. Check the solution in the words of the original problem. The necessary formulas are given in an appendix in the back of the book. See Examples 2–4. Use 3.14 as an approximation for π. (Some of the figures have been provided.)

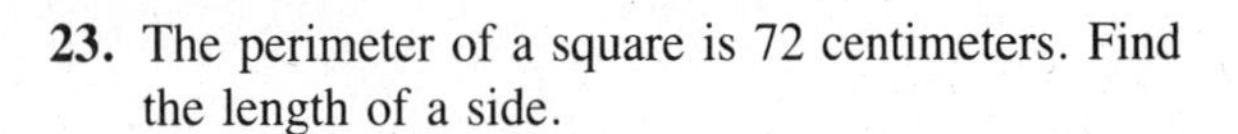

23. The perimeter of a square is 72 centimeters. Find the length of a side.

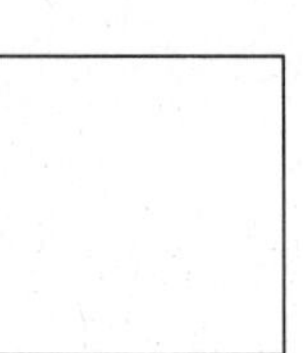

24. The area of a rectangle is 65 square meters, and the width is 5 meters. Find the length.

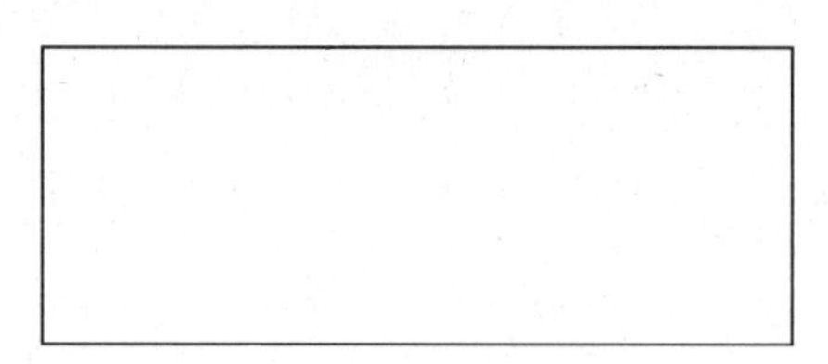

25. The length of a rectangle is 12 inches, and the perimeter is 40 inches. Find the width.

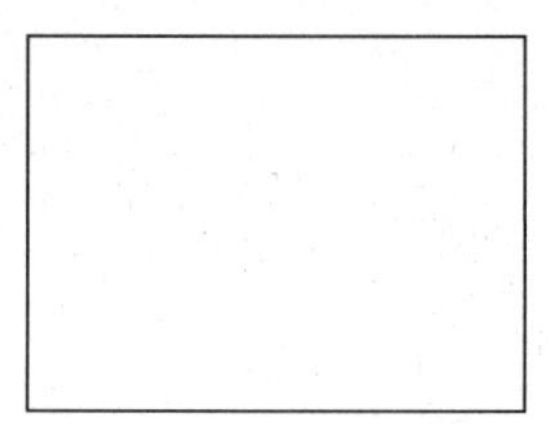

26. The radius of a circle is 8 feet. Find the circumference.

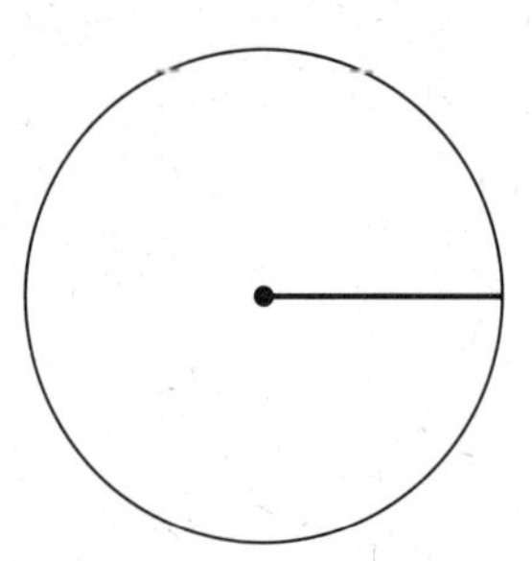

27. The shorter base of a trapezoid is 8 centimeters long, and the longer base is 12 centimeters long. The height is 4 centimeters. Find the area.

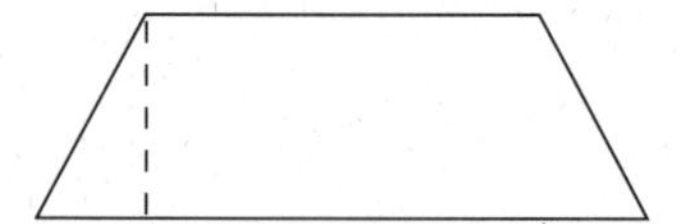

28. The perimeter of a triangle is 40 meters. One side is 16 meters long, and another side is 6 meters long. Find the length of the third side.

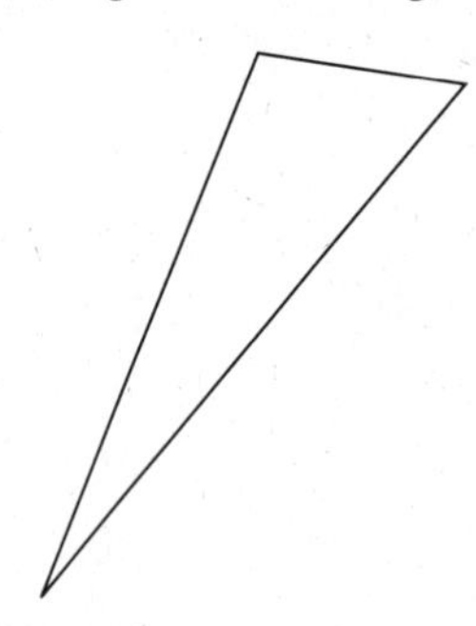

name date hour

29. The area of a triangle is 60 square centimeters. The base is 24 centimeters long. Find the height.

30. Find the radius of a circle with a circumference of 15.072 inches.

31. The volume of a right circular cylinder with a radius of 4 centimeters is 502.4 cubic centimeters. Find the height of the cylinder.

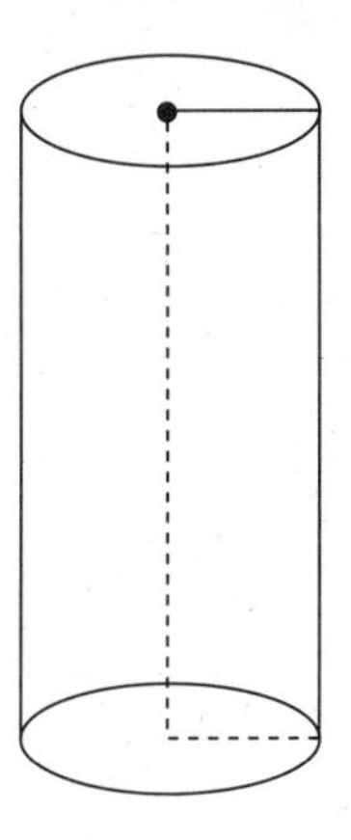

32. The area of a trapezoid is 105 square meters. One base has a length of 7 meters, and the height is 3 meters. Find the length of the other base.

33. Find the height of a right circular cylinder with a radius of 3 inches and a surface area of 339.12 square inches.

34. Find the area of the base of a right pyramid with a height of 12 feet and a volume of 200 cubic feet.

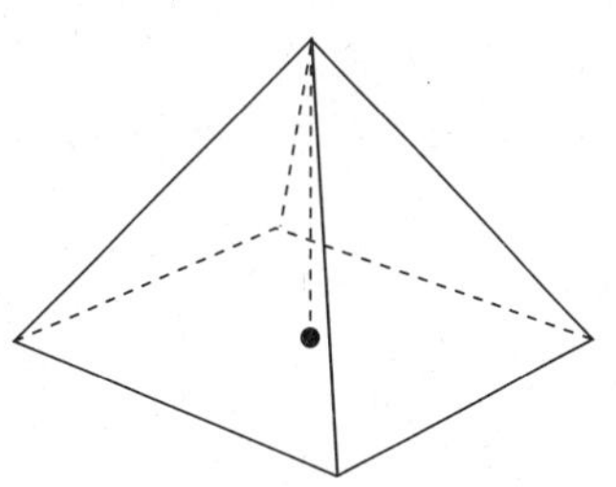

Solve the given formula for the indicated variable. See Examples 5 and 6.

35. $A = LW$ for L

36. $d = rt$ for r

37. $d = rt$ for t

38. $V = LWH$ for W

39. $V = LWH$ for H

40. $I = prt$ for p

41. $I = prt$ for t

42. $C = 2\pi r$ for r

43. $c^2 = a^2 + b^2$ for a^2

44. $a + b + c = P$ for b

45. $A = \frac{1}{2}bh$ for b

46. $A = \frac{1}{2}bh$ for h

47. $V = \pi r^2 h$ for r^2

48. $V = \frac{1}{3}\pi r^2 h$ for r^2

49. $P = 2L + 2W$ for W

50. $A = p + prt$ for r

51. $P = A - Art$ for t

52. $y = ax + b$ for a

53. $d = gt^2 + vt$ for v

54. $F = \frac{9}{5}C + 32$ for C

55. $C = \frac{5}{9}(F - 32)$ for F

SKILL SHARPENERS

Solve each equation. See Section 3.2.

56. $8x = 16$

57. $9y = 54$

58. $7q = 15$

59. $9t = 22$

60. $-\frac{2}{5}m = 12$

61. $-\frac{4}{7}r = 16$

3.6 RATIO AND PROPORTION

1 Ratios provide a way of comparing two numbers or quantities using division. A **ratio** is a quotient of two quantities with the same units.

The ratio of the number a to the number b is written

$$a \text{ to } b, \quad a:b, \quad \text{or} \quad \frac{a}{b}.$$

This last way of writing a ratio is most common in algebra.

EXAMPLE 1 Writing a Word Phrase as a Ratio
Write a ratio for each word phrase.

(a) The ratio of 5 hours to 3 hours is

$$\frac{5}{3}.$$

(b) To find the ratio of 6 hours to 3 days, first convert 3 days to hours.

$$3 \text{ days} = 3 \cdot 24$$
$$= 72 \text{ hours}$$

The ratio of 6 hours to 3 days is thus

$$\frac{6}{72} = \frac{1}{12}. \quad ■$$

WORK PROBLEM 1 AT THE SIDE.

2 A ratio is used to compare two numbers or amounts. A **proportion** says that two ratios are equal. For example,

$$\frac{3}{4} = \frac{15}{20}$$

is a proportion that says that the ratios $\frac{3}{4}$ and $\frac{15}{20}$ are equal. In the proportion

$$\frac{a}{b} = \frac{c}{d}$$

a, b, c, and d are the **terms** of the proportion. Beginning with the proportion

$$\frac{a}{b} = \frac{c}{d}$$

and multiplying both sides by the common denominator, bd, gives

$$bd \cdot \frac{a}{b} = bd \cdot \frac{c}{d}$$

$$\frac{b}{b}(d \cdot a) = \frac{d}{d}(b \cdot c) \qquad \text{Associative and commutative properties}$$

$$ad = bc. \qquad \text{Commutative property}$$

OBJECTIVES

1. Write ratios.
2. Decide whether proportions are true.
3. Solve proportions.
4. Solve word problems using proportions.

1. Write each ratio.

(a) 9 women to 5 women

(b) 4 inches to 1 foot

(c) 3 days to 2 weeks

(d) \$6 to 10 quarters

ANSWERS
1. (a) $\frac{9}{5}$ (b) $\frac{4}{12} = \frac{1}{3}$ (c) $\frac{3}{14}$
(d) $\frac{24}{10} = \frac{12}{5}$

2. Decide whether each proportion is true or false.

(a) $\frac{7}{12} = \frac{35}{60}$

(b) $\frac{21}{15} = \frac{62}{45}$

(c) $\frac{58}{25} = \frac{638}{275}$

The products ad and bc are found by multiplying diagonally, as shown below.

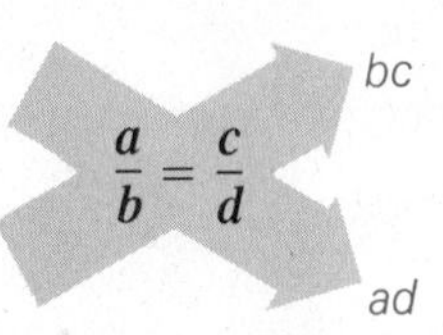

For this reason, ad and bc are called **cross products.**

If $\frac{a}{b} = \frac{c}{d}$, then the cross products ad and bc are equal.

Also, if $ad = bc$, then $\frac{a}{b} = \frac{c}{d}$.

From the rule given above, if $\frac{a}{b} = \frac{c}{d}$ then $ad = bc$. However, if $\frac{a}{c} = \frac{b}{d}$, then $ad = cb$, or $ad = bc$. This means that the two proportions are equivalent, and

$$\text{the proportion } \frac{a}{b} = \frac{c}{d} \text{ can always be written as } \frac{a}{c} = \frac{b}{d}.$$

Sometimes one form is more convenient to work with than the other.

EXAMPLE 2 Deciding Whether a Proportion Is True

Decide whether the following proportions are true or false.

(a) $\frac{3}{4} = \frac{15}{20}$

Check to see whether the cross products are equal.

$4 \cdot 15 = 60$

$\frac{3}{4} = \frac{15}{20}$

$3 \cdot 20 = 60$

The cross products are equal, so the proportion is true.

(b) $\frac{6}{7} = \frac{30}{32}$

The cross products are $6 \cdot 32 = 192$ and $7 \cdot 30 = 210$. The cross products are not equal, so the proportion is false. ■

CAUTION The cross product method cannot be used directly if there is more than one term on either side.

WORK PROBLEM 2 AT THE SIDE.

ANSWERS

2. (a) true (b) false (c) true

3 Four numbers are used in a proportion. If any three of these numbers are known, the fourth can be found.

EXAMPLE 3 Finding an Unknown in a Proportion

(a) Find x in the proportion

$$\frac{5}{9} = \frac{x}{63}.$$

The cross products must be equal.

$$5 \cdot 63 = 9 \cdot x \quad \text{Cross products}$$
$$315 = 9x \quad \text{Multiply.}$$
$$35 = x \quad \text{Divide by 9.}$$

(b) Solve for r in the proportion

$$\frac{8}{5} = \frac{12}{r}.$$

Set the cross products equal to each other.

$$8r = 5 \cdot 12 \quad \text{Cross products}$$
$$8r = 60 \quad \text{Multiply.}$$
$$r = \frac{60}{8} = \frac{15}{2} \quad \text{Divide by 8.} \quad ■$$

WORK PROBLEM 3 AT THE SIDE.

EXAMPLE 4 Solving an Equation Using Proportions

Solve the equation

$$\frac{m-2}{m+1} = \frac{5}{3}.$$

Find the cross products.

$$3(m-2) = 5(m+1) \quad \text{Cross products}$$
$$3m - 6 = 5m + 5 \quad \text{Distributive property}$$
$$3m = 5m + 11 \quad \text{Add 6.}$$
$$-2m = 11 \quad \text{Subtract } 5m.$$
$$m = -\frac{11}{2} \quad \text{Divide by } -2. \quad ■$$

WORK PROBLEM 4 AT THE SIDE.

4 Proportions occur in many practical applications, as the next example shows.

3. Solve each equation.

(a) $\frac{25}{11} = \frac{125}{k}$

(b) $\frac{y}{6} = \frac{35}{42}$

(c) $\frac{24}{a} = \frac{16}{15}$

4. Solve each equation.

(a) $\frac{z}{z+1} = \frac{2}{3}$

(b) $\frac{p+3}{p-5} = \frac{3}{4}$

(c) $\frac{a+6}{2a} = \frac{1}{5}$

ANSWERS

3. (a) 55 (b) 5 (c) $\frac{45}{2}$

4. (a) 2 (b) -27 (c) -10

EXAMPLE 5 Applying Proportions

A hospital charges a patient \$7.80 for 12 capsules. How much should it charge for 18 capsules?

Let x be the cost of 18 capsules. Set up a proportion; one ratio in the proportion can involve the number of capsules, and the other ratio can use the costs. Make sure that corresponding numbers appear in the numerator and the denominator.

$$\begin{array}{cc} \text{Cost} & \text{Number} \\ \dfrac{\text{Cost of 18}}{\text{Cost of 12}} = & \dfrac{18}{12} \\ \dfrac{x}{7.80} = & \dfrac{18}{12} \end{array}$$

As shown earlier, this proportion could also be written as

$$\frac{18}{\text{Cost of 18}} = \frac{12}{\text{Cost of 12}}$$
$$\frac{18}{x} = \frac{12}{7.80}.$$

Solve either equation. Find the cross products, and set them equal.

$$12x = 18(7.80) \quad \text{Cross products}$$
$$12x = 140.40 \quad \text{Multiply.}$$
$$x = 11.70 \quad \text{Divide by 12.}$$

The 18 capsules should cost \$11.70. ■

WORK PROBLEM 5 AT THE SIDE.

5. Solve each word problem.

(a) On a map, 12 inches represents 500 miles. How many miles would be represented by 30 inches?

(b) If 7 shirts cost \$87.50, find the cost of 11 shirts.

ANSWERS
5. (a) 1250 miles (b) \$137.50

name date hour

3.6 EXERCISES

Write the following ratios. Write each ratio in lowest terms. See Example 1.

1. 80 feet to 50 feet

2. 70 miles to 40 miles

3. 140 people to 70 people

4. 180 dollars to 90 dollars

5. 12 yards to 15 feet

6. 10 feet to 15 yards

7. 50 inches to 10 yards

8. 50 inches to 12 feet

9. 15 minutes to 3 hours

10. 7 quarts to 3 pints

11. 55 dimes to 3 dollars

12. 7 dollars to 8 quarters

13. 3 days to 16 hours

14. 40 hours to 10 days

15. 75¢ to $3

Decide whether the following proportions are true. See Example 2.

16. $\frac{4}{7} = \frac{12}{21}$

17. $\frac{9}{10} = \frac{18}{20}$

18. $\frac{6}{8} = \frac{15}{20}$

19. $\frac{12}{18} = \frac{8}{12}$

20. $\frac{5}{8} = \frac{35}{56}$

21. $\frac{12}{7} = \frac{36}{20}$

22. $\frac{7}{10} = \frac{82}{120}$

23. $\frac{18}{20} = \frac{56}{60}$

24. $\frac{16}{40} = \frac{22}{55}$

25. $\frac{19}{30} = \frac{57}{90}$

26. $\frac{110}{18} = \frac{160}{27}$

27. $\frac{420}{600} = \frac{14}{20}$

28. $\frac{7.7}{2.1} = \frac{4.4}{1.2}$

29. $\frac{6.3}{4.5} = \frac{3.5}{2.5}$

30. $\frac{6.3}{5.1} = \frac{7.2}{6.4}$

31. $\frac{8.5}{3.5} = \frac{10.6}{4.8}$

Solve each of the following equations. See Examples 3 and 4.

32. $\frac{35}{4} = \frac{k}{20}$

33. $\frac{z}{56} = \frac{7}{8}$

34. $\frac{m}{32} = \frac{3}{24}$

35. $\frac{6}{x} = \frac{4}{18}$

36. $\frac{z - 2}{80} = \frac{20}{100}$

37. $\frac{25}{100} = \frac{8}{m + 6}$

38. $\frac{2}{3} = \frac{y + 2}{7}$

39. $\frac{5}{8} = \frac{m - 1}{5}$

40. $\frac{5}{9} = \frac{2z + 2}{15}$

41. $\frac{3}{4} = \frac{3n - 1}{10}$

42. $\frac{6}{2k + 3} = \frac{8}{5}$

43. $\frac{5}{4p - 5} = \frac{3}{4}$

44. $\frac{m}{m - 3} = \frac{5}{3}$

45. $\frac{r + 1}{r} = \frac{1}{3}$

46. $\frac{3k - 1}{k} = \frac{6}{7}$

47. $\frac{y}{6y - 5} = \frac{5}{11}$

48. $\frac{p + 7}{p - 1} = \frac{3}{4}$

49. $\frac{r + 8}{r - 9} = \frac{7}{3}$

name date hour

Solve the following applications involving proportions. See Example 5.

50. A certain lawn mower uses 3 tanks of gas to cut 10 acres of lawn. How many tanks of gas would be needed for 30 acres?

51. A bus ride of 80 miles costs $5. Find the charge to ride 180 miles.

52. José can assemble 12 car parts in 40 minutes. How many minutes would he need to assemble 15 car parts?

53. A Hershey bar contains 200 calories. How many bars would you need to eat to get 500 calories?

54. The tax on a $20 item is $1. Find the tax on a $110 item.

55. If 2 pounds of fertilizer will cover 50 square feet of garden, how many pounds would be needed for 225 square feet?

56. A garden service charges $30 to install 50 square feet of sod. Find the charge to install 125 square feet.

57. On a road map, 3 inches represents 8 miles. How many inches would represent a distance of 24 miles?

58. Suppose that 7 sacks of fertilizer cover 3325 square feet of lawn. Find the number of sacks needed for 7125 square feet.

59. If 9 pairs of jeans cost $121.50, find the cost of 5 pairs.

60. Twelve yards of material are needed for 5 dresses. How much material would be needed for 8 dresses?

61. The distance between two cities on a road map is 11 inches. Actually, the cities are 308 miles apart. The distance between two other cities is 15 inches. How far apart are these cities?

62. If 8 ounces of medicine must be mixed with 20 ounces of water, how many ounces of medicine must be mixed with 50 ounces of water?

63. The charge to move a load of freight 750 miles is $90. Find the charge to move the freight 1200 miles.

SKILL SHARPENERS

Place $<$ or $>$ in each blank to make a true statement. See Section 2.1.

64. -8 ____ 5

65. -10 ____ 7

66. -12 ____ -5

67. -9 ____ -14

68. -7 ____ -10

69. -6 ____ -4

70. 7 ____ -12

71. 7 ____ -2

3.7 THE ADDITION AND MULTIPLICATION PROPERTIES OF INEQUALITY

Inequalities are statements with algebraic expressions related in the following ways:

$<$ "is less than"
$\leq$ "is less than or equal to"
$>$ "is greater than"
$\geq$ "is greater than or equal to."

An inequality is solved by finding all real number solutions for it. For example, the solutions of $x \leq 2$ include all *real numbers* that are less than or equal to 2, and not just the *integers* less than or equal to 2.

1 Graphing is a good way to show the solutions of an inequality. To graph all real numbers satisfying $x \leq 2$, place a dot at 2 on a number line and draw an arrow extending from the dot to the left (to represent the fact that all numbers less than 2 are also part of the graph). The graph is shown in Figure 6.

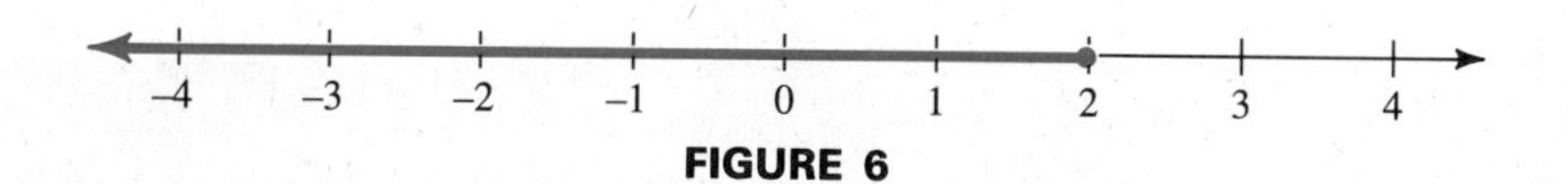

FIGURE 6

EXAMPLE 1 Graphing the Solutions of an Inequality

Graph $x > -5$.

The statement $x > -5$ says that x can take any value greater than -5, but x cannot equal -5 itself. Show this on a graph by placing an open circle at -5 and drawing an arrow to the right, as in Figure 7. The open circle at -5 shows that -5 is not part of the graph. ■

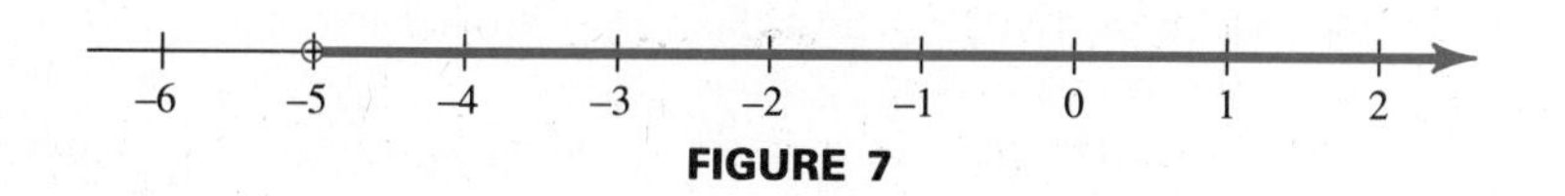

FIGURE 7

EXAMPLE 2 Graphing the Solutions of an Inequality

Graph $3 > x$.

The statement $3 > x$ means the same as $x < 3$. The graph of $x < 3$ is shown in Figure 8. ■

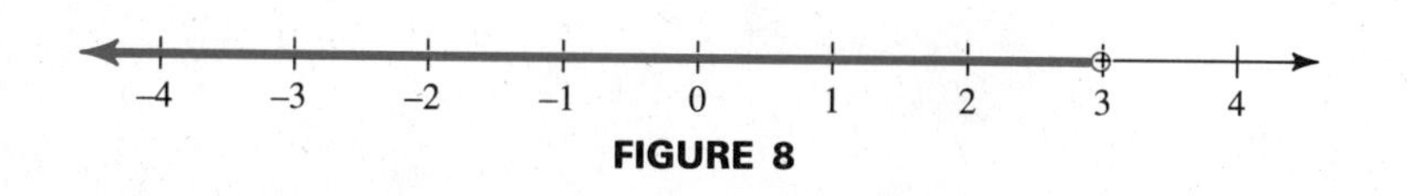

FIGURE 8

WORK PROBLEM 1 AT THE SIDE.

OBJECTIVES

1. Graph inequalities on a number line.
2. Use the addition property of inequality.
3. Use the multiplication property of inequality.
4. Solve inequalities using both properties of inequality.
5. Use inequalities to solve word problems.
6. Solve inequalities about numbers between other numbers.

1. Graph each of the following.

(a) $x \leq 3$

(b) $x > -4$

(c) $-4 \geq x$

(d) $0 < x$

ANSWERS

1. (a)

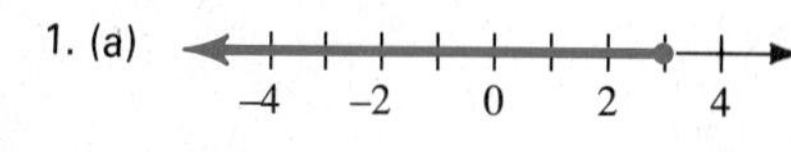

(b)

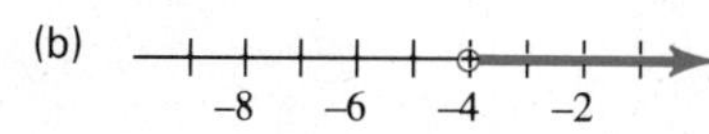

(c)

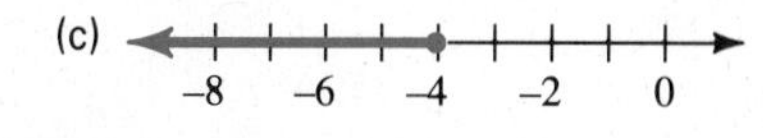

(d) -4 -2 0 2 4

2. Graph the solutions of each inequality.

(a) $4 < x \leq 8$

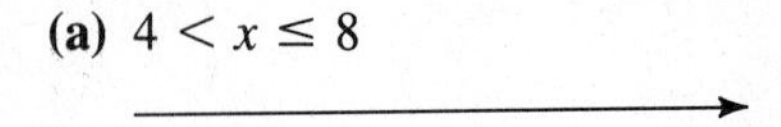

(b) $-7 < x < -2$

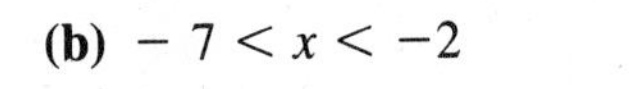

(c) $-6 < x \leq -4$

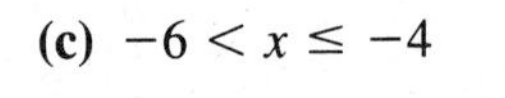

ANSWERS

2. (a) 0 4 8 12 16

(b) −8 −6 −4 −2 0

(c) −8 −6 −4 −2 0

EXAMPLE 3 Graphing the Solutions of an Inequality

Graph $-3 \leq x < 2$.

The statement $-3 \leq x < 2$ is read "-3 is less than or equal to x *and* x is less than 2." Graph the solutions of this inequality by placing a solid dot at -3 (because -3 is part of the graph) and an open circle at 2 (because 2 is not part of the graph), then drawing a line segment between the two circles. See Figure 9. ■

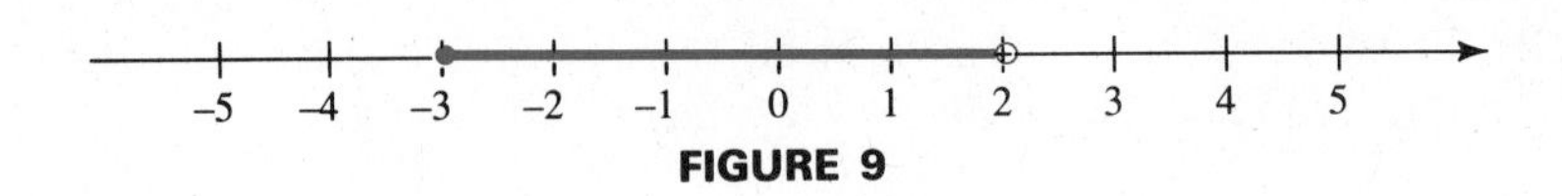

FIGURE 9

WORK PROBLEM 2 AT THE SIDE.

2 Inequalities such as $x + 4 \leq 9$ are solved in much the same way as equations. First, we can use the **addition property of inequality,** which states that the same number can be added to each side of an inequality.

ADDITION PROPERTY OF INEQUALITY

For any real numbers A, B, and C, the inequalities

$$A < B \quad \text{and} \quad A + C < B + C$$

have exactly the same solutions.

In other words, the same number may be added to each side of an inequality without changing the solutions.

The addition property of inequality also works with $>$, $\leq$, or $\geq$. Also, as with the addition property of equality, the same number may be *subtracted* on each side of an inequality.

The following examples show how the addition property is used to solve inequalities.

EXAMPLE 4 Using the Addition Property of Inequality

Solve the inequality $7 + 3k > 2k - 5$.

Use the addition property of inequality twice, once to get the terms containing k on one side of the inequality and a second time to get the integers together on the other side.

$$7 + 3k > 2k - 5$$

$$7 + 3k - 2k > 2k - 5 - 2k \qquad \text{Subtract } 2k.$$

$$7 + k > -5 \qquad \text{Combine terms.}$$

$$7 + k - 7 > -5 - 7 \qquad \text{Subtract 7.}$$

$$k > -12 \qquad \text{Combine terms.}$$

A graph of the solutions, $k > -12$, is shown in Figure 10. ■

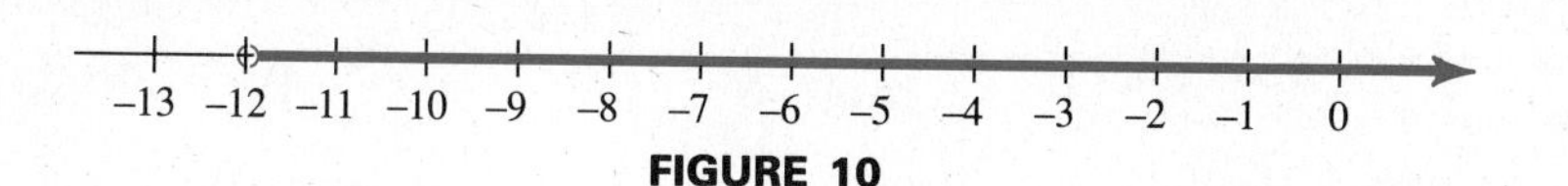

FIGURE 10

WORK PROBLEM 3 AT THE SIDE.

3 The addition property of inequality alone cannot be used to solve inequalities such as $4y \geq 28$. These inequalities require the *multiplication property of inequality*. To see how this property works, it will be helpful to look at some examples.

First, write the inequality $3 < 7$ and then multiply each side by the positive number 2.

$$3 < 7$$
$$2(3) < 2(7) \qquad \text{Multiply each side by 2.}$$
$$6 < 14 \qquad \text{True}$$

Now multiply each side of $3 < 7$ by the negative number -5.

$$3 < 7$$
$$-5(3) < -5(7) \qquad \text{Multiply each side by } -5.$$
$$-15 < -35 \qquad \text{False}$$

Getting a true statement when multiplying each side by -5 requires reversing the direction of the inequality symbol.

$$3 < 7$$
$$-5(3) > -5(7) \qquad \text{Multiply by } -5\text{; reverse the symbol.}$$
$$-15 > -35 \qquad \text{True}$$

Take the inequality $-6 < 2$ as another example. Multiply each side by the positive number 4.

$$-6 < 2$$
$$4(-6) < 4(2) \qquad \text{Multiply by 4.}$$
$$-24 < 8 \qquad \text{True}$$

Multiplying each side of $-6 < 2$ by -5 *and at the same time reversing the direction of the inequality symbol* gives

$$-6 < 2$$
$$(-5)(-6) > (-5)(2) \qquad \text{Multiply by } -5\text{; reverse the symbol.}$$
$$30 > -10. \qquad \text{True}$$

WORK PROBLEM 4 AT THE SIDE.

In summary, the two parts of the multiplication property of inequality are as stated here.

3. Solve each of the following inequalities. Graph the solutions.

(a) $x - 1 < 6$

(b) $2m \geq m - 4$

(c) $-1 + 8r < 7r + 2$

4. (a) Multiply each side of $-2 < 8$ by 6 and then by -5. Reverse the direction of the inequality symbol if necessary.

(b) Multiply each side of $-4 > -9$ by 2 and then by -8. Reverse the direction of the inequality symbol if necessary.

ANSWERS

3. (a) $x < 7$

0 2 4 6 7 8

(b) $m \geq -4$

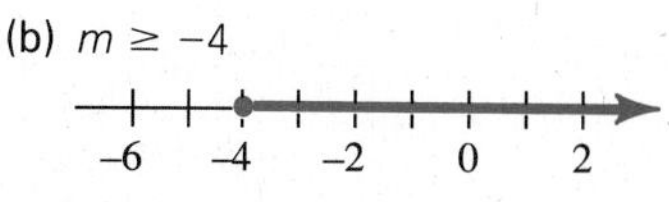

(c) $r < 3$

−4 −2 0 2 3 4

4. (a) $-12 < 48$; $10 > -40$
(b) $-8 > -18$; $32 < 72$

MULTIPLICATION PROPERTY OF INEQUALITY

For any real numbers A, B, and C ($C \neq 0$),

(1) if C is *positive*, then the inequalities

$$A < B \quad \text{and} \quad AC < BC$$

have exactly the same solutions;

(2) if C is *negative*, then the inequalities

$$A < B \quad \text{and} \quad AC > BC$$

have exactly the same solutions.

In other words, each side of an inequality may be multiplied by the same positive number without changing the solutions. If the number is negative, we must reverse the direction of the inequality symbol.

The multiplication property of inequality also works with $>$, $\leq$, or $\geq$. As with the multiplication property of equality, the same nonzero number may be divided into each side.

It is important to remember the differences in the multiplication property for positive and negative numbers.

1. When each side of an inequality is multiplied or divided by a positive number, the direction of the inequality symbol *does not change.* Adding or subtracting terms on each side also does not change the symbol.
2. When each side of an inequality is multiplied or divided by a negative number, the direction of the symbol *does change. Reverse the direction of the symbol of inequality only when multiplying or dividing by a negative number.*

The next examples show how to use the multiplication property to solve inequalities.

EXAMPLE 5 Using the Multiplication Property of Inequality

Solve the inequality $3r < -18$.

Use the multiplication property of inequality and divide each side by 3. Since 3 is a positive number, the direction of the inequality symbol *does not* change. *It does not matter that the number on the right side of the inequality is negative.*

$$3r < -18$$

$$\frac{3r}{3} < \frac{-18}{3} \qquad \text{Divide by 3.}$$

$$r < -6 \qquad \text{Reduce.}$$

The graph of the solutions is shown in Figure 11. ■

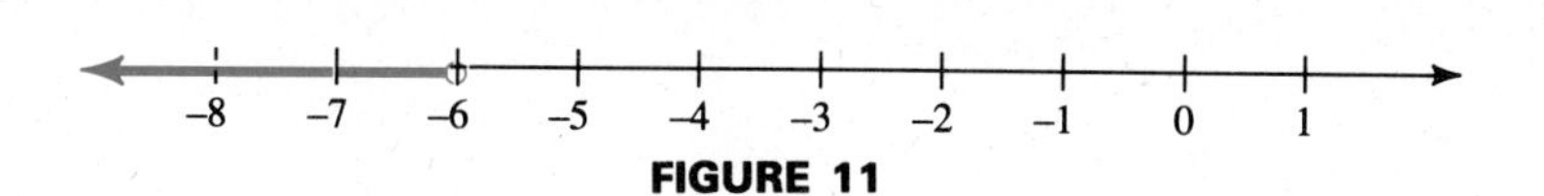

FIGURE 11

EXAMPLE 6 Using the Multiplication Property of Inequality
Solve the inequality $-4t \geq 8$.

Here each side of the inequality must be divided by -4, a negative number, which *does* change the direction of the inequality symbol.

$$-4t \geq 8$$

$$\frac{-4t}{-4} \leq \frac{8}{-4} \quad \text{Divide by } -4\text{; symbol reversed.}$$

$$t \leq -2 \quad \text{Reduce.}$$

The solutions are graphed in Figure 12. ■

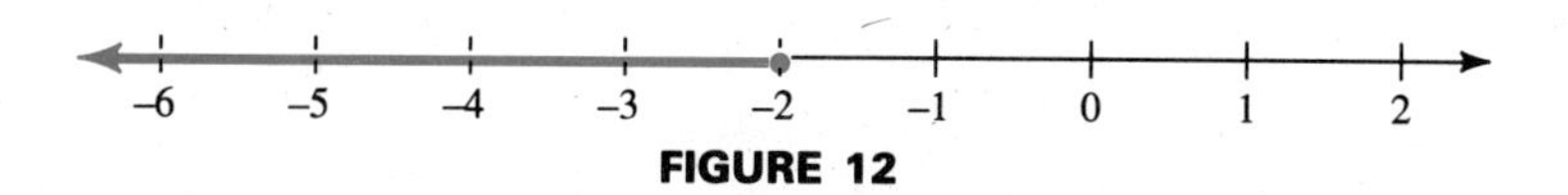

FIGURE 12

WORK PROBLEM 5 AT THE SIDE.

4 The steps in solving an inequality are summarized below. (Remember that $<$ can be replaced with $>$, $\leq$, or $\geq$ in this summary.)

SOLVING INEQUALITIES

Step 1 Use the associative, commutative, and distributive properties to combine like terms on each side of the inequality.

Step 2 Use the addition property of inequality to simplify the inequality to the form $ax < b$, where a and b are real numbers.

Step 3 Use the multiplication property of inequality to simplify further to an inequality of the form $x < c$ or $x > c$, where c is a real number.

Notice how these steps are used in the next example.

EXAMPLE 7 Solving an Inequality
Solve $5(k - 3) - 7k \geq 4(k - 3) + 9$.

Step 1 Use the distributive property; then combine like terms.

$$5(k - 3) - 7k \geq 4(k - 3) + 9$$

$$5k - 15 - 7k \geq 4k - 12 + 9 \quad \text{Distributive property}$$

$$-2k - 15 \geq 4k - 3 \quad \text{Combine like terms.}$$

5. Solve each inequality. Graph your solutions.

(a) $5x \geq 30$

(b) $9y < -18$

(c) $-4p \leq 32$

(d) $-2r > -12$

(e) $-5p \leq 0$

ANSWERS
5. (a) $x \geq 6$
4 5 6 7 8 9 10
(b) $y < -2$
-5 -4 -3 -2 -1 0 1
(c) $p \geq -8$
-11 -10 -9 -8 -7 -6 -5
(d) $r < 6$
3 4 5 6 7 8 9
(e) $p \geq 0$
-3 -2 -1 0 1 2 3

6. Solve. Graph your solutions.

(a) $2m - 4 \geq 3m - 1$

(b) $5r - r + 2 < 7r - 5$

7. Solve
$4(y - 1) - 3y > -15 - (2y + 1)$.
Graph your solutions.

ANSWERS
6. (a) $m \leq -3$

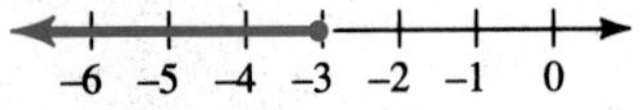

(b) $r > \frac{7}{3}$

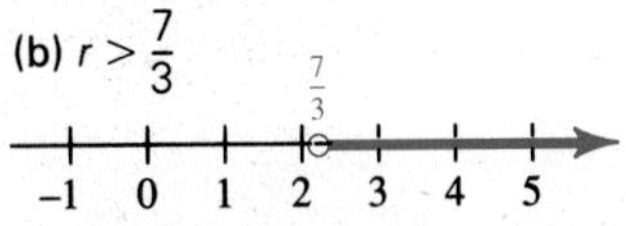

7. $y > -4$

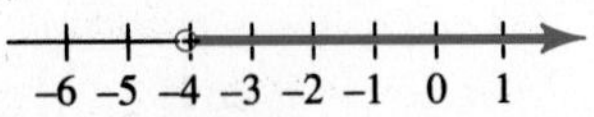

Step 2 Use the addition property.

$-2k - 15 - 4k \geq 4k - 3 - 4k$	Subtract $4k$.
$-6k - 15 \geq -3$	Combine terms.
$-6k - 15 + 15 \geq -3 + 15$	Add 15.
$-6k \geq 12$	Combine terms.

Step 3 Divide each side by -6, a negative number. Change the direction of the inequality symbol.

$$\frac{-6k}{-6} \leq \frac{12}{-6} \qquad \text{Divide by } -6; \text{ reverse the symbol.}$$

$$k \leq -2$$

A graph of the solutions is shown in Figure 13. ■

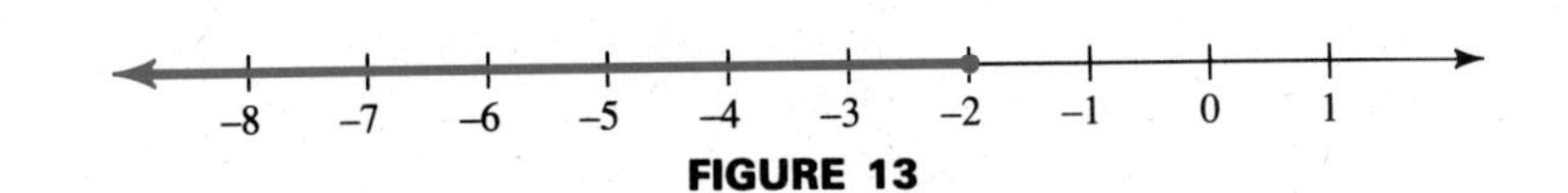

FIGURE 13

WORK PROBLEMS 6 AND 7 AT THE SIDE.

5 Inequalities can be used to solve word problems involving phrases that suggest inequality. The chart below gives some of the more common such phrases, along with examples and translations.

Phrase	*Example*	*Inequality*
Is more than	A number *is more than* 4	$x > 4$
Is less than	A number *is less than* -12	$x < -12$
Is at least	A number *is at least* 6	$x \geq 6$
Is at most	A number *is at most* 8	$x \leq 8$

CAUTION Do not confuse statements like "5 more than a number" and "5 *is* more than a number." The first of these is expressed as "$x + 5$" while the second is expressed as "$5 > x$."

EXAMPLE 8 Finding an Average Test Score
Brent has test grades of 86, 88, and 78 on his first three tests in geometry. If he wants an average of at least 80 after his fourth test, what score must he make on his fourth test?

Let x represent Brent's score on his fourth test. To find the average of the four scores, add them and find $\frac{1}{4}$ of the sum.

Average ↓ is at least 80. ↓ ↓

$$\frac{1}{4}(86 + 88 + 78 + x) \geq 80$$

$$4 \cdot \frac{1}{4}(252 + x) \geq 4 \cdot 80 \quad \text{Multiply by 4.}$$
$$252 + x \geq 320 \quad \text{Multiply.}$$
$$252 - 252 + x \geq 320 - 252 \quad \text{Subtract 252.}$$
$$x \geq 68 \quad \text{Combine terms.}$$

He must score 68 or more on the fourth test to have an average of *at least* 80. ■

WORK PROBLEM 8 AT THE SIDE.

6 The next example shows how to solve inequalities in which one expression is *between* two other ones.

EXAMPLE 9 Solving an Inequality with Three Parts

Solve $4 \leq 3x - 5 < 6$.

First, add 5 to each part.

$$4 \leq 3x - 5 < 6$$
$$4 + 5 \leq 3x - 5 + 5 < 6 + 5 \quad \text{Add 5.}$$
$$9 \leq 3x < 11 \quad \text{Combine terms.}$$

Now divide each part by the positive number 3.

$$\frac{9}{3} \leq \frac{3x}{3} < \frac{11}{3} \quad \text{Divide by 3.}$$
$$3 \leq x < \frac{11}{3} \quad \text{Reduce.}$$

The solutions are graphed in Figure 14. ■

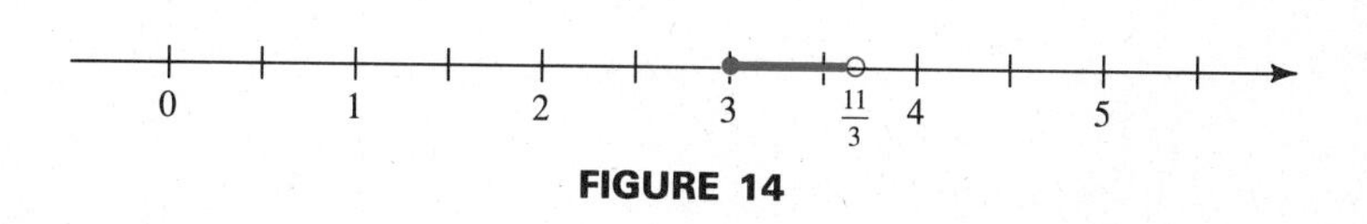

FIGURE 14

WORK PROBLEM 9 AT THE SIDE.

8. Solve the following problem.

Maggie has scores of 98, 86, and 88 on her first three tests in algebra. If she wants an average of at least 90 after her fourth test, what score should she make on her fourth test?

9. Solve $2 \leq 3p - 1 \leq 8$. Graph the solutions.

ANSWERS
8. 88 or more
9. $1 \leq p \leq 3$

–1 0 1 2 3 4 5

Historical Reflections

Historical Reflections

Charles Dodgson (1832–1898)

Lewis Carroll was the pen name of the Reverend Charles Lutwidge Dodgson, a mathematics lecturer at Oxford University in England. When Queen Victoria told Dodgson how much she enjoyed his *Alice's Adventures in Wonderland,* he supposedly sent her *Symbolic Logic,* his most famous mathematical work. In the book, he makes some strong claims for the subject:

> (Symbolic logic) . . . will give you clearness of thought—the ability to *see your way* through a puzzle—the habit of arranging your ideas in an orderly and get-at-able form—and, more valuable than all, the power to detect *fallacies,* and to tear to pieces the flimsy illogical arguments which you will so continually encounter in books, in newspapers, in speeches, and even in sermons . . . ."

Art: Library of Congress

name date hour

3.7 EXERCISES

Graph each inequality on the given number line. See Examples 1–3.

1. $x \leq 4$

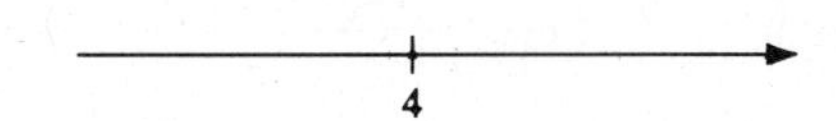

2. $x \leq -3$

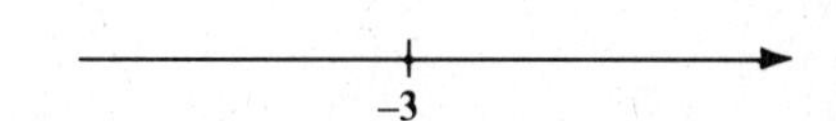

3. $a < 3$

4. $p > 4$

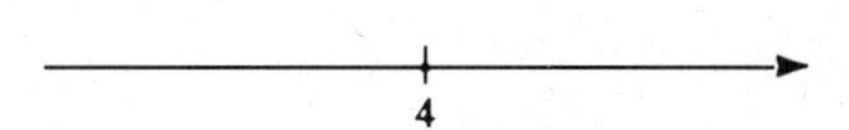

5. $-2 \leq x \leq 5$

6. $8 \leq m \leq 10$

7. $3 \leq y < 5$

8. $0 < y \leq 10$

Solve each inequality. See Examples 4–6.

9. $a + 6 < 8$

10. $k - 4 < 2$

11. $z - 3 \geq -2$

12. $p + 2 \geq -6$

13. $-3 + k \geq 2$

14. $x + 5 > 5$

15. $3x < 27$

16. $5h \geq 20$

17. $-2k \leq 12$

18. $-3v > 6$

19. $-8y < -72$

20. $-9a \geq -63$

Solve each inequality and then graph your solutions. See Example 7.

21. $3n + 5 \leq 2n - 6$

22. $5x - 2 < 4x - 5$

23. $-6(k + 2) + 3 \geq -7(k - 5)$

24. $-3(m - 5) + 8 < -4(m + 2)$

25. $4k + 1 \geq 2k - 9$

26. $5y + 3 < 2y + 12$

27. $4q + 1 - 5 < 8q + 4$

28. $5x - 2 \leq 2x + 6 - x$

29. $3 - 2z + 2 > 4 - z$

30. $3 - 5t < 8 + 2t - 2$

31. $12 - w \leq 4 + 4w - 7$

32. $10 - 4k + 8 \geq 6 - 2k + 10$

33. $-k + 4 + 7k \leq -1 + 3k + 5$

34. $6y - 2y - 4 + 7y > 3y - 4 + 7y$

35. $2(x - 5) + 3x < 4(x - 6) + 3$

36. $5(t + 3) - 6t \leq 3(2t + 1) - 4t$

name date hour

Solve each inequality and then graph your solutions. See Example 9.

37. $8 \le p + 2 \le 15$

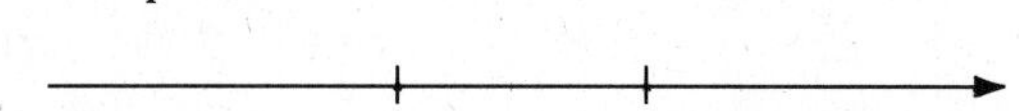

38. $4 \le k - 1 \le 10$

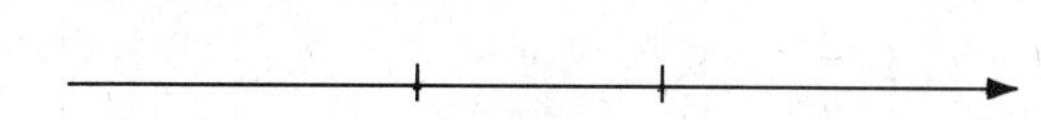

39. $-3 < y - 8 < 4$

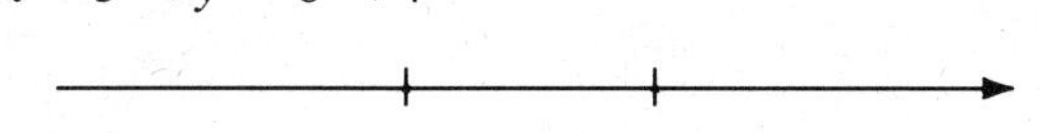

40. $-6 < r - 1 < -2$

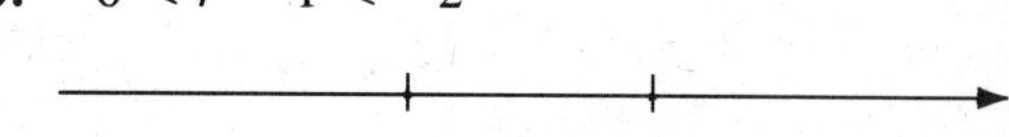

41. $-5 \le 2x - 3 \le 9$

42. $-7 \le 3x - 4 \le 8$

43. $-12 \le \frac{1}{2}z + 1 \le 4$

44. $-6 \le \frac{1}{3}a - 3 \le 5$

Solve each word problem. See Example 8.

45. If half a number is added to 5, the result is greater than or equal to -3. Find all such numbers.

46. When four times a number is subtracted from 8, the result is less than 15. Find all numbers that satisfy this condition.

47. Inkie has grades of 75 and 82 on her first two computer science tests. What must she score on a third test to have an average of at least 80?

48. Mabimi Pampo has grades of 94 and 88 on his first two calculus tests. What must he score on a third test to have an average of at least 90?

49. Audrey earned \$200 at odd jobs during July, \$300 during August, and \$225 during September. If her average salary for the four months from July through October is to be at least \$250, how much must she earn during October?

50. In order to qualify for a company pension plan, an employee must average at least \$1000 per month in earnings. During the first four months of the year, an employee made \$900, \$1200, \$1040, and \$760. What amount of earnings during the fifth month will qualify the employee for the pension plan?

51. One side of a triangle is twice as long as a second side. The third side of the triangle is 17 feet long. The perimeter of the triangle cannot be more than 50 feet. Find the longest possible values for the other two sides of the triangle.

52. The perimeter of a rectangle must be no greater than 120 meters. The width of the rectangle must be 22 meters. Find the greatest possible value for the length of the rectangle.

SKILL SHARPENERS

Evaluate each expression. See Section 1.4.

53. $2 \cdot 2 \cdot 2 \cdot 2 \cdot 2$

54. $3 \cdot 3 \cdot 3$

55. $5 \cdot 5 \cdot 5 \cdot 5$

56. $4 \cdot 4 \cdot 4 \cdot 4$

57. $\frac{2}{3} \cdot \frac{2}{3} \cdot \frac{2}{3}$

58. $\frac{5}{8} \cdot \frac{5}{8}$

Simplify. See Sections 2.2 and 2.3.

59. $8 + (-5)$

60. $9 + (-12)$

61. $3 - (-2)$

62. $5 - (-7)$

63. $8 - 4 - (-9)$

64. $-6 - (-8)$

65. $-2 - [-4 - (3 - 2)]$

66. $1 - [-(8 - 9) - 2]$

67. $3 - [2 + (-4 - 1)]$

Historical Reflections

Historical Reflections

TWEEDLOGIC

"I know what you're thinking about," said Tweedledum, "but it isn't so, nohow." "Contrariwise," continued Tweedledee, "if it was so, it might be; and if it were so, it would be, but as it isn't, it ain't. That's logic." (From *Alice's Adventures in Wonderland*)

Art: John Tenneil illustration from Lewis Carroll's *Through the Looking Glass*

CHAPTER 3 SUMMARY

KEY TERMS

3.1 **linear equation** A linear equation is an equation that can be written in the form $ax + b = c$, for real numbers a, b, and c, with $a \neq 0$.

3.6 **ratio** A ratio is a quotient of two quantities with the same units.

proportion A proportion is a statement that two ratios are equal.

cross products The method of cross products provides a way of determining whether a proportion is true.

3.7 **inequality** An inequality is a statement with algebraic expressions related by $<$, $\leq$, $>$, or $\geq$.

NEW SYMBOLS

***a* to *b*, *a*:*b*, or $\frac{a}{b}$** the ratio of a to b

QUICK REVIEW

Section	Concepts	Examples
3.1 The Addition Property of Equality	The same number may be added to (or subtracted from) each side of an equation without changing the solution.	$x - 6 = 12$ $x - 6 + 6 = 12 + 6$ Add 6. $x = 18$ Combine terms
3.2 The Multiplication Property of Equality	Each side of an equation may be multiplied (or divided) by the same nonzero number without changing the solution.	$\frac{3}{4}x = -9$ $\frac{4}{3} \cdot \frac{3}{4}x = \frac{4}{3}(-9)$ Multiply by $\frac{4}{3}$. $x = -12$
3.3 Solving Linear Equations		Solve the equation $2x + 3x + 3 = 38$.
	1. Combine like terms to simplify each side.	**1.** $2x + 3x + 3 = 38$ $5x + 3 = 38$ Combine like terms.
	2. Get the variable term on one side, a number on the other.	**2.** $5x + 3 - 3 = 38 - 3$ Subtract 3. $5x = 35$ Combine terms.
	3. Get the equation into the form $x =$ a number.	**3.** $\frac{5x}{5} = \frac{35}{5}$ Divide by 5. $x = 7$ Reduce.
	4. Check by substituting the result into the original equation.	**4.** $2x + 3x + 3 = 38$ Check. $2(7) + 3(7) + 3 = 38$ Let $x = 7$. $14 + 21 + 3 = 38$ Multiply. $38 = 38$ True

Section	Concepts	Examples
3.4 An Introduction to Word Problems		One number is 5 more than another. Their sum is 21. Find both numbers.
	1. Choose a variable to represent the unknown.	**1.** Let x be the smaller number.
	2. Determine expressions for any other unknown quantities, using the variable. Draw figures or diagrams if they apply.	**2.** Let $x + 5$ be the larger number.
	3. Translate the problem into an equation.	**3.** $x + (x + 5) = 21$
	4. Solve the equation.	**4.** $2x + 5 = 21$ Combine terms. $2x + 5 - 5 = 21 - 5$ Subtract 5. $2x = 16$ Combine terms. $\frac{2x}{2} = \frac{16}{2}$ Divide by 2. $x = 8$ Reduce.
	5. Answer the question asked in the problem.	**5.** The numbers are 8 and 13.
	6. Check your solution by using the original words of the problem. Be sure that the answer is appropriate and makes sense.	**6.** 13 is 5 more than 8, and $8 + 13 = 21$. It checks.
3.5 Formulas	Given the values of all but one of the variables in a formula, the value of the remaining variable can be found.	Find L if $A = LW$, given that $A = 24$ and $W = 3$. $24 = L \cdot 3$ $A = 24, W = 3$ $\frac{24}{3} = \frac{L \cdot 3}{3}$ Divide by 3. $8 = L$ Reduce.
3.6 Ratio and Proportion	To write a ratio, express quantities in the same units.	4 feet to 8 inches = 48 inches to 8 inches $= \frac{48}{8} = \frac{6}{1}$
	To solve a proportion, use the method of cross products.	Solve $\frac{x}{12} = \frac{35}{60}$. $60x = 12 \cdot 35$ Cross products $60x = 420$ Multiply. $\frac{60x}{60} = \frac{420}{60}$ Divide by 60. $x = 7$ Reduce.

Section	Concepts	Examples
3.7 The Addition and Multiplication Properties of Inequality	The same number may be added or subtracted on each side of an inequality without changing the solutions.	$x - 8 < -4$ $x - 8 + 8 < -4 + 8$ Add 8. $x < 4$ Combine terms. 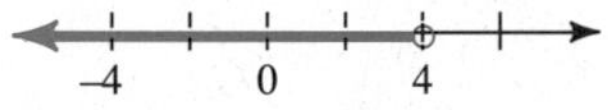
	Each side of an inequality may be multiplied or divided by the same nonzero number. **(a)** If the number is positive, the direction of the inequality symbol *does not* change.	$5x \geq 10$ $\frac{5x}{5} \geq \frac{10}{5}$ Divide by 5. $x \geq 2$ Reduce. 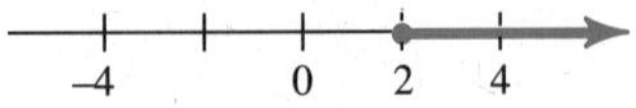
	(b) If the number is negative, the direction of the inequality symbol *must be reversed.*	$-\frac{1}{4}y < 3$ $(-4)\left(-\frac{1}{4}y\right) > -4(3)$ Multiply by −4; reverse sign. $y > -12$ Multiply. 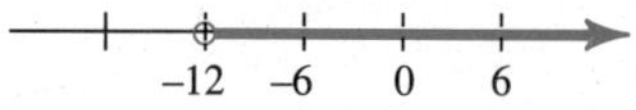
	To solve an inequality: **1.** Combine like terms to simplify each side.	$4x - 3 + 2x - 5 > 10 - 6$ $6x - 8 > 4$ Combine terms. $6x - 8 + 8 > 4 + 8$ Add 8.
	2. Get the variable term on one side, a number on the other.	$6x > 12$ Combine terms. $\frac{6x}{6} > \frac{12}{6}$ Divide by 6.
	3. Get the inequality into the form $x < a$ or $x > a$, where a is a number. (The symbols $\leq$ or $\geq$ may also appear.)	$x > 2$ Reduce. 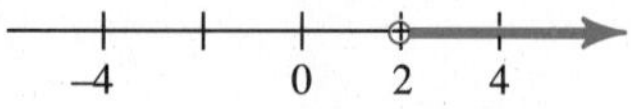
	To solve an inequality such as $4 < 2x + 6 < 8$ work with all three expressions at the same time.	$4 < 2x + 6 < 8$ $4 - 6 < 2x + 6 - 6 < 8 - 6$ Subtract 6. $-2 < 2x < 2$ Combine terms. $\frac{-2}{2} < \frac{2x}{2} < \frac{2}{2}$ Divide by 2. $-1 < x < 1$ Reduce. 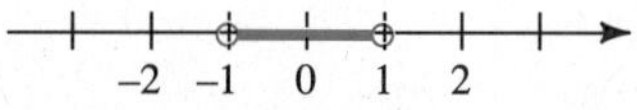

name date hour

CHAPTER 3 REVIEW EXERCISES

[3.1–3.3] *Solve each equation. Check your solutions.*

1. $m - 5 = 1$

2. $y + 8 = -4$

3. $3k + 1 = 2k + 8$

4. $5k = 4k + \frac{2}{3}$

5. $(4r - 2) - (3r + 1) = 8$

6. $3(2y - 5) = 2 + 5y$

7. $7k = 35$

8. $12r = -48$

9. $2p - 7p + 8p = 15$

10. $\frac{m}{12} = -1$

11. $\frac{5}{8}k = 8$

12. $2k - 5 = 4k + 7$

13. $2 - 3(y - 5) = 4 + y$

14. $2(5m - 1) = 10m - 2$

15. $3 - 3(6 - r) = 3(r - 4) + 1$

[3.4] *Solve the following problems. Check your answers.*

16. If 4 is subtracted from twice a number, the result is 8. Find the number.

17. The sum of a number and 5 is multiplied by 6, giving 72 as a result. Find the number.

18. Pedro has 15 more college units than Felix. Altogether, the two men have 95 units. How many units does Felix have?

19. On a test in geometry, the highest grade was 35 points more than the lowest. The sum of the highest and lowest grades was 157. Find the lowest and the highest scores.

20. In a marathon, Susan ran $\frac{2}{3}$ as far as Linda. In all, the two people ran 30 miles. How many miles did Susan run?

21. Pat has $\frac{5}{2}$ as many golf balls as John. Together they have 70 golf balls. How many golf balls does Pat have?

[3.5] *A formula is given in the following exercises, along with the values of some of the variables. Find the value of the variable that is not given.*

22. $A = \frac{1}{2}bh$; $A = 22$, $b = 4$

23. $A = \frac{1}{2}(b + B)h$; $b = 9$, $B = 12$, $h = 8$

24. $C = 2\pi r$; $C = 12.56$, $\pi = 3.14$

25. $V = \frac{4}{3}\pi r^3$; $\pi = 3.14$, $r = 1$

Solve each formula for the specified variable.

26. $A = LW$ for W

27. $I = prt$ for r

28. $m + a = x^2 - m$ for m

29. $A = \frac{1}{2}(b + B)h$ for h

Solve the word problems. Check your answers.

30. The area of a triangle is 25 square meters. The base is 10 meters in length. Find the height.

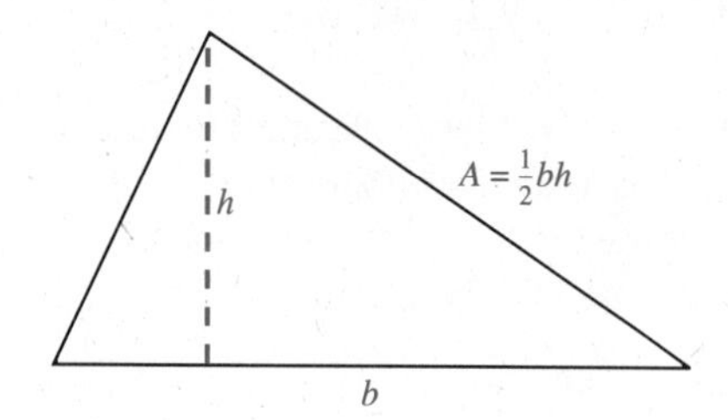

31. The shorter base of a trapezoid is 42 centimeters long, and the longer base is 48 centimeters long. The height of the trapezoid is 8 centimeters. Find the area of the trapezoid.

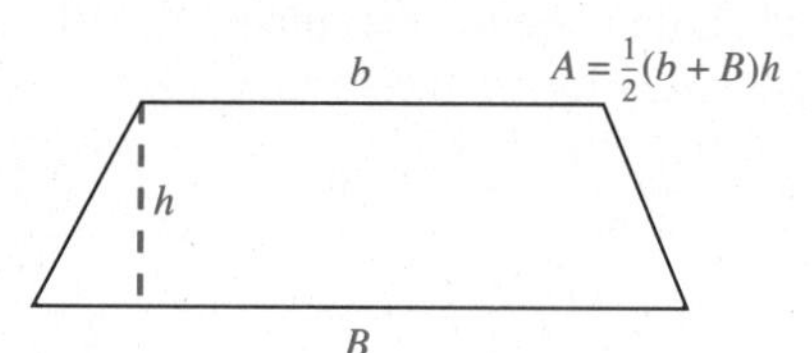

[3.6] *Write the following ratios in lowest terms.*

32. 50 centimeters to 30 centimeters

33. 6 days to 1 week

34. 45 inches to 5 feet

35. 2 months to 3 years

Decide whether the following proportions are true or false.

36. $\frac{15}{18} = \frac{45}{54}$

37. $\frac{11}{19} = \frac{55}{105}$

Solve each equation. Check your answers.

38. $\frac{p}{5} = \frac{21}{30}$

39. $\frac{3}{5} = \frac{k}{12}$

40. $\frac{y}{y - 5} = \frac{7}{2}$

41. $\frac{2 + m}{2 - m} = \frac{3}{4}$

Solve the word problems. Check your answers.

42. If 1 quart of oil must be mixed with 24 quarts of gasoline, how much oil would be needed for 192 quarts of gasoline?

43. Lupe can paint 3 walls in 7 hours. How long would it take her to paint 27 walls?

44. If 15 yards of cloth are needed for 18 nurses' smocks, how much cloth would be needed for 90 smocks?

45. The distance between two cities on a road map is 16 centimeters. The two cities are actually 150 kilometers apart. The distance on the map between two other cities is 40 centimeters. How far apart are these cities?

[3.7] *Graph each inequality on a number line.*

46. $m \geq -2$

47. $-5 \leq p < 6$

Solve each inequality. Graph the solutions.

48. $y + 5 \geq 2$

49. $5y > 4y + 8$

50. $9(k - 5) - (3 + 8k) \geq 5$

51. $3(2z + 5) + 4(8 + 3z) \leq 5(3z + 2) + 2z$

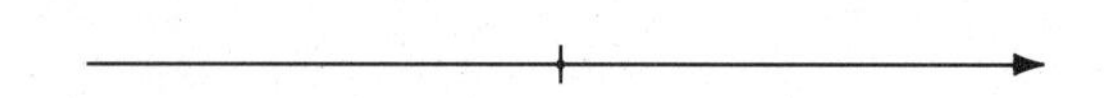

52. $-6 \leq x + 2 \leq 0$

53. $3 < y - 4 < 5$

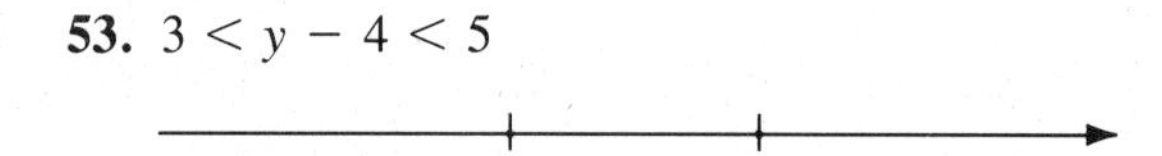

54. $6k \geq -18$

55. $-11y < 22$

56. $-z > -4$

57. $5m - 9 \geq 7m + 3$

58. $6p - 5p > 2 - 4p + 7p + 8$

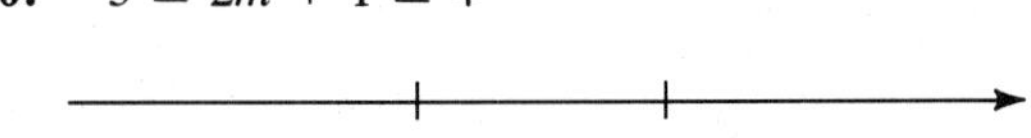

59. $-(y + 2) + 3(2y - 7) \leq 4 - 5y$

60. $-3 \leq 2m + 1 \leq 4$

61. $9 < 3m + 5 \leq 20$

MIXED REVIEW EXERCISES

Solve each of the following.

62. $\dfrac{k + 7}{2k + 5} = -4$

63. $2x + t = 3t - 5x$ for t

64. $-2z > -6$

65. $5x - 6 = 3x + 12$

66. $7m + 3 \leq 5m - 9$

67. $a + b + c = P$ for a

68. $7 - 8(x + 2) = -10x - 3$

69. $-6 \leq 2z + 4 \leq 12$

Solve each problem.

70. Two-thirds of a number added to the number is 10. What is the number?

71. If three-fourths of a number is subtracted from twice the number, the result is 15. Find the number.

72. Buddy defeated Bob in an election. Buddy had twice as many votes as Bob, and together they had 1800 votes. How many votes did each of the candidates receive?

73. John and Gwen commute to work. John travels three times as far as Gwen each day, and together they travel 112 miles. How far does each travel?

74. On a recent diet, Duc lost 18 pounds more than Hoa. Their total weight loss was 42 pounds. How much did Hoa lose?

75. Rick and Steve drove from different towns to a family reunion. Rick drove 43 miles farther than Steve. The two men drove a total of 293 miles. How far did Steve drive to the reunion?

76. A teacher noted that one week he had graded 32 more algebra tests than geometry tests. He had graded 102 tests altogether. How many geometry tests did he grade that week?

77. A parking lot attendant parked 63 cars one day. He parked 11 fewer large cars than small cars. How many small cars did he park?

78. A bird-watcher in the North Atlantic counted 23 more fulmars than shearwaters. Altogether she counted 53 birds of the two species. How many were shearwaters?

79. A recipe calls for $1\frac{1}{3}$ cups of milk. How much milk would be needed to make half the recipe?

80. The perimeter of a rectangle is 288 feet. The length is 4 feet longer than the width. Find the width.

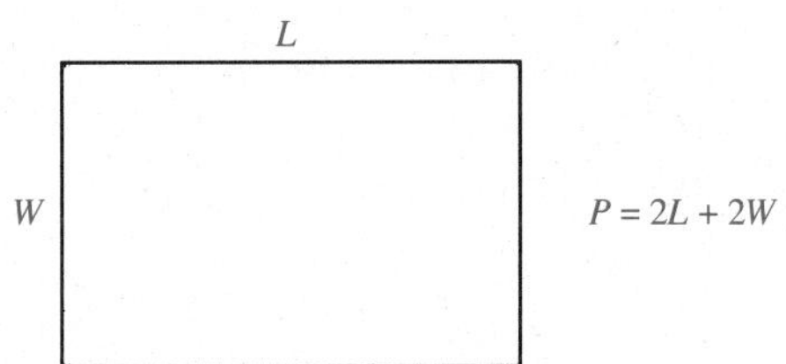

81. The perimeter of a triangle is 96 meters. One side is twice as long as another, and the third side is 30 meters long. What is the length of the longest side?

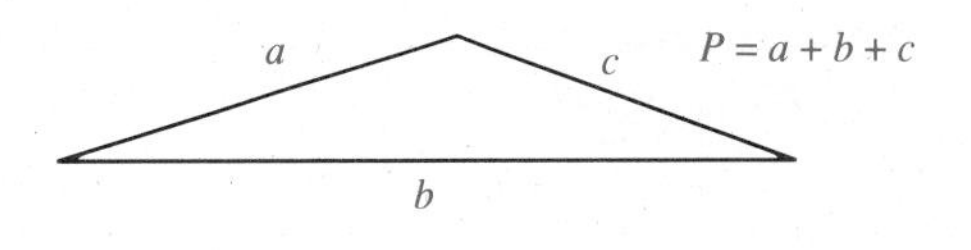

82. The area of a triangle is 182 square inches. The height is 14 inches. Find the length of the base.

83. The perimeter of a rectangle is 86 inches. The width is 17 inches. What is the length?

84. The Celsius temperature is 22°C. What is the corresponding Fahrenheit temperature?

85. The Fahrenheit temperature is 122°. What is the corresponding Celsius temperature?

86. On a certain map, 2 inches represents 100 kilometers. How many kilometers would 5 inches represent?

87. The tax on an $18 item is $1.17. How much tax would be charged on a $215 item?

88. It costs $4.50 to plant 2 square yards with a certain ground cover. How much would it cost to plant $6\frac{1}{2}$ square yards?

89. If $7\frac{3}{4}$ yards of fabric are needed to make 2 dresses, how many yards would be needed to make 9 dresses?

90. Tommy, Mike, and Chuck all have summer jobs. During June, Tommy earned $480 and Mike earned $540. If the average earnings of the three men were at least $550, what is the smallest amount that Chuck could have earned?

91. Nalima has grades of 82 and 96 on her first two English tests. What must she make on her third test so that her average will be at least 90?

name date hour

CHAPTER 3 TEST

Solve each equation and check the solution.

1. $2m - 5 = 3$ **1.** ______

2. $6v + 3 = 8v - 7$ **2.** ______

3. $3(a + 12) = 1 - 2(a - 5)$ **3.** ______

4. $4k - 6k + 8(k - 3) = -2(k + 12)$ **4.** ______

5. $\frac{m}{5} = 2$ **5.** ______

6. $6x - 4 = 12$ **6.** ______

7. $4 - (3 - m) = 12 + m$ **7.** ______

8. $-(r + 4) = 2 + r$ **8.** ______

9. Vern paid $57 more to tune up his Bronco than his Oldsmobile. He paid $257 in all. How much did he pay for the tune-up on the Oldsmobile? **9.** ______

10. Each month, Janet's allowance is five times as much as Louis's allowance. Together, their allowances total $72. What is Louis's monthly allowance? **10.** ______

11. The formula for the perimeter of a rectangle is $P = 2L + 2W$. If $P = 58$ and $L = 20$, find W. **11.** ______

12. Solve the formula $I = prt$ for p. **12.** ______

13. ________

14. ________

15. ________

16. ________

17. ________

18. ________

19. ________

20. ________

21. ________

22. ________

23. ________

13. Solve the formula $A = \frac{1}{2}(b + B)h$ for h.

14. The area of a rectangle is 48 square meters. The width is 12 meters. Find the length.

15. Is the proportion $\frac{15}{79} = \frac{465}{2449}$ true or false?

Solve.

16. $\frac{z}{16} = \frac{3}{48}$

17. $\frac{y + 5}{y - 2} = \frac{1}{4}$

18. If 11 hamburgers cost \$6.05, find the cost of 32 hamburgers.

Solve the following inequalities. Graph the solutions.

19. $x - 1 \leq 3$

20. $-2m < -14$

21. $-4r + 2(r - 3) \geq 5r - (3 + 6r) + 1 - 8$

22. $-8 < 3k - 2 \leq 12$

Solve the following problem.

23. On the first two days of their vacation to Florida, Wally and Joanna drove 430 miles and 470 miles. If they must average at least 450 miles per day, what is the shortest distance they can drive on the third day?

name date hour

CUMULATIVE REVIEW EXERCISES: CHAPTERS 1–3

Write each fraction in lowest terms.

1. $\frac{15}{40}$

2. $\frac{108}{144}$

Work the following problems.

3. $\frac{5}{6} + \frac{1}{4} + \frac{7}{15}$

4. $16\frac{7}{8} - 3\frac{1}{10}$

5. $\frac{9}{8} \cdot \frac{16}{3}$

6. $\frac{3}{4} \div \frac{5}{8}$

7. One dog weighs $8\frac{1}{3}$ pounds, and another dog weighs $12\frac{5}{8}$ pounds. Find the total weight of both dogs.

8. In making dresses, Earth Works uses $\frac{5}{8}$ yard of trim per dress. How many yards of trim would be used to make 56 dresses?

9. A cook wants to increase a recipe that serves 6 to make enough for 20 people. The recipe calls for $1\frac{1}{4}$ cups of cheese. How much cheese will be needed to serve 20?

Work the following problems.

10. $4.8 + 12.5 + 16.73$

11. $56.3 - 28.99$

12. $67.8(.45)$

13. $236.46 \div 4.2$

14. Find 12% of 180.

15. Find 11.6% of 1500.

16. What percent is 24 of 64?

17. What percent of 72 is 18?

18. A purchasing agent bought 3 desks at $211.40 each and 3 chairs for $195, $189.95, and $168.50. What was the final bill (without tax)?

19. The purchasing agent in Exercise 18 paid a sales tax of $6\frac{1}{4}\%$ on his purchase. What was the final bill, including tax?

20. A car has a price of $5000. For trading in her old car, Rosalie will get 25% off. Find the price of the car with the trade-in.

CUMULATIVE REVIEW EXERCISES: CHAPTERS 1–3

Tell whether each of the following is true or false.

21. $\dfrac{8(7) - 5(6 + 2)}{3 \cdot 5 + 1} \geq 1$

22. $\dfrac{4(9 + 3) - 8(4)}{2 + 3 - 3} \geq 2$

Perform the indicated operations.

23. $-11 + 20 + (-2)$

24. $13 + (-19) + 7$

25. $9 - (-4)$

26. $-2(-5)(-4)$

27. $\dfrac{4 \cdot 9}{-3}$

28. $\dfrac{8}{7 - 7}$

29. $(-5 + 8) + (-2 - 7)$

30. $(-7 - 1)(-4) + (-4)$

31. $\dfrac{-3 - (-5)}{1 - (-1)}$

32. $\dfrac{6(9) + 3}{6(-4) + 8(2) - 4(-2)}$

33. $\dfrac{(-3)^2 - (-4)(2^4)}{5 \cdot 2 - (-2)^3}$

34. $\dfrac{-2(5^3) - 6}{4^2 + 2(-5) + (-2)}$

Circle the smaller number in each pair.

35. $-7, \quad -11$

36. $-8.23, \quad -|-7|$

37. $|7.23|, \quad -3$

Find the value of each expression when $x = -2$, $y = -4$, and $z = 3$.

38. xyz

39. $(x + 4y)3z$

40. $2x^2 - y^2$

41. $xz^3 - 5y^2$

42. $\dfrac{3x - y^3}{-4z}$

name date hour

Find the solution for each equation from the domain $\{-3, -2, -1, 0, 1, 2, 3\}$.

43. $-5 + x = -2$

44. $\frac{x}{2} = -1$

Write an equation using x as the variable, then find the solution of the equation from the domain $\{-3, -2, -1, 0, 1, 2, 3\}$.

45. Half a number increased by 3 is 2.

46. When the sum of a number and 6 is divided by the square of the number, the result is 1.

47. The additive inverse of a number plus 7 is 4.

48. The square of a number multiplied by 4 equals 0.

Name the property illustrated by each of the following examples.

49. $5 + 11 = 11 + 5$

50. $18 \cdot 1 = 18$

51. $7(k + m) = 7k + 7m$

52. $3 + (5 + 2) = 3 + (2 + 5)$

53. $7 + (-7) = 0$

54. $\frac{6}{5} \cdot \frac{5}{6} = 1$

55. $3 + 0 = 3$

Simplify the following expressions by combining terms.

56. $4p - 6 + 3p - 8$

57. $-4(k + 2) + 3(2k - 1)$

Solve the following equations and check each solution.

58. $2r - 6 = 8$

59. $2(p - 1) = 3p + 2$

60. $4 - 5(a + 2) = 3(a + 1) - 1$

61. $2 - 6(z + 1) = 4(z - 2) + 10$

62. $-(m - 1) = 3 - 2m$

63. $-(k + 2) = 4 - 2k$

64. $\dfrac{y - 2}{3} = \dfrac{2y + 1}{5}$

65. $\dfrac{2x + 3}{5} = \dfrac{x - 4}{2}$

Solve for x.

66. $3y - x = 2$

67. $\dfrac{2x - a}{2} = 5$

Solve the following inequalities. Graph your solutions.

68. $-5z \geq 4z - 18$

69. $6(r - 1) + 2(3r - 5) \leq -4$

70. $-3 < x + 4 < 9$

71. $-1 \leq 2x + 3 \leq 5$

Solve each word problem. Check the solution in the words of the original problem.

72. Three less than a number is multiplied by 5. The result is 45. Find the number.

73. The ratio of a number to the sum of the number and 4 is $\frac{2}{3}$. Find the number.

74. Ed Calvin bought textbooks at the college bookstore for $52.47, including 6% sales tax. What did the books cost?

75. Louise Palla received a bill from her credit card company for $104.93. The bill included interest at $1\frac{1}{2}$% per month for one month and a $5.00 late charge. How much did her purchases amount to?

76. The perimeter of a rectangle is 98 centimeters. The width is 19 centimeters. Find the length.

77. The area of a triangle is 104 square inches. The base is 13 inches. Find the height.

4

EXPONENTS AND POLYNOMIALS

4.1 EXPONENTS

OBJECTIVES

1. Use exponents.
2. Use the product rule for exponents.
3. Use the rule $(a^m)^n = a^{mn}$.
4. Use the rule $(ab)^m = a^m b^m$.
5. Use the rule $\left(\frac{a}{b}\right)^m = \frac{a^m}{b^m}$.
6. Use combinations of the rules for exponents.

1 In Chapter 1 we used exponents to write repeated products. Recall that in the expression 5^2, the number 5 is called the *base* and 2 is called the *exponent,* The expression 5^2 is called an **exponential expression.** Usually we do not write a quantity with an exponent of 1; however, sometimes it is convenient to do so. In general, for any quantity a, $a = a^1$.

EXAMPLE 1 Using Exponents

Write $3 \cdot 3 \cdot 3 \cdot 3 \cdot 3$ in exponential form, and evaluate the exponential expression.

Since 3 occurs as a factor five times, the base is 3 and the exponent is 5. The exponential expression is 3^5. The value is

$$3^5 = 3 \cdot 3 \cdot 3 \cdot 3 \cdot 3 = 243. \blacksquare$$

WORK PROBLEM 1 AT THE SIDE.

EXAMPLE 2 Evaluating an Exponential Expression

Evaluate each exponential expression. Name the base and the exponent.

	Base	Exponent
(a) $5^4 = 5 \cdot 5 \cdot 5 \cdot 5 = 625$	5	4
(b) $-5^4 = -1 \cdot 5^4 = -1 \cdot (5 \cdot 5 \cdot 5 \cdot 5) = -625$	5	4
(c) $(-5)^4 = (-5)(-5)(-5)(-5) = 625$	-5	4 $\blacksquare$

CAUTION It is important to understand the difference between parts (b) and (c) of Example 2. In -5^4 the lack of parentheses shows that the exponent 4 refers only to the base 5, and not -5; in $(-5)^4$ the parentheses show that the exponent 4 refers to the base -5. In summary, $-a^n$ and $(-a)^n$ are not always the same.

Expression	*Base*	*Exponent*	*Example*
$-a^n$	a	n	$-3^2 = -(3 \cdot 3) = -9$
$(-a)^n$	$-a$	n	$(-3)^2 = (-3)(-3) = 9$

WORK PROBLEM 2 AT THE SIDE.

1. Write each product in exponential form and evaluate.

(a) $2 \cdot 2 \cdot 2 \cdot 2$

(b) $7 \cdot 7 \cdot 7$

2. Evaluate each exponential expression. Name the base and the exponent.

(a) $(-2)^5$ **(b)** -2^5

(c) -4^2 **(d)** $(-4)^2$

ANSWERS

1. (a) $2^4 = 16$ (b) $7^3 = 343$
2. (a) -32; -2; 5 (b) -32; 2; 5 (c) -16; 4; 2 (d) 16; -4; 2

3. Find each product by the product rule, if possible.

(a) $8^2 \cdot 8^5$

(b) $3^5 \cdot 2^6$

(c) $(-7)^5 \cdot (-7)^3$

(d) $y^3 \cdot y$

(e) $6^4 + 6^2$

ANSWERS
3. (a) 8^7 (b) cannot use product rule (c) $(-7)^8$ (d) y^4 (e) cannot use product rule

2 One reason for the importance of exponents is that we can use them to develop several useful rules as shortcuts for working many problems. For example, by the definition of exponential expressions,

$$\begin{aligned} 2^4 \cdot 2^3 &= (2 \cdot 2 \cdot 2 \cdot 2)(2 \cdot 2 \cdot 2) \\ &= 2 \cdot 2 \cdot 2 \cdot 2 \cdot 2 \cdot 2 \cdot 2 \\ &= 2^7. \end{aligned}$$

Generalizing from this example, $2^4 \cdot 2^3 = 2^{4+3} = 2^7$, suggests the product rule for exponents.

PRODUCT RULE FOR EXPONENTS

For any positive integers m and n, $\quad a^m \cdot a^n = a^{m+n}$.

Example: $6^2 \cdot 6^5 = 6^{2+5} = 6^7$

CAUTION The bases must be the same before the product rule for exponents can be applied.

EXAMPLE 3 Using the Product Rule

Use the product rule for exponents to find each product, if possible.

(a) $6^3 \cdot 6^5 = 6^{3+5} = 6^8$ by the product rule.

(b) $(-4)^5(-4)^3 = (-4)^{5+3} = (-4)^8$ by the product rule.

(c) $x^2 \cdot x = x^2 \cdot x^1 = x^{2+1} = x^3$

(d) $m^4 \cdot m^3 = m^{4+3} = m^7$

(e) The product rule does not apply to the product $2^3 \cdot 3^2$, since the bases are different.

(f) The product rule does not apply to $2^3 + 2^4$, since the terms are *added*, not *multiplied*. ■

WORK PROBLEM 3 AT THE SIDE.

EXAMPLE 4 Using the Product Rule

Multiply $2x^3$ and $3x^7$.

Since $2x^3$ means $2 \cdot x^3$ and $3x^7$ means $3 \cdot x^7$, we use the associative and commutative properties and the product rule to get

$$2x^3 \cdot 3x^7 = 2 \cdot 3 \cdot x^3 \cdot x^7 = 6x^{10}. \quad ■$$

CAUTION Be sure you understand the difference between *adding* and *multiplying* exponential expressions. For example,

$$8x^3 + 5x^3 = 13x^3$$

but $$(8x^3)(5x^3) = 8 \cdot 5 \cdot x^{3+3} = 40x^6.$$

WORK PROBLEM 4 AT THE SIDE.

3 We can simplify an expression such as $(8^3)^2$ with the product rule for exponents, as follows.

$$(8^3)^2 = (8^3)(8^3) = 8^{3+3} = 8^6$$

The exponents in $(8^3)^2$, 3 and 2, are multiplied to give the exponent in 8^6. This example suggests power rule (a) for exponents.

POWER RULE (A) FOR EXPONENTS

For any positive integers m and n, $\quad (a^m)^n = a^{mn}$.

Example: $(3^2)^4 = 3^{2 \cdot 4} = 3^8$

EXAMPLE 5 Using Power Rule (a)

Use power rule (a) for exponents to simplify each expression.

(a) $(2^5)^3 = 2^{5 \cdot 3} = 2^{15}$

(b) $(5^7)^2 = 5^{7 \cdot 2} = 5^{14}$

(c) $(x^2)^5 = x^{2 \cdot 5} = x^{10}$

(d) $(n^3)^2 = n^{3 \cdot 2} = n^6$ ■

WORK PROBLEM 5 AT THE SIDE.

4 The properties studied in Chapter 2 can be used to develop two more rules for exponents. Using the definition of an exponential expression and the commutative and associative properties, we can evaluate the expression $(4 \cdot 8)^3$ as shown below.

$(4 \cdot 8)^3 = (4 \cdot 8)(4 \cdot 8)(4 \cdot 8)$	Definition of exponent
$= 4 \cdot 4 \cdot 4 \cdot 8 \cdot 8 \cdot 8$	Commutative and associative properties
$= 4^3 \cdot 8^3$	Definition of exponent

This example suggests power rule (b) for exponents.

POWER RULE (B) FOR EXPONENTS

For any positive integer m, $\quad (ab)^m = a^m b^m$.

Example: $(2p)^5 = 2^5p^5$

4. Multiply.

(a) $5m^2 \cdot 2m^6$

(b) $3p^5 \cdot 9p^4$

(c) $-7p^5 \cdot (3p^8)$

5. Simplify each expression.

(a) $(5^3)^4$

(b) $(6^2)^5$

(c) $(3^2)^4$

(d) $(a^6)^5$

ANSWERS

4. (a) $10m^8$ (b) $27p^9$ (c) $-21p^{13}$
5. (a) 5^{12} (b) 6^{10} (c) 3^8 (d) a^{30}

6. Simplify.

(a) $5(mn)^3$

(b) $(3a^2b^4)^5$

(c) $(5m^2)^3$

7. Simplify. Assume all variables represent nonzero real numbers.

(a) $\left(\frac{5}{2}\right)^4$

(b) $\left(\frac{p}{q}\right)^2$

(c) $\left(\frac{r}{t}\right)^3$

ANSWERS

6. (a) $5m^3n^3$ (b) $3^5a^{10}b^{20}$ (c) 5^3m^6

7. (a) $\frac{5^4}{2^4}$ (b) $\frac{p^2}{q^2}$ (c) $\frac{r^3}{t^3}$

EXAMPLE 6 Using Power Rule (b)

Use power rule (b) to simplify each expression.

(a) $(3xy)^2 = 3^2x^2y^2$
$= 9x^2y^2$

(b) $9(pq)^2 = 9(p^2q^2)$ Power rule (b)
$= 9p^2q^2$ Multiply.

(c) $5(2m^2p^3)^4 = 5[2^4(m^2)^4(p^3)^4]$ Power rule (b)
$= 5(2^4m^8p^{12})$ Power rule (a)
$= 5 \cdot 2^4m^8p^{12}$
$= 80m^8p^{12}$ ■

WORK PROBLEM 6 AT THE SIDE.

5 Since the quotient $\frac{a}{b}$ can be written as $a \cdot \frac{1}{b}$, we can use power rule (b), together with some of the properties of real numbers, to get power rule (c) for exponents.

POWER RULE (C) FOR EXPONENTS

For any integer m, $\left(\frac{a}{b}\right)^m = \frac{a^m}{b^m}$ $(b \neq 0)$.

Example: $\left(\frac{5}{3}\right)^2 = \frac{5^2}{3^2}$

EXAMPLE 7 Using Power Rule (c)

Simplify each expression.

(a) $\left(\frac{2}{3}\right)^5 = \frac{2^5}{3^5}$

(b) $\left(\frac{a}{b}\right)^4 = \frac{a^4}{b^4}$, $b \neq 0$ ■

WORK PROBLEM 7 AT THE SIDE.

The rules for exponents discussed in this section are summarized below. These rules are basic to the study of algebra and should be *memorized.*

RULES FOR EXPONENTS

For positive integers m and n:			*Examples*
Product rule		$a^m \cdot a^n = a^{m+n}$	$6^2 \cdot 6^5 = 6^{2+5} = 6^7$
Power rules	(a)	$(a^m)^n = a^{mn}$	$(3^2)^4 = 3^{2 \cdot 4} = 3^8$
	(b)	$(ab)^m = a^mb^m$	$(2p)^5 = 2^5p^5$
	(c)	$\left(\frac{a}{b}\right)^m = \frac{a^m}{b^m}$ $(b \neq 0)$	$\left(\frac{5}{3}\right)^2 = \frac{5^2}{3^2}$

6 The next example shows that more than one rule may be needed to simplify an expression.

EXAMPLE 8 Using Combinations of Rules

Simplify each expression.

(a) $\left(\frac{2}{3}\right)^2 \cdot 2^3$

$$\left(\frac{2}{3}\right)^2 \cdot 2^3 = \frac{2^2}{3^2} \cdot \frac{2^3}{1} \quad \text{Power rule (c)}$$

$$= \frac{2^2 \cdot 2^3}{3^2 \cdot 1} \quad \text{Multiplying fractions}$$

$$= \frac{2^5}{3^2} \quad \text{Product rule}$$

(b) $(5x)^3(5x)^4$

$$(5x)^3(5x)^4 = (5x)^7 \quad \text{Product rule}$$

$$= 5^7x^7 \quad \text{Power rule (b)}$$

(c) $(2x^2y^3)^4(3xy^2)^3$

$$(2x^2y^3)^4(3xy^2)^3 = 2^4(x^2)^4(y^3)^4 \cdot 3^3x^3(y^2)^3 \quad \text{Power rule (b)}$$

$$= 2^4 \cdot 3^3x^8y^{12}x^3y^6 \quad \text{Power rule (a)}$$

$$= 16 \cdot 27x^{11}y^{18} \quad \text{Product rule}$$

$$= 432x^{11}y^{18} \quad ■$$

CAUTION Refer to Example 8(c). Notice that

$$(2x^2y^3)^4 \neq (2 \cdot 4)x^{2 \cdot 4}y^{3 \cdot 4}.$$

Do not multiply the coefficient 2 and the exponent 4.

WORK PROBLEM 8 AT THE SIDE.

8. Simplify.

(a) $(2m)^3(2m)^4$

(b) $\left(\frac{5k^3}{3}\right)^2$

(c) $\left(\frac{1}{5}\right)^4(2x)^2$

(d) $(3xy^2)^3(x^2y)^4$

ANSWERS

8. (a) 2^7m^7 (b) $\frac{5^2k^6}{3^2}$ (c) $\frac{2^2x^2}{5^4}$ (d) $3^3x^{11}y^{10}$

Historical Reflections

Historical Reflections

Mathematics and Music

Musical tones are related to the lengths of stretched strings by the ratios of the counting numbers. This idea can be tested on a cello, for example. Stop any string halfway, so that the ratio of the whole string to the part is 2/1. If you pick the free half of the string, you get the octave above the fundamental tone of the whole string. The ratio 3/2 gives you the fifth above the octave, and so on.

Pythagoras noted that the simple ratios of 1, 2, 3, and 4 give the most harmonious musical intervals. He claimed that the intervals between the planets must also be ratios of counting numbers. The idea came to be called "music of the spheres." (The planets were believed to orbit around the earth on crystal spheres.)

name date hour

4.1 EXERCISES

Identify the base and exponent for each exponential expression. See Example 2.

1. 5^{12} **2.** a^6 **3.** $(3m)^4$ **4.** -2^4

5. -125^3 **6.** $(-1)^8$ **7.** $(-2x)^2$ **8.** $-(-p)^5$

9. $3m^2$ **10.** $5y^3$ **11.** $-r^5$ **12.** $(-y)^5$

Write each expression using exponents. See Example 1.

13. $3 \cdot 3 \cdot 3 \cdot 3 \cdot 3$ **14.** $4 \cdot 4 \cdot 4$ **15.** $(-2)(-2)(-2)(-2)(-2)$

16. $(-1)(-1)(-1)(-1)$ **17.** $p \cdot p \cdot p \cdot p \cdot p$ **18.** $\frac{1}{4 \cdot 4 \cdot 4 \cdot 4 \cdot 4}$

19. $\frac{1}{(-2)(-2)(-2)}$ **20.** $\frac{1}{3 \cdot 3 \cdot 3 \cdot 3}$ **21.** $\frac{1}{2 \cdot 2 \cdot 2 \cdot 2 \cdot 2}$

22. $\frac{1}{y \cdot y \cdot y \cdot y}$ **23.** $(-2z)(-2z)(-2z)$ **24.** $(-3m)(-3m)(-3m)(-3m)$

Evaluate each expression. For example, $5^2 + 5^3 = 25 + 125 = 150$.

25. $3^2 + 3^4$ **26.** $2^8 - 2^6$ **27.** $4^2 + 4^3$ **28.** $3^3 + 3^4$

29. $2^2 + 2^5$ **30.** $4^2 + 4^1$ **31.** $(-4)^2 - (-2)^2$ **32.** $(-2)^3 - (-3)^2$

Use the product rule to simplify each expression. Write each answer in exponential form. See Examples 3 and 4.

33. $4^2 \cdot 4^3$ **34.** $3^5 \cdot 3^4$ **35.** $3^4 \cdot 3^7$ **36.** $2^5 \cdot 2^{15}$

37. $4^3 \cdot 4^5 \cdot 4^{10}$ **38.** $2^3 \cdot 2^4 \cdot 2^6$ **39.** $(-3)^3(-3)^2$ **40.** $(-4)^5(-4)^3$

41. $y^3 \cdot y^4 \cdot y^7$ **42.** $a^8 \cdot a^5 \cdot a^2$ **43.** $r \cdot r^5 \cdot r^4 \cdot r^7$ **44.** $m^9 \cdot m \cdot m^5 \cdot m^8$

45. $(-9r^3)(7r^6)$ **46.** $(8a^9)(-3a^{14})$ **47.** $(2p^4)(5p^9)$ **48.** $(3q^8)(7q^5)$

In each of the following exercises, first add *the given terms; then start over and* multiply *them.*

49. $4m^3,\ 9m^3$ **50.** $8y^2,\ 7y^2$ **51.** $-12p,\ 11p$ **52.** $3q^4,\ 5q^4$

53. $7r,\ 3r,\ 5r$ **54.** $9a^3,\ 2a^3,\ 3a^3$ **55.** $-5a^2,\ 3a^2$ **56.** $6r^4,\ -8r^4$

Use the power rules for exponents to simplify each expression. Write each answer in exponential form. See Examples 5–7.

57. $(6^3)^2$ **58.** $(8^4)^6$ **59.** $(9^3)^2$ **60.** $(2^3)^4$

61. $(-5^2)^4$ **62.** $(-2^3)^2$ **63.** $(-4^2)^3$ **64.** $(-3^5)^3$

65. $(5m)^3$ **66.** $(2xy)^4$ **67.** $(-2pq)^4$ **68.** $(-3ab)^5$

69. $\left(\frac{-3x^5}{4}\right)^2$ **70.** $\left(\frac{4m^3n^2}{5}\right)^4$ **71.** $\left(\frac{5a^2b}{c^4}\right)^3$ **72.** $\left(\frac{2y^2z^4}{w^3}\right)^5$

Simplify each expression. See Example 8.

73. $\left(\frac{4}{3}\right)^5 \cdot (4)^3$ **74.** $\left(\frac{2}{5}\right)^2 \cdot \left(\frac{2}{5}\right)^3$ **75.** $\left(\frac{3}{4}x\right)^5$ **76.** $\left(\frac{5}{3}y\right)^2$

77. $(3m)^2(3m)^5$ **78.** $(-5p)^4(-5p)^2$ **79.** $(8z)^6(8z)^3$ **80.** $(2p)^4(2p)$

81. $(2m^2n)^3(mn^2)$ **82.** $(3p^2q^2)^3(q^4)$ **83.** $(5ab^2)^5(5a^3b)^2$ **84.** $(-r^3s)^4(r^2s)^3$

SKILL SHARPENERS

Evaluate each expression when **(a)** $x = 2$ *and* **(b)** $x = -3$. *See Section 1.5.*

85. $-x^2 + 4x - 7$ **86.** $-x^2 - 6x + 2$ **87.** $-x^3 + x^2 - 4$ **88.** $8 - 6x - x^3$

4.2 THE QUOTIENT RULE AND INTEGER EXPONENTS

1 In the previous section we studied the product rule for exponents. The rule for division with exponents is similar to the product rule for exponents. For example,

$$\frac{6^5}{6^2} = \frac{6 \cdot 6 \cdot 6 \cdot 6 \cdot 6}{6 \cdot 6} = 6 \cdot 6 \cdot 6 = 6^3.$$

The difference between the exponents, $5 - 2$, gives the exponent in the quotient, 3.

If the exponents in the numerator and denominator are equal, then, for example,

$$\frac{6^5}{6^5} = \frac{6 \cdot 6 \cdot 6 \cdot 6 \cdot 6}{6 \cdot 6 \cdot 6 \cdot 6 \cdot 6} = 1.$$

If, however, the exponents are subtracted as above.

$$\frac{6^5}{6^5} = 6^{5-5} = 6^0.$$

This means that $6^0 = 1$.

DEFINITION OF ZERO EXPONENT

For any nonzero real number a, $\quad a^0 = 1 \quad (a \neq 0)$.

Example: $17^0 = 1$

EXAMPLE 1 Using Zero Exponents

Evaluate each exponential expression.

(a) $60^0 = 1$

(b) $(-60)^0 = 1$

(c) $-60^0 = -(1) = -1$

(d) $y^0 = 1, \quad$ if $y \neq 0$

(e) $-r^0 = -1, \quad$ if $r \neq 0$ ■

CAUTION Notice the difference between parts (b) and (c) of Example 1. In Example 1(b) the base is -60 and the exponent is 0. Any nonzero base raised to a zero exponent is 1. But in Example1(c), the base is 60. Then $60^0 = 1$, and $-60^0 = -1$.

WORK PROBLEM 1 AT THE SIDE.

2 In the discussion above, we found that

$$\frac{6^5}{6^2} = 6^3$$

where the bottom exponent was subtracted from the top exponent. If the bottom exponent were larger than the top exponent, subtracting would result in a negative exponent.

OBJECTIVES

1. Use zero as an exponent.
2. Use negative numbers as exponents.
3. Use the quotient rule for exponents.
4. Use combinations of rules.

1. Evaluate.

(a) 28^0

(b) $(-16)^0$

(c) -7^0

(d) $m^0, \quad m \neq 0$

(e) $-p^0, \quad p \neq 0$

ANSWERS

1. (a) 1 (b) 1 (c) -1 (d) 1 (e) -1

For example,

$$\frac{6^2}{6^5} = 6^{2-5} = 6^{-3}.$$

On the other hand,

$$\frac{6^2}{6^5} = \frac{6 \cdot 6}{6 \cdot 6 \cdot 6 \cdot 6 \cdot 6} = \frac{1}{6 \cdot 6 \cdot 6} = \frac{1}{6^3}$$

so

$$6^{-3} = \frac{1}{6^3}.$$

This example suggests that negative exponents be defined as follows.

DEFINITION OF NEGATIVE EXPONENTS

For any nonzero real number a and any integer n,

$$a^{-n} = \frac{1}{a^n} \quad (a \neq 0).$$

Example: $3^{-2} = \frac{1}{3^2}$

CAUTION A negative exponent does not indicate a negative number; negative exponents lead to reciprocals.

Expression	*Example*	
a^{-n}	$3^{-2} = \frac{1}{3^2} = \frac{1}{9}$	Not negative
$-a^{-n}$	$-3^{-2} = -\frac{1}{3^2} = -\frac{1}{9}$	Negative

By definition, a^{-n} and a^n are reciprocals, since

$$a^n \cdot a^{-n} = a^n \cdot \frac{1}{a^n} = 1.$$

Since $1^n = 1$, the definition of a^{-n} also can be written

$$a^{-n} = \frac{1}{a^n} = \left(\frac{1}{a}\right)^n.$$

For example,

$$6^{-3} = \left(\frac{1}{6}\right)^3 \quad \text{and} \quad \left(\frac{1}{3}\right)^{-2} = 3^2.$$

EXAMPLE 2 Using Negative Exponents

Simplify by using the definition of negative exponents.

(a) $3^{-2} = \frac{1}{3^2} = \frac{1}{9}$

(b) $5^{-3} = \frac{1}{5^3} = \frac{1}{125}$

(c) $\left(\frac{1}{2}\right)^{-3} = 2^3 = 8$

As shown above, we can change the base to its reciprocal if we also change the sign of the exponent.

(d) $\left(\frac{2}{5}\right)^{-4} = \left(\frac{5}{2}\right)^4$

(e) $4^{-1} - 2^{-1} = \frac{1}{4} - \frac{1}{2} = \frac{1}{4} - \frac{2}{4} = -\frac{1}{4}$

Apply the exponents first, then subtract.

(f) $p^{-2} = \frac{1}{p^2}, \quad p \neq 0$

(g) $\frac{1}{x^{-4}}, \quad x \neq 0$

Since any real-number power of 1 is 1,

$$\frac{1}{x^{-4}} = \frac{1^{-4}}{x^{-4}} = \left(\frac{1}{x}\right)^{-4} = x^4. \blacksquare$$

WORK PROBLEM 2 AT THE SIDE.

The definition of negative exponents allows us to move factors in a fraction if we also change the signs of the exponents. For example,

$$\frac{2^{-3}}{3^{-4}} = \frac{\frac{1}{2^3}}{\frac{1}{3^4}} = \frac{1}{2^3} \cdot \frac{3^4}{1} = \frac{3^4}{2^3}$$

so that

$$\frac{2^{-3}}{3^{-4}} = \frac{3^4}{2^3}.$$

CHANGING FROM NEGATIVE TO POSITIVE EXPONENTS

For any nonzero numbers a and b, and any integers m and n,

$$\frac{a^{-m}}{b^{-n}} = \frac{b^n}{a^m}.$$

Example: $\frac{3^{-5}}{2^{-4}} = \frac{2^4}{3^5}$

2. Simplify by using the definition of negative exponents.

(a) 4^{-3}

(b) 6^{-2}

(c) $\left(\frac{2}{3}\right)^{-2}$

(d) $2^{-1} + 5^{-1}$

(e) $m^{-5}, \quad m \neq 0$

(f) $\frac{1}{z^{-4}}, \quad z \neq 0$

ANSWERS

2. (a) $\frac{1}{4^3}$ (b) $\frac{1}{6^2}$ (c) $\left(\frac{3}{2}\right)^2$ (d) $\frac{7}{10}$ (e) $\frac{1}{m^5}$ (f) z^4

3. Write with only positive exponents. Assume all variables represent nonzero real numbers.

(a) $\dfrac{7^{-1}}{5^{-4}}$

(b) $\dfrac{x^{-3}}{y^{-2}}$

(c) $\dfrac{4h^{-5}}{m^{-2}k}$

(d) p^2q^{-5}

ANSWERS

3. (a) $\dfrac{5^4}{7}$ (b) $\dfrac{y^2}{x^3}$ (c) $\dfrac{4m^2}{h^5k}$ (d) $\dfrac{p^2}{q^5}$

EXAMPLE 3 Changing from Negative to Positive Exponents

Write with only positive exponents. Assume all variables represent nonzero real numbers.

(a) $\dfrac{4^{-2}}{5^{-3}} = \dfrac{5^3}{4^2}$

(b) $\dfrac{m^{-5}}{p^{-1}} = \dfrac{p^1}{m^5} = \dfrac{p}{m^5}$

(c) $\dfrac{a^{-2}b}{3d^{-3}} = \dfrac{bd^3}{3a^2}$

(d) $x^3y^{-4} = \dfrac{x^3y^{-4}}{1} = \dfrac{x^3}{y^4}$ ■

WORK PROBLEM 3 AT THE SIDE.

3 Now that zero and negative exponents have been defined, we can state the quotient rule for exponents.

QUOTIENT RULE FOR EXPONENTS

For any nonzero real number a and any integers m and n,

$$\frac{a^m}{a^n} = a^{m-n} \quad (a \neq 0).$$

Example: $\dfrac{5^8}{5^4} = 5^{8-4} = 5^4$

EXAMPLE 4 Using the Quotient Rule for Exponents

Simplify, using the quotient rule for exponents. Write answers with positive exponents.

(a) $\dfrac{5^8}{5^6} = 5^{8-6} = 5^2$

(b) $\dfrac{4^2}{4^9} = 4^{2-9} = 4^{-7} = \dfrac{1}{4^7}$

(c) $\dfrac{5^{-3}}{5^{-7}} = 5^{-3-(-7)} = 5^4$

(d) $\dfrac{q^5}{q^{-3}} = q^{5-(-3)} = q^8, \quad q \neq 0$

(e) $\dfrac{3^2x^5}{3^4x^3} = \dfrac{3^2}{3^4} \cdot \dfrac{x^5}{x^3} = 3^{2-4} \cdot x^{5-3}$

$= 3^{-2}x^2 = \dfrac{x^2}{3^2}, \quad x \neq 0$

Sometimes numerical expressions with small exponents, such as 3^2, are evaluated. Doing that would give the result as $\frac{x^2}{9}$. ■

WORK PROBLEM 4 AT THE SIDE.

Since exponential expressions with negative exponents can be written with positive exponents, the rules for exponents also are true for negative exponents.

The definitions and rules for exponents given in this section and the previous one are summarized below.

DEFINITIONS AND RULES FOR EXPONENTS

For any integers m and n:

			Examples
Product rule	$a^m \cdot a^n = a^{m+n}$		$7^4 \cdot 7^3 = 7^7$
Zero exponent	$a^0 = 1$	$(a \neq 0)$	$(-3)^0 = 1$
Negative exponent	$a^{-n} = \frac{1}{a^n}$	$(a \neq 0)$	$5^{-3} = \frac{1}{5^3}$
Quotient rule	$\frac{a^m}{a^n} = a^{m-n}$	$(a \neq 0)$	$\frac{2^2}{2^5} = 2^{-3} = \frac{1}{2^3}$
Power rules (a)	$(a^m)^n = a^{mn}$		$(4^2)^3 = 4^6$
(b)	$(ab)^m = a^m b^m$		$(3k)^4 = 3^4 k^4$
(c)	$\left(\frac{a}{b}\right)^m = \frac{a^m}{b^m}$	$(b \neq 0)$	$\left(\frac{2}{3}\right)^{-2} = \frac{2^{-2}}{3^{-2}} = \frac{3^2}{2^2}$

4 As shown in the next example we may sometimes need to use more than one rule to simplify an expression.

EXAMPLE 5 Using a Combination of Rules

Use a combination of the rules for exponents to simplify each expression. Assume all variables represent nonzero real numbers.

(a) $\frac{(4^2)^3}{4^5}$

Use power rule (a) and then the quotient rule.

$$\frac{(4^2)^3}{4^5} = \frac{4^6}{4^5} = 4^{6-5} = 4^1 = 4$$

(b) $(2x)^3(2x)^2$

Use the product rule first. Then use power rule (b).

$$(2x)^3(2x)^2 = (2x)^5 = 2^5x^5 \quad \text{or} \quad 32x^5$$

(c) $\left(\frac{2x^3}{5}\right)^{-4}$

By the definition of a negative exponent and the power rules,

$$\left(\frac{2x^3}{5}\right)^{-4} = \left(\frac{5}{2x^3}\right)^4$$ Change the base to its reciprocal, and change the sign of the exponent.

$$= \frac{5^4}{2^4x^{12}}.$$ Power rules

4. Simplify. Write answers with positive exponents.

(a) $\frac{5^{11}}{5^8}$

(b) $\frac{4^7}{4^{10}}$

(c) $\frac{6^{-5}}{6^{-2}}$

(d) $\frac{8^4 \cdot m^9}{8^5 \cdot m^{10}}$

ANSWERS

4. (a) 5^3 (b) $\frac{1}{4^3}$ (c) $\frac{1}{6^3}$ (d) $\frac{1}{8m}$

5. Simplify. Assume all variables represent nonzero real numbers.

(a) $12^5 \cdot 12^{-7} \cdot 12^6$

(b) $y^{-2} \cdot y^5 \cdot y^{-8}$

(c) $\dfrac{(6x)^{-1}}{(3x^2)^{-2}}$

(d) $\dfrac{3^9 \cdot (x^2y)^{-2}}{3^3 \cdot x^{-4}y}$

ANSWERS

5. (a) 12^4 (b) $\dfrac{1}{y^5}$ (c) $\dfrac{3x^3}{2}$

(d) $\dfrac{3^6}{y^3}$

(d) $\left(\dfrac{3x^{-2}}{4^{-1}y^3}\right)^{-3} = \dfrac{3^{-3}x^6}{4^3y^{-9}}$ Power rules

$= \dfrac{\frac{1}{3^3} \cdot x^6}{4^3 \cdot \frac{1}{y^9}}$ Definition of negative exponent

$= \dfrac{\frac{x^6}{3^3}}{\frac{4^3}{y^9}}$ Multiply.

$= \dfrac{x^6}{3^3} \cdot \dfrac{y^9}{4^3}$ Definition of division of fractions

$= \dfrac{x^6y^9}{3^3 \cdot 4^3}$ Multiplication of fractions ■

NOTE Since the steps can be done in several different orders, there are many equally good ways to simplify a problem like Example 5(d).

WORK PROBLEM 5 AT THE SIDE.

name date hour

4.2 EXERCISES

Evaluate each expression. See Examples 1 and 2.

1. $4^0 + 5^0$

2. $3^0 + 8^0$

3. $(-9)^0 + 9^0$

4. $(-8)^0 + (-8)^0$

5. 3^{-3}

6. 2^{-5}

7. $(-12)^{-1}$

8. $(-6)^{-2}$

9. $\left(\frac{1}{2}\right)^{-5}$

10. $\left(\frac{1}{5}\right)^{-2}$

11. $\left(\frac{1}{2}\right)^{-1}$

12. $\left(\frac{5}{4}\right)^{-2}$

Use the quotient rule to simplify each expression. Write each answer with positive exponents. Assume that all variables represent nonzero real numbers. See Examples 3 and 4.

13. $\frac{4^7}{4^2}$

14. $\frac{11^5}{11^3}$

15. $\frac{8^3}{8^9}$

16. $\frac{5^4}{5^{10}}$

17. $\frac{6^{-4}}{6^2}$

18. $\frac{7^{-5}}{7^3}$

19. $\frac{x^6}{x^{-9}}$

20. $\frac{y^2}{y^{-5}}$

21. $\frac{1}{2^{-5}}$

22. $\frac{1}{4^{-3}}$

23. $\frac{2}{k^{-2}}$

24. $\frac{5}{p^{-5}}$

Simplify each expression. Give answers with only positive exponents. Assume that all variables represent nonzero numbers. See Examples 3–5.

25. $\frac{4^3 \cdot 4^{-5}}{4^7}$

26. $\frac{2^5 \cdot 2^{-4}}{2^{-1}}$

27. $\frac{5^4}{5^{-3} \cdot 5^{-2}}$

28. $\frac{8^6}{8^{-2} \cdot 8^5}$

29. $\frac{64^6}{32^6}$

30. $\frac{81^5}{9^5}$

31. $(3x)^5$

32. $(2m)^8$

33. $(5m^{-2})^2$ **34.** $(4x^3)^{-1}$ **35.** $x^7(x^8)^{-2}$ **36.** $(m^3)^2(m^{-2})^4$

37. $\frac{3^{-1}a^{-2}}{3^2a^{-4}}$ **38.** $\frac{2^{-5}p^{-3}}{2^{-7}p^5}$ **39.** $\frac{4k^{-3}m^5}{4^{-1}k^{-7}m^{-3}}$ **40.** $\frac{6^{-1}y^{-2}z^5}{6^2y^{-1}z^{-2}}$

Simplify each expression. Give answers with only positive exponents. Assume all variables are nonzero real numbers.

41. $\frac{(4a^2b^3)^{-2}}{(a^3b)^{-4}}$ **42.** $\frac{(m^6n)^{-2}}{m^{-1}n^{-2}}$ **43.** $\frac{(2y^{-1}z^2)^2}{(y^3z^2)^{-1}}$

44. $\frac{(3p^{-2}q^3)^2}{(p^2q^{-2})^{-3}}$ **45.** $\frac{(9^{-1}z^{-2}x)^{-1}}{(5z^{-2}x^{-3})^2}$ **46.** $\frac{(5a^{-3}b^4)^{-2}}{(3a^{-3}b^{-5})^2}$

SKILL SHARPENERS

Evaluate each of the following.

47. 10(6427) **48.** 100(72.69) **49.** 1000(1.23) **50.** 10,000(26.94)

51. 34 ÷ 10 **52.** 6501 ÷ 100 **53.** 237 ÷ 1000 **54.** 42 ÷ 10,000

4.3 AN APPLICATION OF EXPONENTS: SCIENTIFIC NOTATION

1 One example of the use of exponents comes from science. The numbers occurring in science are often extremely large (such as the distance from the earth to the sun, 93,000,000 miles) or extremely small (the wavelength of yellow-green light is approximately .0000006 meters). Because of the difficulty of working with many zeros, scientists often express such numbers with exponents. Each number is written as $a \times 10^n$, where $1 \leq |a| < 10$ and n is an integer. This form is called **scientific notation.** There is always one nonzero digit before the decimal point. For example, 35 is written 3.5×10^1, or 3.5×10; 56,200 is written 5.62×10^4, since

$$56{,}200 = 5.62 \times \mathbf{10{,}000} = 5.62 \times \mathbf{10^4};$$

and .09 is written as 9×10^{-2}.

The steps involved in writing a number in scientific notation are given below.

OBJECTIVES

1. Express numbers in scientific notation.
2. Convert numbers in scientific notation to numbers without exponents.
3. Use scientific notation in calculations.

WRITING A NUMBER IN SCIENTIFIC NOTATION

Step 1 Move the decimal point to the right of the first nonzero digit.

Step 2 Count the number of places you moved the decimal point.

Step 3 The number of places in Step 2 is the absolute value of the exponent on 10.

Step 4 The exponent on 10 is positive if you made the number smaller in Step 1; the exponent is negative if you made the number larger in Step 1.

EXAMPLE 1 Using Scientific Notation

Write each number in scientific notation.

(a) 93,000,000

The number will be written in scientific notation as 9.3×10^n. To find the value of n, first compare 9.3 with the original number, 93,000,000. Did the number get larger or smaller? Here the number 9.3 is *smaller than* 93,000,000. Therefore, we must multiply by a *positive* power of 10 so the product 9.3×10^n will equal the larger number.

Move the decimal point after the first nonzero digit. Count the number of places the decimal point was moved.

9.3 000 000   7 places

Since the decimal point was moved 7 places, and since n is positive, $93{,}000{,}000 = 9.3 \times 10^7$.

(b) $463{,}000{,}000{,}000{,}000 = 4.63\ 000\ 000\ 000\ 000$ 14 places

$= 4.63 \times 10^{14}$

(c) $302{,}100 = 3.021 \times 10^5$

(d) .00462

Move the decimal point to the right of the first nonzero digit and count the number of places the decimal point was moved.

$$.004.62 \qquad \text{3 places}$$

Since 4.62 is *larger* than .00462, the exponent must be *negative*.

$$.00462 = 4.62 \times 10^{-3}$$

(e) $.0000762 = 7.62 \times 10^{-5}$ ■

WORK PROBLEM 1 AT THE SIDE.

2 To convert a number written in scientific notation to a number without exponents, work in reverse. Multiplying a number by a positive power of 10 will make the number larger; multiplying by a negative power of 10 will make the number smaller.

EXAMPLE 2 Writing Numbers Without Exponents

Write each number without exponents.

(a) 6.2×10^3

Since the exponent is positive, make 6.2 larger by moving the decimal point 3 places to the right.

$$6.2 \times 10^3 = 6.200. = 6200.$$

(b) $4.283 \times 10^5 = 4.28300. = 428{,}300$

(c) $9.73 \times 10^{-2} = .09.73 = .0973$ ■

As these examples show, the exponent tells the number of places that the decimal point is moved.

WORK PROBLEM 2 AT THE SIDE.

3 The next example shows how scientific notation can be used with products and quotients.

EXAMPLE 3 Multiplying and Dividing with Scientific Notation

Write each product or quotient without exponents.

(a) $(6 \times 10^3)(5 \times 10^{-4})$

$$\begin{aligned} &(6 \times 10^3)(5 \times 10^{-4}) \\ &= (6 \times 5)(10^3 \times 10^{-4}) && \text{Commutative and associative properties} \\ &= 30 \times 10^{-1} && \text{Product rule for exponents} \\ &= 3.0 && \text{Write without exponents.} \end{aligned}$$

(b) $\dfrac{6 \times 10^{-5}}{2 \times 10^3} = \dfrac{6}{2} \times \dfrac{10^{-5}}{10^3} = 3 \times 10^{-8} = .00000003$ ■

WORK PROBLEM 3 AT THE SIDE.

1. Write each number in scientific notation.

(a) 63,000

(b) 5,870,000

(c) .0571

(d) .000062

2. Write without exponents.

(a) 4.2×10^3

(b) 8.7×10^5

(c) 6.42×10^{-3}

3. Simplify, and write without exponents.

(a) $(2.6 \times 10^4)(2 \times 10^{-6})$

(b) $\dfrac{4.8 \times 10^2}{2.4 \times 10^{-3}}$

ANSWERS

1. (a) 6.3×10^4 (b) 5.87×10^6 (c) 5.71×10^{-2} (d) 6.2×10^{-5}
2. (a) 4200 (b) 870,000 (c) .00642
3. (a) .052 (b) 200,000

4.3 EXERCISES

Write each number in scientific notation. See Example 1

1. 6,835,000,000

2. 6850

3. 25,000

4. 110,000,000

5. .0101

6. .000012

7. .000000982

8. .834

9. .0069

Write each number without exponents. See Examples 2 and 3.

10. 8.1×10^9

11. 3.5×10^2

12. 9.132×10^6

13. 3.24×10^8

14. 3.2×10^{-4}

15. 5.76×10^{-5}

16. $(2 \times 10^8) \times (4 \times 10^{-3})$

17. $(5 \times 10^4) \times (3 \times 10^{-2})$

18. $(4 \times 10^{-1}) \times (1 \times 10^{-5})$

19. $(6 \times 10^{-5}) \times (2 \times 10^4)$

20. $(7 \times 10^3) \times (2 \times 10^2) \times (3 \times 10^{-4})$

21. $(3 \times 10^{-5}) \times (3 \times 10^2) \times (5 \times 10^{-2})$

22. $\frac{9 \times 10^5}{3 \times 10^{-1}}$

23. $\frac{12 \times 10^{-4}}{4 \times 10^4}$

24. $\frac{8 \times 10^{-3}}{2 \times 10^{-2}}$

25. $\frac{5 \times 10^{-1}}{1 \times 10^{-5}}$

26. $\frac{2.6 \times 10^5}{2 \times 10^2}$

27. $\frac{9.5 \times 10^{-1}}{5 \times 10^3}$

28. $\dfrac{7.2 \times 10^{-3} \times 1.6 \times 10^{5}}{4 \times 10^{-2} \times 3.6 \times 10^{9}}$

29. $\dfrac{8.7 \times 10^{-2} \times 1.2 \times 10^{-6}}{3 \times 10^{-4} \times 2.9 \times 10^{11}}$

Write the numbers in each statement in scientific notation. See Example 1.

30. Light visible to the human eye has a wavelength between .0004 millimeters and .0008 millimeters.

31. In the ocean, the amount of oxygen per cubic mile of water is 4,037,000,000 tons and the amount of radium is .0003 tons.

32. Each tide in the Bay of Fundy carries more than 3,680,000,000,000,000 cubic feet of water into the bay.

33. The mean (average) diameter of the sun is about 865,000 miles.

Write the numbers in each statement without exponents. See Example 2.

34. There are 1×10^{3} cubic millimeters in 6.102×10^{-2} cubic inches.

35. In the food chain that links the largest sea creature (the whale) to the smallest (the diatom), 4×10^{14} diatoms sustain a medium-sized whale for only a few hours.

36. Many ocean trenches have a depth of 3.5×10^{4} feet.

37. The average life span of a human is 1×10^{9} seconds.

SKILL SHARPENERS

Simplify. See Section 2.7.

38. $7m + 8m$

39. $9y + 11y$

40. $12p - 9p + p$

41. $8x - 9x + 10x$

42. $5(2m - 1) - 3m$

43. $-7 + 2(2x + 5)$

4.4 POLYNOMIALS

1 Recall that in an expression such as

$$4x^3 + 6x^2 + 5x$$

the quantities that are added, $4x^3$, $6x^2$, and $5x$, are called **terms.** In the term $4x^3$, the number 4 is called the **numerical coefficient,** or simply the **coefficient,** of x^3. In the same way, 6 is the coefficient of x^2 in the term $6x^2$, and 5 is the coefficient of x in the term $5x$.

EXAMPLE 1 Identifying Coefficients

Name the coefficient of each term in these expressions.

(a) $4x^3$

The coefficient is 4.

(b) $x - 6x^4$

The coefficient of x is 1 because $x = 1 \cdot x$. The coefficient of x^4 is -6, since $x - 6x^4$ can be written as the sum $x + (-6x^4)$.

(c) $5 - v^3$

The coefficient of the term 5 is 5 since $5 = 5v^0$. By writing $5 - v^3$ as a sum, $5 + (-v^3)$, or $5 + (-1v^3)$, the coefficient of v^3 can be identified as -1. ■

WORK PROBLEM 1 AT THE SIDE.

2 Recall that **like terms** have exactly the same combination of variables, with the same exponents on the variables. Only the coefficients may be different. Examples of like terms are

$$19m^5 \quad \text{and} \quad 14m^5;$$
$$6y^9, \quad -37y^9, \quad \text{and} \quad y^9;$$
$$3pq, \quad -2pq, \quad \text{and} \quad 4pq.$$

We add like terms with the distributive property.

EXAMPLE 2 Adding Like Terms

Simplify each expression by adding like terms.

(a) $-4x^3 + 6x^3 = (-4 + 6)x^3$ Distributive property
$= 2x^3$

(b) $3x^4 + 5x^4 = (3 + 5)x^4 = 8x^4$

(c) $9x^6 - 14x^6 + x^6 = (9 - 14 + 1)x^6 = -4x^6$

(d) $12m^2 + 5m + 4m^2 = (12 + 4)m^2 + 5m$
$= 16m^2 + 5m$

(e) $3x^2y + 4x^2y - x^2y = (3 + 4 - 1)x^2y = 6x^2y$ ■

Example 2(d) shows that it is not possible to add $16m^2$ and $5m$. These two terms are unlike because the exponents on the variables are different. **Unlike terms** have different variables or different exponents on the same variables.

WORK PROBLEM 2 AT THE SIDE.

OBJECTIVES

1. Identify terms and coefficients.
2. Add like terms.
3. Know the vocabulary for polynomials.
4. Add polynomials.
5. Subtract polynomials.

1. Name the coefficient of each term in these expressions.

(a) $3m^2$

(b) $2x^3 - x$

(c) $x + 8$

2. Add like terms.

(a) $5x^4 + 7x^4$

(b) $9pq + 3pq - 2pq$

(c) $r^2 + 3r + 5r^2$

(d) $8t + 6w$

ANSWERS
1. (a) 3 (b) 2; -1 (c) 1; 8
2. (a) $12x^4$ (b) $10pq$ (c) $6r^2 + 3r$
(d) cannot be added—unlike terms

3 *Polynomials* are basic to algebra. A **polynomial in x** is the sum of a finite number of terms of the form ax^n, for any real number a and any whole number n. For example,

$$16x^8 - 7x^6 + 5x^4 - 3x^2 + 4$$

is a polynomial in x (here 4 can be written as $4x^0$). This polynomial is written in **descending powers** of the variable, since the exponents on x decrease from left to right. On the other hand,

$$2x^3 - x^2 + \frac{4}{x}$$

is not a polynomial in x, since $\frac{4}{x} = 4x^{-1}$ is not a *product,* ax^n, for a whole number n. (Of course, a polynomial could be defined using any variable, or variables, and not just x.)

WORK PROBLEM 3 AT THE SIDE.

The **degree of a term** with one variable is the exponent on the variable. For example, $3x^4$ has degree 4, $6x^{17}$ has degree 17, $5x$ has degree 1, and -7 has degree 0 (since -7 can be written as $-7x^0$). The **degree of a polynomial** in one variable is the highest exponent found in any nonzero term of the polynomial. For example, $3x^4 - 5x^2 + 6$ is of degree 4, while $5x$ is of degree 1, and 3 (or $3x^0$) is of degree 0.

Three types of polynomials are very common and are given special names. A polynomial with exactly three terms is called a **trinomial.** (*Tri-* means "three," as in *tri*angle.) Examples are

$$9m^3 - 4m^2 + 6, \qquad 19y^2 + 8y + 5, \qquad \text{and} \qquad -3m^5 - 9m^2 + 2.$$

A polynomial with exactly two terms is called a **binomial.** (*Bi-* means "two," as in *bi*cycle.) Examples are

$$-9x^4 + 9x^3, \qquad 8m^2 + 6m, \qquad \text{and} \qquad 3m^5 - 9m^2.$$

A polynomial with only one term is called a **monomial.** (*Mon(o)-* means "one," as in *mono*rail.) Examples are

$$9m, \qquad -6y^5, \qquad a^2, \qquad \text{and} \qquad 6.$$

EXAMPLE 3 Classifying Polynomials

For each polynomial, first simplify if possible by combining like terms. Then give the degree and tell whether the polynomial is a monomial, a binomial, a trinomial, or none of these.

(a) $2x^3 + 5$

The polynomial cannot be simplified. The degree is 3. The polynomial is a binomial.

(b) $4x - 5x + 2x$

Add like terms to simplify: $4x - 5x + 2x = x$. The degree is 1. The simplified polynomial is a monomial. ■

WORK PROBLEM 4 AT THE SIDE.

4 Polynomials may be added, subtracted, multiplied, and divided. Polynomial addition and subtraction are explained in the rest of this section.

3. Choose one or more of the following descriptions for each of the expressions in parts (a)–(d).
(1) Polynomial
(2) Polynomial written in descending order
(3) Not a polynomial

(a) $3m^5 + 5m^2 - 2m + 1$

(b) $2p^4 + p^6$

(c) $\frac{1}{x} + 2x^2 + 3$

(d) $x - 3$

4. For each polynomial, first simplify if possible. Then give the degree and tell whether the polynomial is a monomial, binomial, trinomial, or none of these.

(a) $3x^2 + 2x - 4$

(b) $x^3 + 4x^3$

(c) $x^8 - x^7 + 2x^8$

ANSWERS
3. (a) 1 and 2 (b) 1 (c) 3 (d) 1 and 2
4. (a) degree 2; trinomial (b) degree 3; monomial (simplify to $5x^3$) (c) degree 8; binomial (simplify to $3x^8 - x^7$)

ADDING POLYNOMIALS

To add two polynomials, add like terms.

EXAMPLE 4 Adding Polynomials Vertically
Add $6x^3 - 4x^2 + 3$ and $-2x^3 + 7x^2 - 5$.
Write like terms in columns.

$$\begin{array}{r} 6x^3 - 4x^2 + 3 \\ \underline{-2x^3 + 7x^2 - 5} \end{array}$$

Now add, column by column.

$$\begin{array}{ccc} 6x^3 & -4x^2 & 3 \\ \underline{-2x^3} & \underline{7x^2} & \underline{-5} \\ \mathbf{4x^3} & \mathbf{3x^2} & \mathbf{-2} \end{array}$$

Add the three sums together.

$$\mathbf{4x^3} + \mathbf{3x^2} + (\mathbf{-2}) = 4x^3 + 3x^2 - 2 \quad ■$$

WORK PROBLEM 5 AT THE SIDE.

The polynomials in Example 4 also could be added horizontally, as shown in the next example.

EXAMPLE 5 Adding Polynomials Horizontally
Add $6x^3 - 4x^2 + 3$ and $-2x^3 + 7x^2 - 5$.
Write the sum.

$$(6x^3 - 4x^2 + 3) + (-2x^3 + 7x^2 - 5)$$

Rewrite this sum with the parentheses removed and with subtractions changed to addition of inverses.

$$6x^3 + (-4x^2) + 3 + (-2x^3) + 7x^2 + (-5)$$

Place like terms together.

$$6x^3 + (-2x^3) + (-4x^2) + 7x^2 + 3 + (-5)$$

Combine like terms to get

$$4x^3 + 3x^2 + (-2) \qquad \text{or} \qquad 4x^3 + 3x^2 - 2$$

the same answer found in Example 4. ■

WORK PROBLEM 6 AT THE SIDE.

5 Earlier, the difference $x - y$ was defined as $x + (-y)$. (Find the difference $x - y$ by adding x and the opposite of y.) For example,

$$\begin{aligned} 7 - 2 &= 7 + (-2) \\ &= 5 \end{aligned}$$

and

$$-8 - (-2) = -8 + 2 = -6.$$

A similar method is used to subtract polynomials.

5. Add each pair of polynomials.

(a) $\begin{array}{r} 4x^3 - 3x^2 + 2x \\ \underline{6x^3 + 2x^2 - 3x} \end{array}$

(b) $\begin{array}{r} x^2 - 2x + 5 \\ \underline{4x^2 + 3x - 2} \end{array}$

6. Find each sum.

(a) $(2x^4 - 6x^2 + 7)$
$+ (-3x^4 + 5x^2 + 2)$

(b) $(3x^3 + 4x + 2)$
$+ (6x^3 - 5x - 7)$

ANSWERS
5. (a) $10x^3 - x^2 - x$ (b) $5x^2 + x + 3$
6. (a) $-x^4 - x^2 + 9$ (b) $9x^3 - x - 5$

7. Subtract, and check your answers by addition.

(a) $(14y^3 - 6y^2 + 2y - 5) - (2y^3 - 7y^2 - 4y + 6)$

(b) $(7y^2 - 11y + 8) - (-3y^2 + 4y + 6)$

8. Subtract, using the method of subtracting by columns.

(a) $(14y^3 - 6y^2 + 2y) - (2y^3 - 7y^2 + 6)$

(b) $(6p^4 - 8p^3 + 2p - 1) - (-7p^4 + 6p^2 - 12)$

ANSWERS
7. (a) $12y^3 + y^2 + 6y - 11$
(b) $10y^2 - 15y + 2$
8. (a) $12y^3 + y^2 + 2y - 6$
(b) $13p^4 - 8p^3 - 6p^2 + 2p + 11$

EXAMPLE 6 Subtracting Polynomials

Subtract: $(5x - 2) - (3x - 8)$.

By the definition of subtraction,

$$(5x - 2) - (3x - 8) = (5x - 2) + [-(3x - 8)].$$

From Chapter 2,

$$-(3x - 8) = -1(3x - 8) = -3x + 8$$

so

$$(5x - 2) - (3x - 8) = (5x - 2) + (-3x + 8) = 2x + 6. \blacksquare$$

In summary, polynomials are subtracted as follows.

SUBTRACTING POLYNOMIALS

To subtract polynomials, change all the signs of the second polynomial and add the result to the first polynomial.

EXAMPLE 7 Subtracting Polynomials Horizontally

Subtract $6x^3 - 4x^2 + 2$ from $11x^3 + 2x^2 - 8$.

Start with

$$(11x^3 + 2x^2 - 8) - (6x^3 - 4x^2 + 2).$$

Change all the signs on the second polynomial and add like terms.

$$(11x^3 + 2x^2 - 8) + (-6x^3 + 4x^2 - 2) = 5x^3 + 6x^2 - 10$$

We can check a subtraction problem by using the fact that if $a - b = c$, then $a = b + c$. For example, $6 - 2 = 4$. Check by writing $6 = 2 + 4$, which is correct. Check the polynomial subtraction above by adding $6x^3 - 4x^2 + 2$ and $5x^3 + 6x^2 - 10$. Since the sum is $11x^3 + 2x^2 - 8$, the subtraction was performed correctly. ■

WORK PROBLEM 7 AT THE SIDE.

Subtraction also can be done in columns.

EXAMPLE 8 Subtracting Polynomials Vertically

Use the method of subtracting by columns to find $(14y^3 - 6y^2 + 2y - 5) - (2y^3 - 7y^2 - 4y + 6)$.

Step 1 Arrange like terms in columns.

$$\begin{array}{r} 14y^3 - 6y^2 + 2y - 5 \\ \underline{2y^3 - 7y^2 - 4y + 6} \end{array}$$

Step 2 Change all signs in the second row, and then add *(Step 3*

$$\begin{array}{rl} 14y^3 - 6y^2 + 2y - 5 & \\ \underline{-2y^3 + 7y^2 + 4y - 6} & \text{All signs changed} \\ 12y^3 + y^2 + 6y - 11 & \text{Add.} \blacksquare \end{array}$$

Either the horizontal or the vertical method may be used for adding and subtracting polynomials.

WORK PROBLEM 8 AT THE SIDE.

name date hour

4.4 EXERCISES

In each polynomial, add like terms whenever possible. Write the result in descending powers of the variable. See Examples 2 and 3.

1. $3m^5 + 5m^5$

2. $-4y^3 + 3y^3$

3. $2r^5 + (-3r^5)$

4. $-19y^2 + 9y^2$

5. $2m^5 - 5m^2$

6. $-9y + 9y^2$

7. $3x^5 + 2x^5 - 4x^5$

8. $6x^3 - 8x^7 - 9x^3$

9. $-4p^7 + 8p^7 - 5p^7$

10. $-3a^8 + 4a^8 - 3a^8 + 2a^8$

11. $4y^2 + 3y^2 - 2y^2 + y^2$

12. $3r^5 - 8r^5 + r^5 - 2r^5$

13. $-5p^5 + 8p^5 - 2p^5 - p^5$

14. $6k^3 - 9k^3 + 8k^3 - 2k^3$

15. $y^4 + 8y^4 - 9y^2 + 6y^2 + 10y^2$

16. $11a^2 - 10a^2 + 2a^2 - a^6 + 2a^6$

17. $4z^5 - 9z^3 + 8z^2 + 10z^5$

18. $-9m^3 + 2m^3 - 11m^3 + 15m^2 - 9m$

19. $2p^7 - 8p^6 + 5p^4 - 9p$

20. $7y^3 - 8y^2 + 6y + 2$

Tell whether each statement is true always, sometimes, *or* never.

21. A binomial is a polynomial.

22. A polynomial is a trinomial.

23. A trinomial is a binomial.

24. A monomial has no coefficient.

25. A binomial is a trinomial.

26. A polynomial of degree 4 has 4 terms.

For each polynomial, first simplify, if possible; then give the degree of the polynomial and tell whether it is a monomial, a binomial, a trinomial, or none of these. See Example 3.

	Simplified form	*Degree*	*Kind of polynomial*
27. $5x^4 - 8x$			
28. $4y - 8y$			
29. $23x^9 - \frac{1}{2}x^2 + x$			
30. $2m^7 - 3m^6 + 2m^5 + m$			
31. $x^8 + 3x^7 - 5x^4$			
32. $2x - 2x^2$			
33. $\frac{3}{5}x^5 + \frac{2}{5}x^5$			
34. $\frac{9}{11}x^2$			
35. -8			
36. $2m^8 - 5m^9$			

SKILL SHARPENERS

Add or subtract as indicated. See Examples 4 and 8.

37. Add.

$$\begin{array}{r} 3m^2 + 5m \\ \underline{2m^2 - 2m} \end{array}$$

38. Add.

$$\begin{array}{r} 4a^3 - 4a^2 \\ \underline{6a^3 + 5a^2} \end{array}$$

39. Subtract.

$$\begin{array}{r} 12x^4 - x^2 \\ \underline{8x^4 + 3x^2} \end{array}$$

40. Subtract.

$$\begin{array}{r} 2a + 5d \\ \underline{3a - 6d} \end{array}$$

41. Subtract.

$$\begin{array}{r} 2n^5 - 5n^3 + 6 \\ \underline{3n^5 + 7n^3 + 8} \end{array}$$

42. Subtract.

$$\begin{array}{r} 3r^2 - 4r + 2 \\ \underline{7r^2 + 2r - 3} \end{array}$$

name date hour

43. Add.

$$\begin{array}{r} 9m^3 - 5m^2 + 4m - 8 \\ \underline{3m^3 + 6m^2 + 8m - 6} \end{array}$$

44. Add.

$$\begin{array}{r} 12r^5 + 11r^4 - 7r^3 - 2r^2 - 5r - 3 \\ \underline{-8r^5 - 10r^4 + 3r^3 + 2r^2 - 5r + 7} \end{array}$$

45. Add.

$$\begin{array}{r} 12m^2 - 8m + 6 \\ \underline{3m^2 + 5m - 2} \end{array}$$

46. Subtract.

$$\begin{array}{r} 5a^4 - 3a^3 + 2a^2 \qquad\quad \\ \underline{a^3 - a^2 + a - 1} \end{array}$$

47. Add.

$$\begin{array}{r} 5b^2 + 6b + 2 \\ \underline{3b^2 - 4b + 5} \end{array}$$

48. Add.

$$\begin{array}{r} 3w^2 - 5w + 2 \\ 4w^2 + 6w - 5 \\ \underline{8w^2 + 7w - 2} \end{array}$$

Perform the indicated operations. See Examples 5-7.

49. $(2r^2 + 3r) - (3r^2 + 5r)$

50. $(3r^2 + 5r - 6) + (2r - 5r^2)$

51. $(8m^2 - 7m) - (3m^2 + 7m)$

52. $(x^2 + x) - (3x^2 + 2x - 1)$

53. $8 - (6s^2 - 5s + 7)$

54. $2 - [3 - (4 + s)]$

55. $(8s - 3s^2) + (-4s + 5s^2)$

56. $(3x^2 + 2x + 5) + (8x^2 - 5x - 4)$

57. $(16x^3 - x^2 + 3x) + (-12x^3 + 3x^2 + 2x)$

58. $(-2b^6 + 3b^4 - b^2) - (b^6 + 2b^4 + 2b^2)$

59. $(7y^4 + 3y^2 + 2y) - (18y^4 - 5y^2 - y)$

60. $(3x^2 + 2x + 5) + (-7x^2 - 8x + 2) + (3x^2 - 4x + 7)$

61. $(9a^4 - 3a^2 + 2) + (4a^4 - 4a^2 + 2) + (-12a^4 + 6a^2 - 3)$

62. $(4m^2 - 3m + 2) + (5m^2 + 13m - 4) - (16m^2 + 4m - 3)$

63. $[(8m^2 + 4m - 7) - (2m^2 - 5m + 2)] - (m^2 + m + 1)$

64. $[(9b^3 - 4b^2 + 3b + 2) - (-2b^3 - 3b^2 + b)] - (8b^3 + 6b + 4)$

Write each statement as an equation or an inequality. Do not solve.

65. When $4 + x^2$ is added to $-9x + 2$, the result is larger than 8.

66. When $6 + 3x$ is subtracted from $5 + 2x$, the difference is larger than $8x + x^2$.

67. The sum of $5 + x^2$ and $3 - 2x$ is not equal to 5.

68. The sum of $3 - 2x + x^2$ and $8 - 9x + 3x^2$ is negative.

SKILL SHARPENERS

Multiply. See Section 4.1.

69. $p(2p)$

70. $3k(5k)$

71. $5x^2(2x)$

72. $9r^3(2r)$

73. $7m^3(8m^2)$

74. $4y^5(7y^2)$

75. $6p^5(5p^4)$

76. $2z^8(5z^3)$

4.5 MULTIPLICATION OF POLYNOMIALS

OBJECTIVES

1 Multiply a monomial and a polynomial.

2 Multiply two polynomials.

3 Multiply vertically.

1 As shown earlier, we find the product of two monomials by using the rules for exponents and the commutative and associative properties. For example,

$$(6x^3)(4x^4) = 6 \cdot 4 \cdot x^3 \cdot x^4 = 24x^7.$$

Also,

$$(-8m^6)(-9n^6) = (-8)(-9)(m^6)(n^6) = 72m^6n^6.$$

To find the product of a monomial and a polynomial with more than one term, we use the distributive property and then the method shown above.

EXAMPLE 1 Multiplying a Monomial and a Polynomial
Use the distributive property to find each product.

(a) $4x^2(3x + 5)$

$$\mathbf{4x^2}(3x + 5) = (\mathbf{4x^2})(3x) + (\mathbf{4x^2})(5) \qquad \text{Distributive property}$$
$$= 12x^3 + 20x^2 \qquad \text{Multiply monomials.}$$

(b) $-8m^3(4m^3 + 3m^2 + 2m - 1)$

$$= (-8m^3)(4m^3) + (-8m^3)(3m^2) + (-8m^3)(2m) + (-8m^3)(-1) \qquad \text{Distributive property}$$
$$= -32m^6 - 24m^5 - 16m^4 + 8m^3 \qquad \text{Multiply monomials.}$$ ■

WORK PROBLEM 1 AT THE SIDE.

1. Find each product.

(a) $5m^3(2m + 7)$

(b) $2x^4(3x^2 + 2x - 5)$

(c) $-4y^2(3y^3 + 2y^2 - 4y + 8)$

2 The distributive property is used also to find the product of any two polynomials. For example, to find the product of the polynomials $x + 1$ and $x - 4$, think of $x + 1$ as a single quantity and use the distributive property as follows.

$$\mathbf{(x + 1)}(x - 4) = \mathbf{(x + 1)}x + \mathbf{(x + 1)}(-4)$$

Now use the distributive property to find $(x + 1)x$ and $(x + 1)(-4)$.

$$(x + 1)\mathbf{x} + (x + 1)\mathbf{(-4)} = x\mathbf{(x)} + 1\mathbf{(x)} + x\mathbf{(-4)} + 1\mathbf{(-4)}$$
$$= x^2 + x + (-4x) + (-4)$$
$$= x^2 - 3x - 4$$

EXAMPLE 2 Multiplying Two Binomials
Find the product $(2x + 1)(3x + 5)$.

$$(2x + 1)(3x + 5) = (2x + 1)(3x) + (2x + 1)(5)$$
$$= (2x)(3x) + (1)(3x) + (2x)(5) + (1)(5)$$
$$= 6x^2 + 3x + 10x + 5$$
$$= 6x^2 + 13x + 5$$ ■

WORK PROBLEM 2 AT THE SIDE.

2. Find each product.

(a) $(4x + 3)(2x + 1)$

(b) $(3k - 2)(2k + 1)$

(c) $(m + 5)(3m - 4)$

ANSWERS
1. (a) $10m^4 + 35m^3$
(b) $6x^6 + 4x^5 - 10x^4$
(c) $-12y^5 - 8y^4 + 16y^3 - 32y^2$
2. (a) $8x^2 + 10x + 3$
(b) $6k^2 - k - 2$
(c) $3m^2 + 11m - 20$

3. Multiply.

(a) $(m^3 - 2m + 1) \cdot (2m^2 + 4m + 3)$

(b) $(6p^2 + 2p - 4)(3p^2 - 5)$

(c) $(2k^2 - k)(k^3 + 3k + 4)$

4. Find each product.

(a) $\begin{array}{r} 4k - 6 \\ \underline{2k + 5} \end{array}$

(b) $\begin{array}{r} 3x^2 + 4x - 5 \\ \underline{x + 4} \end{array}$

ANSWERS

3. (a) $2m^5 + 4m^4 - m^3 - 6m^2 - 2m + 3$
   (b) $18p^4 + 6p^3 - 42p^2 - 10p + 20$
   (c) $2k^5 - k^4 + 6k^3 + 5k^2 - 4k$
4. (a) $8k^2 + 8k - 30$
   (b) $3x^3 + 16x^2 + 11x - 20$

EXAMPLE 3 Multiplying Any Two Polynomials

Find the product $(3y^2 + 2y - 1)(y^3 + y^2 - 5)$.

Multiply each term in the first polynomial by each term in the second polynomial, then add any like terms.

$$\begin{aligned}
&(3y^2 + 2y - 1)(y^3 + y^2 - 5) \\
&\quad = 3y^2(y^3) + 3y^2(y^2) + 3y^2(-5) + 2y(y^3) + \\
&\quad\quad 2y(y^2) + 2y(-5) - 1(y^3) - 1(y^2) - 1(-5) \\
&\quad = 3y^5 + 3y^4 - 15y^2 + 2y^4 + 2y^3 - 10y - \\
&\quad\quad y^3 - y^2 + 5 \\
&\quad = 3y^5 + 5y^4 + y^3 - 16y^2 - 10y + 5 \quad \blacksquare
\end{aligned}$$

WORK PROBLEM 3 AT THE SIDE.

3 When at least one of the factors in a product of polynomials has three or more terms, the multiplication can be simplified by writing one polynomial above the other.

EXAMPLE 4 Multiplying Polynomials Vertically

Multiply $2x^2 + 4x + 1$ by $3x + 5$.

Start with

$$\begin{array}{r} 2x^2 + 4x + 1 \\ \underline{3x + 5}. \end{array}$$

It is not necessary to line up terms in columns, because any terms may be multiplied (not just like terms). Begin by multiplying each of the terms in the top row by 5.

Step 1

$$\begin{array}{rl} \mathbf{2x^2 + 4x + 1} & \\ 3x + \mathbf{5} & \\ \hline 10x^2 + 20x + 5 & \leftarrow \text{Product of 5 and } 2x^2 + 4x + 1 \end{array}$$

Notice how this process is similar to multiplication of whole numbers. Now multiply each term in the top row by $3x$. Be careful to place the like terms in columns, since the final step will involve addition (as in multiplying two whole numbers).

Step 2

$$\begin{array}{rl} \mathbf{2x^2 + 4x + 1} & \\ \mathbf{3x} + 5 & \\ \hline 10x^2 + 20x + 5 & \\ 6x^3 + 12x^2 + 3x & \leftarrow \text{Product of } 3x \text{ and } 2x^2 + 4x + 1 \end{array}$$

Step 3 Add like terms.

$$\begin{array}{r} 2x^2 + 4x + 1 \\ 3x + 5 \\ \hline 10x^2 + 20x + 5 \\ 6x^3 + 12x^2 + 3x \\ \hline 6x^3 + 22x^2 + 23x + 5 \end{array}$$

The product is $6x^3 + 22x^2 + 23x + 5$. $\blacksquare$

WORK PROBLEM 4 AT THE SIDE.

EXAMPLE 5 Multiplying Polynomials Vertically

Find the product of $3p - 5q$ and $2p + 7q$.

$$\begin{array}{rrrl} & 3p & -\ 5q & \\ & 2p & +\ 7q & \\ \hline & 21pq & -\ 35q^2 & \leftarrow 7q(3p - 5q) \\ 6p^2 & -\ 10pq & & \leftarrow 2p(3p - 5q) \\ \hline 6p^2 & +\ 11pq & -\ 35q^2 & \end{array}$$

■

EXAMPLE 6 Multiplying Polynomials Vertically

Find the product of $4m^3 - 2m^2 + 4m$ and $m^2 + 5$.

$$\begin{array}{rrrrrl} & & 4m^3 & -\ 2m^2 & +\ 4m & \\ & & & m^2 & +\ 5 & \\ \hline & & 20m^3 & -\ 10m^2 & +\ 20m & \text{Terms of top row multiplied by 5} \\ 4m^5 & -\ 2m^4 & +\ 4m^3 & & & \text{Terms of top row multiplied by } m^2 \\ \hline 4m^5 & -\ 2m^4 & +\ 24m^3 & -\ 10m^2 & +\ 20m & \end{array}$$

■

WORK PROBLEM 5 AT THE SIDE.

EXAMPLE 7 Squaring a Polynomial

Find $(3n^2 + 5n - 1)^2$.

By definition, $(3n^2 + 5n - 1)^2 = (3n^2 + 5n - 1)(3n^2 + 5n - 1)$. Use the vertical method of multiplication.

$$\begin{array}{rrrrr} & & 3n^2 & +\ 5n & -\ 1 \\ & & 3n^2 & +\ 5n & -\ 1 \\ \hline & & -3n^2 & -\ 5n & +\ 1 \\ & 15n^3 & +\ 25n^2 & -\ 5n & \\ 9n^4 & +\ 15n^3 & -\ 3n^2 & & \\ \hline 9n^4 & +\ 30n^3 & +\ 19n^2 & -\ 10n & +\ 1 \end{array}$$

■

EXAMPLE 8 Cubing a Polynomial

Find $(x + 5)^3$.

Since $(x + 5)^3 = (x + 5)(x + 5)(x + 5)$, the first step is to find the product $(x + 5)(x + 5)$.

$$\begin{aligned} (x + 5)(x + 5) &= x^2 + 5x + 5x + 25 \\ &= x^2 + 10x + 25 \end{aligned}$$

Now multiply this result by $x + 5$.

$$\begin{aligned} (x + 5)(x^2 + 10x + 25) &= x^3 + 10x^2 + 25x + 5x^2 + 50x + 125 \\ &= x^3 + 15x^2 + 75x + 125 \end{aligned}$$

■

WORK PROBLEM 6 AT THE SIDE.

5. Find each product.

(a) $\begin{array}{r} 2m + 3p \\ 5m - 4p \\ \hline \end{array}$

(b) $\begin{array}{r} k^3 - k^2 + k + 1 \\ k + 1 \\ \hline \end{array}$

(c) $\begin{array}{r} a^3 + 3a - 4 \\ 2a^2 + 6a + 5 \\ \hline \end{array}$

6. Find each product.

(a) $(3k - 2)^2$

(b) $(2x^2 - 3x + 4)^2$

(c) $(m + 1)^3$

ANSWERS

5. (a) $10m^2 + 7mp - 12p^2$
   (b) $k^4 + 2k + 1$
   (c) $2a^5 + 6a^4 + 11a^3 + 10a^2 - 9a - 20$
6. (a) $9k^2 - 12k + 4$
   (b) $4x^4 - 12x^3 + 25x^2 - 24x + 16$
   (c) $m^3 + 3m^2 + 3m + 1$

Historical Reflections

Historical Reflections

WOMEN IN MATHEMATICS: Amalie ("Emmy") Noether (1882–1935)

Emmy Noether was an outstanding mathematician in the field of abstract algebra. She studied and worked in Germany at a time when it was very difficult for a woman to do so. At the University of Erlangen in 1900, she was one of only two women. Although she could attend classes, professors could and did deny her the right to take exams for their courses. Not until 1904 was she allowed to register officially. She completed her doctorate four years later.

In 1916 she went to Göttingen to work, but it was not until three years later that she become a *Privatdozent*, a member of the lowest rank in the faculty. In 1922 she was made an unofficial professor (or assistant). She received no pay for this post, although she was given a small stipend to lecture in algebra. She left Germany in 1933 and accepted a position at Bryn Mawr College in Pennsylvania. She died two years later at the age of fifty-three.

Art: Courtesy Prof. Gottfried Noether

name date hour

4.5 EXERCISES

Find each product. See Example 1.

1. $(-4x^5)(8x^2)$

2. $(-3x^7)(2x^5)$

3. $(5y^4)(3y^7)$

4. $(10p^2)(5p^3)$

5. $(15a^4)(2a^5)$

6. $(-3m^6)(-5m^4)$

7. $2m(3m + 2)$

8. $-5p(6 - 3p)$

9. $3p(-2p^3 + 4p^2)$

10. $4x(3 + 2x + 5x^3)$

11. $-8z(2z + 3z^2 + 3z^3)$

12. $7y(3 + 5y^2 - 2y^3)$

13. $2y(3 + 2y + 5y^4)$

14. $-2m^4(3m^2 + 5m + 6)$

15. $4z^3(8z^2 + 5xz - 3x^2)$

Find each binomial product. See Examples 2 and 5.

16. $(n - 1)(n + 4)$

17. $(x + 5)(x - 5)$

18. $(y + 8)(y - 8)$

19. $(3r - 1)(r - 4)$

20. $(2k + 5)(k + 2)$

21. $(6p + 5)(p - 1)$

22. $(2x + 3)(6x - 4)$

23. $(4m + 3)(4m + 3)$

24. $(3x - 2)(3x - 2)$

25. $(b + 8)(6b - 2)$

26. $(5a + 1)(2a + 7)$

27. $(8b - 3a)(2b + a)$

28. $(6p - 5m)(2p + 3m)$

29. $(-4h + k)(2h - k)$

30. $(5y - 3x)(4y + x)$

Find each product. See Examples 3, 4, and 6.

31. $(6x + 1)(2x^2 + 4x + 1)$

32. $(9y - 2)(8y^2 - 6y + 1)$

33. $(4m + 3)(5m^3 - 4m^2 + m - 5)$

34. $(y + 4)(3y^4 - 2y^2 + 1)$

35. $(2x - 1)(3x^5 - 2x^3 + x^2 - 2x + 3)$

36. $(2a + 3)(a^4 - a^3 + a^2 - a + 1)$

37. $(5x^2 + 2x + 1)(x^2 - 3x + 5)$

38. $(2m^2 + m - 3)(m^2 - 4m + 5)$

Find each product. See Examples 7 and 8.

39. $(x + 7)^2$

40. $(m + 6)^2$

41. $(2p - 5)^2$

42. $(3m + 1)^2$

43. $(5k + 8)^2$

44. $(8m - 3)^2$

45. $(m - 5)^3$

46. $(p + 3)^3$

47. $(k + 1)^4$

48. $(r - 1)^4$

49. $(3r - 2s)^3$

50. $(2z + 5y)^3$

SKILL SHARPENERS

Find two numbers having the given product and sum. See Sections 2.2 and 2.4.

	Product	*Sum*
51.	6	5
52.	8	6
53.	−21	−4
54.	−35	−2
55.	−54	3
56.	−96	4

4.6 PRODUCTS OF BINOMIALS

1 We can use the methods introduced in the last section to find the product of any two polynomials. They are the only practical methods for multiplying polynomials with three or more terms. However, many of the polynomials to be multiplied are binomials, with only two terms, so in this section we discuss a shortcut that eliminates the need to write all the steps. To develop this shortcut, let us first multiply $x + 3$ and $x + 5$ using the distributive property.

$$\begin{aligned}(x + 3)(x + 5) &= (x + 3)x + (x + 3)5 \\ &= (x)(x) + (3)(x) + (x)(5) + (3)(5) \\ &= x^2 + 3x + 5x + 15 \\ &= x^2 + 8x + 15\end{aligned}$$

The first term in the second line, $(x)(x)$, is the product of the first terms of the two binomials.

$(x + 3)(x + 5)$ Multiply the first terms: $(x)(x)$.

The term $(x)(5)$ is the product of the first term of the first binomial and the last term of the second binomial. This is the **outer product.**

$(x + 3)(x + 5)$ Multiply the outer terms: $(x)(5)$.

The term $(3)(x)$ is the product of the last term of the first binomial and the first term of the second binomial. The product of these middle terms is called the **inner product.**

$(x + 3)(x + 5)$ Multiply the inner terms: $(3)(x)$.

Finally, $(3)(5)$ is the product of the last terms of the two binomials.

$(x + 3)(x + 5)$ Multiply the last terms: $(3)(5)$.

In the third step of the multiplication above, the inner product and the outer product are added. This step should be performed mentally, so that the three terms of the answer can be written without extra steps as

$$(x + 3)(x + 5) = x^2 + 8x + 15.$$

WORK PROBLEM 1 AT THE SIDE.

A summary of these steps is given below. This procedure is sometimes called the **FOIL method,** which comes from the initial letters of the words *First, Outer, Inner,* and *Last.*

MULTIPLYING BINOMIALS BY THE FOIL METHOD

Step 1 Multiply the two first terms of the binomials to get the first term of the answer.

Step 2 Find the outer product and the inner product and add them (mentally, if possible) to get the middle term of the answer.

Step 3 Multiply the two last terms of the binomials to get the last term of the answer.

OBJECTIVES

1. Multiply binomials by the FOIL method.
2. Square binomials.
3. Find the product of the sum and difference of two terms.

1. For the product $(2p - 5)(3p + 7)$, find the following.

(a) Product of first terms

(b) Outer product

(c) Inner product

(d) Product of last terms

(e) Product in simplified form

ANSWERS

1. (a) $2p(3p) = 6p^2$
(b) $2p(7) = 14p$
(c) $-5(3p) = -15p$
(d) $-5(7) = -35$
(e) $6p^2 - p - 35$

2. Use the FOIL method to find each product.

(a) $(m + 4)(m - 3)$

(b) $(y + 7)(y + 2)$

(c) $(r - 8)(r - 5)$

3. Find each product.

(a) $(4k - 1)(2k + 3)$

(b) $(6m + 5)(m - 4)$

(c) $(8y + 3)(2y + 1)$

(d) $(3r + 2t)(3r + 4t)$

ANSWERS
2. (a) $m^2 + m - 12$
(b) $y^2 + 9y + 14$
(c) $r^2 - 13r + 40$
3. (a) $8k^2 + 10k - 3$
(b) $6m^2 - 19m - 20$
(c) $16y^2 + 14y + 3$
(d) $9r^2 + 18rt + 8t^2$

EXAMPLE 1 Using the FOIL Method

Use the FOIL method to find the product $(x + 8)(x - 6)$.

Step 1 **F** Multiply the first terms.

$$x(x) = x^2$$

Step 2 **O** Find the product of the outer terms.

$$x(-6) = -6x$$

I Find the product of the inner terms.

$$8(x) = 8x$$

Add the outer and inner products mentally.

$$-6x + 8x = 2x$$

Step 3 **L** Multiply the last terms.

$$8(-6) = -48$$

The product of $x + 8$ and $x - 6$ is found by adding the terms found in the three steps above, so

$$(x + 8)(x - 6) = x^2 + 2x - 48. \blacksquare$$

As a shortcut, this product can be found in the following manner.

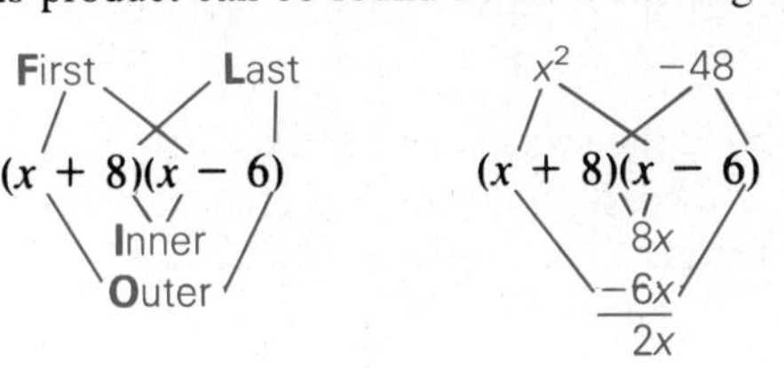

WORK PROBLEM 2 AT THE SIDE.

EXAMPLE 2 Using the FOIL Method

Multiply $9x - 2$ and $3x + 1$.

First	$(9x - 2)(3x + 1)$	$27x^2$
Outer	$(9x - 2)(3x + 1)$	$9x$
Inner	$(9x - 2)(3x + 1)$	$-6x$
Last	$(9x - 2)(3x + 1)$	-2

$$\begin{aligned}(9x - 2)(3x + 1) &= \overset{F}{27x^2} + \overset{O}{9x} - \overset{I}{6x} - \overset{L}{2} \\ &= 27x^2 + 3x - 2 \quad \blacksquare\end{aligned}$$

EXAMPLE 3 Using the FOIL Method

Find the following products.

$$\begin{aligned}\textbf{(a)}\ (2k + 5y)(k + 3y) &= \overset{F}{(2k)(k)} + \overset{O}{(2k)(3y)} + \overset{I}{(5y)(k)} + \overset{L}{(5y)(3y)} \\ &= 2k^2 + 6ky + 5ky + 15y^2 \\ &= 2k^2 + 11ky + 15y^2\end{aligned}$$

(b) $(7p + 2q)(3p - q) = 21p^2 - pq - 2q^2$ ■

WORK PROBLEM 3 AT THE SIDE.

2 Certain special types of binomial products occur so often that the form of the answers should be memorized. For example, to find the square of a binomial quickly, use the method shown in Example 4.

EXAMPLE 4 Squaring a Binomial

Find $(2m + 3)^2$.

Squaring $2m + 3$ by the FOIL method gives

$$(2m + 3)(2m + 3) = 4m^2 + 12m + 9.$$

The result has the square of both the first and the last terms of the binomial:

$$4m^2 = (2m)^2 \quad \text{and} \quad 9 = 3^2.$$

The middle term is twice the product of the two terms of the binomial, that is,

$$12m = 2(2m)(3). \quad \blacksquare$$

This example suggests the following rule.

SQUARE OF A BINOMIAL

The square of a binomial is a trinomial made up of the square of the first term, plus twice the product of the two terms, plus the square of the last term of the binomial. For a and b,

$$(a + b)^2 = a^2 + 2ab + b^2.$$

Also,
$$(a - b)^2 = a^2 - 2ab + b^2.$$

EXAMPLE 5 Squaring Binomials

Use the formula to square each binomial.

(a) $(5z - 1)^2 = (5z)^2 - 2(5z)(1) + (1)^2$
$= 25z^2 - 10z + 1$

Recall that $(5z)^2 = 5^2z^2 = 25z^2$.

(b) $(3b + 5r)^2 = (3b)^2 + 2(3b)(5r) + (5r)^2$
$= 9b^2 + 30br + 25r^2$

(c) $(2a - 9x)^2 = 4a^2 - 36ax + 81x^2$

(d) $\left(4m + \frac{1}{2}\right)^2 = (4m)^2 + 2(4m)\left(\frac{1}{2}\right) + \left(\frac{1}{2}\right)^2$
$= 16m^2 + 4m + \frac{1}{4} \quad \blacksquare$

CAUTION A common error in squaring a binomial is forgetting the middle term of the product. In general,

$$(a + b)^2 \neq a^2 + b^2.$$

WORK PROBLEM 4 AT THE SIDE.

4. Find each square.

(a) $(t + u)^2$

(b) $(2m - p)^2$

(c) $(4p + 3q)^2$

(d) $(5r - 6s)^2$

(e) $\left(3k - \frac{1}{2}\right)^2$

ANSWERS

4. (a) $t^2 + 2tu + u^2$
(b) $4m^2 - 4mp + p^2$
(c) $16p^2 + 24pq + 9q^2$
(d) $25r^2 - 60rs + 36s^2$
(e) $9k^2 - 3k + \frac{1}{4}$

3 Binomial products of the form $(a + b)(a - b)$ also occur frequently. In these products, one binomial is the sum of two terms, and the other is the difference of the same two terms. As an example, the product of $x + 2$ and $x - 2$ is

$$\begin{aligned}(x + 2)(x - 2) &= x^2 - 2x + 2x - 4\\ &= x^2 - 4.\end{aligned}$$

As shown with the FOIL method, the product of $a + b$ and $a - b$ is the difference between two squares.

PRODUCT OF THE SUM AND DIFFERENCE OF TWO TERMS

$$(a + b)(a - b) = a^2 - b^2$$

EXAMPLE 6 Finding the Product of the Sum and Difference of Two Terms

Find each product.

(a) $(5m + 3)(5m - 3)$

Use the pattern for the sum and difference of two terms.

$$\begin{aligned}(5m + 3)(5m - 3) &= (5m)^2 - 3^2\\ &= 25m^2 - 9\end{aligned}$$

(b) $\begin{aligned}(4x + y)(4x - y) &= (4x)^2 - y^2\\ &= 16x^2 - y^2\end{aligned}$

(c) $\left(z - \frac{1}{4}\right)\left(z + \frac{1}{4}\right) = z^2 - \frac{1}{16}$ ■

WORK PROBLEM 5 AT THE SIDE.

The product formulas of this section will be very useful in later work, particularly in Chapter 5. Therefore, it is important to memorize these formulas and practice using them.

5. Find each product by using the pattern for the sum and difference of two terms.

(a) $(6a + 3)(6a - 3)$

(b) $(10m + 7)(10m - 7)$

(c) $(7p + 2q)(7p - 2q)$

(d) $(8r + 5s)(8r - 5s)$

(e) $\left(3r - \frac{1}{2}\right)\left(3r + \frac{1}{2}\right)$

ANSWERS

5. (a) $36a^2 - 9$ (b) $100m^2 - 49$ (c) $49p^2 - 4q^2$ (d) $64r^2 - 25s^2$ (e) $9r^2 - \frac{1}{4}$

name date hour

4.6 EXERCISES

Find each product. See Examples 1–3.

1. $(r - 1)(r + 3)$

2. $(x + 2)(x - 5)$

3. $(x - 7)(x - 3)$

4. $(r + 3)(r + 6)$

5. $(2x - 1)(3x + 2)$

6. $(4y - 5)(2y + 1)$

7. $(6z + 5)(z - 3)$

8. $(8a + 3)(6a + 1)$

9. $(a + 4)(2a + 1)$

10. $(3x - 1)(2x + 3)$

11. $(2r - 1)(4r + 3)$

12. $(5m + 2)(3m - 4)$

13. $(2a + 4)(3a - 2)$

14. $(11m - 10)(10m + 11)$

15. $(4 + 5x)(5 - 4x)$

16. $(8 + 3x)(2 - x)$

17. $(-3 + 2r)(4 + r)$

18. $(-5 + 6z)(2 - z)$

19. $(-3 + a)(-5 - 2a)$

20. $(-6 - 3y)(1 - 4y)$

21. $(p + 3q)(p + q)$

22. $(2r - 3s)(3r + s)$

23. $(5y + z)(2y - z)$

24. $(9m + 4k)(2m - 3k)$

25. $(8y - 9z)(y + 5z)$

26. $(3a + 7b)(-4a + b)$

27. $(4r + 9s)(-2r + 5s)$

Find each square. See Examples 4 and 5.

28. $(m + 2)^2$

29. $(x + 8)^2$

30. $(x - 2y)^2$

31. $(3m - n)^2$

32. $\left(2z - \frac{5}{2}x\right)^2$

33. $\left(6a - \frac{3}{2}b\right)^2$

34. $(5p + 2q)^2$

35. $(8a - 3b)^2$

36. $\left(4a - \frac{5}{4}b\right)^2$

Find the following products. See Example 6.

37. $(p + 2)(p - 2)$

38. $(a + 8)(a - 8)$

39. $(2b + 5)(2b - 5)$

40. $(3x + 4)(3x - 4)$

41. $(6a - p)(6a + p)$

42. $(5y + 3x)(5y - 3x)$

43. $\left(2m - \frac{5}{3}\right)\left(2m + \frac{5}{3}\right)$

44. $\left(3a - \frac{4}{5}\right)\left(3a + \frac{4}{5}\right)$

45. $(7y^2 + 10z)(7y^2 - 10z)$

Write each statement as an equation or an inequality, using x to represent the unknown number. Do not solve.

46. The square of 3 more than a number is 5.

47. The square of the sum of a number and 6 is less than 3.

48. When 3 plus a number is multiplied by the number less 4, the result is greater than 7.

49. Twice a number plus 4, multiplied by 6 times the number, less 5, gives 8.

SKILL SHARPENERS

Simplify. Write answers with positive exponents. See Section 4.2.

50. $\frac{9y^4}{3y}$

51. $\frac{8p^2}{2p}$

52. $\frac{5m^7}{m^3}$

53. $\frac{-10p^4}{2p^3}$

54. $\frac{-8z^5}{10z^7}$

55. $\frac{20k^3}{50k^5}$

56. $\frac{36r^5s^7}{24r^9s}$

57. $\frac{-18p^7q^{10}}{32p^2q^{12}}$

4.7 DIVIDING A POLYNOMIAL BY A MONOMIAL

OBJECTIVE

1 Divide a polynomial by a monomial.

1 The quotient rule for exponents is used to divide a monomial by another monomial. For example,

$$\frac{12x^2}{6x} = 2x, \quad \frac{25m^5}{5m^2} = 5m^3, \quad \text{and} \quad \frac{30a^2b^8}{15a^3b^3} = \frac{2b^5}{a}.$$

This suggests the following rule.

DIVIDING A POLYNOMIAL BY A MONOMIAL

To divide a polynomial by a monomial, divide each term of the polynomial by the monomial:

$$\frac{a+b}{c} = \frac{a}{c} + \frac{b}{c} \quad (c \neq 0).$$

EXAMPLE 1 Dividing a Polynomial by a Monomial

Divide $5m^5 - 10m^3$ by $5m^2$.

Use the rule above, with + replaced by −.

$$\frac{5m^5 - 10m^3}{5m^2} = \frac{5m^5}{5m^2} - \frac{10m^3}{5m^2} = m^3 - 2m$$

Check by multiplication.

$$5m^2(m^3 - 2m) = 5m^5 - 10m^3$$

Since division by 0 is undefined, the quotient

$$\frac{5m^5 - 10m^3}{5m^2}$$

is undefined if $m = 0$. In the rest of the chapter, we assume that no denominators are 0. ■

WORK PROBLEM 1 AT THE SIDE.

1. Divide.

(a) $\dfrac{6p^4 + 18p^7}{3p^2}$

(b) $\dfrac{12m^6 + 18m^5 + 30m^4}{6m^2}$

(c) $(18r^7 - 9r^2) \div (3r)$

EXAMPLE 2 Dividing a Polynomial by a Monomial

Divide: $\dfrac{16a^5 - 12a^4 + 8a^2}{4a^3}$.

Divide each term of $16a^5 - 12a^4 + 8a^2$ by $4a^3$.

$$\frac{16a^5 - 12a^4 + 8a^2}{4a^3} = \frac{16a^5}{4a^3} - \frac{12a^4}{4a^3} + \frac{8a^2}{4a^3}$$

$$= 4a^2 - 3a + \frac{2}{a}$$

The result is not a polynomial because of the expression $\frac{2}{a}$, which has a variable in the denominator. While the sum, difference, and product of two polynomials are always polynomials, the quotient of two polynomials may not be.

ANSWERS

1. (a) $2p^2 + 6p^5$
(b) $2m^4 + 3m^3 + 5m^2$
(c) $6r^6 - 3r$

2. Divide.

(a) $\dfrac{20x^4 - 25x^3 + 5x}{5x^2}$

(b) $\dfrac{50m^4 - 30m^3 + 20m}{10m^3}$

3. Divide.

(a) $\dfrac{8y^7 - 9y^6 - 11y - 4}{y^2}$

(b) $\dfrac{12p^5 + 8p^4 + 6p^3 - 5p^2}{3p^3}$

(c) $\dfrac{45x^4 + 30x^3 - 60x^2}{-15x^2}$

ANSWERS

2. (a) $4x^2 - 5x + \dfrac{1}{x}$

(b) $5m - 3 + \dfrac{2}{m^2}$

3. (a) $8y^5 - 9y^4 - \dfrac{11}{y} - \dfrac{4}{y^2}$

(b) $4p^2 + \dfrac{8p}{3} + 2 - \dfrac{5}{3p}$

(c) $-3x^2 - 2x + 4$

Again, check by multiplying.

$$4a^3\left(4a^2 - 3a + \frac{2}{a}\right) = 4a^3(4a^2) - 4a^3(3a) + 4a^3\left(\frac{2}{a}\right)$$
$$= 16a^5 - 12a^4 + 8a^2 \quad ■$$

WORK PROBLEM 2 AT THE SIDE.

EXAMPLE 3 Dividing a Polynomial by a Monomial

Divide.

$$\frac{12x^4 - 7x^3 + x - 4}{4x} = \frac{12x^4}{4x} - \frac{7x^3}{4x} + \frac{x}{4x} - \frac{4}{4x}$$
$$= 3x^3 - \frac{7x^2}{4} + \frac{1}{4} - \frac{1}{x}$$

Check by multiplication. ■

EXAMPLE 4 Dividing a Polynomial by a Monomial

Divide the polynomial

$$180y^{10} - 150y^8 + 120y^6 - 90y^4 + 100y$$

by the monomial $-30y^2$.

Using the methods of this section,

$$\frac{180y^{10} - 150y^8 + 120y^6 - 90y^4 + 100y}{-30y^2}$$
$$= \frac{180y^{10}}{-30y^2} - \frac{150y^8}{-30y^2} + \frac{120y^6}{-30y^2} - \frac{90y^4}{-30y^2} + \frac{100y}{-30y^2}$$
$$= -6y^8 + 5y^6 - 4y^4 + 3y^2 - \frac{10}{3y}.$$

To check, multiply this result and $-30y^2$. ■

WORK PROBLEM 3 AT THE SIDE.

name date hour

4.7 EXERCISES

Divide.

1. $\frac{4x^2}{2x}$

2. $\frac{8m^7}{2m}$

3. $\frac{10a^3}{5a}$

4. $\frac{36p^8}{4p^3}$

5. $\frac{27k^4m^5}{3km^6}$

6. $\frac{18x^5y^6}{3x^2y^2}$

Divide each polynomial by 2m. See Examples 1–4.

7. $60m^4 - 20m^2$

8. $120m^6 - 60m^3 + 80m^2$

9. $10m^5 - 16m^2 + 8m^3$

10. $6m^5 - 4m^3 + 2m^2$

11. $8m^5 - 4m^3 + 4m^2$

12. $8m^3 - 4m^2 + 6m$

13. $2m^5 - 4m^2 + 8m$

14. $m^2 + m + 1$

15. $2m^2 - 2m + 5$

Divide each polynomial by $3x^2$. See Examples 1–4.

16. $15x^2 - 9x^3$

17. $12x^4 - 3x^3 + 3x^2$

18. $45x^3 + 15x^2 - 9x^5$

19. $27x^3 - 9x^4 + 18x^5$

20. $-18x^6 + 6x^5 + 3x^4 - 9x^3$

21. $36x + 24x^2 + 3x^3$

22. $4x^4 - 3x^3 + 2x$

23. $x^3 + 6x^2 - x$

24. $6x^5 - 3x^4 + 9x^2 + 27$

Perform each division. See Examples 1–4.

25. $\frac{8k^4 - 12k^3 - 2k^2 + 7k - 3}{2k}$

26. $\frac{27r^4 - 36r^3 - 6r^2 + 26r - 2}{3r}$

27. $\dfrac{100p^5 - 50p^4 + 30p^3 - 30p}{-10p^2}$

28. $\dfrac{2m^5 - 6m^4 + 8m^2}{-2m^3}$

29. $(16y^5 - 8y^2 + 12y) \div (4y^2)$

30. $(20a^4 - 15a^5 + 25a^3) \div (15a^4)$

31. $(120x^{11} - 60x^{10} + 140x^9 - 100x^8) \div (10x^{12})$

32. $(5 + x + 6x^2 + 8x^3) \div (3x^4)$

Solve each problem.

33. What polynomial, when divided by $3x^2$, yields $4x^3 + 3x^2 - 4x + 2$ as a quotient?

34. The quotient of a certain polynomial and $-7y^2$ is $9y^2 + 3y + 5 - \frac{2}{y}$. Find the polynomial.

SKILL SHARPENERS

Find each product. See Section 4.5.

35. $-4k(5k^3 - 3k^2 + 2k)$

36. $-7z(z^2 - 5z + 6)$

37. $3m^5(2m^5 - 4m^3 + m^2)$

Subtract. See Section 4.4.

38. $\begin{array}{r} 4k^2 - 4k + 1 \\ \underline{4k^2 + 7k - 5} \end{array}$

39. $\begin{array}{r} 2x^2 - 4x - 7 \\ \underline{2x^2 - 6x - 8} \end{array}$

40. $\begin{array}{r} 3m^2 - 5m - 1 \\ \underline{3m^2 + 2m - 5} \end{array}$

4.8 THE QUOTIENT OF TWO POLYNOMIALS

OBJECTIVE

1 Divide a polynomial by a polynomial.

1 A method of "long division" is used to divide a polynomial by a polynomial (other than a monomial). This method is similar to the method of long division used for two whole numbers. For comparison, the division of whole numbers is shown alongside the division of polynomials.

Division of Whole Numbers	**Division of Polynomials**
Step 1	
Divide 27 into 6696.	Divide $2x + 3$ into $8x^3 - 4x^2 - 14x + 15$.
$27\overline{)6696}$	$2x + 3\overline{)8x^3 - 4x^2 - 14x + 15}$
Step 2	
27 divides into 66 **2** times; $2 \cdot 27 = 54$.	$2x$ divides into $8x^3$ $\mathbf{4x^2}$ times; $4x^2(2x + 3) = 8x^3 + 12x^2$.
$\begin{array}{r} \mathbf{2} \\ 27\overline{)6696} \\ \underline{54} \end{array}$	$\begin{array}{r} \mathbf{4x^2} \\ 2x + 3\overline{)8x^3 - 4x^2 - 14x + 15} \\ \underline{8x^3 + 12x^2} \end{array}$
Step 3	
Subtract; then bring down the next term.	Subtract; then bring down the next term.
$\begin{array}{r} 2 \\ 27\overline{)6696} \\ \underline{54}\downarrow \\ 12\mathbf{9} \end{array}$	$\begin{array}{r} 4x^2 \\ 2x + 3\overline{)8x^3 - 4x^2 - 14x + 15} \\ \underline{8x^3 + 12x^2} \downarrow \\ -16x^2 \mathbf{- 14x} \end{array}$
	(To subtract two polynomials, change the sign of the second and then add.)
Step 4	
27 divides into 129 **4** times; $4 \cdot 27 = 108$.	$2x$ divides into $-16x^2$ $\mathbf{-8x}$ times; $-8x(2x + 3) = -16x^2 - 24x$.
$\begin{array}{r} 2\mathbf{4} \\ 27\overline{)6696} \\ \underline{54} \\ 129 \\ \underline{108} \end{array}$	$\begin{array}{r} 4x^2 \mathbf{- 8x} \\ 2x + 3\overline{)8x^3 - 4x^2 - 14x + 15} \\ \underline{8x^3 + 12x^2} \\ -16x^2 - 14x \\ \underline{-16x^2 - 24x} \end{array}$
Step 5	
Subtract; then bring down the next term.	Subtract; then bring down the next term.
$\begin{array}{r} 24 \\ 27\overline{)6696} \\ \underline{54} \\ 129 \\ \underline{108}\downarrow \\ 21\mathbf{6} \end{array}$	$\begin{array}{r} 4x^2 - 8x \\ 2x + 3\overline{)8x^3 - 4x^2 - 14x + 15} \\ \underline{8x^3 + 12x^2} \\ -16x^2 - 14x \\ \underline{-16x^2 - 24x} \downarrow \\ 10x \mathbf{+ 15} \end{array}$

1. Divide.

(a) $(2y^2 - y - 21) \div (y + 3)$

(b) $(x^3 + x^2 + 4x - 6) \div (x - 1)$

(c) $\dfrac{p^3 - 2p^2 - 5p + 9}{p + 2}$

ANSWERS
1. (a) $2y - 7$ (b) $x^2 + 2x + 6$
(c) $p^2 - 4p + 3 + \dfrac{3}{p+2}$

Step 6

27 divides into 216 **8** times; $8 \cdot 27 = 216$.

$$\begin{array}{r} 248 \\ 27\overline{)6696} \\ \underline{54} \\ 129 \\ \underline{108} \\ 216 \\ \underline{216} \end{array}$$

6696 divided by 27 is 248. There is no remainder.

$2x$ divides into $10x$ **5** times; $5(2x + 3) = 10x + 15$.

$$\begin{array}{r} 4x^2 - 8x + 5 \\ 2x + 3\overline{)8x^3 - 4x^2 - 14x + 15} \\ \underline{8x^3 + 12x^2} \qquad\qquad \\ -16x^2 - 14x \qquad \\ \underline{-16x^2 - 24x} \qquad \\ 10x + 15 \\ \underline{10x + 15} \end{array}$$

$8x^3 - 4x^2 - 14x + 15$ divided by $2x + 3$ is $4x^2 - 8x + 5$. There is no remainder.

Step 7

Check by multiplication.

$27 \cdot 248 = 6696$

Check by multiplication.

$(2x + 3)(4x^2 - 8x + 5)$
$= 8x^3 - 4x^2 - 14x + 15$

EXAMPLE 1 Dividing a Polynomial by a Polynomial

Divide $4x^3 - 4x^2 + 5x - 8$ by $2x - 1$.

$$\begin{array}{r} 2x^2 - x + 2 \\ 2x - 1\overline{)4x^3 - 4x^2 + 5x - 8} \\ \underline{4x^3 - 2x^2} \qquad\qquad \\ -2x^2 + 5x \qquad \\ \underline{-2x^2 + x} \qquad \\ 4x - 8 \\ \underline{4x - 2} \\ -6 \end{array}$$

Step 1 $2x$ divides into $4x^3$ $\mathbf{2x^2}$ times; $2x^2(2x - 1) = 4x^3 - 2x^2$.

Step 2 Subtract; bring down the next term.

Step 3 $2x$ divides into $-2x^2$ $\mathbf{-x}$ times; $-x(2x - 1) = -2x^2 + x$.

Step 4 Subtract; bring down the next term.

Step 5 $2x$ divides into $4x$ **2** times; $2(2x - 1) = 4x - 2$.

Step 6 Subtract. The remainder is -6. Thus $2x - 1$ divides into $4x^3 - 4x^2 + 5x - 8$ with a quotient of $2x^2 - x + 2$ and a remainder of -6. Write the remainder as the numerator of a fraction that has $2x - 1$ as its denominator. The result is not a polynomial because of the remainder.

$$\frac{4x^3 - 4x^2 + 5x - 8}{2x - 1} = 2x^2 - x + 2 + \frac{-6}{2x - 1}$$

Step 7 Check by multiplication.

$$(2x - 1)\left(2x^2 - x + 2 + \frac{-6}{2x - 1}\right)$$
$$= 4x^3 - 4x^2 + 5x - 8 \quad ■$$

WORK PROBLEM 1 AT THE SIDE.

EXAMPLE 2 Dividing into a Polynomial with Missing Terms
Divide $x^3 - 1$ by $x - 1$.

Here the polynomial $x^3 - 1$ is missing the x^2 term and the x term. When terms are missing, use 0 as the coefficient for the missing terms. (Zero acts as a placeholder here, just as it does in our number system.)

$$x^3 - 1 = x^3 + \mathbf{0x^2} + \mathbf{0x} - 1$$

Now divide.

$$\begin{array}{r} x^2 + x + 1 \\ x - 1\overline{)x^3 + 0x^2 + 0x - 1} \\ \underline{x^3 - x^2} \qquad\qquad\quad \\ x^2 + 0x \qquad \\ \underline{x^2 - x} \qquad \\ x - 1 \\ \underline{x - 1} \\ 0 \end{array}$$

The remainder is 0. The quotient is $x^2 + x + 1$. Check by multiplication.

$$(x^2 + x + 1)(x - 1) = x^3 - 1 \quad ■$$

WORK PROBLEM 2 AT THE SIDE.

EXAMPLE 3 Dividing by a Polynomial with Missing Terms
Divide $x^4 + 2x^3 + 2x^2 - x - 1$ by $x^2 + 1$.

Since $x^2 + 1$ has a missing x term, write it as $x^2 + 0x + 1$. Then proceed through the division process as follows.

$$\begin{array}{r} x^2 + 2x + 1 \\ x^2 + \mathbf{0x} + 1\overline{)x^4 + 2x^3 + 2x^2 - x - 1} \\ \underline{x^4 + 0x^3 + x^2} \qquad\qquad\quad \\ 2x^3 + x^2 - x \qquad \\ \underline{2x^3 + 0x^2 + 2x} \qquad \\ x^2 - 3x - 1 \\ \underline{x^2 + 0x + 1} \\ -3x - 2 \end{array}$$

When the result of subtracting ($-3x - 2$, in this case) is a polynomial of smaller degree than the divisor ($x^2 + 0x + 1$), that polynomial is the remainder. Write the result as

$$x^2 + 2x + 1 + \frac{-3x - 2}{x^2 + 1}. \quad ■$$

WORK PROBLEM 3 AT THE SIDE.

2. Divide.

(a) $\dfrac{r^2 - 5}{r + 4}$

(b) $(x^3 - 8) \div (x - 2)$

3. Divide.

(a) $(2x^4 + 3x^3 - x^2 + 6x + 5) \div (x^2 - 1)$

(b) $\dfrac{2m^5 + m^4 + 6m^3 - 3m^2 - 18}{m^2 + 3}$

ANSWERS

2. (a) $r - 4 + \dfrac{11}{r + 4}$
(b) $x^2 + 2x + 4$

3. (a) $2x^2 + 3x + 1 + \dfrac{9x + 6}{x^2 - 1}$
(b) $2m^3 + m^2 - 6$

Historical Reflections

Historical Reflections

Blaise Pascal (1623–1662)

The French mathematician Blaise Pascal began searching for mathematical truth early in his life. As a child prodigy he worked out by himself the fundamental ideas of Euclidean geometry without the aid of books. By the age of 16 he had formulated a theorem in projective geometry. A few years later he invented a calculating machine that employed geared wheels. He summed up his experiments on air and water pressure in a treatise on atmospheric weight in 1647, then applied its principles in the invention of a hydraulic press.

During his lifetime, Pascal alternated his studies between mathematics and religion. He studied the theory of probability, and in one of his works he developed the "rule of the wager": If you bet God exists and live accordingly, you will have gained much even if God does not exist; if you bet the opposite and God does exist, you will have lost the reason for living right—hence, everything.

A mathematical array of numbers known as Pascal's Triangle is an important part of several branches of mathematics. The first few rows are shown below.

$$
\begin{array}{ccccccccc}
 & & & & 1 & & & & \\
 & & & 1 & & 1 & & & \\
 & & 1 & & 2 & & 1 & & \\
 & 1 & & 3 & & 3 & & 1 & \\
1 & & 4 & & 6 & & 4 & & 1
\end{array}
$$

Can you determine the next row of the triangle?

Answer: The next row is 1 5 10 10 5 1. Notice that the first and last entry of any row is always 1. To obtain any other entry, add the entries from the previous row that are just to the left and just to the right of the entry being determined. Thus, $5 = 1 + 4$, $10 = 4 + 6$, $10 = 6 + 4$, and $5 = 4 + 1$.

name date hour

4.8 EXERCISES

Perform each division. See Example 1.

1. $\dfrac{x^2 - x - 6}{x - 3}$

2. $\dfrac{m^2 - 2m - 24}{m + 4}$

3. $\dfrac{2y^2 + 9y - 35}{y + 7}$

4. $\dfrac{y^2 + 2y + 1}{y + 1}$

5. $\dfrac{p^2 + 2p - 20}{p + 6}$

6. $\dfrac{x^2 + 11x + 16}{x + 8}$

7. $\dfrac{r^2 - 8r + 15}{r - 3}$

8. $\dfrac{t^2 - 3t - 10}{t - 5}$

9. $\dfrac{12m^2 - 20m + 3}{2m - 3}$

10. $\dfrac{2y^2 - 5y - 3}{2y + 1}$

11. $\dfrac{2a^2 - 11a + 16}{2a + 3}$

12. $\dfrac{9w^2 + 6w + 10}{3w - 2}$

13. $\dfrac{15m^2 + 34m + 28}{5m + 3}$

14. $\dfrac{2x^3 - x^2 + 3x + 2}{2x + 1}$

15. $\dfrac{12t^3 - 11t^2 + 9t + 18}{4t + 3}$

16. $\dfrac{8k^4 - 12k^3 - 2k^2 + 7k - 6}{2k - 3}$

17. $\dfrac{27r^4 - 36r^3 - 6r^2 + 26r - 24}{3r - 4}$

Perform each division. See Examples 2 and 3.

18. $\dfrac{3y^3 + y^2 + 2}{y + 1}$

19. $\dfrac{2r^3 - 6r - 36}{r - 3}$

20. $\dfrac{3k^3 - 4k^2 - 6k + 10}{k^2 - 2}$

21. $\dfrac{5z^3 - z^2 + 10z + 2}{z^2 + 2}$

22. $\dfrac{x^4 - x^2 - 2}{x^2 - 2}$

23. $\dfrac{r^4 - 2r^2 + 5}{r^2 - 1}$

24. $\dfrac{6p^4 - 15p^3 + 14p^2 - 5p + 10}{3p^2 + 1}$

25. $\dfrac{6r^4 - 10r^3 - r^2 + 15r - 8}{2r^2 - 3}$

26. $\dfrac{4m^5 - 11m^4 - 6m^3 + 5m^2 - 11m + 6}{4m^2 + m - 3}$

27. $\dfrac{2x^5 + 9x^4 + 8x^3 + 10x^2 + 14x + 5}{2x^2 + 3x + 1}$

28. $\dfrac{y^3 + 1}{y + 1}$

29. $\dfrac{x^4 - 1}{x^2 - 1}$

SKILL SHARPENERS

List all positive integer factors of each number. See Section 1.1.

30. 12 **31.** 18 **32.** 20 **33.** 36

34. 50 **35.** 70 **36.** 29 **37.** 41

CHAPTER 4 SUMMARY

KEY TERMS

4.3 **scientific notation** A number written as $a \times 10^n$, where $1 \le |a| < 10$ and n is an integer, is in scientific notation.

4.4 **polynomial** A polynomial is the sum of a finite number of terms.

degree of a term The degree of a term with one variable is the exponent on the variable.

degree of a polynomial The degree of a polynomial in one variable is the highest exponent found in any term of the polynomial.

trinomial A trinomial is a polynomial with three terms.

binomial A binomial is a polynomial with two terms.

monomial A monomial is a polynomial with one term.

NEW SYMBOLS

x^{-n} x to the negative n

QUICK REVIEW

Section	Concepts	Examples
4.1 Exponents	For any integers m and n: **Product rule** $a^m \cdot a^n = a^{m+n}$ **Power rules (a)** $(a^m)^n = a^{mn}$ **(b)** $(ab)^m = a^m b^m$ **(c)** $\left(\frac{a}{b}\right)^m = \frac{a^m}{b^m}$ $(b \ne 0)$	$2^4 \cdot 2^5 = 2^9$ $(3^4)^2 = 3^8$ $(6a)^5 = 6^5 a^5$ $\left(\frac{2}{3}\right)^4 = \frac{2^4}{3^4}$
4.2 The Quotient Rule and Integer Exponents	If $a \ne 0$, for integers m and n: **Zero exponent** $a^0 = 1$ **Negative exponent** $a^{-n} = \frac{1}{a^n}$ **Quotient rule** $\frac{a^m}{a^n} = a^{m-n}$	$15^0 = 1$ $5^{-2} = \frac{1}{5^2} = \frac{1}{25}$ $\frac{4^8}{4^3} = 4^5$
4.3 An Application of Exponents: Scientific Notation	To write a number in scientific notation (as $a \times 10^n$), move the decimal point to follow the first nonzero digit. The number of digits the decimal point is moved is the absolute value of n. If moving the decimal point makes the number smaller, n is positive. Otherwise, n is negative. If the decimal point is not moved, n is 0.	$247 = 2.47 \times 10^2$ $0.0051 = 5.1 \times 10^{-3}$ $4.8 = 4.8 \times 10^0$

Section	Concepts	Examples
4.4 Polynomials	**Addition:** Add like terms.	$\begin{array}{r} 2x^2 + 5x - 3 \\ \underline{5x^2 - 2x + 7} \\ 7x^2 + 3x + 4 \end{array}$
	Subtraction: Change the signs of the terms in the second polynomial and add to the first polynomial.	$(2x^2 + 5x - 3) - (5x^2 - 2x + 7)$ $= (2x^2 + 5x - 3) + (-5x^2 + 2x - 7)$ $= -3x^2 + 7x - 10$
4.5 Multiplication of Polynomials	Multiply each term of the first polynomial by each term of the second polynomial. Then add like terms.	$\begin{array}{rrrrr} & 3x^3 & - 4x^2 & + 2x & - 7 \\ & & & 4x & + 3 \\ \hline & 9x^3 & - 12x^2 & + 6x & - 21 \\ 12x^4 & - 16x^3 & + 8x^2 & - 28x & \\ \hline 12x^4 & - 7x^3 & - 4x^2 & - 22x & - 21 \end{array}$
4.6 Products of Binomials	**FOIL Method** *Step 1* Multiply the two first terms to get the first term of the answer.	Find $(2x + 3)(5x - 4)$. $2x(5x) = 10x^2$
	Step 2 Find the outer product and the inner product and mentally add them, when possible, to get the middle term of the answer.	$2x(-4) + 3(5x) = 7x$
	Step 3 Multiply the two last terms to get the last term of the answer.	$3(-4) = -12$ The product of $(2x + 3)$ and $(5x - 4)$ is $10x^2 + 7x - 12$.
	Square of a Binomial $(a + b)^2 = a^2 + 2ab + b^2$ $(a - b)^2 = a^2 - 2ab + b^2$	$(3x + 1)^2 = 9x^2 + 6x + 1$ $(2m - 5n)^2 = 4m^2 - 20mn + 25n^2$
	Product of the Sum and Difference of Two Squares $(a + b)(a - b) = a^2 - b^2$	$(4a + 3)(4a - 3) = 16a^2 - 9$
4.7 Dividing a Polynomial by a Monomial	Divide each term of the polynomial by the monomial: $\frac{a + b}{c} = \frac{a}{c} + \frac{b}{c}$	$\frac{4x^3 - 2x^2 + 6x - 8}{2x}$ $= 2x^2 - x + 3 - \frac{4}{x}$
4.8 The Quotient of Two Polynomials	Use "long division."	$\begin{array}{r} 2x - 5 + \frac{-1}{3x + 4} \\ 3x + 4 \overline{) 6x^2 - 7x - 21} \\ \underline{6x^2 + 8x} \\ -15x - 21 \\ \underline{-15x - 20} \\ -1 \end{array}$

name date hour

CHAPTER 4 REVIEW EXERCISES

[4.1] *Evaluate each expression.*

1. $4^2 + 4^3$

2. $5^0 + 7^0$

3. 2^{-4}

4. 6^{-1}

5. $\left(\frac{5}{8}\right)^{-2}$

6. $3^{-2} - 3^0$

Simplify. Write each answer in exponential form, using only positive exponents. Assume all variables are positive.

7. $5^4 \cdot 5^7$

8. $9^3 \cdot 9^{-5}$

9. $(-4)^5 \cdot (-4)^3$

10. $\frac{15^{17}}{15^{12}}$

11. $\frac{5^8}{5^{19}}$

12. $\frac{6^{-3}}{6^{-5}}$

13. $\frac{x^{-7}}{x^{-9}}$

14. $\frac{p^{-8}}{p^4}$

15. $\frac{r^{-2}}{r^{-6}}$

[4.2]

16. $(2^4)^2$

17. $(9^3)^{-2}$

18. $(5^{-2})^{-4}$

19. $(8^{-3})^4$

20. $\frac{(m^2)^3}{(m^4)^2}$

21. $\frac{y^4 \cdot y^{-2}}{y^{-5}}$

22. $\frac{r^9 \cdot r^{-5}}{r^{-2} \cdot r^{-7}}$

23. $(-5m^3)^2$

24. $(2y^{-4})^{-3}$

25. $\frac{ab^{-3}}{a^4b^2}$

26. $\frac{(6r^{-1})^2 \cdot (2r^{-4})}{r^{-5}(r^2)^{-3}}$

27. $\frac{(2m^{-5}n^2)^3(3m^2)^{-1}}{m^{-2}n^{-4}(m^{-1})^2}$

[4.3] *Write each number in scientific notation.*

28. 64,000

29. 15,800,000

30. 26,954,000,000

31. .0004251

32. .0000976

33. .784

Write each number without exponents.

34. 1.2×10^4

35. 6.89×10^8

36. 4.253×10^{-4}

37. 8.77×10^{-1}

38. $(6 \times 10^4) \times (1.5 \times 10^3)$

39. $(2 \times 10^{-3}) \times (4 \times 10^5)$

40. $\dfrac{9 \times 10^{-2}}{3 \times 10^2}$

41. $\dfrac{8 \times 10^4}{2 \times 10^{-2}}$

42. $\dfrac{12 \times 10^{-5} \times 5 \times 10^4}{4 \times 10^3 \times 6 \times 10^{-2}}$

43. $\dfrac{2.5 \times 10^5 \times 4.8 \times 10^{-4}}{7.5 \times 10^8 \times 1.6 \times 10^{-5}}$

[4.4] *Combine terms where possible in the following polynomials. Write answers in descending powers of the variable. Give the degree of the answer.*

	Simplified form	*Degree*
44. $9m^2 + 11m^2 + 2m^2$		
45. $-4p + p^3 - p^2 + 8p + 2$		
46. $2r^4 - r^3 + 8r^4 + r^3 - 6r^4$		
47. $12a^5 - 9a^4 + 8a^3 + 2a^2 - a + 3$		
48. $-7x^5 - 8x - x^5 + x + 9x^3$		
49. $-5z^3 + 7 - 6z^2 + 8z$		

Add or subtract as indicated.

50. Add.

$$\begin{array}{r} -2a^3 + 5a^2 \\ \underline{3a^3 - a^2} \end{array}$$

51. Add.

$$\begin{array}{r} 4r^3 - 8r^2 + 6r \\ \underline{-2r^3 + 5r^2 - 3r} \end{array}$$

52. Subtract.

$$\begin{array}{r} 6y^2 - 8y + 2 \\ \underline{5y^2 + 2y - 7} \end{array}$$

53. Subtract.

$$\begin{array}{r} -12k^4 - 8k^2 + 7k - 5 \\ \underline{k^4 + 7k^2 - 11k + 1} \end{array}$$

54. $(2m^3 - 8m^2 + 4) + (3m^3 + 2m^2 - 7)$

55. $(-5y^2 + 3y - 11) + (4y^2 - 7y + 15)$

name date hour

56. $(6p^2 - p - 8) - (-4p^2 + 2p - 3)$

57. $(12r^4 - 7r^3 + 2r^2) - (5r^4 - 3r^3 + 2r^2 - 1)$

[4.5] *Find each product.*

58. $5x(2x - 11)$

59. $-3p^3(2p^2 - 5p)$

60. $(m - 9)(m + 2)$

61. $(3k - 6)(2k + 1)$

62. $(3r - 2)(2r^2 + 4r - 3)$

63. $(5p + 3)(p + 2)(p - 1)$

64. $(r + 2)^3$

[4.6] *Find each product.*

65. $(3k + 1)(2k - 3)$

66. $(a + 3b)(2a - b)$

67. $(6k + 5q)(2k - 7q)$

68. $(a + 4)^2$

69. $(3p - 2)^2$

70. $(2r + 5s)^2$

71. $(8z - 3y)^2$

72. $(6m - 5)(6m + 5)$

73. $(2z + 7)(2z - 7)$

74. $(5a + 6b)(5a - 6b)$

75. $(9y + 8z)(9y - 8z)$

[4.7] *Perform each division.*

76. $\dfrac{-15y^4}{9y^2}$

77. $\dfrac{12x^3y^2}{6xy}$

78. $\dfrac{6y^4 - 12y^2 + 18y}{6y}$

79. $\dfrac{2p^3 - 6p^2 + 5p}{2p^2}$

80. $(-10m^4n^2 + 5m^3n^3 + 6m^2n^4) \div (5m^2n)$

[4.8] *Perform each division.*

81. $(2r^2 + 3r - 14) \div (r - 2)$

82. $\dfrac{12m^2 - 11m - 10}{3m - 5}$

83. $\dfrac{10a^3 + 9a^2 - 14a + 9}{5a - 3}$

84. $\dfrac{2k^4 + 3k^3 + 9k^2 - 8}{2k + 1}$

85. $\dfrac{2m^4 + 4m^3 - 4m^2 - 12m + 6}{2m^2 - 3}$

MIXED REVIEW EXERCISES

Simplify. Perform the indicated operations. Write with positive exponents. Assume all variables represent nonzero real numbers.

86. $(3p)^4(3p^{-7})$

87. 7^{-2}

88. $(-7 + 2k)^2$

89. $\dfrac{2y^3 + 17y^2 + 37y + 7}{2y + 7}$

90. $\left(\dfrac{6r^2s}{5}\right)^3$

91. $-m^5(8m^2 - 10m + 6)$

92. $\left(\dfrac{1}{2}\right)^{-5}$

93. $(25x^2y^3 - 8xy^2 + 15x^3y) \div (10x^2y^3)$

94. $(6r^{-2})^{-1}$

95. $-8 - [6 - (3 + p)]$

96. $2^{-1} + 4^{-1}$

97. $(a + 2)(a^2 - 4a + 1)$

98. $(5y^3 - 8y^2 + 7) - (-3y^3 + y^2 + 2)$

99. $(2r + 5)(5r - 2)$

100. $(12a + 1)(12a - 1)$

101. $2y^2(-11y^2 + 2y + 9)$

name date hour

CHAPTER 4 TEST

Evaluate each expression.

1. 3^{-4} — 1. ______

2. -17^0 — 2. ______

Simplify. Write each answer using only positive exponents. Assume all variables are positive.

3. $5^{-3} \cdot 5^2$ — 3. ______

4. $\dfrac{8^{-4}}{8^{-9}}$ — 4. ______

5. $\dfrac{(p^{-2})^{-3}(p^4)^2}{(p^{-5})^2}$ — 5. ______

6. $\left(\dfrac{a^2b^3}{a^{-3}b}\right)^{-2}$ — 6. ______

Write each number in scientific notation.

7. 245,000,000 — 7. ______

8. .000379 — 8. ______

Simplify, writing without exponents.

9. 4.8×10^{-3} — 9. ______

10. $\dfrac{8 \times 10^{-4}}{2 \times 10^{-6}}$ — 10. ______

For each polynomial, add like terms and then give the degree of the polynomial. Finally, select the most specific description from this list: trinomial, binomial, monomial, or none of these.

11. $3x^2 + 6x - 4x^2$ — 11. ______

12. $11m^3 - m^2 + m^4 + m^4 - 7m^2$ — 12. ______

13. $5x^3 - 4x^2 + 2x - 1$ — 13. ______

Perform the indicated operations.

14. ______

14. $(2x^5 - 4x + 7) - (x^5 + x^2 - 2x - 5)$

15. ______

15. $(y^2 - 5y - 3) + (3y^2 + 2y) - (y^2 - y - 1)$

16. ______

16. $6m^2(m^3 + 2m^2 - 3m + 7)$

17. ______

17. $(r - 5)(r + 2)$

18. ______

18. $(3t + 4w)(2t - 3w)$

19. ______

19. $(5r - 3s)^2$

20. ______

20. $(6p - 8q)(6p + 8q)$

21. ______

21. $(2x - 3)(x^2 + 2x - 5)$

22. ______

22. $\dfrac{9y^3 - 15y^2 + 6y}{3y}$

23. ______

23. $(10r^3 + 25r^2 - 15r + 8) \div (5r)$

24. ______

24. $\dfrac{12y^2 - 15y - 11}{4y + 3}$

25. ______

25. $\dfrac{3x^3 - 2x^2 - 6x - 4}{x - 2}$

FACTORING

5.1 FACTORS: THE GREATEST COMMON FACTOR

OBJECTIVES

1. List the factors of an integer.
2. Identify prime numbers.
3. Find the greatest common factor of a list of integers, or a list of terms.
4. Factor out the greatest common factor.
5. Factor by grouping.

1 Prime factors of whole numbers were discussed in Section 1.1. Now the idea of factoring is extended to the set of integers. Recall that 6 and 2 are **factors** of 12 because the product of 6 and 2 is 12. The expression $6 \cdot 2$ is a **factored form** of 12. To **factor** means to write a quantity as a product. That is, factoring is the opposite of multiplication, as shown in the following example.

Multiplication		*Factoring*
$6 \cdot 2 = 12$	and	$12 = 6 \cdot 2$
↑ ↑ Factors ↑ Product		↑ Product ↑ ↑ Factors

Other factored forms of 12 are $(-6)(-2)$, $3 \cdot 4$, $(-3)(-4)$, $12 \cdot 1$, and $(-12)(-1)$. More than two factors may be used, so another factored form of 12 is $2 \cdot 2 \cdot 3$. The positive integer factors of 12 are

$$1, 2, 3, 4, 6, 12.$$

DEFINITION OF FACTOR

The integer a is a **factor** of b if b can be divided by a with a remainder of zero.

EXAMPLE 1 Listing Positive Integer Factors

(a) The positive integer factors of 36 are 1, 2, 3, 4, 6, 9, 12, 18, and 36.

(b) The positive integer factors of 11 are 1 and 11. ■

WORK PROBLEM 1 AT THE SIDE.

2 As shown in Example 1(b), the only positive integer factors of 11 are 11 and 1. Recall that a positive integer (or natural number) greater than 1 having only itself and 1 as factors is a **prime number.** The first few prime numbers are

$$2, 3, 5, 7, 11, 13, 17, 19, 23, 29, 31, 37, 41, 43.$$

A positive integer (other than 1) that is not prime is called a **composite number.** The number 1 is neither composite nor prime.

1. List the positive integer factors of each number.

(a) 24

(b) 40

(c) 7

(d) 19

ANSWERS
1. (a) 1, 2, 3, 4, 6, 8, 12, 24
(b) 1, 2, 4, 5, 8, 10, 20, 40
(c) 1,7 (d) 1, 19

2. Tell whether each number is prime or composite.

(a) 12

(b) 13

(c) 27

(d) 59

(e) 1,806,954

ANSWERS
2. (b) and (d) are prime; the others are composite.

EXAMPLE 2 Distinguishing Between Prime and Composite Numbers

Tell whether each number is prime or composite.

(a) 33

This number has factors of 3 and 11, as well as 1 and 33, so it is composite.

(b) 53

Try dividing 53 by various integers. It is divisible only by itself and 1, so it is prime.

(c) 14,976,083,922

This number is even and therefore divisible by 2. It is composite. ■

WORK PROBLEM 2 AT THE SIDE.

Each composite number may be expressed as a product of primes. For example,

$$30 = 2 \cdot 3 \cdot 5 \qquad 55 = 5 \cdot 11 \qquad 72 = 2^3 \cdot 3^2$$

and so on. A number written as a product of prime factors is in **prime factored form.** (There is exactly one way to do this for any composite number.)

To decide whether the first three prime numbers are factors of a given number, the following divisibility tests are helpful. (The tests for larger prime numbers are not as simple, so we do not include them here.)

DIVISIBILITY TESTS

Number	*Test*	*Examples*
2	The last digit is even.	2 is a factor of 126, 9432, and 680.
3	The sum of the digits is divisible by 3.	3 is a factor of 414 because $4 + 1 + 4 = 9$ and 3 is a factor of 9.
5	The last digit is 0 or 5.	5 is a factor of 620, 125, and 935.

EXAMPLE 3 Writing a Number in Prime Factored Form

Write each number in prime factored form.

(a) 50

Divide 50 by the first prime, 2.

$$50 = 2 \cdot 25$$

We cannot divide 25 by 2, or by the next prime, 3, but we can divide it by 5.

$$50 = 2 \cdot 5 \cdot 5$$

$$50 = 2 \cdot 5^2 \qquad \text{Prime factored form}$$

(b) $300 = 2 \cdot \mathbf{150}$
$= 2 \cdot 2 \cdot \mathbf{75}$
$= 2 \cdot 2 \cdot 3 \cdot \mathbf{25}$
$= 2 \cdot 2 \cdot 3 \cdot 5 \cdot \mathbf{5}$
$= 2^2 \cdot 3 \cdot 5^2$ Prime factored form

(c) 71

Since 71 is prime, its prime factored form is 71. ■

WORK PROBLEM 3 AT THE SIDE.

A table giving the prime factored form of all positive integers from 2 through 100 follows.

PRIME FACTORED FORM OF THE NUMBERS 2 THROUGH 100

$2 = 2$	$26 = 2 \cdot 13$	$51 = 3 \cdot 17$	$76 = 2^2 \cdot 19$
$3 = 3$	$27 = 3^3$	$52 = 2^2 \cdot 13$	$77 = 7 \cdot 11$
$4 = 2^2$	$28 = 2^2 \cdot 7$	$53 = 53$	$78 = 2 \cdot 3 \cdot 13$
$5 = 5$	$29 = 29$	$54 = 2 \cdot 3^3$	$79 = 79$
$6 = 2 \cdot 3$	$30 = 2 \cdot 3 \cdot 5$	$55 = 5 \cdot 11$	$80 = 2^4 \cdot 5$
$7 = 7$	$31 = 31$	$56 = 2^3 \cdot 7$	$81 = 3^4$
$8 = 2^3$	$32 = 2^5$	$57 = 3 \cdot 19$	$82 = 2 \cdot 41$
$9 = 3^2$	$33 = 3 \cdot 11$	$58 = 2 \cdot 29$	$83 = 83$
$10 = 2 \cdot 5$	$34 = 2 \cdot 17$	$59 = 59$	$84 = 2^2 \cdot 3 \cdot 7$
$11 = 11$	$35 = 5 \cdot 7$	$60 = 2^2 \cdot 3 \cdot 5$	$85 = 5 \cdot 17$
$12 = 2^2 \cdot 3$	$36 = 2^2 \cdot 3^2$	$61 = 61$	$86 = 2 \cdot 43$
$13 = 13$	$37 = 37$	$62 = 2 \cdot 31$	$87 = 3 \cdot 29$
$14 = 2 \cdot 7$	$38 = 2 \cdot 19$	$63 = 3^2 \cdot 7$	$88 = 2^3 \cdot 11$
$15 = 3 \cdot 5$	$39 = 3 \cdot 13$	$64 = 2^6$	$89 = 89$
$16 = 2^4$	$40 = 2^3 \cdot 5$	$65 = 5 \cdot 13$	$90 = 2 \cdot 3^2 \cdot 5$
$17 = 17$	$41 = 41$	$66 = 2 \cdot 3 \cdot 11$	$91 = 7 \cdot 13$
$18 = 2 \cdot 3^2$	$42 = 2 \cdot 3 \cdot 7$	$67 = 67$	$92 = 2^2 \cdot 23$
$19 = 19$	$43 = 43$	$68 = 2^2 \cdot 17$	$93 = 3 \cdot 31$
$20 = 2^2 \cdot 5$	$44 = 2^2 \cdot 11$	$69 = 3 \cdot 23$	$94 = 2 \cdot 47$
$21 = 3 \cdot 7$	$45 = 3^2 \cdot 5$	$70 = 2 \cdot 5 \cdot 7$	$95 = 5 \cdot 19$
$22 = 2 \cdot 11$	$46 = 2 \cdot 23$	$71 = 71$	$96 = 2^5 \cdot 3$
$23 = 23$	$47 = 47$	$72 = 2^3 \cdot 3^2$	$97 = 97$
$24 = 2^3 \cdot 3$	$48 = 2^4 \cdot 3$	$73 = 73$	$98 = 2 \cdot 7^2$
$25 = 5^2$	$49 = 7^2$	$74 = 2 \cdot 37$	$99 = 3^2 \cdot 11$
	$50 = 2 \cdot 5^2$	$75 = 3 \cdot 5^2$	$100 = 2^2 \cdot 5^2$

3 An integer that is a factor of two or more integers is a **common factor** of those integers. For example, 6 is a common factor of 18 and 24 since 6 is a factor of both 18 and 24. Other common factors of 18 and 24 are 1, 2, and 3. The **greatest common factor** of a list of integers is the largest common factor of those integers. This means 6 is the greatest common factor of 18 and 24, since it is the largest of their common factors.

Find the greatest common factor of a list of numbers as follows.

3. Write each number in prime factored form.

(a) 70

(b) 180

(c) 400

(d) 97

ANSWERS
3. (a) $2 \cdot 5 \cdot 7$ (b) $2^2 \cdot 3^2 \cdot 5$ (c) $2^4 \cdot 5^2$ (d) 97

4. Find the greatest common factor for each group of numbers.

(a) 30, 20, 15

(b) 42, 28, 35

(c) 12, 18, 26, 32

(d) 10, 15, 21

FINDING THE GREATEST COMMON FACTOR

Step 1 Write each number in prime factored form.

Step 2 List each prime number that is a factor of every number in the list.

Step 3 Use as exponents on the prime factors the *smallest* exponent from the prime factored forms. (If a prime does not appear in one of the prime factored forms, it cannot appear in the greatest common factor.)

Step 4 Multiply together the primes from Step 3. If there are no primes left after Step 3, the greatest common factor is 1.

EXAMPLE 4 Finding the Greatest Common Factor for Numbers

Find the greatest common factor for each group of numbers.

(a) 30, 45

First write each number in prime factored form.

$$30 = 2 \cdot 3 \cdot 5 \qquad 45 = 3^2 \cdot 5$$

Use each prime the *least* number of times it appears in all the factored forms. There is no 2 in the prime factored form of 45, so there will be no 2 in the greatest common factor. The least number of times 3 appears in all the factored forms is 1; the least number of times 5 appears is also 1. From this, the greatest common factor is

$$3^1 \cdot 5^1 = 3 \cdot 5 = 15.$$

(b) 72, 120, 432

Find the prime factored form of each number.

$$72 = 2^3 \cdot 3^2 \qquad 120 = 2^3 \cdot 3 \cdot 5 \qquad 432 = 2^4 \cdot 3^3$$

The least number of times 2 appears in all the factored forms is 3, and the least number of times 3 appears is 1. There is no 5 in the prime factored form of either 72 or 432, so the greatest common factor is

$$2^3 \cdot 3 = 24.$$

(c) 10, 11, 14

Write the prime factored form of each number.

$$10 = 2 \cdot 5 \qquad 11 = 11 \qquad 14 = 2 \cdot 7$$

There are no primes common to all three numbers, so the greatest common factor is 1. ■

WORK PROBLEM 4 AT THE SIDE.

The greatest common factor can also be found for a list of terms. For example, the terms x^4, x^5, x^6, and x^7 have x^4 as the greatest common factor, because 4 is the smallest exponent on x.

CAUTION The exponent on a variable in the greatest common factor is the *smallest* exponent that appears on that variable.

ANSWERS

4. (a) 5 (b) 7 (c) 2 (d) 1

EXAMPLE 5 Finding Greatest Common Factors for Variable Terms

Find the greatest common factor for each list of terms.

(a) $21m^7, \ -18m^6, \ 45m^8, \ -24m^5$

First, 3 is the greatest common factor of the coefficients 21, -18, 45, and -24. The smallest exponent on m is 5, so the greatest common factor of the terms is $3m^5$.

(b) $x^2y^2, \ x^7y^5, \ x^3y^7, \ y^{15}$

There is no x in the last term, y^{15}, so x will not appear in the greatest common factor. There is a y in each term, however, and 2 is the smallest exponent of y. The greatest common factor is y^2.

(c) $-a^2b, \ -ab^2$

Write $-a^2b$ as $-1a^2b$ and $-ab^2$ as $-1ab^2$. The factors of -1 are -1 and 1. Since $1 > -1$, the greatest common factor is $1ab$ or ab. ■

WORK PROBLEM 5 AT THE SIDE.

4 The idea of a greatest common factor can be used to write a polynomial in factored form. For example, the polynomial

$$3m + 12$$

consists of the two terms $3m$ and 12. The greatest common factor for these two terms is 3. Write $3m + 12$ so that each term is a product with 3 as one factor.

$$3m + 12 = \mathbf{3} \cdot m + \mathbf{3} \cdot 4$$

Now use the distributive property.

$$3m + 12 = \mathbf{3} \cdot m + \mathbf{3} \cdot 4 = \mathbf{3}(m + 4)$$

The factored form of $3m + 12$ is $3(m + 4)$. This process is called **factoring out the greatest common factor.**

EXAMPLE 6 Factoring Out the Greatest Common Factor.

Factor out the greatest common factor.

(a) $20m^5 + 10m^4 + 15m^3$

Look for the greatest common factor of the numbers first, then look at the variables. The greatest common factor for the terms of this polynomial is $5m^3$.

$$\begin{aligned} 20m^5 + 10m^4 + 15m^3 &= (\mathbf{5m^3})(4m^2) + (\mathbf{5m^3})(2m) + (\mathbf{5m^3})3 \\ &= \mathbf{5m^3}(4m^2 + 2m + 3) \end{aligned}$$

Check this work by multiplying $5m^3$ and $4m^2 + 2m + 3$. You should get the original polynomial as your answer.

(b) $\begin{aligned} 48y^{12} - 36y^{10} + 12y^7 &= (12y^7)(4y^5) - (12y^7)(3y^3) + (12y^7)1 \\ &= 12y^7(4y^5 - 3y^3 + 1) \end{aligned}$

(c) $x^5 + x^3 = (x^3)x^2 + (x^3)1 = x^3(x^2 + 1)$

(d) $20m^7p^2 - 36m^3p^4 = 4m^3p^2(5m^4 - 9p^2)$ ■

5. Find the greatest common factor for each group of terms.

(a) $6m^4, \ 9m^2, \ 12m^5$

(b) $25y^{11}, \ 30y^7$

(c) $-12p^5, \ -18q^4$

(d) $-11r^9, \ -10r^{15}, \ -8r^{12}$

(e) $y^4z^2, \ y^6z^8, \ z^9$

(f) $12p^{11}, \ 17q^5$

ANSWERS
5. (a) $3m^2$ (b) $5y^7$ (c) 6 (d) r^9 (e) z^2 (f) 1

6. Factor out the greatest common factor.

(a) $32p^2 + 16p + 48$

(b) $10y^5 - 8y^4 + 6y^2$

(c) $27a^5 + 9a^4$

(d) $m^7 + m^9$

(e) $8p^5q^2 + 16p^6q^3 - 12p^4q^7$

7. Factor by grouping.

(a) $pq + 5q + 2p + 10$

(b) $2mn - 8n + 3m - 12$

(c) $6y^2 - 16y - 15y + 40$

ANSWERS

6. (a) $16(2p^2 + p + 3)$
   (b) $2y^2(5y^3 - 4y^2 + 3)$
   (c) $9a^4(3a + 1)$ (d) $m^7(1 + m^2)$
   (e) $4p^4q^2(2p + 4p^2q - 3q^5)$
7. (a) $(p + 5)(q + 2)$
   (b) $(m - 4)(2n + 3)$
   (c) $(3y - 8)(2y - 5)$

CAUTION Be careful to avoid the common error of leaving out the 1 in a problem like Example 6(b). Always be sure that the factored form can be multiplied out to yield the original polynomial.

WORK PROBLEM 6 AT THE SIDE.

5 Common factors are used in **factoring by grouping,** explained in the next example.

EXAMPLE 7 Factoring by Grouping

Factor by grouping.

(a) $2x + 6 + ax + 3a$

The first two terms have a common factor of 2, and the last two terms have a common factor of a.

$$(2x + 6) + (ax + 3a) = \mathbf{2}(x + 3) + \mathbf{a}(x + 3)$$

Note that there are now only *two* terms, $2(x + 3)$ and $a(x + 3)$. These two terms have a common factor of $x + 3$.

$$\begin{aligned} 2x + 6 + ax + 3a &= 2(\mathbf{x + 3}) + a(\mathbf{x + 3}) \\ &= (\mathbf{x + 3})(2 + a) \end{aligned}$$

The expression $2(x + 3) + a(x + 3)$ is *not* the factored form, since it is a *sum* (as indicated by the + sign).

(b) $m^2 + 6m + 2m + 12 = m(m + 6) + 2(m + 6)$
$= (m + 6)(m + 2)$

(c) $6y^2 - 21y - 8y + 28 = 3y(2y - 7) - 4(2y - 7)$
$= (2y - 7)(3y - 4)$

Since the quantities in parentheses in the second step must be the same, it was necessary in part (c) to factor out -4 rather than 4. ■

WORK PROBLEM 7 AT THE SIDE.

Use these steps when factoring by grouping.

FACTORING BY GROUPING

Step 1 Write the four terms so that the first two have a common factor and the last two have a common factor.

Step 2 Use the distributive property to factor each group of two terms.

Step 3 If possible, factor a common binomial factor from the results of Step 2.

Step 4 If Step 2 does not result in a common binomial factor, try grouping the terms of the original polynomial in a different way.

name date hour

5.1 EXERCISES

Find all positive integer factors of each number. See Example 1.

1. 27 **2.** 35 **3.** 60 **4.** 72

5. 100 **6.** 130 **7.** 29 **8.** 37

Find the prime factored form for each of the following. See Example 3.

9. 120 **10.** 150 **11.** 180 **12.** 225

13. 275 **14.** 350 **15.** 475 **16.** 650

Find the greatest common factor for each set of terms. See Examples 4 and 5.

17. $12y$, 24 **18.** $72m$, 12 **19.** $30p^2$, $20p^3$, $40p^5$

20. $14r^5$, $28r^2$, $56r^8$ **21.** $18m^2n^2$, $36m^4n^5$, $12m^3n$ **22.** $50p^5r^2$, $25p^4r^7$, $30p^7r^8$

Complete the factoring.

23. $12 = 6(\quad)$ **24.** $18 = 9(\quad)$ **25.** $3x^2 = 3x(\quad)$

26. $8x^3 = 8x(\quad)$ **27.** $9m^4 = 3m^2(\quad)$ **28.** $12p^5 = 6p^3(\quad)$

29. $-8z^9 = -4z^5(\quad)$ **30.** $-15k^{11} = -5k^8(\quad)$ **31.** $6m^4n^5 = 3m^3n(\quad)$

32. $27a^3b^2 = 9a^2b(\quad)$ **33.** $14x^4y^3 = 2xy(\quad)$ **34.** $-16m^3n^3 = 4mn^2(\quad)$

Factor out the greatest common factor for each expression. See Example 6.

35. $12x + 24$ **36.** $18m - 9$ **37.** $9a^2 - 18a$

38. $21m^5 - 14m^4$ **39.** $65y^9 - 35y^5$ **40.** $100a^4 + 16a^2$

41. $11z^2 - 100$ **42.** $12z^2 - 11y^4$ **43.** $19y^3p^2 + 38y^2p^3$

44. $4mn^2 - 12m^2n$ **45.** $13y^6 + 26y^5 - 39y^3$ **46.** $5x^4 + 25x^3 - 20x^2$

47. $45q^4p^5 - 36qp^6 + 81q^2p^3$ **48.** $a^5 + 2a^5b + 3a^5b^2 - 4a^5b^3$ **49.** $125z^5a^3 - 60z^4a^5 + 85z^3a^4$

Factor by grouping. See Example 7.

50. $p^2 + 4p + 3p + 12$ **51.** $m^2 + 2m + 5m + 10$ **52.** $a^2 - 2a + 5a - 10$

53. $y^2 - 6y + 4y - 24$ **54.** $7z^2 + 14z - az - 2a$ **55.** $8k^2 + 6kq + 12kq + 9q^2$

56. $5m^2 + 15mp - 2mp - 6p^2$ **57.** $18r^2 + 12ry - 3xr - 2xy$ **58.** $3a^3 + 3ab^2 + 2a^2b + 2b^3$

59. $16m^3 - 4m^2p^2 - 4mp + p^3$ **60.** $1 - a + ab - b$ **61.** $2pq^2 - 8q^2 + p - 4$

SKILL SHARPENERS

Find each product. See Section 4.6.

62. $(x + 2)(x + 5)$ **63.** $(y + 7)(y - 3)$ **64.** $(y - 4)(y - 8)$

65. $(q - 5)(q - 7)$ **66.** $(p + 10)(p - 10)$ **67.** $(a - 7)(a + 7)$

5.2 FACTORING TRINOMIALS

1 The product of the polynomials $k - 3$ and $k + 1$ is

$$(k - 3)(k + 1) = k^2 - 2k - 3.$$

The polynomial $k^2 - 2k - 3$ can be rewritten as the product $(k - 3)(k + 1)$. The product is called the *factored form* of $k^2 - 2k - 3$. The discussion of factoring in this section is limited to trinomials like $x^2 - 2x - 24$ or $y^2 + 2y - 15$, where the coefficient of the squared term is 1.

When factoring polynomials with only integer coefficients, use only integers for numerical factors. For example, we can factor $x^2 + 5x + 6$ by finding integers a and b such that

$$x^2 + 5x + 6 = (x + \boxed{a})(x + \boxed{b}).$$

To find these integers a and b, first find the product of the two terms on the right hand side:

$$(x + a)(x + b) = x^2 + ax + bx + ab.$$

Since $ax + bx = (a + b)x$ by the distributive property,

$$x^2 + ax + bx + ab = x^2 + (a + b)x + ab.$$

By this result, we can factor $x^2 + 5x + 6$ by finding integers a and b having a sum of 5 and a product of 6.

Product of a and b is 6.

$$x^2 + \boxed{5}x + \boxed{6} = x^2 + \boxed{(a + b)}x + \boxed{ab}$$

Sum of a and b is 5.

Since many pairs of integers have a sum of 5, it is best to begin by listing those pairs of integers whose product is 6. Both 5 and 6 are positive, so only pairs in which both integers are positive need be considered.

WORK PROBLEM 1 AT THE SIDE.

As in Problem 1 at the side, the numbers 1 and 6 and the numbers 2 and 3 both have a product of 6, but only the pair 2 and 3 has a sum of 5. So 2 and 3 are the needed integers, and

$$x^2 + 5x + 6 = (x + \boxed{2})(x + \boxed{3}).$$

Check by multiplying the binomials. Make sure that the sum of the outer and inner products produces the correct middle term.

$$(x + 2)(x + 3) = x^2 + \boxed{5x} + 6$$

$2x$
$3x$
$5x$

This method of factoring can be used only for trinomials having the coefficient of the squared term equal to 1. Methods for factoring other trinomials will be given in the next section.

OBJECTIVES

1 Factor trinomials with a coefficient of 1 for the squared term.

2 Factor such polynomials after factoring out the greatest common factor.

1. (a) List all pairs of positive integers whose product is 6.

(b) Find the pair from part (a) whose sum is 5.

ANSWERS
1. (a) 1, 6; 2, 3 (b) 2, 3

2. Complete the given lists of numbers; then factor the given trinomial.

(a) $m^2 + 11m + 30$

$ab = 30,$	$a + b = 11$
30, 1	$30 + 1 = 31$
15, 2	$15 + 2 =$ ___
10, 3	$10 + 3 =$ ___
6, 5	$6 + 5 =$ ___

(b) $y^2 + 12y + 20$

$ab = 20,$	$a + b = 12$
20, 1	$20 + 1 = 21$
10, ___	$10 +$ ___ $=$ ___
5, ___	$5 +$ ___ $=$ ___

3. Factor each trinomial.

(a) $p^2 + 7p + 6$

(b) $y^2 + 4y + 3$

(c) $a^2 - 9a - 22$

(d) $r^2 - 6r - 16$

ANSWERS
2. (a) 17; 13; 11; $(m + 6)(m + 5)$
(b) 2; 2; 12; 4; 4; 9; $(y + 10)(y + 2)$
3. (a) $(p + 6)(p + 1)$
(b) $(y + 3)(y + 1)$
(c) $(a - 11)(a + 2)$
(d) $(r - 8)(r + 2)$

EXAMPLE 1 Factoring a Trinomial with Only Positive Terms

Factor $m^2 + 9m + 14$.

Look for two integers whose product is 14 and whose sum is 9. List the pairs of integers whose products are 14. Then examine the sums. Again, only positive integers are needed because all signs in $m^2 + 9m + 14$ are positive.

14, 1	$14 + 1 = 15$	
7, 2	$7 + 2 = \mathbf{9}$	Sum is 9.

From the list, 7 and 2 are the required integers, since $7 \cdot 2 = 14$ and $7 + 2 = 9$. Thus

$$m^2 + 9m + 14 = (m + 2)(m + 7). \quad \blacksquare$$

NOTE In Example 1, the answer also could have been written $(m + 7)(m + 2)$. Because of the commutative property of multiplication, the order of the factors does not matter.

WORK PROBLEM 2 AT THE SIDE.

EXAMPLE 2 Factoring a Trinomial with Two Negative Terms

Factor $p^2 - 2p - 15$.

Find two integers whose product is -15 and whose sum is -2. If these numbers do not come to mind right away, find them (if they exist) by listing all the pairs of integers whose product is -15. Because the last term is negative, the pairs must include one positive integer and one negative integer (to give a negative product).

$15, -1$	$15 + (-1) = 14$	
$5, -3$	$5 + (-3) = 2$	
$-15, 1$	$-15 + 1 = -14$	
$-5, 3$	$-5 + 3 = \mathbf{-2}$	Sum is -2.

The required integers are -5 and 3.

$$p^2 - 2p - 15 = (p - 5)(p + 3) \quad \blacksquare$$

WORK PROBLEM 3 AT THE SIDE.

As shown in the next example, some trinomials cannot be factored using only integer coefficients. Such trinomials are called **prime polynomials.**

EXAMPLE 3 Factoring a Trinomial with One Negative Term

Factor each trinomial.

(a) $x^2 - 5x + 12$

Since the middle term is negative, the sum must be negative. The last term is positive, so both factors must be negative to give a positive product and a negative sum. First, list all pairs of negative integers whose product is 12. Then examine the sums.

$-12, -1$	$-12 + (-1) = -13$
$-6, -2$	$-6 + (-2) = -8$
$-3, -4$	$-3 + (-4) = -7$

None of the pairs of integers has a sum of -5. Because of this, the trinomial $x^2 - 5x + 12$ *cannot be factored using only integer coefficients,* showing that it is a *prime polynomial.*

(b) $k^2 - 8k + 11$

There is no pair of integers whose product is 11 and whose sum is -8, so $k^2 - 8k + 11$ is a prime polynomial. ■

WORK PROBLEM 4 AT THE SIDE.

EXAMPLE 4 Factoring a Trinomial with Two Variables

Factor $z^2 - 2bz - 3b^2$.

Look for two expressions whose product is $-3b^2$ and whose sum is $-2b$. The expressions are $-3b$ and b, so

$$z^2 - 2bz - 3b^2 = (z - 3b)(z + b). \quad ■$$

WORK PROBLEM 5 AT THE SIDE.

2 The trinomial in the next example does not fit the pattern used above. A preliminary step is necessary before we can use the pattern.

EXAMPLE 5 Factoring a Trinomial with a Common Factor

Factor $4x^5 - 28x^4 + 40x^3$.

First, factor out the greatest common factor, $4x^3$.

$$4x^5 - 28x^4 + 40x^3 = \mathbf{4x^3}(x^2 - 7x + 10)$$

Now factor $x^2 - 7x + 10$. The integers -5 and -2 have a product of 10 and a sum of -7. The complete factored form is

$$4x^5 - 28x^4 + 40x^3 = 4x^3(x - 5)(x - 2). \quad ■$$

CAUTION In factoring, always remember to look for a common factor first. Do not forget to include the common factor as part of the answer.

WORK PROBLEM 6 AT THE SIDE.

4. Factor each trinomial, where possible.

(a) $x^2 + x + 1$

(b) $r^2 - 3r - 4$

(c) $m^2 - 2m + 5$

5. Factor each trinomial.

(a) $b^2 - 3ab - 4a^2$

(b) $p^2 + 6pq + 5q^2$

(c) $r^2 - 6rs + 8s^2$

6. Factor each trinomial as completely as possible.

(a) $2p^3 + 6p^2 - 8p$

(b) $3x^4 - 15x^3 + 18x^2$

ANSWERS

4. (a) prime (b) $(r - 4)(r + 1)$ (c) prime
5. (a) $(b - 4a)(b + a)$ (b) $(p + 5q)(p + q)$ (c) $(r - 4s)(r - 2s)$
6. (a) $2p(p + 4)(p - 1)$ (b) $3x^2(x - 3)(x - 2)$

Historical Reflections

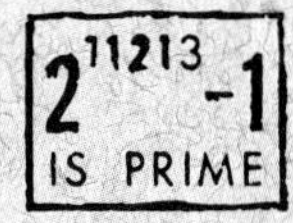

How Many Prime Numbers Are There?

There are infinitely many prime numbers. This fact was proved by Euclid thousands of years ago, and his proof remains a classic; he assumed that there was a largest prime number, and this assumption led to a contradiction. He concluded that there was no largest prime number (or, in other words, infinitely many prime numbers).

At one time,

$$2^{11,213} - 1$$

was the largest known prime. To honor its discovery, the post office at Urbana, Illinois, used the cancellation seen here. But the search for large primes continues. According to the September, 16, 1989, issue of *Science News,* a team of researchers at the Amdahl Corporation's key Computer Laboratories in Fremont, California, discovered that

$$391,581 \times 2^{216,193} - 1$$

is prime. At this writing it is the largest known prime, with 65,087 digits!

name date hour

5.2 EXERCISES

Complete the factoring.

1. $x^2 + 10x + 21 = (x + 7)(\qquad)$

2. $p^2 + 11p + 30 = (p + 5)(\qquad)$

3. $r^2 + 15r + 56 = (r + 7)(\qquad)$

4. $x^2 + 15x + 44 = (x + 4)(\qquad)$

5. $t^2 - 14t + 24 = (t - 2)(\qquad)$

6. $x^2 - 9x + 8 = (x - 1)(\qquad)$

7. $x^2 - 12x + 32 = (x - 4)(\qquad)$

8. $y^2 - 2y - 15 = (y + 3)(\qquad)$

9. $m^2 + 2m - 24 = (m - 4)(\qquad)$

10. $x^2 + 9x - 22 = (x - 2)(\qquad)$

11. $p^2 + 7p - 8 = (p + 8)(\qquad)$

12. $y^2 - 7y - 18 = (y + 2)(\qquad)$

13. $x^2 - 7xy + 10y^2 = (x - 2y)(\qquad)$

14. $k^2 - 3kh - 28h^2 = (k - 7h)(\qquad)$

Factor completely. If a polynomial cannot be factored, write prime. *See Examples 1-3.*

15. $y^2 + 9y + 8$

16. $a^2 + 9a + 20$

17. $b^2 + 8b + 15$

18. $x^2 - 6x - 7$

19. $m^2 + m - 20$

20. $p^2 + 4p + 5$

21. $n^2 + 4n - 12$

22. $y^2 - 6y + 8$

23. $r^2 - r - 30$

24. $s^2 + 2s - 35$

25. $h^2 + 11h + 12$

26. $n^2 - 12n - 35$

27. $a^2 - 2a - 99$

28. $b^2 - 11b + 24$

29. $x^2 - 9x + 20$

30. $k^2 - 10k + 25$

31. $z^2 - 14z + 49$

32. $y^2 - 12y - 45$

Factor completely. See Examples 4 and 5.

33. $x^2 + 4ax + 3a^2$

34. $x^2 - mx - 6m^2$

35. $y^2 - by - 30b^2$

36. $z^2 + 2zx - 15x^2$

37. $x^2 + xy - 30y^2$

38. $a^2 - ay - 56y^2$

39. $r^2 - 2rs + s^2$

40. $m^2 - 2mn - 3n^2$

41. $p^2 - 3pq - 10q^2$

42. $c^2 - 5cd + 4d^2$

43. $3m^3 + 12m^2 + 9m$

44. $3y^5 - 18y^4 + 15y^3$

45. $6a^2 - 48a - 120$

46. $h^7 - 5h^6 - 14h^5$

47. $3j^3 - 30j^2 + 72j$

48. $2x^6 - 8x^5 - 42x^4$

49. $3x^4 - 3x^3 - 90x^2$

50. $2y^3 - 8y^2 - 10y$

51. $a^5 + 3a^4b - 4a^3b^2$

52. $m^3n - 2m^2n^2 - 3mn^3$

53. $y^3z + y^2z^2 - 6yz^3$

54. Use the FOIL method from Section 4.6 to show that $(2x + 4)(x - 3) = 2x^2 - 2x - 12$. Why, then, is it incorrect to completely factor $2x^2 - 2x - 12$ as $(2x + 4)(x - 3)$?

55. Why is it incorrect to completely factor $3x^2 + 9x - 12$ as the product $(x - 1)(3x + 12)$?

56. What polynomial can be factored to give $(y - 7)(y + 6)$?

57. What polynomial can be factored to give $(a + 9)(a + 6)$?

SKILL SHARPENERS

Find each product. See Section 4.6.

58. $(2y - 7)(y + 4)$

59. $(3a + 2)(2a + 1)$

60. $(5z + 2)(3z - 2)$

61. $(4m - 3)(2m + 5)$

62. $(4p + 1)(2p - 3)$

63. $(6r - 2)(3r + 1)$

5.3 MORE ON FACTORING TRINOMIALS

1 Trinomials such as $2x^2 + 7x + 6$, in which the coefficient of the squared term is *not* 1, are factored with an extension of the method presented in the last section. Recall that a trinomial such as $m^2 + 3m + 2$ is factored by finding two numbers whose product is 2 and whose sum is 3.

To factor $2x^2 + 7x + 6$, find integers a, b, c, and d such that

$$2x^2 + 7x + 6 = (ax + b)(cx + d)$$

where, using FOIL, $ac = 2$, $bd = 6$, and $ad + bc = 7$. The possible factors of $2x^2$ are $2x$ and x, or $-2x$ and $-x$. Since the polynomial has only positive coefficients, use the factors with positive coefficients. Then the factored form of $2x^2 + 7x + 6$ can be set up as

$$2x^2 + 7x + 6 = (2x \quad\;)(x \quad\;).$$

The product 6 can be factored as $6 \cdot 1$, $1 \cdot 6$, $2 \cdot 3$, or $3 \cdot 2$. Try each pair to find the correct choices for b and d.

WORK PROBLEM 1 AT THE SIDE.

In part (b) of the problem at the side, the factor $2x + 6 = 2(x + 3)$, so the binomial $2x + 6$ has a common factor of 2. However, $2x^2 + 7x + 6$ does not have a common factor other than 1. The product $(2x + 6)(x + 1)$ cannot be correct.

NOTE If the original polynomial has no common factor, then none of its binomial factors will either.

Now try 2 and 3 as factors of 6. Because of the common factor of 2 in $2x + 2$, the factored form $(2x + 2)(x + 3)$ will not work. Try $(2x + 3)(x + 2)$.

$$(2x + 3)(x + 2) = 2x^2 + \mathbf{7x} + 6 \qquad \text{Correct}$$

$3x$
$4x$
$\overline{7x}$

Finally, $2x^2 + 7x + 6$ factors as

$$2x^2 + 7x + 6 = (2x + 3)(x + 2).$$

Check by multiplying $2x + 3$ and $x + 2$.

EXAMPLE 1 Factoring a Trinomial

Factor $8p^2 + 14p + 5$.

The number 8 has several possible pairs of factors, but 5 has only 1 and 5 or -1 and -5. For this reason, it is easier to begin by considering the factors of 5. Ignore the negative factors since all coefficients in the trinomial are positive. If $8p^2 + 14p + 5$ can be factored, it will be factored as

$$(\quad\; + 5)(\quad\; + 1).$$

The possible pairs of factors of $8p^2$ are $8p$ and p, or $4p$ and $2p$. Try various combinations.

OBJECTIVES

1. Use trial and error to factor trinomials with the coefficient of the squared term not 1.
2. Factor trinomials using factoring by grouping.

1. Decide whether each factored form is correct or incorrect for $2x^2 + 7x + 6$.

(a) $(2x + 1)(x + 6)$

(b) $(2x + 6)(x + 1)$

ANSWERS
1. (a) incorrect (b) incorrect

2. Factor each trinomial.

(a) $2p^2 + 9p + 9$

(b) $8a^2 + 6a + 1$

(c) $6p^2 + 19p + 10$

(d) $2m^2 + 13m + 21$

3. Factor each trinomial.

(a) $6x^2 + 5x - 4$

(b) $3x^2 - 7x - 6$

(c) $5p^2 + 13p - 6$

(d) $6m^2 - 11m - 10$

ANSWERS
2. (a) $(2p + 3)(p + 3)$
(b) $(4a + 1)(2a + 1)$
(c) $(3p + 2)(2p + 5)$
(d) $(m + 3)(2m + 7)$
3. (a) $(3x + 4)(2x - 1)$
(b) $(3x + 2)(x - 3)$
(c) $(5p - 2)(p + 3)$
(d) $(2m - 5)(3m + 2)$

$(8p + 5)(p + 1) = 8p^2 + \mathbf{13p} + 5$ Incorrect

$5p + 8p = 13p$

$(p + 5)(8p + 1) = 8p^2 + \mathbf{41p} + 5$ Incorrect

$40p + p = 41p$

$(4p + 5)(2p + 1) = 8p^2 + \mathbf{14p} + 5$ Correct

$10p + 4p = 14p$

Finally, $8p^2 + 14p + 5$ factors as $(4p + 5)(2p + 1)$. ■

WORK PROBLEM 2 AT THE SIDE.

EXAMPLE 2 Factoring a Trinomial

Factor $6x^2 - 11x + 3$.

There are several possible pairs of factors for 6, but 3 has only 1 and 3 or -1 and -3, so it is better to begin by factoring 3. The middle term of $6x^2 - 11x + 3$ has a negative coefficient, so negative factors must be considered. Try -3 and -1 as factors of 3:

$$(\quad - 3)(\quad - 1).$$

The factors of $6x^2$ are either $6x$ and x, or $2x$ and $3x$. Let us try $2x$ and $3x$.

$(2x - 3)(3x - 1) = 6x^2 \mathbf{- 11x} + 3$ Correct

$-9x + (-2x) = -11x$

Finally, $6x^2 - 11x + 3 = (2x - 3)(3x - 1)$. ■

EXAMPLE 3 Factoring a Trinomial

Factor $8x^2 + 6x - 9$.

The integer 8 has several possible pairs of factors, as does -9. Since the coefficient of the middle term is small, it is wise to avoid large factors such as 8 or 9. Let us begin by trying 4 and 2 as factors of 8, and 3 and -3 as factors of -9.

$(4x + 3)(2x - 3) = 8x^2 \mathbf{- 6x} - 9$ Incorrect

$6x + (-12x) = -6x$

When the wrong sign is obtained, simply exchange the constants, 3 and -3.

$(4x - 3)(2x + 3) = 8x^2 \mathbf{+ 6x} - 9$ Correct ■

$-6x + 12x = 6x$

WORK PROBLEM 3 AT THE SIDE.

EXAMPLE 4 Factoring a Trinomial with Two Variables
Factor $12a^2 - ab - 20b^2$.

There are several possible pairs of factors of $12a^2$, including $12a$ and a, $6a$ and $2a$, and $3a$ and $4a$, just as there are many possible pairs of factors of $-20b^2$, including $-20b$ and b, $10b$ and $-2b$, $-10b$ and $2b$, $4b$ and $-5b$, and $-4b$ and $5b$. Once again, since the desired middle term is small, it is better to avoid the larger factors. Let us try as factors $6a$ and $2a$ and $4b$ and $-5b$.

$$(6a + 4b)(2a - 5b)$$

This cannot be correct since $6a + 4b$ has a common factor of 2 while the given trinomial has none. Let us try $3a$ and $4a$ with $4b$ and $-5b$.

$$(3a + 4b)(4a - 5b) = 12a^2 + ab - 20b^2 \quad \text{Incorrect}$$

Here the middle term has the wrong sign, so we reverse the middle signs of the factors.

$$(3a - 4b)(4a + 5b) = 12a^2 - ab - 20b^2 \quad \text{Correct} \ \blacksquare$$

WORK PROBLEM 4 AT THE SIDE.

EXAMPLE 5 Factoring a Trinomial with a Common Factor
Factor $28x^5 - 58x^4 - 30x^3$.

First factor out the greatest common factor, $2x^3$.

$$28x^5 - 58x^4 - 30x^3 = 2x^3(14x^2 - 29x - 15)$$

Now try to factor $14x^2 - 29x - 15$. Let us try $7x$ and $2x$ as factors of $14x^2$ and -3 and 5 as factors of -15.

$$(7x - 3)(2x + 5) = 14x^2 + 29x - 15 \quad \text{Incorrect}$$

The middle term differs only in sign, so reverse the middle signs of the two factors.

$$(7x + 3)(2x - 5) = 14x^2 - 29x - 15 \quad \text{Correct}$$

Finally, the factored form of $28x^5 - 58x^4 - 30x^3$ is

$$28x^5 - 58x^4 - 30x^3 = 2x^3(7x + 3)(2x - 5). \ \blacksquare$$

CAUTION Do not forget to include the common factor in the final result.

WORK PROBLEM 5 AT THE SIDE.

2 The rest of this section shows an alternative method of factoring trinomials in which the coefficient of the squared term is not 1. This method uses factoring by grouping, introduced in Section 5.1. In the next example, we use the alternative method to factor $2x^2 + 7x + 6$, the same trinomial factored at the beginning of this section.

Recall that we factor a trinomial such as $m^2 + 3m + 2$ by finding two numbers whose product is 2 and whose sum is 3. To factor $2x^2 + 7x + 6$ by

4. Factor each trinomial.

(a) $2x^2 - 5xy - 3y^2$

(b) $8a^2 + 2ab - 3b^2$

(c) $3r^2 + 8rs + 5s^2$

(d) $6m^2 + 11mn - 10n^2$

5. Factor each polynomial as completely as possible.

(a) $4x^2 - 2x - 30$

(b) $15y^3 + 55y^2 + 30y$

(c) $18p^4 + 63p^3 + 27p^2$

(d) $16m^5 - 8m^4 - 168m^3$

ANSWERS
4. (a) $(2x + y)(x - 3y)$
(b) $(4a + 3b)(2a - b)$
(c) $(3r + 5s)(r + s)$
(d) $(3m - 2n)(2m + 5n)$
5. (a) $2(2x + 5)(x - 3)$
(b) $5y(3y + 2)(y + 3)$
(c) $9p^2(2p + 1)(p + 3)$
(d) $8m^3(2m - 7)(m + 3)$

6. Find two integers whose product is 12 and whose sum is 7.

7. Factor each trinomial by grouping.

(a) $2m^2 + 7m + 3$

(b) $5p^2 - 2p - 3$

(c) $3p^2 - 4p + 1$

(d) $3r^2 + 11r - 4$

(e) $15k^2 - k - 2$

ANSWERS
6. 3 and 4
7. (a) $(2m + 1)(m + 3)$
(b) $(5p + 3)(p - 1)$
(c) $(3p - 1)(p - 1)$
(d) $(3r - 1)(r + 4)$
(e) $(5k - 2)(3k + 1)$

the alternative method, look for two integers whose product is 12 (that is, $2 \cdot 6$) and whose sum is 7.

Sum is 7.

$$2x^2 + 7x + 6$$

Product is $2 \cdot 6 = 12$.

WORK PROBLEM 6 AT THE SIDE.

As the problem at the side shows, the necessary integers are 3 and 4. Use these integers to write the middle term, $7x$, as $7x = 3x + 4x$. With this, the trinomial $2x^2 + 7x + 6$ becomes

$$2x^2 + 7x + 6 = 2x^2 + \underbrace{3x + 4x}_{7x = 3x + 4x} + 6.$$

Factor the new polynomial by grouping:

$$2x^2 + 3x + 4x + 6 = x(2x + 3) + 2(2x + 3).$$

Factor out the common factor $2x + 3$:

$$= (2x + 3)(x + 2)$$

so $$2x^2 + 7x + 6 = (2x + 3)(x + 2).$$

In the example above, we could have written $7x$ as $4x + 3x$. Check that the resulting factoring by grouping would give the same answer.

EXAMPLE 6 Factoring Trinomials by Grouping

Factor each trinomial.

(a) $6r^2 + r - 1$

Find two integers whose product is $6(-1) = -6$ and whose sum is 1.

Sum is 1.

$$6r^2 + r - 1 = 6r^2 + 1r - 1$$

Product is -6.

The integers are -2 and 3. Write the middle term, $+r$, as $-2r + 3r$, so that

$$6r^2 + r - 1 = 6r^2 - 2r + 3r - 1.$$

Factor by grouping on the right-hand side.

$$\begin{aligned} 6r^2 + r - 1 &= 6r^2 - 2r + 3r - 1 \\ &= 2r(3r - 1) + 1(3r - 1) \\ &= (3r - 1)(2r + 1) \end{aligned}$$

(b) $12z^2 - 5z - 2$

Look for two integers whose product is $12(-2) = -24$ and whose sum is -5. The required integers are 3 and -8, and

$$\begin{aligned} 12z^2 - 5z - 2 &= 12z^2 + 3z - 8z - 2 \\ &= 3z(4z + 1) - 2(4z + 1) \\ &= (4z + 1)(3z - 2). \quad \blacksquare \end{aligned}$$

WORK PROBLEM 7 AT THE SIDE.

name date hour

5.3 EXERCISES

Complete the factoring.

1. $2x^2 - x - 1 = (2x + 1)(\qquad)$

2. $3a^2 + 5a + 2 = (3a + 2)(\qquad)$

3. $5b^2 - 16b + 3 = (5b - 1)(\qquad)$

4. $2x^2 + 11x + 12 = (2x + 3)(\qquad)$

5. $4y^2 + 17y - 15 = (y + 5)(\qquad)$

6. $7z^2 + 10z - 8 = (z + 2)(\qquad)$

7. $15x^2 + 7x - 4 = (3x - 1)(\qquad)$

8. $12c^2 - 7c - 12 = (4c + 3)(\qquad)$

9. $2m^2 + 19m - 10 = (2m - 1)(\qquad)$

10. $6x^2 + x - 12 = (2x + 3)(\qquad)$

11. $6a^2 + 7ab - 20b^2 = (2a + 5b)(\qquad)$

12. $9m^2 - 3mn - 2n^2 = (3m - 2n)(\qquad)$

13. $4k^2 + 13km + 3m^2 = (4k + m)(\qquad)$

14. $6x^2 - 13xy - 5y^2 = (3x + y)(\qquad)$

15. $4x^3 - 10x^2 - 6x = 2x(\qquad) = 2x(2x + 1)(\qquad)$

16. $15r^3 - 39r^2 - 18r = 3r(\qquad\qquad) = 3r(5r + 2)(\qquad\qquad)$

17. $6m^6 + 7m^5 - 20m^4 = m^4(\qquad\qquad) = m^4(3m - 4)(\qquad\qquad)$

18. $16y^5 - 4y^4 - 6y^3 = 2y^3(\qquad\qquad) = 2y^3(4y - 3)(\qquad\qquad)$

Factor completely. Use either method. See Examples 1–6.

19. $2x^2 + 7x + 3$

20. $3y^2 + 13y + 4$

21. $3a^2 + 10a + 7$

22. $7r^2 + 8r + 1$

23. $4r^2 + r - 3$

24. $3p^2 + 2p - 8$

25. $15m^2 + m - 2$

26. $6x^2 + x - 1$

27. $8m^2 - 10m - 3$

28. $2a^2 - 17a + 30$

29. $5a^2 - 7a - 6$

30. $12s^2 + 11s - 5$

name date hour

5.3 EXERCISES

31. $3r^2 + r - 10$

32. $20x^2 - 28x - 3$

33. $4y^2 + 69y + 17$

34. $21m^2 + 13m + 2$

35. $38x^2 + 23x + 2$

36. $20y^2 + 39y - 11$

37. $10x^2 + 11x - 6$

38. $6b^2 + 7b + 2$

39. $6w^2 + 19w + 10$

40. $20q^2 - 41q + 20$

41. $6q^2 + 23q + 21$

42. $8x^2 + 47x - 6$

43. $10m^2 - 23m + 12$

44. $4t^2 - 5t - 6$

45. $8k^2 + 2k - 15$

46. $16a^2 + 30a + 9$

47. $24x^2 - 42x + 9$

48. $48b^2 - 74b - 10$

49. $40m^2q + mq - 6q$

50. $15a^2b + 22ab + 8b$

51. $2m^3 + 2m^2 - 40m$

52. $15n^4 - 39n^3 + 18n^2$

53. $24a^4 + 10a^3 - 4a^2$

54. $18x^5 + 15x^4 - 75x^3$

55. $32z^5 - 20z^4 - 12z^3$

56. $15x^2y^2 - 7xy^2 - 4y^2$

57. $12p^2 + 7pq - 12q^2$

58. $6m^2 - 5mn - 6n^2$

59. $25a^2 + 25ab + 6b^2$

60. $6x^2 - 5xy - y^2$

61. $6a^2 - 7ab - 5b^2$

62. $25g^2 - 5gh - 2h^2$

63. $6m^6n + 7m^5n^2 + 2m^4n^3$

64. $12k^3q^4 - 4k^2q^5 - kq^6$

SKILL SHARPENERS

Find each product. See Sections 4.5 and 4.6.

65. $(3r - 1)(3r + 1)$

66. $(5p - 3q)(5p + 3q)$

67. $(a + 6b)(a - 6b)$

68. $(3x + 2)^2$

69. $(2z - 3)^2$

70. $(3k - 1)^2$

5.4 SPECIAL FACTORIZATIONS

OBJECTIVES

1. Factor the difference of two squares.
2. Factor a perfect square trinomial.

1 Recall from the last chapter that

$$(a + b)(a - b) = a^2 - b^2.$$

Based on this product, a **difference of two squares** can be factored as follows.

FACTORING A DIFFERENCE OF TWO SQUARES

$$a^2 - b^2 = (a + b)(a - b)$$

For example,

$$m^2 - 16 = m^2 - 4^2 = (m + 4)(m - 4).$$

EXAMPLE 1 Factoring a Difference of Squares
Factor each binomial that is the difference of two squares.

(a) $x^2 - 49 = (x + 7)(x - 7)$

(b) $z^2 - \frac{9}{16} = z^2 - \left(\frac{3}{4}\right)^2 = \left(z + \frac{3}{4}\right)\left(z - \frac{3}{4}\right)$

(c) $y^2 - m^2 = (y + m)(y - m)$

(d) $p^2 + 16$

Since $p^2 + 16$ is the *sum* of two squares, it is not equal to $(p + 4)(p - 4)$. Also, using FOIL,

$$(p - 4)(p - 4) = p^2 - 8p + 16 \neq p^2 + 16$$

and

$$(p + 4)(p + 4) = p^2 + 8p + 16 \neq p^2 + 16$$

so $p^2 + 16$ is a prime polynomial. ■

CAUTION As Example 1(d) suggests, the sum of two squares usually cannot be factored.

EXAMPLE 2 Factoring a Difference of Squares
Factor each binomial that is the difference of two squares.

(a) $25m^2 - 16$

This is the difference of two squares, since

$$25m^2 - 16 = (5m)^2 - 4^2.$$

Factor this as

$$(5m + 4)(5m - 4).$$

(b) $49z^2 - 64 = (7z)^2 - (8)^2 = (7z + 8)(7z - 8)$ ■

WORK PROBLEM 1 AT THE SIDE.

1. Factor.

(a) $p^2 - 100$

(b) $9m^2 - 49$

(c) $64a^2 - 25$

ANSWERS
1. (a) $(p + 10)(p - 10)$
(b) $(3m + 7)(3m - 7)$
(c) $(8a + 5)(8a - 5)$

2. Factor.

(a) $50r^2 - 32$

(b) $27y^2 - 75$

(c) $k^4 - 49$

(d) $9r^4 - 100$

(e) $4z^2 - \frac{25}{49}$

ANSWERS
2. (a) $2(5r + 4)(5r - 4)$
(b) $3(3y + 5)(3y - 5)$
(c) $(k^2 + 7)(k^2 - 7)$
(d) $(3r^2 + 10)(3r^2 - 10)$
(e) $\left(2z + \frac{5}{7}\right)\left(2z - \frac{5}{7}\right)$

EXAMPLE 3 Factoring More Complex Differences of Squares

Factor completely.

(a) $9a^2 - 4b^2 = (3a)^2 - (2b)^2 = (3a + 2b)(3a - 2b)$

(b) $81y^2 - 36$

First factor out the common factor of 9.

$$81y^2 - 36 = 9(9y^2 - 4)$$
$$= 9(3y + 2)(3y - 2) \qquad \text{Difference of squares}$$

(c) $p^4 - 36 = (p^2)^2 - 6^2 = (p^2 + 6)(p^2 - 6)$

Neither $p^2 + 6$ nor $p^2 - 6$ can be factored further.

(d) $m^4 - 16 = (m^2)^2 - 4^2$
$$= (m^2 + 4)(m^2 - 4) \qquad \text{Difference of squares}$$
$$= (m^2 + 4)(m + 2)(m - 2) \qquad \text{Difference of squares}$$ ■

WORK PROBLEM 2 AT THE SIDE.

2 The expressions 144, $4x^2$, and $81m^6$ are called perfect squares, since

$$144 = 12^2, \qquad 4x^2 = (2x)^2, \qquad \text{and} \qquad 81m^6 = (9m^3)^2.$$

A **perfect square trinomial** is a trinomial that is the square of a binomial. As an example, $x^2 + 8x + 16$ is a perfect square trinomial since it is the square of the binomial $x + 4$:

$$x^2 + 8x + 16 = (x + 4)^2.$$

For a trinomial to be a perfect square, two of its terms must be perfect squares. For this reason, $16x^2 + 4x + 15$ is not a perfect square trinomial since only the term $16x^2$ is a perfect square.

On the other hand, even if two of the terms are perfect squares, the trinomial may not be a perfect square trinomial. For example, $x^2 + 6x + 36$ has two perfect square terms, but it is not a perfect square trinomial. (Try to find a binomial that can be squared to give $x^2 + 6x + 36$.)

Multiply to see that the square of a binomial gives the following perfect square trinomials.

FACTORING PERFECT SQUARE TRINOMIALS

$$a^2 + 2ab + b^2 = (a + b)^2$$
$$a^2 - 2ab + b^2 = (a - b)^2$$

The middle term of a perfect square trinomial is always twice the product of the two terms in the squared binomial. (This was shown in Section 4.6.) Use this to check any attempt to factor a trinomial that appears to be a perfect square.

EXAMPLE 4 Factoring a Perfect Square Trinomial

Factor $x^2 + 10x + 25$.

The term x^2 is a perfect square, and so is 25. Try to factor the trinomial as

$$x^2 + 10x + 25 = (x + 5)^2.$$

Check this by finding twice the product of the two terms in the squared binomial.

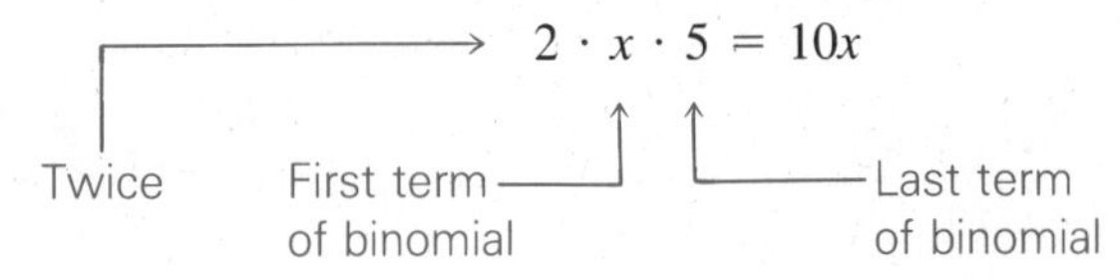

Since $10x$ is the middle term of the trinomial, the trinomial is a perfect square and can be factored as $(x + 5)^2$. ■

WORK PROBLEM 3 AT THE SIDE.

EXAMPLE 5 Factoring Perfect Square Trinomials

Factor each perfect square trinomial.

(a) $x^2 - 22x + 121$

The first and last terms are perfect squares ($121 = 11^2$). Check to see whether the middle term of $x^2 - 22x + 121$ is twice the product of the first and last terms of the binomial $(x - 11)$.

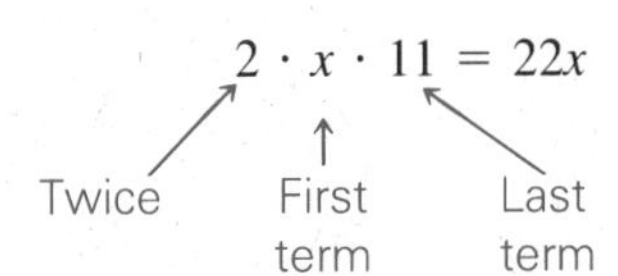

Since twice the product of the first and last terms of the binomial is the middle term, $x^2 - 22x + 121$ is a perfect square trinomial and

$$x^2 - 22x + 121 = (x - 11)^2.$$

(b) $9m^2 - 24m + 16 = (3m)^2 - 2(3m)(4) + 4^2 = (3m - 4)^2$

Twice First term Last term

(c) $25y^2 + 20y + 16$

The first and last terms are perfect squares.

$$25y^2 = (5y)^2 \quad \text{and} \quad 16 = 4^2$$

Twice the product of the first and last terms of the binomial $5y + 4$ is

$$2 \cdot 5y \cdot 4 = 40y$$

which is not the middle term of $25y^2 + 20y + 16$. This polynomial is not a perfect square. In fact, the polynomial cannot be factored even with the methods of Section 5.3; it is a prime polynomial. ■

3. Factor each trinomial that is a perfect square.

(a) $p^2 + 14p + 49$

(b) $m^2 + 8m + 16$

(c) $k^2 + 20k + 100$

(d) $z^2 + 10z + 16$

ANSWERS

3. (a) $(p + 7)^2$ (b) $(m + 4)^2$ (c) $(k + 10)^2$ (d) not a perfect square trinomial

4. Factor each trinomial that is a perfect square.

(a) $p^2 - 8p + 16$

(b) $4m^2 + 20m + 25$

(c) $16a^2 + 56a + 49$

(d) $121p^2 + 110p + 100$

NOTE The sign of the second term in the squared binomial is always the same as the sign of the middle term in the trinomial. Also, the first and last terms of a perfect square trinomial must be *positive,* since they are squares. For example, the polynomial $x^2 - 2x - 1$ cannot be a perfect square because the last term is negative.

WORK PROBLEM 4 AT THE SIDE.

The methods of factoring discussed in this section are summarized here. These rules should be memorized.

SPECIAL FACTORIZATIONS

Difference of two squares	$a^2 - b^2 = (a + b)(a - b)$
Perfect square trinomials	$a^2 + 2ab + b^2 = (a + b)^2$
	$a^2 - 2ab + b^2 = (a - b)^2$

ANSWERS
4. (a) $(p - 4)^2$ (b) $(2m + 5)^2$ (c) $(4a + 7)^2$ (d) not a perfect square trinomial

name date hour

5.4 EXERCISES

Factor each binomial completely. The table of squares and square roots in the back of the book may be helpful. See Examples 1–3.

1. $x^2 - 16$

2. $m^2 - 25$

3. $p^2 - 4$

4. $r^2 - 9$

5. $9m^2 - 1$

6. $16y^2 - 9$

7. $25m^2 - \frac{16}{49}$

8. $25y^2 - \frac{9}{16}$

9. $36t^2 - 16$

10. $9 - 36a^2$

11. $25a^2 - 16r^2$

12. $100k^2 - 49m^2$

13. $81x^2 + 16$

14. $49m^2 + 100$

15. $p^4 - 36$

16. $r^4 - 9$

17. $a^4 - 1$

18. $x^4 - 16$

19. $m^4 - 81$

20. $p^4 - 256$

21. $16k^4 - 1$

Factor any expressions that are perfect square trinomials. See Examples 4 and 5.

22. $a^2 + 4a + 4$

23. $p^2 + 2p + 1$

24. $x^2 - 10x + 25$

25. $y^2 - 8y + 16$

26. $a^2 + 14a + 49$

27. $m^2 - 20m + 100$

28. $k^2 + k + \frac{1}{4}$

29. $r^2 + \frac{2}{3}r + \frac{1}{9}$

30. $b^2 - \frac{2}{5}b + \frac{1}{25}$

31. $9y^2 + 14y + 25$

32. $16m^2 + 42m + 49$

33. $16a^2 - 40ab + 25b^2$

34. $36y^2 - 60yp + 25p^2$

35. $100m^2 + 100m + 25$

36. $100a^2 - 140ab + 49b^2$

37. $49x^2 + 28xy + 4y^2$

38. $64y^2 - 48ya + 9a^2$

39. $4c^2 + 12cd + 9d^2$

40. $9t^2 - 24tr + 16r^2$

41. $25h^2 - 20hy + 4y^2$

42. $9x^2 + 24xy + 16y^2$

43. Find a value of b so that $x^2 + bx + 25 = (x + 5)^2$.

44. For what value of c is $4m^2 - 12m + c = (2m - 3)^2$?

45. Find a so that $ay^2 - 12y + 4 = (3y - 2)^2$.

46. Find b so that $100a^2 + ba + 9 = (10a + 3)^2$.

SKILL SHARPENERS

Solve each equation. *See Section 3.3.*

47. $m - 2 = 0$

48. $r + 1 = 0$

49. $3k - 2 = 0$

50. $4z + 5 = 0$

51. $7a + 9 = 0$

52. $3x + 7 = 0$

53. $8y + 5 = 0$

54. $12k - 11 = 0$

SUMMARY EXERCISES ON FACTORING

As you factor a polynomial, ask yourself these questions.

FACTORING A POLYNOMIAL

1. Is there a common factor? If so, factor it out.
2. How many terms are in the polynomial?

   Two terms: Check to see whether it is the difference of two squares.

   Three terms: Is it a perfect square trinomial? If the trinomial is not a perfect square, check to see whether the coefficient of the squared term is 1. If so, use the method of Section 5.2. If the coefficient of the squared term of the trinomial is not 1, use the general factoring methods of Section 5.3.

   Four terms: Try to factor the polynomial by grouping.
3. Can any factors be factored further? If so, factor them.

Factor as completely as possible.

1. $32m^9 + 16m^5 + 24m^3$

2. $2m^2 - 10m - 48$

3. $14k^3 + 7k^2 - 70k$

4. $9z^2 + 64$

5. $6z^2 + 31z + 5$

6. $m^2 - 3mn - 4n^2$

7. $49z^2 - 16y^2$

8. $100n^2r^2 + 30nr^3 - 50n^2r$

9. $16x + 20$

10. $2m^2 + 5m - 3$

11. $10y^2 - 7yz - 6z^2$

12. $y^4 - 16$

13. $m^2 + 2m - 15$

14. $6y^2 - 5y - 4$

15. $32z^3 + 56z^2 - 16z$

16. $15y + 5$

17. $z^2 - 12z + 36$

18. $9m^2 - 64$

19. $y^2 - 4yk - 12k^2$

20. $16z^2 - 8z + 1$

21. $6y^2 - 6y - 12$

22. $72y^3z^2 + 12y^2 - 24y^4z^2$

23. $p^2 - 17p + 66$

24. $a^2 + 17a + 72$

25. $k^2 + 9$

26. $108m^2 - 36m + 3$

27. $z^2 - 3za - 10a^2$

28. $45a^3b^5 - 60a^4b^2 + 75a^6b^4$

29. $4k^2 - 12k + 9$

30. $a^2 - 3ab - 28b^2$

31. $16r^2 + 24rm + 9m^2$

32. $3k^2 + 4k - 4$

33. $3k^3 - 12k^2 - 15k$

34. $a^4 - 625$

35. $16k^2 - 48k + 36$

36. $8k^2 - 10k - 3$

37. $36y^6 - 42y^5 - 120y^4$

38. $8p^2 + 23p - 3$

39. $5z^3 - 45z^2 + 70z$

40. $8k^2 - 2kh - 3h^2$

41. $54m^2 - 24z^2$

42. $4k^2 - 20kz + 25z^2$

43. $6a^2 + 10a - 4$

44. $15h^2 + 11hg - 14g^2$

45. $m^2 - 81$

46. $10z^2 - 7z - 6$

47. $125m^4 - 400m^3n + 195m^2n^2$

48. $9y^2 + 12y - 5$

49. $m^2 - 4m + 4$

50. $27p^{10} - 45p^9 - 252p^8$

51. $24k^4p + 60k^3p^2 + 150k^2p^3$

52. $10m^2 + 25m - 60$

53. $12p^2 + pq - 6q^2$

54. $k^2 - 64$

55. $64p^2 - 100m^2$

56. $2m^2 + 7mn - 15n^2$

57. $100a^2 - 81y^2$

58. $8a^2 + 23ab - 3b^2$

59. $a^2 + 8a + 16$

60. $4y^2 - 25$

5.5 SOLVING QUADRATIC EQUATIONS BY FACTORING

OBJECTIVES

1. Solve quadratic equations by factoring.
2. Solve other equations by factoring.

1 In this section we introduce **quadratic equations,** which are equations that contain a squared term and no terms of higher degree.

QUADRATIC EQUATIONS

Quadratic equations can be written in the form

$$ax^2 + bx + c = 0$$

where a, b, and c are real numbers, with $a \neq 0$.

For example,

$$x^2 + 5x + 6 = 0, \qquad 2a^2 - 5a = 3, \qquad \text{and} \qquad y^2 = 4$$

are all quadratic equations.

Some quadratic equations can be solved by factoring. A more general method for solving those equations that cannot be solved by factoring is given in Chapter 10.

We use the **zero-factor property** to solve a quadratic equation by factoring.

ZERO-FACTOR PROPERTY

If a and b are real numbers and if $ab = 0$, then $a = 0$ or $b = 0$.

In other words, if the product of two numbers is zero, then at least one of the numbers must be zero.

EXAMPLE 1 Using the Zero-Factor Property

Solve the equation $(x + 3)(2x - 1) = 0$.

The product $(x + 3)(2x - 1)$ is equal to zero. By the zero-factor property, the only way that the product of these two factors can be zero is if at least one of the factors is zero. Therefore, either $x + 3 = 0$ or $2x - 1 = 0$. Solve each of these two linear equations as in Chapter 3.

$$\mathbf{x + 3 = 0} \qquad\qquad \mathbf{2x - 1 = 0}$$
$$x = -3 \qquad\qquad 2x = 1$$
$$x = \frac{1}{2}$$

The given equation $(x + 3)(2x - 1) = 0$ has two solutions, $x = -3$ and $x = \frac{1}{2}$. Check these answers by substituting -3 for x in the original equation, $(x + 3)(2x - 1) = 0$. Then start over and substitute $\frac{1}{2}$ for x.

If $x = -3$, then

$$(-3 + 3)[2(-3) - 1] = 0$$
$$0(-7) = 0. \quad \text{True}$$

If $x = \frac{1}{2}$, then

$$\left(\frac{1}{2} + 3\right)\left(2 \cdot \frac{1}{2} - 1\right) = 0$$
$$\frac{7}{2}(1 - 1) = 0$$
$$\frac{7}{2} \cdot 0 = 0. \quad \text{True}$$

1. Solve each equation. Check your answers.

(a) $(x - 5)(x + 2) = 0$

(b) $(3x - 2)(x + 6) = 0$

(c) $(5x + 7)(2x + 3) = 0$

2. Solve each equation.

(a) $m^2 - 3m - 10 = 0$

(b) $y^2 - y = 6$

(c) $r^2 = 8 - 2r$

ANSWERS

1. (a) 5, −2 (b) $\frac{2}{3}$, −6 (c) $-\frac{7}{5}, -\frac{3}{2}$
2. (a) 5, −2 (b) 3, −2 (c) −4, 2

Both −3 and $\frac{1}{2}$ result in true equations, so they are solutions to the original equation. ■

WORK PROBLEM 1 AT THE SIDE.

In Example 1 the equation to be solved was presented with the polynomial in factored form. If the polynomial in an equation is not already factored, first make sure that all terms are on one side of the equals sign, with 0 alone on the other side. Then factor.

EXAMPLE 2 Solving a Quadratic Equation

Solve each equation.

(a) $x^2 - 5x = -6$

First, rewrite the equation with all terms on one side by adding 6 to both sides.

$$x^2 - 5x + 6 = -6 + 6 \qquad \text{Add 6.}$$
$$x^2 - 5x + 6 = 0$$

Now factor $x^2 - 5x + 6$. Find two numbers whose product is 6 and whose sum is −5. These two numbers are −2 and −3, so the equation becomes

$$(x - 2)(x - 3) = 0.$$

Proceed as in Example 1. Set each factor equal to 0.

$$x - 2 = 0 \quad \text{or} \quad x - 3 = 0$$

Solve the equation on the left by adding 2 to each side. In the equation on the right, add 3 to each side. Doing this gives

$$x = 2 \quad \text{or} \quad x = 3.$$

Check both solutions by substituting first 2 and then 3 for x in the original equation.

(b) $y^2 = y + 20$

To get 0 alone on one side of the equation, subtract y and 20 from each side.

$$y^2 = y + 20$$
$$y^2 - y - 20 = 0 \qquad \text{Subtract } y \text{ and 20.}$$
$$(y - 5)(y + 4) = 0 \qquad \text{Factor.}$$
$$y - 5 = 0 \quad \text{or} \quad y + 4 = 0 \qquad \text{Zero-factor property}$$

Solve each of these two equations to get the solutions

$$y = 5 \quad \text{or} \quad y = -4.$$

Check these solutions by substituting in the original equation. ■

NOTE The word "or" as used in Example 2 means "one or the other or both."

WORK PROBLEM 2 AT THE SIDE.

In summary, go through the following steps to solve quadratic equations by factoring.

SOLVING A QUADRATIC EQUATION BY FACTORING

Step 1 Get all terms on one side of the equals sign, with 0 on the other side.

Step 2 Factor completely.

Step 3 Set each factor containing a variable equal to 0, and solve the resulting equations.

Step 4 Check each solution in the original equation.

Remember: Not all quadratic equations can be solved by factoring.

EXAMPLE 3 Solving a Quadratic Equation

Solve the equation $2p^2 - 13p + 20 = 0$.

Factor $2p^2 - 13p + 20$ as $(2p - 5)(p - 4)$, giving

$$(2p - 5)(p - 4) = 0.$$

Set each of these two factors equal to 0.

$$2p - 5 = 0 \quad \text{or} \quad p - 4 = 0$$

Solve the equation on the left by adding 5 to each side of the equation. Then divide each side by 2. Solve the equation on the right by adding 4 to each side.

$$2p - 5 = 0 \quad \text{or} \quad p - 4 = 0$$
$$2p = 5$$
$$p = \frac{5}{2} \quad \text{or} \quad p = 4$$

The solutions of $2p^2 - 13p + 20 = 0$ are $\frac{5}{2}$ and 4; check them by substituting in the original equation. ■

WORK PROBLEM 3 AT THE SIDE.

EXAMPLE 4 Solving a Quadratic Equation

Solve each equation.

(a) $16m^2 - 25 = 0$

Factor the left-hand side of the equation as the difference of two squares.

$$(4m + 5)(4m - 5) = 0$$

Set each factor equal to 0.

$$4m + 5 = 0 \quad \text{or} \quad 4m - 5 = 0$$

Solve each equation.

$$4m = -5 \quad \text{or} \quad 4m = 5$$
$$m = -\frac{5}{4} \quad \text{or} \quad m = \frac{5}{4}$$

The two solutions, $-\frac{5}{4}$ and $\frac{5}{4}$, should be checked in the original equation.

3. Solve each equation.

(a) $2a^2 - a - 3 = 0$

(b) $3x^2 = 11x + 4$

(c) $12m^2 - 17m = 5$

ANSWERS

3. (a) $\frac{3}{2}, -1$ (b) $-\frac{1}{3}, 4$ (c) $-\frac{1}{4}, \frac{5}{3}$

4. Solve each equation.

(a) $p^2 - 36 = 0$

(b) $49m^2 - 9 = 0$

(c) $p(4p + 7) = 2$

(d) $k(6k + 11) = 10$

ANSWERS

4. (a) 6, −6 (b) $\frac{3}{7}, -\frac{3}{7}$ (c) $\frac{1}{4}, -2$ (d) $\frac{2}{3}, -\frac{5}{2}$

(b) $k(2k + 5) = 3$

Multiply on the left-hand side and then get all terms on one side.

$$k(2k + 5) = 3$$
$$2k^2 + 5k = 3 \quad \text{Multiply.}$$
$$2k^2 + 5k - 3 = 0 \quad \text{Subtract 3.}$$
$$(2k - 1)(k + 3) = 0 \quad \text{Factor.}$$

Place each factor equal to 0 and solve the equations.

$$2k - 1 = 0 \quad \text{or} \quad k + 3 = 0$$
$$2k = 1$$
$$k = \frac{1}{2} \quad \text{or} \quad k = -3$$

Check that the two solutions are $\frac{1}{2}$ and -3. ■

CAUTION In Example 4(b) the zero-factor property could not be used to solve the original equation because of the 3 on the right. Remember that the zero-factor property applies only to a product that equals 0.

WORK PROBLEM 4 AT THE SIDE.

2 The zero-factor property can also be used to solve equations that result in more than two factors, as shown in Example 5. (These equations are not quadratic equations. Why not?)

EXAMPLE 5 Solving Equations with More Than Two Factors

Solve the equation $6z^3 - 6z = 0$.

First, factor out the greatest common factor in $6z^3 - 6z$.

$$6z^3 - 6z = 0$$
$$6z(z^2 - 1) = 0$$

Now factor $z^2 - 1$ as $(z + 1)(z - 1)$ to get

$$6z(z + 1)(z - 1) = 0.$$

By an extension of the zero-factor property, this product can equal 0 only if at least one of the factors is 0. Write three equations, one for each factor with a variable.

$$6z = 0 \quad \text{or} \quad z + 1 = 0 \quad \text{or} \quad z - 1 = 0$$

Solving these three equations gives the three solutions

$$z = 0 \quad \text{or} \quad z = -1 \quad \text{or} \quad z = 1.$$

Check by substituting, in turn, 0, −1, and 1 in the original equation. ■

WORK PROBLEM 5 AT THE SIDE.

EXAMPLE 6 Solving Equations with More Than Two Factors

Solve the equation $(2x - 1)(x^2 - 9x + 20) = 0$.

Factor $x^2 - 9x + 20$ as $(x - 5)(x - 4)$. Then rewrite the original equation as

$$(2x - 1)(x - 5)(x - 4) = 0.$$

Set each of these three factors equal to 0.

$$2x - 1 = 0 \quad \text{or} \quad x - 5 = 0 \quad \text{or} \quad x - 4 = 0$$

Solving these three equations gives

$$x = \frac{1}{2} \quad \text{or} \quad x = 5 \quad \text{or} \quad x = 4$$

as the solutions of the original equation. Check each solution. ■

WORK PROBLEM 6 AT THE SIDE.

5. Solve each equation. Check each solution.

(a) $2a(a - 1)(a + 3) = 0$

(b) $r^3 - 16r = 0$

6. Solve each equation. Check each solution.

(a) $(m + 3)(m^2 - 11m + 10) = 0$

(b) $(2x + 1)(2x^2 + 7x - 15) = 0$

ANSWERS

5. **(a)** 0, 1, −3 **(b)** 0, 4, −4

6. **(a)** −3, 1, 10 **(b)** $-\frac{1}{2}, \frac{3}{2}, -5$

Historical Reflections

Historical Reflections

François Viète (1540–1603)

François Viète, a lawyer at the court of Henri IV of France, studied methods of solving equations. Viète simplified the notation of algebra and was among the first to use letters to represent numbers. For centuries, algebra and arithmetic were expressed in a cumbersome way with words and occasional symbols.

Since the time of Viète, algebra has gone far beyond equation solving. The abstract nature of higher algebra depends on its symbolic language.

Art: Library of Congress

name date hour

5.5 EXERCISES

Solve each equation and check your answer. See Example 1.

1. $(x - 2)(x + 4) = 0$

2. $(y - 3)(y + 5) = 0$

3. $(3k - 8)(k + 7) = 0$

4. $(2a + 3)(a - 2) = 0$

5. $(5p + 1)(2p - 1) = 0$

6. $(3x + 5)(2x - 1) = 0$

Solve each equation and check your answer. See Examples 2-4.

7. $x^2 + 5x + 6 = 0$

8. $y^2 - 3y + 2 = 0$

9. $m^2 + 3m - 28 = 0$

10. $p^2 - p - 6 = 0$

11. $a^2 = 24 - 5a$

12. $r^2 = 2r + 15$

13. $x^2 = 3 + 2x$

14. $m^2 = 3m + 4$

15. $z^2 = -2 - 3z$

16. $p^2 = 2p + 3$

17. $m^2 + 8m + 16 = 0$

18. $b^2 - 6b + 9 = 0$

19. $3a^2 + 5a - 2 = 0$

20. $6r^2 - r - 2 = 0$

21. $2k^2 - k - 10 = 0$

22. $6x^2 - 7x - 5 = 0$

23. $6p^2 = 4 - 5p$

24. $6x^2 - 5x = 4$

25. $6a^2 = 5 - 13a$

26. $9s^2 + 12s = -4$

27. $2z^2 + 3z = 20$

28. $r^2 - 9 = 0$

29. $m^2 - 36 = 0$

30. $16r^2 - 25 = 0$

31. $4k^2 - 9 = 0$

In each of the following exercises, first rewrite each equation so that one side is zero, and then solve. Check each solution. See Example 4(b).

32. $m(m - 7) = -10$

33. $z(2z + 7) = 4$

34. $b(2b + 3) = 9$

35. $5b^2 = 4(2b + 1)$

36. $2(x^2 - 66) = -13x$

37. $3(m^2 + 4) = 20m$

38. $3r(r + 1) = (2r + 3)(r + 1)$

39. $(3k + 1)(k + 1) = 2k(k + 3)$

40. $12k(k - 4) = 3(k - 4)$

41. $y^2 = 4(y - 1)$

Solve each equation and check your answer. See Examples 5 and 6.

42. $(2r - 5)(3r^2 - 16r + 5) = 0$

43. $(3m - 4)(6m^2 + m - 2) = 0$

44. $(2x + 7)(x^2 - 2x - 3) = 0$

45. $(x - 1)(6x^2 + x - 12) = 0$

46. $9y^3 - 49y = 0$

47. $16r^3 - 9r = 0$

48. $r^3 - 2r^2 - 8r = 0$

SKILL SHARPENERS

Solve each problem. See Section 3.4.

49. An animal lover has 3 more cats than dogs. She has a total of 7 cats and dogs. How many cats does she have?

50. A small motorboat engine requires $\frac{1}{5}$ as much oil as gasoline. How much oil would be used in 6 gallons of the mixture?

5.6 APPLICATIONS OF QUADRATIC EQUATIONS

1 Quadratic equations can be applied in solving different kinds of word problems by using the techniques discussed in this section.

EXAMPLE 1 Solving an Area Problem
The width of a rectangular box is 4 centimeters less than the length. The area of the bottom is 96 square centimeters. Find the length and width of the box.

Let x = the length of the box
$x - 4$ = the width (the width is 4 less than the length).

See Figure 1. The area of a rectangle is given by the formula

$$\text{Area} = LW = \text{Length} \times \text{Width}.$$

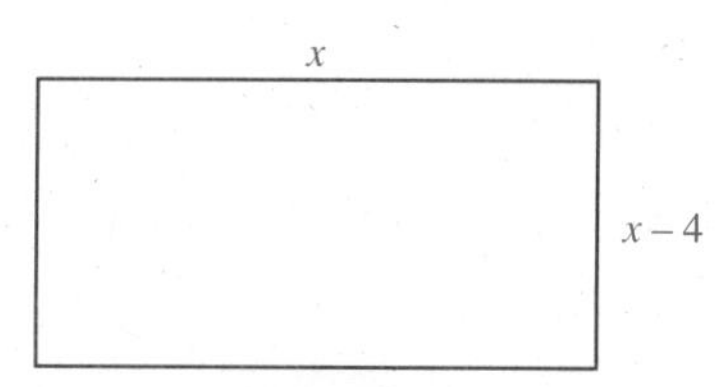

FIGURE 1

Substitute 96 for the area, x for the length, and $x - 4$ for the width in the formula.

$A = LW$

$96 = x(x - 4)$ Let $A = 96$, $L = x$, $W = x - 4$.

Multiply on the right.

$96 = x^2 - 4x$

$0 = x^2 - 4x - 96$ Subtract 96 from both sides.

$0 = (x - 12)(x + 8)$ Factor.

Set each factor equal to 0.

$x - 12 = 0$ or $x + 8 = 0$

Solve each equation.

$x = 12$ or $x = -8$

The solutions of the equations are $x = 12$ or $x = -8$. Since a rectangle cannot have a negative length, discard the solution -8. Then 12 centimeters is the length of the rectangle, and $12 - 4 = 8$ centimeters is the width. As a check, the width is 4 less than the length, and the area is $8 \cdot 12 = 96$ square centimeters. ■

CAUTION In a word problem, always be careful to check solutions against physical facts.

WORK PROBLEM 1 AT THE SIDE.

2 The next word problem involves **perimeter,** the distance around a figure, as well as area.

OBJECTIVES

1. Solve word problems about area.
2. Solve word problems about perimeter.
3. Solve word problems about consecutive integers.
4. Solve word problems using the Pythagorean formula.

1. Solve each word problem.

(a) The length of a room is 2 meters more than the width. The area of the floor is 48 square meters. Find the length and width of the room.

(b) The length of a hall is five times the width. The area of the floor is 45 square meters. Find the length and width of the hall.

ANSWERS
1. (a) 8 meters, 6 meters
(b) 15 meters, 3 meters

2. Solve each problem.

(a) The length of a rectangular floor is 5 feet more than the width. The area is numerically 32 more than the perimeter. Find the length and width of the floor.

(b) The width of a rectangular yard is 4 meters less than the length. The area is numerically 92 more than the perimeter. Find the length and width of the yard.

ANSWERS
2. (a) 11 feet, 6 feet
(b) 14 meters, 10 meters

EXAMPLE 2 Solving an Area and Perimeter Problem

The length of a rectangular rug is 4 feet more than the width. The area of the rug is numerically 1 more than the perimeter. See Figure 2. Find the length and width of the rug.

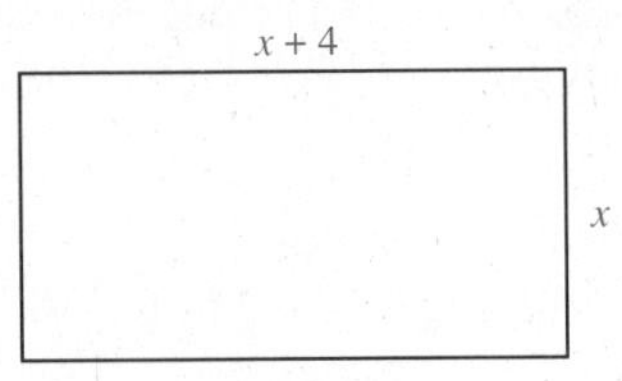

FIGURE 2

Let $\quad x =$ the width of the rug

$x + 4 =$ the length of the rug.

The area is the product of the length and width, so

$$A = LW.$$

Substituting $x + 4$ for the length and x for the width gives

$$A = (x + 4)x.$$

Now substitute into the formula for perimeter,

$$P = 2L + 2W$$
$$P = 2(x + 4) + 2x.$$

According to the information given in the problem, the area is numerically 1 more than the perimeter.

The area	is	1	more than	the perimeter.
↓	↓	↓	↓	↓
$x(x + 4)$	$=$	1	$+$	$2(x + 4) + 2x$

Simplify and solve this equation.

$x^2 + 4x = 1 + 2x + 8 + 2x$	Distributive property
$x^2 + 4x = 9 + 4x$	Combine terms.
$x^2 = 9$	Subtract $4x$ from both sides.
$x^2 - 9 = 0$	Subtract 9 from both sides.
$(x + 3)(x - 3) = 0$	Factor.
$x + 3 = 0 \quad \text{or} \quad x - 3 = 0$	Zero-factor property
$x = -3 \quad \text{or} \quad x = 3$	

A rectangle cannot have a negative width, so ignore -3. The only valid solution is 3, so the width is 3 feet and the length is $3 + 4 = 7$ feet. Check to see that the area is numerically 1 more than the perimeter. The rug is 3 feet wide and 7 feet long. ■

WORK PROBLEM 2 AT THE SIDE.

3 **Consecutive integers** are integers that are next to each other on a number line, such as 5 and 6, or -11 and -10. **Consecutive odd integers** are odd integers that are next to each other, such as 21 and 23, or -17 and -15.

The following list may be helpful in working with consecutive integers. Here x represents the first of the integers.

CONSECUTIVE INTEGERS

Two consecutive integers	$x, x + 1$
Three consecutive integers	$x, x + 1, x + 2$
Two consecutive odd integers	$x, x + 2$
Two consecutive even integers	$x, x + 2$

The next example shows how quadratic equations can occur in work with consecutive integers.

EXAMPLE 3 Solving a Consecutive Integer Problem

The product of two consecutive odd integers is 1 less than five times their sum. Find the integers.

Let s = the smaller integer

$s + 2$ = the next larger odd integer.

Since the problem mentions consecutive *odd* integers, use $s + 2$ for the larger of the two integers. According to the problem, the product is 1 less than five times the sum.

The product	is	five times the sum	less 1.
↓	↓	↓	↓
$s(s + 2)$	$=$	$5(s + s + 2)$	-1

Simplify this equation and solve it.

$$s^2 + 2s = 5s + 5s + 10 - 1 \quad \text{Distributive property}$$

$$s^2 + 2s = 10s + 9 \quad \text{Combine terms.}$$

$$s^2 - 8s - 9 = 0 \quad \text{Subtract } 10s \text{ and } 9.$$

$$(s - 9)(s + 1) = 0 \quad \text{Factor.}$$

$$s - 9 = 0 \quad \text{or} \quad s + 1 = 0$$

$$s = 9 \quad \text{or} \quad s = -1$$

We need to find two consecutive odd integers.

If $s = 9$ is the first, then $s + 2 = 11$ is the second.

If $s = -1$ is the first, then $s + 2 = 1$ is the second.

Check that two pairs of integers satisfy the problem: 9 and 11 or -1 and 1. ■

WORK PROBLEM 3 AT THE SIDE.

4 The next example requires the **Pythagorean formula** from geometry.

3. Solve each problem.

(a) The product of two consecutive even integers is 4 more than two times their sum. Find the integers.

(b) Find three consecutive integers such that the product of the first two is 2 more than six times the third.

ANSWERS

3. (a) 4, 6 or $-2, 0$

(b) 7, 8, 9 or $-2, -1, 0$

4. The hypotenuse of a right triangle is 3 inches longer than the longer leg. The shorter leg is 3 inches shorter than the longer leg. Find the lengths of the sides of the triangle.

PYTHAGOREAN FORMULA

If a right triangle (a triangle with a 90° angle) has longest side of length c and two other sides of lengths a and b, then

$$a^2 + b^2 = c^2.$$

The longest side is called the **hypotenuse** and the two shorter sides are the **legs** of the triangle.

EXAMPLE 4 Using the Pythagorean Formula

The hypotenuse of a right triangle is 2 feet longer than the shorter leg. The longer leg is 1 foot longer than the shorter leg. Find the lengths of the sides of the triangle.

Let x be the length of the shorter leg. Then

$$x = \text{shorter leg}$$
$$x + 1 = \text{longer leg}$$
$$x + 2 = \text{hypotenuse.}$$

Place these on a right triangle, as in Figure 3.

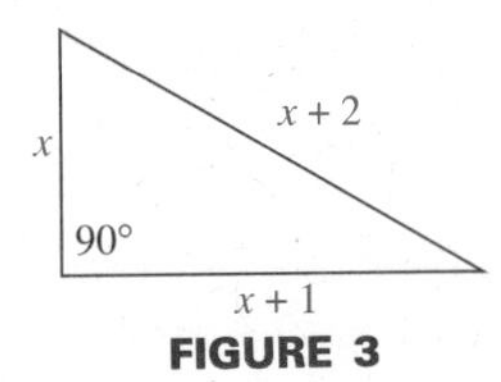

FIGURE 3

Substitute into the Pythagorean formula.

$$a^2 + b^2 = c^2$$
$$x^2 + (x + 1)^2 = (x + 2)^2$$

Since $(x + 1)^2 = x^2 + 2x + 1$, and since $(x + 2)^2 = x^2 + 4x + 4$, the equation becomes

$$x^2 + x^2 + 2x + 1 = x^2 + 4x + 4.$$

Get 0 on one side of the equation.

$$x^2 - 2x - 3 = 0$$

Factor. $$(x - 3)(x + 1) = 0$$

Set each factor equal to 0.

$$x - 3 = 0 \quad \text{or} \quad x + 1 = 0$$
$$x = 3 \quad \text{or} \quad x = -1$$

Since -1 cannot be the length of a side of a triangle, 3 is the only possible answer. The triangle has a shorter leg of length 3 feet, a longer leg of length $3 + 1 = 4$ feet, and a hypotenuse of length $3 + 2 = 5$ feet. Check that $3^2 + 4^2 = 5^2$. ■

WORK PROBLEM 4 AT THE SIDE.

ANSWERS
4. 9 inches, 12 inches, 15 inches

5.6 EXERCISES

Solve each problem. Check your answer. See Examples 1 and 2.

1. The length of a rectangular picture is 5 centimeters more than the width. The area is 66 square centimeters. Find the length and width of the picture.

2. The length of the floor of a rectangular closet is 1 foot more than the width. The area of the floor is 56 square feet. Find the length and width of the floor.

3. The width of a rectangular box is $\frac{1}{2}$ the length. The area of the bottom of the box is 162 square inches. Find the length and width of the box.

4. The length of a book is 1.5 times the width. The area is 37.5 square inches. Find the length of the book.

5. The length of a rectangle is 3 feet more than the width. The area is numerically 4 less than the perimeter. Find the dimensions of the rectangle.

6. The width of a rectangle is 5 meters less than the length. The area is numerically 10 more than the perimeter. Find the dimensions of the rectangle.

7. The length of a rectangular card is twice its width. If the width were increased by 2 inches while the length remained the same, the resulting rectangle would have an area of 48 square inches. Find the dimensions of the original card.

8. The length of a rectangular label is three times its width. If the length were decreased by 1 while the width stayed the same, the area of the new label would be 44 square centimeters. Find the length and width of the original label.

Solve Problems 9 and 10 using the formula for the area of a triangle, Area $= \frac{1}{2}bh$.

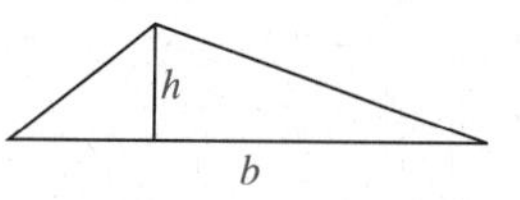

9. The area of a triangle is 25 square centimeters. The base is twice the height. Find the length of the base and the height of the triangle.

10. The height of a triangle is 3 inches more than the base. The area of the triangle is 27 square inches. Find the length of the base and the height of the triangle.

Solve Problems 11 and 12 using the formula for the volume of a pyramid,
$V = \frac{1}{3}Bh$, where B is the area of the base.

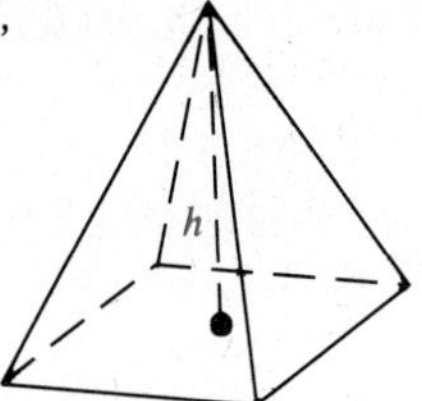

11. The volume of a pyramid is 32 cubic meters. Suppose the numerical value of the height is 10 meters less than the numerical value of the area of the base. Find the height and the area of the base.

12. Suppose a pyramid has a rectangular base whose width is 3 centimeters less than the length. If the height is 8 centimeters and the volume is 144 cubic centimeters, find the dimensions of the base.

Work the following problems. Check your answers.

13. One square has sides 1 foot shorter than the length of the sides of a second square. If the difference between the areas of the two squares is 37 square feet, find the lengths of the sides of the two squares.

14. The sides of one square have a length 2 meters more than the sides of another square. If the area of the larger square is subtracted from three times the area of the smaller square, the answer is 12 square meters. Find the lengths of the sides of each square.

15. Thuy wishes to build a box to hold his tools. The box is to be 4 feet high, and the width of the box is to be 1 foot less than the length. The volume of the box will be 120 cubic feet. Find the length and width of the box. (*Hint:* The formula for the volume of a box is $V = LWH$.)

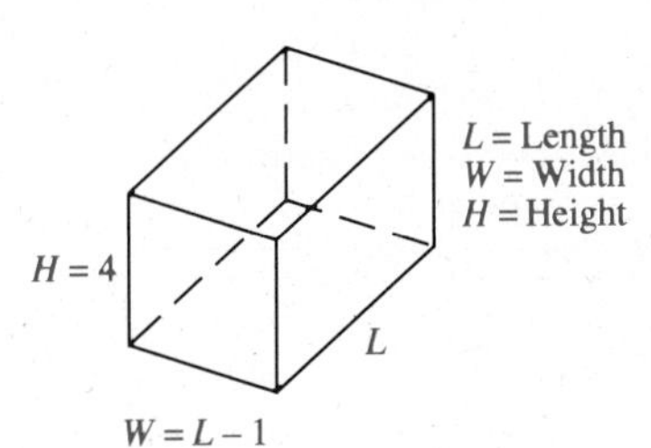

16. The volume of a box must be 315 cubic meters. The length of the box is to be 7 meters, and the height is to be 4 meters more than the width. Find the width and height of the box.

name date hour

Work the following problems. Check your answers. See Example 3.

17. The product of two consecutive integers is 2 more than twice their sum. Find the integers.

18. The product of two consecutive even integers is 60 more than twice the larger. Find the integers.

19. Find three consecutive odd integers such that four times the sum of all three equals 13 more than the product of the smaller two.

20. One number is 4 more than another. The square of the smaller increased by three times the larger is 66. Find the numbers.

21. If the square of the sum of two consecutive integers is reduced by three times their product, the result is 31. Find the integers.

22. If the square of the larger of two numbers is reduced by six times the smaller, the result is five times the larger. The larger is twice the smaller. Find the numbers.

Solve Problems 23–26 using the Pythagorean formula. See Example 4.

23. The hypotenuse of a right triangle is 1 centimeter longer than the longer leg. The shorter leg is 7 centimeters shorter than the longer leg. Find the length of the longer leg of the triangle.

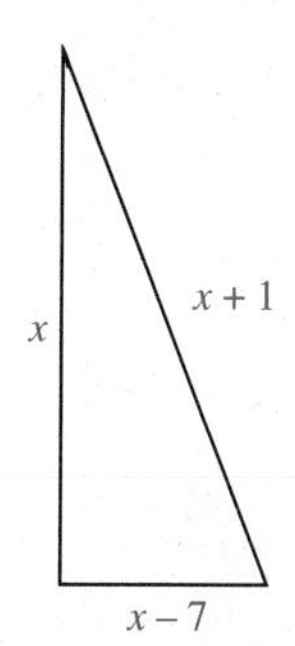

24. The longer leg of a right triangle is 1 meter longer than the shorter leg. The hypotenuse is 1 meter shorter than twice the shorter leg. Find the length of the shorter leg of the triangle.

25. The hypotenuse of a right triangle is 1 foot longer than twice the shorter leg. The longer leg is 1 foot shorter than twice the shorter leg. Find the length of the shorter leg of the triangle.

26. The length of the shorter leg of a right triangle is tripled and 4 inches is added to the result, giving the length of the hypotenuse. The longer leg is 10 inches longer than twice the shorter leg. Find the length of the shorter leg of the triangle.

Work the following problems.

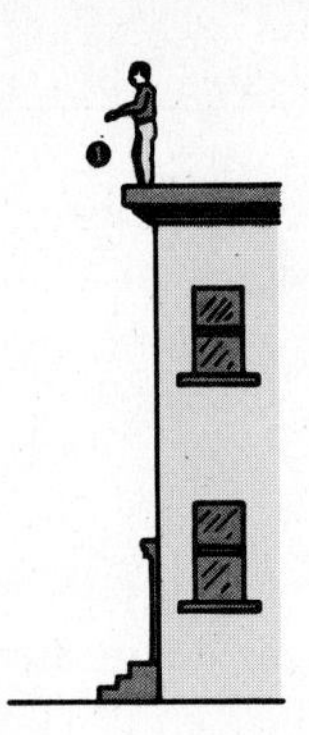

If an object is dropped, the distance d it falls in t seconds (disregarding air resistance) is given by

$$d = \frac{1}{2}gt^2$$

where g is approximately 32 feet per second per second. Find the distance an object would fall in the following periods of time.

27. 4 seconds

28. 8 seconds

How long will it take an object to fall the following distances?

29. 1600 feet

30. 2304 feet

Find a polynomial representing the area of each shaded region.

31.

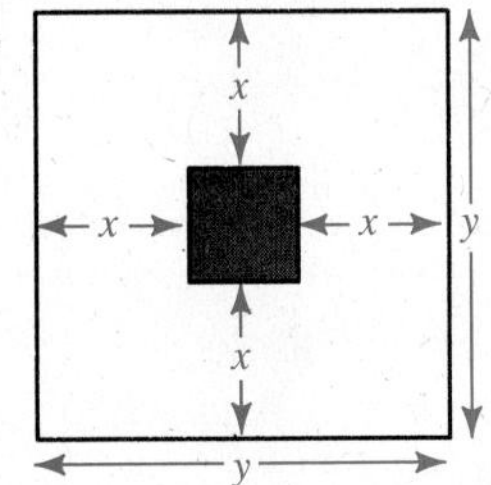

32.

SKILL SHARPENERS

Decide whether the following inequalities are true or false when x is replaced with −2. See Sections 2.1–2.4.

33. $x^2 - x + 4 \leq 0$

34. $x^2 + 2x - 3 < 0$

35. $2x^2 + 3x + 2 > 0$

36. $3x^2 - 5x + 1 < 0$

37. $x^2 + 3x + 7 \leq 0$

38. $x^2 - 5x + 1 \geq 0$

CHAPTER 5 SUMMARY

KEY TERMS

5.1	**factored form**	An expression is in factored form when it is written as a product.
	factor	An expression A is a factor of an expression B if B can be divided by A with zero remainder.
	prime number	A natural number greater than 1 is prime if it has only itself and 1 as factors.
	greatest common factor	The greatest common factor is the largest quantity that is a factor of each of a group of quantities.
5.2	**prime polynomial**	A prime polynomial is a polynomial that cannot be factored.
5.4	**perfect square trinomial**	A perfect square trinomial is a trinomial that can be factored as the square of a binomial.
5.5	**quadratic equation**	A quadratic equation is an equation that can be written in the form $ax^2 + bx + c = 0$, with $a \neq 0$.
5.6	**perimeter**	The perimeter of a geometric figure is the distance around the figure.

QUICK REVIEW

Section	Concepts	Examples
5.1 Factors: The Greatest Common Factor	**Finding the Greatest Common Factor** *Step 1* Write each number in prime factored form. *Step 2* List each prime number that is a factor of every number in the list. *Step 3* Use as exponents on the prime factors the *smallest* exponent from the prime factored forms. (If a prime does not appear in one of the prime factored forms, it cannot appear in the greatest common factor.) *Step 4* Multiply together the primes from Step 3. If there are no primes left after Step 3, the greatest common factor is 1.	$4x^2y, \quad -6x^2y^3, \quad 2xy^2$ $4x^2y = 2^2 \cdot x^2 \cdot y$ $-6x^2y^3 = -1 \cdot 2 \cdot 3 \cdot x^2 \cdot y^3$ $2xy^2 = 2 \cdot x \cdot y^2$ The greatest common factor is $2xy$.
5.2 Factoring Trinomials	To factor $x^2 + mx + n$, find integers a and b so that $a + b = m$ and $ab = n$. Then $x^2 + mx + n = (x + a)(x + b)$.	$x^2 + 6x + 8$ $a + b = 6$ and $ab = 8$ $a = 2$ and $b = 4$ $x^2 + 6x + 8 = (x + 2)(x + 4)$

Section	Concepts	Examples
5.3 More on Factoring Trinomials	To factor $px^2 + mx + n$, find integers a, b, c, and d so that $ac = p$, $bd = n$, and $ad + bc = m$. Then $px^2 + mx + n = (ax + b)(cx + d)$.	$3x^2 + 14x - 5$ $ac = 3, \quad bd = -5$ $ad + bc = 14$ By trial and error or by grouping, $3x^2 + 14x - 5 = (3x - 1)(x + 5)$.
5.4 Special Factorizations	$a^2 - b^2 = (a + b)(a - b)$ $a^2 + 2ab + b^2 = (a + b)^2$ $a^2 - 2ab + b^2 = (a - b)^2$	$4x^2 - 9 = (2x + 3)(2x - 3)$ $9x^2 + 6x + 1 = (3x + 1)^2$ $4x^2 - 20x + 25 = (2x - 5)^2$
5.5 Solving Quadratic Equations by Factoring	**Zero-Factor Property** If a and b are real numbers and if $ab = 0$, then $a = 0$ or $b = 0$.	If $(x - 2)(x + 3) = 0$, then $x - 2 = 0$ or $x + 3 = 0$.
	Solving a Quadratic Equation by Factoring	Solve $2x^2 = 7x + 15$.
	Step 1 Get all terms on one side of the equals sign, with 0 on the other side.	**1.** $2x^2 - 7x - 15 = 0$
	Step 2 Factor completely.	**2.** $(2x + 3)(x - 5) = 0$
	Step 3 Set each factor containing a variable equal to 0, and solve the resulting equations.	**3.** $2x + 3 = 0$ or $x - 5 = 0$ $2x = -3 \qquad x = 5$ $x = -\frac{3}{2}$
	Step 4 Check each solution in the original equation.	**4.** Both solutions satisfy the original equation.
	Remember: Not all quadratic equations can be solved by factoring.	
5.6 Applications of Quadratic Equations	**Pythagorean Formula** In a right triangle, the square of the hypotenuse equals the sum of the squares of the legs. $a^2 + b^2 = c^2$ (right triangle with legs a, b and hypotenuse c)	In a right triangle with legs measuring 3 meters and 5 meters, the square of the hypotenuse is equal to $3^2 + 5^2 = 34$.

name date hour

CHAPTER 5 REVIEW EXERCISES

[5.1] *Find the prime factored form for each number.*

1. 12

2. 48

3. 180

Factor out the greatest common factor.

4. $6p + 12$

5. $40r^2 + 20$

6. $25m^4 - 50m^3$

7. $6 - 18r^5 + 12r^3$

8. $100y^6 - 50y^3 + 300y^4$

9. $6p^2 + 9p + 4p + 6$

10. $4y^2 + 3y + 8y + 6$

[5.2] *Factor completely.*

11. $m^2 + 5m + 6$

12. $y^2 + 13y + 40$

13. $r^2 - 6r - 27$

14. $p^2 + p - 30$

15. $z^2 - 7z - 44$

16. $r^2 - 4rs - 96s^2$

17. $p^2 + 2pq - 120q^2$

18. $4p^3 - 12p^2 - 40p$

19. $3z^4 - 30z^3 + 48z^2$

20. $m^2 - 3mn - 18n^2$

21. $y^2 - 8yz + 15z^2$

22. $p^7 - p^6q - 2p^5q^2$

23. $3r^5 - 6r^4s - 45r^3s^2$

24. $5p^6 - 45p^5 + 70p^4$

25. $6m^9 - 84m^8 + 270m^7$

[5.3]

26. $2k^2 - 5k + 2$

27. $3z^2 + 11z - 4$

28. $6r^2 - 5r - 6$

29. $10p^2 - 3p - 1$

30. $8y^2 + 17y - 21$

31. $24k^5 - 20k^4 + 4k^3$

32. $7m^2 + 19mn - 6n^2$

33. $10r^3s + 17r^2s^2 + 6rs^3$

34. $2z^3 + 9z^2 - 5z$

[5.4]

35. $m^2 - 25$

36. $25p^2 - 121$

37. $49y^2 - 25z^2$

38. $144p^2 - 36q^2$

39. $m^2 + 9$

40. $z^2 - 8z + 16$

41. $9r^2 - 42r + 49$

42. $16m^2 + 40mn + 25n^2$

43. $54x^3 - 72x^2 + 24x$

[5.5] *Solve each equation. Check each solution.*

44. $(3k - 1)(k + 2) = 0$

45. $(5m - 2)(m + 4) = 0$

46. $(2a + 5)(3a - 7) = 0$

47. $z^2 + 4z + 3 = 0$

48. $m^2 - 5m + 4 = 0$

49. $k^2 = 8k - 15$

50. $y^2 = 11y - 30$

51. $3k^2 - 11k - 20 = 0$

52. $100b^2 - 49 = 0$

53. $m(m - 5) = 6$

54. $p^2 = 12(p - 3)$

55. $(3z - 1)(z^2 + 3z + 2) = 0$

56. $(2p + 3)(p^2 - 4p + 3) = 0$

57. $x^2 - 9 = 0$

58. $y^2 - 81 = 0$

[5.6] *Solve each word problem. Check your answer.*

59. The length of a rectangle is 6 meters more than the width. The area is 40 square meters. Find the length and width of the rectangle.

60. The length of a rectangle is three times the width. If the width were increased by 3 meters while the length remained the same, the new rectangle would have an area of 30 square meters. Find the length and width of the original rectangle.

61. The length of a rectangle is 2 centimeters more than the width. The area is numerically 44 more than the perimeter. Find the length and width of the rectangle.

62. The volume of a box is to be 120 cubic meters. The width of the box is to be 4 meters, and the height 1 meter less than the length. Find the length and height of the box.

63. The product of two consecutive even integers is 4 more than twice their sum. Find the integers.

64. One number is 2 more than another. If the square of the smaller is added to the square of the larger, the result is 100. Find the two numbers.

65. The length of a rectangle is 4 meters more than the width. The area is numerically 1 more than the perimeter. Find the dimensions of the rectangle.

66. The width of a rectangle is 5 meters less than the length. The area is numerically 10 more than the perimeter. Find the dimensions of the rectangle.

67. The sum of twice the square of an integer and three times the integer is 5. Find the integer.

68. When five times an integer is subtracted from the square of the integer, the result is -6. Find the integer.

69. The difference of the squares of two consecutive even integers is 28 less than the square of the smaller integer. Find the two integers.

70. When 10 times the larger of two consecutive even integers is subtracted from the square of the smaller, the result is 76. Find the integers.

71. Two cars left an intersection at the same time. One traveled north. The other traveled 14 miles farther, but to the west. How far apart were they then if the distance between them was 4 miles more than the distance traveled west?

72. A ladder is leaning against a building. The distance from the bottom of the laddder to the building is 4 feet less than the length of the ladder. How high up the side of the building is the top of the ladder if that distance is 2 feet less than the length of the ladder?

If an object is thrown straight up with an initial velocity of 128 *feet per second, its height h after t seconds is*

$$h = 128t - 16t^2.$$

Find the height of the object after the following periods of time.

73. 1 second

74. 2 seconds

75. 4 seconds

76. When does the object hit the ground?

77. A 9-inch by 12-inch picture is to be placed on a cardboard mat so that there is an equal border around the picture. The area of the finished mat and picture is to be 208 square inches. How wide will the border be?

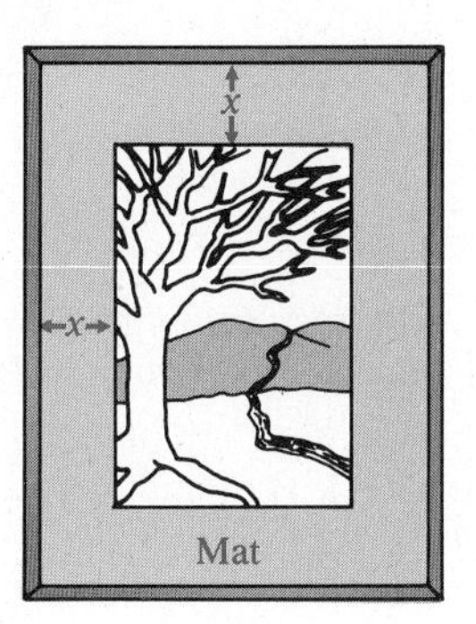

78. A box is made from a 12-centimeter by 10-centimeter piece of cardboard by cutting a square from each corner and folding up the sides. The area of the bottom of the box is to be 48 square centimeters. Find the length of a side of the cutout squares.

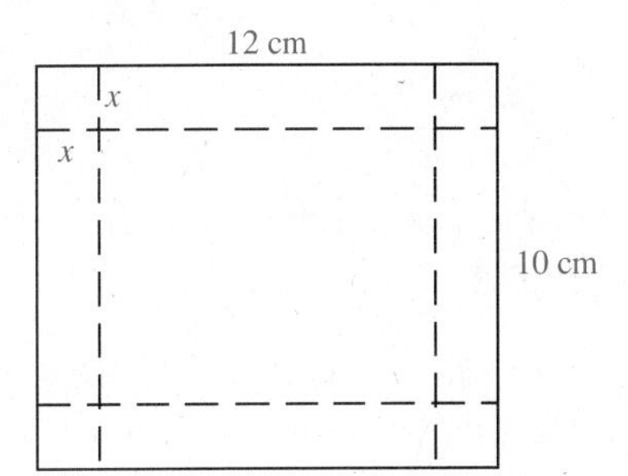

MIXED REVIEW EXERCISES

Factor completely.

79. $z^2 - 11zx + 10x^2$

80. $3k^2 + 11k + 10$

81. $15m^2 + 20mp - 12mp - 16p^2$

82. $y^4 - 625$

83. $6m^3 - 21m^2 - 45m$

84. $24ab^3c^2 - 56a^2bc^3 + 72a^2b^2c$

85. $25a^2 + 15ab + 9b^2$

86. $12x^2yz^3 + 12xy^2z - 30x^3y^2z^4$

87. $2a^5 - 8a^4 - 24a^3$

88. $12r^2 + 18rq - 10rq - 15q^2$

89. $100a^2 - 9$

90. $k^2 + 2k - 63$

Solve.

91. $r^2 + 3r = 10$

92. $25m^2 - 20m + 4 = 0$

93. $8p^2 = 10p + 3$

94. A bicyclist heading east and a motorist traveling south left an intersection at the same time. When the motorcyclist had gone 17 miles farther than the bicyclist, the distance between them was 1 mile more than the distance traveled by the motorist. How far apart were they then? (*Hint:* Draw a sketch.)

95. A pyramid has a rectangular base with a length that is 2 meters more than the width. The height of the pyramid is 6 meters, and its volume is 48 cubic meters. Find the length and width of the base.

96. The product of the smaller two of three consecutive integers equals the largest plus 23. Find the integers.

97. A lot is shaped like a right triangle. The hypotenuse is 3 meters longer than the longer leg. The longer leg is 6 meters longer than twice the length of the shorter leg. Find the lengths of the sides of the lot.

98. If 3 times the square of a number is subtracted from the square of the sum of the number and 2, the result is -2. Find the number.

99. The area of a triangle is 12 square meters. The base is 2 meters longer than the height. Find the base and height of the triangle.

name date hour

CHAPTER 5 TEST

Factor as completely as possible.

1. $16m^2 - 24m$ — 1. ______
2. $6xy + 12y^2$ — 2. ______
3. $28pq + 14p + 56p^2$ — 3. ______
4. $3m + 9n + km + 3kn$ — 4. ______
5. $12p + 11r$ — 5. ______
6. $x^2 + 11x + 30$ — 6. ______
7. $p^2 + 6p - 7$ — 7. ______
8. $2y^2 - 7y - 15$ — 8. ______
9. $4m^2 + 4m - 3$ — 9. ______
10. $3x^2 + 13x - 10$ — 10. ______
11. $10z^2 + 7z + 1$ — 11. ______
12. $10a^2 - 23a - 5$ — 12. ______
13. $12r^2 - 15r + 4r - 5$ — 13. ______
14. $m^2 + 7m - 2m - 14$ — 14. ______
15. $a^2 + 3ab - 10b^2$ — 15. ______
16. $6r^2 - rs - 2s^2$ — 16. ______
17. $x^2 - 25$ — 17. ______

18. ______________ 18. $25m^2 - 49$

19. ______________ 19. $4p^2 + 12p + 9$

20. ______________ 20. $25z^2 - 10z + 1$

21. ______________ 21. $4p^3 + 16p^2 + 16p$

22. ______________ 22. $10m^4 + 55m^3 + 25m^2$

Solve each equation. Check each solution.

23. ______________ 23. $y^2 + 3y + 2 = 0$

24. ______________ 24. $3x^2 + 5x = 2$

25. ______________ 25. $2p^2 + 3p = 20$

26. ______________ 26. $x^2 - 4 = 0$

27. ______________ 27. $z^3 = 16z$

Write an equation for each problem and solve it. Check your answer.

28. ______________ 28. The length of a certain rectangle is 1 inch less than twice the width. The area is 15 square inches. Find the length and width of the rectangle.

29. ______________ 29. Find two consecutive integers such that the square of the sum of the two integers is the negative of the smaller integer.

30. ______________ 30. The length of the hypotenuse of a right triangle is twice the length of the shorter leg, plus 3 meters. The longer leg is 7 meters longer than the shorter leg. Find the lengths of the sides.

RATIONAL EXPRESSIONS

6.1 THE FUNDAMENTAL PROPERTY OF RATIONAL EXPRESSIONS

The quotient of two integers (with denominator not zero) is called a rational number. In the same way, the quotient of two polynomials with denominator not equal to zero is called a *rational expression*.

OBJECTIVES

1. Find the values of the variable for which a rational expression is undefined.
2. Find the numerical value of a rational expression.
3. Write rational expressions in lowest terms.

A **rational expression** is an expression of the form

$$\frac{P}{Q},$$

where P and Q are polynomials, with $Q \neq 0$.

Examples of rational expressions include

$$\frac{-6x}{x^3 + 8}, \qquad \frac{9x}{y + 3}, \qquad \text{and} \qquad \frac{2m^3}{8}.$$

1 A number with a zero denominator is *not* a rational expression, since division by zero is undefined. For that reason, be careful when substituting a number in the denominator of a rational expression. For example, in

$$\frac{8x^2}{x - 3}$$

the variable x can take on any value except 3. When $x = 3$ the denominator becomes $3 - 3 = 0$, making the quotient undefined.

EXAMPLE 1 Finding Values That Make a Rational Expression Undefined

Find any values of the variable for which the following are undefined.

(a) $\dfrac{p + 5}{3p + 2}$

This rational expression is undefined for any value of p that makes the denominator equal to zero. Find these values by solving the equation $3p + 2 = 0$ to get

$$3p = -2 \qquad \text{or} \qquad p = -\frac{2}{3}.$$

Since $p = -\frac{2}{3}$ will make the denominator zero, the given expression is undefined for $-\frac{2}{3}$.

1. Find all values for which the following rational expressions are undefined.

(a) $\dfrac{x+2}{x-5}$

(b) $\dfrac{3r}{r^2+6r+8}$

(c) $\dfrac{-5m}{m^2+4}$

2. Find the value of each rational expression when $x = 3$.

(a) $\dfrac{x}{2x+1}$

(b) $\dfrac{2x+6}{x-3}$

ANSWERS

1. (a) 5 (b) -4, -2 (c) never undefined

2. (a) $\frac{3}{7}$ (b) undefined

(b) $\dfrac{9m^2}{m^2-5m+6}$

Find the numbers that make the denominator zero by solving the equation $m^2 - 5m + 6 = 0$.

$(m-2)(m-3) = 0$ Factor.

$m - 2 = 0$ or $m - 3 = 0$ Set each factor equal to 0.

$m = 2$ or $m = 3$

The original fraction is undefined for 2 and for 3.

(c) $\dfrac{2r}{r^2+1}$

This denominator cannot equal zero for any value of r, since $r^2 + 1$ is always greater than zero. There are no values for which the rational expression is undefined. ■

WORK PROBLEM 1 AT THE SIDE.

2 The next example shows how to find the numerical value of a rational expression for a given value of the variable.

EXAMPLE 2 Evaluating a Rational Expression

Find the numerical value of $\dfrac{3x+6}{2x-4}$ for each of the following values of x.

(a) $x = 1$

Find the value of the rational expression by substituting 1 for x.

$$\frac{3x+6}{2x-4} = \frac{3(1)+6}{2(1)-4} \quad \text{Let } x = 1.$$

$$= \frac{9}{-2} = -\frac{9}{2}$$

(b) $x = 2$

Substitute 2 for x.

$$\frac{3x+6}{2x-4} = \frac{3(2)+6}{2(2)-4} = \frac{6+6}{0}$$

Substituting 2 for x makes the denominator zero, so the rational expression is undefined when $x = 2$. ■

WORK PROBLEM 2 AT THE SIDE.

3 A rational expression represents a number for each value of the variable that does not make the denominator zero. For this reason, the properties of rational numbers also apply to rational expressions. For example, the **fundamental property of rational expressions** permits rational expressions to be written in lowest terms. A rational expression is in **lowest terms** when there are no common factors in the numerator and the denominator (except 1).

FUNDAMENTAL PROPERTY OF RATIONAL EXPRESSIONS

If $\frac{P}{Q}$ is a rational expression and if K represents any factor, where $K \neq 0$, then

$$\frac{PK}{QK} = \frac{P}{Q}.$$

This property is based on the identity property of multiplication:

$$\frac{PK}{QK} = \frac{P}{Q} \cdot \frac{K}{K} = \frac{P}{Q} \cdot 1 = \frac{P}{Q}.$$

The next example shows how to write both a rational number and a rational expression in lowest terms. Notice the similarity in the procedures.

EXAMPLE 3 Writing in Lowest Terms

Write in lowest terms.

(a) $\frac{30}{72}$

Begin by factoring.

$$\frac{30}{72} = \frac{2 \cdot 3 \cdot 5}{2 \cdot 2 \cdot 2 \cdot 3 \cdot 3}$$

(b) $\frac{14k^2}{2k^3}$

Write k^2 as $k \cdot k$.

$$\frac{14k^2}{2k^3} = \frac{2 \cdot 7 \cdot k \cdot k}{2 \cdot k \cdot k \cdot k}$$

Group any factors common to the numerator and denominator.

$$\frac{30}{72} = \frac{5 \cdot (2 \cdot 3)}{2 \cdot 2 \cdot 3 \cdot (2 \cdot 3)} \qquad \frac{14k^2}{2k^3} = \frac{7(2 \cdot k \cdot k)}{k(2 \cdot k \cdot k)}$$

Use the fundamental property.

$$\frac{30}{72} = \frac{5}{2 \cdot 2 \cdot 3} = \frac{5}{12} \qquad \frac{14k^2}{2k^3} = \frac{7}{k} \quad ■$$

WORK PROBLEM 3 AT THE SIDE.

EXAMPLE 4 Writing in Lowest Terms

Write $\frac{3x - 12}{5x - 20}$ in lowest terms.

Begin by factoring both numerator and denominator. Then use the fundamental property.

$$\frac{3x - 12}{5x - 20} = \frac{3(x - 4)}{5(x - 4)} = \frac{3}{5} \quad ■$$

WORK PROBLEM 4 AT THE SIDE.

EXAMPLE 5 Writing in Lowest Terms

Write $\frac{m^2 + 2m - 8}{2m^2 - m - 6}$ in lowest terms.

Always begin by factoring both numerator and denominator, if possible. Then use the fundamental property.

$$\frac{m^2 + 2m - 8}{2m^2 - m - 6} = \frac{(m + 4)(m - 2)}{(2m + 3)(m - 2)} = \frac{m + 4}{2m + 3} \quad ■$$

WORK PROBLEM 5 AT THE SIDE.

3. Use the fundamental property to write the following rational expressions in lowest terms.

(a) $\frac{5x^4}{15x^2}$

(b) $\frac{6p^3}{2p^2}$

4. Write each rational expression in lowest terms.

(a) $\frac{8p + 8q}{5p + 5q}$

(b) $\frac{4y + 2}{6y + 3}$

5. Write each rational expression in lowest terms.

(a) $\frac{a^2 - b^2}{a^2 + 2ab + b^2}$

(b) $\frac{x^2 + 4x + 4}{4x + 8}$

ANSWERS

3. (a) $\frac{x^2}{3}$ (b) $3p$

4. (a) $\frac{8}{5}$ (b) $\frac{2}{3}$

5. (a) $\frac{a - b}{a + b}$ (b) $\frac{x + 2}{4}$

6. Write each rational expression in lowest terms.

(a) $\dfrac{5-y}{y-5}$

(b) $\dfrac{m-n}{n-m}$

(c) $\dfrac{x-3}{3-x}$

(d) $\dfrac{9-k}{9+k}$

(e) $\dfrac{z^2-5}{5-z^2}$

ANSWERS
6. (a) -1 (b) -1 (c) -1
(d) cannot be written in simpler form (e) -1

In Example 4, the rational expression $\dfrac{3x-12}{5x-20}$ is restricted to values of x not equal to 4. For this reason

$$\frac{3x-12}{5x-20}=\frac{3}{5} \text{ for } x \neq 4.$$

Similarly, in Example 5,

$$\frac{m^2+2m-8}{2m^2-m-6}=\frac{m+4}{2m+3} \text{ for } m \neq -\frac{3}{2} \text{ or } 2.$$

From now on we will assume that such restrictions are understood without actually writing them down.

EXAMPLE 6 Writing in Lowest Terms

Write $\dfrac{x-y}{y-x}$ in lowest terms.

At first glance, there does not seem to be any way in which $x-y$ and $y-x$ can be factored to get a common factor. However, $y-x$ can be factored as

$$y-x=-1(-y+x)=-1(x-y).$$

With these factors, use the fundamental property to simplify the rational expression.

$$\frac{x-y}{y-x}=\frac{1(x-y)}{-1(x-y)}=\frac{1}{-1}=-1 \quad ■$$

Example 6 suggests the following rule.

The quotient of two nonzero expressions that differ only in sign is -1.

CAUTION Although x and y appear in both the numerator and denominator in Example 6, it is not possible to use the fundamental property right away since they are *terms,* not *factors.* Terms are *added,* while factors are *multiplied.*

EXAMPLE 7 Writing in Lowest Terms

Write each rational expression in lowest terms.

(a) $\dfrac{2-m}{m-2}$

Since $2-m$ and $m-2$ (or $-2+m$) differ only in sign,

$$\frac{2-m}{m-2}=-1.$$

(b) $\dfrac{3+r}{3-r}$

The quantities $3+r$ and $3-r$ *do not* differ only in sign. This rational expression cannot be written in simpler form. ■

WORK PROBLEM 6 AT THE SIDE.

name date hour

6.1 EXERCISES

Find all values for which the following rational expressions are undefined. See Example 1.

1. $\frac{3}{4x}$

2. $\frac{5}{2x}$

3. $\frac{8}{x - 4}$

4. $\frac{6}{x + 3}$

5. $\frac{x^2}{x + 5}$

6. $\frac{3x^2}{2x - 1}$

7. $\frac{a + 4}{a^2 - 8a + 15}$

8. $\frac{p + 6}{p^2 - p - 12}$

9. $\frac{8r + 2}{2r^2 - r - 3}$

10. $\frac{7k + 2}{3k^2 - k - 10}$

11. $\frac{9y}{y^2 + 16}$

12. $\frac{12z}{z^2 + 100}$

Find the numerical value of each rational expression when **(a)** $x = 2$ *and* **(b)** $x = -3$. *See Example 2.*

13. $\frac{4x - 2}{3x}$

(a) **(b)**

14. $\frac{-5x + 1}{2x}$

(a) **(b)**

15. $\frac{4x^2 - 2x}{3x}$

(a) **(b)**

16. $\frac{x^2 - 1}{x}$

(a) **(b)**

17. $\frac{(-8x)^2}{3x + 9}$

(a) **(b)**

18. $\frac{2x^2 + 5}{3 + x}$

(a) **(b)**

19. $\frac{x + 8}{x^2 - 4x + 2}$

(a) **(b)**

20. $\frac{2x - 1}{x^2 - 7x + 3}$

(a) **(b)**

21. $\frac{5x^2}{6 - 3x - x^2}$

(a) **(b)**

22. $\frac{-2x^2}{8 + x - x^2}$

(a) **(b)**

23. $\frac{2x + 5}{x^2 + 3x - 10}$

(a) **(b)**

24. $\frac{3x - 7}{2x^2 - 3x - 2}$

(a) **(b)**

Write each rational expression in lowest terms. See Examples 3–5.

25. $\frac{12k^2}{6k}$

26. $\frac{9m^3}{3m}$

27. $\frac{6a^2b^3}{24a^3b^2}$

28. $\frac{8r - 12}{4}$

29. $\frac{9z + 6}{3}$

30. $\frac{12m^2 - 9}{3}$

31. $\frac{15p^2 - 10}{5}$

32. $\frac{9m + 18}{5m + 10}$

33. $\frac{x^2 - 1}{(x + 1)^2}$

34. $\frac{3t + 15}{t^2 + 4t - 5}$

35. $\frac{5m^2 - 5m}{10m - 10}$

36. $\frac{3y^2 - 3y}{2(y - 1)}$

37. $\frac{16r^2 - 4s^2}{4r - 2s}$

38. $\frac{11s^2 - 22s^3}{6 - 12s}$

39. $\frac{m^2 - 4m + 4}{m^2 + m - 6}$

40. $\frac{a^2 - a - 6}{a^2 + a - 12}$

41. $\frac{x^2 + 3x - 4}{x^2 - 1}$

42. $\frac{8m^2 + 6m - 9}{16m^2 - 9}$

Write each rational expression in lowest terms. See Examples 6 and 7.

43. $\frac{m - 5}{5 - m}$

44. $\frac{3 - p}{p - 3}$

45. $\frac{x^2 - 1}{1 - x}$

46. $\frac{p^2 - q^2}{q - p}$

47. $\frac{m^2 - 4m}{4m - m^2}$

48. $\frac{s^2 - r^2}{r^2 - s^2}$

SKILL SHARPENERS

Multiply or divide as indicated. See Section 1.1.

49. $\frac{3}{4} \cdot \frac{5}{8}$

50. $\frac{7}{10} \cdot \frac{3}{5}$

51. $\frac{8}{15} \cdot \frac{20}{3}$

52. $\frac{15}{4} \div \frac{5}{12}$

53. $\frac{6}{5} \div \frac{3}{10}$

54. $\frac{27}{8} \div \frac{5}{12}$

6.2 MULTIPLICATION AND DIVISION OF RATIONAL EXPRESSIONS

1 The product of two fractions is found by multiplying the numerators and multiplying the denominators. Rational expressions are multiplied in the same way.

MULTIPLYING RATIONAL EXPRESSIONS

The product of the rational expressions $\frac{P}{Q}$ and $\frac{R}{S}$ is

$$\frac{P}{Q} \cdot \frac{R}{S} = \frac{PR}{QS}.$$

In words: Multiply the numerators and multiply the denominators.

The next example shows the multiplication of both two rational numbers and two rational expressions. This parallel discussion lets you compare the steps.

EXAMPLE 1 Multiplying Rational Expressions

Multiply. Write answers in lowest terms.

(a) $\frac{3}{10} \cdot \frac{5}{9}$ | **(b)** $\frac{6}{x} \cdot \frac{x^2}{12}$

Find the product of the numerators and the denominators.

$$\frac{3}{10} \cdot \frac{5}{9} = \frac{3 \cdot 5}{10 \cdot 9} \qquad \frac{6}{x} \cdot \frac{x^2}{12} = \frac{6 \cdot x^2}{x \cdot 12}$$

Factor the numerator and denominator to identify any common factors. Then use the fundamental property to write each product in lowest terms.

$$\frac{3}{10} \cdot \frac{5}{9} = \frac{3 \cdot 5}{2 \cdot 5 \cdot 3 \cdot 3} = \frac{1}{6} \qquad \frac{6}{x} \cdot \frac{x^2}{12} = \frac{6 \cdot x \cdot x}{2 \cdot 6 \cdot x} = \frac{x}{2}$$

Notice in the second step above that the products were left in factored form, since common factors are needed to write the product in lowest terms. ■

WORK PROBLEM 1 AT THE SIDE.

EXAMPLE 2 Multiplying Rational Expressions

Find the product of $\frac{x + y}{2x}$ and $\frac{x^2}{(x + y)^2}$.

Use the definition of multiplication.

$$\frac{x + y}{2x} \cdot \frac{x^2}{(x + y)^2} = \frac{(x + y)x^2}{2x(x + y)^2} \qquad \text{Multiply numerators. Multiply denominators.}$$

$$= \frac{(x + y)x \cdot x}{2x(x + y)(x + y)} \qquad \text{Factor.}$$

$$= \frac{x}{2(x + y)} \qquad \text{Lowest terms}$$

The product was written in lowest terms by factoring and using the fundamental property of rational expressions. ■

WORK PROBLEM 2 AT THE SIDE.

OBJECTIVES

1. Multiply rational expressions.
2. Divide rational expressions.

1. Multiply.

(a) $\frac{3m^2}{2} \cdot \frac{10}{m}$

(b) $\frac{8p^2q}{3} \cdot \frac{9}{pq^2}$

2. Multiply.

(a) $\frac{a + b}{5} \cdot \frac{30}{2(a + b)}$

(b) $\frac{3(p - q)}{p} \cdot \frac{q}{2(p - q)}$

ANSWERS

1. (a) $15m$ (b) $\frac{24p}{q}$
2. (a) 3 (b) $\frac{3q}{2p}$

3. Multiply.

(a) $\dfrac{x^2 + 7x + 10}{3x + 6} \cdot \dfrac{6x - 6}{x^2 + 2x - 15}$

(b) $\dfrac{m^2 + 4m - 5}{m + 5} \cdot \dfrac{m^2 + 8m + 15}{m - 1}$

ANSWERS

3. (a) $\dfrac{2(x - 1)}{x - 3}$ (b) $(m + 5)(m + 3)$

EXAMPLE 3 Multiplying Rational Expressions

Find the product of $\dfrac{x^2 + 3x}{x^2 - 3x - 4}$ and $\dfrac{x^2 - 5x + 4}{x^2 + 2x - 3}$.

Before multiplying, factor the numerators and denominators whenever possible. Then use the fundamental property to write the product in lowest terms.

$$\frac{x^2 + 3x}{x^2 - 3x - 4} \cdot \frac{x^2 - 5x + 4}{x^2 + 2x - 3} = \frac{x(x + 3)}{(x - 4)(x + 1)} \cdot \frac{(x - 4)(x - 1)}{(x + 3)(x - 1)}$$

$$= \frac{x(x + 3)(x - 4)(x - 1)}{(x - 4)(x + 1)(x + 3)(x - 1)}$$

$$= \frac{x}{x + 1}$$ ■

WORK PROBLEM 3 AT THE SIDE.

2 The fraction $\frac{a}{b}$ is divided by the nonzero fraction $\frac{c}{d}$ by multiplying $\frac{a}{b}$ and the reciprocal of $\frac{c}{d}$, which is $\frac{d}{c}$. Division of rational expressions is defined in the same way.

DIVIDING RATIONAL EXPRESSIONS

If $\frac{P}{Q}$ and $\frac{R}{S}$ are any two rational expressions, with $\frac{R}{S} \neq 0$, then

$$\frac{P}{Q} \div \frac{R}{S} = \frac{P}{Q} \cdot \frac{S}{R} = \frac{PS}{QR}.$$

In words: Multiply the first rational expression by the reciprocal of the second rational expression.

The next example shows the division of two rational numbers and the division of two rational expressions.

EXAMPLE 4 Dividing Rational Expressions

Divide. Write answers in lowest terms.

(a) $\dfrac{5}{8} \div \dfrac{7}{16}$

(b) $\dfrac{y}{y + 3} \div \dfrac{4y}{y + 5}$

Multiply the first expression and the reciprocal of the second.

(a)

$$\frac{5}{8} \div \frac{7}{16} = \frac{5}{8} \cdot \frac{16}{7}$$

Reciprocal of $\frac{7}{16}$

$$= \frac{5 \cdot 16}{8 \cdot 7}$$

$$= \frac{5 \cdot 8 \cdot 2}{8 \cdot 7}$$

$$= \frac{10}{7}$$

(b)

$$\frac{y}{y + 3} \div \frac{4y}{y + 5}$$

$$= \frac{y}{y + 3} \cdot \frac{y + 5}{4y}$$

Reciprocal of $\frac{4y}{y + 5}$

$$= \frac{y(y + 5)}{(y + 3)(4y)}$$

$$= \frac{y + 5}{4(y + 3)}$$ ■

WORK PROBLEM 4 AT THE SIDE.

EXAMPLE 5 Dividing Rational Expressions

Divide: $\frac{(3m)^2}{(2p)^3} \div \frac{6m^3}{16p^2}$.

$$\frac{(3m)^2}{(2p)^3} \div \frac{6m^3}{16p^2} = \frac{9m^2}{8p^3} \div \frac{6m^3}{16p^2} \quad \text{Properties of exponents}$$

$$= \frac{9m^2}{8p^3} \cdot \frac{16p^2}{6m^3} \quad \text{Multiply by reciprocal.}$$

$$= \frac{9 \cdot 16m^2p^2}{8 \cdot 6p^3m^3} \quad \text{Multiply numerators. Multiply denominators.}$$

$$= \frac{3}{mp} \quad \text{Lowest terms} \blacksquare$$

WORK PROBLEM 5 AT THE SIDE.

EXAMPLE 6 Dividing Rational Expressions

Divide: $\frac{x^2 - 4}{(x + 3)(x - 2)} \div \frac{(x + 2)(x + 3)}{-2x}$.

First, use the definition of division.

$$\frac{x^2 - 4}{(x + 3)(x - 2)} \div \frac{(x + 2)(x + 3)}{-2x}$$

$$= \frac{x^2 - 4}{(x + 3)(x - 2)} \cdot \frac{-2x}{(x + 2)(x + 3)} \quad \text{Reciprocal of second expression}$$

Next, be sure all numerators and all denominators are factored.

$$\frac{x^2 - 4}{(x + 3)(x - 2)} \div \frac{(x + 2)(x + 3)}{-2x}$$

$$= \frac{(x + 2)(x - 2)}{(x + 3)(x - 2)} \cdot \frac{-2x}{(x + 2)(x + 3)}$$

Now multiply numerators and denominators and simplify.

$$\frac{(x + 2)(x - 2)}{(x + 3)(x - 2)} \cdot \frac{-2x}{(x + 2)(x + 3)}$$

$$= \frac{-(x + 2)(x - 2)(2x)}{(x + 3)(x - 2)(x + 2)(x + 3)}$$

$$= -\frac{2x}{(x + 3)^2} \quad \text{Lowest terms} \blacksquare$$

WORK PROBLEM 6 AT THE SIDE.

4. Divide.

(a) $\frac{r}{r - 1} \div \frac{3r}{r + 4}$

(b) $\frac{9p^2}{3p + 4} \div \frac{6p^3}{3p + 4}$

(c) $\frac{6x - 4}{3} \div \frac{15x - 10}{9}$

5. Divide.

(a) $\frac{5a^2b}{2} \div \frac{10ab^2}{8}$

(b) $\frac{(3t)^2}{w} \div \frac{3t^2}{5w^4}$

6. Divide.

(a) $\frac{y^2 + 4y + 3}{y + 3} \div \frac{y^2 - 4y - 5}{y - 3}$

(b) $\frac{4x(x + 3)}{2x + 1} \div \frac{-x^2(x + 3)}{4x^2 - 1}$

ANSWERS

4. (a) $\frac{r + 4}{3(r - 1)}$ (b) $\frac{3}{2p}$ (c) $\frac{6}{5}$

5. (a) $\frac{2a}{b}$ (b) $15w^3$

6. (a) $\frac{y - 3}{y - 5}$ (b) $-\frac{4(2x - 1)}{x}$

7. Divide.

(a) $\dfrac{x^2 - y^2}{x^2 - 1} \div \dfrac{x^2 + 2xy + y^2}{x^2 + x}$

(b) $\dfrac{ab - a^2}{a^2 - 1} \div \dfrac{b - a}{a^2 + 2a + 1}$

ANSWERS

7. (a) $\dfrac{x(x - y)}{(x - 1)(x + y)}$ (b) $\dfrac{a(a + 1)}{a - 1}$

EXAMPLE 7 Dividing Rational Expressions

Divide: $\dfrac{m^2 - 4}{m^2 - 1} \div \dfrac{2m^2 + 4m}{1 - m}$.

$$\frac{m^2 - 4}{m^2 - 1} \div \frac{2m^2 + 4m}{1 - m} = \frac{m^2 - 4}{m^2 - 1} \cdot \frac{1 - m}{2m^2 + 4m} \qquad \text{Definition of division}$$

$$= \frac{(m + 2)(m - 2)}{(m + 1)(m - 1)} \cdot \frac{1 - m}{2m(m + 2)} \qquad \text{Factor.}$$

As shown in Section 6.1, $\dfrac{1 - m}{m - 1} = -1$, so

$$\frac{(m + 2)(m - 2)}{(m + 1)\mathbf{(m - 1)}} \cdot \frac{\mathbf{1 - m}}{2m(m + 2)} = \frac{\mathbf{-1}(m - 2)}{2m(m + 1)}$$

$$= \frac{2 - m}{2m(m + 1)}. \quad ■$$

WORK PROBLEM 7 AT THE SIDE.

name date hour

6.2 EXERCISES

Multiply or divide. Write each answer in lowest terms. See Examples 1 and 5.

1. $\frac{9m^2}{16} \cdot \frac{4}{3m}$

2. $\frac{21z^4}{8} \cdot \frac{12}{7z^3}$

3. $\frac{4p^2}{8p} \cdot \frac{3p^3}{16p^4}$

4. $\frac{6x^3}{9x} \cdot \frac{12x}{x^2}$

5. $\frac{8a^4}{12a^3} \cdot \frac{9a^5}{3a^2}$

6. $\frac{14p^5}{2p^2} \cdot \frac{8p^6}{28p^3}$

7. $\frac{3r^2}{9r^3} \div \frac{8r^4}{6r^5}$

8. $\frac{25m^{10}}{9m^5} \div \frac{15m^6}{10m^4}$

9. $\frac{3m^2}{(4m)^3} \div \frac{9m^3}{32m^4}$

10. $\frac{5x^3}{(4x)^2} \div \frac{15x^2}{8x^4}$

11. $\frac{-6r^4}{3r^5} \div \frac{(2r^2)^2}{-4}$

12. $\frac{-10a^6}{3a^2} \div \frac{(3a)^3}{81a}$

Multiply or divide. Write each answer in lowest terms. See Examples 2, 3, 4, 6, and 7.

13. $\frac{a+b}{2} \cdot \frac{12}{(a+b)^2}$

14. $\frac{3(x-1)}{y} \cdot \frac{2y}{5(x-1)}$

15. $\frac{a-3}{16} \div \frac{a-3}{32}$

16. $\frac{9}{8-2y} \div \frac{3}{4-y}$

17. $\frac{2k+8}{6} \div \frac{3k+12}{2}$

18. $\frac{5m+25}{10} \cdot \frac{12}{6m+30}$

19. $\frac{9y-18}{6y+12} \cdot \frac{3y+6}{15y-30}$

20. $\frac{12p+24}{36p-36} \div \frac{6p+12}{8p-8}$

21. $\frac{3r+12}{8} \cdot \frac{16r}{9r+36}$

22. $\frac{2r+2p}{8z} \div \frac{r^2+rp}{72}$

23. $\frac{9(y-4)^2}{8(z+3)^2} \cdot \frac{16(z+3)}{3(y-4)}$

24. $\frac{6(m+2)}{3(m-1)^2} \div \frac{(m+2)^2}{9(m-1)}$

25. $\frac{2-y}{8} \cdot \frac{7}{y-2}$

26. $\frac{9-2z}{3} \cdot \frac{9}{2z-9}$

27. $\frac{8-r}{8+r} \div \frac{r-8}{r+8}$

28. $\frac{a^2+2a-15}{a^2+3a-10} \div \frac{a-3}{a+1}$

29. $\frac{y^2+y-2}{y^2+3y-4} \div \frac{y+2}{y+3}$

30. $\frac{k^2-k-6}{k^2+k-12} \div \frac{k^2+2k-3}{k^2+3k-4}$

31. $\frac{m^2+3m+2}{m^2+5m+4} \cdot \frac{m^2+10m+24}{m^2+5m+6}$

32. $\frac{z^2-z-6}{z^2-2z-8} \cdot \frac{z^2+7z+12}{z^2-9}$

33. $\frac{p^2-4p+3}{p^2-3p+2} \div \frac{p-3}{p+5}$

34. $\frac{2k^2+3k-2}{6k^2-7k+2} \cdot \frac{4k^2-5k+1}{k^2+k-2}$

35. $\frac{2m^2-5m-12}{m^2-10m+24} \div \frac{4m^2-9}{m^2-9m+18}$

36. $\frac{m^2+2mp-3p^2}{m^2-3mp+2p^2} \div \frac{m^2+4mp+3p^2}{m^2+2mp-8p^2}$

37. $\frac{r^2+rs-12s^2}{r^2-rs-20s^2} \div \frac{r^2-2rs-3s^2}{r^2+rs-30s^2}$

38. $\left(\frac{x^2+10x+25}{x^2+10x} \cdot \frac{10x}{x^2+15x+50}\right) \div \frac{x+5}{x+10}$

39. $\left(\frac{m^2-12m+32}{8m} \cdot \frac{m^2-8m}{m^2-8m+16}\right) \div \frac{m-8}{m-4}$

SKILL SHARPENERS

Write the prime factored form of each number. See Section 5.1.

40. 12 **41.** 50 **42.** 72 **43.** 105

Find the greatest common factor of each set of terms. See Section 5.1.

44. $12m$, $15m^2$, 6 **45.** $45a^3$, $85a^2$, $105a^4$ **46.** $72p^2$, $28p^5$

6.3 LEAST COMMON DENOMINATORS

1 In this section, we demonstrate a preliminary step needed to add or subtract rational expressions with different denominators. Just as with rational numbers, adding or subtracting rational expressions (to be discussed in the next section) often requires a **least common denominator,** the least expression that all denominators divide into without a remainder. For example, the least common denominator for $\frac{2}{9}$ and $\frac{5}{12}$ is 36, since 36 is the smallest number that both 9 and 12 divide into.

Least common denominators often can be found by inspection. For example, the least common denominator for $\frac{1}{6}$ and $\frac{2}{3m}$ is $6m$. In other cases, a least common denominator can be found by a procedure similar to that used in Chapter 5 for finding the greatest common factor.

OBJECTIVES

1. Find least common denominators.
2. Rewrite rational expressions with the least common denominator.

FINDING A LEAST COMMON DENOMINATOR

1. Factor each denominator into prime factors.
2. List each different denominator factor the *greatest* number of times it appears in any denominator.
3. Multiply the denominators from Step 2 to get the least common denominator.

When each denominator is factored into prime factors, every prime factor must divide evenly into the least common denominator. The least common denominator is often abbreviated LCD.

In Example 1, the least common denominator is found both for numerical denominators and algebraic denominators.

EXAMPLE 1 Finding the Least Common Denominator
Find the least common denominator for each pair of fractions.

(a) $\frac{1}{24}, \frac{7}{15}$

(b) $\frac{1}{8x}, \frac{3}{10x}$

Write each denominator in factored form, with numerical coefficients in prime factored form.

(a)
$$24 = 2 \cdot 2 \cdot 2 \cdot 3 = 2^3 \cdot 3$$
$$15 = 3 \cdot 5$$

(b)
$$8x = 2 \cdot 2 \cdot 2 \cdot x = 2^3 \cdot x$$
$$10x = 2 \cdot 5 \cdot x$$

The LCD is found by taking each different factor the greatest number of times it appears as a factor in any of the denominators. That is, each factor must be in the LCD and raised to its highest power.

(a) The factor 2 appears three times in one product and not at all in the other, so the greatest number of times 2 appears is three. The greatest number of times both 3 and 5 appear is one.

$$\text{LCD} = 2 \cdot 2 \cdot 2 \cdot 3 \cdot 5 = 2^3 \cdot 3 \cdot 5 = 120$$

(b) Here 2 appears three times in one product and once in the other, so the greatest number of times the 2 appears is three. The greatest number of times the 5 appears is one, and the greatest number of times x appears is one.

$$\text{LCD} = 2 \cdot 2 \cdot 2 \cdot 5 \cdot x = 2^3 \cdot 5 \cdot x = 40x$$ ■

WORK PROBLEM 1 AT THE SIDE.

1. Find the least common denominator.

(a) $\frac{7}{20p}, \frac{11}{30p}$

(b) $\frac{13}{12m}, \frac{17}{30m}$

(c) $\frac{19}{60r}, \frac{5}{72r}$

(d) $\frac{9}{8m^4}, \frac{11}{12m^6}$

(e) $\frac{8}{15p^8}, \frac{7}{20p^2}$

ANSWERS
1. (a) $60p$ (b) $60m$ (c) $360r$ (d) $24m^6$ (e) $60p^8$

EXAMPLE 2 Finding the LCD

Find the LCD for $\frac{5}{6r^2}$ and $\frac{3}{4r^3}$.

Factor each denominator.

$$6r^2 = 2 \cdot 3 \cdot r^2 \text{ and } 4r^3 = 2 \cdot 2 \cdot r^3 = 2^2 \cdot r^3$$

Look for the greatest number of times each different factor appears, that is for the highest power of each factor. The highest power of 2 is 2, the highest power of 3 is 1, and the highest power of r is 3; therefore,

$$\text{LCD} = 2^2 \cdot 3^1 \cdot r^3 = 12r^3. \quad ■$$

WORK PROBLEM 2 AT THE SIDE.

EXAMPLE 3 Finding the LCD

Find the LCD.

(a) $\frac{6}{5m}, \frac{4}{m^2 - 3m}$

Factor each denominator.

$$5m = 5 \cdot m \qquad m^2 - 3m = m(m - 3)$$

Take each different factor the greatest number of times it appears as a factor.

$$\text{LCD} = 5 \cdot m \cdot (m - 3) = 5m(m - 3)$$

Since m is not a *factor* of $m - 3$, both factors, m and $m - 3$, must appear in the least common denominator.

(b) $\frac{1}{r^2 - 4r - 5}, \frac{3}{r^2 - r - 20}$

Factor each denominator.

$$r^2 - 4r - 5 = (r - 5)(r + 1)$$
$$r^2 - r - 20 = (r - 5)(r + 4)$$

The LCD is

$$(r - 5)(r + 1)(r + 4).$$

(c) $\frac{1}{q - 5}, \frac{3}{5 - q}$

The expression $5 - q$ can be written as $-1(q - 5)$, since

$$-1(q - 5) = -q + 5 = 5 - q.$$

Because of this, either $q - 5$ or $5 - q$ can be used as the LCD. ■

WORK PROBLEM 3 AT THE SIDE.

2. Find the LCD.

(a) $\frac{4}{16m^3n}, \frac{5}{9m^5}$

(b) $\frac{3}{25a^2}, \frac{2}{10a^3b}$

3. Find the LCD.

(a) $\frac{1}{12a}, \frac{5}{a^2 - 4a}$

(b) $\frac{6}{x^2 - 4x}, \frac{3x - 1}{x^2 - 16}$

(c) $\frac{2m}{m^2 - 3m + 2}, \frac{5m - 3}{m^2 + 3m - 10}$

ANSWERS
2. (a) $144m^5n$ (b) $50a^3b$
3. (a) $12a(a - 4)$
(b) $x(x - 4)(x + 4)$
(c) $(m - 1)(m - 2)(m + 5)$

2 Once the least common denominator has been found, the next step in preparing two fractions for addition or subtraction is to use the fundamental property to rewrite the fractions with the least common denominator. The next example shows how to do this with both numerical and algebraic fractions.

EXAMPLE 4 Writing a Fraction with a Given Denominator

Rewrite each rational expression with the indicated denominator.

(a) $\frac{3}{8} = \frac{}{40}$

(b) $\frac{9k}{25} = \frac{}{50k}$

For each example, decide what quantity the denominator on the left must be multiplied by to get the denominator on the right. (Find this quantity by dividing.)

Since $40 \div 8$ is 5, using the multiplicative identity property, multiply by $\frac{5}{5}$ to get a denominator of 40.

$$\frac{3}{8} = \frac{3}{8} \cdot \frac{5}{5} = \frac{15}{40}$$

Since $50k \div 25 = 2k$, using the multiplicative identity property, multiply by $\frac{2k}{2k}$ to get a denominator of $50k$.

$$\frac{9k}{25} = \frac{9k}{25} \cdot \frac{2k}{2k} = \frac{18k^2}{50k}$$ ■

EXAMPLE 5 Writing a Fraction with a Given Denominator

Rewrite the following rational expression with the indicated denominator.

$$\frac{12p}{p^2 + 8p} = \frac{}{p(p + 8)(p - 4)}$$

Factor $p^2 + 8p$ as $p(p + 8)$. Since $\frac{p(p + 8)(p - 4)}{p(p + 8)} = p - 4$, multiply by $\frac{p - 4}{p - 4}$.

$$\frac{12p}{p^2 + 8p} = \frac{12p}{p(p + 8)} \cdot \frac{p - 4}{p - 4} = \frac{12p(p - 4)}{p(p + 8)(p - 4)}$$ ■

WORK PROBLEM 4 AT THE SIDE.

4. Rewrite each rational expression with the indicated denominator.

(a) $\frac{7k}{5} = \frac{}{30}$

(b) $\frac{9}{2a + 5} = \frac{}{6a + 15}$

(c) $\frac{5k + 1}{k^2 + 2k} = \frac{}{k(k + 2)(k - 1)}$

ANSWERS

4. (a) $\frac{42k}{30}$ (b) $\frac{27}{6a + 15}$ (c) $\frac{(5k + 1)(k - 1)}{k(k + 2)(k - 1)}$

Historical Reflections

Babylonian Mathematics: Plimpton 322

Our knowledge of the mathematics of the Babylonians of Mesopotamia is based largely on archeological discoveries of thousands of clay tablets, some of which are strictly mathematical in nature. Probably the most remarkable of all the tablets is the one labelled Plimpton 322. It includes several columns of inscriptions that represent numbers. Researchers have determined that the last column on the right simply serves to number the lines, but two other columns represent lengths of hypotenuses and legs of right triangles with integer-valued sides. For example, the fifth line contains the entries 65 and 97. The number 65 represents the length of the shorter leg of a right triangle with hypotenuse 97 and longer leg 72. This can be verified by the Pythagorean relationship

$$a^2 + b^2 = c^2.$$

Thus, it is apparent that even though the famous theorem relating sides of right triangles is named after the Greek Pythagoras, the relationship was known more than one thousand years prior to his time.

Art: George Arthur Plimpton Collection, Rare Book & Manuscript Library, Columbia University

name date hour

6.3 EXERCISES

Find the least common denominator for each list of rational expressions. See Examples 1, 2, and 3.

1. $\frac{5}{12}, \frac{7}{10}$

2. $\frac{1}{4}, \frac{5}{6}$

3. $\frac{7}{15}, \frac{11}{20}, \frac{5}{24}$

4. $\frac{9}{10}, \frac{12}{25}, \frac{11}{35}$

5. $\frac{17}{100}, \frac{13}{120}, \frac{29}{180}$

6. $\frac{17}{250}, \frac{1}{300}, \frac{127}{360}$

7. $\frac{9}{x^2}, \frac{8}{x^5}$

8. $\frac{2}{m^7}, \frac{3}{m^8}$

9. $\frac{2}{5p}, \frac{5}{6p}$

10. $\frac{4}{15k}, \frac{3}{4k}$

11. $\frac{2}{25m}, \frac{17}{30m}$

12. $\frac{9}{40a}, \frac{11}{60a}$

13. $\frac{7}{15y^2}, \frac{5}{36y^4}$

14. $\frac{4}{25m^3}, \frac{7}{10m^4}$

15. $\frac{3}{50r^5}, \frac{7}{60r^3}$

16. $\frac{9}{25a^5}, \frac{17}{80a^6}$

17. $\frac{5}{32r^2}, \frac{9}{16r - 32}$

18. $\frac{13}{18m^3}, \frac{8}{9m - 36}$

19. $\frac{7}{6r - 12}, \frac{4}{9r - 18}$

20. $\frac{4}{5p - 30}, \frac{5}{6p - 36}$

21. $\frac{5}{12p + 60}, \frac{1}{p^2 + 5p}$

22. $\frac{1}{r^2 + 7r}, \frac{3}{5r + 35}$

23. $\frac{9}{p - q}, \frac{3}{q - p}$

24. $\frac{4}{z - x}, \frac{8}{x - z}$

25. $\frac{6}{a^2 + 6a}, \frac{4}{a^2 + 3a - 18}$

26. $\frac{3}{y^2 - 5y}, \frac{2}{y^2 - 2y - 15}$

27. $\frac{5}{k^2 + 2k - 35}, \frac{3}{k^2 + 3k - 40}$

28. $\frac{9}{z^2 + 4z - 12}, \frac{1}{z^2 + z - 30}$

29. $\frac{3}{2y^2 + 7y - 4}, \frac{7}{2y^2 - 7y + 3}$

30. $\frac{2}{5a^2 + 13a - 6}, \frac{6}{5a^2 - 22a + 8}$

Rewrite each rational expression with the given denominator. See Examples 4 and 5.

31. $\frac{7}{11} = \frac{}{66}$

32. $\frac{5}{8} = \frac{}{56}$

33. $\frac{9}{r} = \frac{}{6r}$

34. $\frac{7}{k} = \frac{}{9k}$

35. $\frac{-11}{m} = \frac{}{8m}$

36. $\frac{-5}{z} = \frac{}{6z}$

37. $\frac{12}{35y} = \frac{}{70y^3}$

38. $\frac{17}{9r} = \frac{}{36r^2}$

39. $\frac{15m^2}{8k} = \frac{}{32k^4}$

40. $\frac{5t^2}{3y} = \frac{}{9y^2}$

41. $\frac{19z}{2z - 6} = \frac{}{6z - 18}$

42. $\frac{2r}{5r - 5} = \frac{}{15r - 15}$

43. $\frac{-2a}{9a - 18} = \frac{}{18a - 36}$

44. $\frac{-5y}{6y + 18} = \frac{}{24y + 72}$

45. $\frac{6}{k^2 - 4k} = \frac{}{k(k - 4)(k + 1)}$

46. $\frac{15}{m^2 - 9m} = \frac{}{m(m - 9)(m + 8)}$

SKILL SHARPENERS

Add or subtract as indicated. See Section 1.1.

47. $\frac{2}{3} + \frac{5}{3}$

48. $\frac{9}{5} - \frac{2}{5}$

49. $\frac{1}{2} + \frac{2}{5}$

50. $\frac{5}{4} + \frac{1}{3}$

51. $\frac{8}{10} - \frac{2}{15}$

52. $\frac{7}{12} - \frac{6}{15}$

53. $\frac{5}{24} + \frac{1}{18}$

54. $\frac{11}{36} - \frac{7}{45}$

6.4 ADDITION AND SUBTRACTION OF RATIONAL EXPRESSIONS

OBJECTIVES

1. Add rational expressions having the same denominator.
2. Add rational expressions having different denominators.
3. Subtract rational expressions.

1 The sum of two rational expressions is found with a procedure similar to that for adding two fractions.

ADDING RATIONAL EXPRESSIONS

If $\frac{P}{Q}$ and $\frac{R}{Q}$ are rational expressions, then

$$\frac{P}{Q} + \frac{R}{Q} = \frac{P + R}{Q}.$$

Again, the first example shows how the addition of rational expressions compares with that of rational numbers.

EXAMPLE 1 Adding Rational Expressions with the Same Denominator

Add.

(a) $\frac{4}{7} + \frac{2}{7}$

(b) $\frac{3x}{x + 1} + \frac{2x}{x + 1}$

The denominators are the same, so the sum is found by adding the two numerators and keeping the same (common) denominator.

$$\frac{4}{7} + \frac{2}{7} = \frac{4 + 2}{7} = \frac{6}{7}$$

$$\frac{3x}{x + 1} + \frac{2x}{x + 1} = \frac{3x + 2x}{x + 1} = \frac{5x}{x + 1}$$ ■

WORK PROBLEM 1 AT THE SIDE.

1. Find each sum.

(a) $\frac{3}{y + 4} + \frac{2}{y + 4}$

(b) $\frac{x}{x + y} + \frac{1}{x + y}$

2 Use the steps given below to add two rational expressions with different denominators. These are the same steps that are used to add fractions with different denominators.

ADDING RATIONAL EXPRESSIONS WITH DIFFERENT DENOMINATORS

Step 1 Find the least common denominator (LCD).

Step 2 Rewrite each rational expression as a fraction with the least common denominator as the denominator.

Step 3 Add the numerators to get the numerator of the sum. The least common denominator is the denominator of the sum.

Step 4 Write the answer in lowest terms.

ANSWERS

1. (a) $\frac{5}{y + 4}$ (b) $\frac{x + 1}{x + y}$

2. Find each sum.

(a) $\dfrac{6}{5x} + \dfrac{9}{2x}$

(b) $\dfrac{m}{3n} + \dfrac{2}{7n}$

3. Find the sums.

(a) $\dfrac{2p}{3p + 3} + \dfrac{5p}{2p + 2}$

(b) $\dfrac{4}{y^2 - 1} + \dfrac{6}{y + 1}$

(c) $\dfrac{-2}{p + 1} + \dfrac{4p}{p^2 - 1}$

ANSWERS

2. (a) $\dfrac{57}{10x}$ (b) $\dfrac{7m + 6}{21n}$

3. (a) $\dfrac{19p}{6(p + 1)}$ (b) $\dfrac{6y - 2}{(y + 1)(y - 1)}$

(c) $\dfrac{2}{p - 1}$

EXAMPLE 2 Adding Rational Expressions with Different Denominators

Add.

(a) $\dfrac{1}{12} + \dfrac{7}{15}$ | **(b)** $\dfrac{2}{3y} + \dfrac{1}{4y}$

First find the LCD, using the methods of the last section.

$\text{LCD} = 2^2 \cdot 3 \cdot 5 = 60$ | $\text{LCD} = 2^2 \cdot 3 \cdot y = 12y$

Now rewrite each rational expression as a fraction with the LCD, either 60 or $12y$, as the denominator.

$$\frac{1}{12} + \frac{7}{15} = \frac{1(5)}{12(5)} + \frac{7(4)}{15(4)} = \frac{5}{\mathbf{60}} + \frac{28}{\mathbf{60}}$$

$$\frac{2}{3y} + \frac{1}{4y} = \frac{2(4)}{3y(4)} + \frac{1(3)}{4y(3)} = \frac{8}{\mathbf{12y}} + \frac{3}{\mathbf{12y}}$$

Since the fractions now have common denominators, add the numerators and use the LCD as the denominator of the sum. Write in lowest terms.

$$\frac{5}{60} + \frac{28}{60} = \frac{5 + 28}{60} = \frac{33}{60} = \frac{11}{20}$$

$$\frac{8}{12y} + \frac{3}{12y} = \frac{8 + 3}{12y} = \frac{11}{12y} \quad ■$$

WORK PROBLEM 2 AT THE SIDE.

EXAMPLE 3 Adding Rational Expressions

Add $\dfrac{2x}{x^2 - 1}$ and $\dfrac{-1}{x + 1}$.

Find the least common denominator by factoring both denominators.

$$x^2 - 1 = (x + 1)(x - 1); \qquad x + 1 \text{ cannot be factored.}$$

Write the sum with denominators in factored form as

$$\frac{2x}{(x + 1)(x - 1)} + \frac{-1}{x + 1}.$$

The LCD is $(x + 1)(x - 1)$. Here only the second fraction must be changed. Multiply the numerator and denominator of the second fraction by $x - 1$.

$$\frac{2x}{(x + 1)(x - 1)} + \frac{-1\,\mathbf{(x - 1)}}{(x + 1)\,\mathbf{(x - 1)}} \qquad \text{Multiply by } \frac{x - 1}{x - 1}.$$

With both denominators now the same, add the numerators and use the LCD as the denominator of the sum.

$$\frac{2x - (x - 1)}{(x + 1)(x - 1)} = \frac{2x - x + 1}{(x + 1)(x - 1)} \qquad \text{Add numerators.}$$

$$= \frac{x + 1}{(x + 1)(x - 1)} \qquad \text{Combine terms.}$$

$$= \frac{1}{x - 1} \qquad \text{Lowest terms} \quad ■$$

WORK PROBLEM 3 AT THE SIDE.

EXAMPLE 4 Adding Rational Expressions

Add $\frac{2x}{x^2 + 5x + 6}$ and $\frac{x + 1}{x^2 + 2x - 3}$.

$$\frac{2x}{(x + 2)(x + 3)} + \frac{x + 1}{(x + 3)(x - 1)}$$ Factor denominators.

The LCD is $(x + 2)(x + 3)(x - 1)$. By the fundamental property,

$$\frac{2x}{(x + 2)(x + 3)} + \frac{x + 1}{(x + 3)(x - 1)}$$
$$= \frac{2x(\mathbf{x - 1})}{(x + 2)(x + 3)(\mathbf{x - 1})} + \frac{(x + 1)(\mathbf{x + 2})}{(x + 3)(x - 1)(\mathbf{x + 2})}.$$

Since the two rational expressions now have the same denominator, add their numerators. The LCD is the denominator of the sum.

$$\frac{2x(x - 1)}{(x + 2)(x + 3)(x - 1)} + \frac{(x + 1)(x + 2)}{(x + 3)(x - 1)(x + 2)}$$
$$= \frac{2x(x - 1) + (x + 1)(x + 2)}{(x + 2)(x + 3)(x - 1)}$$ Add numerators.
$$= \frac{2x^2 - 2x + x^2 + 3x + 2}{(x + 2)(x + 3)(x - 1)}$$ Distributive property
$$= \frac{3x^2 + x + 2}{(x + 2)(x + 3)(x - 1)}$$ Combine terms.

It is usually more convenient to leave the denominator in factored form. ■

WORK PROBLEM 4 AT THE SIDE.

3 Subtract rational expressions as follows.

SUBTRACTING RATIONAL EXPRESSIONS

If $\frac{P}{Q}$ and $\frac{R}{Q}$ are rational expressions, then

$$\frac{P}{Q} - \frac{R}{Q} = \frac{P - R}{Q}.$$

EXAMPLE 5 Subtracting Rational Expressions

Subtract: $\frac{2m}{m - 1} - \frac{2}{m - 1}$.

$$\frac{2m}{m - 1} - \frac{2}{m - 1} = \frac{2m - 2}{m - 1}$$ Subtract numerators.
$$= \frac{2(m - 1)}{m - 1}$$ Factor the numerator.
$$= 2.$$ Lowest terms ■

WORK PROBLEM 5 AT THE SIDE.

4. Add.

(a) $\frac{2k}{k^2 - 5k + 4} + \frac{3}{k^2 - 1}$

(b) $\frac{4m}{m^2 + 3m + 2} + \frac{2m - 1}{m^2 + 6m + 5}$

5. Find each difference. Write answers in lowest terms.

(a) $\frac{5}{2y} - \frac{9}{2y}$

(b) $\frac{3}{m^2} - \frac{2}{m^2}$

ANSWERS

4. (a) $\frac{2k^2 + 5k - 12}{(k - 4)(k - 1)(k + 1)}$
(b) $\frac{6m^2 + 23m - 2}{(m + 2)(m + 1)(m + 5)}$

5. (a) $\frac{-2}{y}$ (b) $\frac{1}{m^2}$

6. Subtract.

(a) $\frac{9}{5p} - \frac{2}{3p}$

(b) $\frac{4}{m^2} - \frac{3}{m}$

(c) $\frac{1}{k+4} - \frac{2}{k}$

(d) $\frac{6}{a+2} - \frac{1}{a-3}$

7. Subtract.

(a) $\frac{4y}{y^2-1} - \frac{5}{y^2+2y+1}$

(b) $\frac{3r}{r^2-5r} - \frac{4}{r^2-10r+25}$

ANSWERS

6. (a) $\frac{17}{15p}$ (b) $\frac{4-3m}{m^2}$
(c) $\frac{-k-8}{k(k+4)}$ (d) $\frac{5a-20}{(a+2)(a-3)}$

7. (a) $\frac{4y^2-y+5}{(y+1)^2(y-1)}$ (b) $\frac{3r-19}{(r-5)^2}$

EXAMPLE 6 Subtracting Rational Expressions

Subtract: $\frac{9}{x-2} - \frac{3}{x}$.

The LCD is $x(x-2)$.

$$\frac{9}{x-2} - \frac{3}{x} = \frac{9\mathbf{x}}{\mathbf{x}(x-2)} - \frac{3\mathbf{(x-2)}}{x\mathbf{(x-2)}} \quad \text{Get least common denominator.}$$

$$= \frac{9x - 3(x-2)}{x(x-2)} \quad \text{Subtract numerators.}$$

$$= \frac{9x - 3x + 6}{x(x-2)} \quad \text{Distributive property}$$

$$= \frac{6x+6}{x(x-2)} \quad \text{Combine terms.}$$

$$= \frac{6(x+1)}{x(x-2)} \quad \text{Factor.} \blacksquare$$

WORK PROBLEM 6 AT THE SIDE.

EXAMPLE 7 Subtracting Rational Expressions

Find $\frac{6x}{x^2-2x+1} - \frac{1}{x^2-1}$.

Begin by factoring the denominators.

$$x^2 - 2x + 1 = (x-1)(x-1) \text{ and } x^2 - 1 = (x-1)(x+1)$$

From the factored denominators, identify the LCD, $(x-1)(x-1)(x+1)$. Use the factor $x - 1$ twice, since it appears twice in the first denominator.

$$\frac{6x}{(x-1)(x-1)} - \frac{1}{(x-1)(x+1)}$$

$$= \frac{6x\mathbf{(x+1)}}{(x-1)(x-1)\mathbf{(x+1)}} - \frac{1\mathbf{(x-1)}}{\mathbf{(x-1)}(x-1)(x+1)} \quad \text{Fundamental property}$$

$$= \frac{6x(x+1) - 1(x-1)}{(x-1)(x-1)(x+1)} \quad \text{Subtract.}$$

$$= \frac{6x^2 + 6x - x + 1}{(x-1)(x-1)(x+1)} \quad \text{Distributive property}$$

$$= \frac{6x^2 + 5x + 1}{(x-1)(x-1)(x+1)} \quad \text{Combine terms.}$$

The result may be written as $\frac{6x^2+5x+1}{(x-1)^2(x+1)}$. ■

WORK PROBLEM 7 AT THE SIDE.

EXAMPLE 8 Subtracting Rational Expressions

Find $\frac{q}{q^2 - 4q - 5} - \frac{3}{2q^2 - 13q + 15}$.

Factor each denominator.

$$q^2 - 4q - 5 = (q + 1)(q - 5)$$
$$2q^2 - 13q + 15 = (q - 5)(2q - 3)$$

Now rewrite each of the two rational expressions with the LCD, $(q + 1)(q - 5)(2q - 3)$. Then subtract numerators.

$$\frac{q}{(q + 1)(q - 5)} - \frac{3}{(q - 5)(2q - 3)}$$

$$= \frac{q(\mathbf{2q - 3})}{(q + 1)(q - 5)(\mathbf{2q - 3})} - \frac{3(\mathbf{q + 1})}{(\mathbf{q + 1})(q - 5)(2q - 3)}$$

$$= \frac{q(2q - 3) - 3(q + 1)}{(q + 1)(q - 5)(2q - 3)} \qquad \text{Subtract.}$$

$$= \frac{2q^2 - 3q - 3q - 3}{(q + 1)(q - 5)(2q - 3)} \qquad \text{Distributive property}$$

$$= \frac{2q^2 - 6q - 3}{(q + 1)(q - 5)(2q - 3)} \qquad \text{Combine terms.} \blacksquare$$

WORK PROBLEM 8 AT THE SIDE.

8. Subtract.

(a) $\frac{2}{p^2 - 5p + 4} - \frac{3}{p^2 - 1}$

(b) $\frac{q}{2q^2 + 5q - 3} - \frac{3q + 4}{3q^2 + 10q + 3}$

ANSWERS

8. (a) $\frac{14 - p}{(p - 4)(p - 1)(p + 1)}$

(b) $\frac{-3q^2 - 4q + 4}{(2q - 1)(q + 3)(3q + 1)}$

Historical Reflections

Historical Reflections

WOMEN IN MATHEMATICS: Ada Augusta, Countess of Lovelace (1815–1852)

Ada Augusta, the daughter of British poet Lord Byron, has been described as the world's first computer programmer. She was one of the few people who understood what Charles Babbage was trying to accomplish with his calculating machines (forerunners of today's computers). She provided some corrections to his work, and on her own, she devised the concept of repetition of instructions known today as a "loop" or "subroutine" in computer programming.

Lady Lovelace met an untimely death at age 36 from cancer of the uterus, and her programming ideas were not implemented for another century. In 1987 the U.S. Department of Defense solidified its insistence that contractors designing programs to govern military aircraft and missile systems use a language called *Ada,* named for Lady Lovelace and first introduced in 1980. As a result of this standardization move, it was expected that the demand for Ada programmers would quadruple by the 1990s.

Art: *From Faster than Thought: A Symposium on Digital Computing Machines*. B.V. Browden, Pitman Publishing Corp. © 1953. Used with permission.

name date hour

6.4 EXERCISES

Find the sums or differences. Write each answer in lowest terms. See Examples 1 and 5.

1. $\frac{2}{p} + \frac{5}{p}$

2. $\frac{3}{r} + \frac{6}{r}$

3. $\frac{9}{k} - \frac{12}{k}$

4. $\frac{p - q}{3} - \frac{p + q}{3}$

5. $\frac{a + b}{2} - \frac{a - b}{2}$

6. $\frac{m^2}{m + 6} + \frac{6m}{m + 6}$

7. $\frac{y^2}{y - 1} + \frac{-y}{y - 1}$

8. $\frac{q^2 - 4q}{q - 2} + \frac{4}{q - 2}$

9. $\frac{z^2 - 10z}{z - 5} + \frac{25}{z - 5}$

Find the sums or differences. Write each answer in lowest terms. See Examples 2, 3, 4, 6, 7, and 8.

10. $\frac{m}{3} + \frac{1}{2}$

11. $\frac{p}{6} - \frac{2}{3}$

12. $\frac{7}{5} - \frac{z}{4}$

13. $\frac{9}{10} - \frac{r}{2}$

14. $-\frac{4}{3} - \frac{1}{y}$

15. $\frac{8}{5} - \frac{2}{a}$

16. $\frac{5m}{6} - \frac{2m}{3}$

17. $\frac{3}{x} + \frac{5}{2x}$

18. $\frac{4 + 2k}{5} + \frac{2 + k}{10}$

19. $\frac{5 - 4r}{8} - \frac{2 - 3r}{6}$

20. $\frac{3m + 5}{3} - \frac{m + 2}{6}$

21. $\frac{5r + 4}{3} - \frac{2r - 3}{9}$

22. $\frac{m + 2}{m} + \frac{m + 3}{4m}$

23. $\frac{2x - 5}{x} + \frac{x - 1}{2x}$

24. $\frac{6}{y^2} - \frac{2}{y}$

25. $\frac{3}{p} + \frac{5}{p^2}$

26. $\frac{9}{2p} - \frac{4}{p^2}$

27. $\frac{15}{4k^2} - \frac{3}{k}$

28. $\dfrac{-1}{x^2} + \dfrac{3}{xy}$

29. $\dfrac{9}{p^2} + \dfrac{4}{px}$

30. $\dfrac{8}{x-2} - \dfrac{4}{x+2}$

31. $\dfrac{6}{m-5} - \dfrac{2}{m+5}$

32. $\dfrac{2x}{x+y} - \dfrac{3x}{2x+2y}$

33. $\dfrac{1}{a-b} - \dfrac{a}{4a-4b}$

34. $\dfrac{1}{m^2-9} + \dfrac{1}{3m+9}$

35. $\dfrac{-6}{y^2-4} - \dfrac{3}{2y+4}$

36. $\dfrac{1}{m^2-1} - \dfrac{1}{m^2+3m+2}$

37. $\dfrac{1}{y^2-4} + \dfrac{3}{y^2+5y+6}$

38. $\dfrac{8}{m-2} + \dfrac{3}{5m} + \dfrac{7}{5m(m-2)}$

39. $\dfrac{-1}{7z} + \dfrac{3}{z+2} + \dfrac{4}{7z(z+2)}$

40. $\dfrac{4}{r^2-r} + \dfrac{6}{r^2+2r} - \dfrac{1}{r^2+r-2}$

41. $\dfrac{6}{k^2+3k} - \dfrac{1}{k^2-k} + \dfrac{2}{k^2+2k-3}$

42. $\dfrac{4y-1}{2y^2+5y-3} - \dfrac{y+3}{6y^2+y-2}$

43. $\dfrac{2q+1}{3q^2+10q-8} - \dfrac{3q+5}{2q^2+5q-12}$

SKILL SHARPENERS

Simplify each expression using the order of operations as necessary. See Sections 1.1 and 1.4.

44. $\dfrac{\frac{2}{3}+\frac{5}{3}}{\frac{3}{4}-\frac{1}{4}}$

45. $\dfrac{\frac{5}{8}-\frac{3}{8}}{\frac{2}{5}+\frac{1}{5}}$

46. $\dfrac{\frac{5}{6}-\frac{2}{3}}{\frac{3}{4}+\frac{1}{12}}$

47. $\dfrac{\frac{9}{7}-\frac{3}{14}}{\frac{1}{3}+\frac{5}{12}}$

6.5 COMPLEX FRACTIONS

A rational expression with fractions in the numerator, denominator, or both, is called a **complex fraction.** Examples of complex fractions include

$$\frac{3 + \frac{4}{x}}{5}, \quad \frac{\frac{3x^2 - 5x}{6x^2}}{2x - \frac{1}{x}}, \quad \text{and} \quad \frac{3 + x}{5 - \frac{2}{x}}.$$

The parts of a complex fraction are named as follows.

$$\frac{\frac{2}{p} - \frac{1}{q}}{\frac{3}{p} + \frac{5}{q}}$$

←Numerator of complex fraction
←Main fraction bar
←Denominator of complex fraction

Complex fractions can always be simplified to rational expressions without fractions in the numerator and denominator. Two methods are commonly used to do this. We show both methods in this section. You should choose the method you like best and stick to it to avoid confusion.

OBJECTIVES

1. Simplify complex fractions by simplifying numerator and denominator (Method 1).
2. Simplify complex fractions by multiplying by least common denominator (Method 2).

1 **Method 1** Since a fraction represents a quotient, one method of simplifying complex fractions is to rewrite both the numerator and denominator as single fractions, and then to perform the indicated division.

EXAMPLE 1 Simplifying Complex Fractions by Method 1

Simplify each complex fraction.

(a) $\dfrac{\frac{2}{3} + \frac{5}{9}}{\frac{1}{4} + \frac{1}{12}}$ **(b)** $\dfrac{6 + \frac{3}{x}}{\frac{x}{4} + \frac{1}{8}}$

First, write each numerator as a single fraction.

$$\frac{2}{3} + \frac{5}{9} = \frac{2(3)}{3(3)} + \frac{5}{9} = \frac{6}{9} + \frac{5}{9} = \frac{11}{9}$$

$$6 + \frac{3}{x} = \frac{6}{1} + \frac{3}{x} = \frac{6x}{x} + \frac{3}{x} = \frac{6x + 3}{x}$$

Do the same thing with each denominator.

$$\frac{1}{4} + \frac{1}{12} = \frac{1(3)}{4(3)} + \frac{1}{12} = \frac{3}{12} + \frac{1}{12} = \frac{4}{12}$$

$$\frac{x}{4} + \frac{1}{8} = \frac{x(2)}{4(2)} + \frac{1}{8} = \frac{2x}{8} + \frac{1}{8} = \frac{2x + 1}{8}$$

The original complex fraction can now be written as follows.

$$\frac{\frac{11}{9}}{\frac{4}{12}} \qquad \frac{\frac{6x + 3}{x}}{\frac{2x + 1}{8}}$$

1. Simplify the complex fractions.

(a) $\dfrac{6 + \dfrac{1}{x}}{5 - \dfrac{2}{x}}$

(b) $\dfrac{9 - \dfrac{4}{p}}{\dfrac{2}{p} + 1}$

(c) $\dfrac{m + \dfrac{1}{2}}{\dfrac{6m + 3}{4m}}$

2. Simplify the complex fractions.

(a) $\dfrac{\dfrac{rs^2}{t}}{\dfrac{r^2s}{t^2}}$

(b) $\dfrac{\dfrac{m^2n^3}{p}}{\dfrac{m^4n}{p^2}}$

ANSWERS

1. (a) $\dfrac{6x + 1}{5x - 2}$ (b) $\dfrac{9p - 4}{2 + p}$ (c) $\dfrac{2m}{3}$

2. (a) $\dfrac{st}{r}$ (b) $\dfrac{n^2p}{m^2}$

Now use the rule for division and the fundamental property.

$$\frac{11}{9} \div \frac{4}{12} = \frac{11}{9} \cdot \frac{12}{4} = \frac{11 \cdot 2 \cdot 2 \cdot 3}{3 \cdot 3 \cdot 2 \cdot 2} = \frac{11}{3}$$

$$\frac{6x + 3}{x} \div \frac{2x + 1}{8} = \frac{6x + 3}{x} \cdot \frac{8}{2x + 1} = \frac{3(2x + 1)}{x} \cdot \frac{8}{2x + 1} = \frac{24}{x} \quad ■$$

WORK PROBLEM 1 AT THE SIDE.

EXAMPLE 2 Simplifying a Complex Fraction by Method 1

Simplify the complex fraction $\dfrac{\dfrac{xp}{q^3}}{\dfrac{p^2}{qx^2}}$.

Here the numerator and denominator are already single fractions, so use the rule for division and then the fundamental property.

$$\frac{xp}{q^3} \div \frac{p^2}{qx^2} = \frac{xp}{q^3} \cdot \frac{qx^2}{p^2} = \frac{x^3}{q^2p} \quad ■$$

WORK PROBLEM 2 AT THE SIDE.

2 **Method 2** As an alternative method, complex fractions may be simplified by multiplying both numerator and denominator by the least common denominator of all the denominators appearing in the complex fraction. In the next example, this second method is used to simplify the same complex fractions as in Example 1.

EXAMPLE 3 Simplifying Complex Fractions by Method 2

Simplify each complex fraction.

(a) $\dfrac{\dfrac{2}{3} + \dfrac{5}{9}}{\dfrac{1}{4} + \dfrac{1}{12}}$

(b) $\dfrac{6 + \dfrac{3}{x}}{\dfrac{x}{4} + \dfrac{1}{8}}$

Find the least common denominator for all the denominators in the complex fraction.

The LCD for 3, 9, 4, and 12 is 36.

The LCD for x, 4, and 8 is $8x$.

Multiply the numerator and denominator of the complex fraction by the LCD.

$$\frac{\frac{2}{3}+\frac{5}{9}}{\frac{1}{4}+\frac{1}{12}} = \frac{36\left(\frac{2}{3}+\frac{5}{9}\right)}{36\left(\frac{1}{4}+\frac{1}{12}\right)} = \frac{36\left(\frac{2}{3}\right)+36\left(\frac{5}{9}\right)}{36\left(\frac{1}{4}\right)+36\left(\frac{1}{12}\right)} = \frac{24+20}{9+3} = \frac{44}{12} = \frac{11}{3}$$

$$\frac{6+\frac{3}{x}}{\frac{x}{4}+\frac{1}{8}} = \frac{8x\left(6+\frac{3}{x}\right)}{8x\left(\frac{x}{4}+\frac{1}{8}\right)} = \frac{8x(6)+8x\left(\frac{3}{x}\right)}{8x\left(\frac{x}{4}\right)+8x\left(\frac{1}{8}\right)} = \frac{48x+24}{2x^2+x} = \frac{24(2x+1)}{x(2x+1)} = \frac{24}{x} \quad ■$$

EXAMPLE 4 Simplifying a Complex Fraction by Method 2

Simplify $\dfrac{\frac{3}{m+5}}{\frac{9}{m+2}}$.

The LCD is $(m+5)(m+2)$. Multiply numerator and denominator of the complex fraction by the LCD.

$$\frac{\frac{3}{m+5}}{\frac{9}{m+2}} \cdot \frac{(m+5)(m+2)}{(m+5)(m+2)} = \frac{3(m+2)}{9(m+5)}$$

$$= \frac{m+2}{3(m+5)} \quad ■$$

WORK PROBLEM 3 AT THE SIDE.

3. Simplify by the second method.

(a) $\dfrac{2-\frac{6}{a}}{3+\frac{4}{a}}$

(b) $\dfrac{\frac{5}{p}-6}{\frac{2p+1}{p}}$

(c) $\dfrac{\frac{-4}{3+x}}{\frac{5}{2-x}}$

ANSWERS

3. (a) $\dfrac{2a-6}{3a+4}$ (b) $\dfrac{5-6p}{2p+1}$ (c) $\dfrac{4(x-2)}{5(3+x)}$

The two methods for simplifying a complex fraction are summarized below.

SIMPLIFYING COMPLEX FRACTIONS

Method 1 Simplify the numerator and denominator of the complex fraction separately. Then divide the simplified numerator by the simplified denominator.

Method 2 Multiply numerator and denominator of the complex fraction by the least common denominator of all the denominators appearing in the complex fraction.

You may want to choose one method of simplifying complex fractions and stick to it. Although either method can be used correctly to simplify any complex fraction, some students prefer to use Method 1 for problems like Example 2 and Example 4, which are the quotient of two fractions. They find Method 2 works best for problems like Example 1 (or Example 3), which has a sum or difference in the numerator or denominator or both.

name date hour

6.5 EXERCISES

Simplify each complex fraction. Use either method. See Examples 1-4.

1. $\dfrac{\frac{5}{8} + \frac{2}{3}}{\frac{7}{3} - \frac{1}{4}}$

2. $\dfrac{\frac{6}{5} - \frac{1}{9}}{\frac{2}{5} + \frac{5}{3}}$

3. $\dfrac{1 - \frac{3}{8}}{2 + \frac{1}{4}}$

4. $\dfrac{3 + \frac{5}{4}}{1 - \frac{7}{8}}$

5. $\dfrac{\frac{m^3p^4}{5m}}{\frac{8mp^5}{p^2}}$

6. $\dfrac{\frac{9z^5x^3}{2x}}{\frac{8z^2x^5}{3z}}$

7. $\dfrac{\frac{x + 1}{4}}{\frac{x - 3}{x}}$

8. $\dfrac{\frac{m + 6}{m}}{\frac{m - 6}{2}}$

9. $\dfrac{\frac{3}{y} + 1}{\frac{3 + y}{2}}$

10. $\dfrac{6 + \frac{2}{r}}{\frac{r + 2}{3}}$

11. $\dfrac{\frac{1}{x} + x}{\frac{x^2 + 1}{8}}$

12. $\dfrac{\frac{3}{m} - m}{\frac{3 - m^2}{4}}$

13. $\dfrac{y - \frac{6}{y}}{y + \frac{2}{y}}$

14. $\dfrac{\frac{p + 3}{p}}{\frac{1}{p} + \frac{1}{5}}$

15. $\dfrac{r + \frac{1}{r}}{\frac{1}{r} - r}$

16. $\dfrac{\frac{2}{p^2} - \frac{3}{5p}}{\frac{4}{p} + \frac{1}{4p}}$

17. $\dfrac{\frac{1}{m^3p} + \frac{2}{mp^2}}{\frac{4}{mp} + \frac{1}{m^2p}}$

18. $\dfrac{\frac{-15}{a - 4}}{\frac{3}{a + 2}}$

19. $\dfrac{\frac{9m}{m - 3}}{\frac{18}{m + 3}}$

20. $\dfrac{\dfrac{1}{z+5}}{\dfrac{4}{z^2-25}}$

21. $\dfrac{\dfrac{a}{a+1}}{\dfrac{2}{a^2-1}}$

22. $\dfrac{\dfrac{1}{m+1}-1}{\dfrac{1}{m+1}+1}$

23. $\dfrac{\dfrac{2}{x-1}+2}{\dfrac{2}{x-1}-2}$

24. $\dfrac{\dfrac{1}{m-1}+\dfrac{2}{m+2}}{\dfrac{2}{m+2}-\dfrac{1}{m-3}}$

25. $\dfrac{\dfrac{5}{r+3}-\dfrac{1}{r-1}}{\dfrac{2}{r+2}+\dfrac{3}{r+3}}$

The following exercises are real "head-scratchers." (Hint: Begin at the lower right and work upward.)

26. $1-\dfrac{1}{1+\dfrac{1}{1+1}}$

27. $3-\dfrac{2}{4+\dfrac{2}{4-2}}$

SKILL SHARPENERS

Solve each equation. See Section 3.3.

28. $3x - 5 = 7x + 3$

29. $9z + 2 = 7z - 6$

30. $8(2q + 5) - 1 = 7q$

31. $6(z - 3) + 5 = 8z$

32. $9 - (5 - 3y) = 6$

33. $-(8 - a) + 5a = -7$

6.6 EQUATIONS INVOLVING RATIONAL EXPRESSIONS

1 When an equation involves fractions, the multiplication property of equality can be used first to clear it of fractions. The equation can then be solved in the usual way. The goal is to get an equivalent equation (one with the same solutions) that does not have fractions. Choose as multiplier the least common denominator of all denominators in the fractions of the equation.

OBJECTIVES

1. Solve equations involving rational expressions.
2. Solve for a specified variable.

EXAMPLE 1 Solving an Equation Involving Rational Expressions

Solve $\frac{x}{3} + \frac{x}{4} = 10 + x$.

Since the least common denominator of the two fractions is 12, begin by multiplying each side of the equation by 12.

$$12\left(\frac{x}{3} + \frac{x}{4}\right) = 12(10 + x)$$

$$12\left(\frac{x}{3}\right) + 12\left(\frac{x}{4}\right) = 12(10) + 12x \qquad \text{Distributive property}$$

$$\frac{12x}{3} + \frac{12x}{4} = 120 + 12x$$

$$4x + 3x = 120 + 12x$$

This equation has no fractions. Solve it using the methods of solving linear equations given earlier.

$$7x = 120 + 12x \qquad \text{Combine terms.}$$

$$-5x = 120 \qquad \text{Subtract } 12x.$$

$$x = -24 \qquad \text{Divide by } -5.$$

Check by substituting -24 for x in the original equation. ■

WORK PROBLEMS 1 AND 2 AT THE SIDE.

CAUTION Note that use of the LCD here is different from its use in the previous section. Here, we multiply each side of an *equation* by the LCD. Earlier, we multiplied a *fraction* by another fraction that had the LCD as both its numerator and denominator.

EXAMPLE 2 Solving an Equation Involving Rational Expressions

Solve $\frac{p}{2} - \frac{p-1}{3} = 1$.

$$6\left(\frac{p}{2} - \frac{p-1}{3}\right) = 6 \cdot 1 \qquad \text{Multiply by the LCD, 6.}$$

$$6\left(\frac{p}{2}\right) - 6\left(\frac{p-1}{3}\right) = 6 \qquad \text{Distributive property}$$

$$3p - 2(p - 1) = 6$$

Be very careful to put parentheses around $p - 1$; otherwise an incorrect solution may be found. Continue simplifying and solve.

$$3p - 2p + 2 = 6 \qquad \text{Distributive property}$$

$$p + 2 = 6 \qquad \text{Combine terms.}$$

$$p = 4 \qquad \text{Subtract 2.}$$

Check to see that 4 is correct by replacing p with 4 in the original equation. ■

1. Check the solution to Example 1 by substituting -24 for x in the original equation,

$$\frac{x}{3} + \frac{x}{4} = 10 + x.$$

Is -24 the solution?

2. Solve each equation and check your answer.

(a) $\frac{x}{5} + 3 = \frac{3}{5}$

(b) $\frac{x}{2} - \frac{x}{3} = \frac{5}{6}$

(c) $\frac{k}{7} - \frac{k}{2} = -5$

ANSWERS
1. yes
2. (a) -12 (b) 5 (c) 14

3. Solve each equation and check your answer.

(a) $\frac{k}{6} - \frac{k+1}{4} = -\frac{1}{2}$

(b) $\frac{2m-3}{5} - \frac{m}{3} = -\frac{6}{5}$

(c) $\frac{r-2}{3} + \frac{r+1}{5} = \frac{3}{5}$

4. Solve $1 - \frac{2}{x+1} = \frac{2x}{x+1}$ and check your answer.

ANSWERS
3. (a) 3 (b) −9 (c) 2
4. When the equation is solved, −1 is found. However, since $x = -1$ leads to a 0 denominator in the original equation, there is no solution.

WORK PROBLEM 3 AT THE SIDE.

In solving equations that have a variable in the denominator, remember that the number 0 cannot be used as a denominator. Therefore, the solution cannot be a number that will make the denominator equal 0.

EXAMPLE 3 Solving an Equation Involving Rational Expressions

Solve $\frac{x}{x-2} = \frac{2}{x-2} + 2$.

The common denominator is $x - 2$. Since $x = 2$ makes a denominator 0, x cannot equal 2. Solve the equation by multiplying each side of the equation by $x - 2$.

$$(x-2)\left(\frac{x}{x-2}\right) = (x-2)\left(\frac{2}{x-2}\right) + (x-2)(2)$$

$$x = 2 + 2x - 4$$

$$x = -2 + 2x \qquad \text{Combine terms.}$$

$$-x = -2 \qquad \text{Subtract } 2x.$$

$$x = 2 \qquad \text{Divide by } -1.$$

The proposed solution is 2. However, 2 cannot be a solution because 2 makes a denominator equal 0. The equation has no solution. (Equations with no solutions are one of the main reasons that it is important to always check proposed solutions.) ■

WORK PROBLEM 4 AT THE SIDE.

In summary, solve equations with rational expressions as follows.

SOLVING EQUATIONS WITH RATIONAL EXPRESSIONS

Step 1 Multiply each side of the equation by the least common denominator. (This clears the equation of fractions.)

Step 2 Solve the resulting equation.

Step 3 Check each proposed solution by substituting it in the original equation.

EXAMPLE 4 Solving an Equation Involving Rational Expressions

Solve $\frac{2m}{m^2-4} + \frac{1}{m-2} = \frac{2}{m+2}$.

Multiply by the LCD, $(m+2)(m-2)$.

$$(m+2)(m-2)\left(\frac{2m}{m^2-4} + \frac{1}{m-2}\right) = (m+2)(m-2)\frac{2}{m+2}$$

$$(m+2)(m-2)\frac{2m}{m^2-4} + (m+2)(m-2)\frac{1}{m-2}$$

$$= (m+2)(m-2)\frac{2}{m+2}$$

$$2m + m + 2 = 2(m-2)$$

$$3m + 2 = 2m - 4 \qquad \text{Distributive property; combine terms.}$$

$$m + 2 = -4 \qquad \text{Subtract } 2m.$$

$$m = -6 \qquad \text{Subtract 2.}$$

Check to see that −6 is indeed a valid solution for the given equation. ■

WORK PROBLEMS 5 AND 6 AT THE SIDE.

EXAMPLE 5 Solving an Equation Involving Rational Expressions

Solve $\frac{2}{x^2 - x} = \frac{1}{x^2 - 1}$.

Begin by finding a least common denominator. Since $x^2 - x = x(x - 1)$, and $x^2 - 1 = (x + 1)(x - 1)$, the LCD is $x(x + 1)(x - 1)$. Proceed as usual.

$$x(x + 1)(x - 1)\frac{2}{x(x - 1)} = x(x + 1)(x - 1)\frac{1}{(x + 1)(x - 1)}$$

$$2(x + 1) = x$$

$$2x + 2 = x \qquad \text{Distributive property}$$

$$x + 2 = 0 \qquad \text{Subtract } x.$$

$$x = -2 \qquad \text{Subtract 2.}$$

To be sure that $x = -2$ is a solution, substitute -2 for x in the original equation. Since -2 satisfies the equation, the solution is -2. ■

WORK PROBLEM 7 AT THE SIDE.

EXAMPLE 6 Solving an Equation Involving Rational Expressions

Solve $\frac{1}{x - 1} + \frac{1}{2} = \frac{2}{x^2 - 1}$.

The least common denominator is $2(x + 1)(x - 1)$. Multiply each side of the equation by the LCD, $2(x + 1)(x - 1)$.

$$2(x + 1)(x - 1)\left(\frac{1}{x - 1} + \frac{1}{2}\right) = 2(x + 1)(x - 1)\frac{2}{(x + 1)(x - 1)}$$

$$2(x + 1)(x - 1)\frac{1}{x - 1} + 2(x + 1)(x - 1)\frac{1}{2}$$

$$= 2(x + 1)(x - 1)\frac{2}{(x + 1)(x - 1)}$$

$$2(x + 1) + (x + 1)(x - 1) = 4$$

$$2x + 2 + x^2 - 1 = 4 \qquad \text{Distributive property}$$

$$x^2 + 2x + 1 = 4 \qquad \text{Combine terms.}$$

$$x^2 + 2x - 3 = 0 \qquad \text{Subtract 4.}$$

Factoring gives

$$(x + 3)(x - 1) = 0.$$

Solving this equation suggests that $x = -3$ or $x = 1$. But 1 makes a denominator of the original equation equal 0, so 1 is not a solution. However, -3 is a solution, as shown by substituting -3 for x in the original equation.

$$\frac{1}{x - 1} + \frac{1}{2} = \frac{2}{x^2 - 1}$$

$$\frac{1}{-3 - 1} + \frac{1}{2} = \frac{2}{(-3)^2 - 1} \qquad \text{Let } x = -3.$$

$$\frac{1}{-4} + \frac{1}{2} = \frac{2}{9 - 1} \qquad \text{Simplify.}$$

$$\frac{1}{4} = \frac{1}{4} \qquad \text{True}$$

The check shows that -3 is a solution. ■

5. Check -6 as a solution to Example 4. Is the solution correct?

6. Solve each equation and check your answer.

(a) $\frac{2p}{p^2 - 1} = \frac{2}{p + 1} - \frac{1}{p - 1}$

(b) $\frac{8r}{4r^2 - 1} = \frac{3}{2r + 1} + \frac{3}{2r - 1}$

7. Solve each equation and check your answer.

(a) $\frac{4}{3m + 3} = \frac{m + 1}{m^2 + m}$

(b) $\frac{2}{p^2 - 2p} = \frac{3}{p^2 - p}$

ANSWERS
5. yes
6. (a) -3 (b) 0
7. (a) 3 (b) 4

8. Solve each equation and check your answer.

(a) $\frac{1}{x-2} + \frac{1}{5} = \frac{2}{5(x^2-4)}$

(b) $\frac{6}{5a+10} - \frac{1}{a-5} = \frac{4}{a^2-3a-10}$

9. Solve each equation for the specified variable.

(a) $z = \frac{x}{x+y}$ for v

(b) $a = \frac{v-w}{t}$ for v

ANSWERS

8. (a) $-4, -1$ (b) 60

9. (a) $y = \frac{x - zx}{z}$ (b) $v = at + w$

EXAMPLE 7 Solving an Equation Involving Rational Expressions

Solve $\frac{1}{k^2+4k+3} + \frac{1}{2k+2} = \frac{3}{4k+12}$.

Factor the three denominators to get the common denominator, $4(k+1)(k+3)$. Multiply each side by this product.

$$4(k+1)(k+3)\left(\frac{1}{(k+1)(k+3)} + \frac{1}{2(k+1)}\right) = 4(k+1)(k+3)\frac{3}{4(k+3)}$$

$$4(k+1)(k+3)\frac{1}{(k+1)(k+3)} + 2 \cdot 2(k+1)(k+3)\frac{1}{2(k+1} = 4(k+1)(k+3)\frac{3}{4(k+3)}$$

$$4 + 2(k+3) = 3(k+1) \quad \text{Simplify.}$$

$$4 + 2k + 6 = 3k + 3 \quad \text{Distributive property}$$

$$2k + 10 = 3k + 3 \quad \text{Combine terms.}$$

$$7 = k$$

Check to see that 7 actually is a solution for the given equation. ■

WORK PROBLEM 8 AT THE SIDE.

2 Solving a formula for a specified variable was discussed in Chapter 3. In the next example this process is applied to formulas with fractions.

EXAMPLE 8 Solving for a Specified Variable

Solve the formula $S = \frac{a(r^n - 1)}{r - 1}$ for a.

$$S = \frac{a(r^n-1)}{r-1}$$

$$(r-1)S = (r-1)\frac{a(r^n-1)}{r-1} \quad \text{Multiply each side by the LCD, } r-1.$$

$$(r-1)S = a(r^n-1)$$

$$\frac{(r-1)S}{r^n-1} = a \quad \text{Divide each side by } (r^n-1).$$

The last step was done to get a alone on one side of the equation. The distributive property can be applied to give

$$a = \frac{rS - S}{r^n - 1}. \quad ■$$

WORK PROBLEM 9 AT THE SIDE.

name date hour

6.6 EXERCISES

Solve each equation and check your answers. See Examples 1 and 2.

1. $\frac{6}{x} - \frac{4}{x} = 5$

2. $\frac{3}{x} + \frac{2}{x} = 5$

3. $\frac{x}{2} - \frac{x}{4} = 6$

4. $\frac{4}{y} + \frac{2}{3} = 1$

5. $\frac{9}{m} = 5 - \frac{1}{m}$

6. $\frac{3x}{5} + 2 = \frac{1}{4}$

7. $\frac{2t}{7} - 5 = t$

8. $\frac{1}{2} + \frac{2}{m} = 1$

9. $\frac{x + 1}{2} = \frac{x + 2}{3}$

10. $\frac{t - 4}{3} = \frac{t + 2}{4}$

11. $\frac{3m}{2} + m = 5$

12. $\frac{2z}{5} + z = \frac{7}{5}$

13. $\frac{2p + 8}{9} = \frac{10p + 4}{27}$

14. $\frac{2k + 3}{k} = \frac{3}{2}$

15. $\frac{5 - y}{y} + \frac{3}{4} = \frac{7}{y}$

16. $\frac{x}{x - 4} = \frac{2}{x - 4} + 5$

17. $\frac{a - 4}{4} = \frac{a}{16} + \frac{1}{2}$

18. $\frac{m - 2}{5} = \frac{m}{10} + \frac{4}{5}$

19. $\frac{y+2}{5} + \frac{y-5}{3} = \frac{7}{5}$

20. $\frac{a+7}{8} - \frac{a-2}{3} = \frac{4}{3}$

21. $\frac{m+2}{5} - \frac{m-6}{7} = 2$

22. $\frac{p}{2} - \frac{p-1}{4} = \frac{5}{4}$

23. $\frac{r}{6} - \frac{r-2}{3} = \frac{-4}{3}$

24. $\frac{5y}{3} - \frac{2y-1}{4} = \frac{1}{4}$

25. $\frac{8k}{5} - \frac{3k-4}{2} = \frac{5}{2}$

26. $\frac{y-1}{2} - \frac{y-3}{4} = 1$

27. $\frac{r+5}{3} - \frac{r-1}{4} = \frac{7}{4}$

Solve each equation and check your answers. See Examples 3–7.

28. $\frac{4}{k-2} + \frac{3}{5k-10} = \frac{23}{5}$

29. $\frac{1}{3p+15} + \frac{5}{4p+20} = \frac{19}{24}$

30. $\frac{m}{2m+2} = \frac{2m-3}{m+1} - \frac{m}{2m+2}$

31. $\frac{5p+1}{3p+3} = \frac{5p-5}{5p+5} + \frac{3p-1}{p+1}$

32. $\frac{2}{y} = \frac{y}{5y-12}$

33. $\frac{8x+3}{x} = 3x$

34. $\dfrac{2}{k-3} - \dfrac{3}{k+3} = \dfrac{12}{k^2-9}$

35. $\dfrac{1}{r+5} - \dfrac{3}{r-5} = \dfrac{-10}{r^2-25}$

36. $\dfrac{3y}{y^2+5y+6} = \dfrac{5y}{y^2+2y-3} - \dfrac{2}{y^2+y-2}$

37. $\dfrac{x+4}{x^2-3x+2} - \dfrac{5}{x^2-4x+3} = \dfrac{x-4}{x^2-5x+6}$

38. $\dfrac{4}{p} - \dfrac{2}{p+1} = 3$

39. $\dfrac{6}{r} + \dfrac{1}{r-2} = 3$

40. $\dfrac{2}{m-1} + \dfrac{1}{m+1} = \dfrac{5}{4}$

41. $\dfrac{5}{z-2} + \dfrac{10}{z+2} = 7$

Solve each formula for the specified variable. See Example 8.

42. $P = \dfrac{kT}{V}$ for T

43. $I = \dfrac{kE}{R}$ for R

44. $N = \dfrac{kF}{d}$ for d

45. $F = \dfrac{k}{d - D}$ for D

46. $I = \dfrac{E}{R + r}$ for R

47. $I = \dfrac{E}{R + r}$ for r

48. $S = \dfrac{a}{1 - r}$ for r

49. $h = \dfrac{2A}{B + b}$ for b

50. $F = \dfrac{GmM}{d^2}$ for M

SKILL SHARPENERS

Write a mathematical expression for each of the following. See Section 3.4.

51. Sharon goes 800 kilometers in p hours; find her rate.

52. Joann drives for 10 hours, going d kilometers. Find her rate.

53. Sam goes 780 kilometers at z kilometers per hour. Find his time.

54. Alejandro can do a job in x hours. What portion of the job is done in 1 hour?

55. Kathy needs h hours to tune up her car. How much of the job does she do in 1 hour?

name date hour

SUMMARY EXERCISES ON RATIONAL EXPRESSIONS

A common error when working with rational expressions is to confuse *operations* on rational expressions with the *solution of equations* with rational expressions. For example, the four possible operations on the rational expressions

$$\frac{1}{x} \quad \text{and} \quad \frac{1}{x-2}$$

can be performed as follows.

Add. $\dfrac{1}{x} + \dfrac{1}{x-2} = \dfrac{x-2}{x(x-2)} + \dfrac{x}{x(x-2)} = \dfrac{x-2+x}{x(x-2)} = \dfrac{2x-2}{x(x-2)}$

Subtract. $\dfrac{1}{x} - \dfrac{1}{x-2} = \dfrac{x-2}{x(x-2)} - \dfrac{x}{x(x-2)} = \dfrac{x-2-x}{x(x-2)} = \dfrac{-2}{x(x-2)}$

Multiply. $\dfrac{1}{x} \cdot \dfrac{1}{x-2} = \dfrac{1}{x(x-2)}$

Divide. $\dfrac{1}{x} \div \dfrac{1}{x-2} = \dfrac{1}{x} \cdot \dfrac{x-2}{1} = \dfrac{x-2}{x}$

On the other hand, the *equation*

$$\frac{1}{x} + \frac{1}{x-2} = \frac{3}{4}$$

is solved by multiplying each side by the least common denominator, $4x(x-2)$, giving an equation with no denominators.

$$4x(x-2)\frac{1}{x} + 4x(x-2)\frac{1}{x-2} = 4x(x-2)\frac{3}{4}$$

$$4x - 8 + 4x = 3x^2 - 6x$$

$$0 = 3x^2 - 14x + 8$$

$$0 = (3x-2)(x-4)$$

$$x = \frac{2}{3} \quad \text{or} \quad x = 4$$

In each of the following exercises, first decide whether the given rational expressions should be added, subtracted, multiplied, or divided; then perform the operation, or solve the given equation.

1. $\dfrac{6}{m} + \dfrac{2}{m}$

2. $\dfrac{b^2c^3}{b^5c^4} \cdot \dfrac{c^5}{b^7}$

3. $\dfrac{2}{x^2 + 2x - 3} \div \dfrac{8x^2}{3x - 3}$

4. $\dfrac{4}{m-2} = 1$

5. $\dfrac{2r^2 - 3r - 9}{2r^2 - r - 6} \cdot \dfrac{r^2 + 2r - 8}{r^2 - 2r - 3}$

6. $\dfrac{1}{m^2 - 3m} + \dfrac{4}{m^2 - 9}$

7. $\dfrac{p + 3}{8} = \dfrac{p - 2}{9}$

8. $\dfrac{4t^2 - t}{6t^2 + 10t} \div \dfrac{8t^2 + 2t - 1}{3t^2 + 11t + 10}$

9. $\dfrac{5}{y - 1} + \dfrac{2}{3y - 3}$

10. $\dfrac{1}{z} + \dfrac{1}{z + 2} = \dfrac{8}{15}$

11. $\dfrac{2}{r - 1} + \dfrac{1}{r} = \dfrac{5}{2}$

12. $\dfrac{2}{y} - \dfrac{7}{5y}$

13. $\dfrac{4}{9z} - \dfrac{3}{2z}$

14. $\dfrac{r - 3}{2} = \dfrac{2r - 5}{5}$

15. $\dfrac{1}{m^2 + 5m + 6} + \dfrac{2}{m^2 + 4m + 3}$

16. $\dfrac{2k^2 - 3k}{20k^2 - 5k} \div \dfrac{2k^2 - 5k + 3}{4k^2 + 11k - 3}$

6.7 APPLICATIONS OF RATIONAL EXPRESSIONS

1 Rational expressions are often useful in solving word problems. Example 1, involving numbers, is included in this section to show the basic steps in solving a word problem with fractions.

EXAMPLE 1 Solving a Problem About Numbers

If the same number is added to both the numerator and denominator of the fraction $\frac{3}{4}$, the result is $\frac{5}{6}$. Find the number.

Let x = the number added to numerator and denominator. Then

$$\frac{3 + x}{4 + x} \quad \text{Same number added to numerator and denominator}$$

represents the result of adding the same number to both the numerator and denominator. Since this result is $\frac{5}{6}$,

$$\frac{3 + x}{4 + x} = \frac{5}{6}.$$

Solve this equation by multiplying each side by the common denominator, $6(4 + x)$.

$$6(4 + x)\frac{3 + x}{4 + x} = 6(4 + x)\frac{5}{6}$$

$$6(3 + x) = 5(4 + x)$$

$$18 + 6x = 20 + 5x \qquad \text{Distributive property}$$

$$x = 2$$

Check this solution in the words of the original problem: if 2 is added to both the numerator and denominator of $\frac{3}{4}$, the result is $\frac{5}{6}$, as needed. ■

WORK PROBLEM 1 AT THE SIDE.

2 The next example shows how to solve word problems involving distance.

EXAMPLE 2 Solving a Problem About Distance, Rate, and Time

The Big Muddy River has a current of 3 miles per hour. A motorboat takes as long to go 12 miles downstream as to go 8 miles upstream. What is the speed of the boat in still water?

This problem requires the distance formula,

$$d = rt \text{ (distance = rate} \cdot \text{time).}$$

Let x = the speed of the boat in still water. Since the current pushes the boat when the boat is going downstream, the speed of the boat downstream will be the sum of the speed of the boat and the speed of the current, $x + 3$ miles per hour. Also, the boat's speed going upstream is given by $x - 3$ miles per hour. This information is summarized in the following chart.

	d	r	t
Downstream	12	$x + 3$	
Upstream	8	$x - 3$	

Fill in the column representing time by solving the formula $d = rt$ for t.

$$d = rt$$

$$\frac{d}{r} = t \qquad \text{Divide by } r.$$

OBJECTIVES

1. Solve word problems about numbers using rational expressions.
2. Solve word problems about distance using rational expressions.
3. Solve word problems about work using rational expressions.
4. Solve word problems about variation using rational expressions.

1. Solve each word problem.

(a) A certain number is added to the numerator and subtracted from the denominator of $\frac{5}{8}$. The new number equals the reciprocal of $\frac{5}{8}$. Find the number.

(b) The denominator of a fraction is 1 more than the numerator. If 6 is added to the numerator and subtracted from the denominator, the result is $\frac{15}{4}$. Find the original fraction.

ANSWERS

1. (a) 3 (b) $\frac{9}{10}$

2. (a) A boat can go 20 miles against a wind in the same time it can go 60 miles with the wind. The wind is blowing at 4 miles per hour. Find the speed of the boat with no wind.

(b) An airplane, maintaining a constant airspeed, takes as long to go 450 miles with the wind as it does to go 375 miles against the wind. If the wind is blowing at 15 miles per hour, what is the speed of the plane?

ANSWERS
2. (a) 8 miles per hour (b) 165 miles per hour

Then the time upstream is the distance divided by the rate, or

$$t = \frac{d}{r} = \frac{8}{x - 3},$$

and the time downstream is also the distance divided by the rate, or

$$t = \frac{d}{r} = \frac{12}{x + 3}.$$

Now complete the chart.

	d	r	t
Downstream	12	$x + 3$	$\frac{12}{x+3}$
Upstream	8	$x - 3$	$\frac{8}{x-3}$

Times are equal.

According to the original problem, the time upstream equals the time downstream. The two times from the chart must therefore be equal, giving the equation

$$\frac{12}{x + 3} = \frac{8}{x - 3}.$$

Solve this equation by multiplying each side by $(x + 3)(x - 3)$.

$$(x + 3)(x - 3)\frac{12}{x + 3} = (x + 3)(x - 3)\frac{8}{x - 3}$$

$$12(x - 3) = 8(x + 3)$$

$$12x - 36 = 8x + 24 \quad \text{Distributive property}$$

$$4x = 60 \quad \text{Subtract } 8x\text{; add 36.}$$

$$x = 15 \quad \text{Divide by 4.}$$

The speed of the boat in still water is 15 miles per hour.

Check the solution by first finding the speed of the boat downstream, which is $15 + 3 = 18$ miles per hour. Traveling 12 miles would take

$$d = rt$$
$$12 = 18t$$
$$t = \frac{2}{3} \text{ hour.}$$

On the other hand, the speed of the boat upstream is $15 - 3 = 12$ miles per hour, and traveling 8 miles would take

$$d = rt$$
$$8 = 12t$$
$$t = \frac{2}{3} \text{ hour.}$$

The time upstream equals the time downstream, as required. ■

WORK PROBLEM 2 AT THE SIDE.

3 This example shows a word problem about the length of time needed to do a job. This type of problem is often called a **work problem.**

EXAMPLE 3 Solving a Problem About Work

With a riding lawn mower, John, the grounds keeper in a large park, can cut the lawn in 8 hours. With a small mower, his assistant Walt needs 14 hours to cut the same lawn. If both John and Walt work on the lawn, how long will it take to cut it?

Let x = the number of hours that it takes John and Walt to cut the lawn, working together. Certainly x will be less than 8, since John alone can cut the lawn in 8 hours. In one hour, John can do $\frac{1}{8}$ of the lawn, and in one hour Walt can do $\frac{1}{14}$ of the lawn. Since it takes them x hours to cut the lawn when working together, in one hour together they can do $\frac{1}{x}$ of the lawn.

Summarize this information in a chart.

Worker	*Number of hours*	*Part of job in 1 hour*
John	8	$\frac{1}{8}$
Walt	14	$\frac{1}{14}$
John and Walt together	x	$\frac{1}{x}$

The amount of the lawn cut by John in one hour plus the amount cut by Walt in one hour must equal the amount they can do together in one hour, or

Amount by Walt ↓

$$\text{Amount by John} \rightarrow \frac{1}{8} + \frac{1}{14} = \frac{1}{x}. \leftarrow \text{Amount by both together}$$

Since $56x$ is the least common denominator for 8, 14, and x, multiply each side of the equation by $56x$.

$$56x\left(\frac{1}{8} + \frac{1}{14}\right) = 56x \cdot \frac{1}{x}$$

$$56x \cdot \frac{1}{8} + 56x \cdot \frac{1}{14} = 56x \cdot \frac{1}{x}$$

$$7x + 4x = 56$$

$$11x = 56$$

$$x = \frac{56}{11}$$

Working together, John and Walt can cut the lawn in $\frac{56}{11}$ hours, or $5\frac{1}{11}$ hours, about 5 hours and 5 minutes. ■

WORK PROBLEM 3 AT THE SIDE.

3. (a) Geraldo and Luisa operate a small roofing company. Luisa can roof an average house alone in 9 hours. Geraldo can roof a house alone in 8 hours. How long will it take them to do the job if they work together?

(b) Wing can do a job in 7 hours, but Tammie needs only 6 hours. How long would it take them working together?

ANSWERS

3. (a) $\frac{72}{17}$ or $4\frac{4}{17}$ hours

(b) $\frac{42}{13}$ or $3\frac{3}{13}$ hours

4 Equations with fractions often result when discussing **variation.** Two variables **vary directly** if one is a constant multiple of the other, as stated below.

4. (a) If z varies directly as t, and $z = 11$ when $t = 4$, find z when $t = 32$.

(b) Suppose q varies directly as m, and $q = \frac{1}{2}$ when $m = 2$. Find q when $m = 5$.

5. (a) Suppose z varies inversely as t, and $z = 8$ when $t = 2$. Find z when $t = 32$.

(b) If p varies inversely as q, and $p = 5$ when $q = 1$, find p when $q = 4$.

ANSWERS

4. (a) 88 (b) $\frac{5}{4}$

5. (a) $\frac{1}{2}$ (b) $\frac{5}{4}$

DIRECT VARIATION

y* varies directly as *x if there exists a constant k such that

$$y = kx.$$

EXAMPLE 4 Using Direct Variation

Suppose y varies directly as x, and $y = 20$ when $x = 4$. Find y when $x = 9$.

Since y varies directly as x, there is a constant k such that $y = kx$. We know that $y = 20$ when $x = 4$. Substituting these values into $y = kx$ gives

$$y = kx$$

$$20 = k \cdot 4,$$

from which $k = 5$.

Since $y = kx$ and $k = 5$,

$$y = 5x. \qquad \text{Let } k = 5.$$

When $x = 9$, $y = 5x = 5 \cdot 9 = 45$. Let $x = 9$.

Thus, $y = 45$ when $x = 9$. ■

WORK PROBLEM 4 AT THE SIDE.

In another common type of variation, the value of one variable increases while the value of another decreases.

INVERSE VARIATION

y* varies inversely as *x if there exists a constant k such that

$$y = \frac{k}{x}.$$

EXAMPLE 5 Using Inverse Variation

Suppose z varies inversely as t, and $z = 8$ when $t = 5$. Find z when $t = 20$.

Since z varies inversely as t, there is a constant k such that

$$z = \frac{k}{t}.$$

Find k by replacing z with 8 and t with 5.

$$8 = \frac{k}{5} \qquad \text{Let } z = 8 \text{ and } t = 5.$$

Multiply both sides by 5 to get

$$k = 40,$$

so that $z = \frac{40}{t}$.

When $t = 20$, $z = \frac{40}{20} = 2$. ■

WORK PROBLEM 5 AT THE SIDE.

name date hour

6.7 EXERCISES

Solve each problem. See Example 1.

1. One-half of a number is 3 more than one-sixth of the same number. What is the number?

2. The numerator of the fraction $\frac{4}{7}$ is increased by an amount such that the value of the resulting fraction is $\frac{27}{21}$. By what amount was the numerator increased?

3. One number is 3 more than another. If the smaller is added to two-thirds the larger, the result is four-fifths the sum of the original numbers. Find the numbers.

4. The sum of a number and its reciprocal is $\frac{5}{2}$. Find the number.

5. If twice the reciprocal of a number is subtracted from the number, the result is $-\frac{7}{3}$. Find the number.

6. The sum of the reciprocals of two consecutive integers is $\frac{5}{6}$. Find the integers.

7. A mother and her daughter worked four days at a job. The daughter earned $\frac{2}{5}$ as much as her mother. They earned a total of $1344. How much did each earn per day?

8. The profits from a carnival are to be given to two scholarships so that one scholarship receives $\frac{3}{2}$ as much money as the other. If the total amount given to the two scholarships is $780, how much goes to the scholarship that receives the lesser amount?

9. An apprentice is paid $\frac{3}{4}$ the salary of an experienced journeyman. If the total wages paid an apprentice and a journeyman are $56,000, find the amount paid to the journeyman.

10. A child takes $\frac{5}{8}$ the number of pills that an adult takes for the same illness. Together the child and the adult use 26 pills. Find the number used by the adult.

Solve each problem. See Example 2.

11. Sam can row 4 miles per hour in still water. It takes as long to row 8 miles upstream as 24 miles downstream. How fast is the current?

	d	r	t
Upstream	8	$4 - x$	
Downstream	24	$4 + x$	

12. Elena flew from Philadelphia to Des Moines at 180 miles per hour and from Des Moines to Philadelphia at 150 miles per hour. The trip at the slower speed took 1 hour longer than the trip at the higher speed. Find the distance between the two cities. (Assume there was no wind in either direction.)

	d	r	t
P to D	x	180	
D to P	x	150	

13. On a business trip, Arlene traveled to her destination at an average speed of 60 miles per hour. Coming home, her average speed was 50 miles per hour and the trip took $\frac{1}{2}$ hour longer. How far did she travel each way?

14. Reynaldo flew his airplane 500 miles against the wind in the same time it took him to fly it 600 miles with the wind. If the speed of the wind was 10 miles per hour, what was the average speed of his plane?

15. The distance from Seattle, Washington, to Victoria, British Columbia, is about 148 miles by ferry. It takes about 4 hours less to travel by ferry from Victoria to Vancouver, British Columbia, a distance of about 74 miles. What is the average speed of the ferry?

16. A boat goes 210 miles downriver in the same time it can go 140 miles upriver. The speed of the current is 5 miles per hour. Find the speed of the boat in still water.

Solve each problem. See Example 3.

17. Paul can tune up his Toyota in 2 hours. His friend Marco can do the job in 3 hours. How long would it take them if they worked together?

18. Jorge can paint a room, working alone, in 8 hours. Caterina can paint the same room, working alone, in 6 hours. How long will it take them if they work together?

19. Machine A can do a certain job in 7 hours, and machine B takes 12 hours. How long will the job take the two machines working together?

20. One pipe can fill a swimming pool in 6 hours, and another pipe can do it in 9 hours. How long will it take the two pipes working together to fill the pool $\frac{3}{4}$ full?

21. Dennis can do a job in 4 days. When Dennis and Sue work together, the job takes $2\frac{1}{3}$ days. How long would the job take Sue if she worked alone?

22. An inlet pipe can fill a swimming pool in 9 hours, and an outlet pipe can empty the pool in 12 hours. Through an error, both pipes are left open. How long will it take to fill the pool?

The next four exercises are more challenging.

23. A cold water faucet can fill a sink in 12 minutes, and a hot water faucet can fill it in 15. The drain can empty the sink in 25 minutes. If both faucets are on and the drain is open, how long will it take to fill the sink?

24. Refer to Exercise 22. Assume the error was discovered after both pipes had been running for 3 hours, and the outlet pipe was then closed. How much more time would then be required to fill the pool? (*Hint:* How much of the job had been done when the error was discovered?)

25. An experienced employee can enter tax data into a computer twice as fast as a new employee. Working together, it takes the employees 2 hours. How long would it take the experienced employee working alone?

26. One roofer can put a new roof on a house three times faster than another. Working together they can roof a house in 4 days. How long would it take the faster roofer working alone?

Solve the following problems about variation. See Examples 4 and 5.

27. If x varies directly as y, and $x = 9$ when $y = 2$, find x when y is 7.

28. If z varies directly as x, and $z = 15$ when $x = 4$, find z when x is 11.

29. If m varies directly as p, and $m = 20$ when $p = 2$, find m when p is 5.

30. If a varies directly as b, and $a = 48$ when $b = 4$, find a when $b = 7$.

31. If y varies inversely as x, and $y = 10$ when $x = 3$, find y when $x = 12$.

32. If r varies inversely as s, and $r = 7$ when $s = 8$, find r when $s = 12$.

33. The circumference of a circle varies directly as the radius. A circle with a radius of 7 centimeters has a circumference of 43.96 centimeters. Find the circumference if the radius changes to 11 centimeters.

34. The pressure exerted by a certain liquid at a given point varies directly as the depth of the point beneath the surface of the liquid. The pressure at 10 feet is 50 pounds per square inch. What is the pressure at 20 feet?

35. The current in a simple electrical circuit varies inversely as the resistance. If the current is 50 amp (an *ampere* is a unit for measuring current) when the resistance is 10 ohm (an *ohm* is a unit for measuring resistance), find the current if the resistance is 5 ohm.

36. The force required to compress a spring varies directly as the change in the length of the spring. If a force of 12 pounds is required to compress a certain spring 3 inches, how much force is required to compress the spring 5 inches?

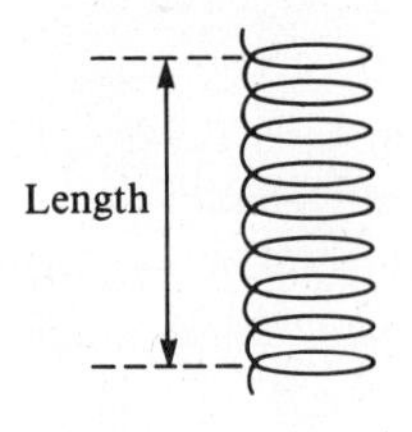

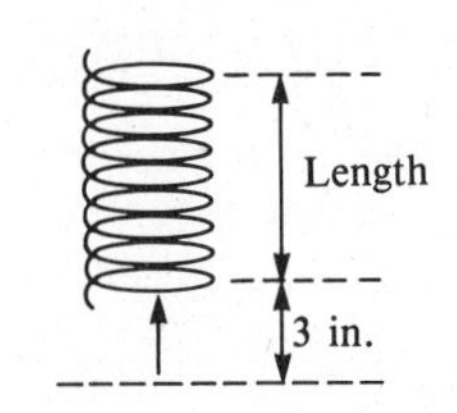

SKILL SHARPENERS

Find the value of y when **(a)** $x = 2$ *and* **(b)** $x = -4$. *See Sections 1.5 and 3.3.*

37. $y = 4x - 7$

38. $y = 3 - 2x$

39. $4x - 1 = y$

40. $3x + 7y = 10$

41. $2x - 3y = 5$

42. $y + 3x = 2$

Historical Reflections

Historical Reflections

Alfred North Whitehead (1861–1947) and Bertrand Russell (1872–1970)

Bertrand Russell was a student of Alfred North Whitehead, and together the two worked on the massive three-volume, 2000+ page treatise *Principia Mathematica*. In this work they attempted to prove that mathematics and logic were essentially one and the same.

Russell became a public figure because of his involvement in social issues. Deeply aware of human loneliness, he was "passionately desirous of finding ways of diminishing this tragic isolation." During World War I he was an antiwar crusader and was briefly imprisoned. Again in the 1960s he championed peace. He wrote many books on social issues, winning the Nobel Prize for Literature in 1950.

Art: (Top) Culver Pictures; (Bottom) AP/Wide World

CHAPTER 6 SUMMARY

KEY TERMS

6.1 **rational expression** The quotient of two polynomials with denominator not zero is called a rational expression.

lowest terms A rational expression is written in lowest terms if there are no common factors in the numerator and denominator (except 1).

6.3 **least common denominator** (LCD) The smallest expression that all denominators divide into without remainder is called the least common denominator.

6.5 **complex fraction** A rational expression with one or more fractions in the numerator, denominator, or both is a complex fraction.

6.7 **direct variation** y varies directly as x if there is a constant k such that $y = kx$.

inverse variation y varies inversely as x if there is a constant k such that $y = \frac{k}{x}$.

QUICK REVIEW

Section	Concepts	Examples
6.1 The Fundamental Property of Rational Expressions	To find the values for which a rational expression is not defined, set the denominator equal to zero and solve the equation.	Given: $\frac{x-4}{x^2-16}$. $x^2 - 16 = 0$; $(x-4)(x+4) = 0$; $x - 4 = 0$ or $x + 4 = 0$; $x = 4$ or $x = -4$. The rational expression is not defined for 4 or -4.
	To write a rational expression in lowest terms, (1) factor; and (2) use the fundamental property to remove common factors from the numerator and denominator.	Given: $\frac{x^2-1}{(x-1)^2}$. $\frac{(x-1)(x+1)}{(x-1)(x-1)} = \frac{x+1}{x-1}$

Section	Concepts	Examples
6.2 Multiplication and Division of Rational Expressions	**Multiplication** 1. Factor. 2. Multiply numerators and multiply denominators. 3. Write in lowest terms.	$\frac{3x+9}{x-5} \cdot \frac{x^2-3x-10}{x^2-9}$ $= \frac{3(x+3)}{x-5} \cdot \frac{(x-5)(x+2)}{(x+3)(x-3)}$ $= \frac{3(x+3)(x-5)(x+2)}{(x-5)(x+3)(x-3)}$ $= \frac{3(x+2)}{x-3}$
	Division 1. Factor. 2. Multiply the first rational expression by the reciprocal of the second rational expression. 3. Write in lowest terms.	$\frac{2x+1}{x+5} \div \frac{6x^2-x-2}{x^2-25}$ $= \frac{2x+1}{x+5} \div \frac{(2x+1)(3x-2)}{(x+5)(x-5)}$ $= \frac{2x+1}{x+5} \cdot \frac{(x+5)(x-5)}{(2x+1)(3x-2)}$ $= \frac{x-5}{3x-2}$
6.3 Least Common Denominators	**Finding the LCD** 1. Factor each denominator into prime factors. 2. List each different denominator factor the greatest number of times it appears in any denominator. 3. Multiply the denominators from Step 2 to get the LCD.	Find the LCD for $\frac{3}{k^2-8k+16}$ and $\frac{1}{4k^2-16k}$. $k^2-8k+16 = (k-4)^2$ $4k^2-16k = 4k(k-4)$ $\text{LCD} = (k-4)^2 \cdot 4 \cdot k$ $= 4k(k-4)^2$
	To write a rational expression with the LCD as denominator: 1. Factor both denominators. 2. Decide what factors the denominator must be multiplied by to equal the LCD. 3. Multiply the rational expression by that factor over itself (multiply by 1).	$\frac{5}{2z^2-6z} = \frac{}{4z^3-12z^2}$ $\frac{5}{2z(z-3)} = \frac{}{4z^2(z-3)}$ $2z(z-3)$ must be multiplied by $2z$. $\frac{5}{2z(z-3)} \cdot \frac{2z}{2z} = \frac{10z}{4z^2(z-3)}$

Section	Concepts	Examples
6.4 Addition and Subtraction of Rational Expressions	**Adding Rational Expressions** 1. Find the LCD. 2. Rewrite each rational expression as a fraction with the LCD as denominator. 3. Add the numerators to get the numerator of the sum. The LCD is the denominator of the sum. 4. Write in lowest terms.	Add $\frac{2}{3m+6} + \frac{m}{m^2-4}$. $\frac{2}{3(m+2)} + \frac{m}{(m+2)(m-2)}$ The LCD is $3(m+2)(m-2)$. $= \frac{2(m-2)}{3(m+2)(m-2)} + \frac{3m}{3(m+2)(m-2)}$ $= \frac{2m-4+3m}{3(m+2)(m-2)}$ $= \frac{5m-4}{3(m+2)(m-2)}$
	Subtracting Rational Expressions Follow the same steps as for addition, but subtract in Step 3.	Subtract $\frac{6}{k+4} - \frac{2}{k}$. The LCD is $k(k+4)$. $\frac{6k}{(k+4)k} - \frac{2(k+4)}{k(k+4)}$ $= \frac{6k-2(k+4)}{k(k+4)}$ $= \frac{6k-2k-8}{k(k+4)} = \frac{4k-8}{k(k+4)}$
6.5 Complex Fractions	**Simplifying Complex Fractions** Method 1 Simplify the numerator and denominator separately. Then divide the simplified numerator by the simplified denominator.	$\frac{\frac{1}{a}-a}{1-a} = \frac{\frac{1}{a}-\frac{a^2}{a}}{1-a} = \frac{\frac{1-a^2}{a}}{1-a}$ $= \frac{1-a^2}{a} \cdot \frac{1}{1-a}$ $= \frac{(1-a)(1+a)}{a(1-a)} = \frac{1+a}{a}$
	Method 2 Multiply numerator and denominator of the complex fraction by the LCD of all the denominators in the complex fraction.	$\frac{\frac{1}{a}-a}{1-a} = \frac{\frac{1}{a}-a}{1-a} \cdot \frac{a}{a} = \frac{\frac{a}{a}-a^2}{(1-a)a}$ $= \frac{1-a^2}{(1-a)a} = \frac{(1+a)(1-a)}{(1-a)a}$ $= \frac{1+a}{a}$

Section	Concepts	Examples
6.6 Equations Involving Rational Expressions	**1.** Using the multiplication property of equality, multiply each side of the equation by the LCD of all denominators in the equation. **2.** Solve the resulting equation which should have no fractions. **3.** Check each proposed solution.	Solve $\frac{2}{x-1} + \frac{3}{4} = \frac{5}{x-1}$. The LCD is $4(x - 1)$. $4(x-1)\left(\frac{2}{x-1} + \frac{3}{4}\right) = 4(x-1)\left(\frac{5}{x-1}\right)$ $4(x-1)\left(\frac{2}{x-1}\right) + 4(x-1)\left(\frac{3}{4}\right) = 4(x-1)\left(\frac{5}{x-1}\right)$ $8 + 3(x - 1) = 20$ $8 + 3x - 3 = 20$ $3x = 15$ $x = 5$ The proposed solution, 5, checks.
6.7 Applications of Rational Expressions	**Solving Problems About Distance** **1.** State what the variable represents. **2.** Use a chart to identify distance, rate, and time. **3.** Solve $d = rt$ for the unknown quantity. **4.** From the wording in the problem, decide what quantities are equal. Use those expressions from the chart to write an equation. **5.** Solve the equation. **6.** Check the solution.	On a trip from Sacramento to Monterey, Marge traveled at an average speed of 60 miles per hour. The return trip, at an average speed of 64 miles per hour, took $\frac{1}{4}$ hour less. How far did she travel between the two cities? (see chart below) Since the time for the return trip was $\frac{1}{4}$ hour less, the time going equals the time returning plus $\frac{1}{4}$. $\frac{x}{60} = \frac{x}{64} + \frac{1}{4}$ The solution of this equation is $x = 240$. She traveled 240 miles.

	d	r	$t = \frac{d}{r}$
Going	x	60	$\frac{x}{60}$
Returning	x	64	$\frac{x}{64}$

Section	Concepts	Examples
	Solving Problems About Work **1.** State what the variable represents. **2.** Put the information from the problem in a chart giving the part of the job done in 1 hour. **3.** Write the equation: the sum of the parts done separately should equal the part done together. **4.** Solve the equation. **5.** Check the solution.	It takes the regular mail carrier 6 hours to cover the route. A substitute took 8 hours to cover the route. How long would it take them together? Let x be the number of hours to cover the route together. (see chart below) $\frac{1}{6} + \frac{1}{8} = \frac{1}{x}$ The solution of this equation is $x = \frac{24}{7} = 3\frac{3}{7}$. It would take them $3\frac{3}{7}$ hours to cover the route together.
	Solving Variation Problems **1.** Write the variation equation using $y = kx$ or $y = \frac{k}{x}$. **2.** Find k by substituting the given values of x and y into the equation. **3.** Write the equation with the value of k from Step 2 and the given value of x or y. Solve for the remaining variable.	If a varies inversely as b, and $a = 4$ when $b = 4$, find a when $b = 6$. The equation is $a = \frac{k}{b}$. Substitute $a = 4$ and $b = 4$. $4 = \frac{k}{4}$ The solution is $k = 16$. Let $k = 16$ and $b = 6$ in the variation equation. $a = \frac{16}{6} = \frac{8}{3}$

	Hours to complete	*Part done in 1 hour*
Regular	6	$\frac{1}{6}$
Sub	8	$\frac{1}{8}$
Together	x	$\frac{1}{x}$

name date hour

CHAPTER 6 REVIEW EXERCISES

[6.1] *Find any values of the variables for which the following rational expressions are undefined.*

1. $\frac{2}{7x}$

2. $\frac{3}{m - 5}$

3. $\frac{r - 3}{r^2 - 2r - 8}$

4. $\frac{3z + 5}{2z^2 + 5z - 3}$

Find the numerical value of each rational expression when **(a)** $x = 3$ *and* **(b)** $x = -1$.

5. $\frac{x^2}{x + 2}$

(a) **(b)**

6. $\frac{5x + 3}{2x - 1}$

(a) **(b)**

7. $\frac{8x}{x^2 - 2}$

(a) **(b)**

8. $\frac{x - 5}{x - 3}$

(a) **(b)**

Write each rational expression in lowest terms.

9. $\frac{6y^2z^3}{9y^4z^2}$

10. $\frac{9x^2 - 16}{6x + 8}$

11. $\frac{m - 5}{5 - m}$

12. $\frac{6k^2 + 11ky - 10y^2}{12k^2 - 11ky + 2y^2}$

[6.2] *Find each product or quotient. Write each answer in lowest terms.*

13. $\frac{12y^2}{8} \cdot \frac{24}{y}$

14. $\frac{6z^2}{9z^4} \cdot \frac{8z^3}{3z}$

15. $\frac{8z^2}{(4z)^3} \div \frac{4z^5}{32z}$

16. $\frac{m - 1}{2} \cdot \frac{5}{2m - 2}$

17. $\frac{7y + 14}{8y - 5} \div \frac{4y + 8}{16y - 10}$

18. $\frac{3k + 5}{k + 2} \cdot \frac{k^2 - 4}{18k^2 - 50}$

19. $\dfrac{2a^2 - a - 6}{a^2 + 4a - 5} \div \dfrac{2a + 3}{a + 5}$

20. $\dfrac{z^2 + z - 2}{z^2 + 7z + 10} \div \dfrac{z - 3}{z + 5}$

21. $\dfrac{2p^2 + 3p - 2}{p^2 + 5p + 6} \div \dfrac{p^2 - 2p - 15}{2p^2 - 7p - 15}$

22. $\dfrac{8r^2 + 23r - 3}{64r^2 - 1} \div \dfrac{r^2 - 4r - 21}{64r^2 + 16r + 1}$

[6.3] *Find the least common denominator for each list of fractions.*

23. $\dfrac{1}{15}, \dfrac{7}{30}, \dfrac{4}{45}$

24. $\dfrac{3}{8y}, \dfrac{7}{12y^2}, \dfrac{1}{16y^3}$

25. $\dfrac{1}{y^2 + 2y}, \dfrac{4}{y^2 + 6y + 8}$

26. $\dfrac{3}{z^2 + z - 6}, \dfrac{2}{z^2 + 4z + 3}$

Rewrite each rational expression with the given denominator.

27. $\dfrac{4}{9}, \dfrac{\quad}{45}$

28. $\dfrac{12}{m}, \dfrac{\quad}{5m}$

29. $\dfrac{3}{8m^2}, \dfrac{\quad}{24m^3}$

30. $\dfrac{12}{y - 4}, \dfrac{\quad}{8y - 32}$

31. $\dfrac{-2k}{3k + 15}, \dfrac{\quad}{15k + 75}$

32. $\dfrac{12y}{y^2 - y - 2}, \dfrac{\quad}{(y - 2)(y + 1)(y - 4)}$

[6.4] *Add or subtract as indicated. Write each answer in lowest terms.*

33. $\dfrac{5}{m} + \dfrac{8}{m}$

34. $\dfrac{11}{3r} - \dfrac{8}{3r}$

35. $\dfrac{4}{p} + \dfrac{1}{3}$

36. $\frac{7}{k} - \frac{2}{5}$

37. $\frac{3 + 5m}{2} - \frac{m}{4}$

38. $\frac{8}{r^2} - \frac{3}{2r}$

39. $\frac{4}{2p - q} + \frac{3}{2p + q}$

40. $\frac{2}{p^2 - 4} - \frac{3}{5p + 10}$

41. $\frac{6}{m^2 + 5m} - \frac{1}{m^2 + 3m - 10}$

42. $\frac{10}{p^2 - 2p} - \frac{2}{p^2 - 5p + 6}$

[6.5] *Simplify each complex fraction.*

43. $\frac{\frac{r^2}{q^4}}{\frac{r^5}{q}}$

44. $\frac{\frac{m - 2}{m}}{\frac{m + 2}{3m}}$

45. $\frac{\frac{5k - 1}{k}}{\frac{4k + 3}{8k}}$

46. $\frac{\frac{1}{m} - \frac{1}{n}}{\frac{1}{n - m}}$

47. $\frac{z + \frac{1}{x}}{z - \frac{1}{x}}$

48. $\frac{\frac{1}{a + b} - 1}{\frac{1}{a + b} + 1}$

[6.6] *Solve each equation. Check your answer.*

49. $\frac{p}{4} - \frac{1}{2} = \frac{1}{3}$

50. $\frac{5 + m}{m} + \frac{3}{4} = \frac{-2}{m}$

51. $\frac{y}{3} - \frac{y - 2}{8} = -1$

52. $\frac{3y - 1}{y - 2} = \frac{5}{y - 2} + 1$

53. $\frac{3}{m - 2} + \frac{1}{m - 1} = \frac{7}{m^2 - 3m + 2}$

Solve for the specified variable.

54. $P = \frac{kT}{V}$ for V

55. $x = \frac{3}{2y + z}$ for y

56. $p^2 = \frac{4}{3m - q}$ for m

[6.7] *Solve each word problem.*

57. The sum of a number and its reciprocal is 2. Find the number.

58. Five times a number is added to three times the reciprocal of the number, giving $\frac{17}{2}$. Find the number.

59. The commission received by a salesperson for selling a small car is $\frac{2}{3}$ that received for selling a large car. On a recent day, Linda sold one of each, earning a commission of \$300. Find the commission for each type of car.

60. A boat goes 7 miles per hour in still water. It takes as long to go 20 miles upstream as 50 miles downstream. Find the speed of the current.

61. When Mary and Sue work together on a job, they can do it in $3\frac{3}{7}$ days. Mary can do the job working alone in 8 days. How long would it take Sue working alone?

62. One painter can paint a house twice as fast as another. Working together, they can paint the house in $1\frac{1}{3}$ days. How long would it take the faster painter working alone?

63. If x varies directly as y, and $x = 12$ when $y = 5$, find x when $y = 3$.

64. If m varies directly as q, and $m = 8$ when $q = 4$, find m when $q = 6$.

MIXED REVIEW EXERCISES

Perform the indicated operations.

65. $\dfrac{\dfrac{2}{p - q} + 3}{5 - \dfrac{4}{p + q}}$

66. $\dfrac{2}{y + 1} - \dfrac{3}{y - 1}$

67. $\dfrac{10p^5}{5} \div \dfrac{3p^7}{20}$

68. $\dfrac{p + 4}{6} \div \dfrac{3p + 12}{2}$

69. $\dfrac{\dfrac{6}{r} - 1}{\dfrac{6 - r}{4r}}$

70. $\dfrac{4}{z^2 - 2z + 1} - \dfrac{3}{z^2 - 1}$

Solve.

71. $F = \dfrac{k}{d - D}$ for d

72. $\dfrac{2}{z} - \dfrac{z}{z + 3} = \dfrac{1}{z + 3}$

73. A man can plant his garden in 5 hours, working alone. His daughter can do the same job in 8 hours. How long would it take them if they worked together?

74. In 1985 the ratio of the amount of aluminum used in the U.S. compared to steel was $\frac{7}{8}$. If the total steel and aluminum used in 1985 was 825 kilograms per capita, how much aluminum was used in 1985?

75. Kerrie flew her plane 400 kilometers with the wind in the same time it took her to go 200 kilometers against the wind. The speed of the wind is 50 kilometers per hour. Find the speed of the plane in still air.

76. Domestic car sales in 1986 were $\frac{3}{4}$ of the sales in 1976. If the total sales for those years were $1.5 billion, what were the sales in 1986?

name date hour

CHAPTER 6 TEST

1. Find any values for which $\frac{8k + 1}{k^2 - 4k + 3}$ is undefined.

1. ____________

Write each rational expression in lowest terms.

2. $\frac{5s^3 - 5s}{2s + 2}$

2. ____________

3. $\frac{4p^2 + pq - 3q^2}{p^2 + 2pq + q^2}$

3. ____________

Multiply or divide. Write all answers in lowest terms.

4. $\frac{x^6y}{x^3} \cdot \frac{y^2}{x^2y^3}$

4. ____________

5. $\frac{8y - 16}{9} \div \frac{3y - 6}{5}$

5. ____________

6. $\frac{6m^2 - m - 2}{8m^2 + 10m + 3} \cdot \frac{4m^2 + 7m + 3}{3m^2 + 5m + 2}$

6. ____________

7. $\frac{5a^2 + 7a - 6}{2a^2 + 3a - 2} \div \frac{5a^2 + 17a - 12}{2a^2 + 5a - 3}$

7. ____________

Rewrite each rational expression with the given denominator.

8. $\frac{11}{7r}, \frac{\quad}{49r^2}$

8. ____________

9. $\frac{5}{8m - 16}, \frac{\quad}{24m - 48}$

9. ____________

Add or subtract as indicated. Write each answer in lowest terms.

10. $\frac{5}{x} - \frac{6}{x}$

10. ____________

11. ______

12. ______

13. ______

14. ______

15. ______

16. ______

17. ______

18. ______

19. ______

20. ______

11. $\dfrac{-3}{a+1} + \dfrac{5}{6a+6}$

12. $\dfrac{3}{2k^2 + 3k - 2} + \dfrac{1}{k^2 + 3k + 2}$

13. $\dfrac{5}{2m^2 - m - 10} - \dfrac{3}{2m^2 + 5m + 2}$

Simplify each complex fraction.

14. $\dfrac{\dfrac{1}{k} - 2}{\dfrac{1}{3} + k}$

15. $\dfrac{\dfrac{1}{p+4} - 2}{\dfrac{1}{p+4} + 2}$

Solve each equation.

16. $\dfrac{3}{2p} + \dfrac{12}{5p} = \dfrac{13}{20}$

17. $\dfrac{2}{p^2 - 2p - 3} = \dfrac{3}{p-3} + \dfrac{2}{p+1}$

For each problem, write an equation and solve it.

18. If four times a number is added to the reciprocal of twice the number, the result is 3. Find the number.

19. A man can paint a room in his house, working alone, in 5 hours. His wife can do the job in 4 hours. How long would it take them working together?

20. If x varies directly as y, and $x = 8$ when $y = 12$, find x when $y = 28$.

name date hour

CUMULATIVE REVIEW EXERCISES: CHAPTERS 4–6

Evaluate each expression.

1. $(-3)^2(-3)$

2. $2^{-3} \cdot 2^2$

3. $\left(\frac{3}{4}\right)^{-2}$

4. $(2^{-3} \cdot 3^2)^2$

5. $\frac{7^{-1}}{7}$

6. $\frac{6^5 \cdot 6^{-2}}{6^3}$

7. $\frac{(4^{-2})^3}{4^6 \cdot 4^{-3}}$

8. $\left(\frac{4^{-3} \cdot 4^4}{4^5}\right)^{-1}$

Simplify each expression. Write with only positive exponents.

9. $\frac{(2x^3)^{-1} \cdot x}{2^3x^5}$

10. $\frac{(p^2)^3p^{-4}}{(p^{-3})^{-1}p}$

11. $\frac{(m^{-2})^3m}{m^5m^{-4}}$

Perform the indicated operations.

12. $(x^3 + 3x^2 - 5x) + (4x^3 - x^2 + 7)$

13. $(2k^2 + 3k) - (5k^2 - 2) - (k^2 + k - 1)$

14. $8x^2y^2(9x^4y^5)$

15. $3m^2(2m^5 - 5m^3 + m)$

16. $(2y + 1)(y - 4)$

17. $(3z - 2w)(4z + 2w)$

18. $(y^2 + 3y + 5)(3y - 1)$

19. $(3p + 2)^2$

20. $(4a - b)^2$

21. $(5k - 4)(5k + 4)$

22. $(2p + 3q)(2p - 3q)$

23. $\dfrac{3a^3 - 9a^2 + 15a}{3a}$

24. $\dfrac{8x^4 + 12x^3 - 6x^2 + 20x}{2x}$

25. $\dfrac{12p^3 + 2p^2 - 12p + 4}{2p - 2}$

26. $\dfrac{15z^3 - 11z^2 + 22z + 8}{5z - 2}$

Factor completely.

27. $8x^2 - 24x$

28. $9rq^5 + 18q^6p$

29. $8r^2 - 9rs + 12s^3$

30. $m^2 + 9m + 14$

31. $r^2 - 3r - 40$

32. $p^2 - 7pq - 18q^2$

33. $2a^2 + 7a - 4$

34. $10m^2 + 19m + 6$

35. $15x^2 - xy - 6y^2$

36. $8t^2 + 10tv + 3v^2$

37. $9x^2 + 6x + 1$

38. $4p^2 - 12p + 9$

39. $16t^2 + 56tz + 49z^2$

40. $25r^2 - 81t^2$

41. $100p^2 - 49q^2$

42. $100y^2 - 169p^4$

43. $6a^2m + am - 2m$

44. $2pq + 6p^3q + 8p^2q$

Solve each of the following equations.

45. $(2p - 3)(p + 2) = 0$

46. $y^2 + y - 6 = 0$

47. $r^2 = 2r + 15$

48. $q^2 = 4q + 77$

49. $6m^2 + m - 2 = 0$

50. $8m^2 = 64m$

51. $9p^2 = 36$

52. $(r - 5)(2r + 1)(3r - 5) = 0$

53. $x^3 = 16x$

Solve each word problem.

54. One number is 4 more than another. The product of the numbers is 2 less than the smaller number. Find the smaller number.

55. The length of a rectangle is 2 meters less than twice the width. The area is 60 square meters. Find the width of the rectangle.

56. The length of a rectangle is 3 centimeters more than twice the width. The area of the rectangle is 44 square centimeters. Find the width of the rectangle.

57. The length of a rectangle is 1 centimeter less than twice the width. The area is 15 square centimeters. Find the width of the rectangle.

Write each expression in lowest terms.

58. $\frac{5m}{20}$

59. $\frac{2p^3}{8p^5}$

60. $\frac{y^2 - 4}{6y - 12}$

61. $\frac{5r - 40}{r^2 - 7r - 8}$

Perform each operation.

62. $\frac{x^6y^2}{y^5x^9} \cdot \frac{x^2}{y^3}$

63. $\frac{r^6s^4}{r^3s^5} \div \frac{r^{14}}{s^{10}}$

64. $\frac{2}{m} + \frac{5}{m}$

65. $\frac{5}{q} - \frac{1}{q}$

66. $\frac{4}{3} - \frac{1}{z}$

67. $\frac{3}{7} + \frac{4}{r}$

68. $\frac{p}{p - 5} - \frac{7}{p - 5}$

69. $\frac{5}{2m + 6} + \frac{2}{m + 3}$

70. $\frac{4}{5q - 20} - \frac{1}{3q - 12}$

71. $\dfrac{1}{a^2 - 2a} + \dfrac{2}{a^2 + a}$

72. $\dfrac{2}{k^2 + k} - \dfrac{3}{k^2 - k}$

73. $\dfrac{2m - 2}{3m^2 + 3m - 6} \cdot \dfrac{5m + 10}{10m - 10}$

74. $\dfrac{7z^2 + 49z + 70}{16z^2 + 72z - 40} \div \dfrac{3z + 6}{4z^2 - 1}$

75. $\dfrac{3x^2 + 5x - 2}{6x^2 + x - 1} \cdot \dfrac{2x^2 + 7x + 3}{x^2 + 3x + 2}$

Simplify each complex fraction.

76. $\dfrac{\frac{5}{x}}{1 + \frac{1}{x}}$

77. $\dfrac{\frac{5z^3y}{y^2z}}{\frac{15y^4}{z^2}}$

78. $\dfrac{\frac{4}{a} + \frac{5}{2a}}{\frac{7}{6a} - \frac{1}{5a}}$

Solve each of the following equations.

79. $\dfrac{r + 2}{5} = \dfrac{r - 3}{3}$

80. $\dfrac{1}{x} = \dfrac{1}{x + 1} + \dfrac{1}{2}$

81. $\dfrac{1}{p} + \dfrac{1}{p - 1} = \dfrac{5}{6}$

Solve each word problem.

82. Juanita can weed the yard in 3 hours. Benito can weed the yard in 2 hours. How long would it take them if they worked together?

83. Awwad can paint a room in 9 hours. Mujeer can paint the same room in 6 hours. How long would it take them if they worked together on the room?

84. When Mary and Joann work together, a job takes 2 days. Joann can do the job alone in 6 days. How long would the job take Mary if she worked alone?

85. When working together, Fred and Alan can do a job in 2 hours. Alan needs 4 hours when working alone. How long would it take Fred to do the job if he worked alone?

7

GRAPHING LINEAR EQUATIONS

7.1 LINEAR EQUATIONS IN TWO VARIABLES

OBJECTIVES

1. Write a solution as an ordered pair.
2. Decide whether a given ordered pair is a solution of a given equation.
3. Complete ordered pairs for a given equation.
4. Complete a table of values.

Most of the equations discussed so far, such as

$$3x + 5 = 12 \quad \text{or} \quad 2x^2 + x + 5 = 0,$$

have had only one variable. Equations in two variables, such as

$$y = 4x + 5 \quad \text{or} \quad 2x + 3y = 6,$$

are discussed in this chapter. Both of these equations are examples of *linear equations*.

A **linear equation** in two variables is an equation that can be put in the form

$$Ax + By = C$$

where A, B and C are real numbers and A and B are not both 0.

1 A solution of a linear equation in two variables requires two numbers, one for each variable. For example, the equation $y = 4x + 5$ is satisfied by replacing x with 2 and y with 13, since

$$13 = 4(2) + 5. \qquad \text{Let } x = 2; y = 13.$$

The pair of numbers $x = 2$ and $y = 13$ gives a solution of the equation $y = 4x + 5$. The phrase "$x = 2$ and $y = 13$" is abbreviated

$$(2, 13)$$

with the x-value, 2, and the y-value, 13, given as a pair of numbers written inside parentheses. The x-value is always given first. A pair of numbers such as (2, 13) is called an **ordered pair.** As the name indicates, the order in which the numbers are written is important. The ordered pairs (2, 13) and (13, 2) are not the same. The second pair indicates that $x = 13$ and $y = 2$.

Of course, letters other than x and y may be used in the equation, with the numbers in the ordered pair usually placed in alphabetical order. For example, one solution to the equation $3p - q = -11$ is $p = -2$ and $q = 5$. This solution is written as the ordered pair $(-2, 5)$.

WORK PROBLEM 1 AT THE SIDE.

1. Write each solution as an ordered pair.

(a) $x = 5$ and $y = 7$

(b) $y = 6$ and $x = -1$

(c) $q = 4$ and $p = -3$

(d) $r = 3$ and $s = 12$

ANSWERS
1. (a) (5, 7) (b) (−1, 6) (c) (−3, 4) (d) (3, 12)

2. Decide whether or not the ordered pairs satisfy the equation $5x + 2y = 20$.

(a) (0, 10)

(b) (2, −5)

(c) (3, 2)

(d) (−4, 20)

3. Complete the given ordered pairs for the equation $y = 2x - 9$.

(a) (5,)

(b) (2,)

(c) (, 7)

(d) (, −13)

ANSWERS
2. (a) yes (b) no (c) no (d) yes
3. (a) (5, 1) (b) (2, −5) (c) (8, 7) (d) (−2, −13)

2 The next example shows how to decide whether an ordered pair is a solution of an equation. An ordered pair that is a solution of an equation is said to *satisfy* the equation.

EXAMPLE 1 Deciding Whether an Ordered Pair Satisfies an Equation

Decide whether the given ordered pair is a solution of the given equation.

(a) (3, 2); $2x + 3y = 12$

To see whether (3, 2) is a solution of the equation $2x + 3y = 12$, substitute 3 for x and 2 for y in the given equation.

$$\begin{aligned} 2x + 3y &= 12 & \\ 2(3) + 3(2) &= 12 & \text{Let } x = 3; \text{ let } y = 2. \\ 6 + 6 &= 12 & \\ 12 &= 12 & \text{True} \end{aligned}$$

This result is true, so (3, 2) satisfies $2x + 3y = 12$.

(b) (−2, −7); $2m + 3n = 12$

$$\begin{aligned} 2(-2) + 3(-7) &= 12 & \text{Let } m = -2; \text{ let } n = -7. \\ -4 + (-21) &= 12 & \\ -25 &= 12 & \text{False} \end{aligned}$$

This result is false, so (−2, −7) does *not* satisfy $2m + 3n = 12$. ■

WORK PROBLEM 2 AT THE SIDE.

3 Choosing a number for one variable in a linear equation makes it possible to find the value of the other, as shown in the next example.

EXAMPLE 2 Completing an Ordered Pair

Complete the given ordered pairs for the equation $y = 4x + 5$.

(a) (7,)

In this ordered pair, $x = 7$. (Remember that x always comes first.) Find the corresponding value of y by replacing x with 7 in the equation $y = 4x + 5$.

$$y = 4(7) + 5 = 28 + 5 = 33$$

The ordered pair is (7, 33).

(b) (, −3)

In this ordered pair $y = -3$. Find the value of x by replacing y with −3 in the equation; then solve for x.

$$\begin{aligned} y &= 4x + 5 & \\ -3 &= 4x + 5 & \text{Let } y = -3. \\ -8 &= 4x & \text{Subtract 5 from each side.} \\ -2 &= x & \text{Divide each side by 4.} \end{aligned}$$

The ordered pair is (−2, −3). ■

WORK PROBLEM 3 AT THE SIDE.

EXAMPLE 3 Completing an Ordered Pair

Complete the given ordered pairs for the equation $5x - y = 24$.

Equation	*Ordered pairs*
$5x - y = 24$	(5,) (−3,) (0,)

Find the y-value for the ordered pair (5,) by replacing x with 5 in the given equation and solving for y.

$$5x - y = 24$$
$$5(5) - y = 24 \quad \text{Let } x = 5.$$
$$25 - y = 24$$
$$-y = -1 \quad \text{Subtract 25 from each side.}$$
$$y = 1$$

The ordered pair is (5, 1).

Complete the ordered pair (−3,) by letting $x = -3$ in the given equation. Also, complete (0,) by letting $x = 0$.

If $x = -3$, then $5x - y = 24$ becomes

$$5(-3) - y = 24$$
$$-15 - y = 24$$
$$-y = 39$$
$$y = -39.$$

If $x = 0$, then $5x - y = 24$ becomes

$$5(0) - y = 24$$
$$0 - y = 24$$
$$-y = 24$$
$$y = -24.$$

The completed ordered pairs are as follows.

Equation	*Ordered pairs*
$5x - y = 24$	(5, 1) (−3, −39) (0, −24) ■

WORK PROBLEM 4 AT THE SIDE.

4 Ordered pairs of an equation often are displayed in a **table of values** as in the next example. The table may be written either vertically or horizontally. We will write these tables horizontally in this section and vertically in Section 7.3.

EXAMPLE 4 Completing a Table of Values

Complete the given table of values for the equation $x - 2y = 8$.

x	2	10		
y			0	−2

Complete the first two ordered pairs by letting $x = 2$ and $x = 10$, respectively.

If $x = 2$, then $x - 2y = 8$ becomes

$$2 - 2y = 8$$
$$-2y = 6$$
$$y = -3.$$

If $x = 10$, then $x - 2y = 8$ becomes

$$10 - 2y = 8$$
$$-2y = -2$$
$$y = 1.$$

4. Complete the given ordered pairs for the equation $x + 2y = 12$.

(a) (0,)

(b) (, 0)

(c) (4,)

(d) (−6,)

ANSWERS

4. (a) (0, 6) (b) (12, 0) (c) (4, 4) (d) (−6, 9)

Now complete the last two ordered pairs by letting $y = 0$ and $y = -2$, respectively.

If	$y = 0,$	If	$y = -2,$
then	$x - 2y = 8$	then	$x - 2y = 8$
becomes	$x - 2(\mathbf{0}) = 8$	becomes	$x - 2(\mathbf{-2}) = 8$
	$x - 0 = 8$		$x + 4 = 8$
	$x = 8.$		$x = 4.$

The completed table of values is as follows.

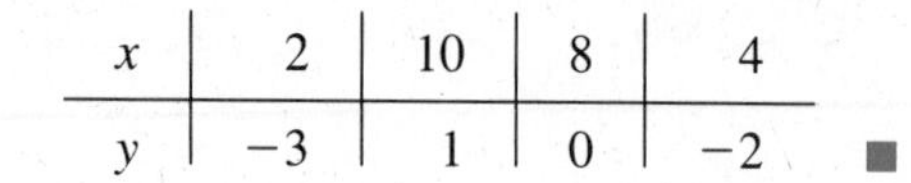

x	2	10	8	4
y	−3	1	0	−2

■

WORK PROBLEM 5 AT THE SIDE.

EXAMPLE 5 Completing a Table of Values

Complete the given table of values for the equation $x = 5$.

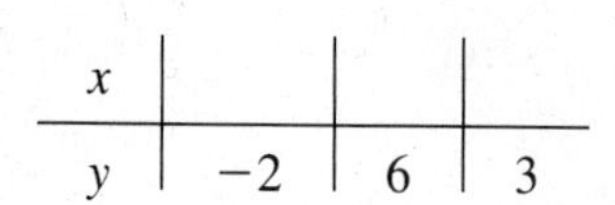

x			
y	−2	6	3

The given equation is $x = 5$. No matter which value of y might be chosen, the value of x is always the same, 5. Each ordered pair can be completed by placing 5 in the first position.

x	5	5	5
y	−2	6	3

■

When an equation such as $x = 5$ is discussed along with equations in two variables, think of $x = 5$ as an equation in two variables by rewriting $x = 5$ as $x + 0y = 5$. This form of the equation shows that for any value of y, the value of x is 5.

Each of the equations in this section has many ordered pairs as solutions. Each choice of a real number for one variable will lead to a particular real number for the other variable. This is true of linear equations in general: linear equations in two variables have an infinite number of solutions.

WORK PROBLEM 6 AT THE SIDE.

5. Complete the table of values for the equation $2x - 3y = 12$.

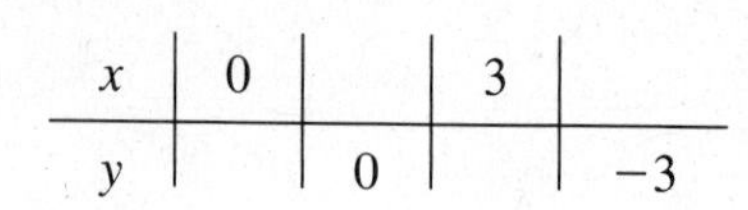

x	0		3	
y		0		−3

6. Complete the table of values for each equation.

(a) $x = 3$

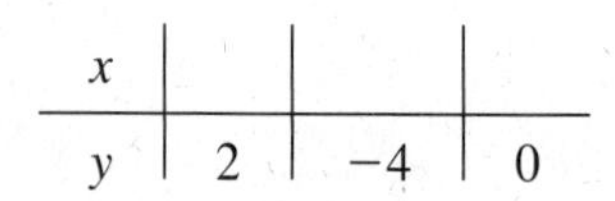

x			
y	2	−4	0

(b) $y = -4$

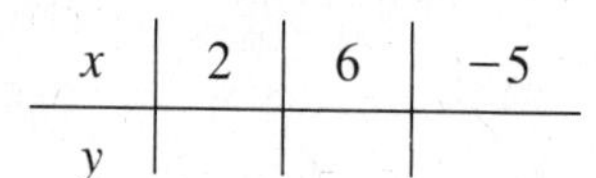

x	2	6	−5
y			

ANSWERS

5.

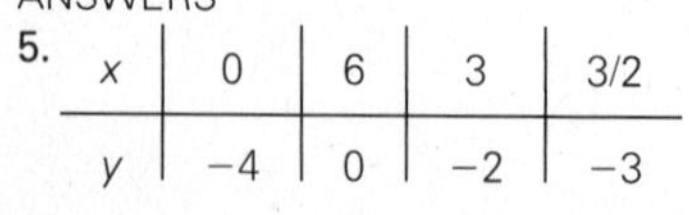

x	0	6	3	3/2
y	−4	0	−2	−3

6. (a)

x	3	3	3
y	2	−4	0

(b)

x	2	6	−5
y	−4	−4	−4

7.1 EXERCISES

Decide whether the given ordered pair is a solution of the given equation. See Example 1.

1. $x + y = 9$; $(2, 7)$

2. $3x + y = 8$; $(0, 8)$

3. $2p - q = 6$; $(2, -2)$

4. $2v + w = 5$; $(2, 1)$

5. $4x - 3y = 6$; $(1, 2)$

6. $5x - 3y = 1$; $(0, 1)$

7. $y = 3x$; $(1, 3)$

8. $x = -4y$; $(8, -2)$

9. $x = -6$; $(-6, 8)$

10. $y = 2$; $(9, 2)$

11. $x + 4 = 0$; $(-5, 1)$

12. $x - 6 = 0$; $(5, -1)$

Complete the given ordered pairs for the equation $y = 3x + 5$. See Example 2.

13. (2,)

14. (5,)

15. (8,)

16. (0,)

17. (−3,)

18. (−4,)

19. (, 14)

20. (, −10)

21. (, 8)

Complete the given ordered pairs for the equation $y = -4x + 8$. See Example 2.

22. (0,)

23. (2,)

24. (,16)

25. (, 24)

26. (,−4)

27. (, −8)

Complete the given ordered pairs or table of values for each equation. See Examples 2–4.

28. $y = 2x + 1$ (3,) (0,) (−1,)

29. $y = 3x - 5$ (2,) (0,) (−3,)

30. $y = 8 - 3x$ (2,) (0,) (−3,)

31. $y = -2 - 5x$ (4,) (0,) (−4,)

32. $2x + y = 9$ (0,) (3,) (12,)

33. $-3m + n = 4$

m	1	0	−2
n			

34. $2p + 3q = 6$

p	0		
q		0	4

35. $4t + 3w = 12$

t	0		
w		0	8

36. $3u - 5v = 15$

u	0		
v		0	−6

37. $4x - 9y = 36$

x		0	
y	0		4

Complete the given table of values for each equation. See Example 5.

38. $x = -4$

x			
y	6	2	−3

39. $x = 2$

x			
y	3	8	0

40. $y = 3$

x	8	4	−2
y			

41. $y = -8$

x	4	0	−4
y			

42. $x + 9 = 0$

x			
y	8	3	0

43. $y + 4 = 0$

x	9	2	0
y			

44. $y - 5 = 0$

x	6	4	−8
y			

SKILL SHARPENERS

Graph each group of numbers on a number line. See Section 2.1.

45. $-4, -2, 1, 2, 5$

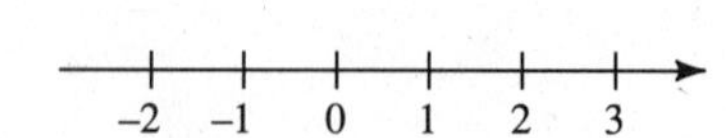

46. $-8, -5, -1, 0, 3$

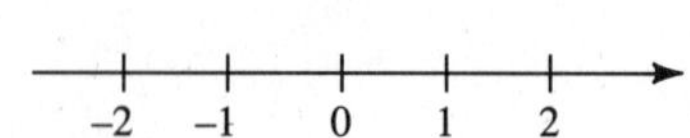

47. $-\frac{1}{2}, \frac{2}{3}, 1, 2, 2.5$

−2 −1 0 1 2 3

48. $-\frac{5}{4}, -\frac{1}{3}, 0, \frac{1}{3}, 2$

−2 −1 0 1 2

7.2 GRAPHING ORDERED PAIRS

OBJECTIVES

1. Decide in which quadrant a point is located.
2. Plot ordered pairs.
3. Given an equation, find and graph ordered pairs.

Earlier we used a number line to graph the solutions of an equation in one variable. Techniques for graphing the solutions of an equation in *two* variables will be shown in this section. The solutions of such an equation are *ordered pairs* of numbers of the form (x, y), so *two* number lines are needed, one for x and one for y. These two number lines are drawn as shown in Figure 1. The horizontal line is called the *x-axis* and the vertical line is called the *y-axis*. Together, the x-axis and y-axis form a **coordinate system.**

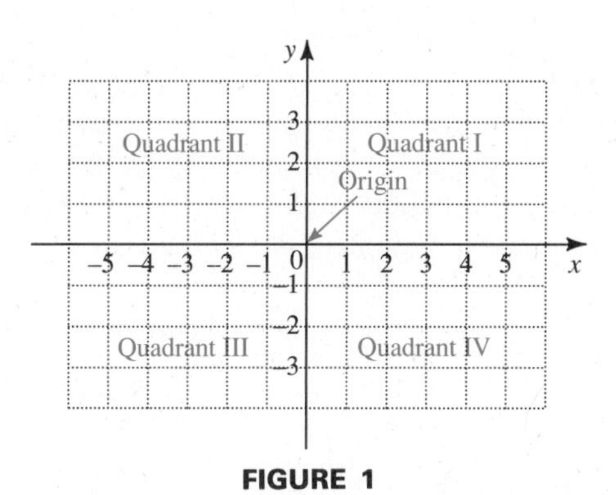

FIGURE 1

1 The coordinate system is divided into four regions, called **quadrants.** These quadrants are numbered counterclockwise, as shown in Figure 1. Points on the axes themselves are not in any quadrant. The point at which the x-axis and y-axis meet is called the **origin.** The origin is labeled 0 in Figure 1.

WORK PROBLEM 1 AT THE SIDE.

1. Name the quadrant in which each point in the figure is located.

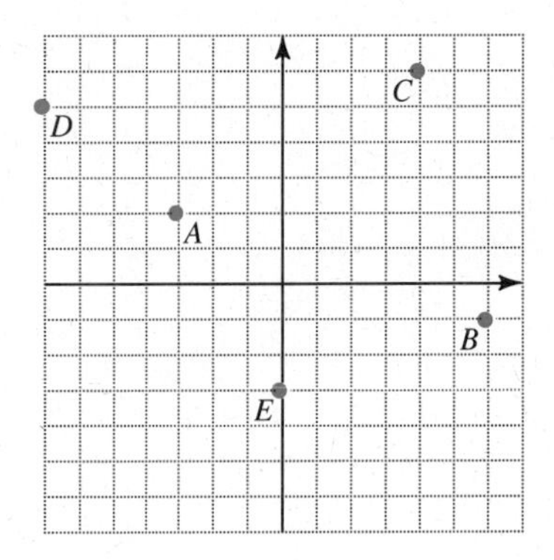

2 By referring to the two axes, every point in the plane can be associated with an ordered pair. The numbers in the ordered pair are called the **coordinates** of the point. For example, locate the point associated with the ordered pair (2, 3) by starting at the origin. Since the x-coordinate is 2, go 2 units to the right along the x-axis. Then, since the y-coordinate is 3, turn and go up 3 units on a line parallel to the y-axis. The point (2, 3) is **plotted** in Figure 2. From now on we will refer to the point with x-coordinate 2 and y-coordinate 3 as the point (2, 3).

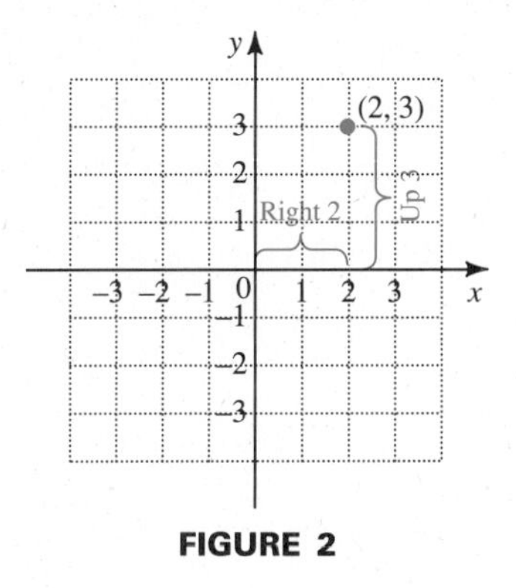

FIGURE 2

ANSWERS
1. *A*, II; *B*, IV; *C*, I; *D*, II; *E*, no quadrant

2. Plot the given ordered pairs on a coordinate system.

(a) (3, 5)

(b) (−2, 6)

(c) (−4, 0)

(d) (−5, −2)

(e) (5, −2)

(f) (0, −6)

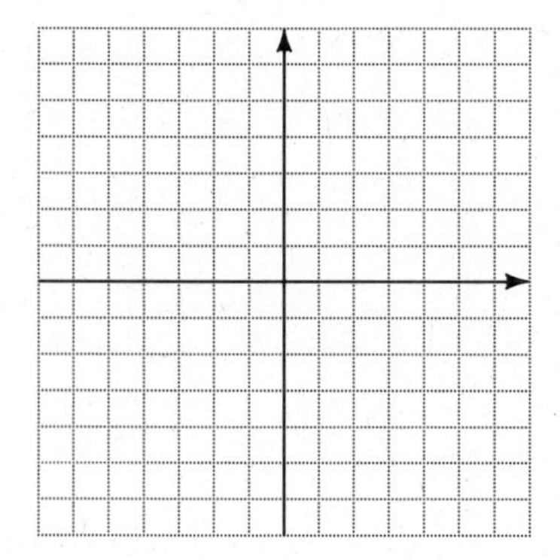

ANSWERS

2.

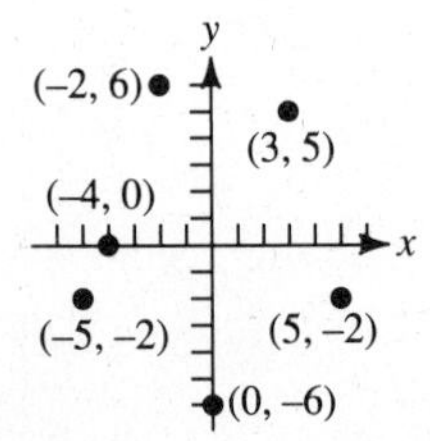

EXAMPLE 1 Plotting Ordered Pairs

Plot the given ordered pairs on a coordinate system.

(a) (1, 5)

(b) (−2, 3)

(c) (−1, −4)

(d) (7, −2)

(e) $\left(\frac{3}{2}, 2\right)$

(f) (5, 0)

(g) (0, −3)

Locate the point (−1, −4), for example, by first going 1 unit to the left along the x-axis. Then turn and go 4 units down, parallel to the y-axis. Plot the point $(\frac{3}{2}, 2)$ by first going $\frac{3}{2}$ (or $1\frac{1}{2}$) units to the right along the x-axis. Then turn and go 2 units up, parallel to the y-axis. Figure 3 shows the graphs of the points in this example. ■

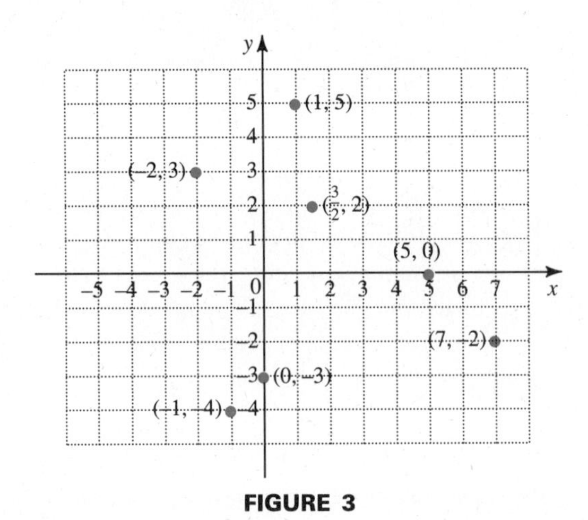

FIGURE 3

WORK PROBLEM 2 AT THE SIDE.

3 The next section shows how to graph a linear equation. The first step in graphing these equations is to find and plot some ordered pairs that satisfy (are solutions of) the equation, as shown in the next example.

EXAMPLE 2 Finding and Plotting Ordered Pairs

Complete the table of values for $x + 2y = 7$. Plot the ordered pairs on a coordinate system.

Equation	*Ordered pairs*
$x + 2y = 7$	(1,) (−3,) (3,) (7,)

Complete the ordered pairs by substituting the given x-values into the equation $x + 2y = 7$.

When $x = \mathbf{1}$,
$$x + 2y = 7$$
$$1 + 2y = 7$$
$$2y = 6$$
$$y = \mathbf{3}.$$

When $x = \mathbf{-3}$,
$$x + 2y = 7$$
$$-3 + 2y = 7$$
$$2y = 10$$
$$y = \mathbf{5}.$$

The ordered pairs are (1, 3) and (−3, 5).

In the same way, if $x = 3$, then $y = 2$, giving (3, 2). Finally, if $x = 7$, then $y = 0$, giving (7, 0). The completed table of values follows.

x	1	−3	3	7
y	3	5	2	0

The graph of these ordered pairs is shown in Figure 4. ■

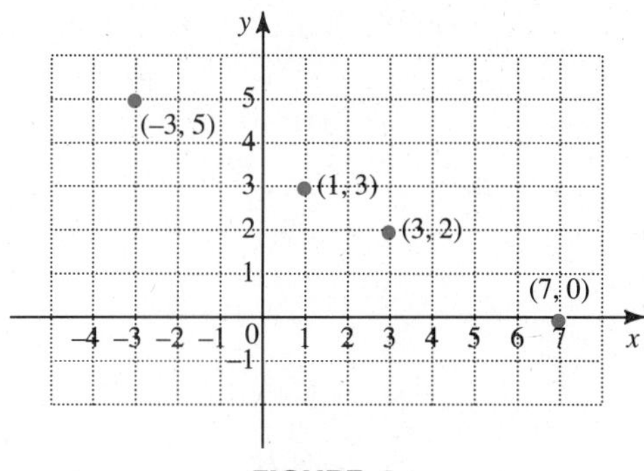

FIGURE 4

WORK PROBLEM 3 AT THE SIDE.

EXAMPLE 3 Completing Ordered Pairs for an Application

A company has found that the cost to produce x small calculators is

$$y = 25x + 250$$

where y represents the cost in cents. Complete the following table of values.

x	1	2	3
y			

To complete the ordered pair (1,), let $x = 1$.

$$y = 25x + 250$$
$$y = 25(1) + 250 \qquad \text{Let } x = 1.$$
$$y = 25 + 250$$
$$y = 275$$

This gives the ordered pair (1, 275), which says that the cost to produce 1 calculator is 275 cents or \$2.75. Complete the ordered pairs (2,) and (3,) as follows.

$$y = 25x + 250 \qquad y = 25x + 250$$
$$y = 25(2) + 250 \qquad y = 25(3) + 250$$
$$y = 50 + 250 \qquad y = 75 + 250$$
$$y = 300 \qquad y = 325$$

This gives the ordered pairs (2, 300) and (3, 325). ■

3. Complete the given ordered pairs for the equation $2x + y = 6$ and graph the results.

(a) (0,)

(b) (2,)

(c) (4,)

(d) (, 1)

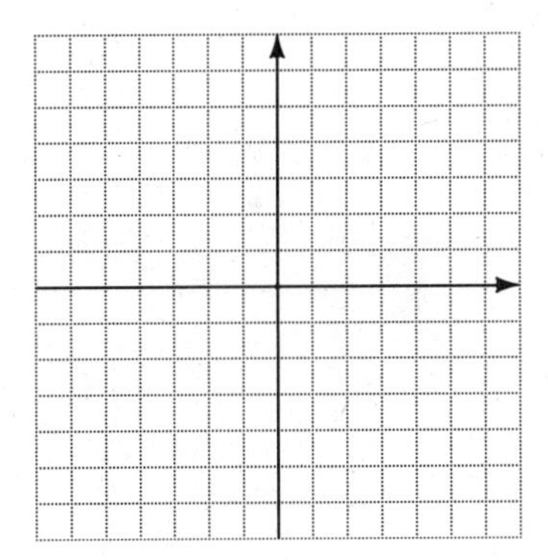

ANSWERS

3. (a) (0, 6) (b) (2, 2) (c) (4, −2)
(d) $\left(\frac{5}{2}, 1\right)$

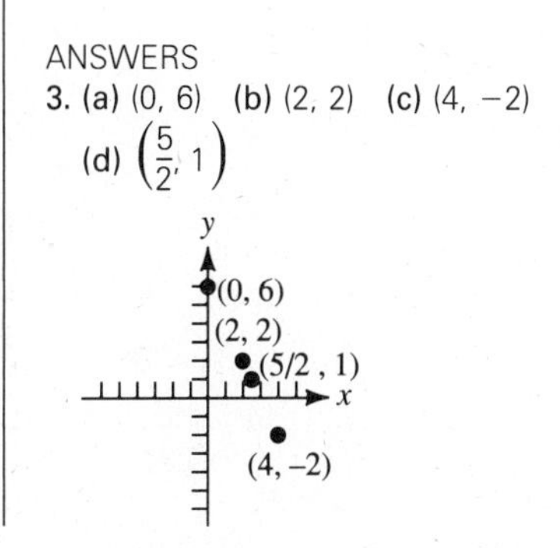

4. Complete the given ordered pairs using the equation $y = 25x + 250$ in Example 3.

(a) (4,)

(b) (5,)

(c) (6,)

(d) (7,)

(e) (8,)

(f) (9,)

ANSWERS
4. (a) (4, 350) (b) (5, 375)
(c) (6, 400) (d) (7, 425)
(e) (8, 450) (f) (9, 475)

WORK PROBLEM 4 AT THE SIDE.

The ordered pairs (2, 300) and (3, 325), along with the ones given in Problem 4, are graphed in Figure 5. In this graph, different scales are used on the two axes, since the y-values in the ordered pairs are much larger than the x-values. Here, each square represents 50 units in the vertical direction.

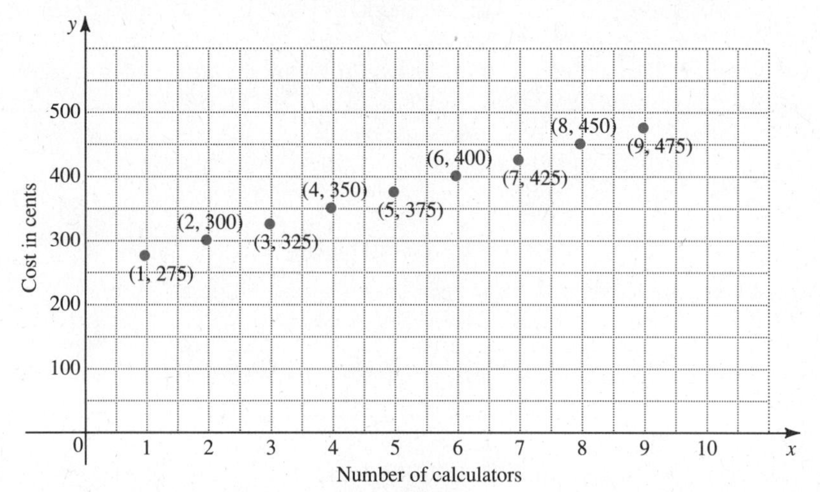

FIGURE 5

name date hour

7.2 EXERCISES

Write the x- and y-coordinates of the points labeled A through F in the figure.

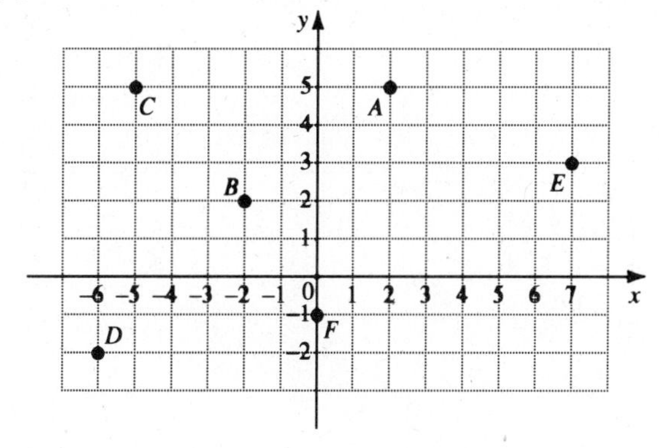

1. *A* **2.** *B*

3. *C* **4.** *D*

5. *E* **6.** *F*

Plot the ordered pairs on the coordinate system provided. See Example 1.

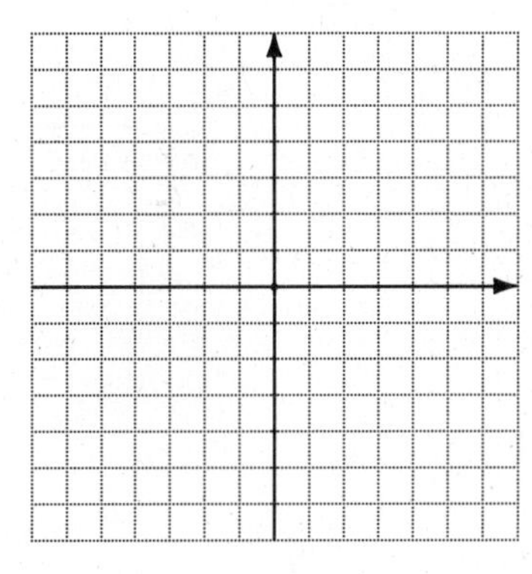

7. (6, 1) **8.** (4, −2) **9.** (3, 5)

10. (−4, −5) **11.** (−2, 4) **12.** (−5, −1)

13. (−3, 5) **14.** (3, −5) **15.** (4, 0)

16. (1, 0) **17.** (−2, 0) **18.** (0, 3)

Without plotting the given point, name the quadrant in which each point lies.

19. (2, 3) **20.** (2, −3) **21.** (−2, 3) **22.** (−2, −3) **23.** (−1, −1)

24. (4, 7) **25.** (−3, 6) **26.** (−2, 0) **27.** (5, −4) **28.** (0, 4)

Complete the table of values for each equation; then plot the ordered pairs. See Example 2.

29. $x + 2y = 6$

x	y
0	
	0
4	
−1	

30. $3x - y = 6$

x	y
0	
	3
	0
	−3

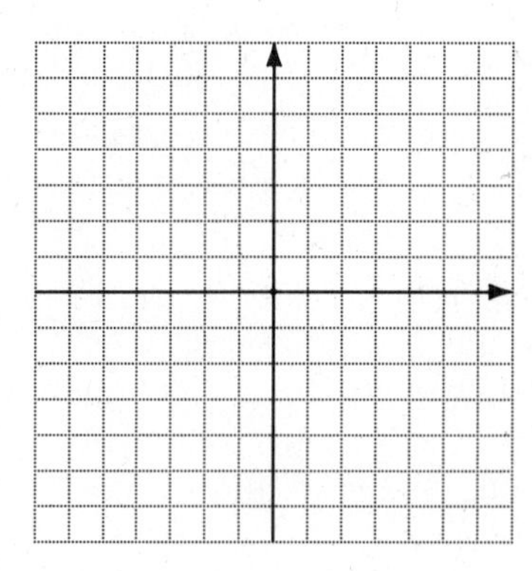

31. $4x - 3y = 12$

x	y
0	
	0
6	
$1\frac{1}{2}$	

32. $3x + 4y = 12$

x	y
0	
	0
2	
	$-1\frac{1}{2}$

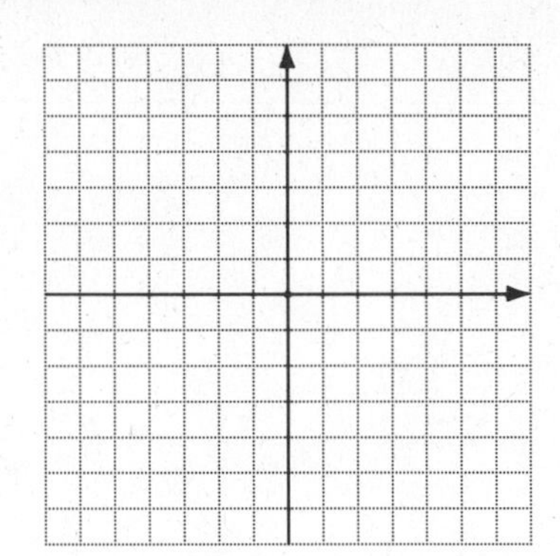

33. $y + 2 = 0$

x	y
5	
0	
−3	
−2	

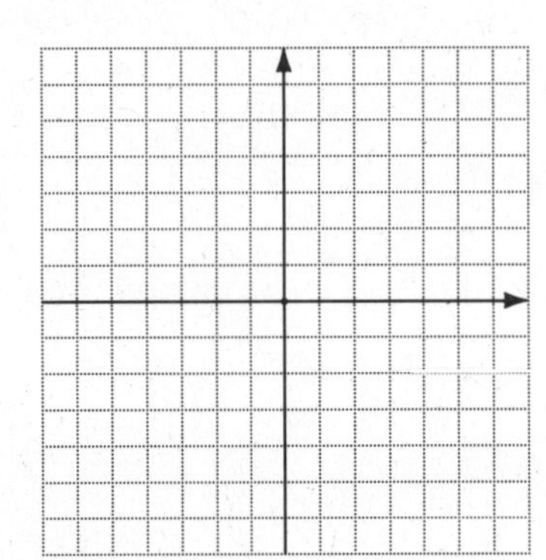

34. $x - 4 = 0$

x	y
	6
	0
	−4
	4

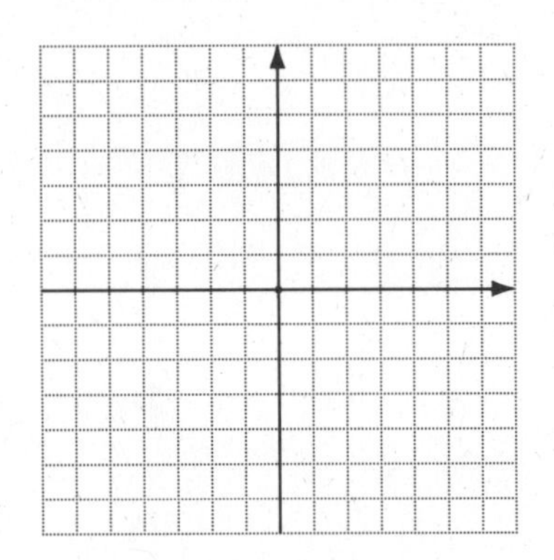

Work the following problem. See Example 3.

35. In statistics we often want to know whether two quantities (such as the height and weight of an individual) are related in such a way that we can predict one given the other. To find out, ordered pairs that give these quantities for a number of individuals are plotted on a graph called a *scatter diagram*. If the points lie approximately in a line, the variables have a linear relationship.

(a) Make a scatter diagram by plotting on the given axes the following pairs of heights and weights for six women: (62, 105), (65, 130), (67, 142), (63, 115), (66, 120), (60, 98). (The break on the horizontal and vertical axes shows that numbers have been skipped.)

(b) Does there seem to be a linear relationship between height and weight?

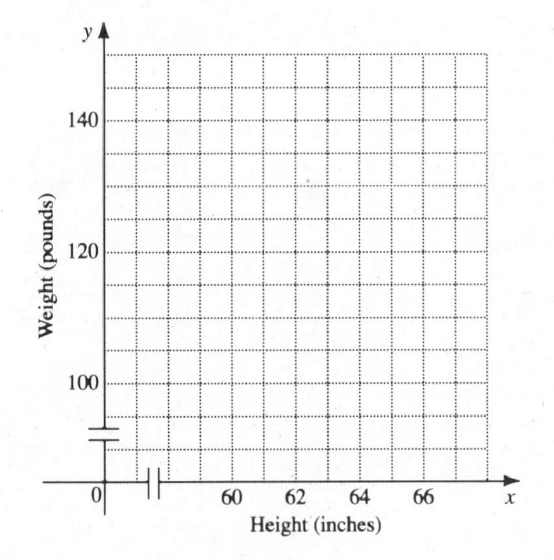

SKILL SHARPENERS

Solve each equation. See Section 3.3.

36. $3m + 5 = 13$

37. $2k - 7 = 9$

38. $8 - q = -7$

39. $-4 - x = 6$

40. $7 + 5p = 12$

41. $-3 + 8z = -19$

7.3 GRAPHING LINEAR EQUATIONS IN TWO VARIABLES

1 There are an infinite number of ordered pairs that satisfy an equation in two variables. For example, we find ordered pairs that are solutions of the equation $x + 2y = 7$ by choosing as many values of x (or y) as we wish and then completing each ordered pair.

For example, if we choose $x = 1$, then $y = 3$, so that the ordered pair (1, 3) is a solution of the equation $x + 2y = 7$.

$$1 + 2(3) = 1 + 6 = 7$$

WORK PROBLEM 1 AT THE SIDE.

Figure 6 shows a graph of all the ordered pairs found for $x + 2y = 7$ in Problem 1 at the side.

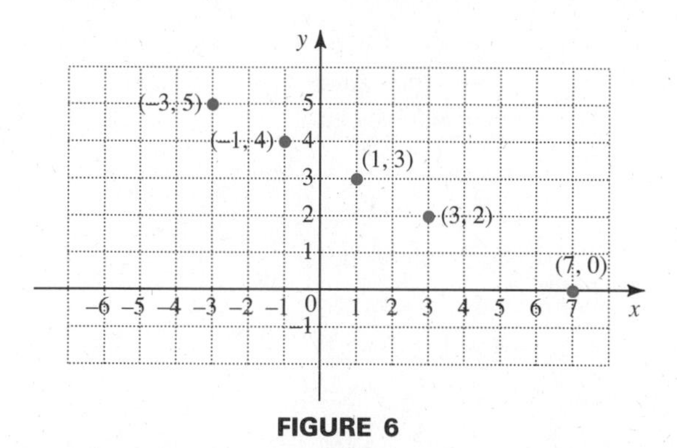

FIGURE 6

Notice that the points plotted in this figure all appear to lie on a straight line. The line that goes through these points is shown in Figure 7. In fact, all ordered pairs satisfying the equation $x + 2y = 7$ correspond to points that lie on this same straight line. This graph gives a "picture" of all the solutions of the equation $x + 2y = 7$. Only a portion of the line is shown here, but it extends indefinitely in both directions, as suggested by the arrowhead on each end of the line.

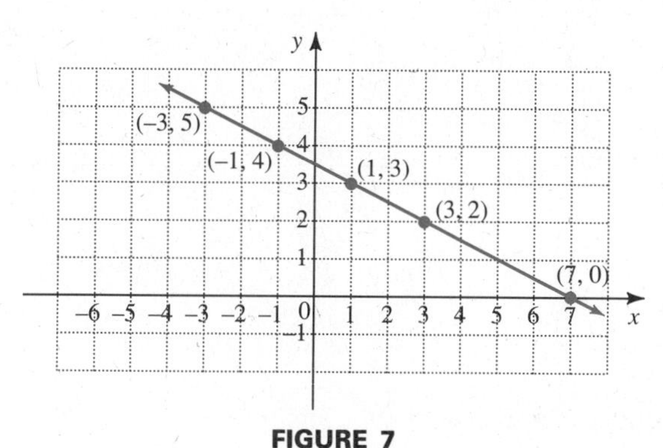

FIGURE 7

It can be shown that this is always the case.

The graph of any linear equation in two variables in a straight line.

(Notice that the word *line* appears in the name "*linear* equation.")

Since two distinct points determine a line, a straight line can be graphed by finding any two different points on the line. However, it is a good idea to plot a third point as a check.

OBJECTIVES

1. Graph linear equations by completing ordered pairs.
2. Find intercepts.
3. Graph linear equations with just one intercept.
4. Graph linear equations of the form $y = k$ or $x = k$.

1. Complete the given ordered pairs for the equation $x + 2y = 7$.

(a) (−3,)

(b) (3,)

(c) (−1,)

(d) (7,)

ANSWERS
1. (a) (−3, 5) (b) (3, 2) (c) (−1, 4) (d) (7, 0)

2. Complete the table of values and then graph the line.

$x + y = 6$

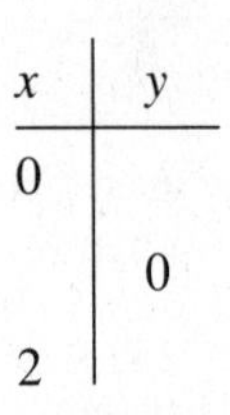

x	y
0	
	0
2	

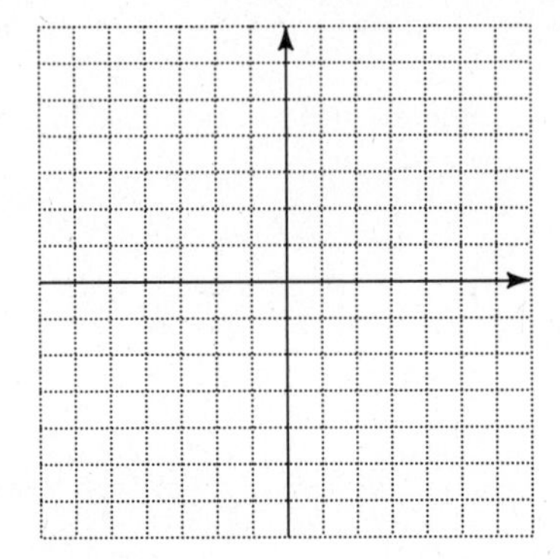

EXAMPLE 1 Graphing a Linear Equation

Graph the linear equation $3x + 2y = 6$.

For most linear equations, two different points on the graph can be found by first letting $x = 0$ and then letting $y = 0$.

If $x = 0$,		If $y = 0$,	
$3x + 2y = 6$		$3x + 2y = 6$	
$3(0) + 2y = 6$	Let $x = 0$.	$3x + 2(0) = 6$	Let $y = 0$.
$0 + 2y = 6$		$3x + 0 = 6$	
$2y = 6$		$3x = 6$	
$y = 3$.		$x = 2$.	

This gives the ordered pairs (0, 3) and (2, 0). Get a third point (as a check) by letting x or y equal some other number. For example, let $x = -2$. (Any number could have been used instead.) Replace x with -2 in the given equation.

$$3x + 2y = 6$$
$$3(-2) + 2y = 6 \qquad \text{Let } x = -2.$$
$$-6 + 2y = 6$$
$$2y = 12$$
$$y = 6$$

These three ordered pairs are shown in the table of values with Figure 8. Plot the corresponding points, then draw a line through them. This line, shown in Figure 8, is the graph of $3x + 2y = 6$. ■

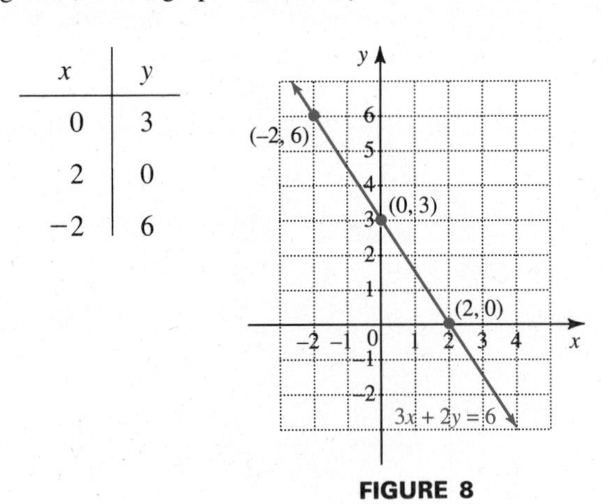

x	y
0	3
2	0
−2	6

FIGURE 8

WORK PROBLEM 2 AT THE SIDE.

EXAMPLE 2 Graphing a Linear Equation

Graph the linear equation $4x = 5y + 20$.

Although this equation is not in the form $Ax + By = C$, it *could* be put in that form, and so is a linear equation. As above, at least two different points are needed to draw the graph. First let $x = 0$ and then let $y = 0$ to complete two ordered pairs.

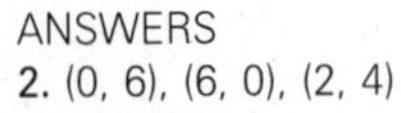

ANSWERS

2. (0, 6), (6, 0), (2, 4)

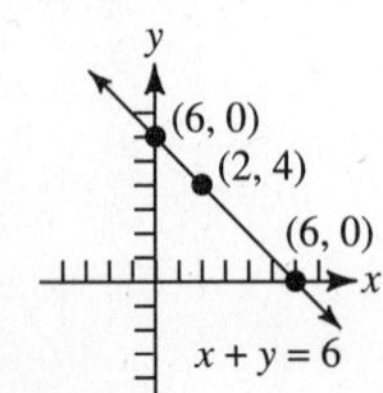

$$4x = 5y + 20 \qquad\qquad 4x = 5y + 20$$
$$4(0) = 5y + 20 \quad \text{Let } x = 0. \qquad 4x = 5(0) + 20 \quad \text{Let } y = 0.$$
$$0 = 5y + 20 \qquad\qquad 4x = 20$$
$$5y = 20$$
$$y = -4 \qquad\qquad x = 5$$

The ordered pairs are $(0, -4)$ and $(5, 0)$. Get a third ordered pair (as a check) by choosing some number other than 0 for x or y. This time, let us choose $y = 2$. Replacing y with 2 in the equation $4x - 5y = 20$ leads to the ordered pair $(\frac{15}{2}, 2)$, or $(7\frac{1}{2}, 2)$.

Plot the three ordered pairs we have found, $(0, -4)$, $(5, 0)$, and $(7\frac{1}{2}, 2)$, and draw a line through them. This line, shown in Figure 9, is the graph of $4x - 5y = 20$. ■

x	y
0	-4
5	0
$7\frac{1}{2}$	2

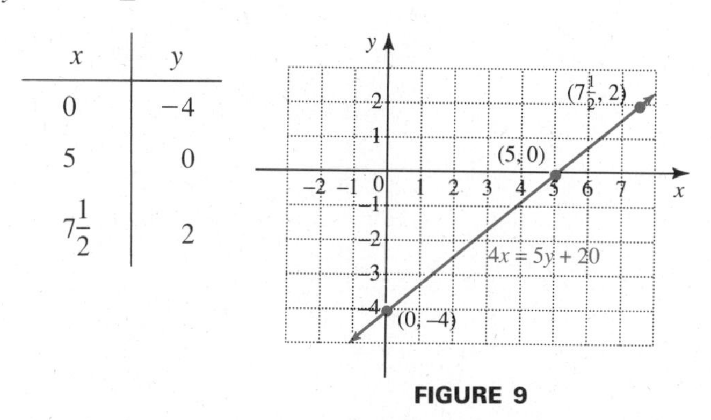

FIGURE 9

WORK PROBLEM 3 AT THE SIDE.

3. Complete three ordered pairs and graph each line.

(a) $2x - 3y = 6$

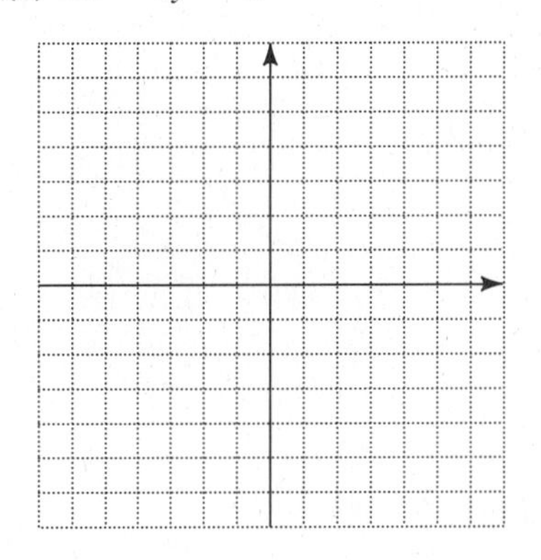

(b) $3y - 2x = 6$

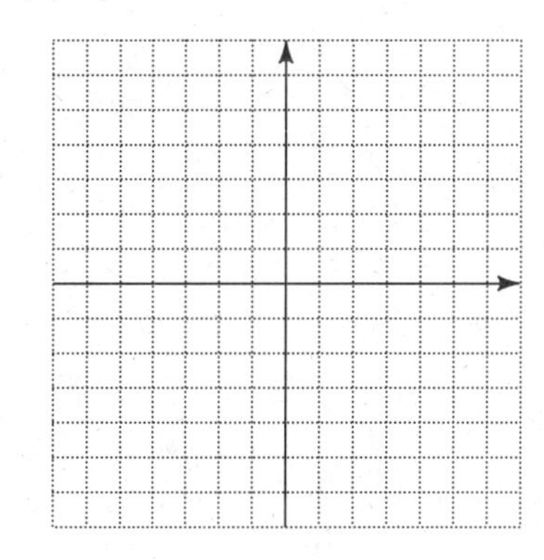

2 In Figure 9 the graph crosses the y-axis at $(0, -4)$ and crosses the x-axis at $(5, 0)$. For this reason, $(0, -4)$ is called the **y-intercept,** and $(5, 0)$ is called the **x-intercept** of the graph. The intercepts are particularly useful for graphing linear equations, as in Examples 1 and 2. The intercepts are found by replacing, in turn, each variable with 0 in the equation and solving for the value of the other variable.

FINDING INTERCEPTS

Find the x-intercept by letting $y = 0$ in the given equation and solving for x.

Find the y-intercept by letting $x = 0$ in the given equation and solving for y.

EXAMPLE 3 Finding Intercepts

Find the intercepts for the graph of $2x + y = 4$. Draw the graph.

Find the y-intercept by letting $x = 0$; find the x-intercept by letting $y = 0$.

$$2x + y = 4 \qquad\qquad 2x + y = 4$$
$$2(0) + y = 4 \quad \text{Let } x = 0. \qquad 2x + 0 = 4 \quad \text{Let } y = 0.$$
$$0 + y = 4 \qquad\qquad 2x = 4$$
$$y = 4 \qquad\qquad x = 2$$

ANSWERS

3. (a) (b)

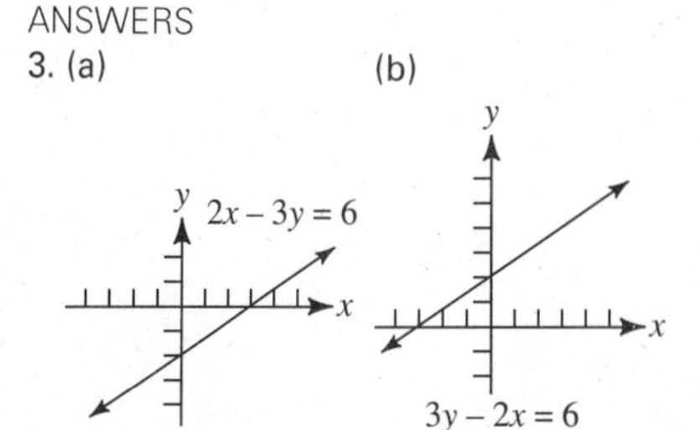

4. Find the intercepts for $5x + 2y = 10$.

The y-intercept is $(0, 4)$. The x-intercept is $(2, 0)$. The graph with the two intercepts shown in color is given in Figure 10. Get a third point as a check. For example, choosing $x = 1$ gives $y = 2$. Plot $(0, 4)$, $(2, 0)$, and $(1, 2)$ and draw a line through them. This line, shown in Figure 10, is the graph of $2x + y = 4$. ■

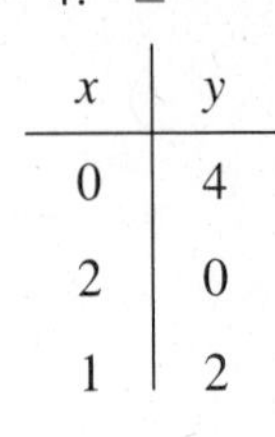

x	y
0	4
2	0
1	2

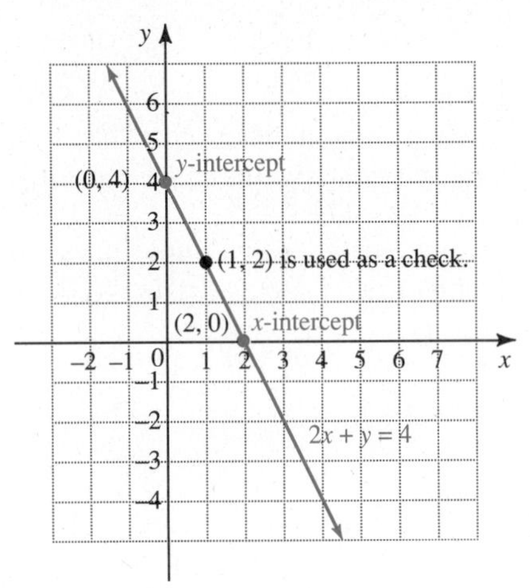

FIGURE 10

WORK PROBLEM 4 AT THE SIDE.

3 In the examples above, the x- and y-intercepts were used to help draw the graphs. This is not always possible, as the following examples show. Example 4 shows what to do when the x- and y-intercepts are the same point.

EXAMPLE 4 Graphing an Equation of the Form $Ax + By = 0$

Graph the linear equation $x - 3y = 0$.

If we let $x = 0$, then $y = 0$, giving the ordered pair $(0, 0)$. Letting $y = 0$ also gives $(0, 0)$. This is the same ordered pair, so choose two additional values for x or y. Choosing 2 for y gives $x - 3 \cdot 2 = 0$, giving the ordered pair $(6, 2)$. For a check point, choose -6 for x getting -2 for y. This ordered pair, $(-6, -2)$, along with $(0, 0)$ and $(6, 2)$, was used to get the graph shown in Figure 11. ■

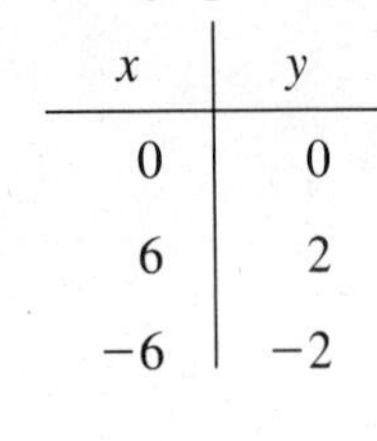

x	y
0	0
6	2
−6	−2

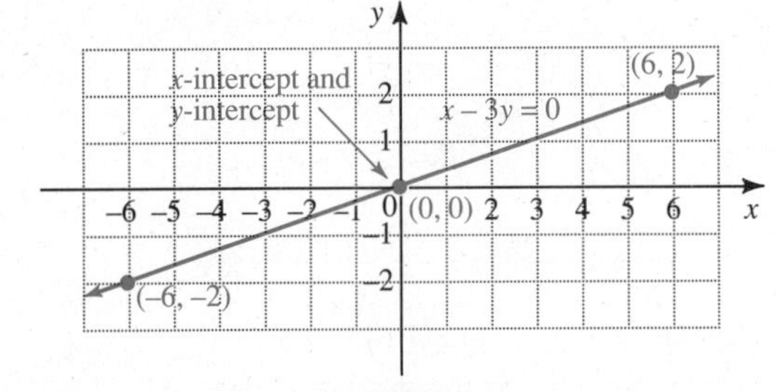

FIGURE 11

Example 4 can be generalized as follows.

If A and B are real numbers, the graph of a linear equation of the form

$$Ax + By = 0$$

goes through the origin $(0, 0)$.

ANSWERS

4. x-intercept $(2, 0)$; y-intercept $(0, 5)$

WORK PROBLEM 5 AT THE SIDE.

4 The equation $y = -4$ is a linear equation with the coefficient of x equal to 0. (To see this, write $y = -4$ as $0x + y = -4$.) Also, $x = 3$ is a linear equation with the coefficient of y equal to 0. These equations lead to horizontal or vertical straight lines, as the next examples show.

EXAMPLE 5 Graphing an Equation of the Form $y = k$

Graph the linear equation $y = -4$.

As the equation states, for any value of x that might be chosen, y is always equal to -4. Get ordered pairs that are solutions of this equation by choosing different numbers for x but always using -4 for y. Three ordered pairs that can be used are shown in the table of values with Figure 12. Drawing a line through these points gives the horizontal line shown in Figure 12. ■

x	y
-2	-4
0	-4
3	-4

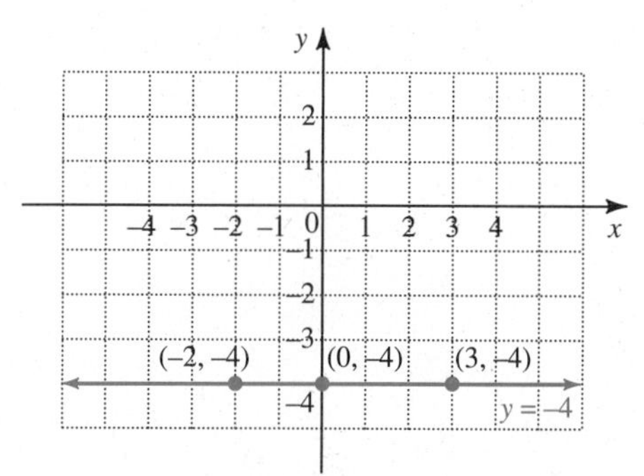

FIGURE 12

HORIZONTAL LINE

The graph of the linear equation $y = k$, where k is a real number, is the horizontal line going through the point $(0, k)$.

EXAMPLE 6 Graphing an Equation of the Form $x = k$

Graph the linear equation $x - 3 = 0$.

First add 3 to both sides of $x - 3 = 0$ to get $x = 3$. All the ordered pairs that satisfy this equation have an x-coordinate of 3. Any number can be used for y. Three ordered pairs that work are shown in the table of values with Figure 13. Drawing a line through these points gives the vertical line shown in Figure 13. ■

x	y
3	3
3	0
3	-2

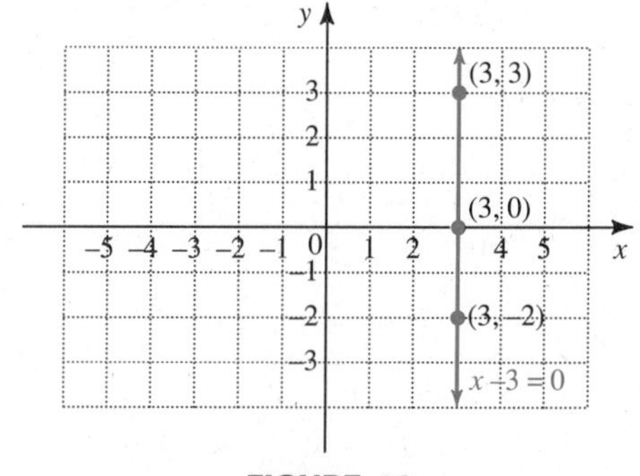

FIGURE 13

5. Graph each equation.

(a) $2x = y$

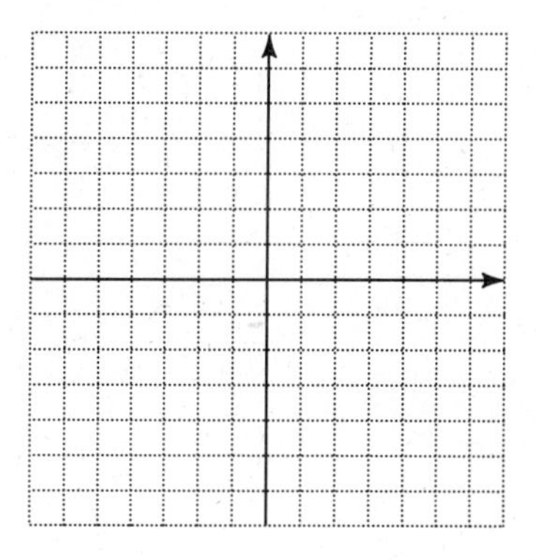

(b) $x = -4y$

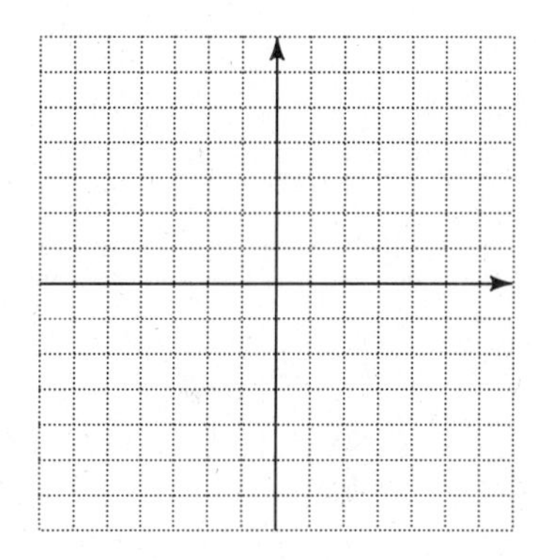

ANSWERS

5. (a) (b)

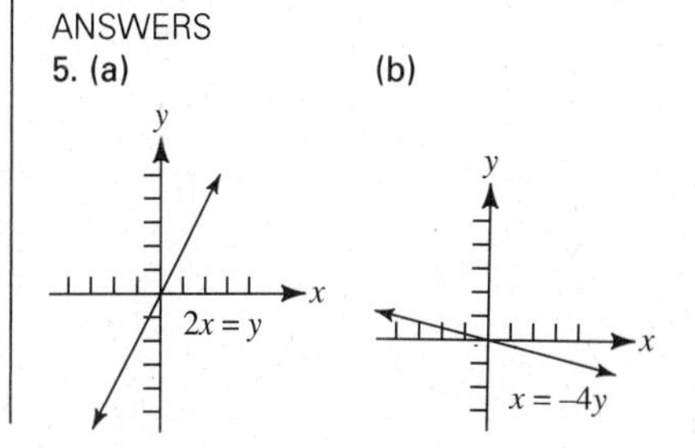

6. Graph each equation.

(a) $y = -5$

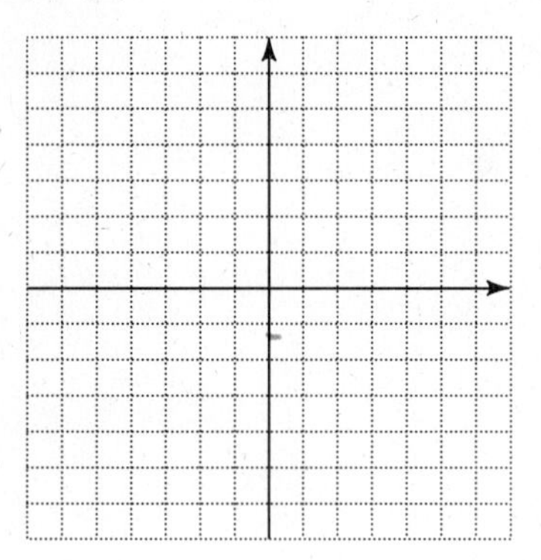

(b) $x = 2$

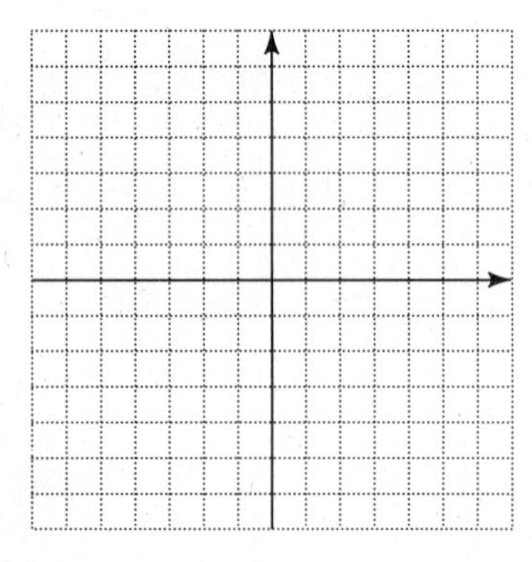

ANSWERS

6. (a) (b)

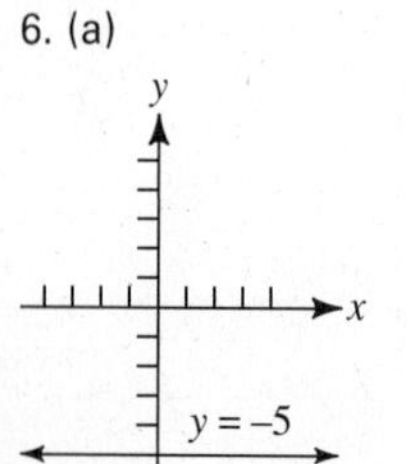

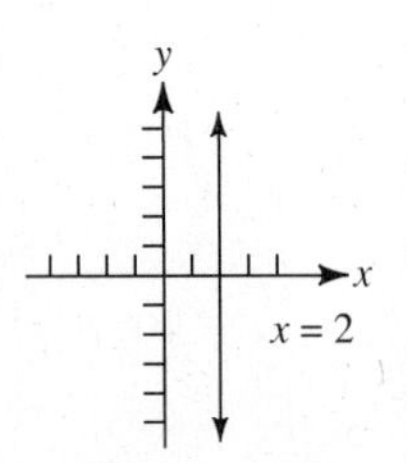

VERTICAL LINE

The graph of the linear equation $x = k$, where k is a real number, is the vertical line going through the point $(k, 0)$.

WORK PROBLEM 6 AT THE SIDE.

The different forms of straight-line equations and the methods of graphing them are summarized below.

GRAPHING LINEAR EQUATIONS

Equation	*Graphing methods*	*Example*
$y = k$	Draw a horizontal line, through $(0, k)$.	$y = -2$ 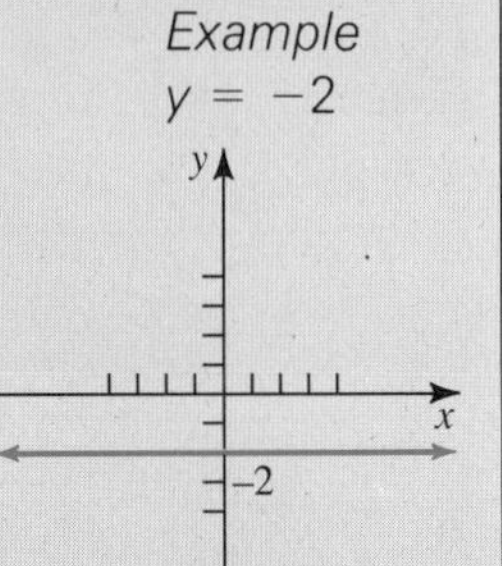
$x = k$	Draw a vertical line, through $(k, 0)$.	$x = 4$ 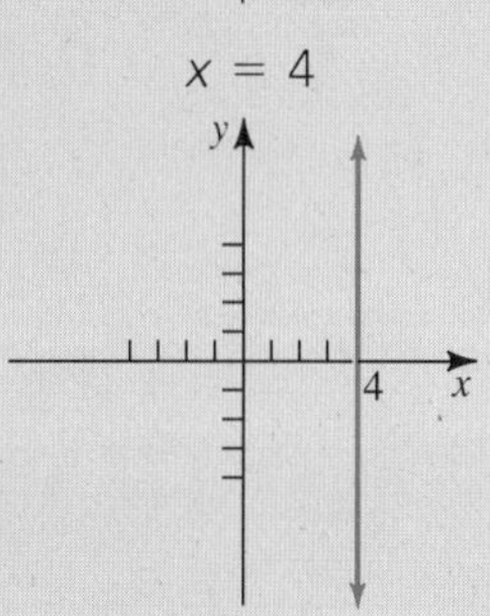
$Ax + By = 0$	Graph goes through $(0, 0)$. To get additional points that lie on the graph, choose any value of x, or y, except 0.	$3x + y = 0$ 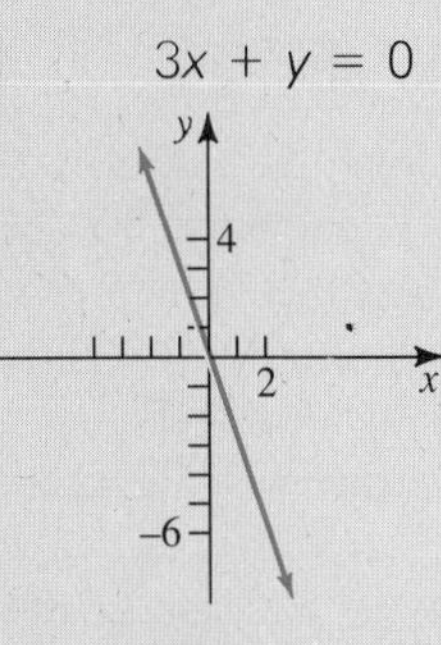
$Ax + By = C$ A, B, and $C \neq 0$	Find any two points the line goes through. A good choice is to find the intercepts: let $x = 0$, and find the corresponding value of y; then let $y = 0$, and find x. As a check, get a third point by choosing a value of x or y that has not yet been used.	$3x - 2y = 6$ 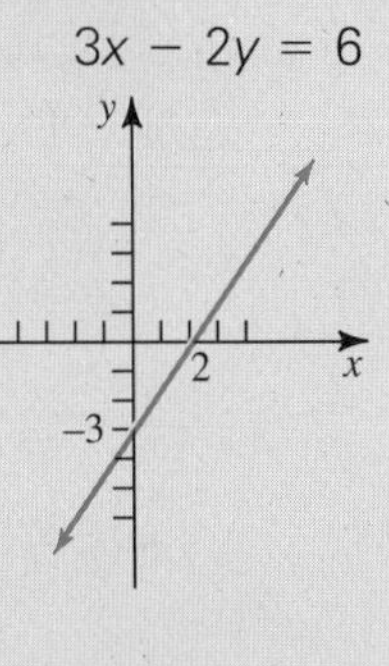

name date hour

7.3 EXERCISES

Complete the given ordered pairs using the given equation. Then graph the equation by plotting the points and drawing a line through them. See Examples 1 and 2.

1. $x + y = 5$
(0,) (, 0) (2,)

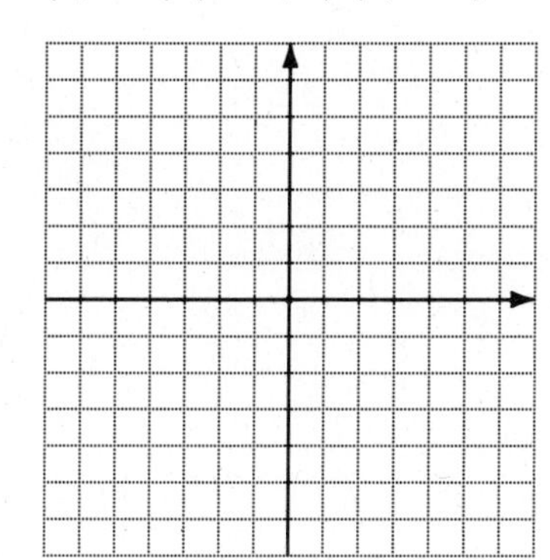

2. $y = x - 3$

(0,) (, 0) (5,)

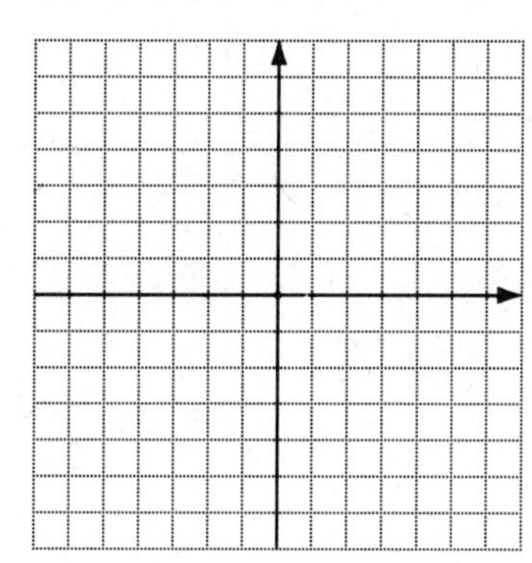

3. $y = x + 4$

(0,) (, 0) (−2,)

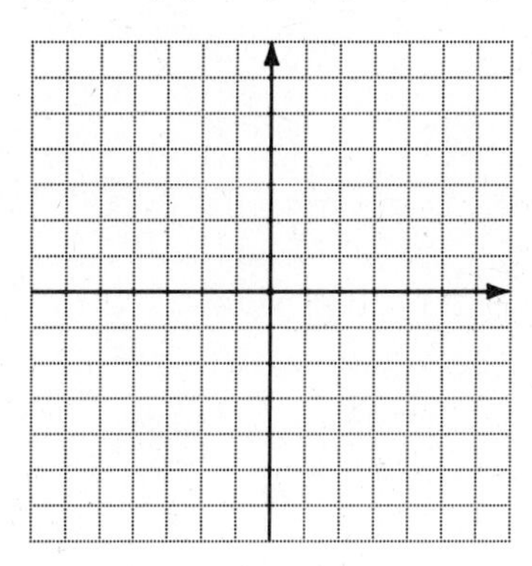

4. $y + 5 = x$
(0,) (, 0) (6,)

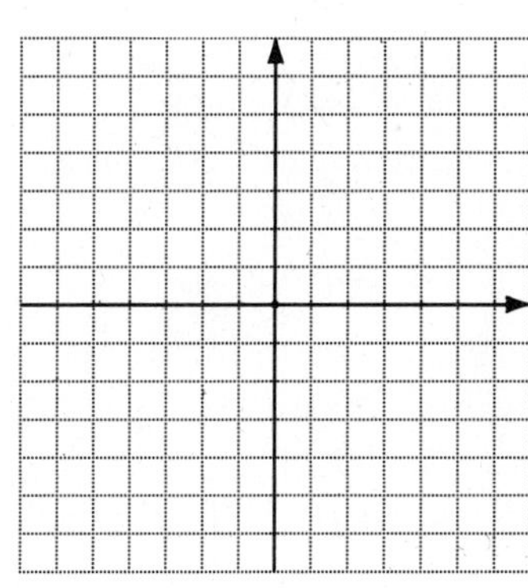

5. $y = 3x - 6$
(0,) (, 0) (3,)

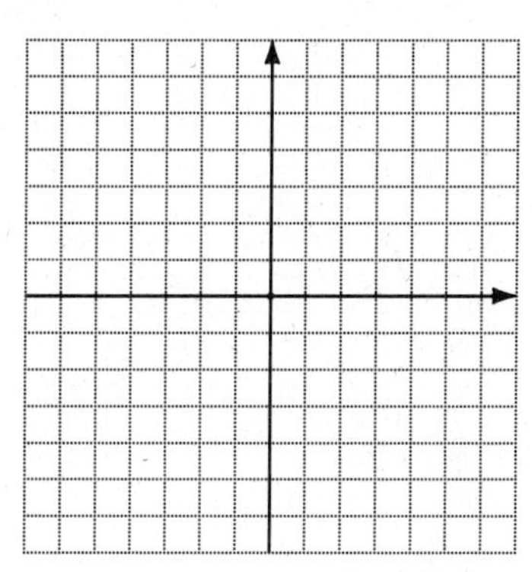

6. $x = 2y + 1$
(, 2) (, 0) (, −3)

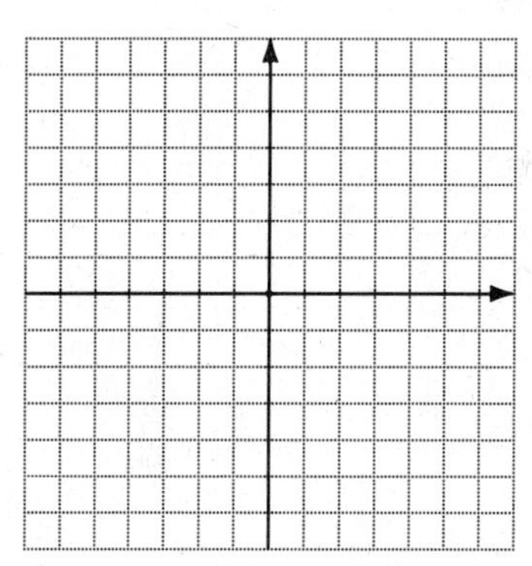

Find the intercepts for each equation. See Example 3.

7. $2x + 3y = 6$

x-intercept:

y-intercept:

8. $7x + 2y = 14$

x-intercept:

y-intercept:

9. $3x - 5y = 9$

x-intercept:

y-intercept:

10. $6x - 5y = 12$

x-intercept:

y-intercept:

11. $2y = 5x$

x-intercept:

y-intercept:

12. $6y = 12$

x-intercept:

y-intercept:

Graph each linear equation. See Examples 1, 2, and 4-6.

13. $x - y = 2$

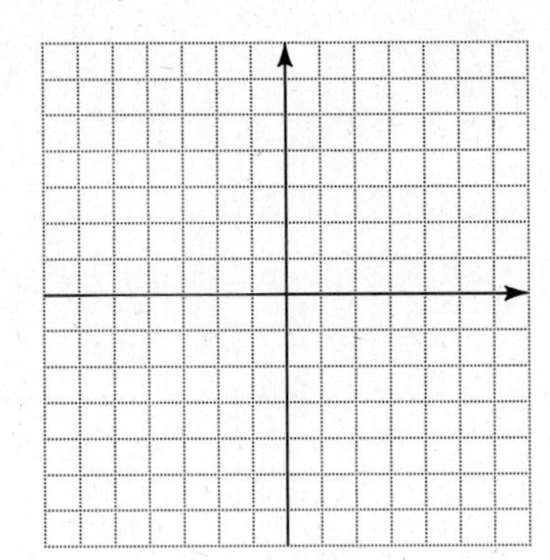

14. $x + y = 6$

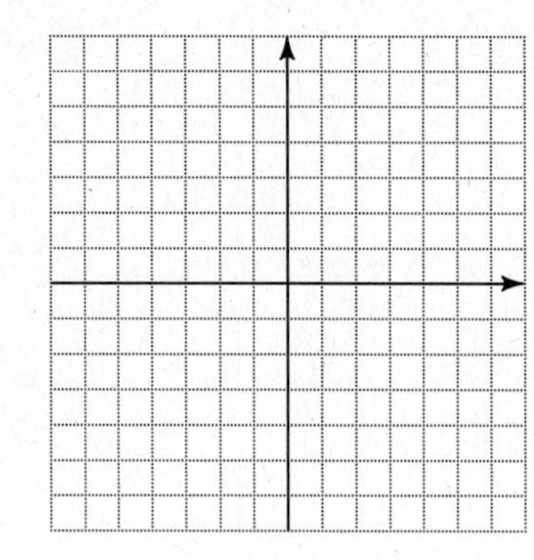

15. $y = x + 4$

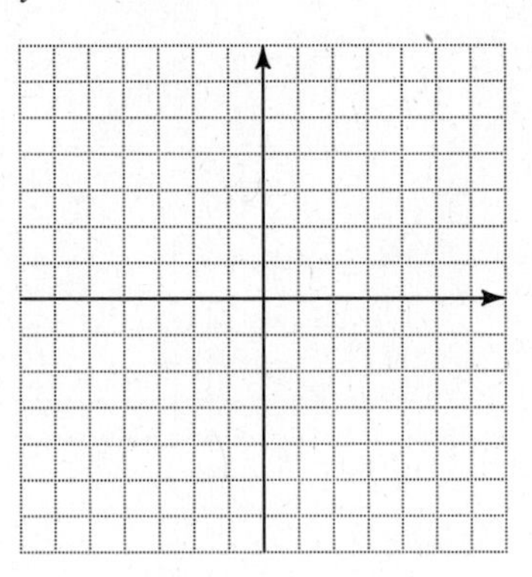

16. $y = x - 5$

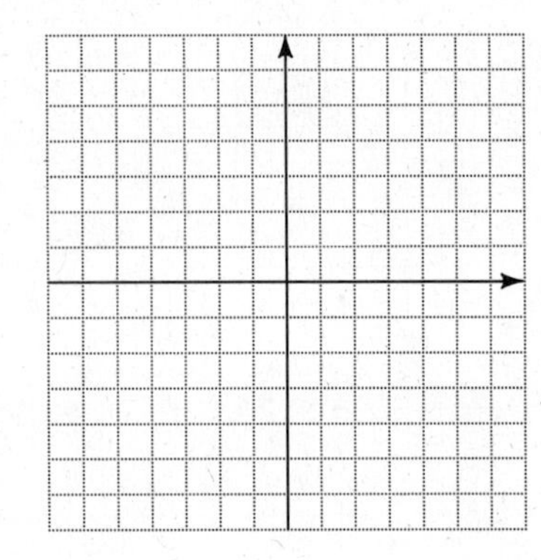

17. $x + 2y = 6$

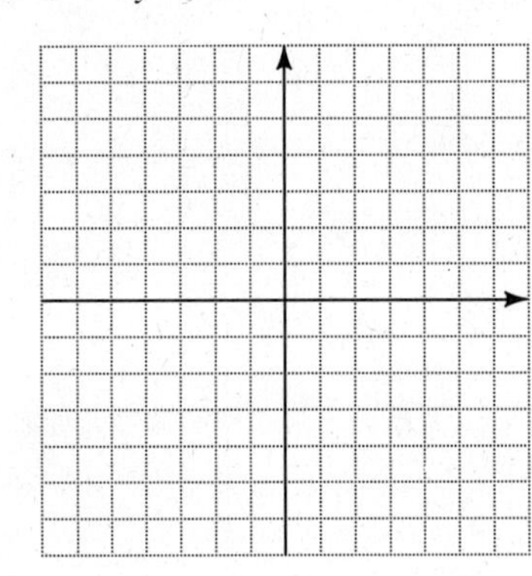

18. $3x - y = 6$

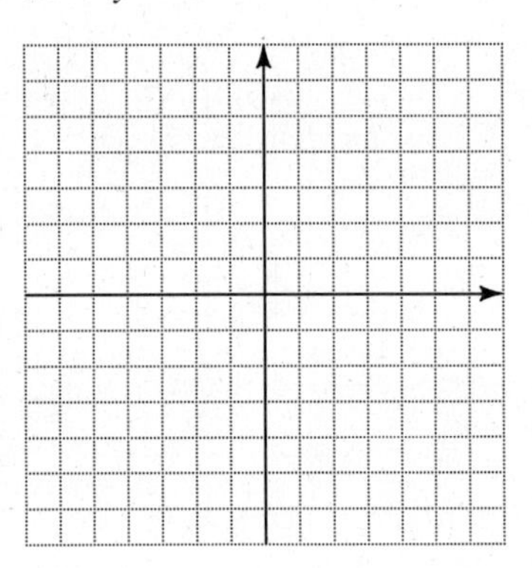

19. $4x = 3y - 12$

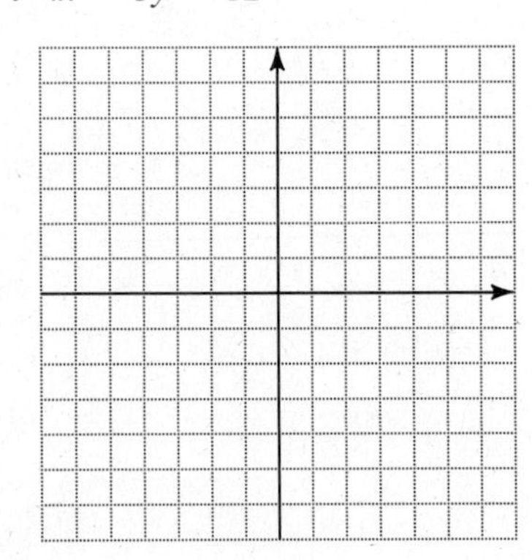

20. $5x = 2y - 10$

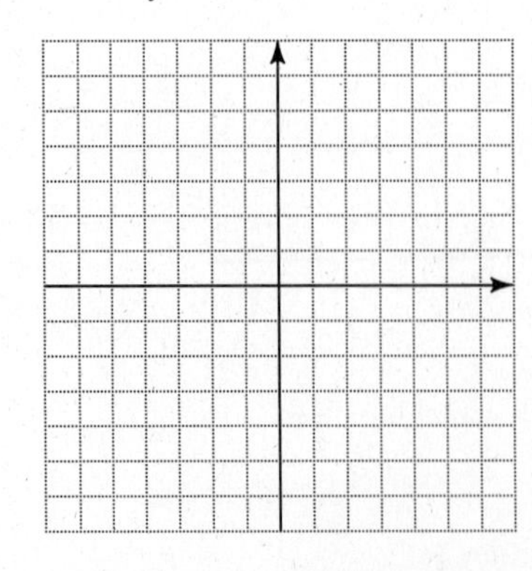

21. $3x = 6 - 2y$

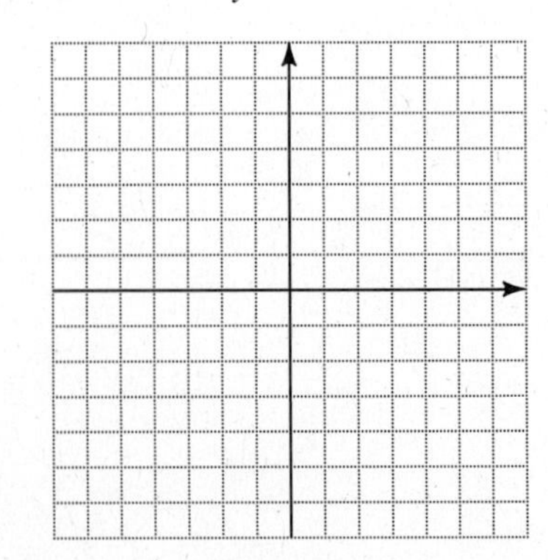

name date hour

22. $3x + 7y = 14$

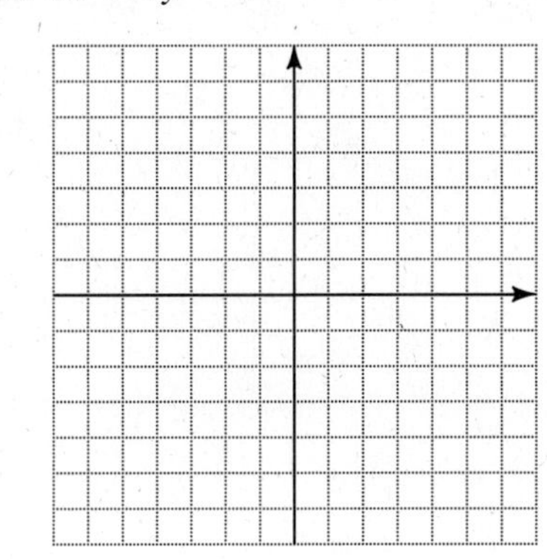

23. $6x - 5y = 18$

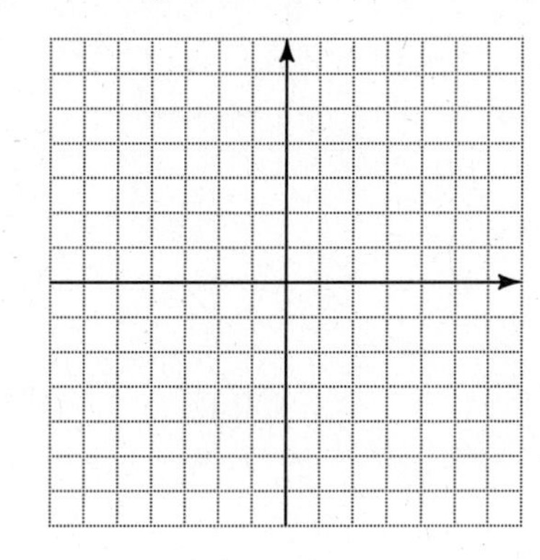

24. $y = 2x$

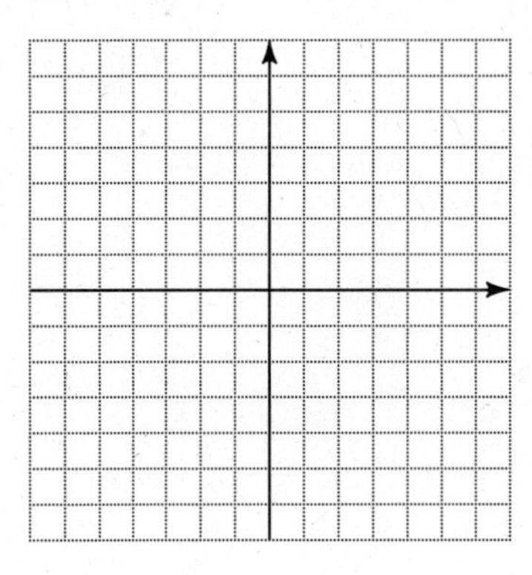

25. $y = -3x$

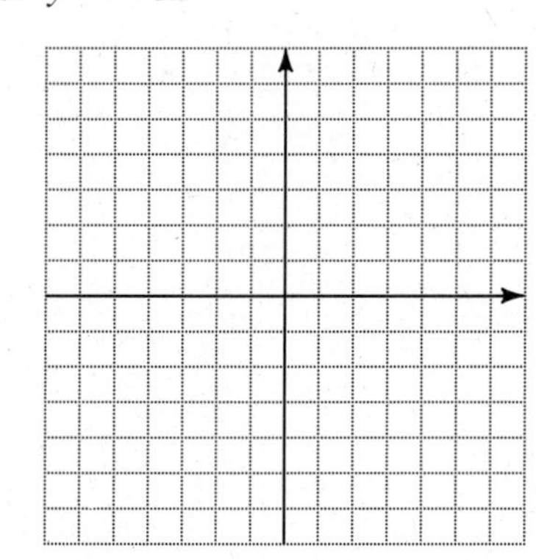

26. $y + 6x = 0$

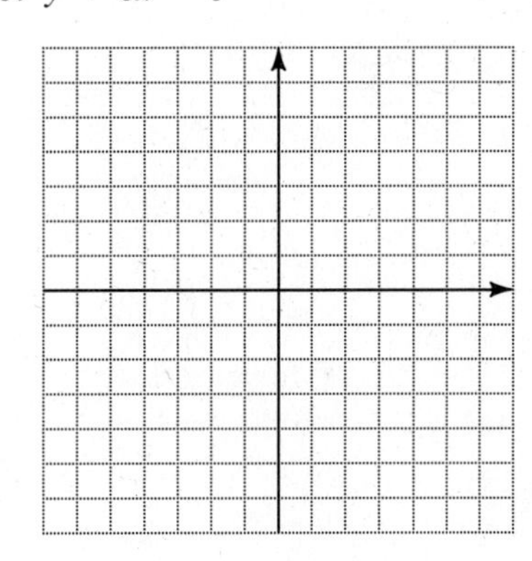

27. $y - 4x = 0$

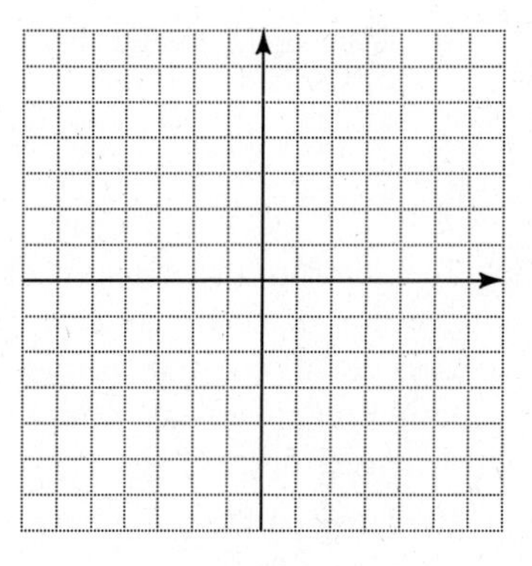

28. $2x + 3y = 0$

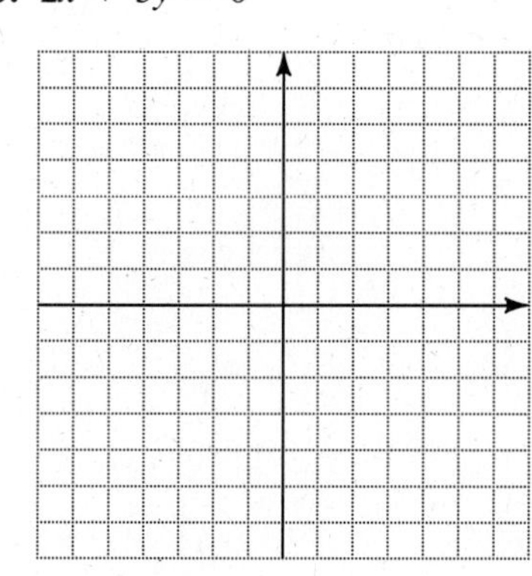

29. $3x - 4y = 0$

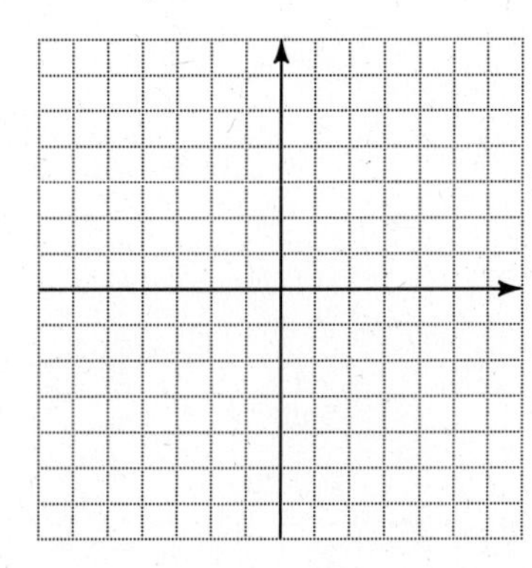

30. $x + 2 = 0$

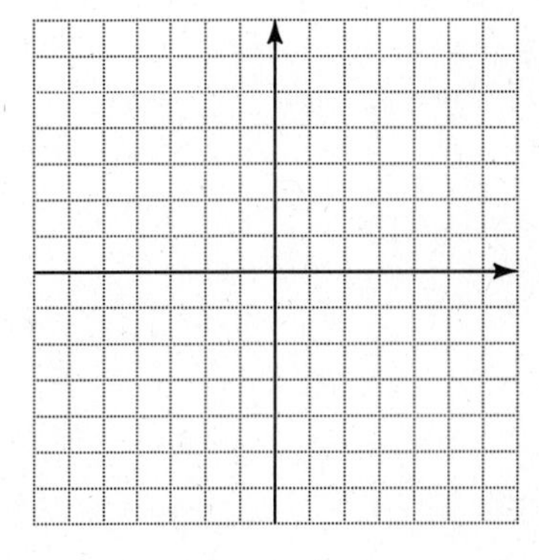

31. $y - 3 = 0$

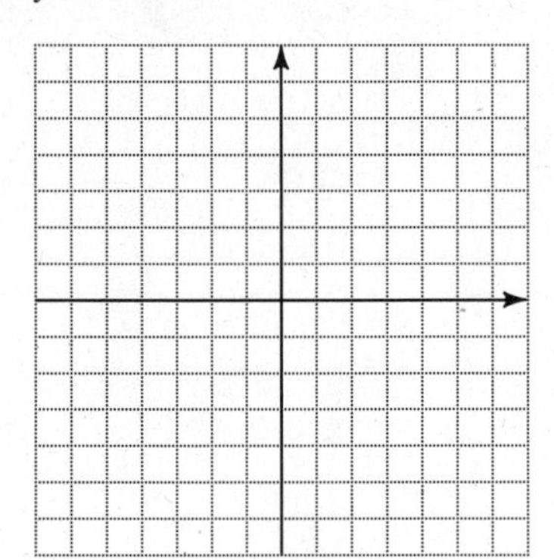

32. $y = 6$

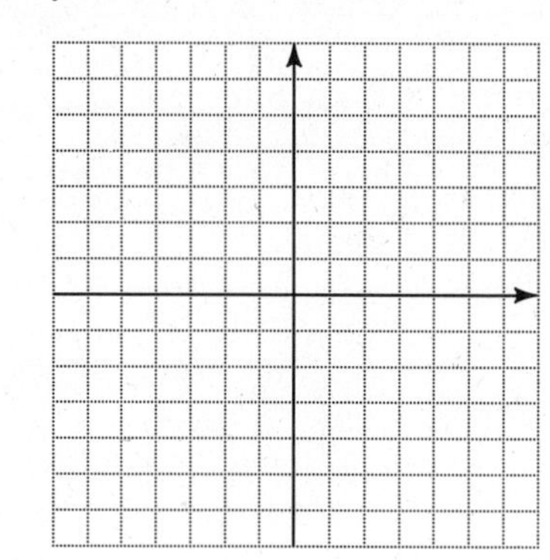

33. $y = 0$

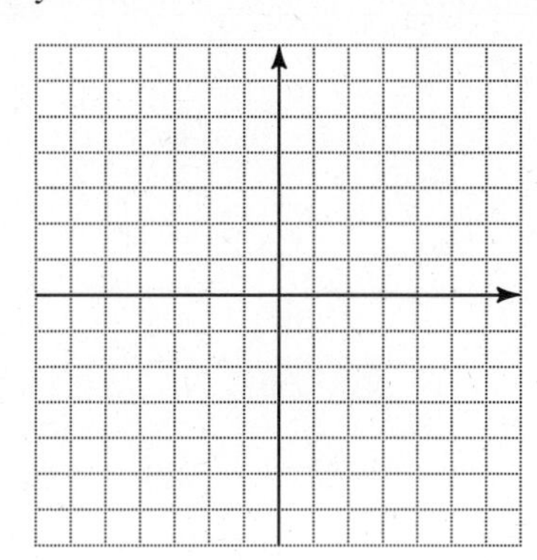

Work the following problems.

34. The graph shows the value of a certain automobile over its first five years. Use the graph to estimate the depreciation (loss in value) during the following years.

(a) First **(b)** Second **(c)** Fifth

(d) What is the total depreciation over the 5-year period?

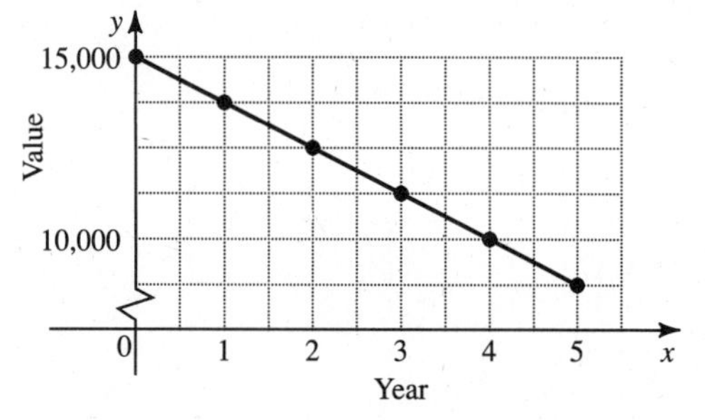

35. The demand for an item is closely related to its price. As price goes up, demand goes down. On the other hand, when price goes down, demand goes up. Suppose the demand for a certain fashionable watch is 1000 when its price is $30 and 8000 when it costs $15.

(a) Let x be the price and y be the demand for the watch. Graph the two given pairs of prices and demands.

(b) Assume the relationship is linear. Draw a line through the two points from part (a). From your graph estimate the demand if the price drops to $10.

(c) Use the graph to estimate the price if the demand is 4000.

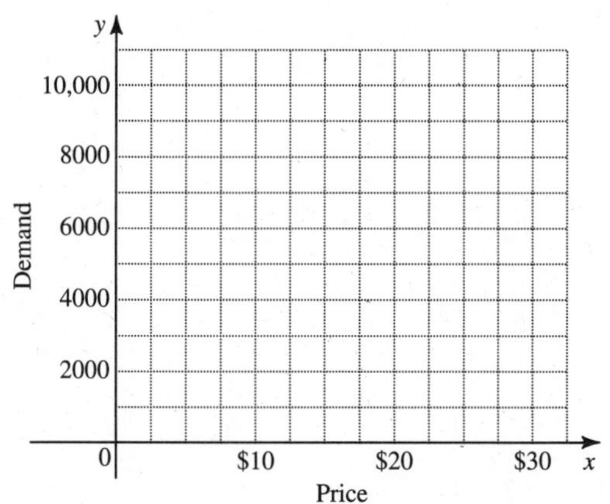

SKILL SHARPENERS

Find each quotient. See Section 2.5.

36. $\dfrac{-3 - 5}{2 - 7}$

37. $\dfrac{-2 - (-4)}{3 - (-1)}$

38. $\dfrac{5 - (-7)}{-4 - (-1)}$

39. $\dfrac{-6 - 3}{4 - (-5)}$

40. $\dfrac{-9 - 4}{-2 - 3}$

41. $\dfrac{12 - (-4)}{3 - 3}$

7.4 THE SLOPE OF A LINE

We can graph a straight line if at least two different points on the line are known. A line can also be graphed by using just one point on the line, along with the "steepness" of the line.

OBJECTIVES

1. Find the slope of a line given two points.
2. Find the slope from the equation of a line.
3. Use slopes to determine whether two lines are parallel, perpendicular, or neither.

1 One way to measure the steepness of a line is to compare the vertical change in the line (the rise) to the horizontal change (the run) while moving along the line from one fixed point to another. This measure of steepness is called the *slope* of the line.

Figure 14 shows a line with the points (x_1, y_1) and (x_2, y_2). (Read x_1 as "x-sub-one" and x_2 as "x-sub-two.") Moving along the line from the point (x_1, y_1) to the point (x_2, y_2) causes y to change by $y_2 - y_1$ units. This is the vertical change. Similarly, x changes by $x_2 - x_1$ units, the horizontal change. The ratio of the change in y to the change in x gives the slope of the line. We usually denote slope with the letter m. The slope of a line is defined as follows.

SLOPE FORMULA

The **slope** of the line through the points (x_1, y_1) and (x_2, y_2) is

$$m = \frac{\text{change in } y}{\text{change in } x} = \frac{y_2 - y_1}{x_2 - x_1} \quad \text{if } x_2 \neq x_1.$$

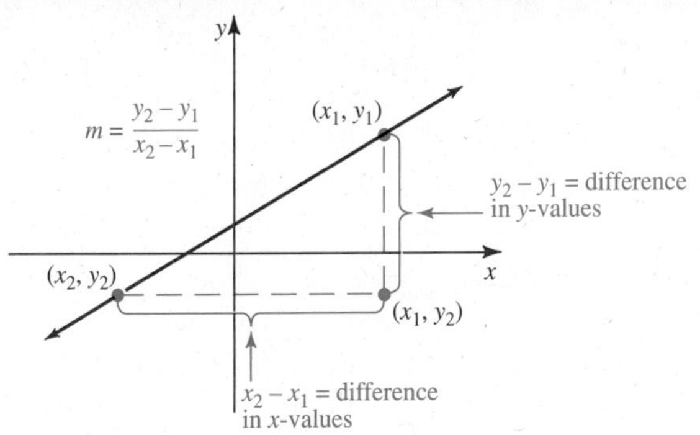

FIGURE 14

WORK PROBLEM 1 AT THE SIDE.

1. Evaluate $\frac{y_2 - y_1}{x_2 - x_1}$ for the following values.

(a) $y_2 = 4$, $y_1 = -1$, $x_2 = 3$, $x_1 = 4$

(b) $x_1 = 3$, $x_2 = -5$, $y_1 = 7$, $y_2 = -9$

(c) $x_1 = 2$, $x_2 = 7$, $y_1 = 4$, $y_2 = 9$

The slope of a line tells how fast y changes for each unit of change in x; that is, the slope gives the rate of change in y for each unit of change in x.

The idea of slope is useful in many everyday situations. For example, a highway with a 10% or $\frac{1}{10}$ grade (or slope) rises 1 meter for every 10 meters horizontally. Architects specify the pitch of a roof by indicating the slope; a $\frac{5}{12}$ roof means that the roof rises 5 feet for every 12 feet in the horizontal direction. The slope of a stairwell also indicates the ratio of the vertical rise to the horizontal run. See Figure 15.

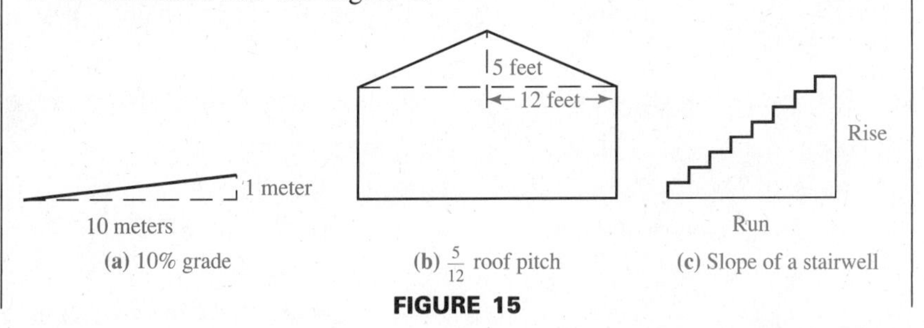

(a) 10% grade (b) $\frac{5}{12}$ roof pitch (c) Slope of a stairwell

FIGURE 15

ANSWERS
1. (a) −5 (b) 2 (c) 1

2. Find the slope of each of the following lines.

(a) Through (6, −2) and (5, 4)

(b) Through (−3, 5) and (−4, 7)

(c) Through (6, −8) and (−2, 4)

ANSWERS

2. (a) −6 (b) −2 (c) $-\frac{3}{2}$

EXAMPLE 1 Finding the Slope of a Line

Find the slope of each of the following lines.

(a) The line through (−4, 7) and (1, −2)

Use the definition given above. Let $(-4, 7) = (x_2, y_2)$ and $(1, -2) = (x_1, y_1)$. Then

$$\text{slope} = m = \frac{y_2 - y_1}{x_2 - x_1} = \frac{7 - (-2)}{-4 - 1} = \frac{9}{-5} = -\frac{9}{5}.$$

See Figure 16.

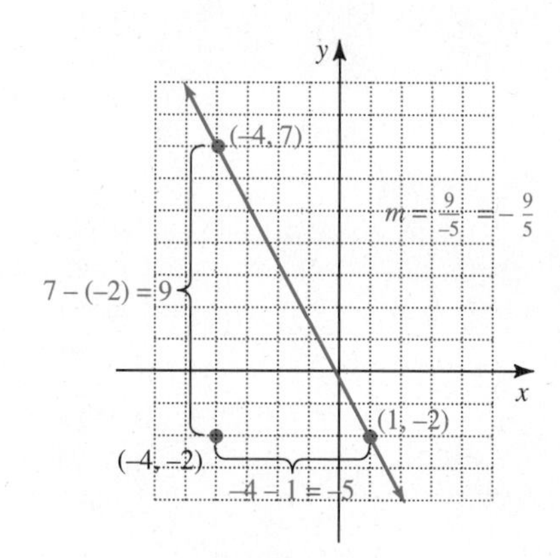

FIGURE 16

(b) The line through (12, 5) and (−9, −2)

$$m = \frac{y_2 - y_1}{x_2 - x_1} = \frac{5 - (-2)}{12 - (-9)} = \frac{7}{21} = \frac{1}{3}$$

The same slope is obtained by subtracting in reverse order.

$$\frac{-2 - 5}{-9 - 12} = \frac{-7}{-21} = \frac{1}{3} \quad \blacksquare$$

CAUTION It makes no difference which point is (x_1, y_1) or (x_2, y_2); however, it is important to be consistent: start with the *x*- and *y*-values of *one* point (either one) and subtract the corresponding values of the *other* point.

WORK PROBLEM 2 AT THE SIDE.

In Example 1(a) the slope was negative and the corresponding line in Figure 16 fell from left to right. The slope in Example 1(b) was positive and the corresponding line, shown in Figure 17, rises from left to right. These facts can be generalized.

A line with positive slope rises from left to right.
A line with negative slope falls from left to right.

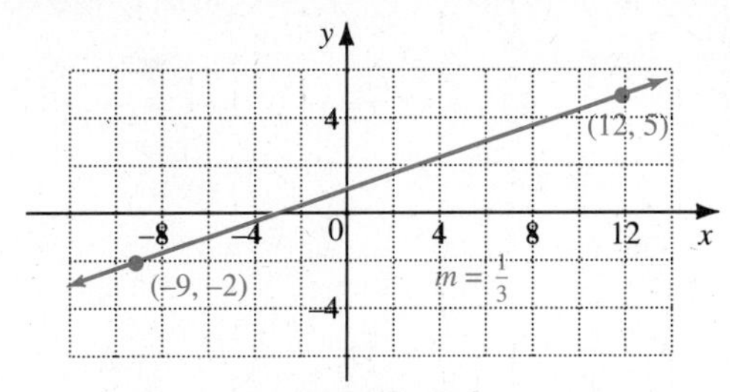

FIGURE 17

The next examples illustrate slopes of horizontal and vertical lines.

EXAMPLE 2 Finding the Slope of a Horizontal Line

Find the slope of the line through (−8, 4) and (2, 4).

Use the definition of slope.

$$m = \frac{y_2 - y_1}{x_2 - x_1} = \frac{4 - 4}{-8 - 2} = \frac{0}{-10} = 0$$

As shown in Figure 18 by a sketch of the graph through these two points, the line through the given points is horizontal. *All horizontal lines have a slope of 0,* since the difference in *y*-values will always be 0. ■

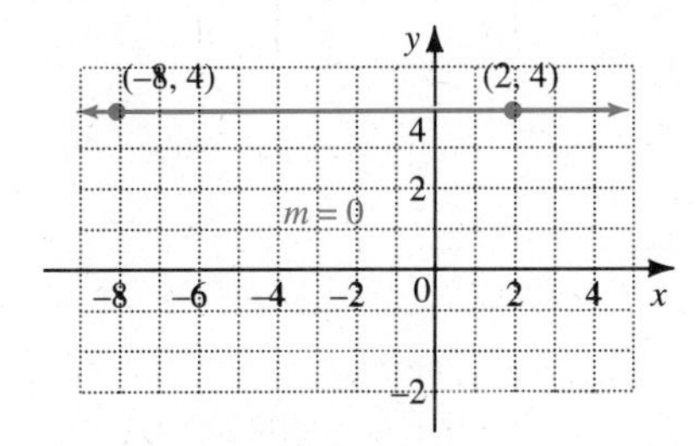

FIGURE 18

EXAMPLE 3 Finding the Slope of a Vertical Line

Find the slope of the line through (6, 2) and (6, −9).

$$m = \frac{y_2 - y_1}{x_2 - x_1} = \frac{2 - (-9)}{6 - 6} = \frac{11}{0} \qquad \text{Undefined}$$

Since division by zero is undefined, this line has an undefined slope. (This is why the formula for slope at the beginning of this section had the restriction $x_1 \neq x_2$.) The graph in Figure 19 shows that this line is vertical. Since all points on a vertical line have the same *x*-value, *vertical lines have undefined slope.* ■

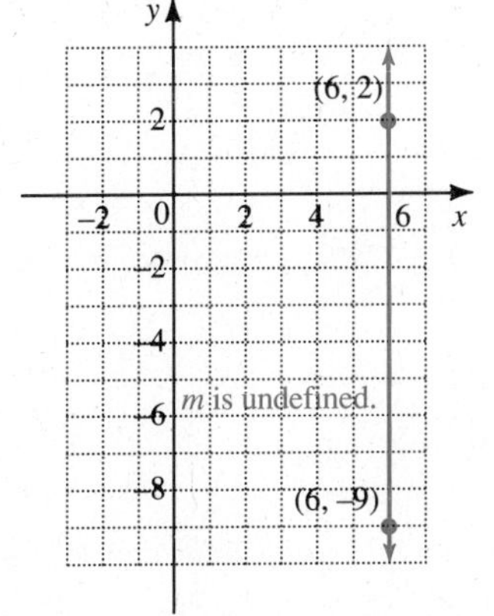

FIGURE 19

3. Find the slope of each line.

(a) Through (2, 5) and (−1, 5)

(b) Through (3, 1) and (3, −4)

(c) With equation $y = -1$

(d) With equation $x - 4 = 0$

ANSWERS
3. (a) 0 (b) undefined slope
(c) 0 (d) undefined slope

SLOPES OF HORIZONTAL AND VERTICAL LINES

Horizontal lines, which have equations of the form $y = k$, have a slope of 0.

Vertical lines, which have equations of the form $x = k$, have undefined slope.

WORK PROBLEM 3 AT THE SIDE.

2 The slope of a line can be found directly from its equation. For example, let us find the slope of the line

$$y = -3x + 5.$$

The definition of slope requires two different points on the line. Get these two points by first choosing two different values of x and then finding the corresponding values of y. Let us choose $x = -2$ and $x = 4$.

$$y = -3(-2) + 5 \quad \text{Let } x = -2.$$
$$y = 6 + 5$$
$$y = 11$$

$$y = -3(4) + 5 \quad \text{Let } x = 4.$$
$$y = -12 + 5$$
$$y = -7$$

The ordered pairs are (−2, 11) and (4, −7). Now find the slope.

$$\text{Slope} = \frac{11 - (-7)}{-2 - 4} = \frac{18}{-6} = -3$$

The slope, −3, is the same number as the coefficient of x in the equation $y = -3x + 5$. It can be shown that this always happens, *as long as the equation is solved for y*. This fact is used to find the slope of a line from its equation.

FINDING THE SLOPE OF A LINE FROM ITS EQUATION

Step 1 Solve the equation for y.

Step 2 The slope is given by the coefficient of x.

EXAMPLE 4 Finding Slope from an Equation

Find the slope of each of the following lines.

(a) $2x - 5y = 4$

Solve the equation for y.

$$2x - 5y = 4$$
$$-5y = -2x + 4 \quad \text{Subtract } 2x \text{ from each side.}$$
$$y = \frac{2}{5}x - \frac{4}{5} \quad \text{Divide each side by } -5.$$

The slope is given by the coefficient of x, so

$$\text{slope} = \frac{2}{5}.$$

(b) $8x + 4y = 1$

Solve for y; you should get

$$y = -2x + \frac{1}{4}.$$

The slope of this line is given by the coefficient of x, -2. ■

WORK PROBLEM 4 AT THE SIDE.

3 Two lines in a plane that never intersect are **parallel.** Slopes can be used to tell whether two lines are parallel. For example, Figure 20 shows the graph of $x + 2y = 4$ and the graph of $x + 2y = -6$. These lines appear to be parallel. Solve for y to find that both $x + 2y = 4$ and $x + 2y = -6$ have a slope of $-\frac{1}{2}$. Nonvertical parallel lines always have equal slopes.

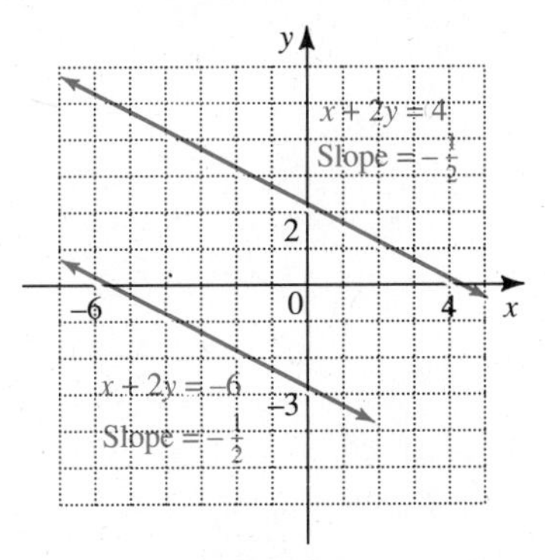

FIGURE 20

Figure 21 shows the graph of $x + 2y = 4$ and the graph of $2x - y = 6$. These lines appear to be **perpendicular** (meet at a 90° angle). Solving for y shows that the slope of $x + 2y = 4$ is $-\frac{1}{2}$, while the slope of $2x - y = 6$ is 2. The product of $-\frac{1}{2}$ and 2 is

$$\left(-\frac{1}{2}\right)(2) = -1.$$

This is true in general; the product of the slopes of two perpendicular lines is always -1.

SLOPES OF PARALLEL AND PERPENDICULAR LINES

Two lines with the same slope are parallel; two lines whose slopes have a product of -1 are perpendicular.

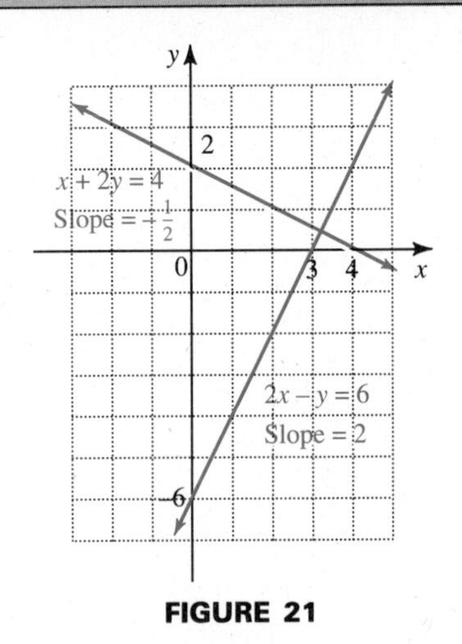

FIGURE 21

4. Find the slope of each line.

(a) $y = 2x - 7$

(b) $y = -\frac{7}{2}x + 1$

(c) $3x + 2y = 9$

(d) $y + 4 = 0$

ANSWERS

4. (a) 2 (b) $-\frac{7}{2}$ (c) $-\frac{3}{2}$ (d) 0

5. Write *parallel, perpendicular,* or *neither* for each pair of lines.

(a) $x + y = 6$
$x + y = 1$

(b) $3x - y = 4$
$x + 3y = 9$

(c) $2x - y = 5$
$2x + y = 3$

ANSWERS
5. (a) parallel (b) perpendicular
(c) neither

EXAMPLE 5 Deciding Whether Lines Are Parallel, Perpendicular, or Neither

Decide whether each pair of lines is *parallel, perpendicular,* or *neither*.

(a) $x + 2y = 7$
$-2x + y = 3$

Find the slope of each line by first solving each equation for y.

$$x + 2y = 7$$
$$2y = -x + 7$$
$$y = -\frac{1}{2}x + \frac{7}{2}$$

Slope is $-\frac{1}{2}$.

$$-2x + y = 3$$
$$y = 2x + 3$$

Slope is 2.

Since the slopes are not equal, the lines are not parallel. Check the product of the slopes: $(-\frac{1}{2})(2) = -1$. The two lines are perpendicular because the product of their slopes is -1. See Figure 22.

(b) $3x - y = 4$
$6x - 2y = -12$

Find the slopes. Both lines have a slope of 3, so the lines are parallel. The graphs of these lines are shown in Figure 23.

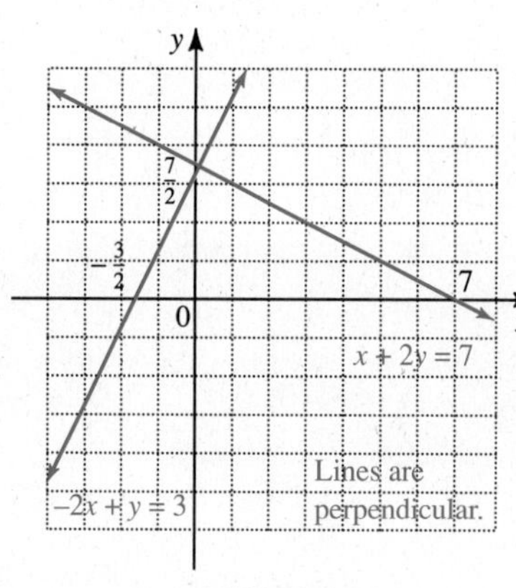

FIGURE 22

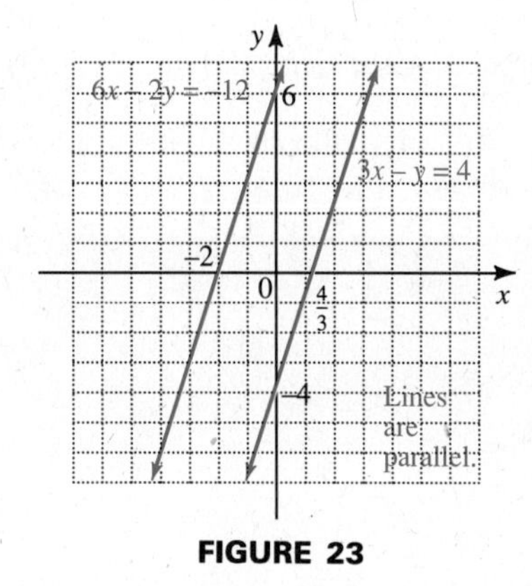

FIGURE 23

(c) $4x + 3y = 6$
$2x - y = 5$

Here the slopes are $-\frac{4}{3}$ and 2. These lines are neither parallel nor perpendicular. ■

WORK PROBLEM 5 AT THE SIDE.

name date hour

7.4 EXERCISES

Use the coordinates of the indicated points to find the slope of each of the following lines. See Example 1.

1.

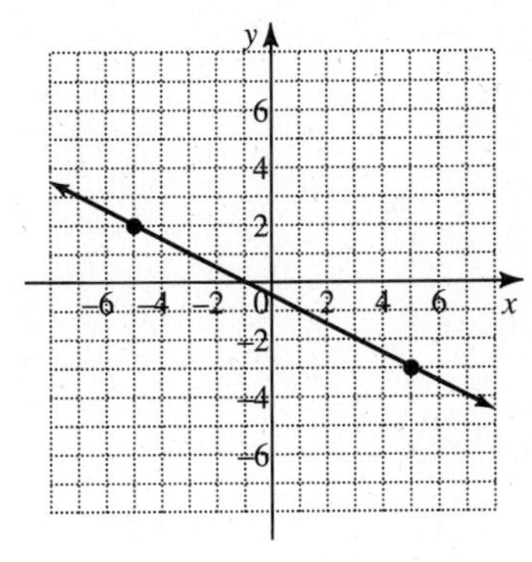

2.

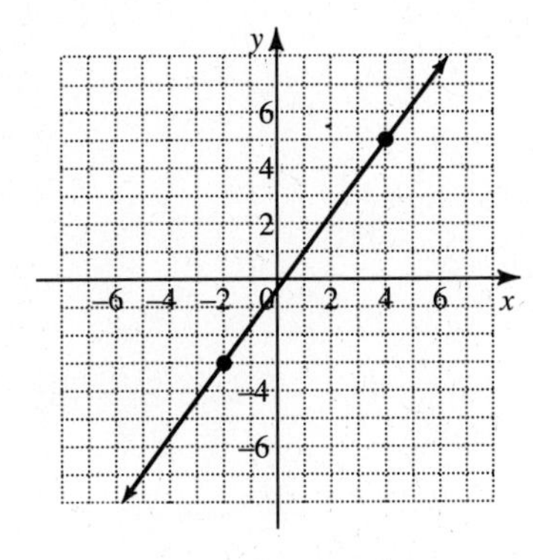

3.

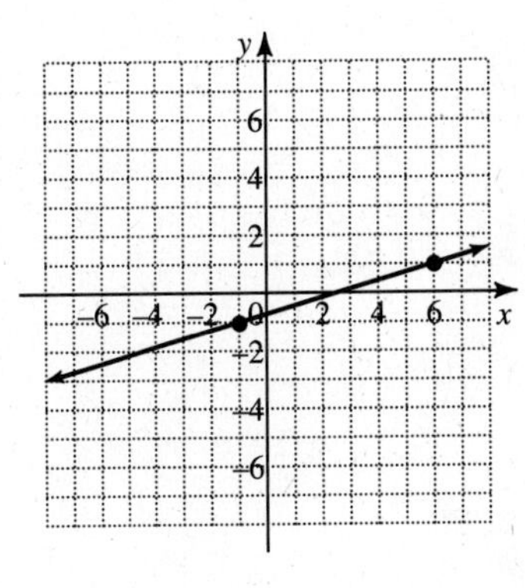

4.

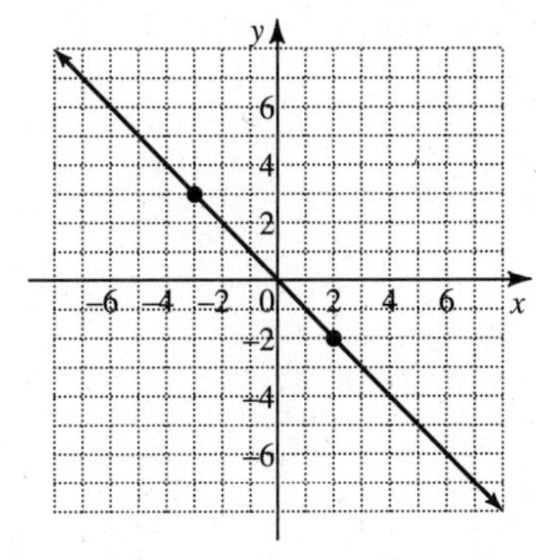

5.

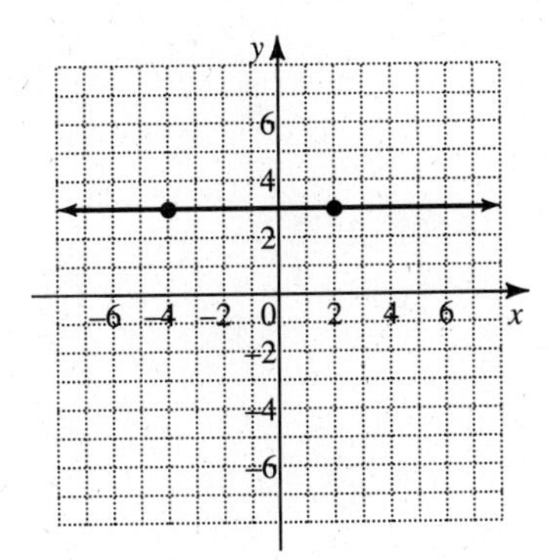

6.

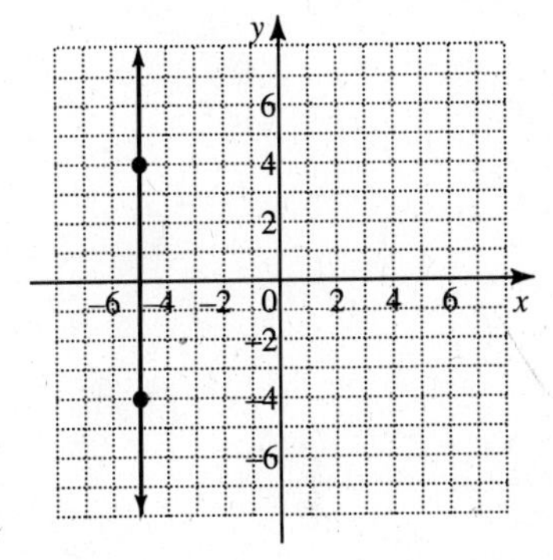

Find the slope of the line going through each pair of points. See Examples 1–3.

7. $(-4, 1)$ $(2, 8)$

8. $(-1, 2)$ $(-3, -7)$

9. $(8, 0)$ $(0, 5)$

10. $(0, -3)\ (2, 0)$

11. $(-1, 6)\ (4, 6)$

12. $\left(5, \frac{2}{3}\right)\left(-1, -\frac{4}{3}\right)$

13. $\left(\frac{7}{5}, -\frac{3}{10}\right)\left(-\frac{1}{5}, \frac{1}{2}\right)$

14. $(4, -11)\ (3, -11)$

15. $(-9, 1)\ (-9, 0)$

Find the slope of each of the following lines. See Example 4.

16. $y = 2x - 1$

17. $y = 5x + 2$

18. $y = -x + 4$

19. $y = x + 1$

20. $y = 6 - 5x$

21. $y = 3 + 9x$

22. $-6x + 4y = 1$

23. $3x - 2y = 5$

24. $5x - 7y = 8$

25. $6x - 9y = 5$

26. $9x + 7y = 5$

27. $y + 4 = 0$

name date hour

7.4 EXERCISES

Give the slope of each of the following lines that has a slope and then decide whether each pair of lines is parallel, perpendicular, *or* neither. *See Example 5.*

28. $y = 4x - 3$
$y = 4x + 1$

29. $y = 5 - 5x$
$y = 4 + \frac{1}{5}x$

30. $y - x = 3$
$y - x = 5$

31. $x + y = 5$
$x - y = 1$

32. $y - x = 4$
$y + x = 3$

33. $2x - 5y = 4$
$4x - 10y = 1$

34. $3x - 2y = 4$
$2x + 3y = 1$

35. $3x - 5y = 2$
$5x + 3y = -1$

36. $4x - 3y = 4$
$8x - 6y = 0$

37. $x - 4y = 2$
$2x + 4y = 1$

38. $8x - 9y = 2$
$8x + 6y = 1$

39. $5x - 3y = 8$
$3x - 5y = 10$

Find the slope (or pitch) of the roofs shown in the figures. Measurements are given in feet.

40.

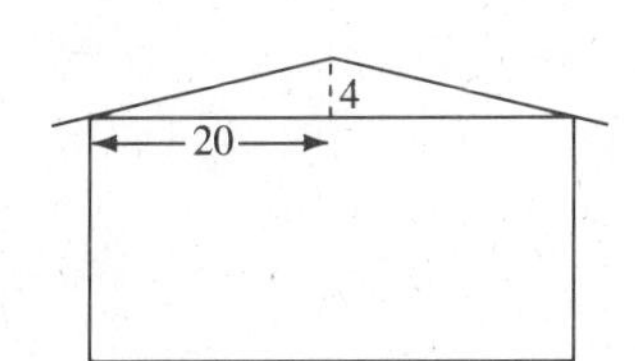

41.

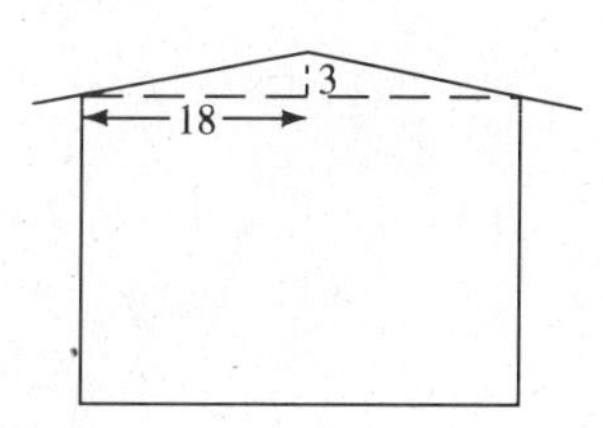

What is the slope (or grade) of the hills shown in the figures? Measurements are given in meters.

42.

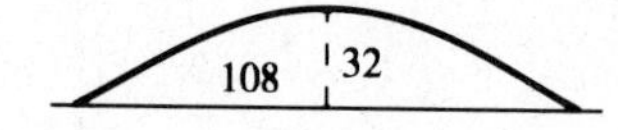

43.

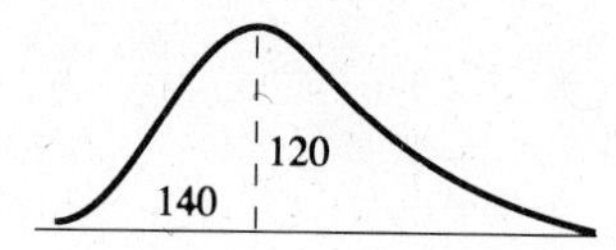

SKILL SHARPENERS

Solve for y. Simplify your answers. See Sections 3.1 and 3.5.

44. $y - 2 = 3(x - 4)$

45. $y + 1 = 2(x - 5)$

46. $y - (-3) = -2(x - 1)$

47. $y - (-4) = -(x + 1)$

48. $y - (-5) = 4(x - 1)$

49. $y - 2 = -3[x - (-1)]$

7.5 EQUATIONS OF A LINE

The last section showed how to find the slope of a line from its equation. For example, the slope of the line having the equation $y = 2x + 3$ is 2, the coefficient of x. What does the number 3 represent? If we let $x = 0$, the equation becomes $y = 2(0) + 3 = 0 + 3 = 3$. Since we found $y = 3$ by letting $x = 0$, (0, 3) is the y-intercept of the graph of $y = 2x + 3$.

An equation solved for y is said to be in **slope-intercept form** because both the slope and the y-intercept of the line can be found from the equation.

OBJECTIVES

1. Write an equation of a line, given its slope and y-intercept.
2. Graph a line, given its slope and a point on the line.
3. Write an equation of a line, given its slope and a point on the line.
4. Write an equation of a line, given two points on the line.

SLOPE-INTERCEPT FORM

The slope-intercept form of the equation of a line with slope m and y-intercept $(0, b)$ is

$$y = mx + b.$$

1 Given the slope and y-intercept of a line, we can use the slope-intercept form to find an equation of the line.

EXAMPLE 1 Finding an Equation of a Line

Find an equation for each of the following lines.

(a) With slope 5 and y-intercept (0, 3)

Use the slope-intercept form. Let $m = 5$ and $b = 3$.

$$y = \mathbf{m}x + \mathbf{b}$$
$$y = \mathbf{5}x + \mathbf{3}$$

(b) With slope $\frac{2}{3}$ and y-intercept $(0, -1)$

Here $m = \frac{2}{3}$ and $b = -1$.

$$y = \frac{2}{3}x - 1 \quad \blacksquare$$

WORK PROBLEM 1 AT THE SIDE.

1. Find the equation of the line with the given slope and value of b.

(a) $m = \frac{1}{2}$; $b = -4$

(b) $m = -1$; $b = 8$

(c) $m = 3$; $b = 0$

(d) $m = 0$; $b = 2$

2 The slope and y-intercept can be used to graph a line. Graph $y = \frac{2}{3}x - 1$ by first locating the y-intercept, $(0, -1)$, on the y-axis. From the definition of slope and the fact that the slope of this line is $\frac{2}{3}$,

$$m = \frac{\text{difference in } y\text{-values}}{\text{difference in } x\text{-values}} = \frac{2}{3}.$$

Another point on the graph can be found by counting from the y-intercept 2 units up and then counting 3 units to the right. The line is then drawn through this point and the y-intercept, as shown in Figure 24. This method can be extended to graph a line given its slope and a point on the line.

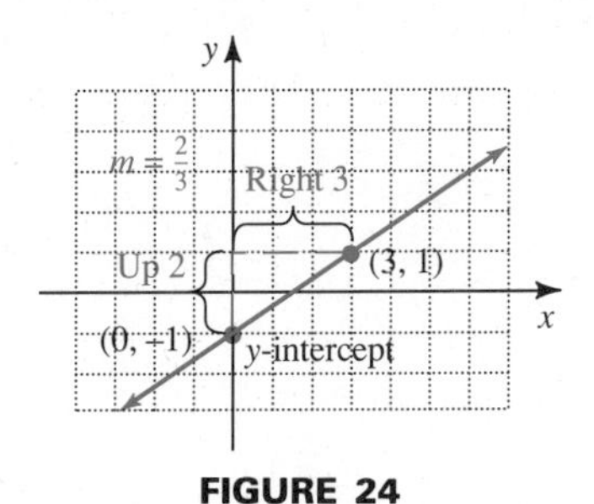

FIGURE 24

ANSWERS

1. (a) $y = \frac{1}{2}x - 4$ (b) $y = -x + 8$ (c) $y = 3x$ (d) $y = 2$

2. Graph each line.

(a) Through $(-1, 4)$, with slope $\frac{5}{2}$

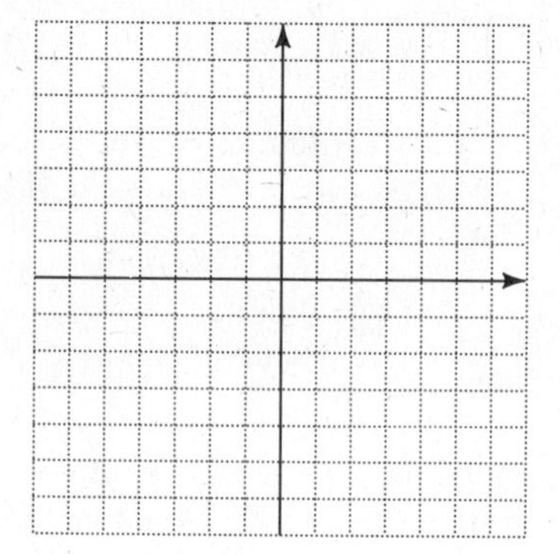

(b) Through $(2, -3)$, with slope $-\frac{1}{3}$

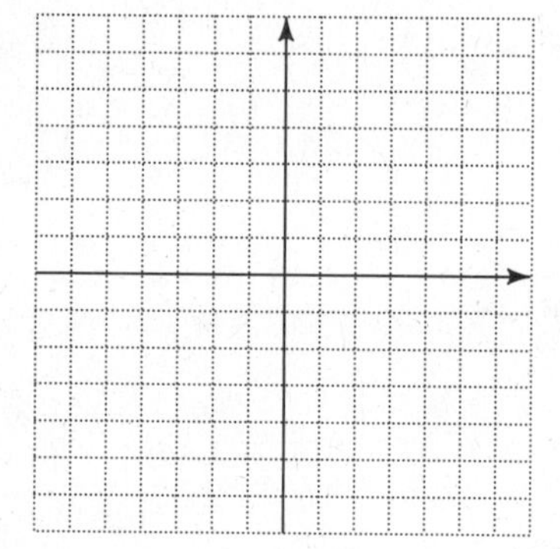

ANSWERS
2. (a) (b)

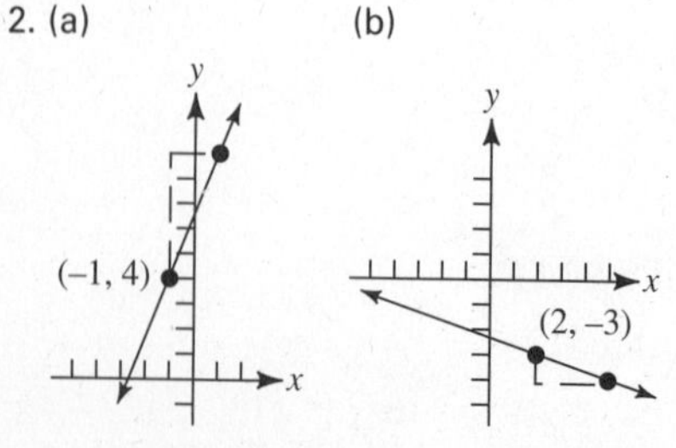

EXAMPLE 2 Graphing a Line Given Its Slope and a Point

Graph the line through $(-2, 3)$ with slope $-\frac{5}{4}$.

First, locate the point $(-2, 3)$. Write the slope as

$$m = \frac{\text{difference in } y\text{-values}}{\text{difference in } x\text{-values}} = \frac{-5}{4}.$$

(We could have used $\frac{5}{-4}$ instead.) Another point on the line is located by counting 5 units *down* (because of the negative sign) and then 4 units to the right. Finally, draw the line through this new point and the given point $(-2, 3)$. See Figure 25. ■

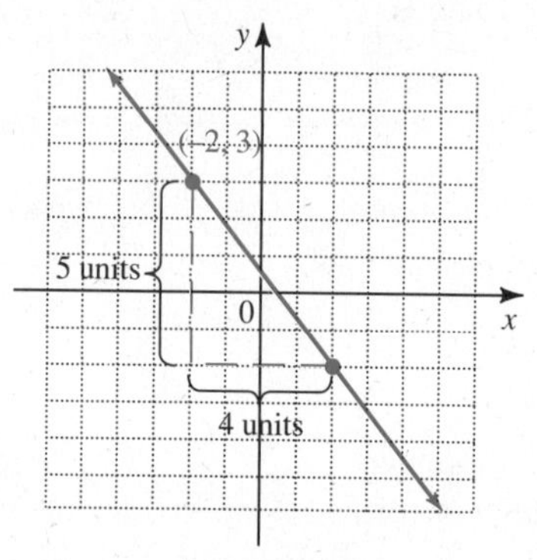

FIGURE 25

WORK PROBLEM 2 AT THE SIDE.

3 An equation for a line can also be found from any point on the line and the slope of the line. Let m represent the slope of the line and let (x_1, y_1) represent the given point on the line. Let (x, y) represent any other point on the line. Then by the definition of slope,

$$\frac{y - y_1}{x - x_1} = m$$

$$\text{or} \quad y - y_1 = m(x - x_1).$$

This result is the **point-slope form** of the equation of a line.

POINT-SLOPE FORM

The point-slope form of the equation of a line with slope m going through (x_1, y_1) is

$$\mathbf{y - y_1 = m(x - x_1).}$$

This very important result should be memorized.

EXAMPLE 3 Using the Point-Slope Form to Write an Equation

Find an equation of each of the following lines. Write the equation in the form $Ax + By = C$.

(a) Through $(-2, 4)$, with slope -3

The given point is $(-2, 4)$ so $x_1 = -2$ and $y_1 = 4$. Also, $m = -3$. Substitute these values into the point-slope form.

$$y - y_1 = m(x - x_1)$$
$$y - 4 = -3[x - (-2)]$$
$$y - 4 = -3(x + 2)$$
$$y - 4 = -3x - 6 \quad \text{Distributive property}$$
$$y = -3x - 2 \quad \text{Add 4.}$$
$$3x + y = -2 \quad \text{Add } 3x.$$

(a) Through $(4, 2)$ with slope $\frac{3}{5}$

Use $x_1 = 4$, $y_1 = 2$, and $m = \frac{3}{5}$ in the point-slope form.

$$y - y_1 = m(x - x_1)$$
$$y - 2 = \frac{3}{5}(x - 4)$$

Multiply each side by 5 to clear the fractions.

$$5(y - 2) = 5 \cdot \frac{3}{5}(x - 4)$$
$$5(y - 2) = 3(x - 4)$$
$$5y - 10 = 3x - 12 \quad \text{Distributive property}$$
$$5y = 3x - 2 \quad \text{Add 10.}$$
$$-3x + 5y = -2 \quad \text{Subtract } 3x.$$ ■

WORK PROBLEM 3 AT THE SIDE.

4 The point-slope form also can be used to find an equation of a line when two different points on the line are known.

EXAMPLE 4 Finding the Equation of a Line from Two Points

Find an equation for the line through the points $(-2, 5)$ and $(3, 4)$. Write the equation in the form $Ax + By = C$.

First find the slope of the line, using the definition of slope.

$$\text{Slope} = \frac{5 - 4}{-2 - 3} = \frac{1}{-5} = -\frac{1}{5}$$

Now use either $(-2, 5)$ or $(3, 4)$ and the point-slope form. Using $(3, 4)$ gives

$$y - y_1 = m(x - x_1)$$
$$y - 4 = -\frac{1}{5}(x - 3)$$
$$5(y - 4) = -1(x - 3) \quad \text{Multiply by 5.}$$
$$5y - 20 = -x + 3 \quad \text{Distributive property}$$
$$5y = -x + 23 \quad \text{Add 20 on each side.}$$
$$x + 5y = 23. \quad \text{Add } x \text{ on each side.}$$

The same result would be found by using $(-2, 5)$ for (x_1, y_1). ■

3. Find an equation for each line. Write answers in the form $Ax + By = C$.

(a) Through $(-1, 3)$, with slope -2

(b) Through $(5, 2)$, with slope $-\frac{1}{3}$

ANSWERS
3. (a) $2x + y = 1$ (b) $x + 3y = 11$

4. Write an equation for the line through the following points. Write answers in the form $Ax + By = C$.

(a) $(-3, 1)$ and $(2, 4)$

(b) $(2, 5)$ and $(-1, 6)$

ANSWERS
4. (a) $3x - 5y = -14$
(b) $x + 3y = 17$

WORK PROBLEM 4 AT THE SIDE.

A summary of the types of *linear equations* is given here.

LINEAR EQUATIONS

$Ax + By = C$	**Standard form** (neither A nor B is 0) Slope is $-\frac{A}{B}$ x-intercept is $\left(\frac{C}{A}, 0\right)$ y-intercept is $\left(0, \frac{C}{B}\right)$.
$x = k$	**Vertical line** Slope is undefined. x-intercept is $(k, 0)$.
$y = k$	**Horizontal line** Slope is 0. y-intercept is $(0, k)$.
$y = mx + b$	**Slope-intercept form** Slope is m. y-intercept is $(0, b)$.
$y - y_1 = m(x - x_1)$	**Point-slope form** Slope is m. Line goes through (x_1, y_1).

CAUTION The above definition of "standard form" is not the same in all texts. Also, a linear equation can be written as $Ax + By = C$ in many different (equally correct) ways. For example, $3x + 4y = 12$, $6x + 8y = 24$, and $9x + 12y = 36$ all represent the same set of ordered pairs. Let us agree that $3x + 4y = 12$ is preferable to the other forms because the greatest common factor of 3, 4, and 12 is 1.

name date hour

7.5 EXERCISES

Write an equation for each line given its slope and y-intercept. See Example 1.

	Slope	*y-intercept*	*Equation*		*Slope*	*y-intercept*	*Equation*
1.	3	(0, 5)		**2.**	−2	(0, 4)	
3.	−1	(0, −6)		**4.**	2	(0, −3)	
5.	$\frac{5}{3}$	$\left(0, \frac{1}{2}\right)$		**6.**	$\frac{2}{5}$	$\left(0, -\frac{1}{4}\right)$	
7.	8	(0, 0)		**8.**	0	(0, −5)	

Graph the line going through the given point and having the given slope. (In Exercises 17–20, recall the type of lines having 0 slope and undefined slope.) See Example 2.

9. $(2, 5), \quad m = \frac{1}{2}$

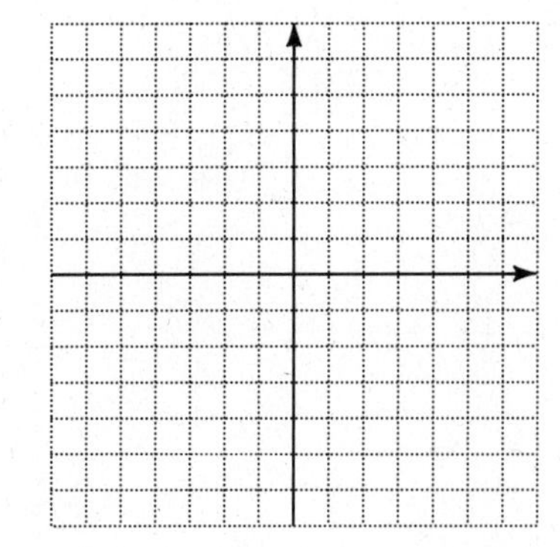

10. $(-4, -3), \quad m = -\frac{2}{5}$

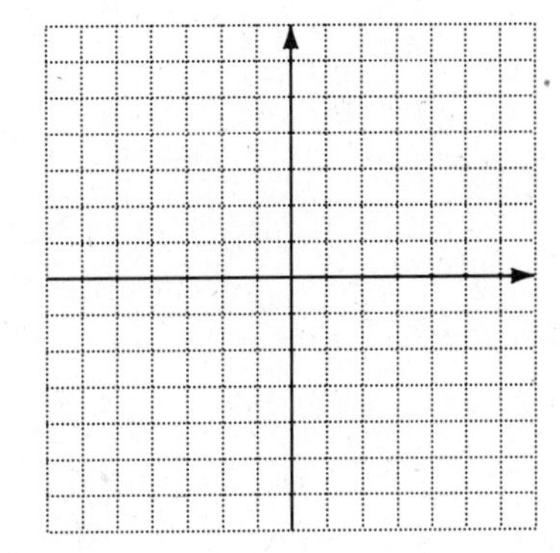

11. $(-1, -1), \quad m = -\frac{3}{8}$

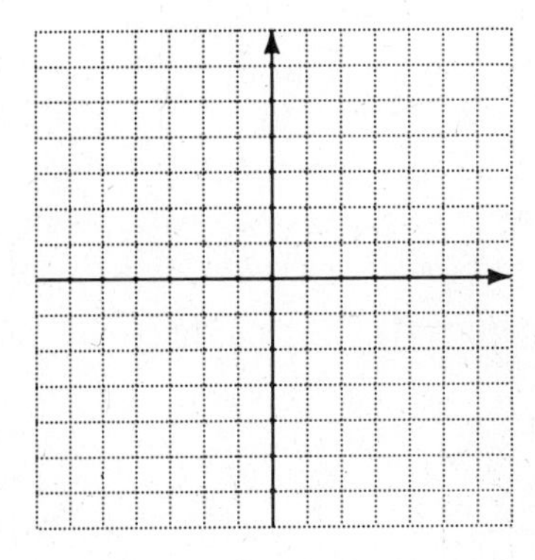

12. (0, 2), $m = \frac{3}{4}$

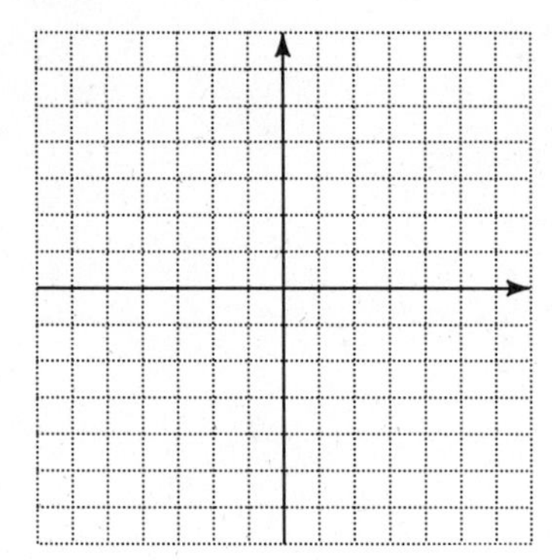

13. (−3, 0), $m = -\frac{5}{4}$

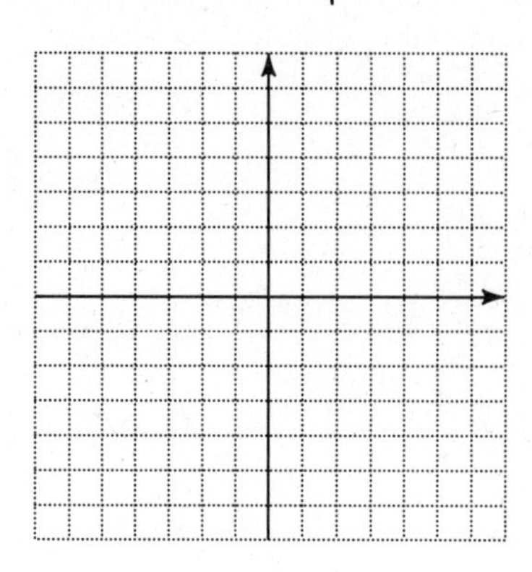

14. (2, −3), $m = -4$

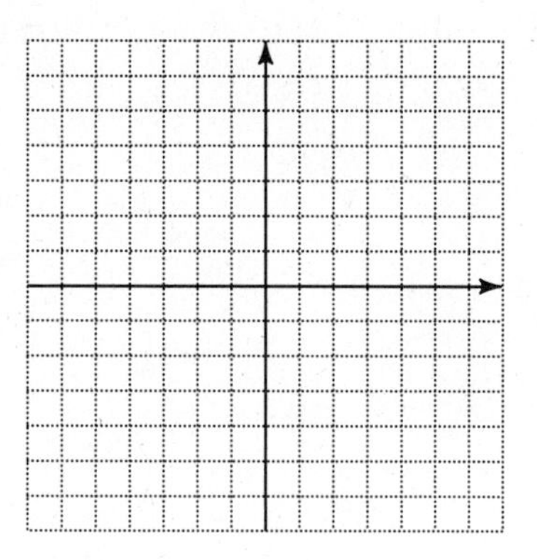

15. (4, −1), $m = 2$

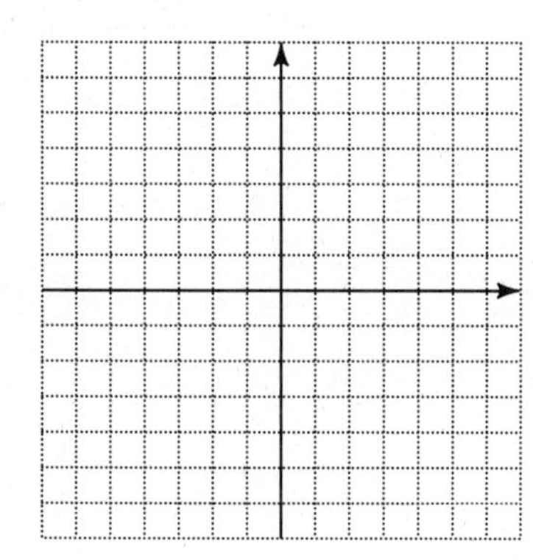

16. (3, −5), $m = 1$

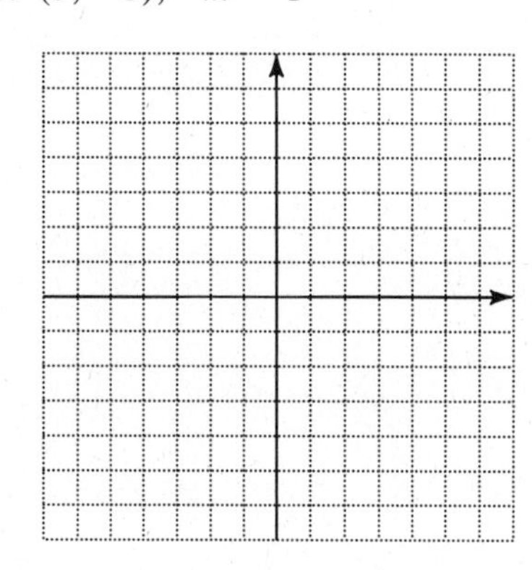

17. (1, 2), $m = 0$

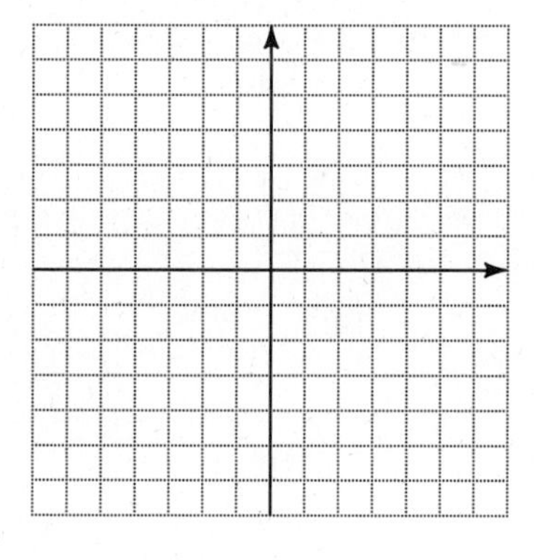

18. (−4, 3), $m = 0$

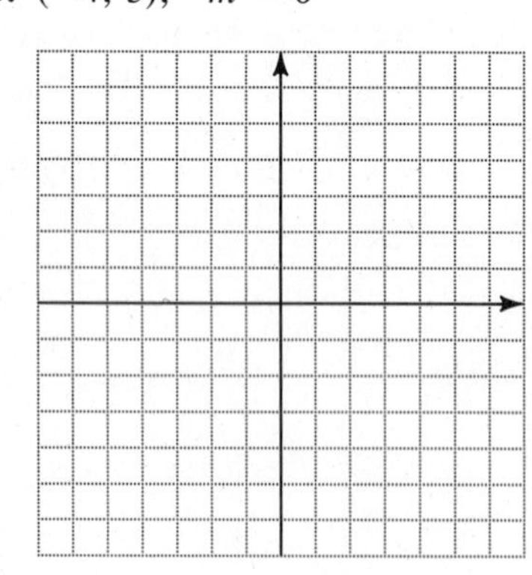

19. (3, 5), undefined slope

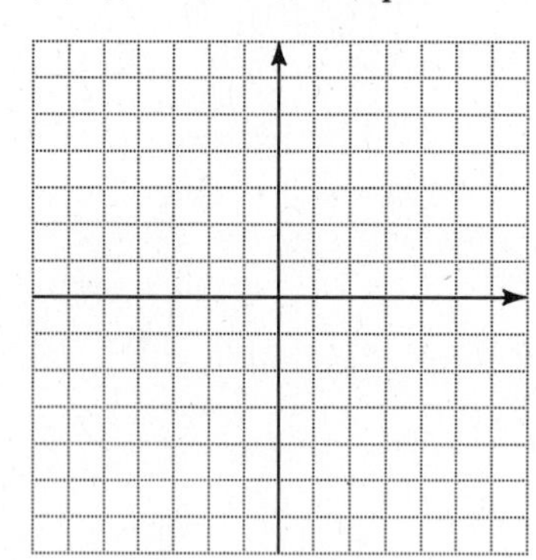

20. (2, 3), undefined slope

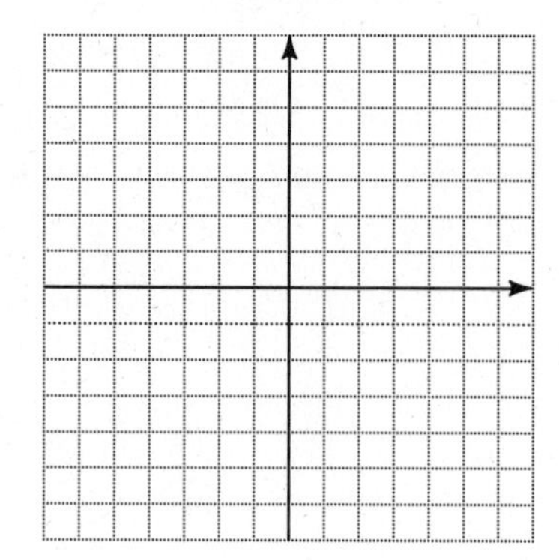

Write an equation for the line passing through the given point and having the given slope. Write the equation in the form $Ax + By = C$. See Example 3.

21. (5, 3), $m = 2$

22. (1, 4), $m = 3$

23. (2, −8), $m = -2$

name date hour

24. $(-1, 7)$, $m = -4$

25. $(3, 5)$, $m = \frac{2}{3}$

26. $(2, -4)$, $m = \frac{4}{5}$

27. $(-3, -2)$, $m = -\frac{3}{4}$

28. $(-8, -2)$, $m = -\frac{5}{9}$

29. $(6, 0)$, $m = -\frac{8}{11}$

Write an equation for the line passing through each of the given pairs of points. Write the equations in the form $Ax + By = C$. See Example 4.

30. $(7, 4)$, $(8, 5)$

31. $(-2, 1)$, $(3, 4)$

32. $(-8, -2)$, $(-1, -7)$

33. $(3, -4)$, $(-2, -1)$

34. $(-7, -5)$, $(-9, -2)$

35. $(0, 2)$, $(3, 0)$

36. $(4, 0)$, $(0, -2)$

37. $(2, -5)$, $(-4, 7)$

38. $(3, -7)$, $(-5, 0)$

39. $\left(\frac{1}{2}, \frac{3}{2}\right)$, $\left(-\frac{1}{4}, \frac{5}{4}\right)$

40. $\left(-\frac{2}{3}, \frac{8}{3}\right)$, $\left(\frac{1}{3}, \frac{7}{3}\right)$

41. $\left(-1, \frac{5}{8}\right)$, $\left(\frac{1}{8}, 2\right)$

The cost to produce x items is often expressed as $y = mx + b$. The number b gives the fixed cost (the cost that is the same no matter how many items are produced), and the number m is the variable cost (the cost to produce an additional item). Write the cost equation for each of the following.

42. Fixed cost 50, variable cost 9

43. Fixed cost 100, variable cost 12

44. Fixed cost 70.5, variable cost 3.5

45. Refer to Exercise 43 and find the total cost to make

(a) 50 items and **(b)** 125 items.

The sales of a company for a given year can be written as an ordered pair in which the first number gives the year (perhaps since the company started business) and the second number gives the sales for that year. If the sales increase at a steady rate, a linear equation for sales can be found. Sales for two years are given for each of two companies below.

Company A

Year x	Sales y
1	24
5	48

Company B

Year x	Sales y
1	18
4	27

46. (a) Write two ordered pairs of the form (year, sales) for Company A.

(b) Write an equation of the line through the two pairs in part (a). This is a sales equation for Company A.

47. (a) Write two ordered pairs of the form (year, sales) for Company B.

(b) Write an equation of the line through the two pairs in part (a). This is a sales equation for Company B.

SKILL SHARPENERS

Graph each inequality. See Section 3.7.

48. $m < 10$

49. $k \geq 10$

50. $a + 3 > 8$

51. $p - 5 \leq 12$

52. $2x + 7 \leq 8$

53. $3p - 9 \geq 4$

7.6 GRAPHING LINEAR INEQUALITIES IN TWO VARIABLES

1 In Section 7.3 we covered methods for graphing linear equations, such as $2x + 3y = 6$. Now this discussion is extended to **linear inequalities in two variables,** such as

$$2x + 3y \leq 6.$$

(Recall that $\leq$ is read "is less than or equal to.")

The inequality $2x + 3y \leq 6$ means that

$$2x + 3y < 6 \quad \text{or} \quad 2x + 3y = 6.$$

As we found at the beginning of this chapter, the graph of $2x + 3y = 6$ is a line. This line divides the plane into two regions. The graph of the solutions of the inequality $2x + 3y < 6$ will include only *one* of these regions. Find the required region by solving the given inequality for y.

$$2x + 3y \leq 6$$

$$3y \leq -2x + 6 \qquad \text{Subtract } 2x.$$

$$y \leq -\frac{2}{3}x + 2 \qquad \text{Divide by 3.}$$

By this last statement, ordered pairs in which y is *less than or equal to* $-\frac{2}{3}x + 2$ will be solutions to the inequality. The ordered pairs in which y is equal to $-\frac{2}{3}x + 2$ are on a line, so the ordered pairs in which y is less than $-\frac{2}{3}x + 2$ will be *below* that line. Indicate the solution by shading the region below the line, as in Figure 26. The shaded region, along with the original line, is the desired graph.

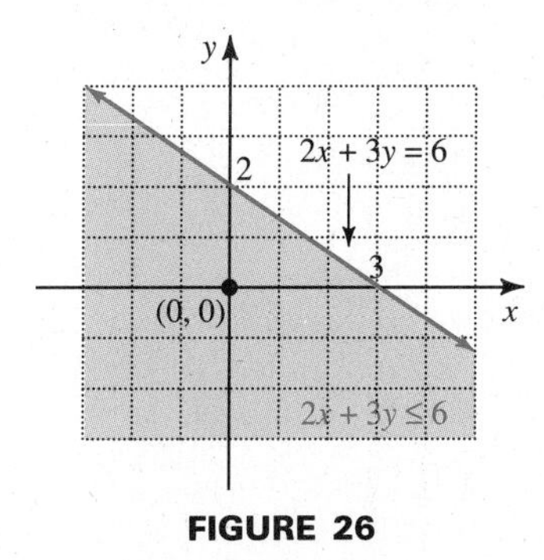

FIGURE 26

WORK PROBLEM 1 AT THE SIDE.

A test point gives a quick way to find the correct region to shade. Choose any point *not* on the line. Because (0, 0) is easy to substitute into an inequality, it is often a good choice, and we will use it here.

Substitute 0 for x and 0 for y in the given inequality to see whether the resulting statement is true or false. In the example above,

$$2\mathbf{x} + 3\mathbf{y} \leq 6$$

$$2(\mathbf{0}) + 3(\mathbf{0}) \leq 6 \qquad \text{Let } x = 0 \text{ and } y = 0.$$

$$0 + 0 \leq 6$$

$$0 \leq 6. \qquad \text{True}$$

Since the last statement is true, shade the region that includes the test point (0,0). This agrees with the result shown in Figure 26.

OBJECTIVES

1. Graph $\leq$ or $\geq$ linear inequalities.
2. Graph $>$ or $<$ linear inequalities.
3. Graph inequalities with a boundary through the origin.

1. Shade the appropriate region for each linear inequality.

(a) $x + 2y \geq 6$

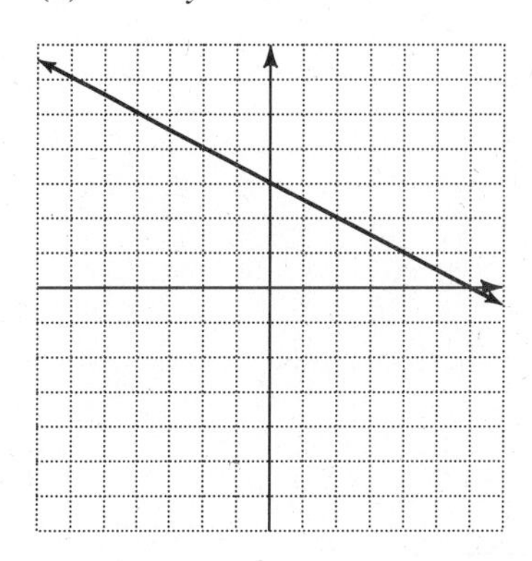

(b) $3x + 4y \leq 12$

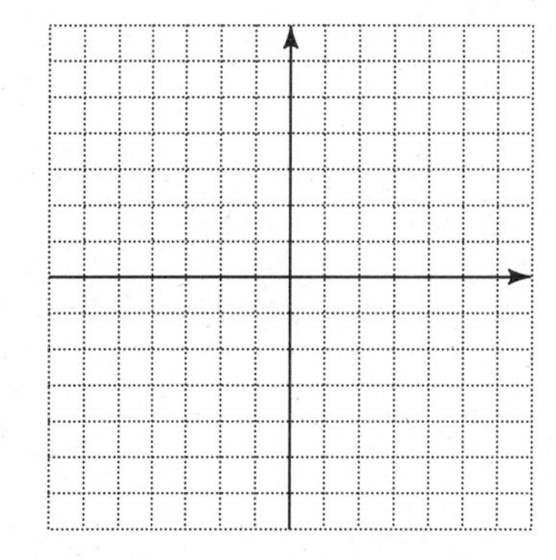

ANSWERS

1. (a) (b)

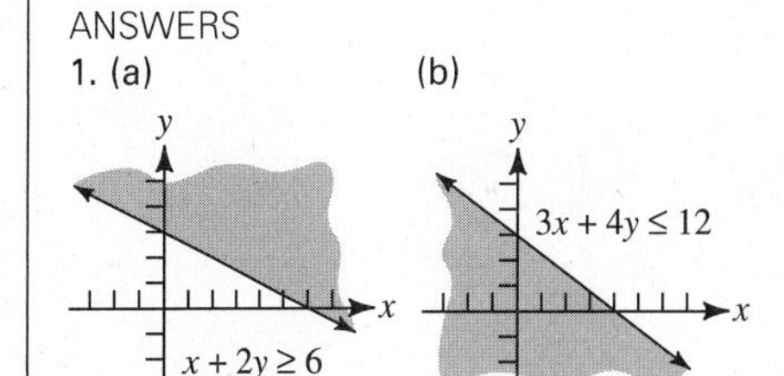

2. Use (0, 0) as a test point to shade the proper region for the inequality $4x - 5y \leq 20$.

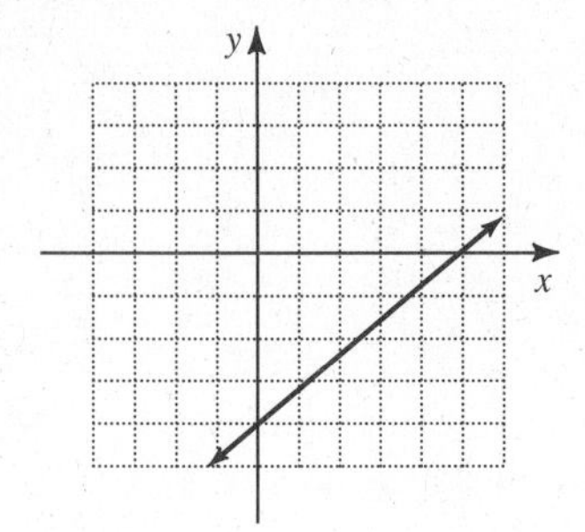

3. Use (1, 2) as a test point to shade the proper region for the inequality $3x + 5y > 15$.

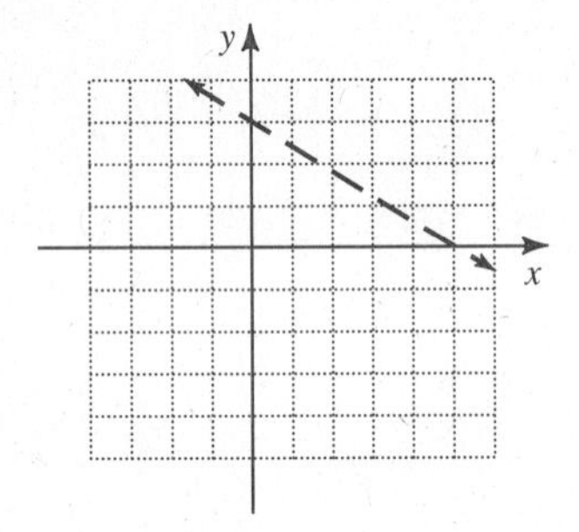

ANSWERS

2.

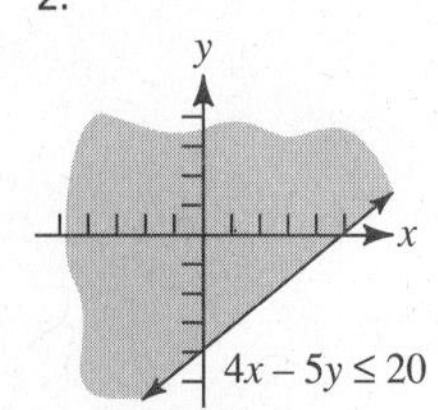

3.

y

x

$3x + 5y > 15$

WORK PROBLEM 2 AT THE SIDE.

2 Inequalities that do not include the equals sign are graphed in a similar way.

EXAMPLE 1 Graphing a Linear Inequality

Graph the inequality $x - y > 5$.

This inequality does not include the equals sign. Therefore, the points on the line

$$x - y = 5$$

do not belong to the graph. However, the line still serves as a boundary for two regions, one of which satisfies the inequality. To graph the inequality, first graph the equation $x - y = 5$. Use a dashed line to show that the points on the line are *not* solutions of the inequality $x - y > 5$. Choose a test point to see which side of the line satisfies the inequality. Let us choose $(1, -2)$ this time.

$x - y > 5$	
$1 - (-2) > 5$	Let $x = 1$ and $y = -2$.
$3 > 5$	False

Since $3 > 5$ is false, the graph of the inequality includes the region that does *not* contain $(1, -2)$. Shade this region, as shown in Figure 27. This shaded region is the desired graph. Check that the proper region is shaded by selecting a point in the shaded region and substituting for x and y in the inequality $x - y > 5$. For example, use $(4, -3)$ from the shaded region, as follows.

$x - y > 5$	
$4 - (-3) > 5$	Let $x = 4$ and $y = -3$.
$7 > 5$	True ■

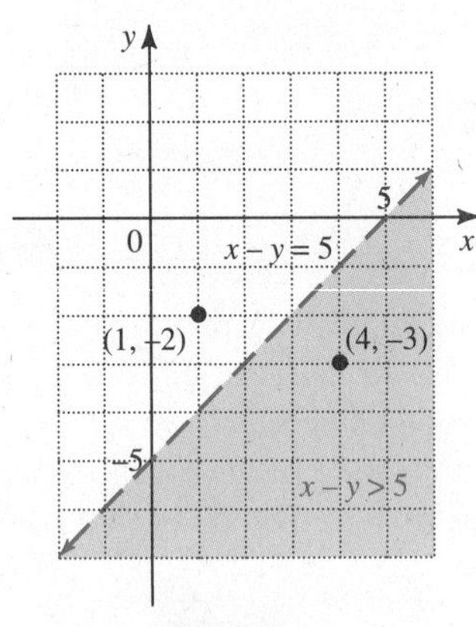

FIGURE 27

WORK PROBLEM 3 AT THE SIDE.

A summary of the steps used to graph a linear inequality is given below.

GRAPHING A LINEAR INEQUALITY IN TWO VARIABLES

Step 1 Graph the line that is the boundary of the region. Use the methods of Section 7.3. Make the line solid if the inequality has $\leq$ or $\geq$; make the line dashed if the inequality has $<$ or $>$.

Step 2 Use any point not on the line as a test point. Substitute for x and y in the *inequality.* If a true statement results, shade the side containing the test point. If a false statement results, shade the other side.

EXAMPLE 2 Graphing a Linear Inequality

Graph the inequality $2x - 5y \geq 10$.

Start by graphing the equation

$$2x - 5y = 10.$$

Use a solid line to show that the points on the line are solutions of the inequality $2x - 5y \geq 10$. Choose any test point not on the line. Again, we choose (0, 0).

$$2x - 5y \geq 10$$

$$2(0) - 5(0) \geq 10 \qquad \text{Let } x = 0 \text{ and } y = 0.$$

$$0 - 0 \geq 10$$

$$0 \geq 10 \qquad \text{False}$$

Since $0 \geq 10$ is false, shade the region *not* containing (0, 0). (See Figure 28.) ■

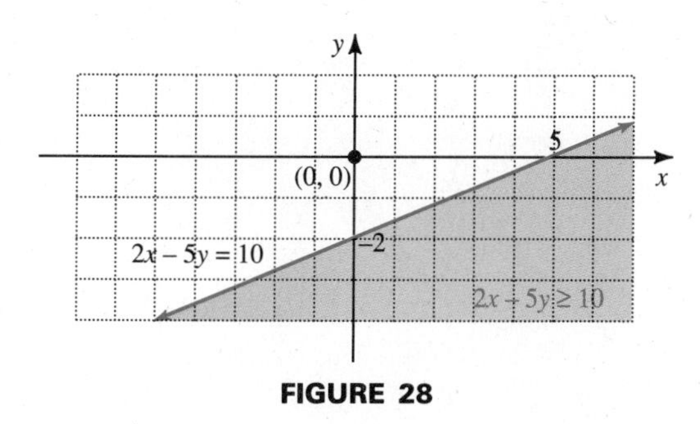

FIGURE 28

WORK PROBLEM 4 AT THE SIDE.

EXAMPLE 3 Graphing a Linear Inequality

Graph the inequality $x \leq 3$.

First graph $x = 3$, a vertical line going through the point (3, 0). Use a solid line. (Why?) Choose (0, 0) as a test point.

$$x \leq 3$$

$$0 \leq 3 \qquad \text{Let } x = 0.$$

$$0 \leq 3 \qquad \text{True}$$

Since $0 \leq 3$ is true, shade the region containing (0, 0), as in Figure 29. ■

4. Graph $2x - y \geq -4$.

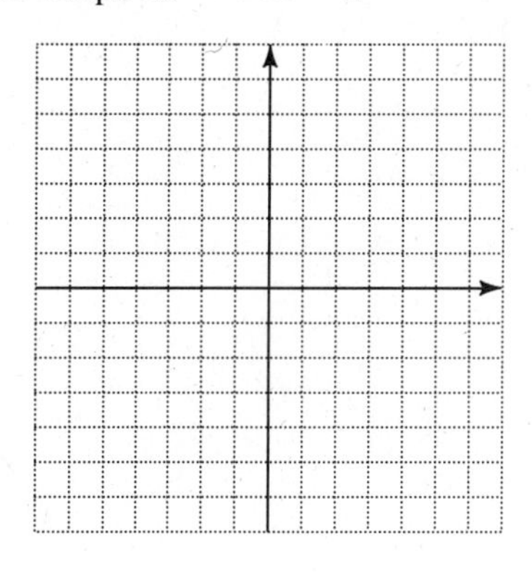

ANSWERS

4.

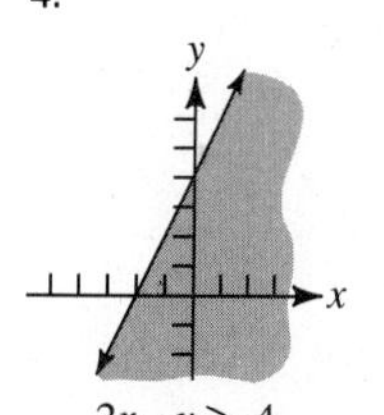

5. Graph $y < 4$.

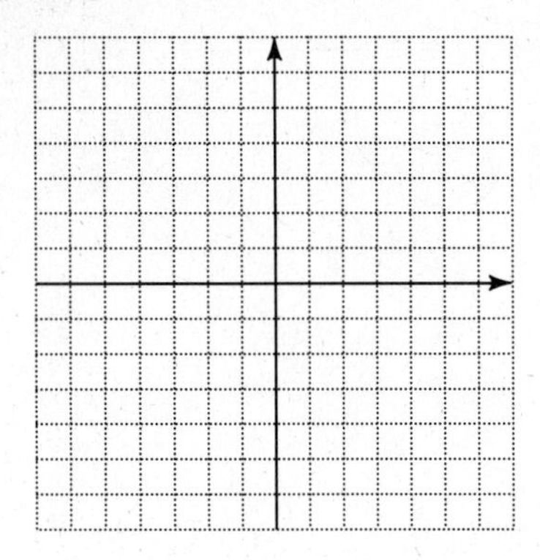

6. Graph each linear inequality.

(a) $3x < y$

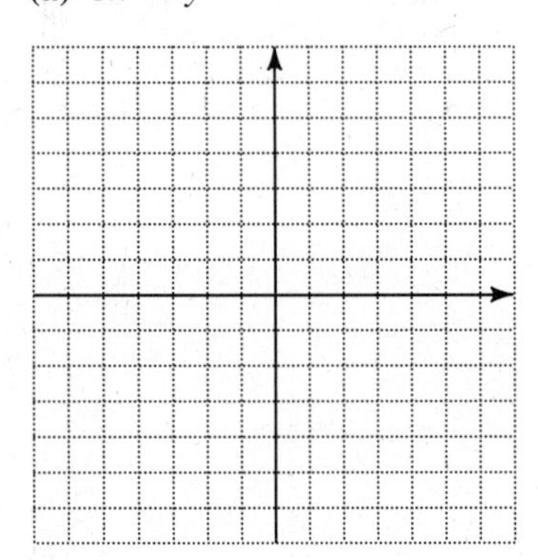

(b) $x \geq -3y$

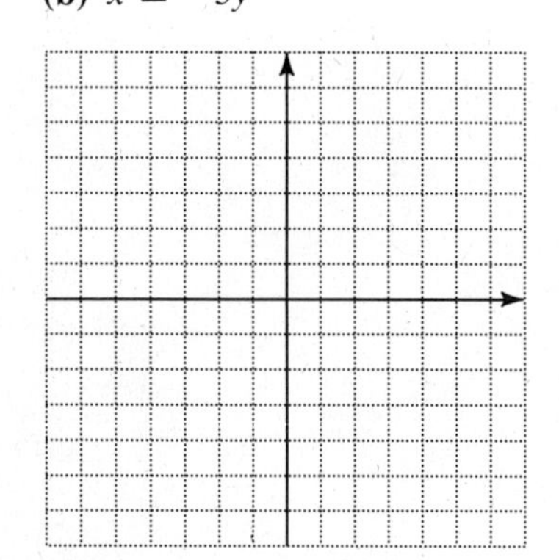

ANSWERS

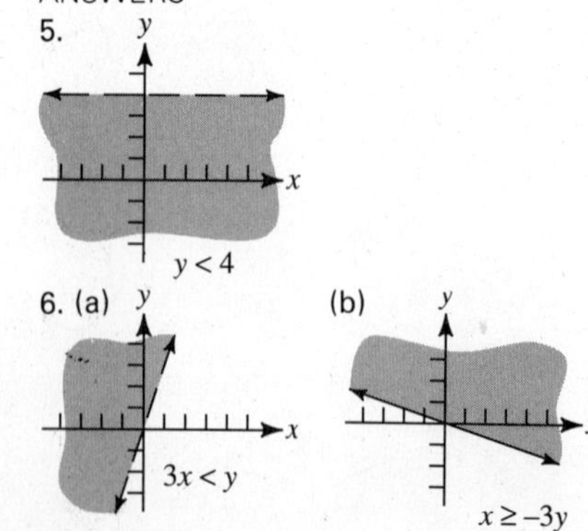

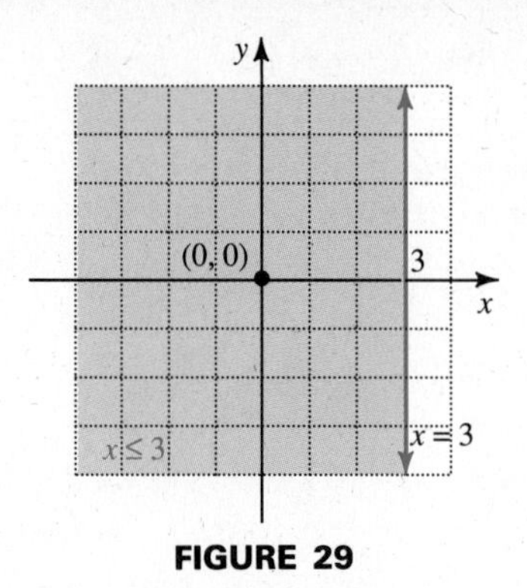

FIGURE 29

WORK PROBLEM 5 AT THE SIDE.

3 The next example shows how to graph an inequality having a boundary line going through the origin, an inequality in which (0, 0) cannot be used as a test point.

EXAMPLE 4 Graphing a Linear Inequality

Graph the inequality $x \leq 2y$.

Begin by graphing $x = 2y$. Some ordered pairs that can be used to graph this line are (0, 0), (6, 3), and (4, 2). Use a solid line. We cannot use (0, 0) as a test point since (0, 0) is on the line $x = 2y$. Instead, choose a test point off the line $x = 2y$. Let us choose (1, 3), which is not on the line.

$x \leq 2y$	Original inequality
$1 \leq 2(3)$	Let $x = 1$ and $y = 3$.
$1 \leq 6$	True

Since $1 \leq 6$ is true, shade the side of the graph containing the test point (1, 3). (See Figure 30.) ■

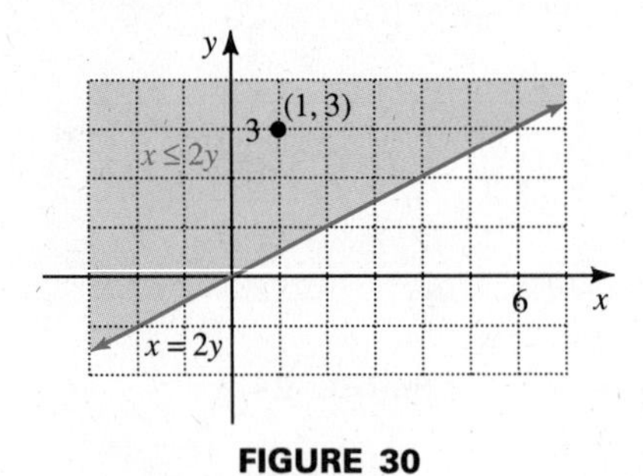

FIGURE 30

WORK PROBLEM 6 AT THE SIDE.

name date hour

7.6 EXERCISES

In Exercises 1–12, the straight line for each inequality has been drawn. Complete each graph by shading the correct region. See Examples 1–4.

1. $x + y \leq 4$

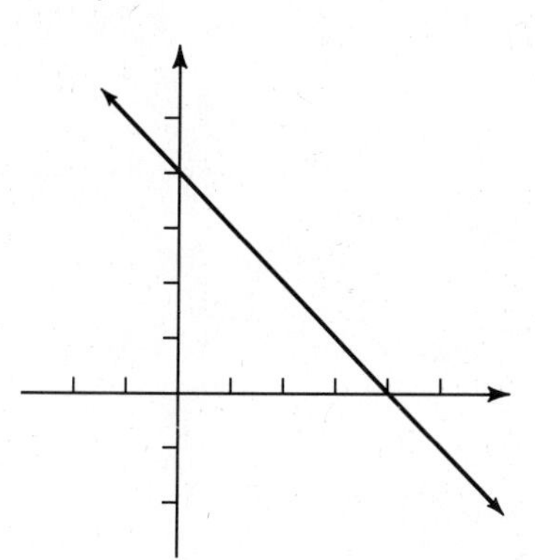

2. $x + y \geq 2$

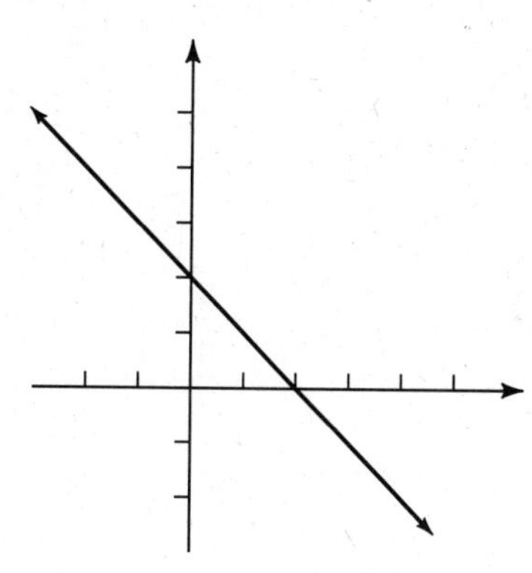

3. $x + 2y \leq 7$

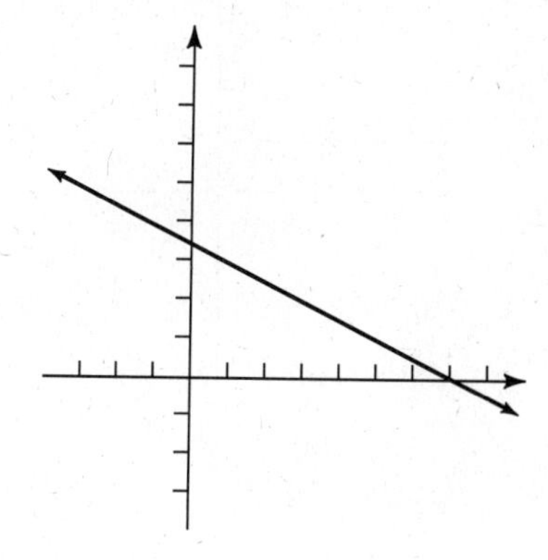

4. $2x + y \leq 5$

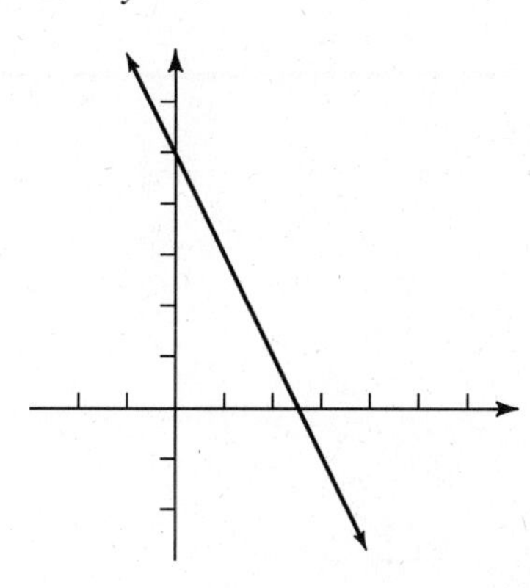

5. $-3x + 4y < 12$

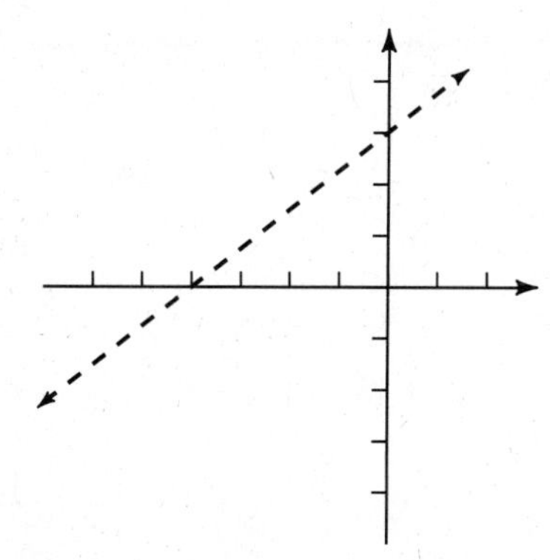

6. $4x - 5y > 20$

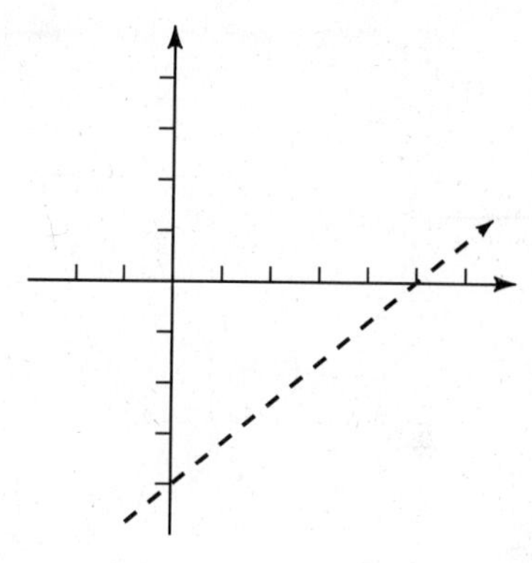

7. $5x + 3y > 15$

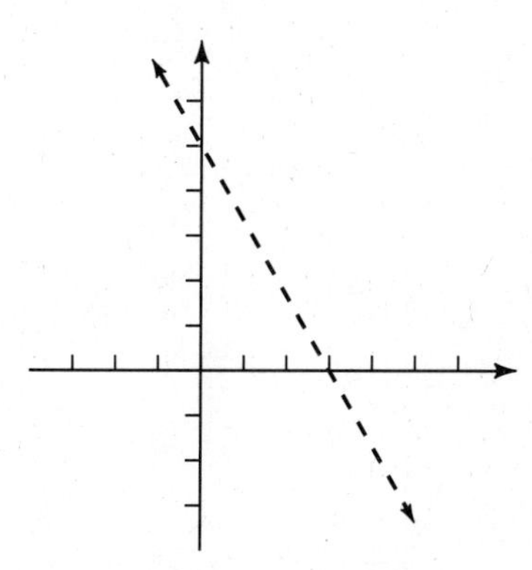

8. $6x - 5y < 30$

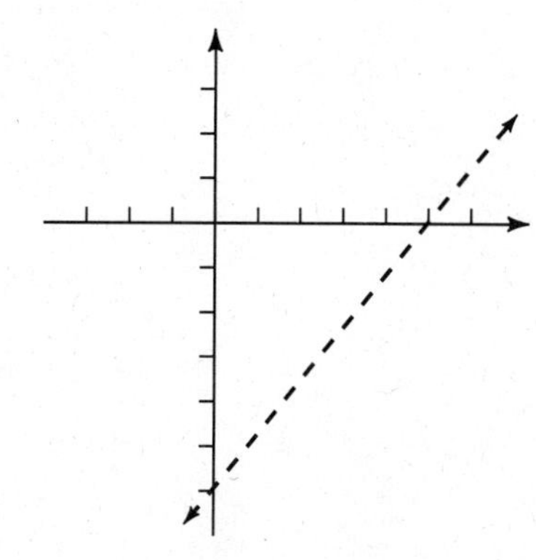

9. $x < 4$

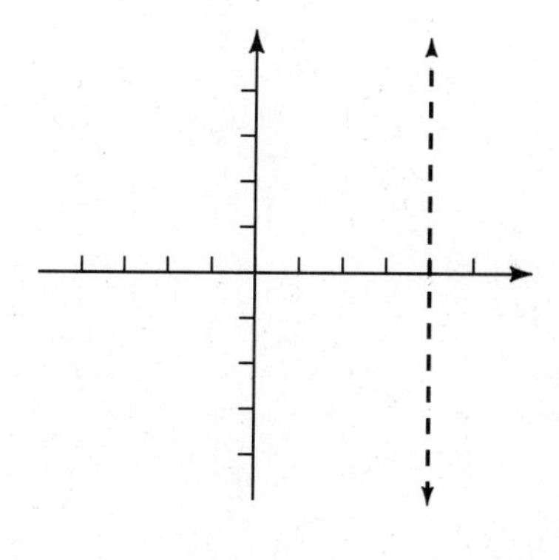

10. $y > -1$

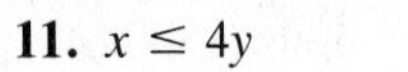

11. $x \leq 4y$

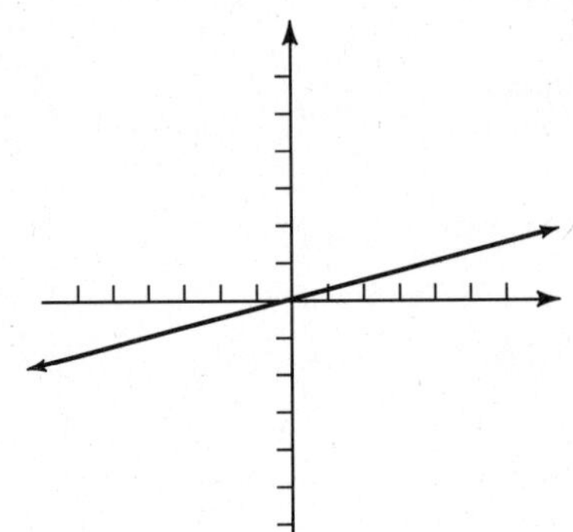

12. $-2x > y$

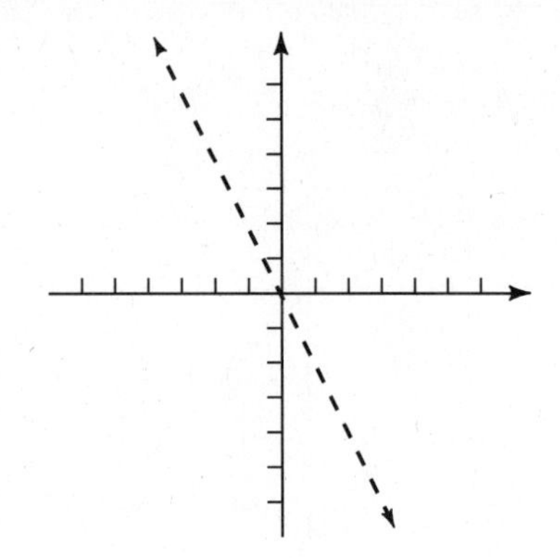

Graph each linear inequality. See Examples 1-4.

13. $x + y \leq 8$

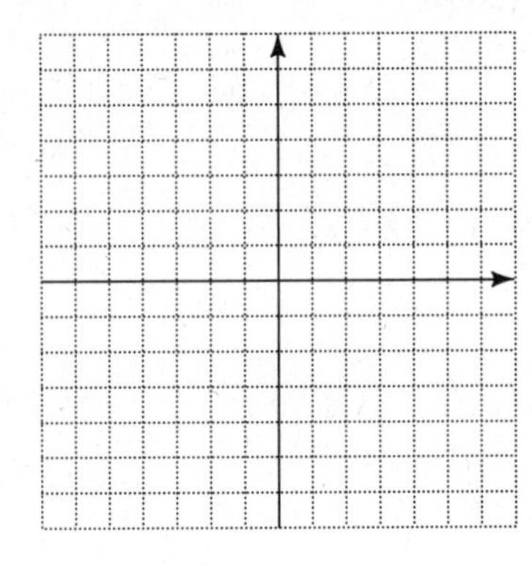

14. $x + y \geq 2$

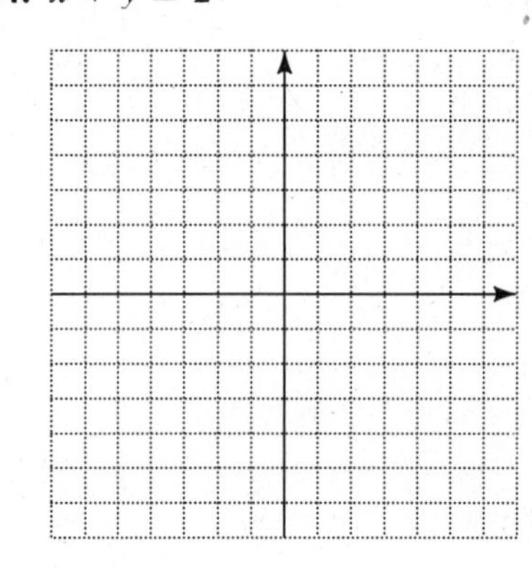

15. $x - y \leq -2$

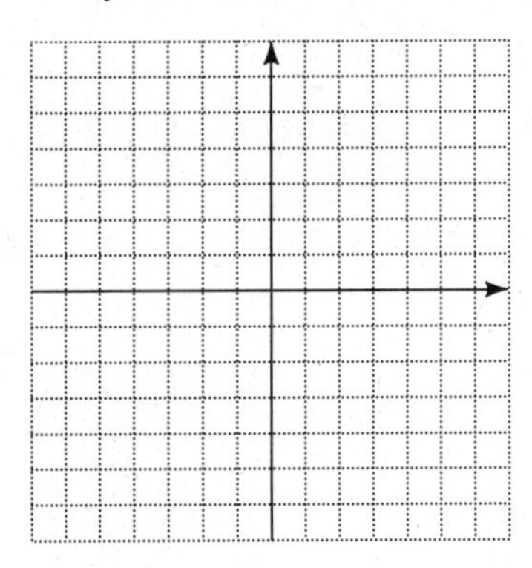

16. $x - y \leq 3$

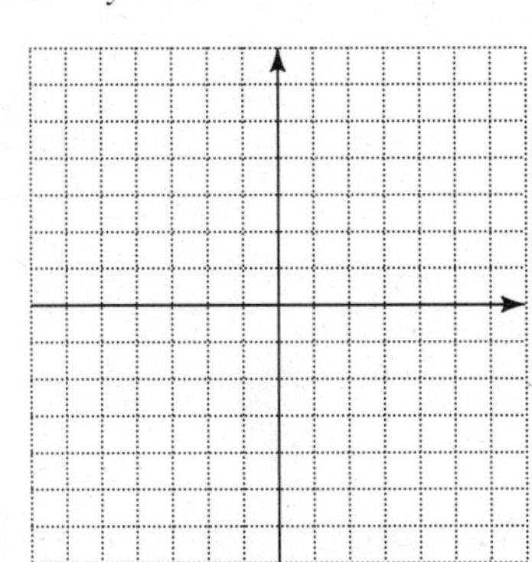

17. $x + 2y \geq 4$

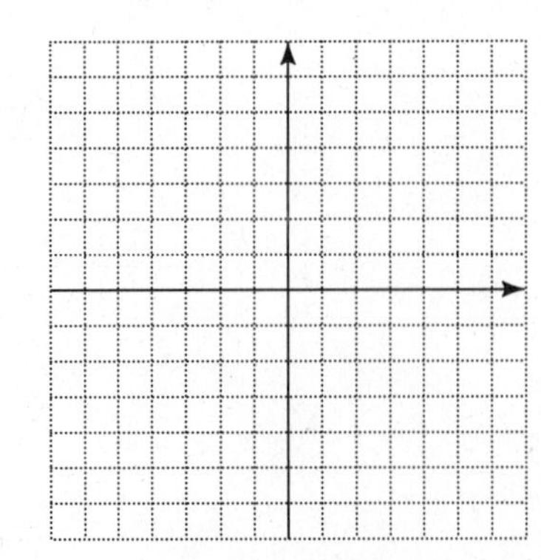

18. $x + 3y \leq 6$

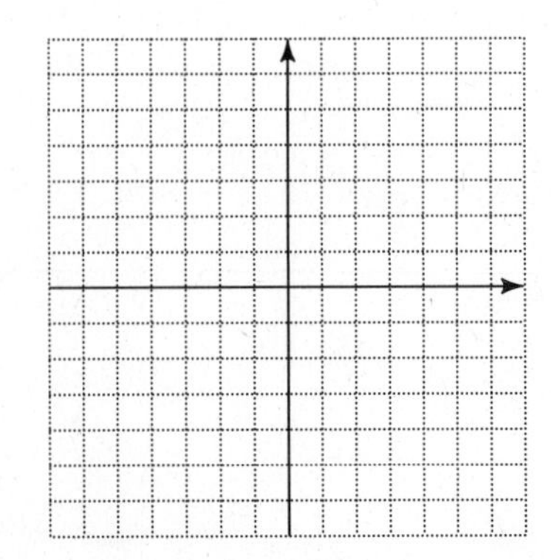

name date hour

19. $2x + 3y > 6$

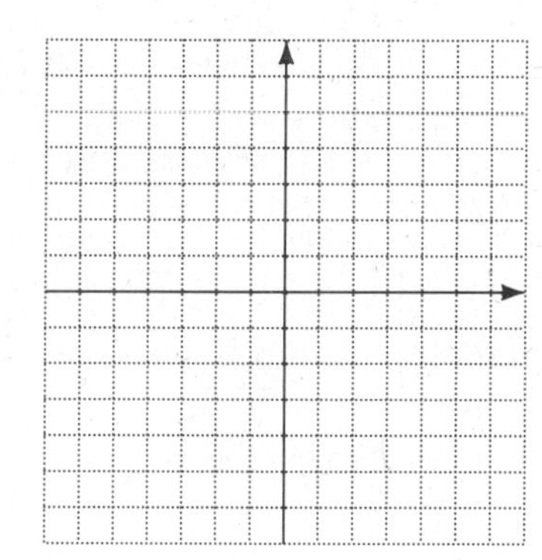

20. $3x + 4y > 12$

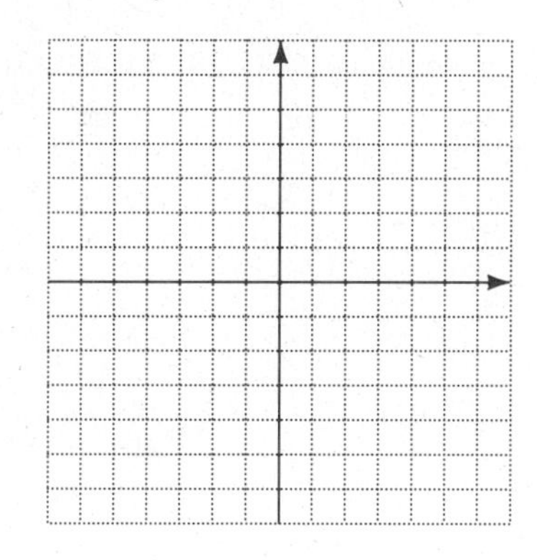

21. $x - 4 < 0$

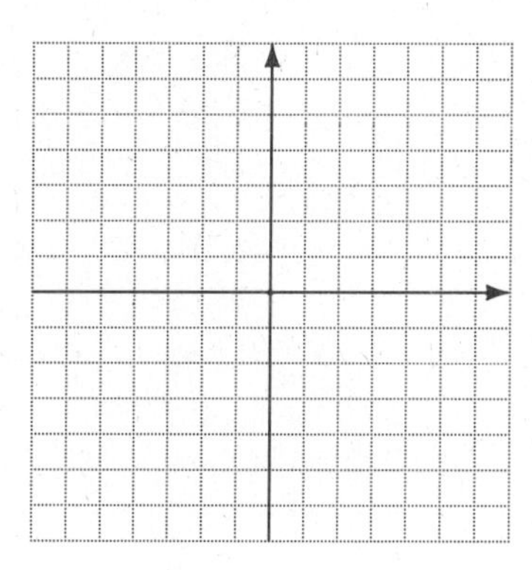

22. $x < -2$

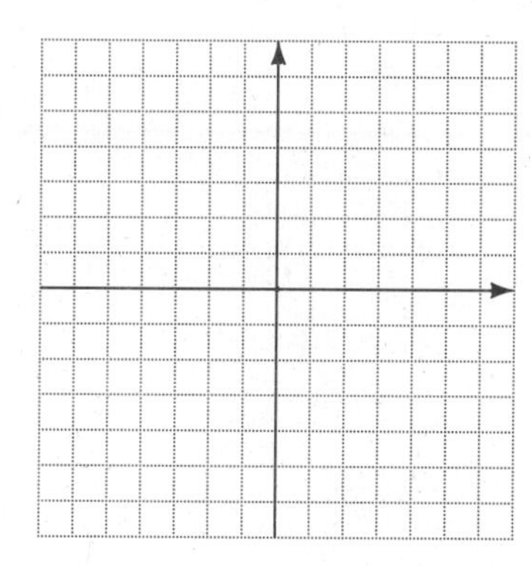

23. $y \leq 2$

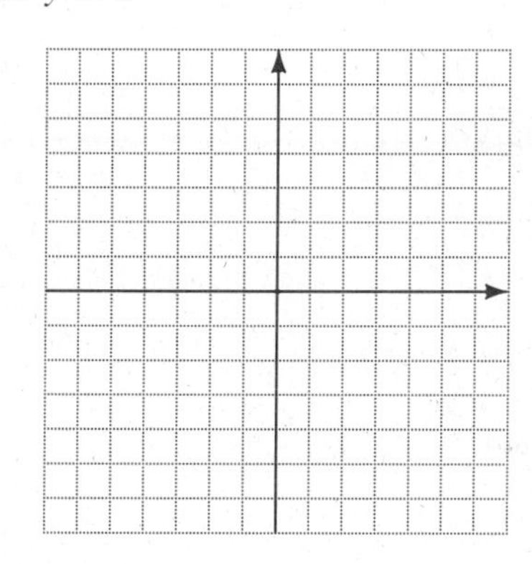

24. $y \leq -3$

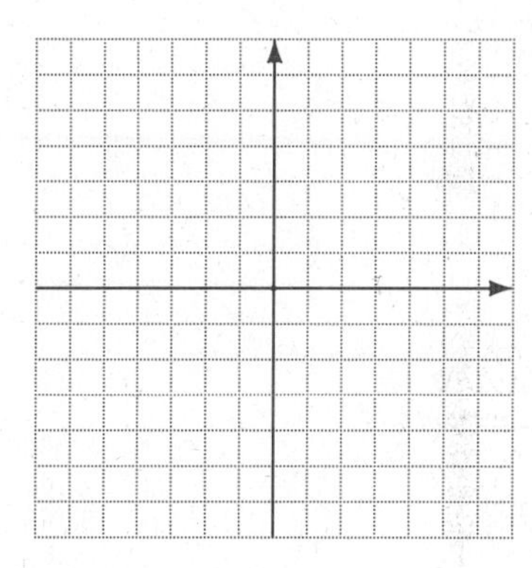

25. $x \leq 3y$

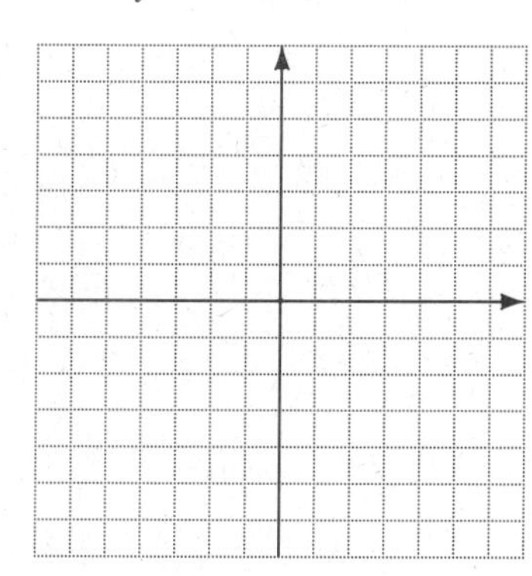

26. $x \leq 5y$

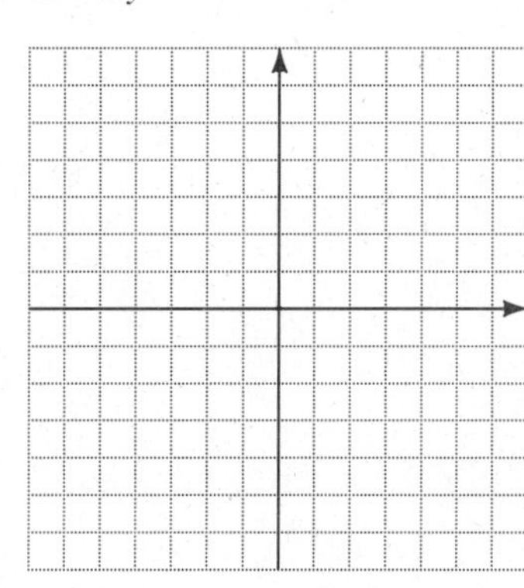

27. $x \geq -2y$

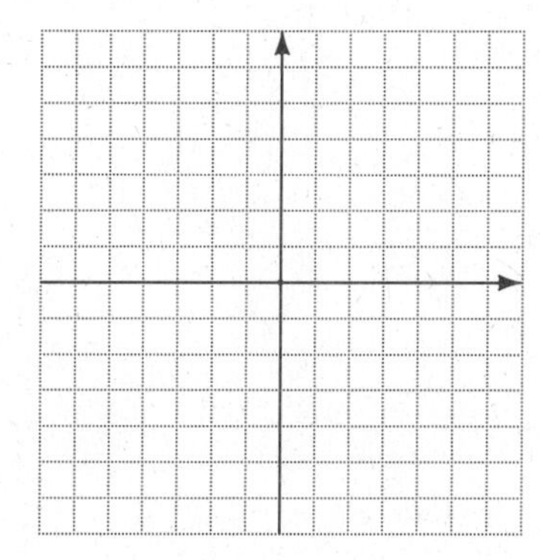

Solve each problem.

28. The Sweet Tooth Candy Company uses x pounds of chocolate for cookies and y pounds of chocolate for cakes. The company has 500 pounds of chocolate available, so that

$$x + y \leq 500.$$

(a) Graph this inequality. (Here $x \geq 0$ and $y \geq 0$.)

(b) Give some sample values of x and y that satisfy the inequality.

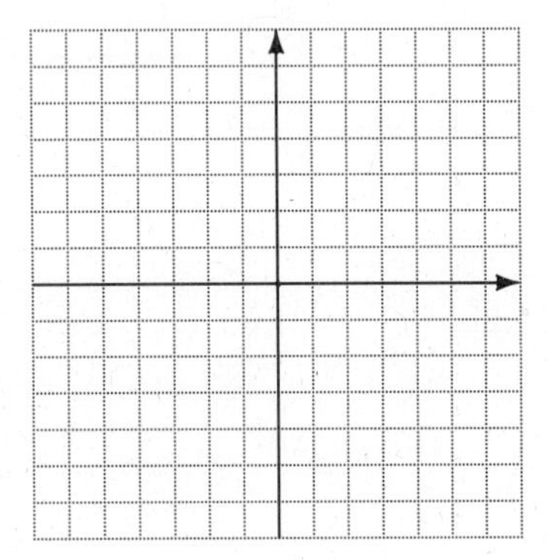

29. Ms. Branson takes x vitamin C tablets each day at a cost of 10¢ each and y multivitamins each day at a cost of 15¢ each. She wants the total cost to be no more than 50¢ a day. Express this as

$$10x + 15y \leq 50.$$

(a) Graph this inequality. (Here $x \geq 0$ and $y \geq 0$.)

(b) Give some sample values of x and y that satisfy the inequality.

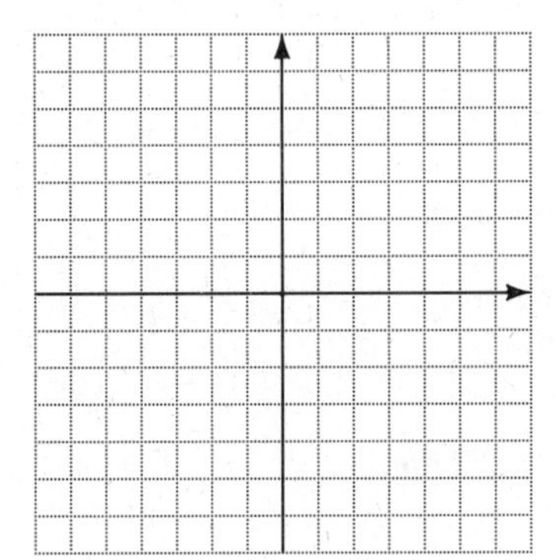

SKILL SHARPENERS

Add the polynomials. *See Section 4.4.*

30. $\begin{array}{r} x - 2y \\ \underline{3x + 2y} \end{array}$

31. $\begin{array}{r} 5x - 7y \\ \underline{12x + 7y} \end{array}$

32. $\begin{array}{r} 9a - 5b \\ \underline{-9a + 7b} \end{array}$

33. $\begin{array}{r} -11p + 8q \\ \underline{11p - 9q} \end{array}$

name date hour

CHAPTER 7 SUMMARY

KEY TERMS

7.1	**linear equation**	An equation that can be written in the form $Ax + By = C$ is a linear equation in two variables.
	ordered pair	A pair of numbers written between parentheses in which the order is important is called an ordered pair.
	table of values	A table showing ordered pairs of numbers is called a table of values.
7.2	**coordinate system**	An x-axis and y-axis at right angles form a coordinate system.
	quadrants	A coordinate system divides the plane into four regions called quadrants.
	origin	The point at which the x-axis and y-axis intersect is called the origin.
	coordinates	The numbers in an ordered pair are called the coordinates of the corresponding point.
	plot	To plot an ordered pair is to find the corresponding point on a coordinate system.
7.3	***y*-intercept**	If a graph crosses the y-axis at k, then the y-intercept is $(0, k)$.
	***x*-intercept**	If a graph crosses the x-axis at k, then the x-intercept is $(k, 0)$.
7.4	**slope**	The slope of a line is the ratio of the change in y compared to the change in x when moving along the line.
	parallel lines	Two lines in a plane that never intersect are parallel.
	perpendicular lines	Perpendicular lines intersect at a 90° angle.

NEW SYMBOLS

(a, b) an ordered pair

m slope

QUICK REVIEW

Section	Concepts	Examples
7.1 Linear Equations in Two Variables	An ordered pair is a solution of an equation if it satisfies the equation.	Is $(2, -5)$ or $(0, -6)$ a solution of $4x - 3y = 18$? $4(2) - 3(-5) = 23 \neq 18$ $(2, -5)$ is not a solution. $4(0) - 3(-6) = 18$ $(0, -6)$ is a solution.
	If a value of either variable in an equation is given, the other variable can be found by substitution.	Complete the ordered pair $(0, \quad)$ for $3x = y + 4$. $3(0) = y + 4$ $0 = y + 4$ $-4 = y$ The ordered pair is $(0, -4)$.

Section	Concepts	Examples
7.2 Graphing Ordered Pairs	Plot the ordered pair $(-2, 4)$ by starting at the origin, going 2 units to the left, then going 4 units up.	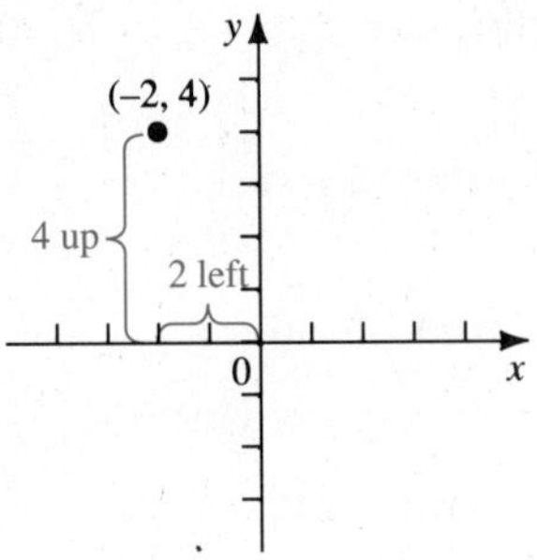
7.3 Graphing Linear Equations in Two Variables	The graph of $y = k$ is a horizontal line through $(0, k)$.	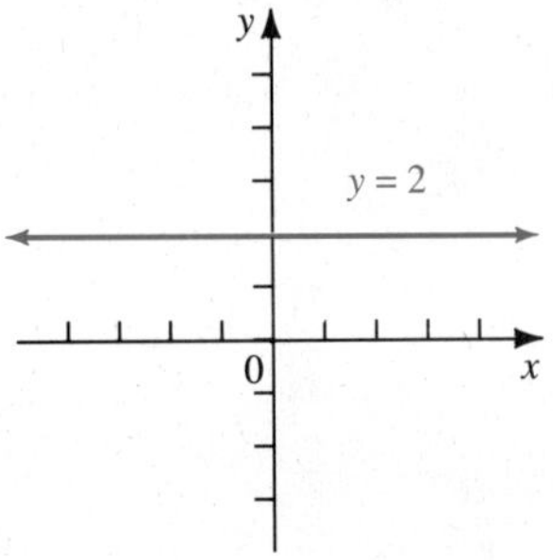
	The graph of $x = k$ is a vertical line through $(k, 0)$.	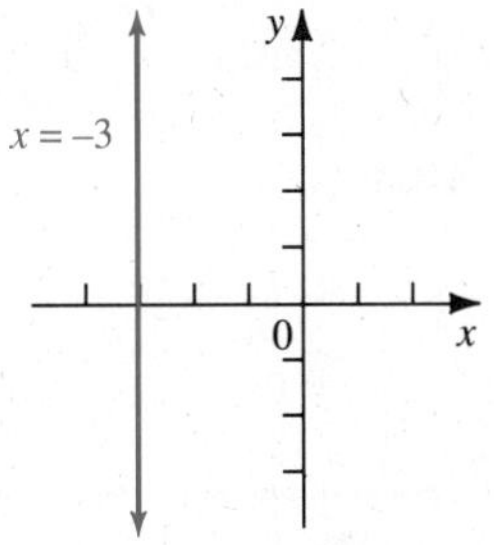
	The graph of $Ax + By = 0$ goes through the origin. Find and plot another point that satisfies the equation. Then draw the line through the two points.	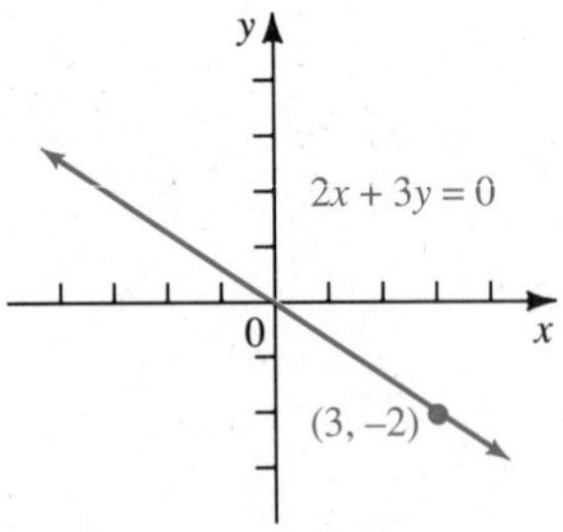

Section	Concepts	Examples
	To graph a linear equation: **1.** Find at least two pairs that satisfy the equation. **2.** Plot the corresponding points. **3.** Draw a straight line through the points.	$2y = x - 4$ $(4, 0)$ $(0, -2)$
7.4 The Slope of a Line	The slope of the line through (x_1, y_1) and (x_2, y_2) is $m = \dfrac{y_2 - y_1}{x_2 - x_1}, \ (x_1 \neq y_2).$ Horizontal lines have slope 0. Vertical lines have undefined slope. To find the slope of a line from its equation, solve for y. The slope is the coefficient of x.	The line through $(-2, 3)$ and $(4, -5)$ has slope $m = \dfrac{-5 - 3}{4 - (-2)} = \dfrac{-8}{6} = -\dfrac{4}{3}.$ The line $y = -2$ has slope 0. The line $x = 4$ has undefined slope. Given: $3x - 4y = 12.$ $-4y = -3x + 12$ $y = \dfrac{3}{4}x - 3$ Slope is $\dfrac{3}{4}$.
7.5 Equations of a Line	**Slope-Intercept Form** $\boldsymbol{y = mx + b}$ m is the slope. $(0, b)$ is the y-intercept. **Point-Slope Form** $\boldsymbol{y - y_1 = m(x - x_1)}$ m is the slope. (x_1, y_1) is a point on the line.	Find an equation of the line with slope 2 and y-intercept $(0, -5)$: $y = 2x - 5$. Find an equation of the line with slope $-\frac{1}{2}$ through $(-4, 5)$: $y - 5 = -\dfrac{1}{2}(x - (-4))$ $2(y - 5) = -(x + 4)$ $2y - 10 = -x - 4$ $x + 2y = 6$

Section	Concepts	Examples
7.6 Graphing Linear Inequalities in Two Variables	**1.** Graph the line that is the boundary of the region. Make it solid if the inequality is $\leq$ or $\geq$; make it dashed if the inequality is $<$ or $>$. **2.** Use any point not on the line as a test point. Substitute for x and y in the inequality. If the result is true, shade the side of the line containing the test point; if the result is false, shade the other side.	Graph $2x + y \leq 5$: Graph the line $2x + y = 5$. Make it solid because of $\leq$. Use $(1, 0)$ as a test point. $2(1) + 0 \leq 5$ $2 \leq 5$ True Shade the side of the line containing $(1, 0)$. 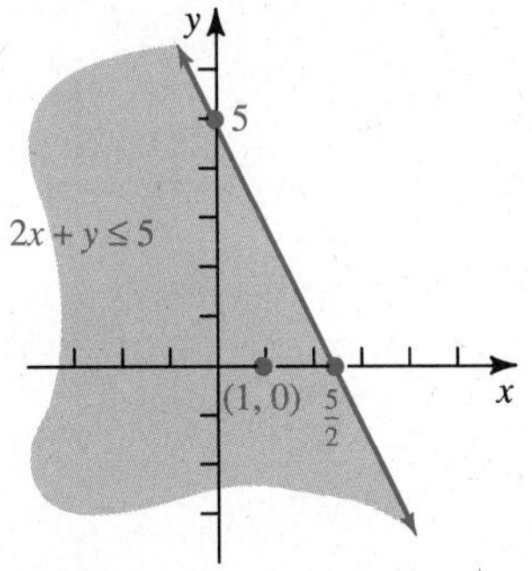 y 5 $2x + y \leq 5$ (1, 0) $\frac{5}{2}$ x

name date hour

CHAPTER 7 REVIEW EXERCISES

[7.1] *Complete the given ordered pairs for each equation.*

1. $y = 3x - 2$ $(-1,\quad)$ $(0,\quad)$ $(\quad, 5)$

2. $4x + 3y = 9$ $(0,\quad)$ $(\quad, 0)$ $(-2,\quad)$

3. $x = 2y$ $(0,\quad)$ $(8,\quad)$ $(\quad, -3)$

4. $x + 4 = 0$ $(\quad, -3)$ $(\quad, 0)$ $(\quad, 5)$

Decide whether the given ordered pair is a solution of the given equation.

5. $x + y = 7$; $(3, 4)$

6. $2x + y = 5$; $(1, 4)$

7. $3x - y = 4$; $(1, -1)$

8. $5x - 3y = 16$; $(2, -2)$

[7.2] *Plot the ordered pairs on the given coordinate system.*

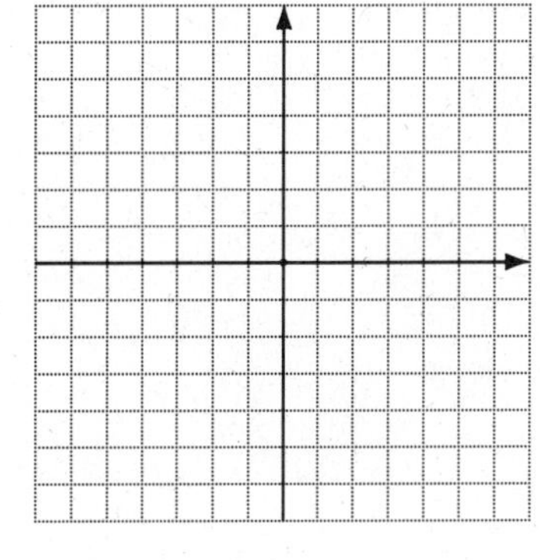

9. $(4, 2)$ **10.** $(-5, 1)$ **11.** $(2, 0)$

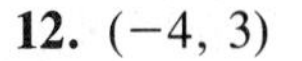

12. $(-4, 3)$ **13.** $(0, 3)$ **14.** $(-2, -5)$

Without plotting the given point, name the quadrant in which each point lies.

15. $(-2, 3)$ **16.** $(-1, -4)$ **17.** $(0, 4)$ **18.** $(0, 0)$

[7.3] *Find the intercepts for each equation.*

19. $y = 2x - 5$

x-intercept:

y-intercept:

20. $2x + y = 7$

x-intercept:

y-intercept:

21. $3x - 2y = 8$

x-intercept:

y-intercept:

Graph each linear equation.

22. $y = 2x + 3$

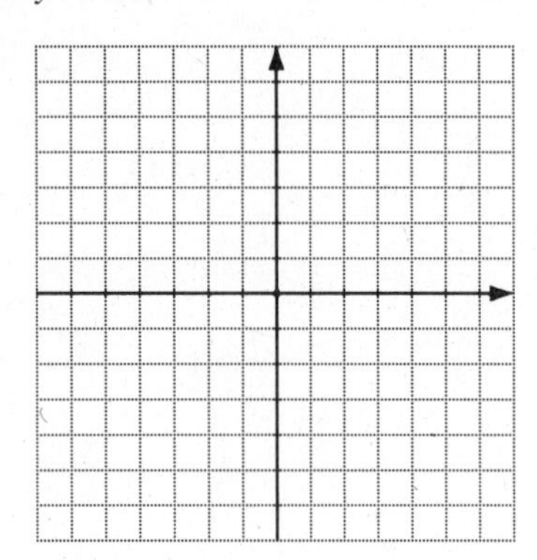

23. $x + y = 5$

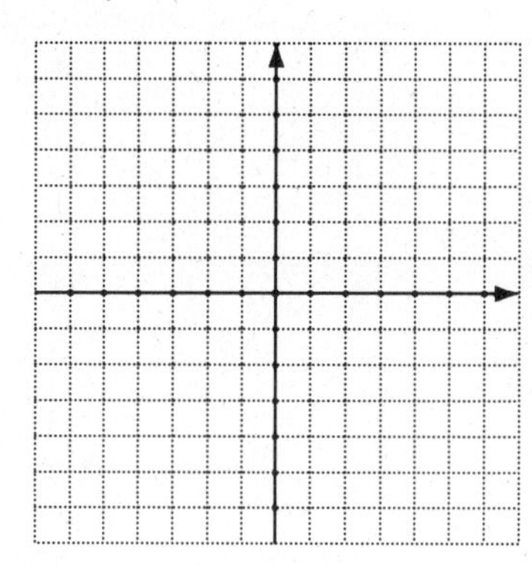

24. $x - y = 0$

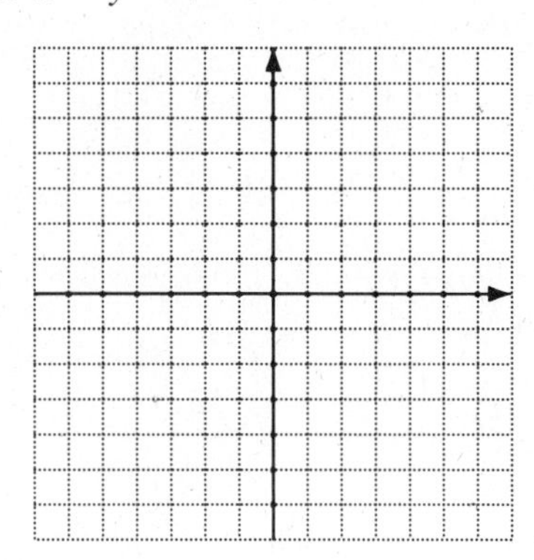

[7.4] *Find the slope of each line.*

25. Through $(2, 3)$ and $(-1, 1)$

26. Through $(0, 0)$ and $(-1, -2)$

27. Through $(0, 6)$ and $(5, 6)$

28. Through $(2, 5)$ and $(2, -2)$

29. $y = 3x - 1$

30. $y = \frac{2}{3}x + 5$

31. $y = 8$

32. $x = 2$

33. $6x + 7y = 9$

Decide whether the lines of each pair are parallel, perpendicular, *or* neither.

34. $3x + 2y = 5$
$6x + 4y = 12$

35. $x - 3y = 8$
$3x + y = 6$

36. $4x + 3y = 10$
$3x - 4y = 12$

37. $x - 2y = 3$
$x + 2y = 3$

[7.5] *Write an equation for each line. Write equations in the form* $Ax + By = C$.

38. $m = 3;\quad b = -2$

39. $m = -1;\quad b = \frac{3}{4}$

40. $m = \frac{2}{3};\quad b = 5$

41. Through (5, −2); $m = 1$

42. Through (−1, 4); $m = \frac{2}{3}$

43. Through (1, −1); $m = -\frac{3}{4}$

44. Through (2, 1) and (−2, 2)

45. Through (−2, 6) with slope 0

46. Through $\left(\frac{1}{2}, -\frac{3}{4}\right)$ with undefined slope

[7.6] *Complete the graph of each linear inequality by shading the correct region.*

47. $x - y \leq 3$

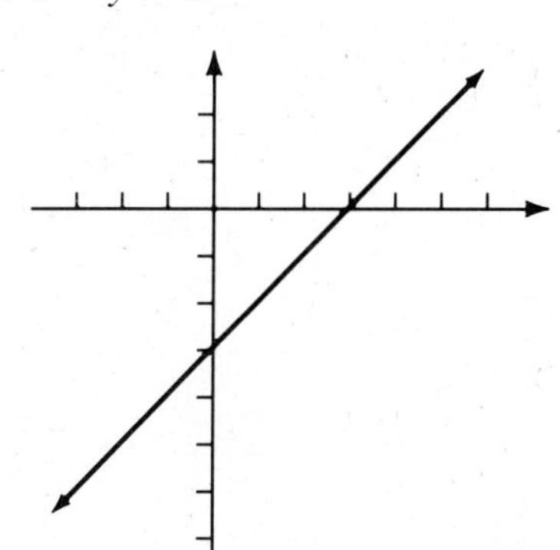

48. $3x - y \geq 5$

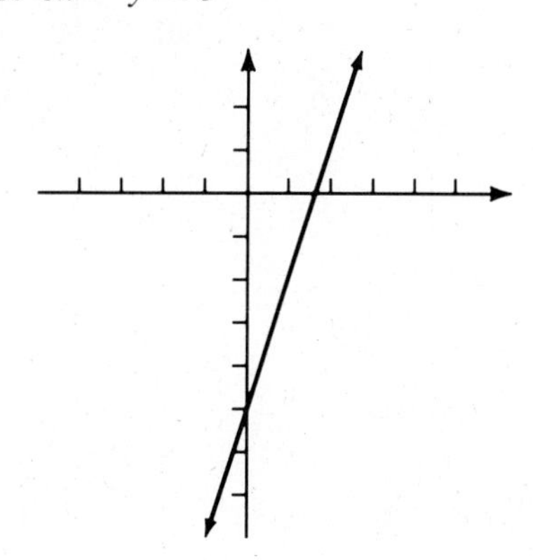

49. $x + 2y > 6$

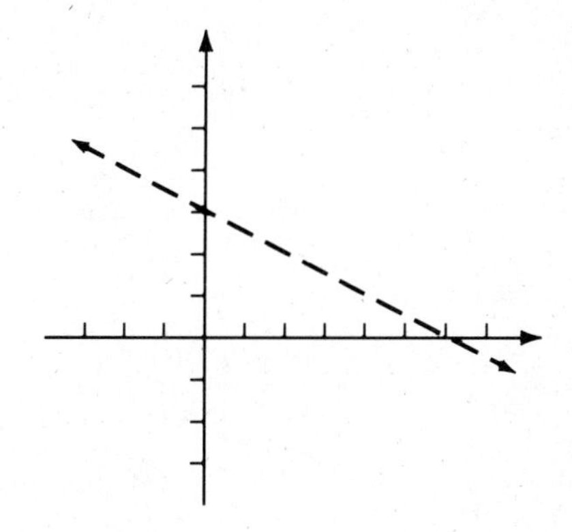

Graph each linear inequality.

50. $3x + 5y \geq 9$

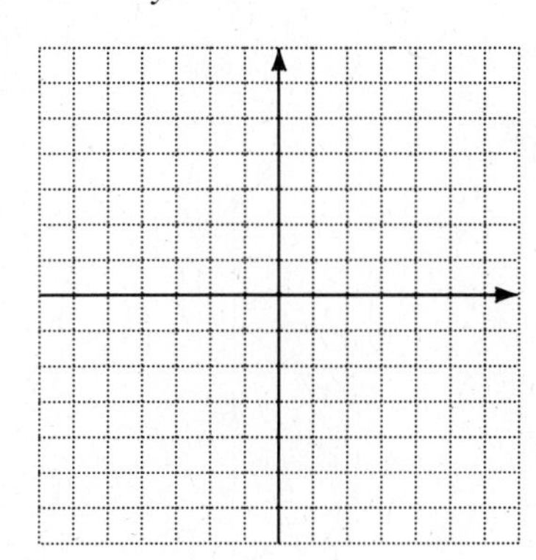

51. $2x - 3y \geq -6$

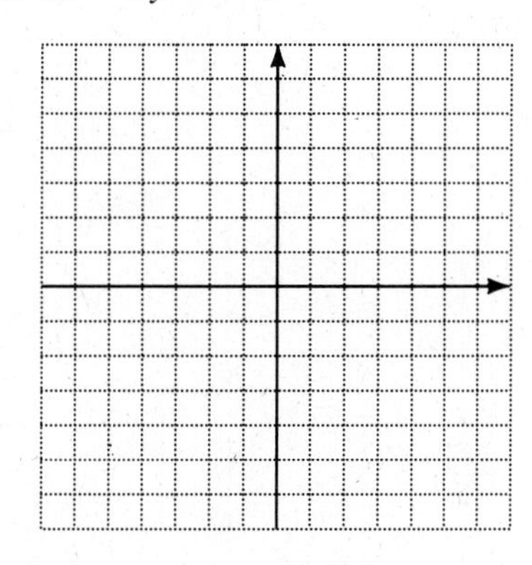

52. $x - 2y > 0$

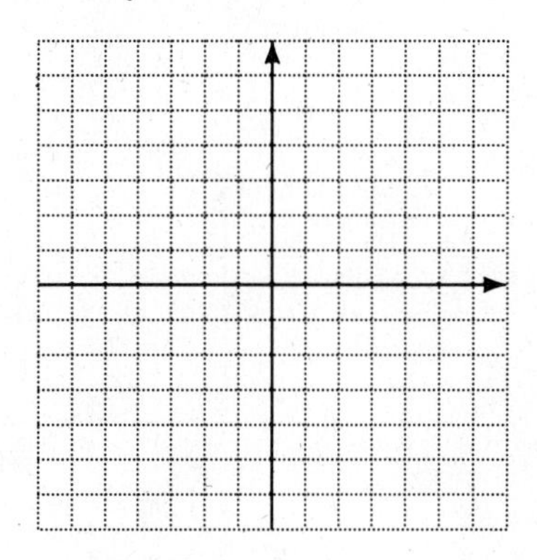

MIXED REVIEW EXERCISES

Find the slope and the intercepts of each line.

53. $11x - 3y = 4$

54. Through $(4, -1)$ and $(-2, -3)$

55. Through $(0, -1)$ and $(9, -5)$

56. $8x = 6 - 3y$

Write an equation in the form $Ax + By = C$ for each of the following lines.

57. Through $(5, 0)$ and $(5, -1)$

58. $m = -\frac{1}{4}$; $b = -\frac{5}{4}$

59. Through $(8, 6)$; $m = -3$

60. Through $(3, -5)$ and $(-4, -1)$

Graph each equation or inequality.

61. $x + 2y \leq 4$

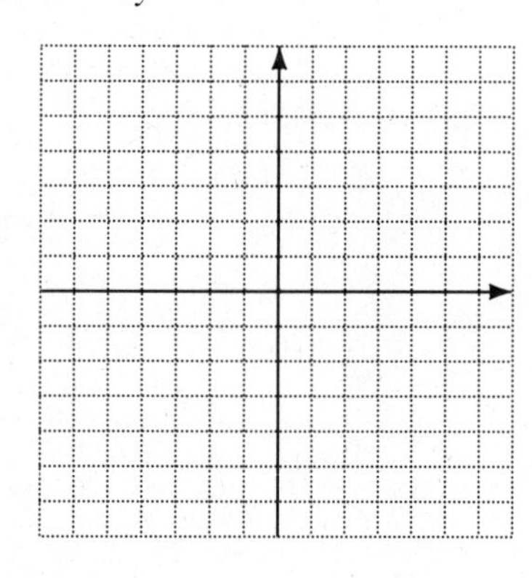

62. $x + 2y = 0$

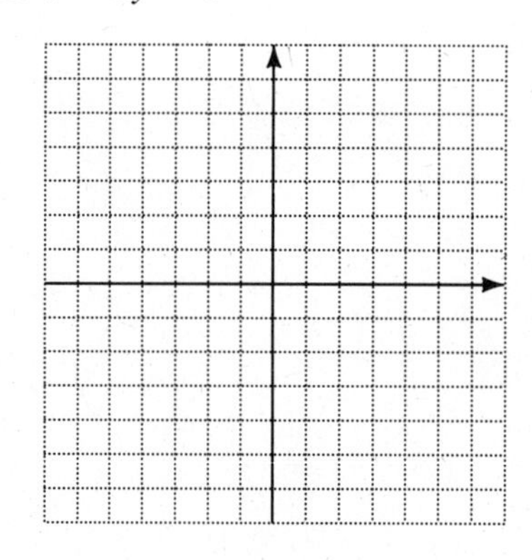

63. $y > 3x$

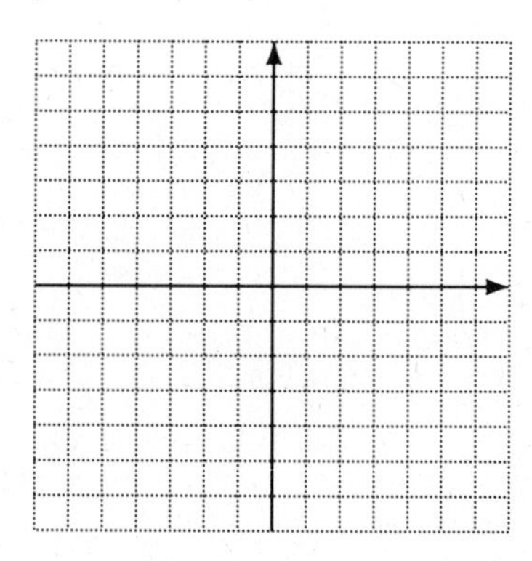

64. $y + 3 = 0$

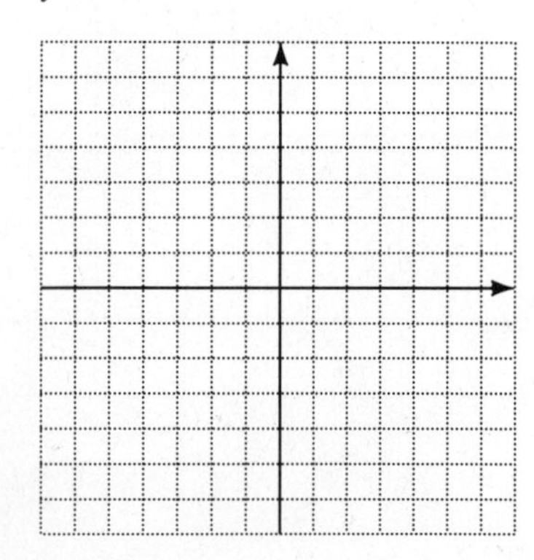

65. $2x - y = 5$

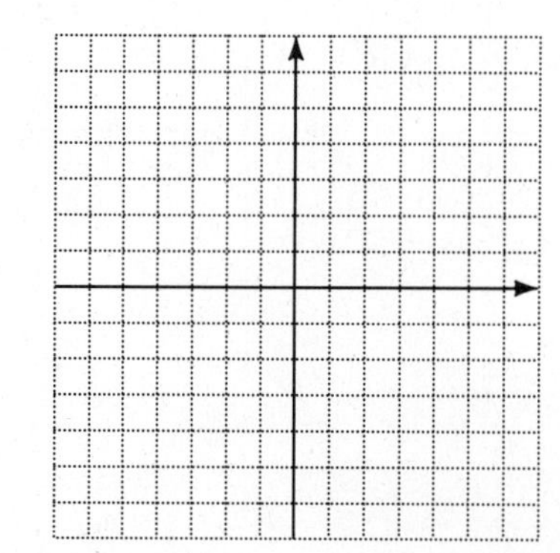

66. $x \leq -2$

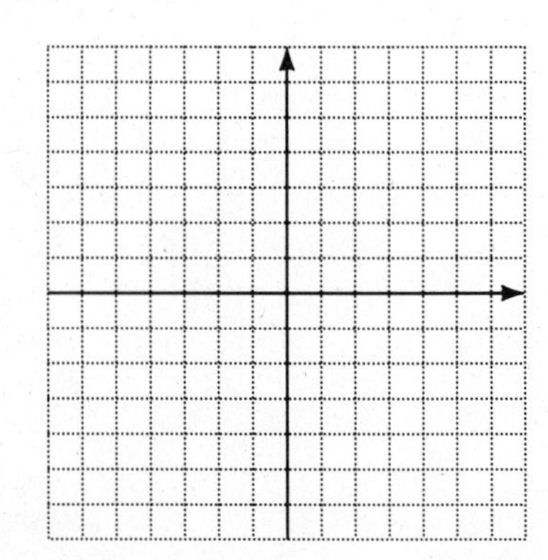

name date hour

CHAPTER 7 TEST

Complete the ordered pairs using the given equations.

	Equation	*Ordered pairs*	
1.	$y = 5x - 6$	(0,) (−2,) (, 14)	**1.** ________
2.	$3x - 5y = 30$	(0,) (, 0) (5,)	**2.** ________
3.	$x = 3y$	(0,) (, 2) (8,) (−12,)	**3.** ________
4.	$y - 2 = 0$	(5,) (4,) (0,) (−3,)	**4.** ________

Graph each linear equation. Give the x- and y-intercepts.

5. $x + y = 4$

5.

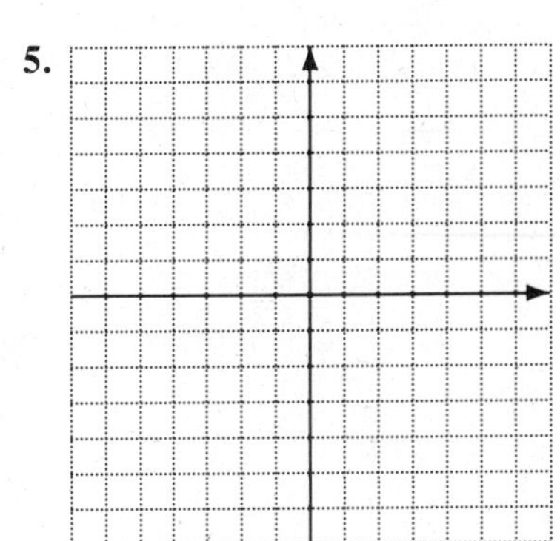

6. $2x + y = 6$

6.

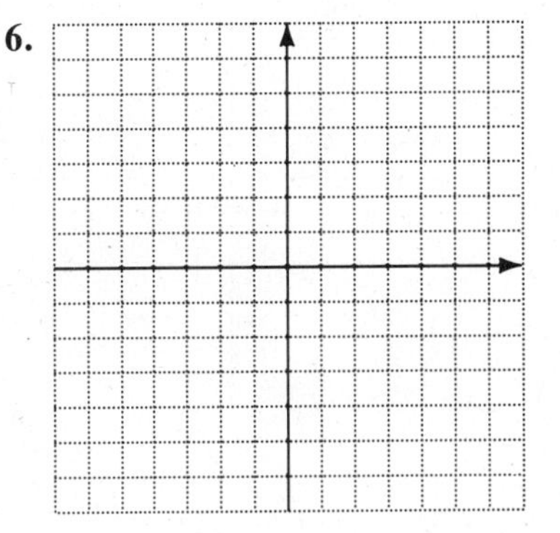

7.

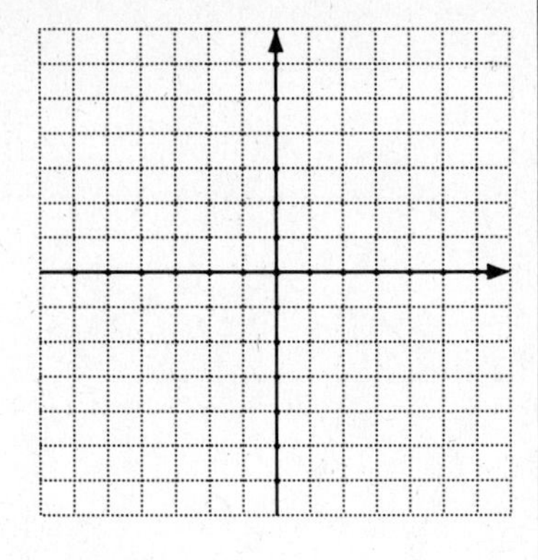

7. $3x - 4y = 18$

8.

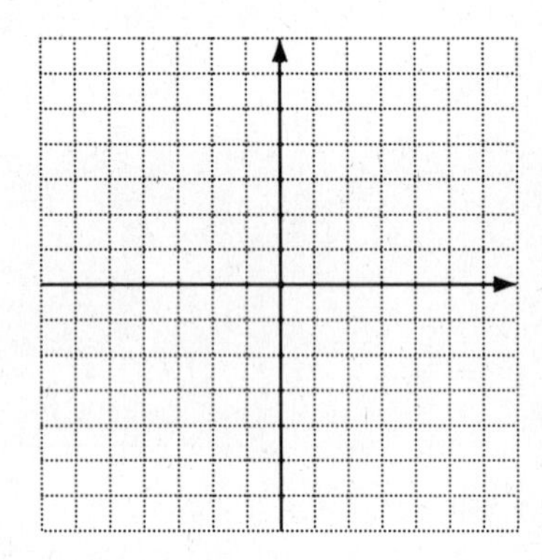

8. $2x + y = 0$

9.

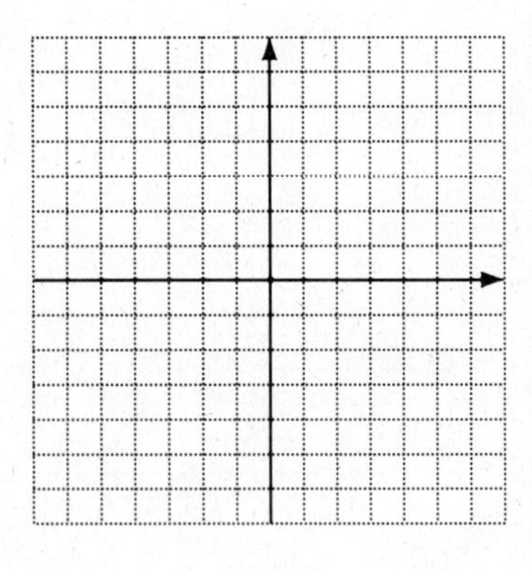

9. $x + 5 = 0$

name date hour

10. $y = 2$

10. 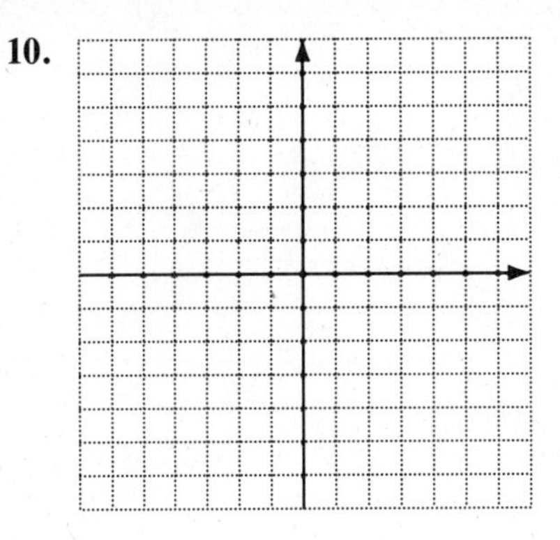

Find the slope of each line.

11. Through $(-2, 4)$ and $(5, 1)$

11. ____________________

12. $y = -\frac{3}{4}x + 6$

12. ____________________

Write an equation for each line. Write the equations in the form $Ax + By = C$.

13. Through $(1, -3)$; $m = -4$

13. ____________________

14. $m = 3$; $b = -1$

14. ____________________

15. $m = -\frac{3}{4}$; $b = 2$

15. ____________________

16. Through $(-2, -6)$ and $(-1, 3)$

16. ____________________

Graph each linear inequality.

17.

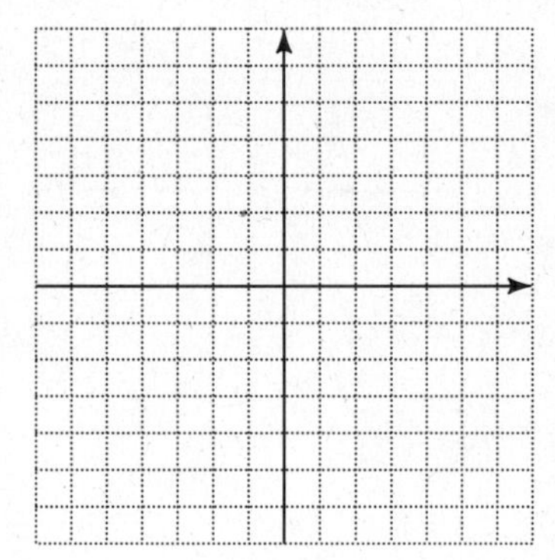

17. $x + y \leq 6$

18.

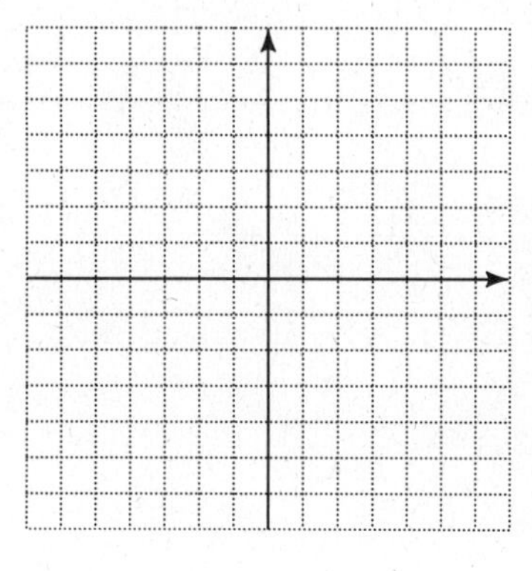

18. $3x - 4y > 12$

19.

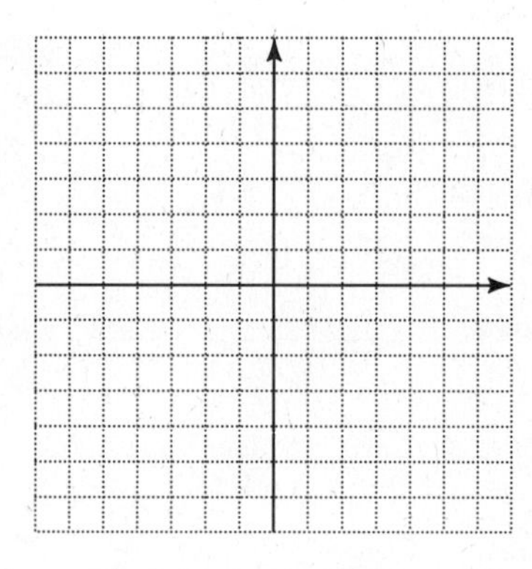

19. $x < 3y$

20.

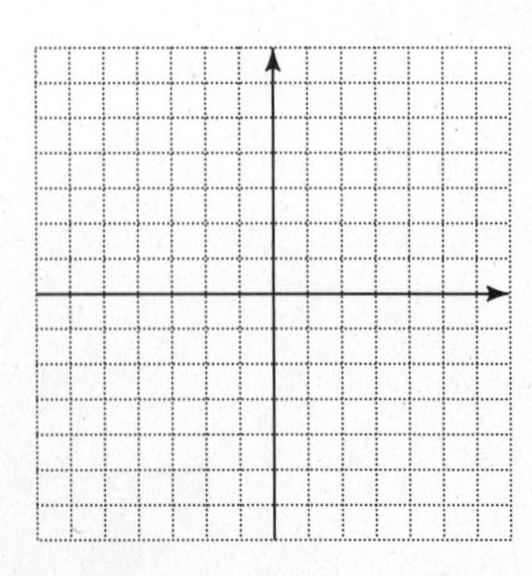

20. $y \geq -3$

LINEAR SYSTEMS

8.1 SOLVING SYSTEMS OF LINEAR EQUATIONS BY GRAPHING

A **system of linear equations** consists of two or more linear equations with the same variables. Examples of systems of two linear equations with two variables are shown below.

$$\begin{aligned} 2x + 3y &= 4 \\ 3x - y &= -5 \end{aligned} \qquad \begin{aligned} x + 3y &= 1 \\ -y &= 4 - 2x \end{aligned} \qquad \begin{aligned} x - y &= 1 \\ y &= 3 \end{aligned}$$

In the system on the right, think of $y = 3$ as an equation in two variables by writing it as $0x + y = 3$.

1 The **solution of a system** of linear equations includes all the ordered pairs that make all the equations of the system true at the same time.

EXAMPLE 1 Determining Whether an Ordered Pair Is a Solution

Is $(4, -3)$ a solution of the following systems?

(a) $x + 4y = -8$
$3x + 2y = 6$

Decide whether or not $(4, -3)$ is a solution of the system by substituting 4 for x and -3 for y in each equation.

$x + 4y = -8$		$3x + 2y = 6$	
$4 + 4(-3) = -8$		$3(4) + 2(-3) = 6$	
$4 + (-12) = -8$	Multiply.	$12 + (-6) = 6$	Multiply.
$-8 = -8$	True	$6 = 6$	True

Since $(4, -3)$ satisfies both equations, it is a solution of the system.

(b) $2x + 5y = -7$
$3x + 4y = 2$

Again, substitute 4 for x and -3 for y in both equations.

$2x + 5y = -7$		$3x + 4y = 2$	
$2(4) + 5(-3) = -7$		$3(4) + 4(-3) = 2$	
$8 + (-15) = -7$	Multiply.	$12 + (-12) = 2$	Multiply.
$-7 = -7$	True	$0 = 2$	False

The ordered pair $(4, -3)$ is not a solution, because it does not satisfy the second equation. ■

WORK PROBLEM 1 AT THE SIDE.

OBJECTIVES

1. Decide whether a given ordered pair is a solution of a system.
2. Solve linear systems by graphing.
3. Identify systems with no solutions or with an infinite number of solutions.

1. Decide whether the given ordered pair is a solution of the system.

(a) $(2, 5)$
$3x - 2y = -4$
$5x + y = 15$

(b) $(1, -2)$
$x - 3y = 7$
$4x + y = 5$

(c) $(4, -1)$
$5x + 6y = 14$
$2x + 5y = 3$

ANSWERS
1. (a) yes (b) no (c) yes

Applications often require solving a system of equations. We now look at ways to solve systems of two linear equations in two variables.

2 One way to find the solution of a system of two linear equations is to graph both equations on the same axes. The graph of each line shows points whose coordinates satisfy the equation of that line. The coordinates of any point where the lines intersect give a solution of the system. Since two different straight lines can intersect at no more than one point, there can never be more than one solution for such a system.

EXAMPLE 2 Solving Systems by Graphing

Solve each system of equations by graphing both equations on the same axes.

(a) $2x + 3y = 4$
$3x - y = -5$

As shown in Chapter 7, these two equations can be graphed by plotting several points for each line.

$2x + 3y = 4$

x	y
0	$\frac{4}{3}$
2	0
-2	$\frac{8}{3}$

$3x - y = -5$

x	y
0	5
$-\frac{5}{3}$	0
-2	-1

The lines in Figure 1 suggest that the graphs intersect at the point $(-1, 2)$. Check this by substituting -1 for x and 2 for y in both equations. Since $(-1, 2)$ satisfies both equations, the solution of this system is $(-1, 2)$.

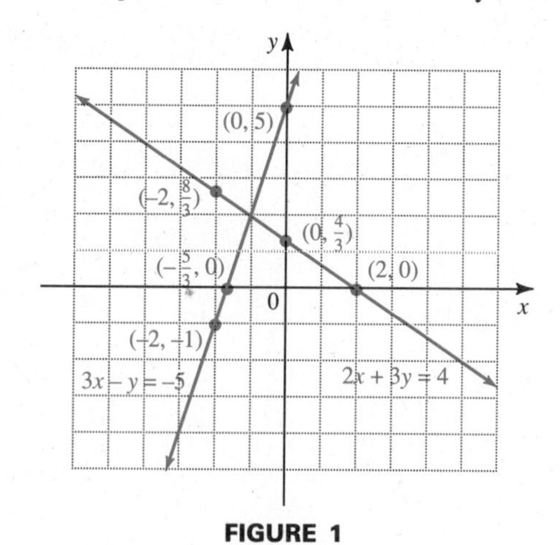

FIGURE 1

(b) $2x + y = 0$
$4x - 3y = 10$

Find the solution of the system by graphing the two lines on the same axes. As suggested by Figure 2, the solution is $(1, -2)$, the point at which the graphs of the two lines intersect. Check by substituting 1 for x and -2 for y in both equations of the system. ■

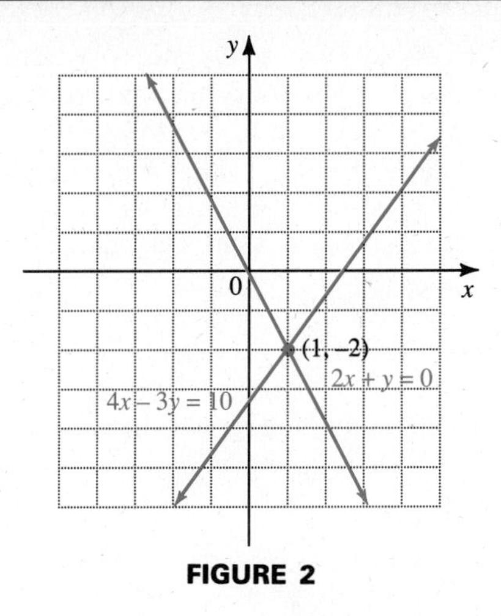

FIGURE 2

WORK PROBLEM 2 AT THE SIDE.

3 Sometimes the graphs of the two equations in a system either do not intersect at all or are the same line, as in the systems in Example 3.

EXAMPLE 3 Solving Special Systems

Solve each system by graphing.

(a) $2x + y = 2$
$2x + y = 8$

The graphs of these lines are shown in Figure 3. The two lines are parallel and have no points in common. For a system whose equations lead to graphs with no points in common, we will write "no solution."

(b) $2x + 5y = 1$
$6x + 15y = 3$

The graphs of these two equations are the same line. See Figure 4. Here the second equation can be obtained by multiplying each side of the first equation by 3. In this case, every point on the line is a solution of the system, and the solution is an infinite number of ordered pairs. We will write "infinite number of solutions" or "same line" to indicate this situation. ■

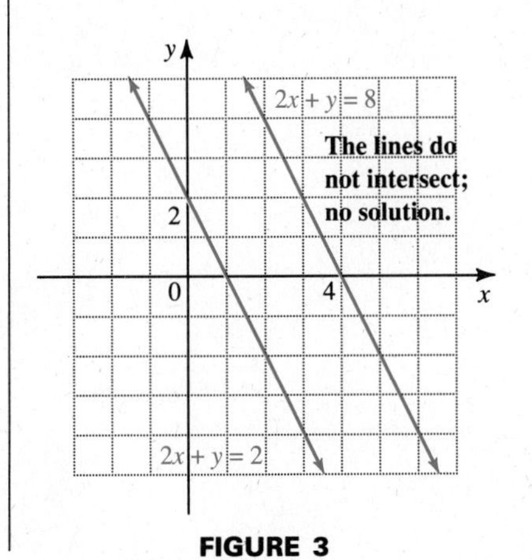

FIGURE 3

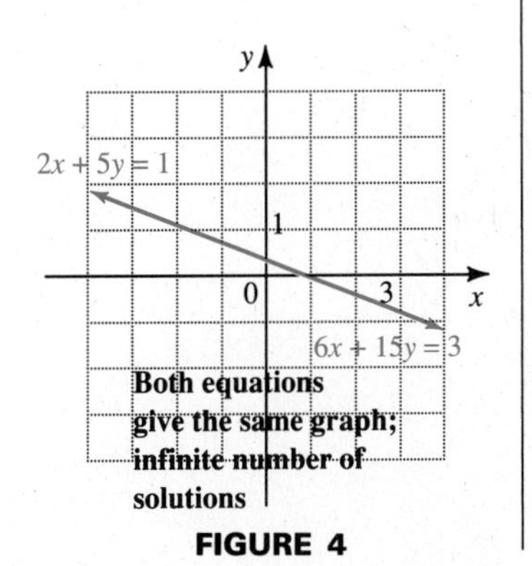

FIGURE 4

2. Solve each system of equations by graphing both equations on the same axes. Check your answers.

(a) $5x - 3y = 9$
$x + 2y = 7$

(One of the lines is already graphed.)

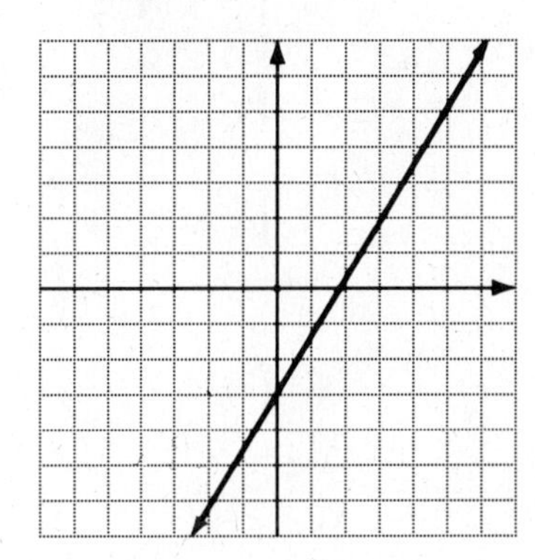

(b) $x + y = 4$
$2x - y = -1$

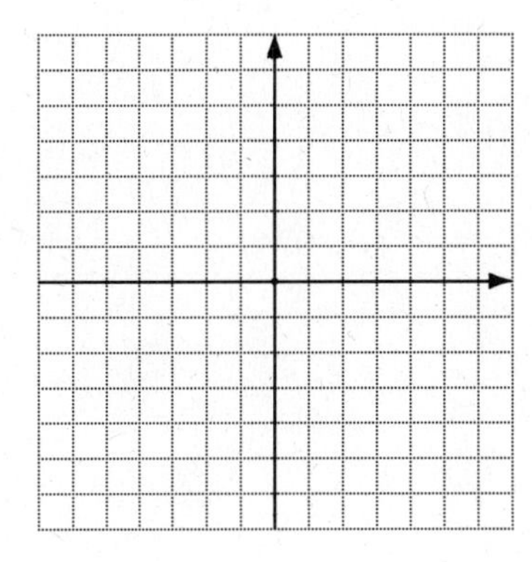

ANSWERS
2. (a) (3, 2) (b) (1, 3)

3. Solve each system of equations by graphing both equations on the same axes.

(a) $3x - y = 4$
$6x - 2y = 12$

(One of the lines is already graphed.)

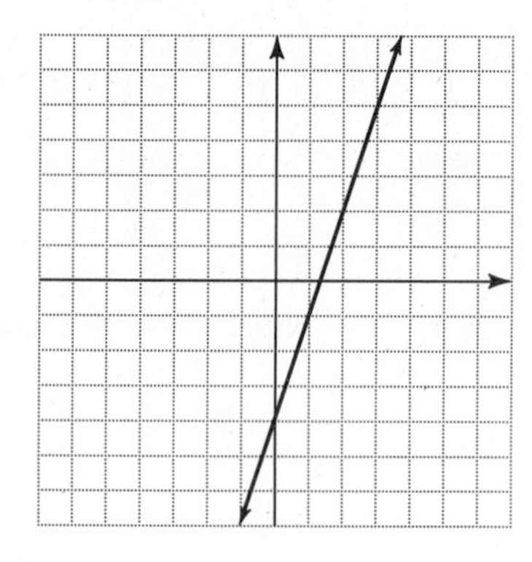

(b) $-x + 3y = 2$
$2x - 6y = -4$

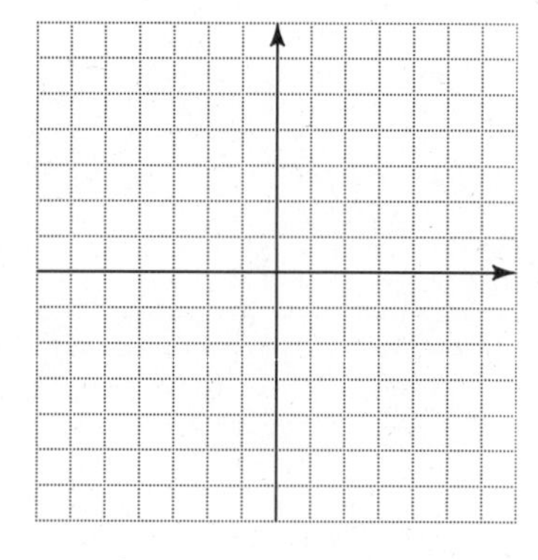

Each system in Example 2 has a solution. A system with a solution is called a **consistent system.** A system of equations with no solutions, such as the one in Example 3(a), is called an **inconsistent system.** The equations in Example 2 are independent equations. **Independent equations** have different graphs. The equations of the system in Example 3(b) have the same graph. Because they are different forms of the same equation, these equations are called **dependent equations.**

WORK PROBLEM 3 AT THE SIDE.

Examples 2 and 3 show the three cases that may occur in a system of two equations with two unknowns.

1. The graphs intersect at exactly one point, which gives the (single) solution of the system. The system is consistent and the equations are independent.

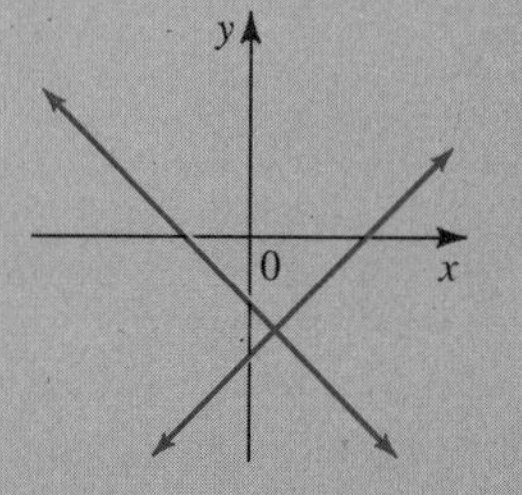

2. The graphs are parallel lines, so there is no solution. The system is inconsistent.

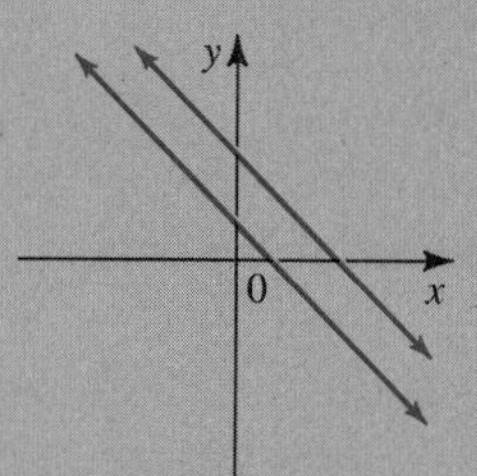

3. The graphs are the same line. The solution is an infinite set of ordered pairs. The equations are dependent.

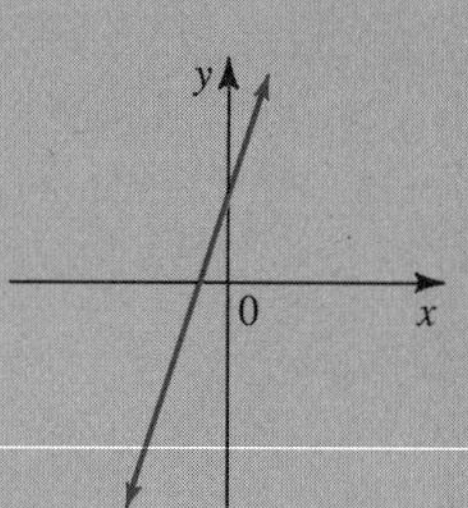

ANSWERS
3. (a) no solution (b) same line

name date hour

8.1 EXERCISES

Decide whether the given ordered pair is the solution of the given system. See Example 1.

1. $(2, -5)$
$3x + y = 1$
$2x + 3y = -11$

2. $(-1, 6)$
$2x + y = 4$
$3x + 2y = 9$

3. $(4, -2)$
$x + y = 2$
$2x + 5y = 2$

4. $(-6, 3)$
$x + 2y = 0$
$3x + 5y = 3$

5. $(2, 0)$
$3x + 5y = 6$
$4x + 2y = 5$

6. $(0, -4)$
$2x - 5y = 20$
$3x + 6y = -20$

7. $(5, 2)$
$4x = 26 - 3y$
$3x = 29 - 7y$

8. $(9, 1)$
$2x = 23 - 5y$
$3x = 24 - 2y$

9. $(6, -8)$
$2y = -x - 10$
$3y = 2x + 30$

10. $(-5, 2)$
$5y = 3x + 20$
$-3y = 2x + 4$

11. $(-3, -6)$
$2x - y = 0$
$2y = x - 9$

12. $(-2, 5)$
$3x + 4y = 8$
$3y = -x + 13$

Solve each system of equations by graphing both equations on the same axes. See Example 2.

13. $x + y = 6$
$x - y = 2$

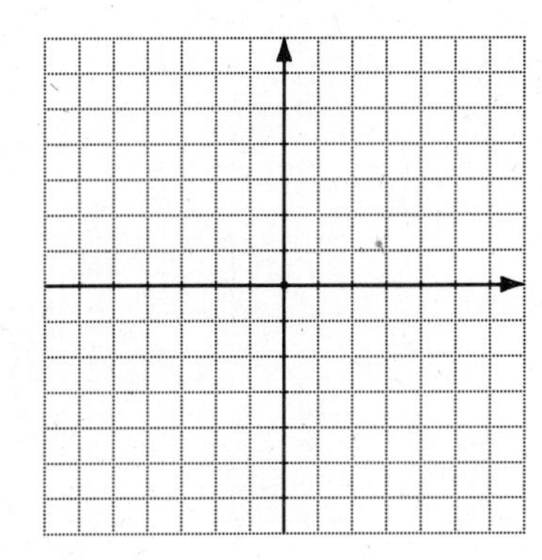

14. $x + y = -1$
$x - y = 3$

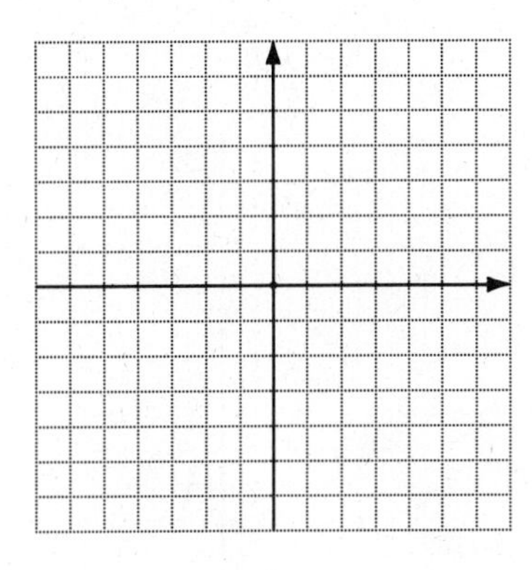

15. $x + y = 4$
$y - x = 4$

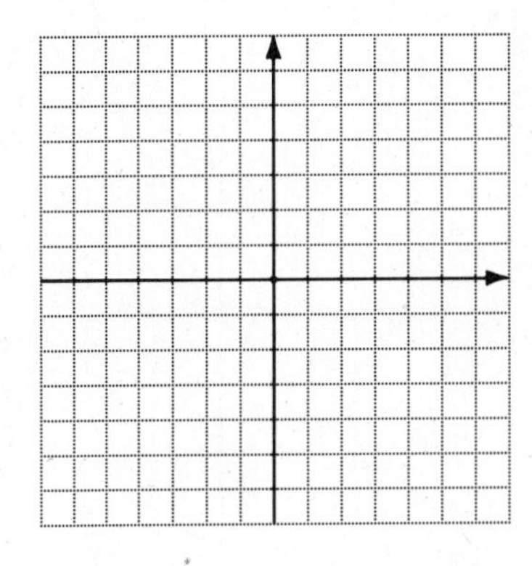

16. $y - x = -5$
$x + y = 1$

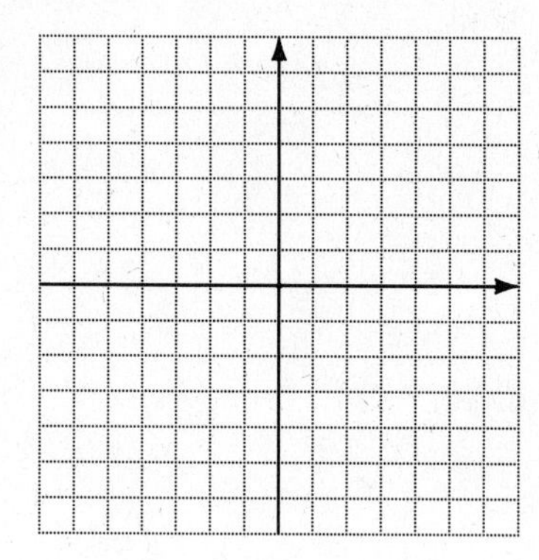

17. $x + 2y = 2$
$x - 2y = 6$

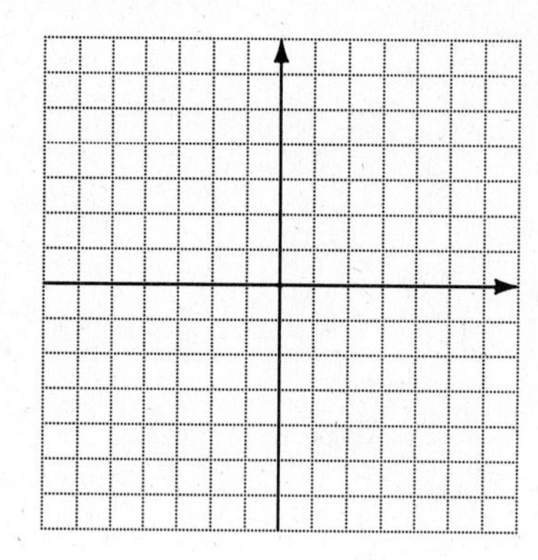

18. $4x + y = 2$
$2x - y = 4$

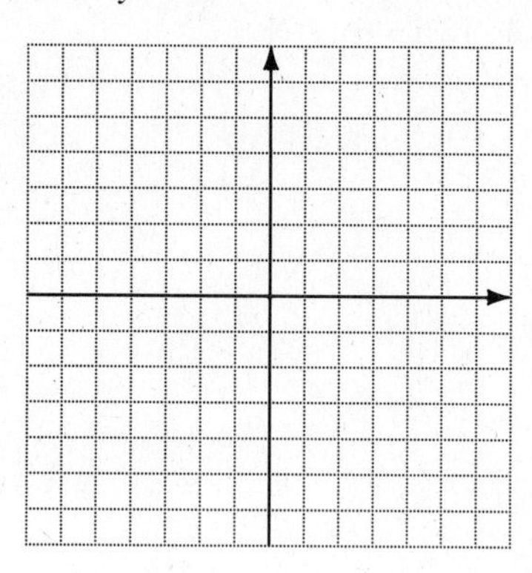

19. $3x - 2y = -3$
$3x + y = 6$

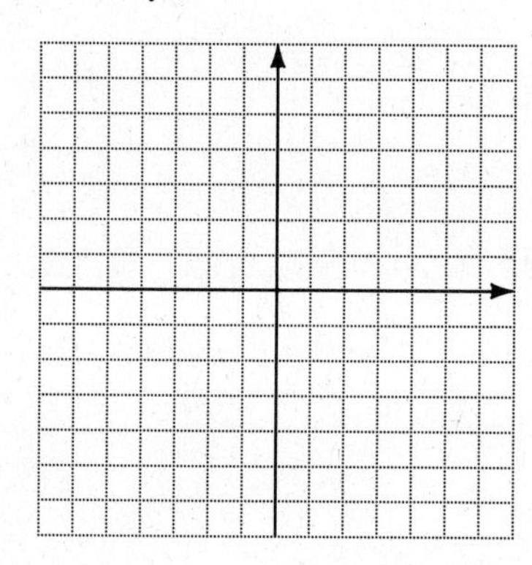

20. $2x + 3y = 12$
$2x - y = 4$

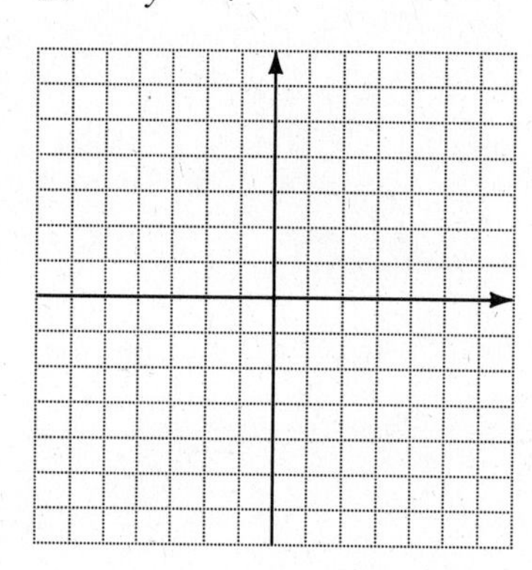

21. $-2x + 3y = 6$
$3x + y = 2$

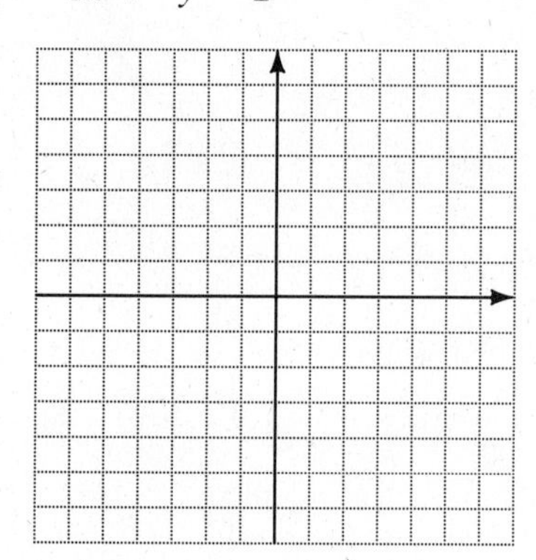

22. $3x - y = 3$
$-x + y = -3$

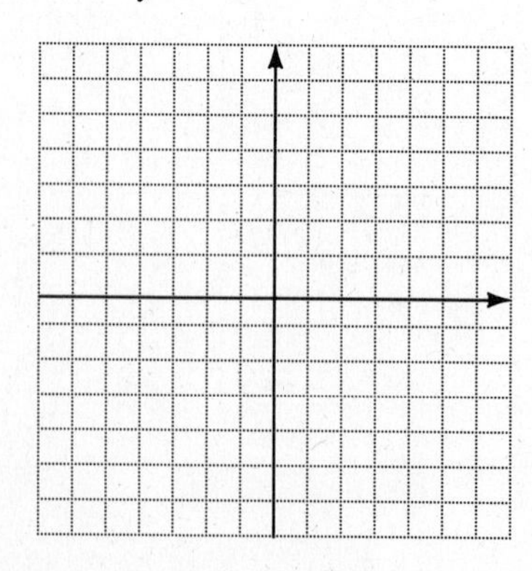

23. $-x + 3y = -4$
$x + 2y = 4$

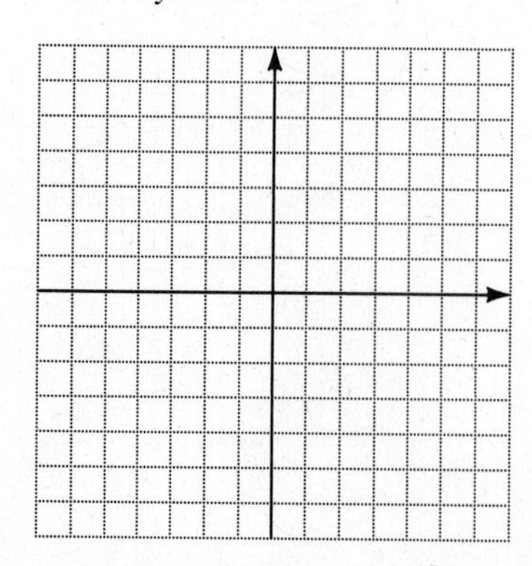

24. $3x + 2y = 6$
$3x - 4y = 24$

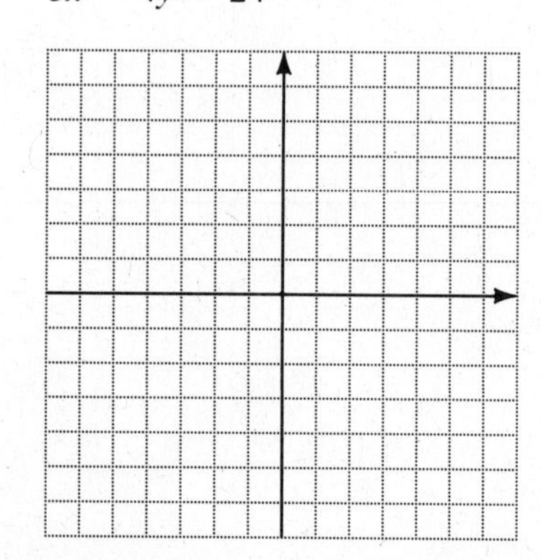

name date hour

25. $3x + 2y = -10$
$x - 2y = -6$

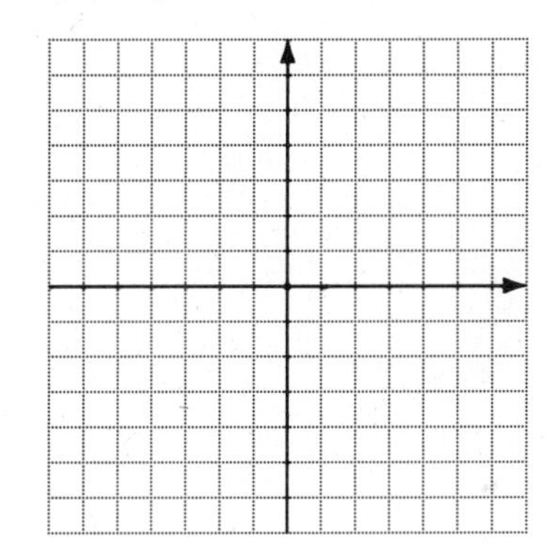

26. $4x + y = 5$
$3x - 2y = 12$

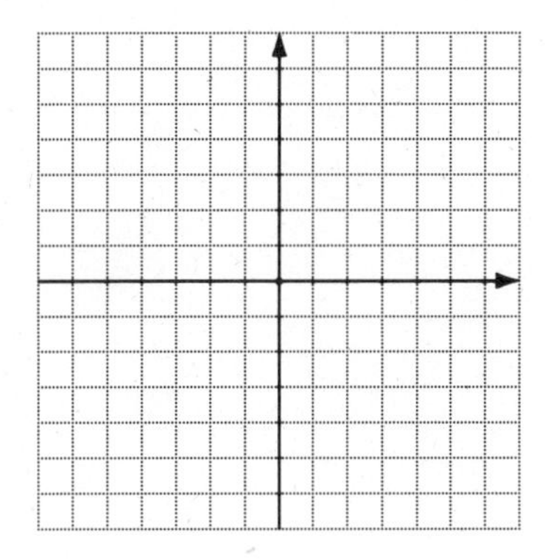

27. $3x - 4y = -24$
$5x + 2y = -14$

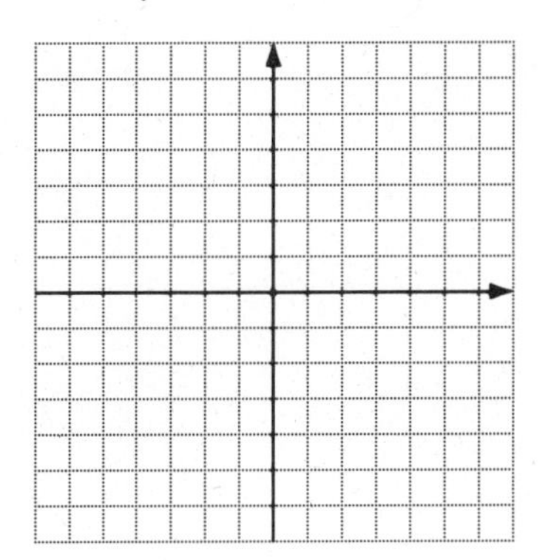

28. $y = x + 1$
$y = 3x - 1$

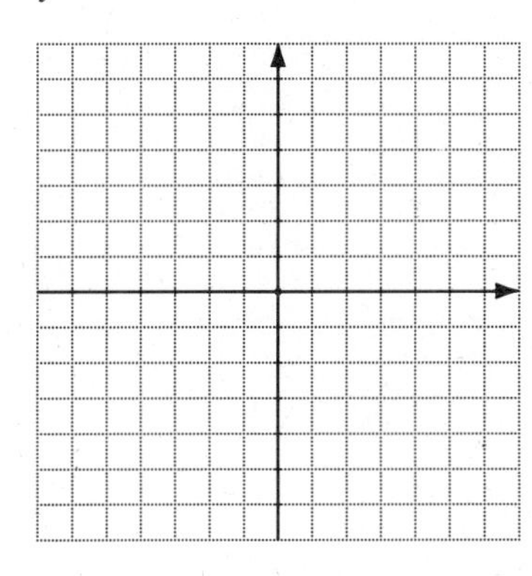

29. $y = x + 6$
$y = -x - 2$

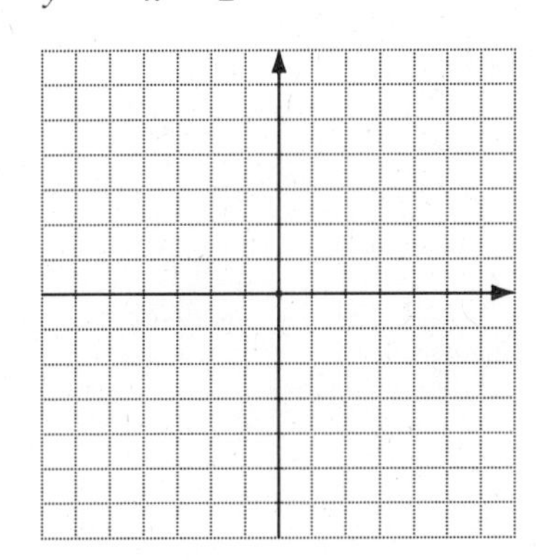

30. $5x + y = 5$
$x + y = 5$

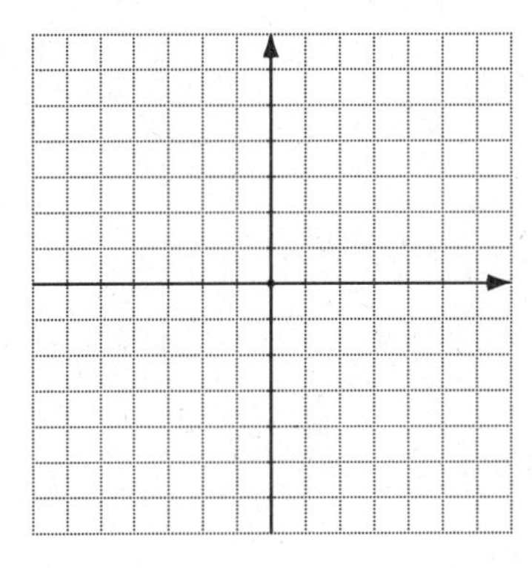

31. $3x - 4y = 8$
$4x + 5y = -10$

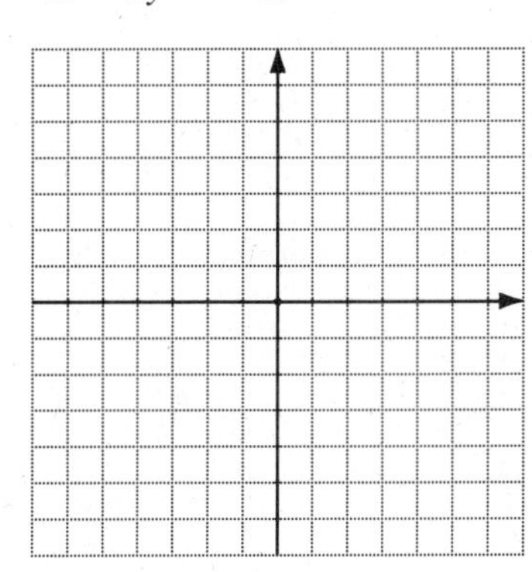

32. $3x + 2y = 10$
$4x - 3y = -15$

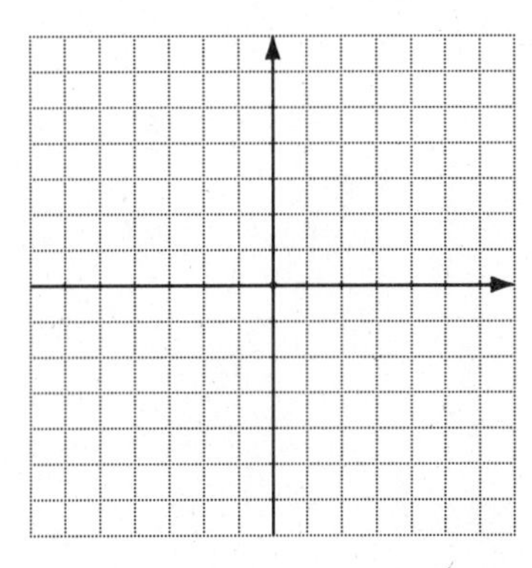

33. $2x + 5y = 12$
$x + y = 3$

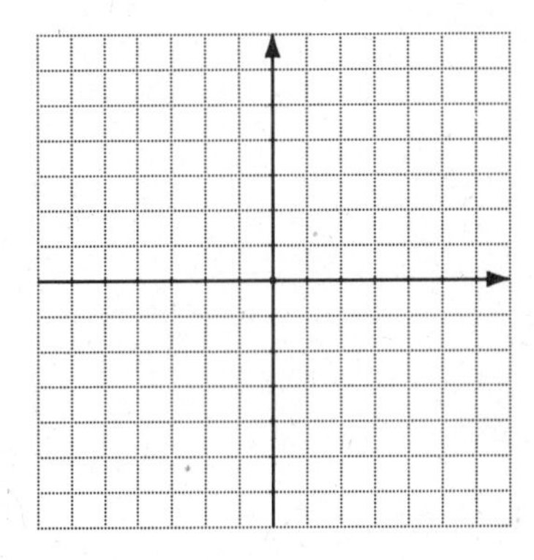

Solve each system by graphing. If the two equations produce parallel lines, write no solution. *If the two equations produce the same line, write* same line. *See Example 3.*

34. $x + 2y = 4$
$2x + 4y = 12$

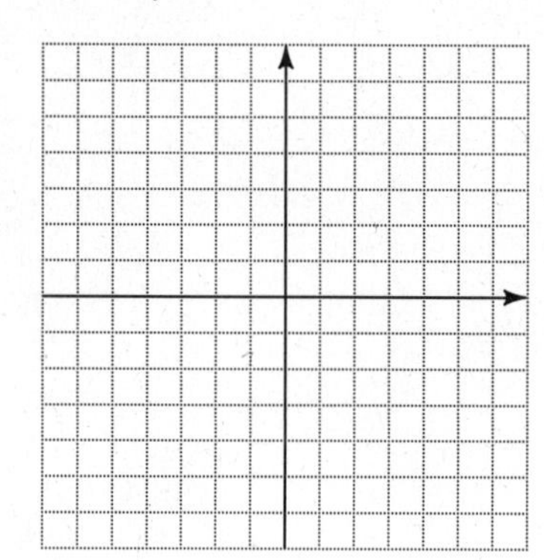

35. $2x - y = 6$
$4x - 2y = 8$

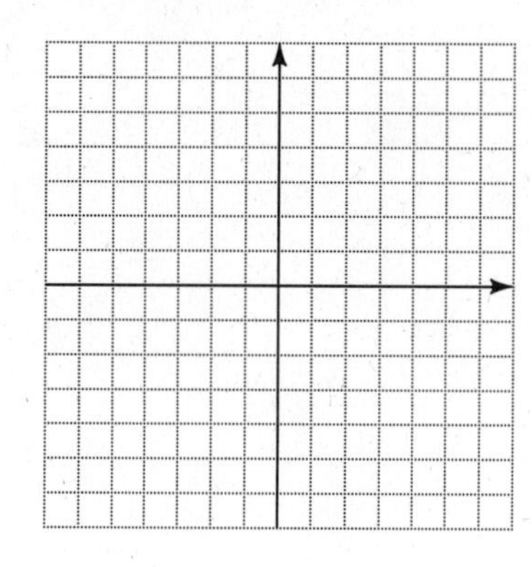

36. $3x = y + 5$
$6x - 2y = 5$

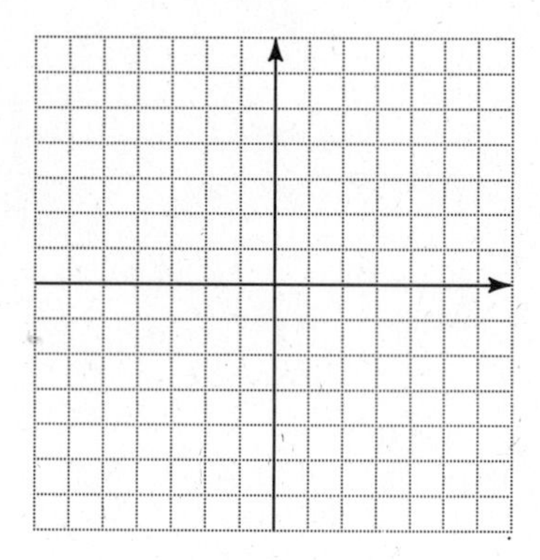

37. $4y + 1 = x$
$2x - 3 = 8y$

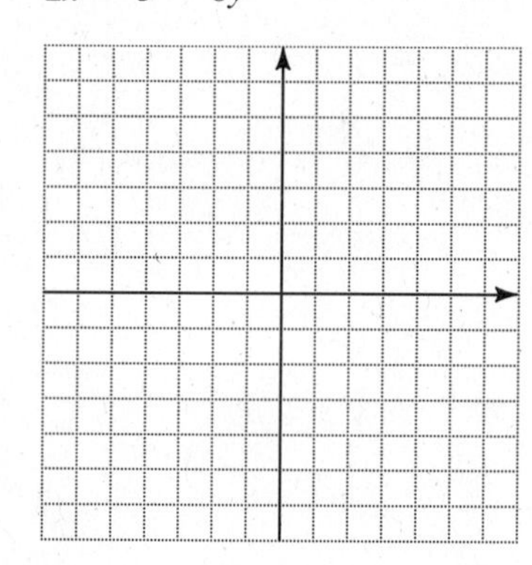

38. $2x - y = 4$
$4x = 2y + 8$

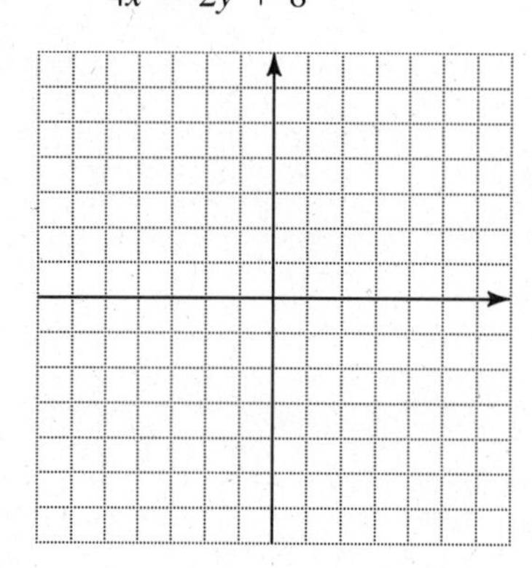

39. $3x = 5 - y$
$6x + 2y = 10$

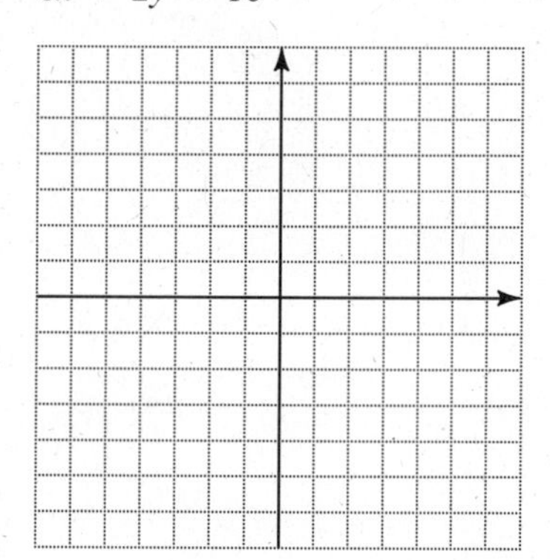

SKILL SHARPENERS

Add. See Section 4.4.

40. $\begin{array}{r} 2m + 3n \\ \underline{5m - n} \end{array}$

41. $\begin{array}{r} 6p + 2q \\ \underline{4p + 5q} \end{array}$

42. $\begin{array}{r} x - 2y \\ \underline{-x + 3y} \end{array}$

43. $\begin{array}{r} 3r - 2t \\ \underline{5r + 2t} \end{array}$

44. $-2x + y$ and $-y + 5x$

45. $4a + 3b$ and $-3b - 2a$

46. $6z + 5w$ and $-5w - 6z$

47. $-3m + 2n$ and $-2n + 3m$

8.2 SOLVING SYSTEMS OF LINEAR EQUATIONS BY ADDITION

Graphing to solve a system of equations has a serious drawback: It is difficult to accurately find a solution such as $(\frac{1}{3}, -\frac{5}{6})$ from a graph.

1 An algebraic method that depends on the addition property of equality can be used to solve systems. As mentioned earlier, adding the same quantity to each side of an equation results in equal sums.

$$\text{If } A = B, \quad \text{then} \quad A + C = B + C.$$

This addition can be taken a step further. Adding *equal* quantities, rather than the *same* quantity, to both sides of an equation also results in equal sums.

$$\text{If } A = B \quad \text{and} \quad C = D, \quad \text{then} \quad A + C = B + D.$$

The use of the addition property to solve systems is called the **addition method** for solving systems of equations. For most systems, this method is more efficient than graphing.

When using the addition method, the idea is to eliminate one of the variables. To do this, one of the variables must have coefficients that are opposites. Keep this in mind throughout the examples in this section.

EXAMPLE 1 Using the Addition Method

Use the addition method to solve the system

$$x + y = 5$$
$$x - y = 3.$$

Each equation in this system is a statement of equality, so, as discussed above, the sum of the right-hand sides equals the sum of the left-hand sides. Adding in this way gives

$$(x + y) + (x - y) = 5 + 3.$$

Combine terms to get

$$2x = 8$$
$$x = 4. \qquad \text{Divide by 2.}$$

The result, $x = 4$, gives the x-value of the solution of the given system. Find the y-value of the solution by substituting 4 for x in either of the two equations in the system.

WORK PROBLEM 1 AT THE SIDE.

The solution found at the side, (4, 1), can be checked by substituting 4 for x and 1 for y into both equations in the given system.

$$x + y = 5 \qquad\qquad x - y = 3$$
$$4 + 1 = 5 \qquad\qquad 4 - 1 = 3$$
$$5 = 5 \quad \text{True} \qquad\qquad 3 = 3 \quad \text{True}$$

Since both results are true, the solution of the given system is (4, 1). The two equations are graphed in Figure 5 on the next page. Notice that the point of intersection is (4, 1), as indicated by the solution of the system using the addition method. ■

OBJECTIVES

1. Solve linear systems by addition.
2. Multiply one or both equations of a system so the addition method can be used.
3. Write equations in the proper form to use the addition method.

1. **(a)** Choose the equation $x + y = 5$ and let $x = 4$ to find the value of y.

(b) Give the solution of the system.

ANSWERS
1. (a) $y = 1$ (b) (4, 1)

2. Solve each system by the addition method. Check each solution.

(a) $x + y = 8$
$x - y = 2$

(b) $3x - y = 7$
$2x + y = 3$

(c) $x - 8y = -8$
$-x + 2y = 2$

3. Solve each system by the addition method. Check each solution.

(a) $2x - y = 2$
$4x + y = 10$

(b) $x + 3y = -2$
$-x - 2y = 0$

(c) $8x - 5y = 32$
$4x + 5y = 4$

ANSWERS
2. (a) (5, 3) (b) (2, −1) (c) (0, 1)
3. (a) (2, 2) (b) (4, −2) (c) $\left(3, -\frac{8}{5}\right)$

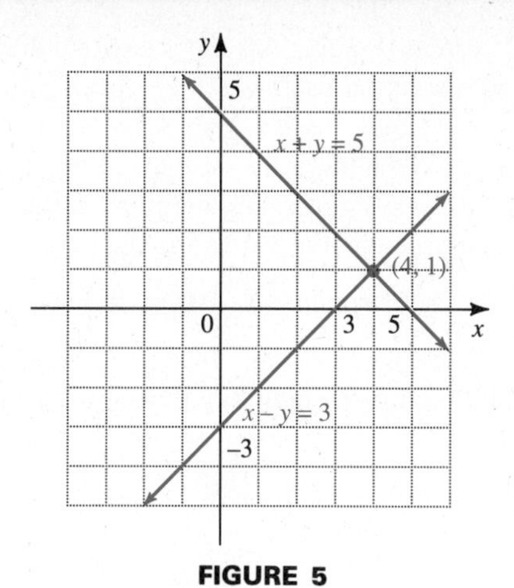

FIGURE 5

WORK PROBLEM 2 AT THE SIDE.

EXAMPLE 2 Using the Addition Method

Solve the system

$$-2x + y = -11$$
$$5x - y = 26.$$

As above, add left-hand sides and add right-hand sides. This may be done most easily by drawing a line under the second equation and adding vertically. (Like terms must be placed in columns.)

$$\begin{array}{rcl} -2x + y &=& -11 \\ 5x - y &=& 26 \\ \hline 3x &=& 15 \end{array} \quad \text{Add in columns.}$$

$$x = 5 \quad \text{Divide by 3.}$$

Substitute 5 for x in either of the original equations. We choose the first.

$$-2x + y = -11$$
$$-2(5) + y = -11 \quad \text{Let } x = 5.$$
$$-10 + y = -11 \quad \text{Multiply.}$$
$$y = -1 \quad \text{Add 10.}$$

The solution is $(5, -1)$. Check this solution by substituting 5 for x and -1 for y in both of the original equations. ■

WORK PROBLEM 3 AT THE SIDE.

2 In both earlier examples, the addition step eliminated a variable. Sometimes one or both equations in a system must be multiplied by some number before the addition step will eliminate a variable.

EXAMPLE 3 Multiplying Before Using the Addition Method

Solve the system

$$x + 3y = 7 \quad \textbf{(1)}$$
$$2x + 5y = 12. \quad \textbf{(2)}$$

Adding the two equations gives $3x + 8y = 19$, which does not help to solve the system. However, if each side of equation (1) is first multiplied by -2 using the multiplication property of equality, the terms with the variable x will drop out after adding.

$$\mathbf{-2}(x + 3y) = \mathbf{-2}(7)$$
$$-2x - 6y = -14 \qquad \textbf{(3)}$$

Now add equations (3) and (2).

$$\begin{aligned} \mathbf{-2x} - 6y &= -14 && \textbf{(3)} \\ \mathbf{2x} + 5y &= 12 && \textbf{(2)} \\ \hline -y &= -2 && \text{Add.} \\ y &= 2 && \text{Multiply by } -1. \end{aligned}$$

Substituting into equation (1) gives

$$\begin{aligned} x + 3\mathbf{y} &= 7 \\ x + 3\mathbf{(2)} &= 7 && \text{Let } y = 2. \\ x + 6 &= 7 && \text{Multiply.} \\ x &= 1. && \text{Subtract 6.} \end{aligned}$$

The solution of this system is (1, 2). Check that this ordered pair satisfies both of the original equations. ■

WORK PROBLEM 4 AT THE SIDE.

EXAMPLE 4 Multiplying Twice Before Using the Addition Method

Solve the system

$$\begin{aligned} 2x + 3y &= -15 && \textbf{(1)} \\ 5x + 2y &= 1. && \textbf{(2)} \end{aligned}$$

Here the multiplication property of equality is used with both equations instead of just one. Multiply by numbers that will cause the coefficients of x (or of y) in the two equations to be additive inverses of each other. For example, multiply each side of equation (1) by 5, and each side of equation (2) by -2.

$$\begin{aligned} \mathbf{10x} + 15y &= -75 && \text{Multiply } \textbf{(1)} \text{ by 5.} \\ \mathbf{-10x} - 4y &= -2 && \text{Multiply } \textbf{(2)} \text{ by } -2. \\ \hline 11y &= -77 && \text{Add.} \\ y &= -7 && \text{Divide by 11.} \end{aligned}$$

Substituting -7 for y in either equation (1) or (2) gives $x = 3$. The solution of the system is $(3, -7)$. Check this solution.

The same result would have been obtained by multiplying each side of equation (1) by 2 and each side of equation (2) by -3. This process would eliminate the y terms so that the value of x would have been found first. ■

WORK PROBLEM 5 AT THE SIDE.

3 Before a system can be solved by the addition method, the two equations of the system must have like terms in the same positions. When this is not the case, the terms should first be rearranged, as the next example shows. This example also shows an alternative way to get the second number when finding the solution of a system.

EXAMPLE 5 Rearranging Terms Before Using the Addition Method

Solve the system

$$\begin{aligned} 4x &= 9 - 3y && \textbf{(1)} \\ 5x - 2y &= 8. && \textbf{(2)} \end{aligned}$$

4. Solve each system by the addition method. Check each solution.

(a) $x - 3y = -7$
$3x + 2y = 23$

(b) $8x + 2y = 2$
$3x - y = 6$

5. Solve each system of equations. Check each solution.

(a) $4x - 5y = -18$
$3x + 2y = -2$

(b) $6x + 7y = 4$
$5x + 8y = -1$

ANSWERS
4. (a) (5, 4) (b) (1, −3)
5. (a) (−2, 2) (b) (3, −2)

6. Solve each system of equations.

(a) $5x = 7 + 2y$
$5y = 5 - 3x$

(b) $3y = 8 + 4x$
$6x = 9 - 2y$

ANSWERS
6. (a) $\left(\frac{45}{31}, \frac{4}{31}\right)$ (b) $\left(\frac{11}{26}, \frac{42}{13}\right)$

Rearrange the terms in equation (1) so that the like terms can be aligned in columns. Add $3y$ to each side to get the following system.

$$4x + 3y = 9 \qquad \textbf{(3)}$$
$$5x - 2y = 8 \qquad \textbf{(2)}$$

Let us eliminate y by multiplying each side of equation (3) by 2 and each side of equation (2) by 3, and then adding.

$$\begin{array}{rcll} 8x + 6y &=& 18 & \text{Multiply by 2.} \\ 15x - 6y &=& 24 & \text{Multiply by 3.} \\ \hline 23x &=& 42 & \text{Add.} \\ x &=& \dfrac{42}{23} & \text{Divide by 23.} \end{array}$$

Substituting $\frac{42}{23}$ for x in one of the given equations would give y, but the arithmetic involved would be messy. Instead, solve for y by starting again with the original equations and eliminating x. Do this by multiplying each side of equation (3) by 5 and each side of equation (2) by -4, and then adding.

$$\begin{array}{rcll} 20x + 15y &=& 45 & \text{Multiply by 5.} \\ -20x + 8y &=& -32 & \text{Multiply by } -4. \\ \hline 23y &=& 13 & \text{Add.} \\ y &=& \dfrac{13}{23} & \text{Divide by 23.} \end{array}$$

The solution is $\left(\frac{42}{23}, \frac{13}{23}\right)$. ■

When the value of the first variable is a fraction, the method used in Example 5 avoids errors that often occur when working with fractions. (Of course, this method could be used in solving any system of equations.)

WORK PROBLEM 6 AT THE SIDE.

The solution of a linear system of equations having exactly one solution can be found by the addition method. A summary of the steps is given below.

SOLVING A LINEAR SYSTEM

Step 1 Write both equations of the system in the form $Ax + By = C$.

Step 2 Multiply one or both equations by appropriate numbers so that the coefficients of x (or y) are negatives of each other.

Step 3 Add the two equations to get an equation with only one variable.

Step 4 Solve the equation from Step 3.

Step 5 Substitute the solution from Step 4 into either of the original equations.

Step 6 Solve the resulting equation from Step 5 for the remaining variable.

Step 7 Check the answer.

name date hour

8.2 EXERCISES

Solve each system by the addition method. Check your answers. See Examples 1 and 2.

1. $x - y = 3$
$x + y = -1$

2. $x + y = 7$
$x - y = -3$

3. $x + y = 2$
$2x - y = 4$

4. $3x - y = 8$
$x + y = 4$

5. $2x + y = 14$
$x - y = 4$

6. $2x + y = 2$
$-x - y = 1$

7. $3x + 2y = 6$
$-3x - y = 0$

8. $5x - y = 9$
$-5x + 2y = -8$

9. $6x - y = 1$
$-6x + 5y = 7$

10. $6x + y = -2$
$-6x + 3y = -14$

11. $2x - y = 5$
$4x + y = 4$

12. $x - 4y = 13$
$-x + 6y = -18$

Solve each system by the addition method. Check your answers. See Example 3.

13. $2x - y = 7$
$3x + 2y = 0$

14. $x + y = 7$
$-3x + 2y = -11$

15. $x + 3y = 16$
$2x - y = 4$

16. $4x - 3y = 8$
$2x + y = 14$

17. $x + 4y = -18$
$3x + 5y = -19$

18. $2x + y = 3$
$5x - 2y = -15$

19. $5x - 3y = 15$
$-3x + 6y = -9$

20. $4x + 3y = -9$
$5x - 6y = 18$

21. $3x - 2y = -6$
$-5x + 4y = 16$

22. $-4x + 3y = 0$
$5x - 6y = 9$

23. $2x - y = -8$
$5x + 2y = -20$

24. $5x + 3y = -9$
$7x + y = -3$

25. $2x + y = 5$
$5x + 3y = 11$

26. $2x + 7y = -53$
$4x + 3y = -7$

name date hour

Solve each system by the addition method. Check your answers. See Examples 4 and 5.

27. $5x - 4y = -1$
$-7x + 5y = 8$

28. $3x + 2y = 12$
$5x - 3y = 1$

29. $3x + 5y = 33$
$4x - 3y = 15$

30. $2x + 5y = 3$
$5x - 3y = 23$

31. $3x + 5y = -7$
$5x + 4y = 10$

32. $2x + 3y = -11$
$5x + 2y = 22$

33. $2x + 3y = -12$
$5x - 7y = -30$

34. $2x + 9y = 16$
$5x - 6y = 40$

35. $4x - 3y = -7$
$6x + 5y = 18$

36. $8x + 3y = -4$
$12x + 7y = 4$

37. $24x + 12y = 19$
$16x - 18y = -9$

38. $9x + 4y = -4$
$6x + 6y = -11$

39. $3x = 3 + 2y$
$-\frac{4}{3}x + y = \frac{1}{3}$

40. $3x = 27 + 2y$
$x - \frac{7}{2}y = -25$

41. $-7y = 6 - 5x$
$3x - 6y = 2$

42. $4y = 2 - 3x$
$4x + 3y = 6$

43. $6.5x + 2.3y = 6.1$
$5.4x - 4.6y = 24.6$

44. $-2.2x + 7.1y = -1.7$
$3.8x + 14.2y = 29.4$

SKILL SHARPENERS

Solve each equation. See Section 3.1.

45. $x + 7 = x + 8$

46. $4x - 9 = 4x - 3$

47. $-2x + 4 = 4 - 2x$

48. $6y + 13 = 13 + 6y$

49. $5x + x + 1 = 2x + 4x + 1$

50. $3x + 3x - 4 = 6x - 3$

8.3 TWO SPECIAL CASES

OBJECTIVES

1. Solve linear systems having parallel lines as their graphs.
2. Solve linear systems having the same line as their graphs.

1 In Section 8.1 some of the systems had equations with graphs that were parallel lines (from an inconsistent system), while the equations of other systems had graphs that were the same line (dependent equations). This section shows how to solve these systems with the addition method discussed in the previous section. The first example shows the solution of an inconsistent system, where the graphs of the equations are parallel lines.

EXAMPLE 1 Using Addition to Solve an Inconsistent System
Solve by the addition method.

$$2x + 4y = 5$$
$$4x + 8y = -9$$

Multiply each side of $2x + 4y = 5$ by -2 and then add.

$$\begin{aligned} -4x - 8y &= -10 \\ 4x + 8y &= -9 \\ \hline 0 &= -19 \end{aligned} \quad \text{False}$$

The false statement $0 = -19$ shows that the given system is self-contradictory. *It has no solution.* This means that the graphs of the equations of this system are parallel lines, as shown in Figure 6. Since this system has no solution, it is inconsistent. ■

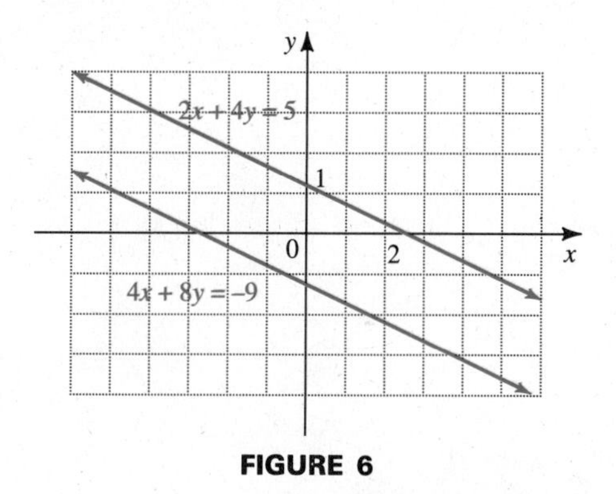

FIGURE 6

WORK PROBLEM 1 AT THE SIDE.

1. Solve each system by the addition method.

(a) $4x + 3y = 10$
$2x + \frac{3}{2}y = 12$

(b) $-2x - 4y = -1$
$5x + 10y = 15$

2 The next example shows the result of using the addition method when the equations of the system are dependent, with the graphs of the equations in the system the same line.

EXAMPLE 2 Using Addition to Solve a Dependent System
Solve by the addition method.

$$3x - y = 4$$
$$-9x + 3y = -12$$

Multiply each side of the first equation by 3 and then add the two equations to get

$$\begin{aligned} 9x - 3y &= 12 \\ -9x + 3y &= -12 \\ \hline 0 &= 0. \end{aligned} \quad \text{True}$$

This means that every solution of one equation is also a solution of the other, so the system has an infinite number of solutions: all the ordered pairs

ANSWERS
1. (a) no solution (b) no solution

2. Solve each system by the addition method.

(a) $6x + 3y = 9$
$-8x - 4y = -12$

(b) $4x - 6y = 10$
$-10x + 15y = -25$

ANSWERS
2. (a) infinite number of solutions, same line
(b) infinite number of solutions, same line

corresponding to points that lie on the common graph. As mentioned in Section 8.1, the equations in this system are dependent. In the answers at the back of this book, a solution of such a system of dependent equations is indicated by the words *same line*. A graph of the equations of this system is shown in Figure 7. ■

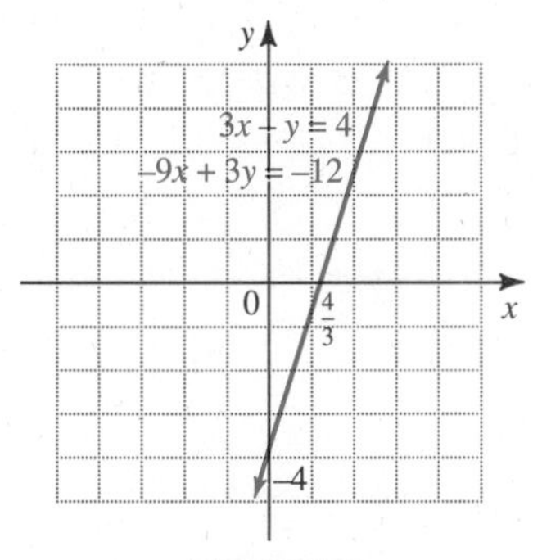

FIGURE 7

WORK PROBLEM 2 AT THE SIDE.

One of three situations may occur when the addition method is used to solve a linear system of equations.

1. The result of the addition step is a statement such as $x = 2$ or $y = -3$. The solution will be exactly one ordered pair. The graphs of the equations of the system will intersect at exactly one point.

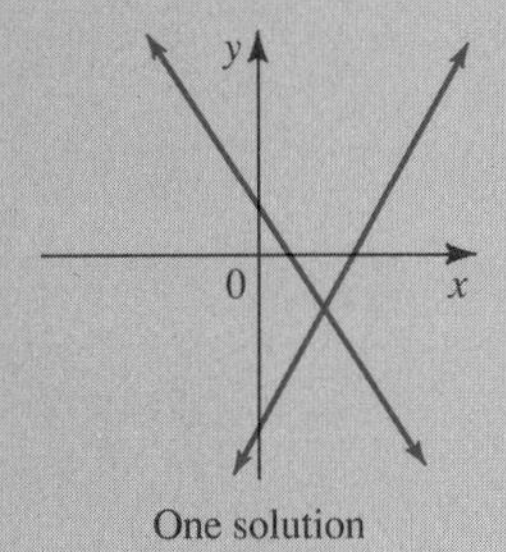

One solution

2. The result of the addition step is a false statement, such as $0 = 4$. In this case, the graphs are parallel lines and there is no solution for the system.

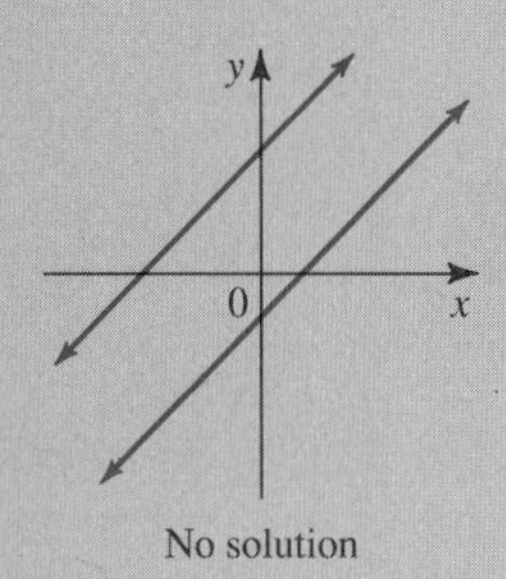

No solution

3. The result of the addition step is a true statement, such as $0 = 0$. The graphs of the equations of the system are the same line, and an infinite number of ordered pairs are solutions.

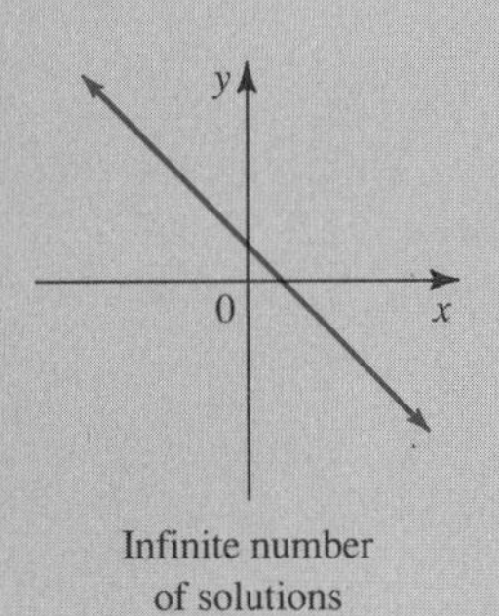

Infinite number of solutions

name date hour

8.3 EXERCISES

Use the addition method to solve each system. See Examples 1 and 2.

1. $x + y = 4$
$x + y = -2$

2. $2x - y = 1$
$2x - y = 4$

3. $5x - 2y = 6$
$10x - 4y = 10$

4. $3x - 5y = 2$
$6x - 10y = 8$

5. $x + 3y = 5$
$2x + 6y = 10$

6. $6x - 2y = 12$
$-3x + y = -6$

7. $2x + 3y = 8$
$4x + 6y = 12$

8. $4x + y = 6$
$-8x - 2y = 21$

9. $5x = y + 4$
$5x = y - 4$

10. $4y = 3x - 2$
$4y = 3x + 5$

11. $6x + 3y = 0$
$-12x - 6y = 0$

12. $3x - 5y = 0$
$6x - 10y = 0$

13. $2x - 3y = 0$
$4x + 5y = 0$

14. $3x - 5y = 0$
$6x + 10y = 0$

15. $3x + 5y = 19$
$4x - 3y = 6$

16. $2x + 5y = 17$
$4x + 3y = -1$

17. $4x - 2y = 1$
$8x - 4y = 1$

18. $-2x + 3y = 5$
$4x - 6y = 5$

19. $3x - 2y = 8$
$-3x + 2y = -8$

20. $4x + y = 4$
$-8x - 2y = -8$

SKILL SHARPENERS

Solve each equation and check the solutions. See Sections 3.3 and 6.6.

21. $-2(y - 2) + 5y = 10$

22. $2m - 3(4 - m) = 8$

23. $p + 4(6 - 2p) = 3$

24. $4\left(\frac{3 - 2k}{2}\right) + 3k = -3$

25. $4x - 2\left(\frac{1 - 3x}{2}\right) = 6$

26. $a + 3\left(\frac{1 - a}{2}\right) = 5$

27. $4\left(\frac{-2 + y}{3}\right) - 2y = 2$

28. $3p + 2\left(\frac{2p - 1}{3}\right) = 4$

29. $4r + 5\left(\frac{2r + 7}{4}\right) = 1$

8.4 SOLVING SYSTEMS OF LINEAR EQUATIONS BY SUBSTITUTION

OBJECTIVES

1. Solve linear systems by substitution.
2. Solve linear systems with fractions as coefficients.

1 As discussed in the preceding sections, the graphing method and the addition method can be used for solving systems of linear equations. A third method, the **substitution method,** is particularly useful for solving systems where one equation can be solved quickly for one of the variables.

EXAMPLE 1 Using the Substitution Method

Solve the system

$$3x + 5y = 26$$
$$y = 2x.$$

The second of these two equations is already solved for y: this equation says that $y = 2x$. Substituting $2x$ for y in the first equation gives

$3x + 5y = 26$	
$3x + 5(2x) = 26$	Let $y = 2x$.
$3x + 10x = 26$	Multiply.
$13x = 26$	Combine terms.
$x = 2.$	Divide by 13.

Since $x = 2$, find y from the equation $y = 2x$ by substituting 2 for x.

$$y = 2(2) = 4 \qquad \text{Let } x = 2.$$

Check that the solution of the given system is (2, 4). ■

WORK PROBLEM 1 AT THE SIDE.

1. Solve by the substitution method.

(a) $3x + 5y = 69$
$y = 4x$

(b) $-x + 4y = 26$
$y = -3x$

(c) $2x + 7y = -12$
$x = -2y$

EXAMPLE 2 Using the Substitution Method

Use substitution to solve the system

$$2x + 5y = 7$$
$$x = -1 - y.$$

The second equation gives x in terms of y. Substitute $-1 - y$ for x in the first equation.

$2x + 5y = 7$	
$2(-1 - y) + 5y = 7$	Let $x = -1 - y$.
$-2 - 2y + 5y = 7$	Distributive property
$-2 + 3y = 7$	Combine terms.
$3y = 9$	Add 2.
$y = 3$	Divide by 3.

To find x, substitute $y = 3$ in the equation $x = -1 - y$ to get $x = -1 - 3 = -4$. Check that the solution of the given system is $(-4, 3)$. ■

EXAMPLE 3 Using the Substitution Method

Use substitution to solve the system

$$x = 5 - 2y$$
$$2x + 4y = 6.$$

Substitute $5 - 2y$ for x in the second equation.

ANSWERS
1. (a) (3, 12) (b) (−2, 6)
(c) (8, −4)

$$2x + 4y = 6$$
$$2(5 - 2y) + 4y = 6 \quad \text{Let } x = 5 - 2y.$$
$$10 - 4y + 4y = 6 \quad \text{Distributive property}$$
$$10 = 6 \quad \text{False}$$

As shown in the last section, this false result means that the equations in the system have graphs that are parallel lines. The system is inconsistent and has no solution. ■

WORK PROBLEM 2 AT THE SIDE.

EXAMPLE 4 Using the Substitution Method

Use substitution to solve the system

$$2x + 3y = 8$$
$$-4x - 2y = 0.$$

To use the substitution method, one of the equations must be solved for one of the variables. Let us choose the first equation of the system, $2x + 3y = 8$, and solve for x. (This procedure was first explained in Section 3.5.)

To get x alone on one side, subtract $3y$ from each side.

$$2x + 3y = 8$$
$$2x = 8 - 3y \quad \text{Subtract } 3y.$$

Now divide each side by 2.

$$x = \frac{8 - 3y}{2} \quad \text{Divide by 2.}$$
$$x = 4 - \frac{3}{2}y \quad \text{Write right side as two terms.}$$

Now substitute this value for x in the second equation of the system.

$$-4x - 2y = 0$$
$$-4\left(4 - \frac{3}{2}y\right) - 2y = 0 \quad \text{Let } x = 4 - \frac{3}{2}y.$$
$$-16 + 6y - 2y = 0 \quad \text{Distributive property}$$
$$-16 + 4y = 0 \quad \text{Combine terms.}$$
$$4y = 16 \quad \text{Add 16.}$$
$$y = 4 \quad \text{Divide by 4.}$$

Find x by letting $y = 4$ in $x = 4 - \frac{3}{2}y$.

$$x = 4 - \frac{3}{2}(4) \quad \text{Let } y = 4.$$
$$x = 4 - 6 \quad \text{Multiply.}$$
$$x = -2 \quad \text{Subtract.}$$

The solution of the given system is $(-2, 4)$. Check this solution in both equations. ■

WORK PROBLEM 3 AT THE SIDE.

2. Solve each system by substitution. Check each solution.

(a) $3x - 4y = -11$
$x = y - 2$

(b) $5x + 2y = -2$
$y = 1 - 2x$

(c) $8x - y = 4$
$y = 8x + 4$

3. Solve each system by substitution. Check each solution.

(a) $x + 4y = -1$
$2x - 5y = 11$

(b) $2x + 5y = 4$
$x + y = -1$

ANSWERS

2. (a) $(3, 5)$ (b) $(-4, 9)$ (c) no solution
3. (a) $(3, -1)$ (b) $(-3, 2)$

EXAMPLE 5 Using the Substitution Method

Use substitution to solve the system

$$2x = 4 - y \qquad (1)$$
$$6 + 3y + 4x = 16 - x. \qquad (2)$$

Start by simplifying the second equation by adding x and subtracting 6 on each side. This gives the simplified system

$$2x = 4 - y \qquad (1)$$
$$5x + 3y = 10. \qquad (3)$$

For the substitution method, one of the equations must be solved for either x or y. Since the coefficient of y in equation (1) is -1, avoid fractions by solving this equation for y.

$$2x = 4 - y \qquad (1)$$
$$2x - 4 = -y \qquad \text{Subtract 4.}$$
$$-2x + 4 = y \qquad \text{Multiply by } -1.$$

Now substitute $-2x + 4$ for y in equation (3).

$$5x + 3\mathbf{y} = 10$$
$$5x + 3(\mathbf{-2x + 4}) = 10 \qquad \text{Let } y = -2x + 4.$$
$$5x - 6x + 12 = 10 \qquad \text{Distributive property}$$
$$-x + 12 = 10 \qquad \text{Combine terms.}$$
$$-x = -2 \qquad \text{Subtract 12.}$$
$$x = 2 \qquad \text{Multiply by } -1.$$

Since $y = -2x + 4$ and $x = 2$,

$$y = -2(2) + 4 = 0,$$

and the solution is (2, 0). Check this solution by substitution in both equations of the given system. ■

WORK PROBLEM 4 AT THE SIDE.

2 When a system includes equations with fractions as coefficients, eliminate the fractions by multiplying each side by a common denominator. Then solve the resulting system.

EXAMPLE 6 Using the Substitution Method with Fractions as Coefficients

Solve the following system by the substitution method.

$$3x + \frac{1}{4}y = 2 \qquad (1)$$
$$\frac{1}{2}x + \frac{3}{4}y = -\frac{5}{2} \qquad (2)$$

4. Solve each system by substitution. First simplify where necessary.

(a) $x = 5 - 3y$
$2x + 3 = 5x - 4y + 14$

(b) $4x + 2x - y + 1 = 30$
$y = -4 + x$

(c) $5x - y = -14 + 2x + y$
$7x + 9y + 4 = 3x + 8y$

ANSWERS
4. (a) $(-1, 2)$ (b) $(5, 1)$ (c) $(-2, 4)$

5. Solve each system by any method. First clear all fractions.

(a) $\frac{2}{3}x + \frac{1}{2}y = 6$

$\frac{1}{2}x - \frac{3}{4}y = 0$

(b) $\frac{3}{5}x + \frac{1}{2}y = 7$

$\frac{7}{10}x - \frac{1}{5}y = 16$

ANSWERS
5. (a) (6, 4) (b) (20, −10)

Begin by eliminating fractions. Clear equation (1) of fractions by multiplying each side by 4.

$$4\left(3x + \frac{1}{4}y\right) = 4(2) \qquad \text{Multiply by 4.}$$

$$4(3x) + 4\left(\frac{1}{4}y\right) = 4(2) \qquad \text{Distributive property}$$

$$12x + y = 8 \qquad \textbf{(3)}$$

Now clear equation (2) of fractions by multiplying each side by the common denominator 4.

$$4\left(\frac{1}{2}x + \frac{3}{4}y\right) = 4\left(-\frac{5}{2}\right) \qquad \text{Multiply by 4.}$$

$$4\left(\frac{1}{2}x\right) + 4\left(\frac{3}{4}y\right) = 4\left(-\frac{5}{2}\right) \qquad \text{Distributive property}$$

$$2x + 3y = -10 \qquad \textbf{(4)}$$

The given system of equations has been simplified as follows.

$$12x + y = 8 \qquad \textbf{(3)}$$

$$2x + 3y = -10 \qquad \textbf{(4)}$$

Let us solve this system by the substitution method. Equation (3) can be solved for y by subtracting $12x$ from each side.

$$12x + y = 8$$

$$y = -12x + 8 \qquad \text{Subtract } 12x.$$

Now substitute the result for y in equation (4).

$$2x + 3(-12x + 8) = -10 \qquad \text{Let } y = -12x + 8.$$

$$2x - 36x + 24 = -10 \qquad \text{Distributive property}$$

$$-34x = -34 \qquad \text{Combine terms; subtract 24.}$$

$$x = 1 \qquad \text{Divide by } -34.$$

Using $y = -12x + 8$ and $x = 1$ gives $y = -12(1) + 8 = -4$. The solution is $(1, -4)$. Check by substituting 1 for x and -4 for y in both of the original equations. Verify that the same solution is obtained if the addition method is used to solve the system of equations (3) and (4). ■

WORK PROBLEM 5 AT THE SIDE.

name date hour

8.4 EXERCISES

Solve each system by the substitution method. Check each solution. See Examples 1–4.

1. $x + y = 6$
$y = 2x$

2. $x + 3y = -11$
$y = -4x$

3. $3x + 2y = 26$
$x = y + 2$

4. $4x + 3y = -14$
$x = y - 7$

5. $x + 5y = 3$
$x - 2y = 10$

6. $5x + 2y = 14$
$2x - y = 11$

7. $3x - 2y = 14$
$2x + y = 0$

8. $2x - 5 = -y$
$x + 3y = 0$

9. $x + y = 6$
$x - y = 4$

10. $3x - 2y = 13$
$x + y = 6$

11. $3x - y = 6$
$y = 3x - 5$

12. $4x - y = 4$
$y = 4x + 3$

13. $2x + 3y = 11$
$2x - y = 7$

14. $3x + 4y = -10$
$2x + y = -4$

15. $4x + y = 5$
$5x + 3y = 1$

16. $5x + 2y = -19$
$4x + y = -17$

17. $6x - 8y = 4$
$3x = 4y + 2$

18. $12x + 18y = 12$
$2x = 2 - 3y$

19. $4x + 5y = 5$
$2x + 3y = 1$

20. $6x + 5y = 13$
$3x + 2y = 4$

Solve each system by either the addition method or the substitution method. First simplify equations where necessary. Check each solution. See Example 5.

21. $x + 4y = 34$
$y = 4x$

22. $3x - y = -14$
$x = -2y$

23. $4 + 4x - 3y = 34 + x$
$4x = -y - 2 + 3x$

24. $5x - 4y = 42 - 8y - 2$
$2x + y = x + 1$

25. $4x - 2y + 8 = 3x + 4y - 1$
$3x + y = x + 8$

26. $5x - 4y - 8x - 2 = 6x + 3y - 3$
$4x - y = -2y - 8$

27. $2x - 8y + 3y + 2 = 5y + 16$
$8x - 2y = 4x + 28$

28. $7x - 9 + 2y - 8 = -3y + 4x + 13$
$4y - 8x = -8 + 9x + 32$

29. $-2x + 3y = 12 + 2y$
$2x - 5y + 4 = -8 - 4y$

30. $2x + 5y = 7 + 4y - x$
$5x + 3y + 8 = 22 - x + y$

31. $y + 9 = 3x - 2y + 6$
$5 - 3x + 24 = -2x + 4y + 3$

32. $5x - 2y = 16 + 4x - 10$
$4x + 3y = 60 + 2x + y$

Solve each system by the addition method or by the substitution method. First clear all fractions. Check each solution. See Example 6.

33. $x + \frac{1}{3}y = y - 2$
$\frac{1}{4}x - y = x - y$

34. $\frac{5}{3}x + 2y = \frac{1}{3} + y$
$2x - 3 + \frac{y}{3} = -2 + x$

35. $\frac{x}{6} + \frac{y}{6} = 1$
$-\frac{1}{2}x - \frac{1}{3}y = -5$

36. $\frac{x}{2} - \frac{y}{3} = \frac{5}{6}$
$\frac{x}{5} - \frac{y}{4} = \frac{1}{10}$

37. $\frac{x}{3} - \frac{3y}{4} = -\frac{1}{2}$
$\frac{2x}{3} + \frac{y}{2} = 3$

38. $\frac{x}{5} + 2y = \frac{8}{5}$
$\frac{3x}{5} + \frac{y}{2} = -\frac{7}{10}$

SKILL SHARPENERS

Solve each word problem. See Section 3.4.

39. If three times a number is added to 6, the result is 69. Find the number.

40. The product of 5 and 1 more than a number is 35. Find the number.

41. The perimeter of a rectangle is 46 feet. The width is 7 feet less than the length. Find the width.

42. The area of a rectangle is numerically 20 more than the width, and the length is 6 centimeters. What is the width?

8.5 APPLICATIONS OF LINEAR SYSTEMS

Many practical problems are more easily translated into equations if two variables are used. With two variables, a system of two equations is needed to find the desired solution. The examples in this section illustrate the method of solving word problems using two equations and two variables.

Recall from Chapter 3 the steps used in solving word problems. The steps presented there can be modified as follows to allow for two variables and two equations.

OBJECTIVES

1. Use linear systems to solve word problems about two numbers.
2. Use linear systems to solve word problems about money.
3. Use linear systems to solve word problems about mixtures.
4. Use linear systems to solve word problems about rate or speed using the distance formula.

SOLVING A WORD PROBLEM WITH TWO VARIABLES

Step 1 Choose a variable to represent each of the unknown values that must be found. Write down what each variable is to represent.

Step 2 Translate the problem into two equations using both variables.

Step 3 Solve the system of two equations.

Step 4 Answer the question or questions asked in the problem.

Step 5 Check the solution in the words of the original problem.

1 The first example shows how to use two variables to solve a problem about two unknown numbers.

EXAMPLE 1 Solving a Problem About Two Numbers

The sum of two numbers is 63. Their difference is 19. Find the two numbers.

Step 1 Let x = one number; y = the other number.

Step 2 Set up a system of equations from the information in the problem.

$$x + y = 63 \quad \text{The sum is 63.}$$
$$x - y = 19 \quad \text{The difference is 19.}$$

Step 3 Solve the system from Step 2.

WORK PROBLEM 1 AT THE SIDE.

Step 4 The numbers required in the problem are 41 and 22.

Step 5 Check: The sum of 41 and 22 is 63, and their difference is 19. The solution satisfies the conditions of the problem. ■

WORK PROBLEM 2 AT THE SIDE.

2 The next example illustrates how to set up and solve a common type of word problem that involves two quantities and their costs.

EXAMPLE 2 Solving a Problem About Quantities and Costs

Admission prices at a football game were $6 for adults and $2 for children. The total value of the tickets sold was $2528, and 454 tickets were sold. How many adults and how many children attended the game?

Step 1 Let a = the number of adults' tickets sold; c = the number of children's tickets sold.

1. Solve the system of equations

$$x + y = 63$$
$$x - y = 19.$$

2. (a) The sum of two numbers is 97. Their difference is 41. What are the numbers?

(b) The sum of two numbers is 38. If twice the first is added to three times the second, the result is 99. Find the numbers.

ANSWERS
1. $x = 41$, $y = 22$
2. (a) 69, 28 (b) 15, 23

Step 2 The information given in the problem is summarized in the table. The entries in the "total value" column were found by multiplying the number of tickets sold by the price per ticket.

Kind of ticket	*Number sold*	*Cost of each (in dollars)*	*Total value (in dollars)*
Adult	a	6	$6a$
Child	c	2	$2c$
Total	454	—	2528

The total number of tickets sold was 454, so

$$a + c = 454.$$

Since the total value was $2528, the right-hand column leads to

$$6a + 2c = 2528.$$

Step 3 These two equations give the following system.

$$a + c = 454 \qquad \textbf{(1)}$$
$$6a + 2c = 2528 \qquad \textbf{(2)}$$

Solve the system of equations with the addition method. First, multiply both sides of equation (1) by -2 to get

$$-2a - 2c = -908.$$

Then add this result to equation (2).

$$-2a - 2c = -908 \qquad \text{Multiply (1) by } -2.$$
$$6a + 2c = 2528$$
$$4a = 1620 \qquad \text{Add.}$$
$$a = 405 \qquad \text{Divide by 4.}$$

Substitute 405 for a in equation (1) to get

$$a + c = 454 \qquad \textbf{(1)}$$
$$\mathbf{405} + c = 454 \qquad \text{Let } a = 405.$$
$$c = 49. \qquad \text{Subtract 405.}$$

Step 4 There were 405 adults and 49 children at the game.

Step 5 Since 405 adults paid $6 each and 49 children paid $2 each, the value of tickets sold should be $405(6) + 49(2) = 2528$, or $2528. This result agrees with the given information. ■

WORK PROBLEM 3 AT THE SIDE.

In the rest of the examples try to identify each step in the solution of the problems.

3 Mixture problems occur in many fields of application of mathematics. One important type of mixture problem occurs in the study of chemistry. This kind of problem can be solved as shown in the next example.

3. The value of the tickets sold for a concert was $1850. The price for a regular ticket was $5.00, and student tickets were $3.50. A total of 400 tickets were sold.

(a) Complete this table.

Kind of ticket	*Number sold*	*Cost of each*	*Total value*
Regular	r		
Student	s		
Totals			

(b) Write a system of equations.

(c) Solve the system.

ANSWERS

3. (a)

Kind of ticket	Number sold	Cost of each	Total value
Regular	r	5	$5r$
Student	s	3.50	$3.50s$
Totals	400	–	1850

(b) $r + s = 400$
$5r + 3.50s = 1850$

(c) 300 regular, 100 student

EXAMPLE 3 Solving a Problem About Mixture (Involving Percent)

A pharmacist needs 100 liters of 50% alcohol solution. She has on hand 30% alcohol solution and 80% alcohol solution, which she can mix. How many liters of each will be required to make the 100 liters of 50% alcohol solution?

A 30% solution means that 30% of the solution is alcohol and the rest is water. The information given above was used for the sketch in Figure 8.

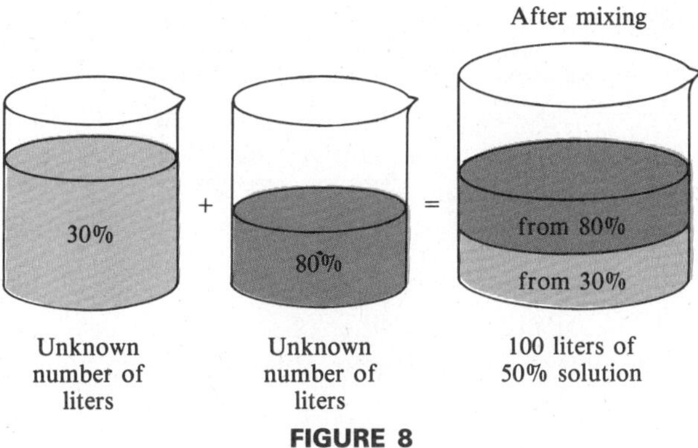

FIGURE 8

Let x represent the number of liters of 30% alcohol needed. Let y represent the number of liters of 80% alcohol needed.

The information given in the problem is summarized in the table below. (Each percent has been changed to its decimal form in the third column.)

Percent	*Liters of solution*	*Liters of pure alcohol*
30	x	$.30x$
80	y	$.80y$
50	100	$.50(100)$

The pharmacist will have $.30x$ liters of alcohol from the x liters of 30% solution and $.80y$ liters of alcohol from the y liters of 80% solution, for a total of $.30x + .80y$ liters of pure alcohol. In the mixture, she wants 100 liters of 50% solution. This 100 liters would contain $.50(100) = 50$ liters of pure alcohol. Since the amounts of pure alcohol must be equal,

$$.30x + .80y = 50.$$

The total number of liters is 100, or

$$x + y = 100.$$

These two equations give the system

$$\begin{aligned} .30x + .80y &= 50 \\ x + \quad y &= 100. \end{aligned}$$

4. Solve the system

$$30x + 80y = 5000$$
$$x + \quad y = 100.$$

5. How many liters of 25% alcohol solution must be mixed with 12% solution to get 13 liters of 15% solution?

(a) Complete the table.

Percent	*Liters*	*Liters of pure alcohol*
25	x	$.25x$
12	y	
15	13	

(b) Write a system of equations and solve it.

(c) Joe needs 100 cc (cubic centimeters) of 20% acid solution for a chemistry experiment. The lab has on hand only 10% and 25% solutions. How much of each should he mix to get the desired amount of 20% solution?

ANSWERS
4. $x = 60$, $y = 40$
5. (a)

Percent	*Liters*	*Liters of pure alcohol*
25	x	$.25x$
12	y	$.12y$
15	13	$.15(13)$

(b) $x + \quad y = 13$
$.25x + .12y = .15(13)$
3 liters of 25%, 10 liters of 12%

(c) $33\frac{1}{3}$ cc of 10%, $66\frac{2}{3}$ cc of 25%

The first of these equations involves decimal fractions. Recall from Example 6 in Section 8.4 that we may clear an equation of fractions by multiplying each side by the least common denominator. When the coefficients are decimals, clear the decimals by multiplying by the power of 10 that is the least common denominator. In the equation $.30x + .80y = 50$, start by multiplying by 100. The system then becomes

$$30x + 80y = 5000$$
$$x + \quad y = 100.$$

WORK PROBLEM 4 AT THE SIDE.

From Problem 4 at the side, $x = 60$ and $y = 40$. The pharmacist should use 60 liters of the 30% solution and 40 liters of the 80% solution. Since $.30(60) + (.80)40 = 50$, this mix will give 100 liters of 50% solution, as required in the original problem. ■

WORK PROBLEM 5 AT THE SIDE.

4 Word problems that use the distance formula, relating distance, rate, and time, often result in a system of two linear equations.

EXAMPLE 4 Solving a Problem About Distance, Rate, and Time
Two cars start from positions 400 miles apart and travel toward each other. They meet after 4 hours. (See Figure 9.) Find the average speed of each car if one travels 20 miles per hour faster than the other.

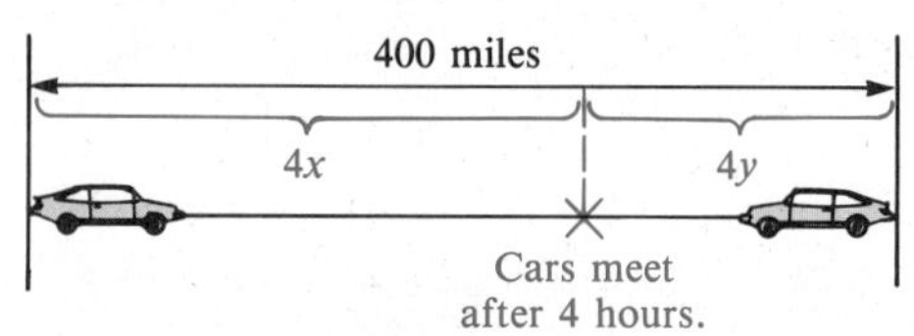

FIGURE 9

Use the formula that relates distance, rate, and time, $d = rt$. Let x be the speed of the faster car and y the speed of the slower car. Since each car travels for 4 hours, the time, t, for each car is 4. This information is shown in the chart. The distance is found by using the formula $d = rt$ and the expressions already entered in the chart.

	r	t	d
Faster car	x	4	$4x$
Slower car	y	4	$4y$

Find d from $d = rt$.

Since the total distance traveled by both cars is 400 miles, $4x + 4y = 400$. The faster car goes 20 miles per hour faster than the slower, giving $x = 20 + y$. This system of equations,

$$4x + 4y = 400$$
$$x = 20 + y,$$

can be solved by substitution.

WORK PROBLEM 6 AT THE SIDE.

Problem 6 at the side gives the solution of the system: $x = 60$, $y = 40$. Thus, the speeds of the two cars were 40 miles per hour and 60 miles per hour. If each car travels for 4 hours, the total distance is

$$4(40) + 4(60) = 160 + 240 = 400$$

miles, as required. ■

WORK PROBLEM 7 AT THE SIDE.

The problems in this section also could be solved using only one variable, but for most of them the solution is simpler with two variables.

CAUTION Be careful: don't forget that two variables require two equations.

6. Use substitution to solve the system

$$4x + 4y = 400$$
$$x = 20 + y.$$

7. (a) In one hour Ann can row 2 miles against the current or 10 miles with the current. Find the speed of the current and Ann's speed in still water.

(b) Two cars that were 450 miles apart traveled toward each other. They met after 5 hours. If one car traveled twice as fast as the other, what were their speeds? Complete this table.

	r	t	d
Faster car	x	5	
Slower car	y	5	

Write a system and solve it.

ANSWERS

6. $x = 60$, $y = 40$

7. (a) 4 miles per hour, 6 miles per hour

(b)

	r	t	d
Faster car	x	5	$5x$
Slower car	y	5	$5y$

$$5x + 5y = 450$$
$$x = 2y$$

$x = 60$ miles per hour, $y = 30$ miles per hour

Historical Reflections

Historical Reflections

René Descartes (1596–1650)

The rectangular coordinate system that we use to plot points is credited to René Descartes, a French mathematician of the seventeenth century. It has been said that he developed the coordinate system while lying in bed, watching an insect move across the bedroom ceiling. He realized that he could locate the position of the insect at any given time by finding its distance from each of two perpendicular walls.

Descartes' attempts to unify algebra and geometry influenced the creation of what is known as analytic geometry. His famous *La geometrie* is an appendix to a philosophical work whose title translates as "A Discourse on the Method of Rightly Conducting the Reason and Seeking the Truth in Sciences."

In 1649 he went to Sweden to tutor Queen Christina. She preferred working in the unheated castle during the early morning hours, while Descartes was accustomed to staying in bed until noon. The rigors of the Swedish winter proved too much for him, and he died less than a year later.

Art: Groninger Museum

name date hour

8.5 EXERCISES

Write a system of equations for each problem and then solve the system. See Example 1.

1. Find two numbers whose sum is 69 and whose difference is 23.

2. The sum of two numbers is 40 and their difference is 8. Find the numbers.

3. The sum of two angles is 180°. The larger angle is 20° more than three times the smaller. Find the measures of the angles.

4. Two angles have a sum of 90°. Their difference is 54°. Find the measures of the angles.

5. The perimeter of a triangle is 42 centimeters. If two sides are of equal length, and the third side measures 6 centimeters more than one of the equal sides, find the lengths of the three sides.

6. A rectangle is three times as long as it is wide. Its perimeter is 40 meters. Find the dimensions of the rectangle.

Write a system of equations for each problem and then solve the system. See Example 2.

7. A store clerk counts her \$1 and \$5 bills at the end of the day. She has a total of 81 bills, and the total value of the money is \$193. How many of each type of bill does she have?

Kind of bill	*Number of bills*	*Total value*
\$1	x	
\$5	y	
Totals	81	\$193

8. Juanita Gomez is a bank teller. At the end of a day she has a total of 86 ten-dollar bills and twenty-dollar bills. The total value of the money is \$1220. How many of each type of bill does she have?

Kind of bill	*Number of bills*	*Total value*
\$10	x	$\$10x$
\$20	y	
Totals		

9. 193 tickets were sold for a soccer game. Student tickets cost \$.50 each and non-student tickets cost \$1.50 each. A total of \$206.50 was collected. How many of each type of ticket were sold?

10. Wally Smart bought some records at \$5 each and some compact discs at \$9 each. He spent \$123 and got a total of 15 records and discs. How many of each type did he buy?

11. Joanna Collins has twice as much money invested at 10% annual simple interest as she has at 8%. If her yearly income from these investments is \$700, how much does she have invested at each rate?

12. Jairo Santanilla has three times as much money invested at 9% interest as he does at 7%. If his yearly income from the investments is \$1360, how much does he have invested at each rate?

13. An artist bought some large canvases at \$7 each and some small ones at \$4 each, paying \$219 in total. Altogether, he bought 39 canvases. How many of each size did he buy?

14. A hospital bought a total of 146 bottles of glucose solution. Small bottles cost \$2 each, and large ones cost \$3 each. The total cost was \$336. How many of each size bottle were bought?

Write a system of equations for each problem and then solve the system. See Example 3.

15. A 40% dye solution is to be mixed with a 70% dye solution to get 60 liters of a 50% solution. How many liters of the 40% and 70% solutions will be needed?

Liters of solution	*Percent (as a decimal)*	*Liters of pure dye*
x	.40	$.40x$
y	.70	
60	.50	

16. A 90% antifreeze solution is to be mixed with a 75% solution to make 40 liters of a 78% solution. How many liters of the 90% and 75% solutions will be used?

Liters of solution	*Percent (as a decimal)*	*Liters of pure antifreeze*
x	.90	
y	.75	
40	.78	

17. An alloy that is 60% copper is to be combined with an alloy that is 80% copper to obtain 40 pounds of a 65% alloy. How many pounds of the 60% and the 80% alloys will be needed?

18. How many liters of a 25% indicator solution and a 55% indicator solution should be mixed to obtain 24 liters of a 45% solution?

19. A grocer wishes to blend candy selling for $1.20 a pound with candy selling for $1.80 a pound to get a mixture that will be sold for $1.40 a pound. How many pounds of the $1.20 and the $1.80 candy should be used to get 90 pounds of the mixture?

Pounds	*Dollars per pound*	*Cost*
x	1.20	
y	1.80	
90		

20. A merchant wishes to mix coffee worth $6 per pound with coffee worth $3 per pound to get 180 pounds of a mixture worth $4 per pound. How many pounds of the $6 and the $3 coffee will be needed?

Pounds	*Dollars per pound*	*Cost*
x	6	
y	3	
180		

21. The owner of a nursery wants to mix some fertilizer worth \$70 per bag with some worth \$90 per bag to obtain 80 bags of mixture worth \$77.50 per bag. How many bags of each type should she use?

22. How many barrels of pickles worth \$40 per barrel and pickles worth \$60 per barrel must be mixed to obtain 100 barrels of a mixture worth \$48 per barrel?

Write a system of equations for each problem and then solve the system. See Example 4.

23. It takes a boat $1\frac{1}{2}$ hours to go 12 miles downstream, and 6 hours to return. Find the speed of the boat in still water and the speed of the current. Let x = the speed of the boat in still water and y = the speed of the current.

	d	r	t
Downstream	12	$x + y$	$\frac{3}{2}$
Upstream	12	$x - y$	6

24. A boat takes 3 hours to go 24 miles upstream. It can go 36 miles downstream in the same time. Find the speed of the current and the speed of the boat in still water if x = the speed of the boat in still water and y = the speed of the current.

	d	r	t
Downstream	36	$x + y$	3
Upstream	24	$x - y$	3

25. A small plane travels 100 miles per hour with the wind and 60 miles per hour against it. Find the speed of the wind and the speed of the plane in still air.

26. If a plane can travel 400 miles per hour into the wind and 540 miles per hour with the wind, find the speed of the wind and the speed of the plane in still air.

27. Two cars start from towns 300 miles apart and travel toward each other. They meet after 3 hours. Find the rate of each car if one car travels 30 miles per hour slower than the other.

28. Two trains start from stations 1000 miles apart and travel toward each other. They meet after 5 hours. Find the rate of each train if one travels 20 miles per hour faster than the other.

29. Mr. Anderson left Farmersville in a plane at noon to travel to Exeter. Mr. Bentley left Exeter in his automobile at 2 P.M. to travel to Farmersville. It is 400 miles from Exeter to Farmersville. If the sum of their speeds was 120 miles per hour, and if they met at 4 P.M., find the speed of each.

30. At the beginning of a walk for charity, John and Harriet are 30 miles apart. If they leave at the same time and walk in the same direction, John overtakes Harriet in 60 hours. If they walk toward each other, they meet in 5 hours. What are their speeds?

SKILL SHARPENERS

Graph each linear inequality. *See Section 7.6.*

31. $x + y \le 5$

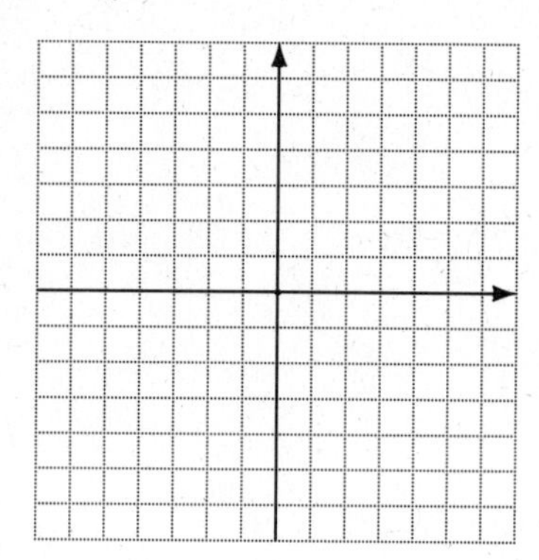

32. $2x - y \ge 7$

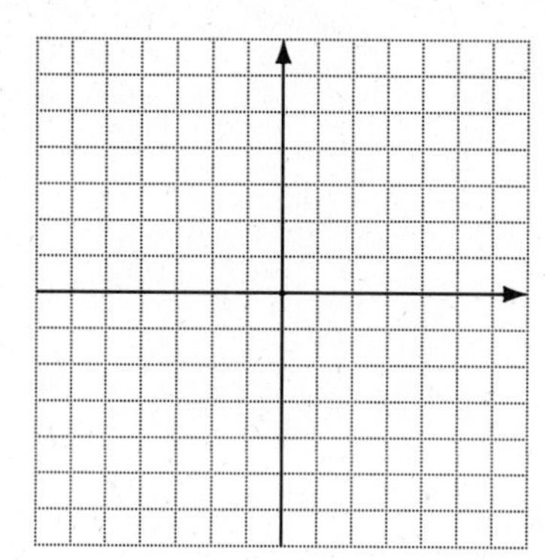

33. $3x + 2y > 6$

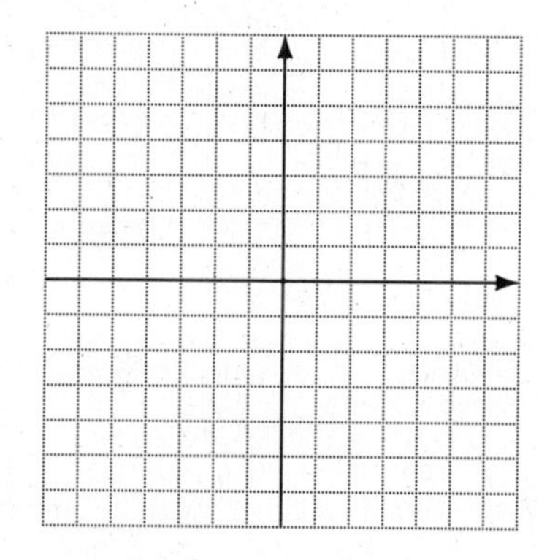

34. $x + 3y < 9$

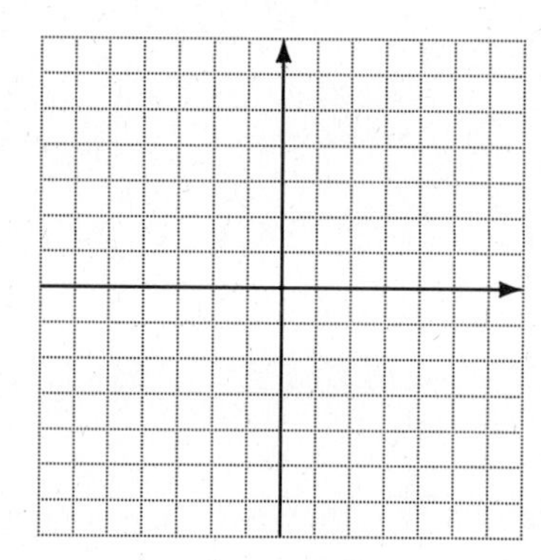

35. $5x + 4y > 0$

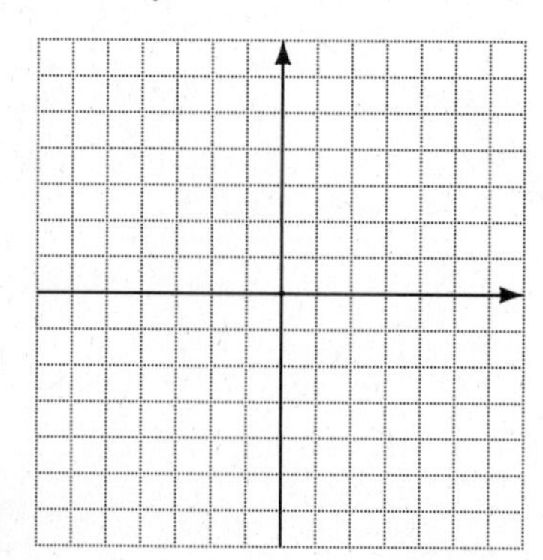

36. $3x - 4y < 0$

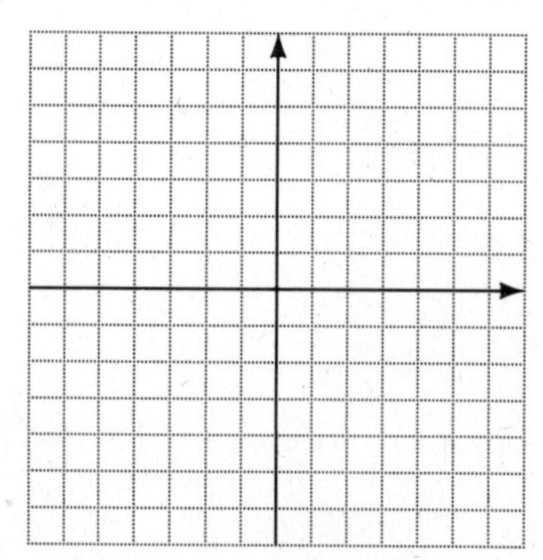

8.6 SOLVING SYSTEMS OF LINEAR INEQUALITIES

OBJECTIVE

1 Solve systems of linear inequalities by graphing.

Graphing the solution of a linear inequality was discussed in Section 7.6. Let us review the method. To graph the solution of $x + 3y > 12$, first graph the line $x + 3y = 12$. (Find a few ordered pairs that satisfy the equation.) Because the points on the line do not satisfy the inequality, use a dashed line. Decide which side of the line should be shaded by choosing any test point not on the line, such as (0, 0). Substitute 0 for x and 0 for y in the given inequality.

$$x + 3y > 12$$
$$0 + 3(0) > 12 \quad \text{Let } x = 0 \text{ and let } y = 0.$$
$$0 > 12 \quad \text{False}$$

The test point does not satisfy the inequality, so shade the region on the side of the line that does not include (0, 0), as shown in Figure 10.

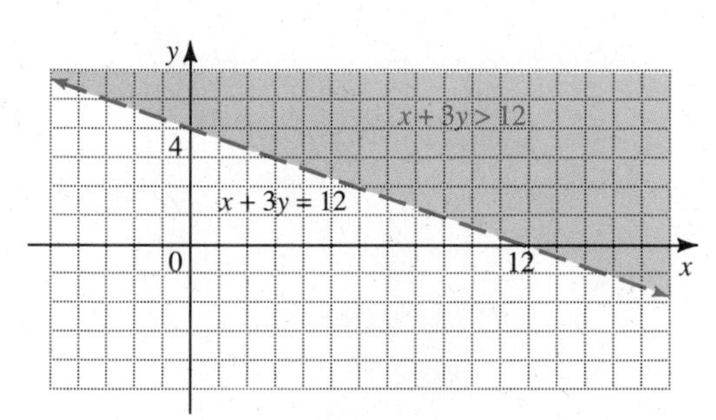

FIGURE 10

WORK PROBLEM 1 AT THE SIDE.

1. Graph each inequality.

(a) $3x + 2y \geq 6$
The line $3x + 2y = 6$ is graphed already.

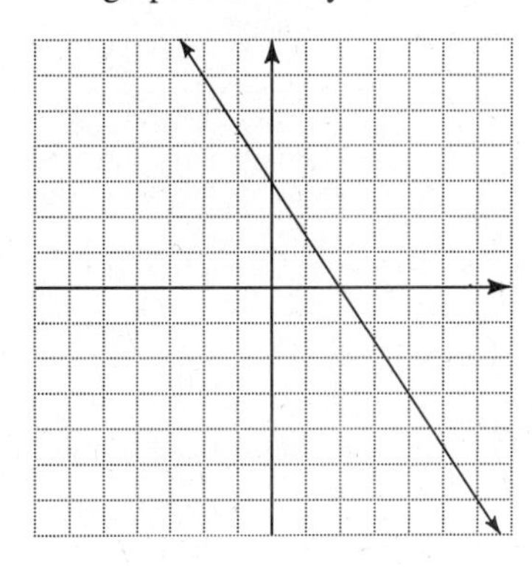

(b) $2x - 5y < 10$

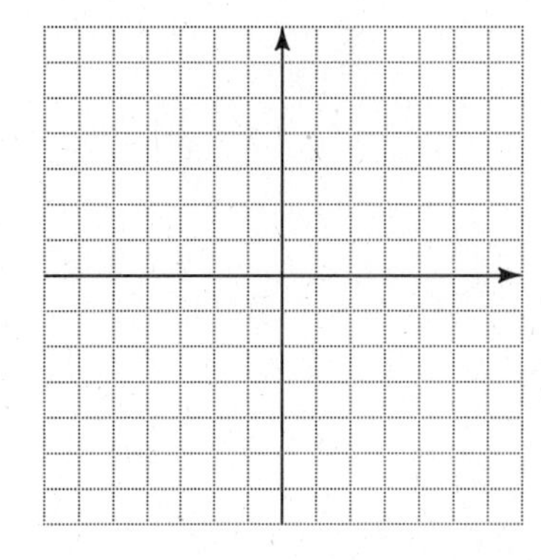

1 The same method is used to find the solution of a system of two linear inequalities, as shown in the following examples. A **system of linear inequalities** contains two or more linear inequalities (and no other kinds of inequalities). The **solution of a system of linear inequalities** consists of all points that make all inequalities of the system true *at the same time*.

EXAMPLE 1 Solving a System of Linear Inequalities

Graph the solution of the system

$$3x + 2y \leq 6$$
$$2x - 5y \geq 10.$$

First graph the inequality $3x + 2y \leq 6$ using the steps described above. Then, on the same axes, graph the second inequality, $2x - 5y \geq 10$. The solution of the system is given by the overlap of the regions of the two graphs. This solution is the region in Figure 11 with the darkest shading; this region includes portions of both boundary lines. ■

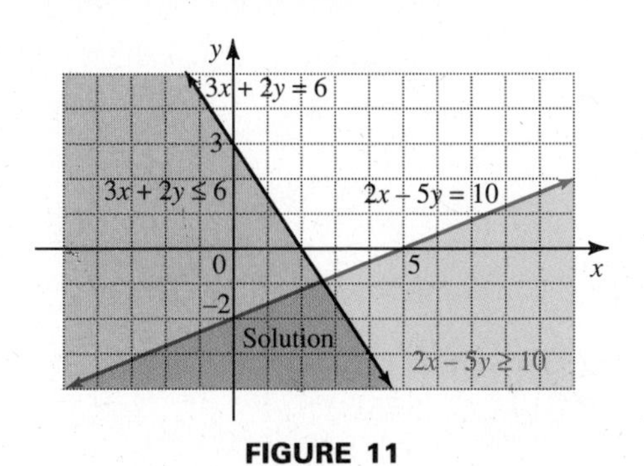

FIGURE 11

ANSWERS

1. (a) (b)

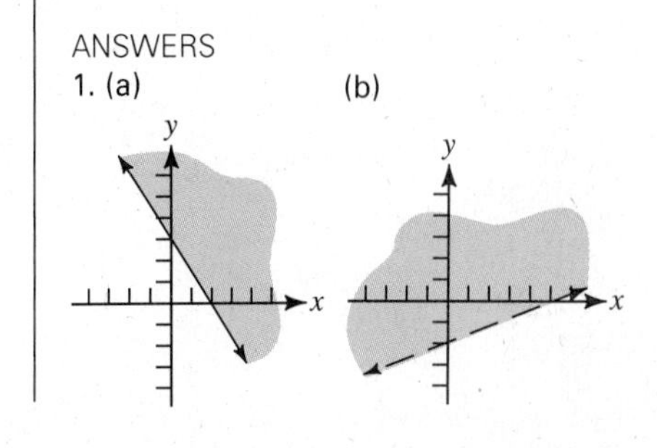

2. Graph the solution of the system

$$x - 2y \leq 8$$
$$3x + y \geq 6.$$

To get started, the graphs of $x - 2y = 8$ and $3x + y = 6$ are included.

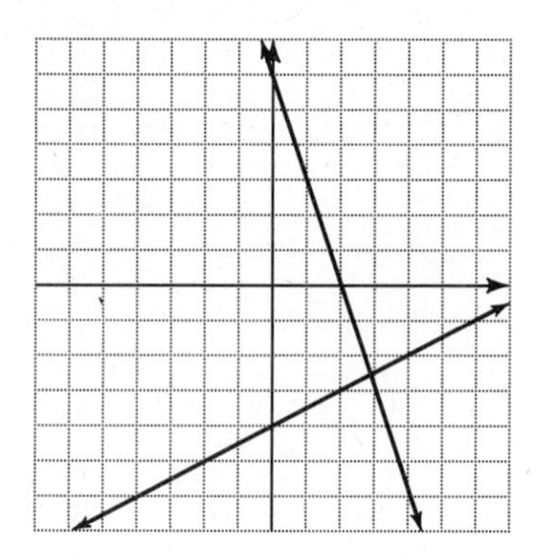

3. Graph the solution of the system

$$x + 2y < 0$$
$$3x - 4y < 12.$$

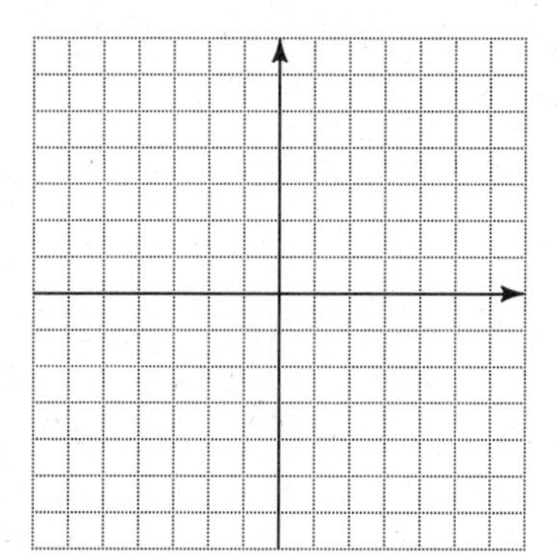

ANSWERS

2.

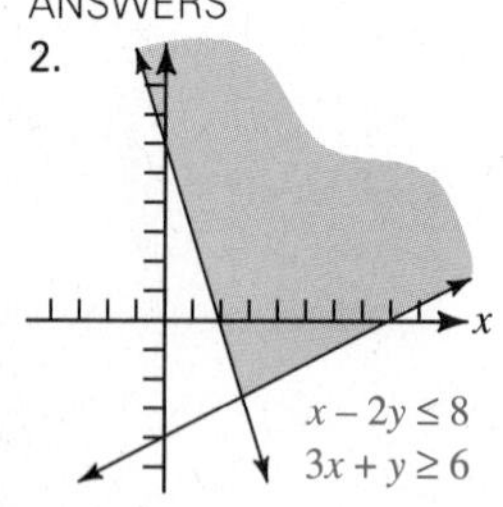

3.

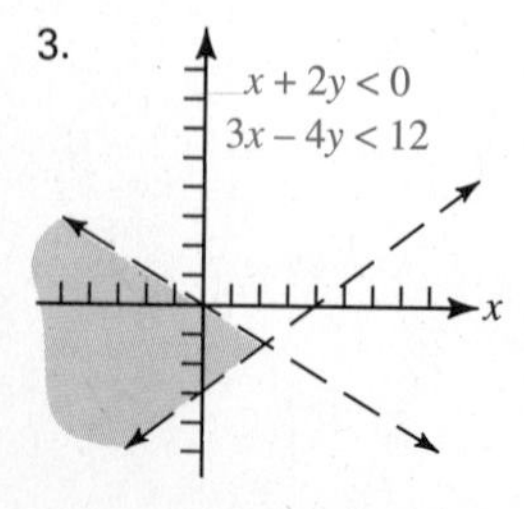

WORK PROBLEM 2 AT THE SIDE.

EXAMPLE 2 Solving a System of Linear Inequalities
Graph the solution of the system

$$x - y > 5$$
$$2x + y < 2.$$

Figure 12 shows the graphs of both $x - y > 5$ and $2x + y < 2$. Dashed lines show that the graphs of the inequalities do not include their boundary lines. The solution of the system is the region with the darkest shading. The solution does not include either boundary line. ■

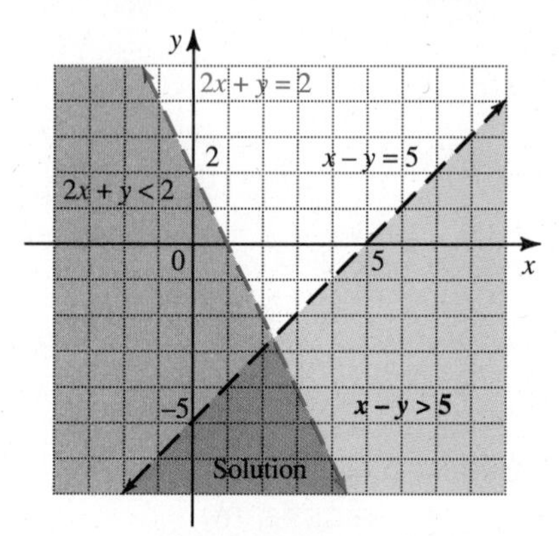

FIGURE 12

EXAMPLE 3 Solving A System of Linear Inequalities
Graph the solution of the system

$$4x - 3y \leq 8$$
$$x \geq 2.$$

Recall that $x = 2$ is a vertical line through the point (2, 0). The graph of the system is the region in Figure 13 with the darkest shading. ■

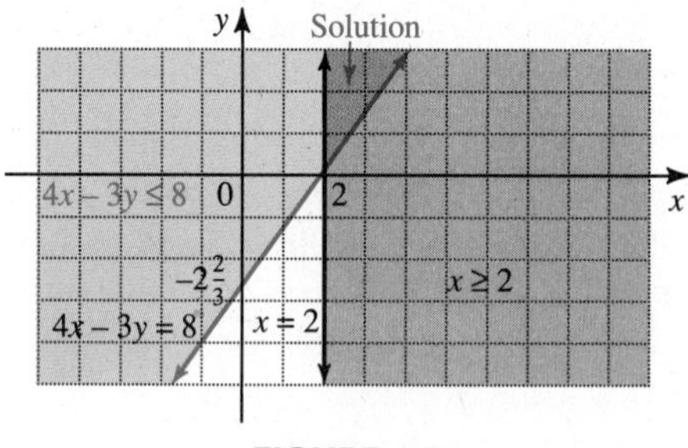

FIGURE 13

WORK PROBLEM 3 AT THE SIDE.

name date hour

8.6 EXERCISES

Graph the solution of each system of linear inequalities. See Examples 1-3.

1. $x + y \le 6$
$x - y \le 1$

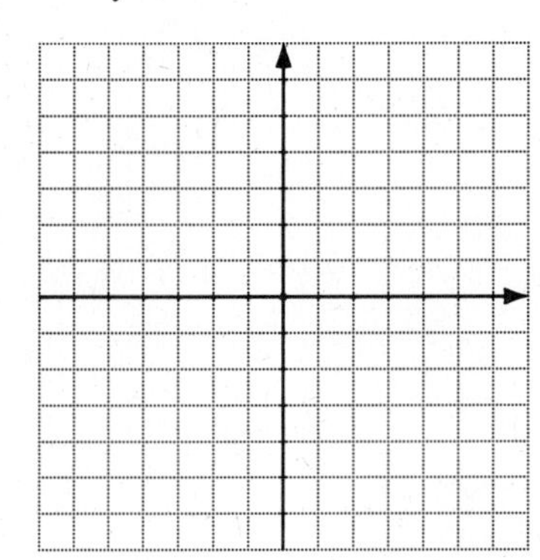

2. $x + y \ge 2$
$x - y \le 3$

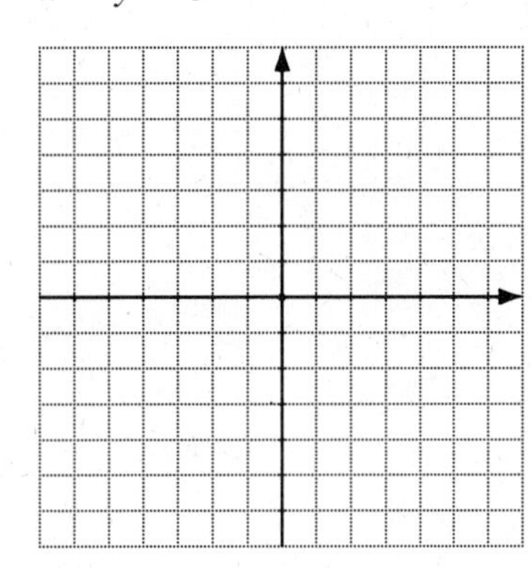

3. $2x - 3y \le 6$
$x + y \ge -1$

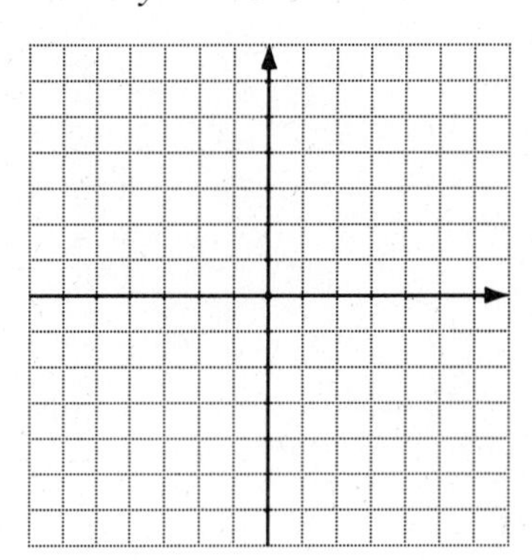

4. $4x + 5y \le 20$
$x - 2y \le 5$

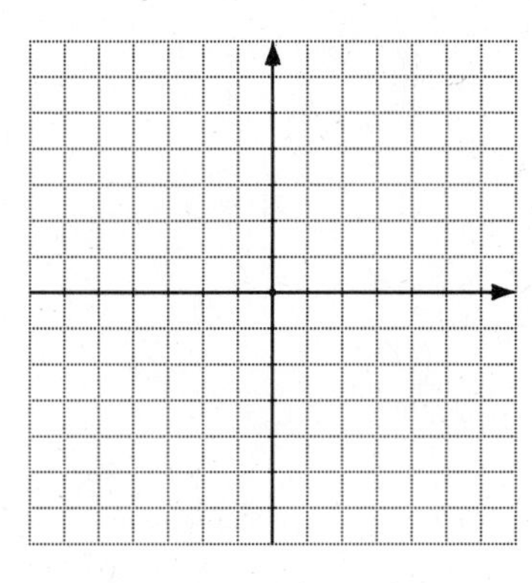

5. $x + 4y \le 8$
$2x - y \le 4$

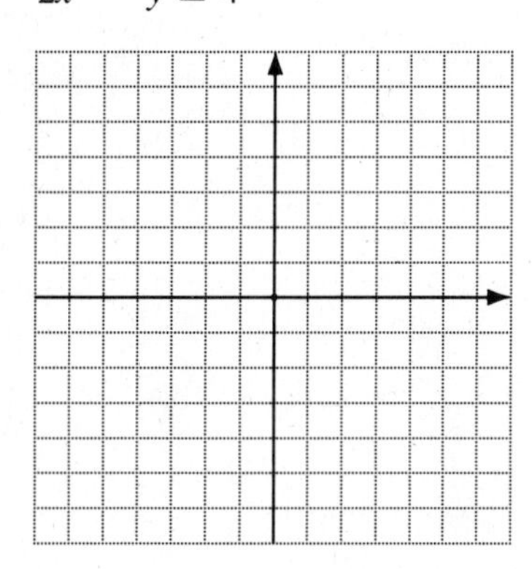

6. $3x + y \le 6$
$2x - y \le 8$

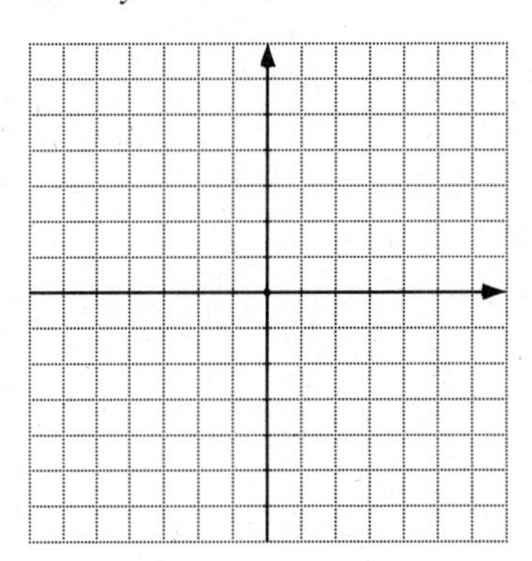

7. $x - 4y \le 3$
$x \ge 2y$

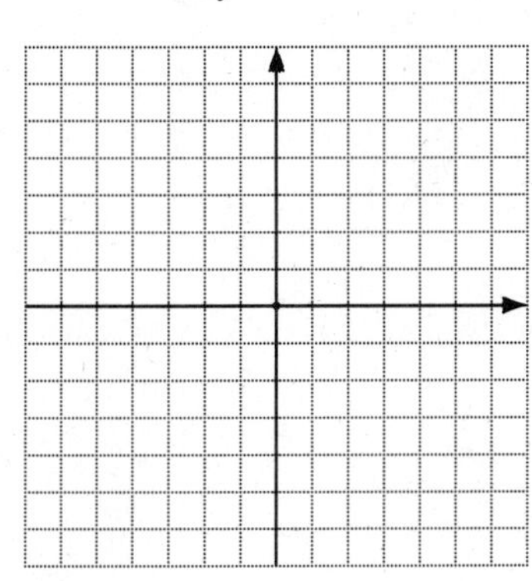

8. $2x + 3y \le 6$
$x - y \ge 5$

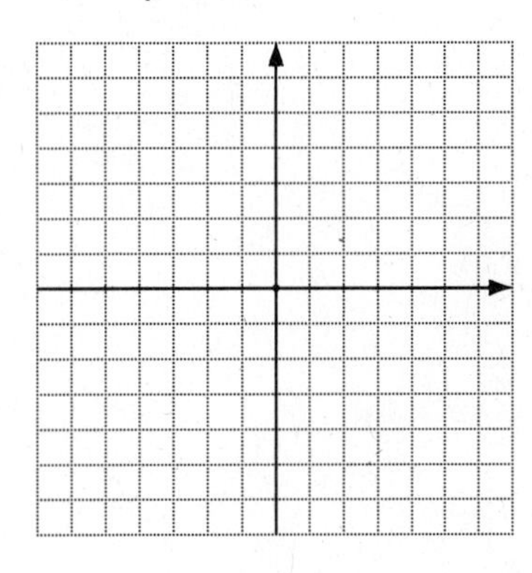

9. $x + 2y \le 4$
$x + 1 \ge y$

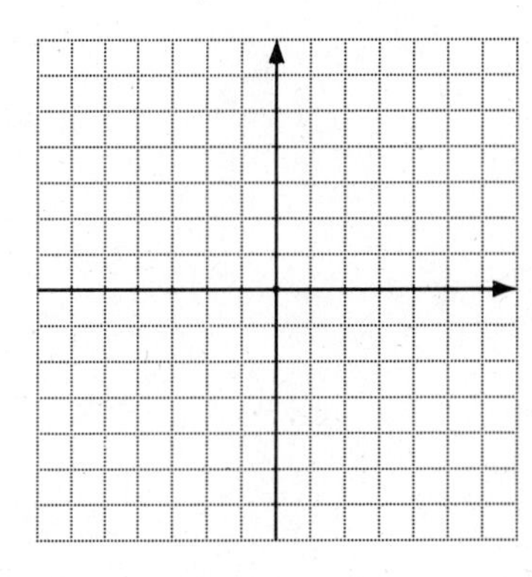

10. $y \leq 2x - 5$
$x - 3y \leq 2$

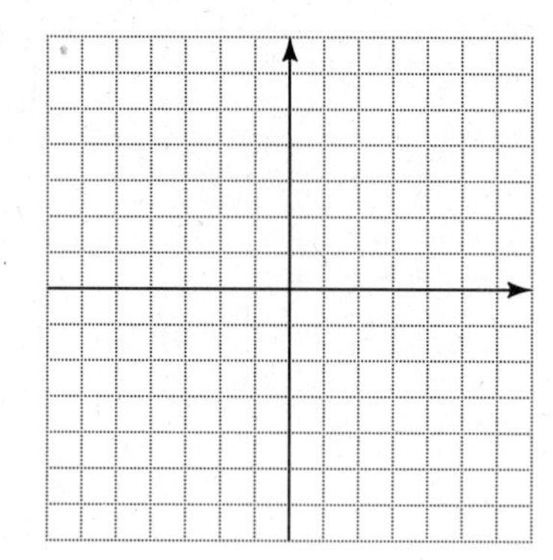

11. $4x + 3y \leq 6$
$x - 2y \geq 4$

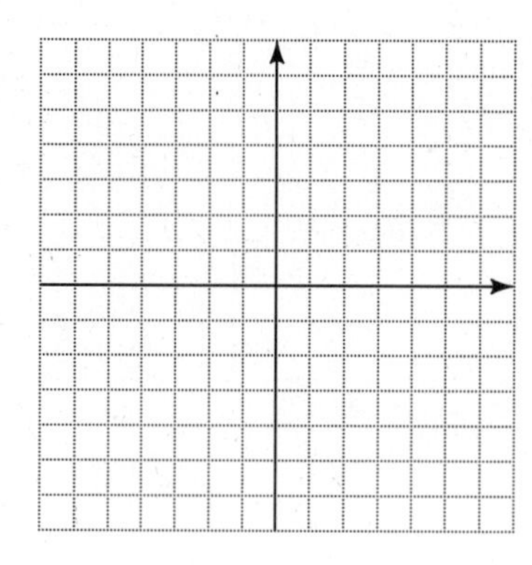

12. $3x - y \leq 4$
$-6x + 2y \leq -10$

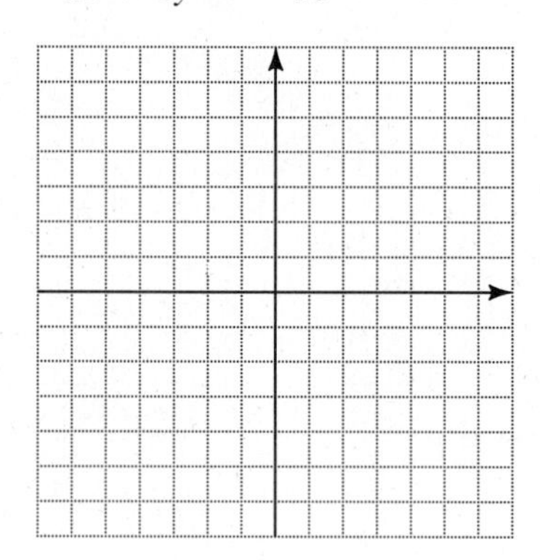

13. $x - 2y > 6$
$2x + y > 4$

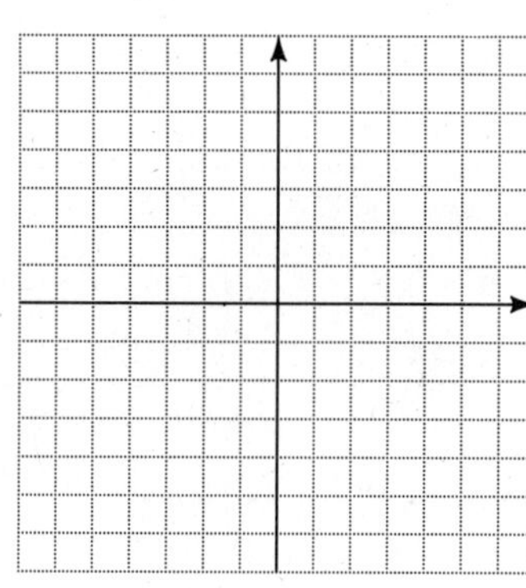

14. $3x + y < 4$
$x + 2y > 2$

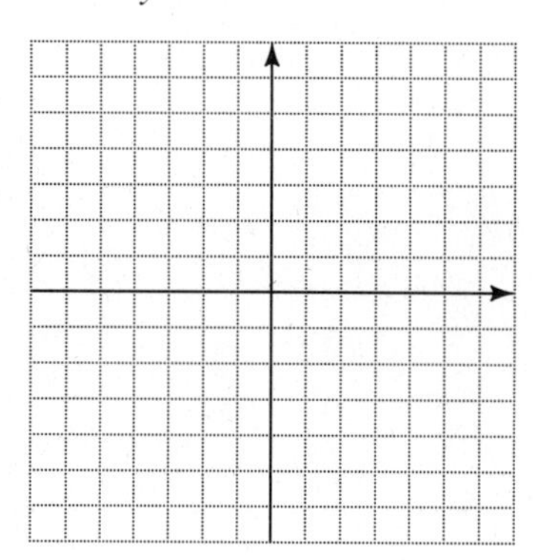

15. $x < 2y + 3$
$x + y > 0$

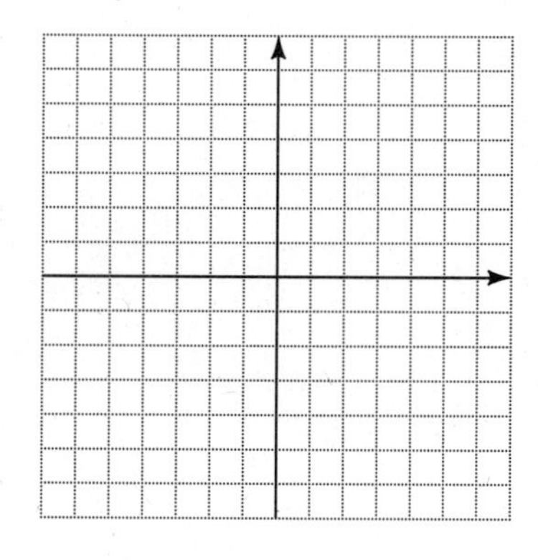

16. $2x + 3y < 6$
$4x + 6y > 18$

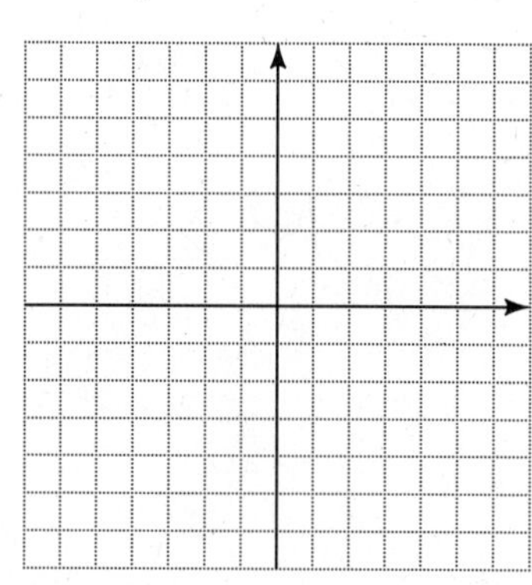

17. $x - 3y \leq 6$
$x \geq -1$

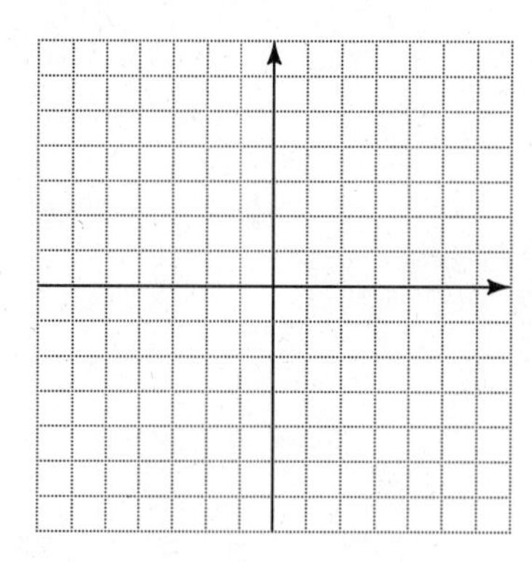

18. $2x + 5y \geq 10$
$x \leq 4$

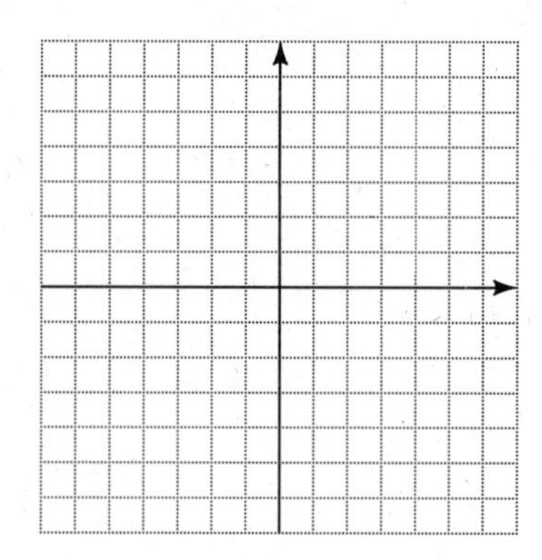

SKILL SHARPENERS

Find each value. See Section 1.4.

19. 5^2

20. 7^2

21. 11^2

22. 15^2

23. 20^2

24. 25^2

Historical Reflections

Historical Reflections

WOMEN IN MATHEMATICS: Sonya Kovalevski (1850–1891)

Sonya Kovalevski was the most widely known Russian mathematician in the late nineteenth century. She did most of her work in the branch of mathematics known as differential equations. These equations are invaluable for expressing rates of change. For example, in biology the rate of growth of a population (of, say, microbes) can be precisely studied by differential equations.

Kovalevski studied privately because during her time public lectures were not open to women. She eventually received a degree (1874) from the University of Göttingen, Germany. In 1884 she became a lecturer at the University of Stockholm and later was appointed professor of higher mathematics.

Kovalevski was well known as a writer. She wrote novels about Russian life, notably *The Sisters Rajevski* and *Vera Vorontzoff,* as well as her *Recollections of Childhood,* which has been translated into English.

Art: TASS from/SOVFOTO

CHAPTER 8 SUMMARY

KEY TERMS

8.1	**system of linear equations**	A system of linear equations consists of two or more linear equations with the same variables.
	solution of a system	The solution of a system of linear equations includes all the ordered pairs that make all the equations of the system true at the same time.
	consistent system	A system of equations with a solution is a consistent system.
	inconsistent system	An inconsistent system of equations is a system with no solutions.
	independent equations	Equations of a system that have different graphs are called independent equations.
	dependent equations	Equations of a system that have the same graph (because they are different forms of the same equation) are called dependent equations.
8.6	**system of linear inequalities**	A system of linear inequalities contains two or more linear inequalities (and no other kinds of inequalities).
	solution of a system of linear inequalities	The solution of a system of linear inequalities consists of all points that make all inequalities of the system true at the same time.

QUICK REVIEW

Section	Concepts	Examples
8.1 Solving Systems of Linear Equations by Graphing	An ordered pair is a solution of a system if it makes all equations of the system true at the same time.	Is $(4, -1)$ a solution of the following system? $$x + y = 3$$ $$2x - y = 9$$ Yes, because $4 + (-1) = 3$, and $2(4) - (-1) = 9$ are both true.
	If the graphs of the equations of a system are both sketched on the same axes, the points of intersection, if any, are solutions of the system.	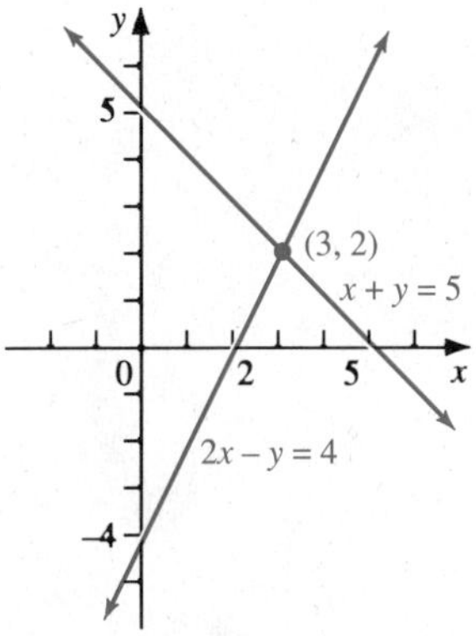 $(3, 2)$ is the solution of the system $$x + y = 5$$ $$2x - y = 4.$$

Section	Concepts	Examples
8.2 Solving Systems of Linear Equations by Addition	*Step 1* Write both equations in the form $Ax + By = C$.	Solve by addition. $x + 3y = 7$ $3x - y = 1$
	Step 2 Multiply one or both equations by appropriate numbers so that the coefficients of x (or y) are negatives of each other.	Multiply the top equation by -3 to eliminate the x terms. $-3x - 9y = -21$ $3x - y = 1$
	Step 3 Add the equations to get an equation with only one variable.	$-10y = -20$ Add.
	Step 4 Solve the equation from Step 3.	$y = 2$ Divide by -10.
	Step 5 Substitute the solution from Step 4 into either of the original equations.	Substitute to get the value of x.
	Step 6 Solve the resulting equation from Step 5 for the remaining variable.	$x + 3y = 7$ $x + 3(2) = 7$ Let $y = 2$. $x + 6 = 7$ Multiply. $x = 1$ Subtract 6.
	Step 7 Check the answer.	The solution, (1, 2), checks.
8.3 Two Special Cases	If the result of the addition step is a false statement, such as $0 = 4$, the graphs are parallel lines and there is *no solution* for the system.	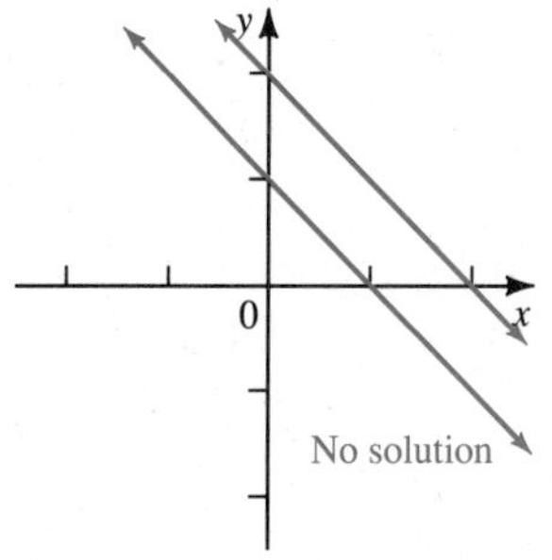 No solution
	If the result is a true statement, such as $0 = 0$, the graphs are the *same line,* and an infinite number of ordered pairs are solutions.	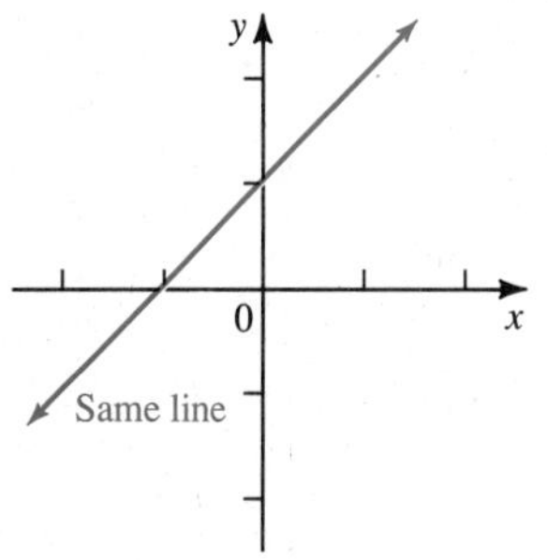 Same line

LINEAR SYSTEMS

Section	Concepts	Examples
8.4 Solving Systems of Linear Equations by Substitution	Solve one equation for one variable, and substitute the expression into the other equation to get an equation in one variable. Solve the equation, and then substitute the solution into either of the original equations to obtain the other variable. Check the answer.	Solve by substitution. $x + 2y = -5$ **(1)** $y = -2x - 1$ **(2)** Substitute $-2x - 1$ for y in equation (1). $x + 2(-2x - 1) = -5$ Solve to get $x = 1$. To find y, let $x = 1$ in equation (2): $y = -2(1) - 1 = -3$. The solution is $(1, -3)$.
8.5 Applications of Linear Systems	*Step 1* Choose a variable to represent each unknown value.	The sum of two numbers is 30. Their difference is 6. Find the numbers.
	Step 2 Translate the problem into two equations using both variables.	Let x represent one number. Let y represent the other number. $x + y = 30$ $x - y = 6$
	Step 3 Solve the system.	$2x = 36$ Add. $x = 18$ Divide by 2.
	Step 4 Answer the question or questions asked.	One number is 18. Let $x = 18$ in the top equation: $18 + y = 30$. Solve to get $y = 12$.
	Step 5 Check the solution in the words of the problem.	$18 + 12 = 30$ and $18 - 12 = 6$, so the solution checks. The numbers are 18 and 12.
8.6 Solving Systems of Linear Inequalities	To solve a system of linear inequalities, graph each inequality on the same axes. (This was explained in Section 7.6.) The solution of the system is the overlap of the regions of the two graphs.	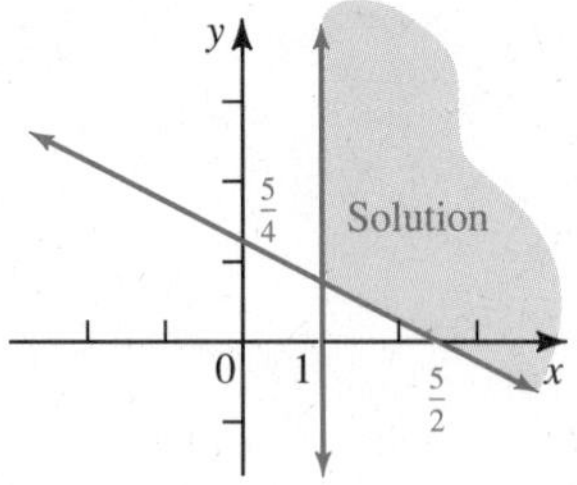 The shaded region is the solution of the system $2x + 4y \geq 5$ $x \geq 1$.

name date hour

CHAPTER 8 REVIEW EXERCISES

[8.1] *Decide whether the given ordered pair is a solution of the given system.*

1. $(3, 4)$
$4x - 2y = 4$
$5x + y = 17$

2. $(-2, 1)$
$5x + 3y = -7$
$2x - 3y = -7$

3. $(-5, 2)$
$x - 4y = -10$
$2x + 3y = -4$

4. $(6, 3)$
$3x + 8y = 42$
$4x - 3y = 15$

Solve each system by graphing.

5. $x + y = 3$
$2x - y = 3$

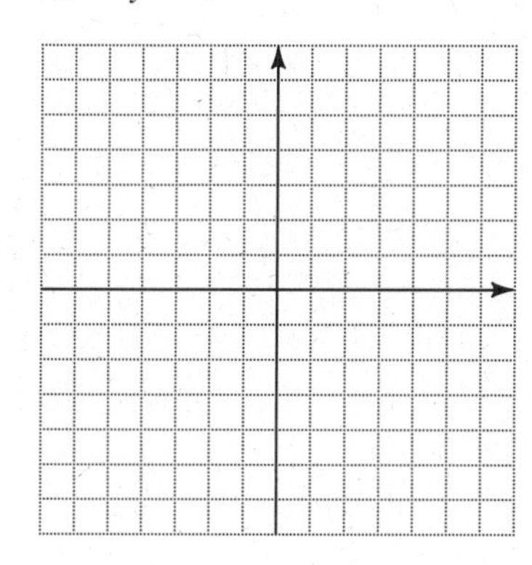

6. $x - 2y = -6$
$2x + y = -2$

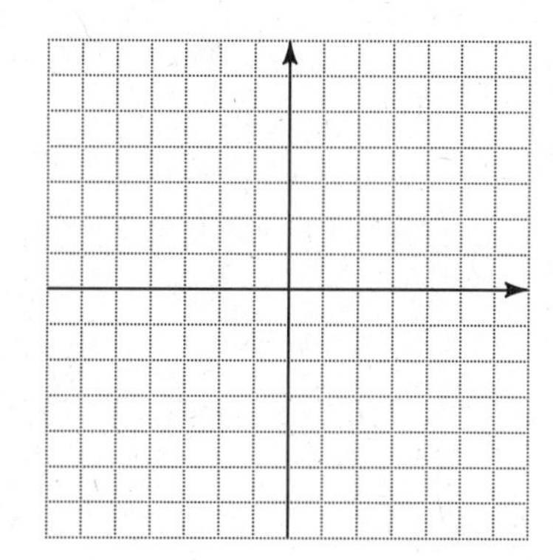

7. $2x + 3y = 1$
$4x - y = -5$

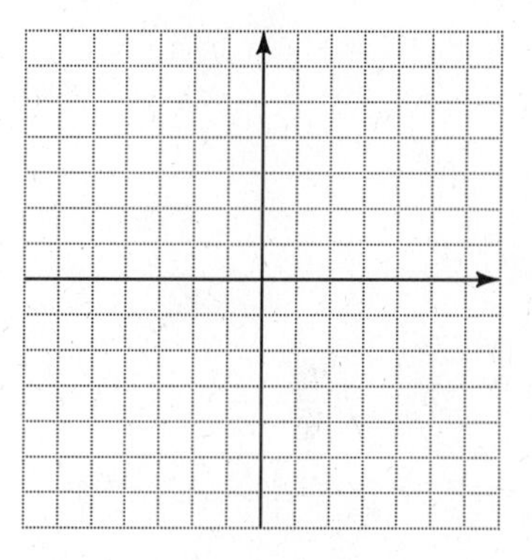

8. $x = 2y + 2$
$2x - 4y = 4$

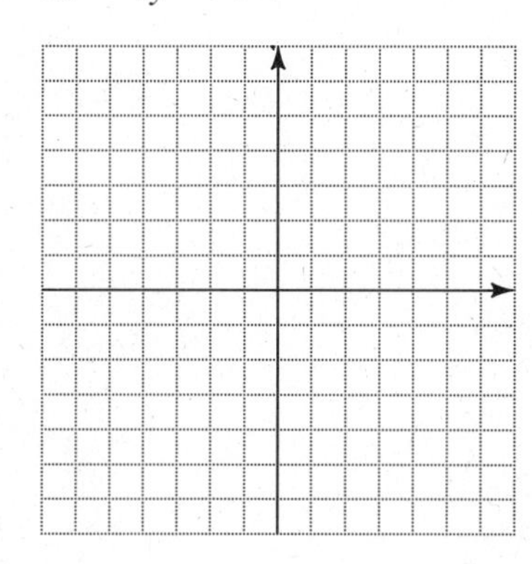

9. $y + 2 = 0$
$3y = 6x$

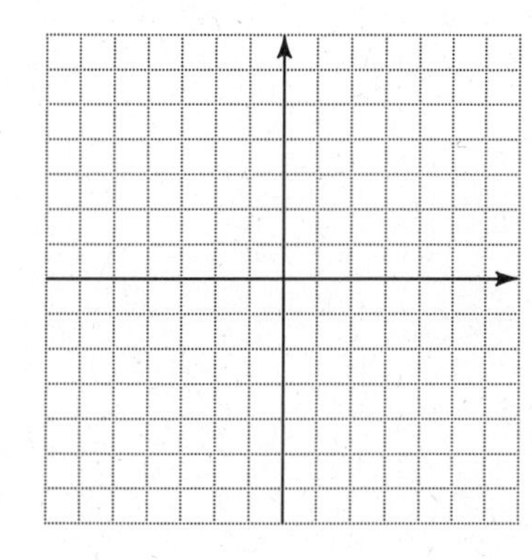

10. $2x + 5y = 10$
$2x + 5y = 8$

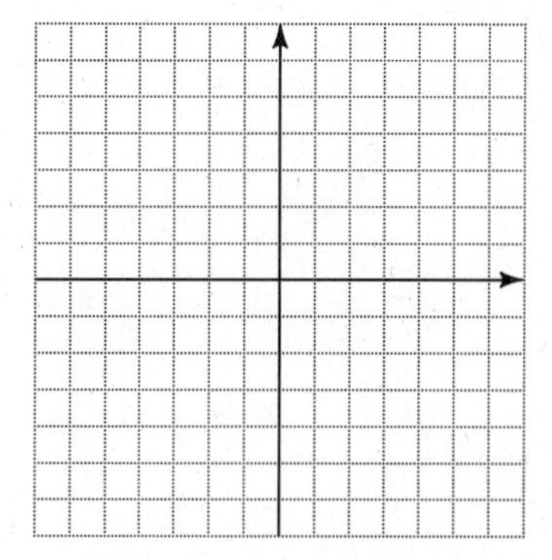

[8.2] *Solve each system by the addition method.*

11. $\begin{aligned} x + y &= 6 \\ 2x + y &= 8 \end{aligned}$

12. $\begin{aligned} 3x - y &= 13 \\ x - 2y &= 1 \end{aligned}$

13. $\begin{aligned} 5x + 4y &= 7 \\ 3x - 4y &= 17 \end{aligned}$

14. $\begin{aligned} -4x + 3y &= -7 \\ 6x - 5y &= 11 \end{aligned}$

15. $\begin{aligned} 3x - 2y &= 14 \\ 5x - 4y &= 24 \end{aligned}$

16. $\begin{aligned} 2x + 6y &= 18 \\ 3x + 5y &= 19 \end{aligned}$

[8.3]

17. $\begin{aligned} 3x - 4y &= 7 \\ 6x - 8y &= 14 \end{aligned}$

18. $\begin{aligned} 2x + y &= 5 \\ 2x + y &= 8 \end{aligned}$

[8.4] *Solve each system by the substitution method.*

19. $\begin{aligned} 3x + y &= 14 \\ x &= 2y \end{aligned}$

20. $\begin{aligned} 2x - 5y &= 5 \\ y &= x + 2 \end{aligned}$

21. $\begin{aligned} 4x + 5y &= 35 \\ x + 1 &= 2y \end{aligned}$

22. $\begin{aligned} 5x + y &= -6 \\ x - 3y &= 2 \end{aligned}$

name date hour

Solve each system by any method. First simplify equations and clear of fractions where necessary.

23. $2x + 3y = 5$
$3x + 4y = 8$

24. $6x + 5y = 9$
$2x - 3y = 17$

25. $2x + y - x = 3y + 5$
$y + 2 = x - 5$

26. $5x - 3 + y = 4y + 8$
$2y + 1 = x - 3$

27. $\frac{x}{2} + \frac{y}{3} = -\frac{8}{3}$
$\frac{x}{4} + 2y = \frac{1}{2}$

28. $\frac{3x}{4} - \frac{y}{3} = \frac{7}{6}$
$\frac{x}{2} + \frac{2y}{3} = \frac{5}{3}$

[8.5] *Solve each word problem by any method. Use two variables.*

29. The sum of two numbers is 42, and their difference is 14. Find the numbers.

30. One number is 2 more than twice as large as another. Their sum is 17. Find the numbers.

31. The perimeter of a rectangle is 40 meters. Its length is $1\frac{1}{2}$ times its width. Find the length and width of the rectangle.

32. A cashier has 20 bills, all of which are ten-dollar or twenty-dollar bills. The total value of the money is \$250. How many of each type does he have?

33. Mr. Dawkins has \$18,000 to invest. He wants the total income from the money to be \$2600. He can invest part of it at 12% interest and the rest at 16%. How much should he invest at each rate?

34. Candy that sells for \$1.30 a pound is to be mixed with candy selling for \$.90 a pound to get 50 pounds of a mix that will sell for \$1 per pound. How much of each type should be used?

35. A 40% antifreeze solution is to be mixed with a 70% solution to get 60 liters of a 50% solution. How many liters of the 40% and 70% solutions will be needed?

36. A certain plane flying with the wind travels 540 miles in 2 hours. Later, flying against the same wind, the plane travels 690 miles in 3 hours. Find the speed of the plane in still air and the speed of the wind.

[8.6] *Graph the solution for each system of linear inequalities.*

37. $x + y \leq 3$
$2x \geq y$

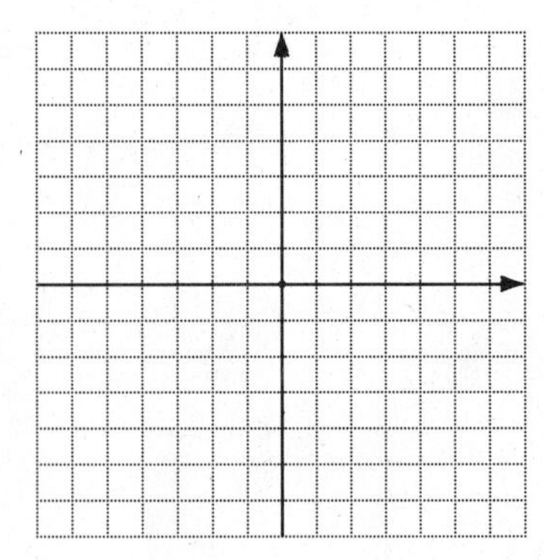

38. $x + y \geq 4$
$x - y \leq 2$

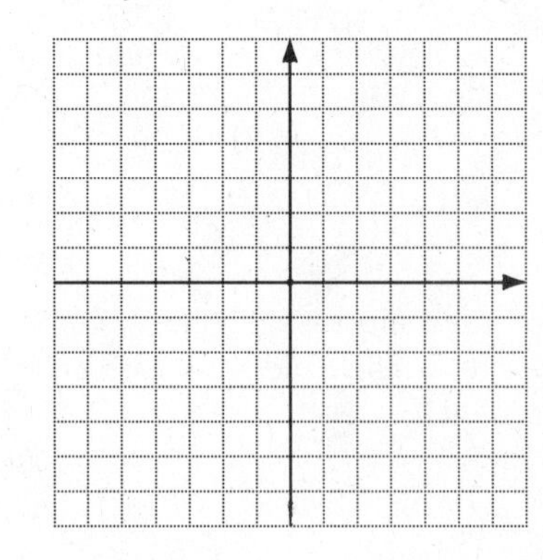

39. $2x + y < 6$
$y - 2x < 6$

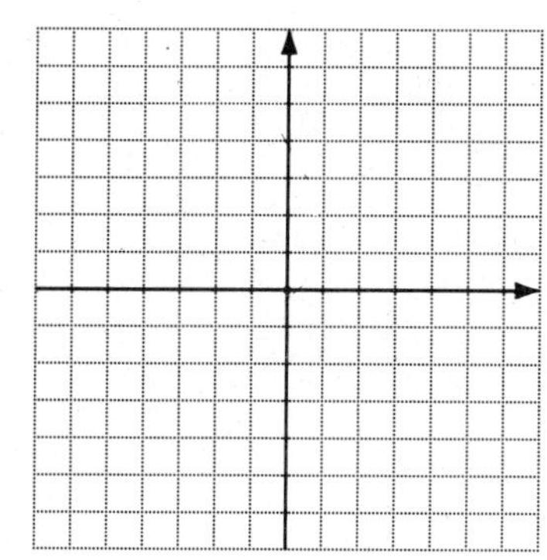

40. $x \geq 2$
$y \leq 5$

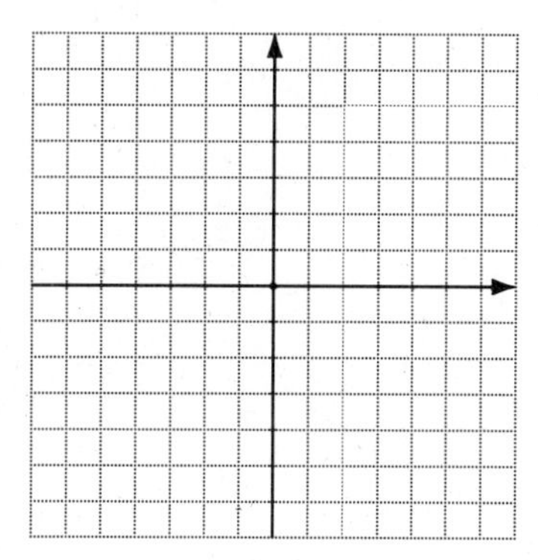

41. $y \geq 3x$
$2x + 3y \leq 4$

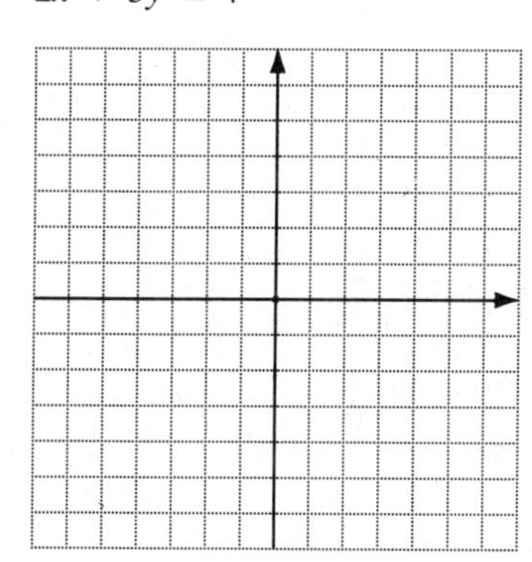

42. $y + 2 < 2x$
$x > 3$

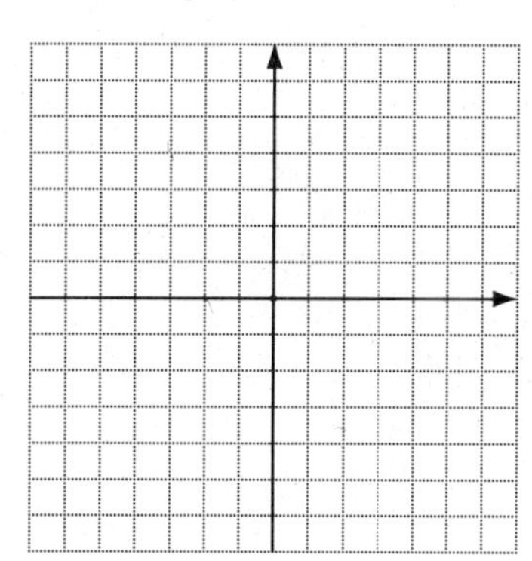

MIXED REVIEW EXERCISES

Solve each of the following exercises using the methods of this chapter.

43. $\frac{2x}{3} - \frac{y}{4} = \frac{2}{3}$
$\frac{x}{2} + \frac{y}{12} = \frac{8}{3}$

44. $3x + y - 2x = 2y + 6$
$y + 8 = x + 2$

45. The perimeter of an isosceles triangle is 58 inches. One side of the triangle is 7 inches longer than each of the two equal sides. Find the lengths of the sides of the triangle.

46. $3x + 4y = 8$
$4x - 5y = -10$

47. $\frac{3x}{2} + \frac{y}{5} = -3$
$4x + \frac{y}{3} = -8$

48. $3x - y < 6$
$y - 4x < -8$

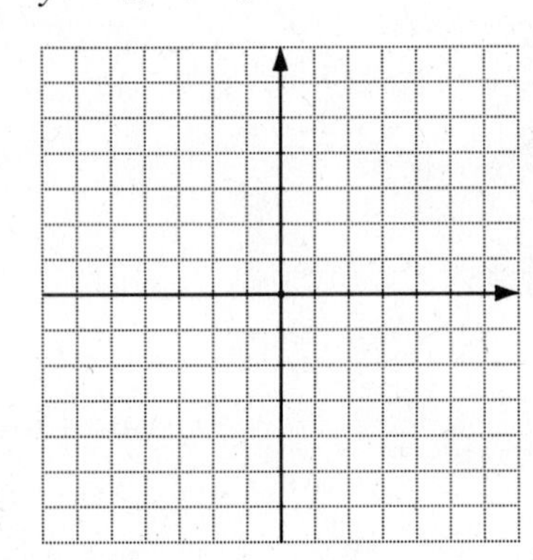

49. $x \geq 3$
$y \leq -1$

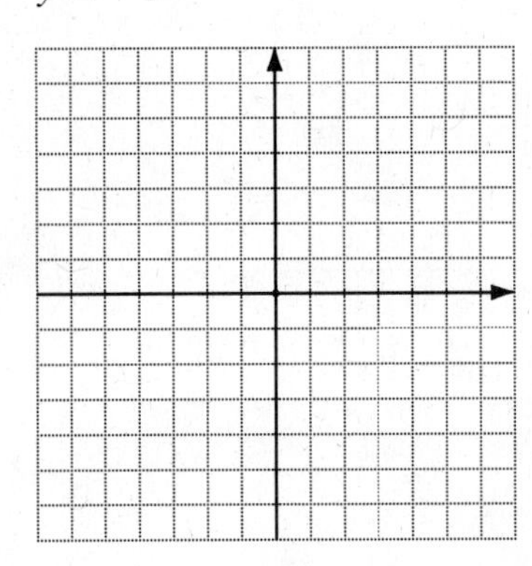

50. $x + 6y = 5$
$2x + 12y = 1$

51. $3x - 2y = 9$
$-6x + 4y = -18$

52. The sum of two numbers is -12, and their difference is -2. Find the numbers.

name date hour

CHAPTER 8 TEST

Solve each system by graphing.

1. $2x + y = 5$
$3x - y = 15$

1.

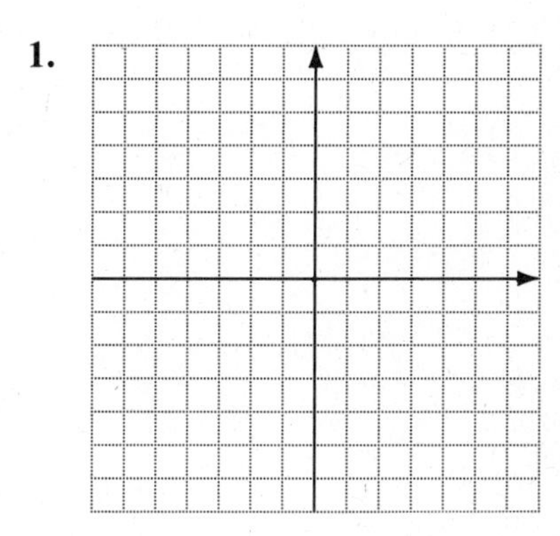

2. $3x + 2y = 8$
$5x + 4y = 10$

2.

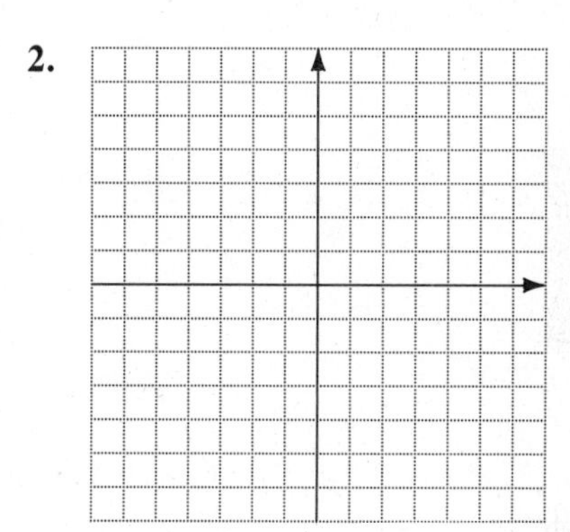

3. $x + 2y = 6$
$2x - y = 7$

3.

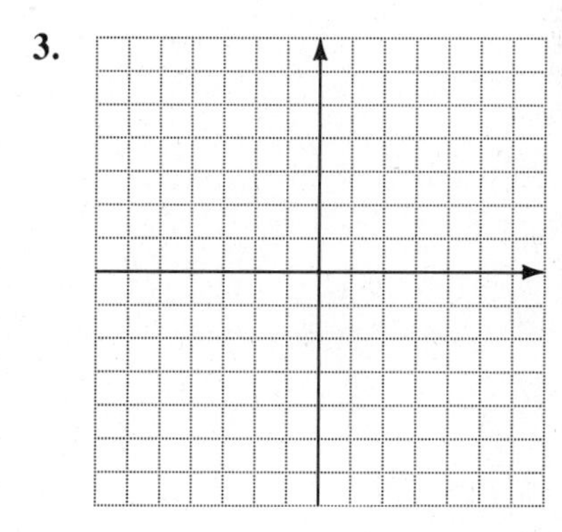

Solve each system by the addition method.

4. $2x - 5y = -13$
$3x + 5y = 43$

4. ______________________

5. ______________

5. $4x + 3y = 26$
$5x + 4y = 32$

6. ______________

6. $6x + 5y = -13$
$3x + 2y = -4$

7. ______________

7. $4x + 5y = 8$
$-8x - 10y = -6$

8. ______________

8. $6x - 5y = 2$
$-2x + 3y = 2$

9. ______________

9. $\frac{6}{5}x - y = \frac{1}{5}$
$-\frac{2}{3}x + \frac{1}{6}y = \frac{1}{3}$

Solve each system by substitution.

10. ______________

10. $2x + y = 1$
$x = 8 + y$

11. ______________

11. $4x + 3y = 0$
$x + y = 2$

Solve each system by any method.

12. ______________

12. $8 + 3x - 4y = 14 - 3y$
$3x + y + 12 = 9x - y$

name date hour

13. $\frac{x}{2} - \frac{y}{4} = -4$

$\frac{2x}{3} + \frac{5y}{4} = 1$

13. ______________________

14. The sum of two numbers is 39. If one number is doubled, the result is 3 less than the other. Find the numbers.

14. ______________________

15. The local music shop is having a sale. Some cassettes cost $2.50 and some cost $3.75. Joe has exactly $20 to spend and wants to buy 6 cassettes. How many can he buy at each price?

15. ______________________

16. A 40% solution of acid is to be mixed with a 60% solution to get 100 liters of a 45% solution. How many liters of each solution should be used?

16. ______________________

17. Two cars leave from the same place and travel in the same direction. One car travels one and one third times as fast as the other. After 3 hours, they are 45 miles apart. What is the speed of each car?

17. ______________________

Graph the solution of each system of inequalities.

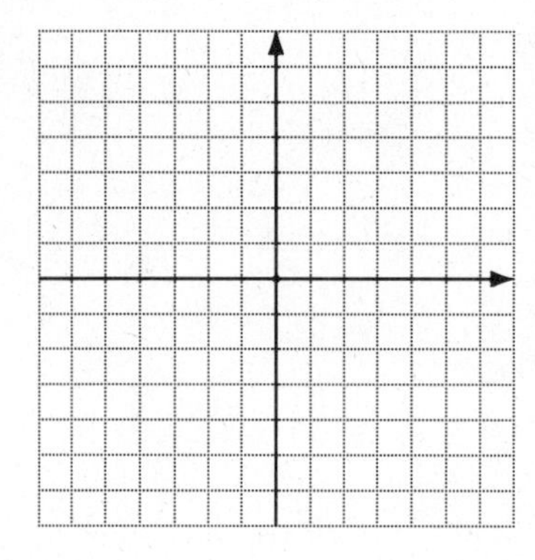

18. $2x + 7y \leq 14$
$x - y \geq 1$

19.

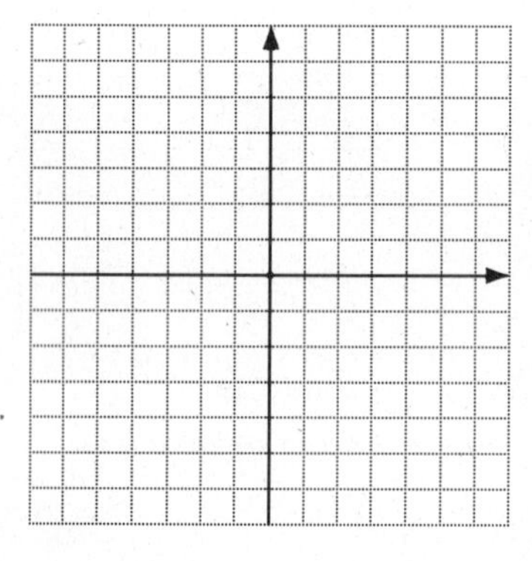

19. $2x - y \leq 6$
$4y + 12 \geq -3x$

20.

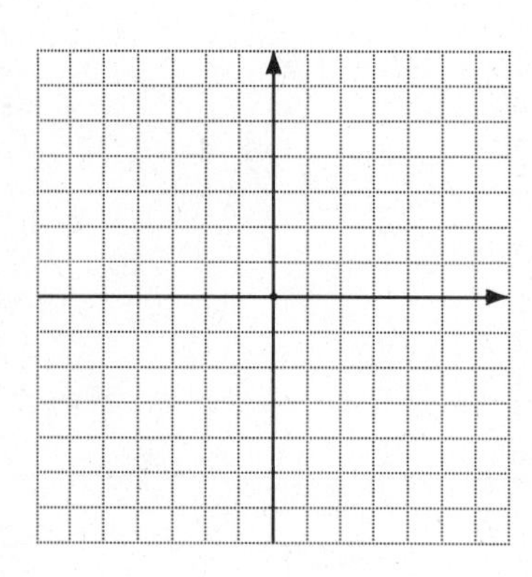

20. $3x - 5y < 15$
$y < 2$

SQUARE ROOTS AND RADICALS

9.1 FINDING SQUARE ROOTS

In Section 1.4, we discussed the idea of the *square* of a number. Recall that squaring a number means multiplying the number by itself.

If $a = 8$, then $a^2 = 8 \cdot 8 = 64$.
If $a = -4$, then $a^2 = (-4)(-4) = 16$.
If $a = -\frac{1}{2}$, then $a^2 = \left(-\frac{1}{2}\right)\left(-\frac{1}{2}\right) = \frac{1}{4}$.

In this chapter, the opposite problem is considered.

If $a^2 = 49$, then $a = ?$
If $a^2 = 100$, then $a = ?$
If $a^2 = 25$, then $a = ?$

1 Finding a in the three statements above requires finding a number that can be multiplied by itself to result in the given number. The number a is called a **square root** of the number a^2.

EXAMPLE 1 Finding the Square Roots of a Number
Find all square roots of 49.

Find a square root of 49 by thinking of a number that multiplied by itself gives 49. One square root is 7, since $7 \cdot 7 = 49$. Another square root of 49 is -7, since $(-7)(-7) = 49$. The number 49 has two square roots, 7 and -7. One is positive and one is negative. ■

WORK PROBLEM 1 AT THE SIDE.

All numbers that have rational number square roots are called **perfect squares.** The first 100 perfect squares that are natural numbers can be found in Table 2 in the back of the book. (They are found in the column headed n^2.)

The positive square root of a number is written with the symbol $\sqrt{}$. For example, the positive square root of 121 is 11, written

$$\sqrt{121} = 11.$$

The symbol $-\sqrt{}$ is used for the negative square root of a number. For example, the negative square root of 121 is -11, written

$$-\sqrt{121} = -11.$$

OBJECTIVES

1. Find square roots.
2. Decide whether a given root is rational, irrational, or not a real number.
3. Find decimal approximations for irrational square roots.
4. Use the Pythagorean formula.
5. Use the Pythagorean formula to solve word problems.

1. Find all square roots.

(a) 100

(b) 25

(c) 36

(d) 64

ANSWERS
1. (a) 10, −10 (b) 5, −5 (c) 6, −6 (d) 8, −8

The symbol $\sqrt{}$ is called a **radical sign** and, used alone, always represents the positive square root (except that $\sqrt{0} = 0$). The number inside the radical sign is called the **radicand** and the entire expression, radical sign and radicand, is called a **radical.** An algebraic expression containing a radical is called a **radical expression.**

If a is a nonnegative real number,

$\sqrt{a}$ is the positive square root of a,

$-\sqrt{a}$ is the negative square root of a.

Also, for nonnegative a,

$$\sqrt{a} \cdot \sqrt{a} = (\sqrt{a})^2 = a \quad \text{and} \quad -\sqrt{a} \cdot -\sqrt{a} = (-\sqrt{a})^2 = a.$$

When the square root of a positive real number is squared, the result is that positive real number. (Also, $(\sqrt{0})^2 = 0$.) This is illustrated in the next example.

EXAMPLE 2 Squaring Radical Expressions

Find the *square* of each radical expression.

(a) $\sqrt{13}$

$(\sqrt{13})^2 = 13$, by the definition of square root.

(b) $-\sqrt{29}$

$(-\sqrt{29})^2 = 29$ The square of a *negative* number is positive.

(c) $\sqrt{p^2 + 1}$

$(\sqrt{p^2 + 1})^2 = p^2 + 1$ ■

WORK PROBLEM 2 AT THE SIDE.

EXAMPLE 3 Finding Square Roots

Find each square root.

(a) $\sqrt{144}$

The radical $\sqrt{144}$ represents the positive square root of 144. Think of a positive number whose square is 144. The number is 12 (see the table of perfect squares), and

$$\sqrt{144} = 12.$$

(b) $-\sqrt{1024}$

This symbol represents the negative square root of 1024. From the table,

$$-\sqrt{1024} = -32.$$

(c) $-\sqrt{\dfrac{16}{49}}$

Since $\left(-\dfrac{4}{7}\right)\left(-\dfrac{4}{7}\right) = \dfrac{16}{49}$, $-\sqrt{\dfrac{16}{49}} = -\dfrac{4}{7}$. ■

WORK PROBLEM 3 AT THE SIDE.

2. Find the *square* of each radical expression.

(a) $\sqrt{41}$

(b) $-\sqrt{39}$

(c) $\sqrt{2x^2 + 3}$

3. Find each square root.

(a) $\sqrt{16}$

(b) $-\sqrt{169}$

(c) $-\sqrt{225}$

(d) $\sqrt{729}$

(e) $\sqrt{\dfrac{36}{25}}$

ANSWERS

2. (a) 41 (b) 39 (c) $2x^2 + 3$

3. (a) 4 (b) -13 (c) -15 (d) 27 (e) $\dfrac{6}{5}$

2 A number that is not a perfect square has a square root that is not a rational number. For example, $\sqrt{5}$ is not a rational number, because it cannot be written as the ratio of two integers. However, $\sqrt{5}$ is a real number and corresponds to a point on the number line. As mentioned in Chapter 2, a real number that is not rational is called an **irrational number.** The number $\sqrt{5}$ is irrational. Many square roots of integers are irrational.

Not every number has a *real number* square root. For example, there is no real number that can be squared to get -36. (The square of a real number can never be negative.) Because of this, $\sqrt{-36}$ is not a real number.

If a is a negative real number, $\sqrt{a}$ is not a real number.

EXAMPLE 4 Identifying Types of Square Roots

Tell whether each square root is rational, irrational, or not a real number.

(a) $\sqrt{17}$

Since 17 is not a perfect square, $\sqrt{17}$ is irrational.

(b) $\sqrt{64}$

The number 64 is a perfect square, 8^2, so $\sqrt{64} = 8$, a rational number.

(c) $\sqrt{-25}$

There is no real number whose square is -25. Therefore, $\sqrt{-25}$ is not a real number. ■

WORK PROBLEM 4 AT THE SIDE.

Not all irrational numbers are square roots of integers. For example, π (approximately 3.14159) is an irrational number that is not a square root of any integer.

3 Even if a number is irrational, a decimal that approximates the number can be found by using a table or certain calculators. For square roots, the square root table can be used, as shown in Example 5.

EXAMPLE 5 Approximating Irrational Square Roots

Find a decimal approximation for each square root. Round answers to the nearest thousandth.

(a) $\sqrt{11}$

Using the square root key of a calculator gives $3.31662479 \approx 3.317$, where $\approx$ means "is approximately equal to." To use the square root table, find 11 at the left. The approximate square root is given in the column having $\sqrt{n}$ at the top. You should find that

$$\sqrt{11} \approx 3.317.$$

(b) $\sqrt{39} \approx 6.245$. Use the table or a calculator.

4. Tell whether each square root is *rational, irrational,* or *not a real number.*

(a) $\sqrt{9}$

(b) $\sqrt{7}$

(c) $\sqrt{\dfrac{4}{9}}$

(d) $\sqrt{72}$

(e) $\sqrt{-43}$

ANSWERS

4. (a) rational (b) irrational (c) rational (d) irrational (e) not a real number

5. Find a decimal approximation for each square root.

(a) $\sqrt{28}$

(b) $\sqrt{63}$

(c) $\sqrt{190}$

(d) $\sqrt{1000}$

6. Find the unknown side in each right triangle.

(a) $a = 7, b = 24$

(b) $c = 15, b = 13$

(c) $c = 11, a = 8$

ANSWERS
5. (a) 5.292 (b) 7.937 (c) 13.784 (d) 31.623
6. (a) 25 (b) $\sqrt{56} \approx 7.483$ (c) $\sqrt{57} \approx 7.550$

(c) $\sqrt{740}$

There is no 740 in the "n" column of the square root table. However, $740 = 74 \times 10$, so $\sqrt{740}$ can be found in the "$\sqrt{10n}$" column. Using the table or a calculator,

$$\sqrt{740} \approx 27.203. \blacksquare$$

WORK PROBLEM 5 AT THE SIDE.

4 One application of square roots comes from the Pythagorean formula. Recall from Section 5.6 that by this formula if c is the length of the hypotenuse of a right triangle, and a and b are the lengths of the two legs, then

$$a^2 + b^2 = c^2.$$

(See Figure 1.)

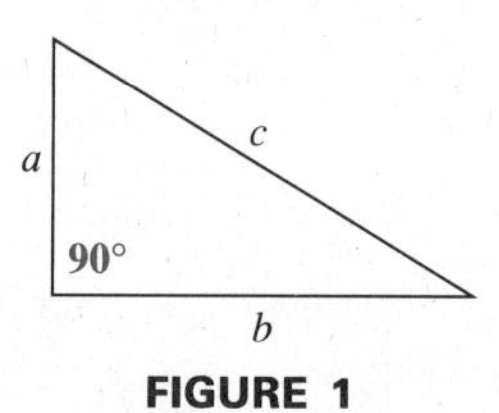

FIGURE 1

EXAMPLE 6 Using the Pythagorean Formula

Find the third side of each right triangle with sides a, b, and c, where c is the hypotenuse.

(a) $a = 3, b = 4$

Use the formula to find c^2 first.

$$c^2 = a^2 + b^2$$

$$c^2 = 3^2 + 4^2 \qquad \text{Let } a = 3 \text{ and } b = 4.$$

$$c^2 = 9 + 16 = 25 \qquad \text{Square and add.}$$

Now find the positive square root of 25 to get c.

$$c = \sqrt{25} = 5$$

(Although -5 is also a square root of 25, the length of a side of a triangle must be a positive number.)

(b) $c = 9, b = 5$

Substitute the given values in the formula, $c^2 = a^2 + b^2$. Then solve for a^2.

$$9^2 = a^2 + 5^2 \qquad \text{Let } c = 9 \text{ and } b = 5.$$

$$81 = a^2 + 25 \qquad \text{Square.}$$

$$56 = a^2$$

Use the table or a calculator to find $a = \sqrt{56} \approx 7.483$. $\blacksquare$

WORK PROBLEM 6 AT THE SIDE.

5 The Pythagorean formula can be used to solve word problems that involve right triangles.

EXAMPLE 7 Using the Pythagorean Formula

A ladder 10 feet long leans against a wall. The foot of the ladder is 6 feet from the base of the wall. How high up the wall does the top of the ladder rest?

As shown in Figure 2, a right triangle is formed with the ladder as the hypotenuse.

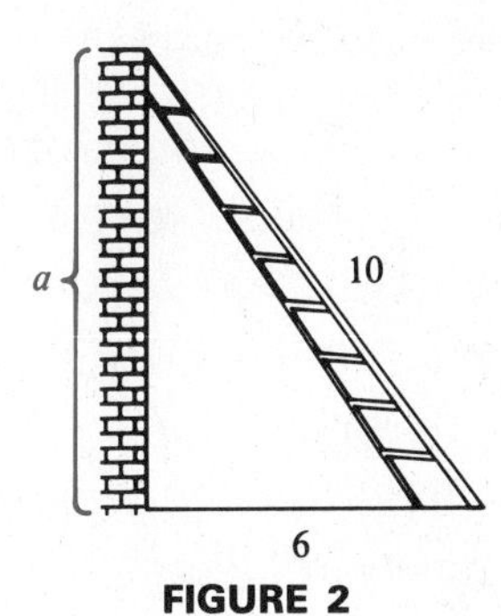

FIGURE 2

Let a represent the height of the top of the ladder.
By the Pythagorean formula,

$$c^2 = a^2 + b^2$$

$$10^2 = a^2 + 6^2 \qquad \text{Let } c = 10 \text{ and } b = 6.$$

$$100 = a^2 + 36 \qquad \text{Square.}$$

$$64 = a^2 \qquad \text{Subtract 36.}$$

$$\sqrt{64} = a$$

$$a = 8. \qquad \sqrt{64} = 8$$

Choose the positive square root of 64 since a represents a length. The top of the ladder rests 8 feet up the wall. ■

WORK PROBLEM 7 AT THE SIDE.

7. A rectangle has dimensions 5 feet by 12 feet. Find the length of its diagonal.

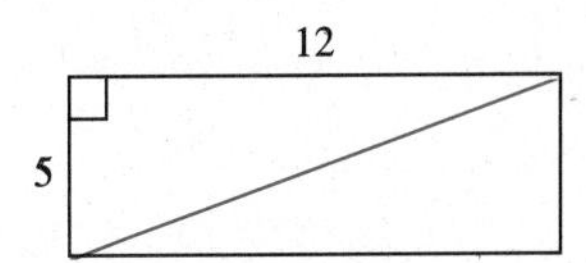

ANSWERS
7. 13 feet

Historical Reflections

Historical Reflections

Sir Isaac Newton (1642–1727)

"Nature and Nature's laws lay hid in night;
God said, 'Let Newton be,' and all was light."

This quote from Alexander Pope emphasizes the importance of the work done by Sir Isaac Newton, pictured on the French stamp here. In his epochal work *Philosophiae Naturalis Principia Mathematica* of 1687, he pictured a kind of "clockwork universe" that ran via his Law of Gravitation. Newton independently invented the branch of mathematics known as calculus, unaware that Gottfried W. Leibniz (1646–1716) had published his own formulation of it earlier. A controversy over their priority continued into the eighteenth century. Today the two are considered co-inventors of calculus.

name date hour

9.1 EXERCISES

Find all square roots of each number. See Example 1.

1. 9 **2.** 16 **3.** 121 **4.** $\frac{196}{25}$

5. $\frac{400}{81}$ **6.** $\frac{900}{49}$ **7.** 625 **8.** 961

Find the square of each radical expression. See Example 2.

9. $\sqrt{16}$ **10.** $\sqrt{25}$ **11.** $-\sqrt{57}$

12. $-\sqrt{97}$ **13.** $\sqrt{4x^2 + 3}$ **14.** $\sqrt{7y^2 + 2}$

Find each square root that is a real number. See Examples 3 and 4.

15. $\sqrt{25}$ **16.** $\sqrt{36}$ **17.** $\sqrt{64}$ **18.** $\sqrt{100}$

19. $-\sqrt{169}$ **20.** $-\sqrt{196}$ **21.** $\sqrt{900}$ **22.** $\sqrt{1600}$

23. $-\sqrt{1681}$ **24.** $-\sqrt{2116}$ **25.** $\sqrt{\frac{36}{49}}$ **26.** $\sqrt{\frac{121}{144}}$

27. $-\sqrt{\frac{100}{169}}$ **28.** $-\sqrt{5625}$ **29.** $\sqrt{-9}$ **30.** $\sqrt{-25}$

Write rational, irrational, *or* not a real number *for each number. If a number is rational, give its exact value. If a number is irrational, give a decimal approximation to the nearest thousandth for the square root. See Examples 4 and 5.*

31. $\sqrt{16}$ **32.** $\sqrt{81}$ **33.** $\sqrt{15}$ **34.** $\sqrt{31}$

35. $\sqrt{47}$ **36.** $\sqrt{53}$ **37.** $-\sqrt{121}$ **38.** $-\sqrt{144}$

39. $\sqrt{-121}$ **40.** $\sqrt{-144}$ **41.** $\sqrt{400}$ **42.** $\sqrt{900}$

43. $-\sqrt{200}$ **44.** $-\sqrt{260}$ **45.** $-\sqrt{-31}$ **46.** $-\sqrt{-29}$

Find the unknown side of each right triangle with sides a, b, and c, where c is the hypotenuse. See Example 6.

47. $a = 6, b = 8$ **48.** $a = 5, b = 12$ **49.** $c = 17, a = 8$ **50.** $c = 26, b = 10$

51. $a = 10, b = 8$ **52.** $a = 9, b = 7$ **53.** $c = 12, b = 7$ **54.** $c = 8, a = 3$

name date hour

Use the Pythagorean formula to solve each word problem. If necessary, round your answer to the nearest thousandth. See Example 7.

55. The diagonal of a rectangle measures 25 inches. The length of the rectangle is 24 inches. Find the width of the rectangle.

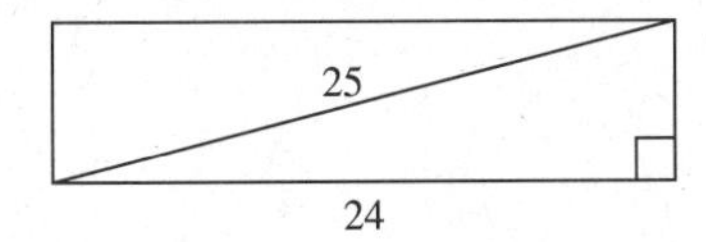

56. The width of a rectangle is 9 meters. The diagonal measures 41 meters. Find the length of the rectangle.

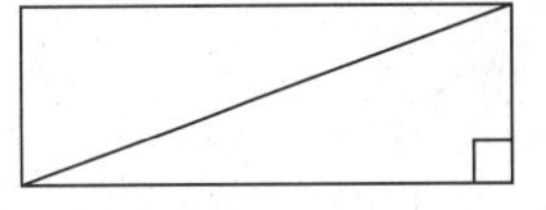

57. A guy wire is attached to the mast of a television antenna. It is attached 48 feet above ground level. If the wire is staked to the ground 36 feet from the base of the mast, how long is the wire?

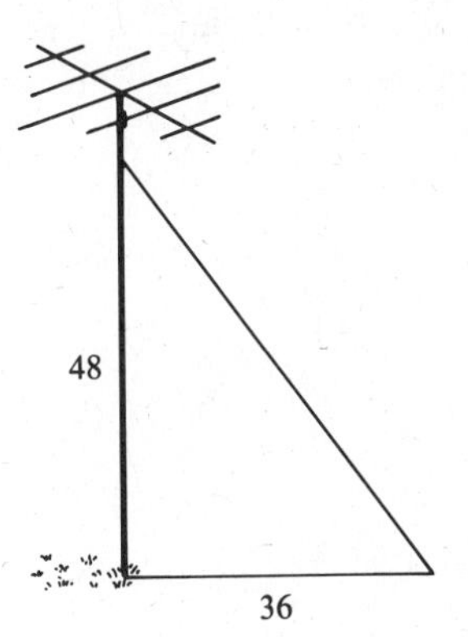

58. Margaret is flying a kite on 50 feet of string. How high is it above her hand if the horizontal distance between Margaret and the kite is 30 feet?

59. A boat is being pulled into a dock with a rope attached at water level. When the boat is 12 feet from the dock, 15 feet of rope is extended. Find the height of the dock above the water.

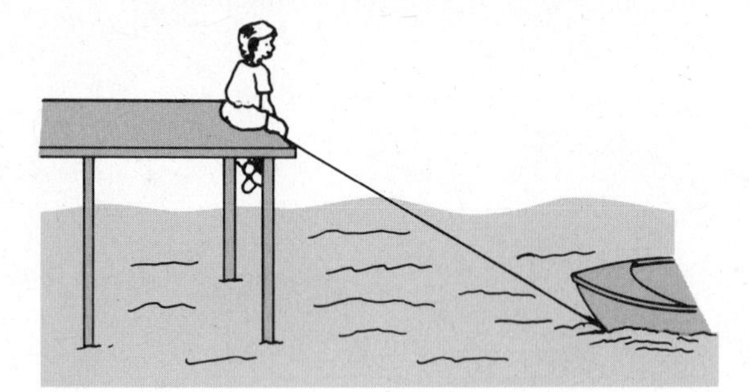

60. Two cars leave Manistee, Michigan, at the same time. One travels south at 25 miles per hour and the other travels east at 60 miles per hour. How far apart are they after 2 hours?

61. The hypotenuse of a right triangle measures 9 inches and one leg measures 3 inches. What is the measure of the other leg?

62. Find the length of the hypotenuse of a right triangle if the two legs measure 7 centimeters and 9 centimeters.

SKILL SHARPENERS

Write each number in prime factored form. See Section 5.1.

63. 24 **64.** 70 **65.** 300 **66.** 48

67. 72 **68.** 75 **69.** 150 **70.** 120

9.2 PRODUCTS AND QUOTIENTS OF RADICALS

OBJECTIVES

1. Multiply radicals.
2. Simplify radical expressions.
3. Simplify radical quotients.

1 Several useful rules for finding products and quotients of radicals are developed in this section. To illustrate the rule for products, notice that

$$\sqrt{4} \cdot \sqrt{9} = 2 \cdot 3 = 6 \quad \text{and} \quad \sqrt{4 \cdot 9} = \sqrt{36} = 6,$$

showing that

$$\sqrt{4} \cdot \sqrt{9} = \sqrt{4 \cdot 9}.$$

This result is a particular case of the more general *product rule for radicals.*

PRODUCT RULE FOR RADICALS

For nonnegative real numbers a and b,

$$\sqrt{a} \cdot \sqrt{b} = \sqrt{a \cdot b} \quad \text{and} \quad \sqrt{a \cdot b} = \sqrt{a} \cdot \sqrt{b}.$$

The radical of a product is the product of the radicals, and the product of two radicals is the radical of the product.

EXAMPLE 1 Using the Product Rule to Multiply Radicals

Use the product rule for radicals to find each product.

(a) $\sqrt{2} \cdot \sqrt{3} = \sqrt{2 \cdot 3} = \sqrt{6}$

(b) $\sqrt{7} \cdot \sqrt{5} = \sqrt{35}$

(c) $\sqrt{11} \cdot \sqrt{a} = \sqrt{11a}$ Assume $a > 0$. ■

WORK PROBLEM 1 AT THE SIDE.

1. Use the product rule for radicals to find each product.

(a) $\sqrt{6} \cdot \sqrt{11}$

(b) $\sqrt{2} \cdot \sqrt{7}$

(c) $\sqrt{17} \cdot \sqrt{3}$

(d) $\sqrt{10} \cdot \sqrt{r}, \quad r > 0$

2 A very important use of the product rule is in simplifying radical expressions: as a first step, a radical expression is *simplified* when no perfect square factor remains under the radical sign. This is accomplished by using the product rule in the form $\sqrt{a \cdot b} = \sqrt{a} \cdot \sqrt{b}$. Example 2 shows how a radical may be simplified using the product rule in this way.

EXAMPLE 2 Using the Product Rule to Simplify Radicals

Simplify each radical.

(a) $\sqrt{20}$

Since 20 has a perfect square factor of 4,

$$\begin{aligned} \sqrt{20} &= \sqrt{4 \cdot 5} && \text{4 is a perfect square.} \\ &= \sqrt{4} \cdot \sqrt{5} && \text{Product rule} \\ &= 2\sqrt{5}. && \sqrt{4} = 2 \end{aligned}$$

Thus, $\sqrt{20} = 2\sqrt{5}$. Since 5 has no perfect square factor (other than 1), $2\sqrt{5}$ is called the *simplified form* of $\sqrt{20}$.

ANSWERS
1. (a) $\sqrt{66}$ (b) $\sqrt{14}$ (c) $\sqrt{51}$ (d) $\sqrt{10r}$

2. Simplify each radical.

(a) $\sqrt{8}$

(b) $\sqrt{27}$

(c) $\sqrt{50}$

(d) $\sqrt{60}$

3. Find each product and simplify.

(a) $\sqrt{3} \cdot \sqrt{15}$

(b) $\sqrt{7} \cdot \sqrt{14}$

(c) $\sqrt{10} \cdot \sqrt{50}$

(d) $\sqrt{12} \cdot \sqrt{12}$

ANSWERS
2. (a) $2\sqrt{2}$ (b) $3\sqrt{3}$ (c) $5\sqrt{2}$ (d) $2\sqrt{15}$
3. (a) $3\sqrt{5}$ (b) $7\sqrt{2}$ (c) $10\sqrt{5}$ (d) 12

(b) $\sqrt{72}$

Begin by looking for the largest perfect square that is a factor of 72. This number is 36, so

$$\begin{aligned} \sqrt{72} &= \sqrt{36 \cdot 2} && \text{36 is a perfect square.} \\ &= \sqrt{36} \cdot \sqrt{2} && \text{Product rule} \\ &= 6\sqrt{2}. && \sqrt{36} = 6 \end{aligned}$$

Notice that 9 is a perfect square factor of 72. We could begin by factoring 72 as $9 \cdot 8$, to get

$$\sqrt{72} = \sqrt{9 \cdot 8} = 3\sqrt{8},$$

but then we would have to factor 8 as $4 \cdot 2$ in order to complete the simplification.

$$\sqrt{72} = 3\sqrt{8} = 3\sqrt{4 \cdot 2} = 3\sqrt{4} \cdot \sqrt{2} = 3 \cdot 2\sqrt{2} = 6\sqrt{2}$$

In either case, we obtain $6\sqrt{2}$ as the simplified form of $\sqrt{72}$; however, our work is simpler if we begin with the largest perfect square factor.

(c)
$$\begin{aligned} \sqrt{300} &= \sqrt{100 \cdot 3} && \text{100 is a perfect square.} \\ &= \sqrt{100} \cdot \sqrt{3} && \text{Product rule} \\ &= 10\sqrt{3} && \sqrt{100} = 10 \end{aligned}$$

(d) $\sqrt{15}$

The number 15 has no perfect square factors (except 1), so $\sqrt{15}$ cannot be simplified further. ■

WORK PROBLEM 2 AT THE SIDE.

Sometimes the product rule can be used to simplify an answer, as Example 3 shows.

EXAMPLE 3 Multiplying and Simplifying Radicals
Find each product and simplify.

(a)
$$\begin{aligned} \sqrt{9} \cdot \sqrt{75} &= 3\sqrt{75} && \sqrt{9} = 3 \\ &= 3\sqrt{25 \cdot 3} && \text{25 is a perfect square.} \\ &= 3\sqrt{25} \cdot \sqrt{3} && \text{Product rule} \\ &= 3 \cdot 5\sqrt{3} && \sqrt{25} = 5 \\ &= 15\sqrt{3} && \text{Multiply.} \end{aligned}$$

Notice that we could have used the product rule to get $\sqrt{9} \cdot \sqrt{75} = \sqrt{675}$, and then simplified. However, the product rule as used above allows us to obtain the final answer without using a number as large as 675.

(b)
$$\begin{aligned} \sqrt{8} \cdot \sqrt{12} &= \sqrt{8 \cdot 12} && \text{Product rule} \\ &= \sqrt{96} && \text{Multiply.} \\ &= \sqrt{16 \cdot 6} && \text{16 is a perfect square.} \\ &= \sqrt{16} \cdot \sqrt{6} && \text{Product rule} \\ &= 4\sqrt{6} && \sqrt{16} = 4 \end{aligned}$$ ■

WORK PROBLEM 3 AT THE SIDE.

3 The quotient rule for radicals is very similar to the product rule. It, too, can be used either way.

QUOTIENT RULE FOR RADICALS

If a and b are nonnegative real numbers and b is not 0,

$$\sqrt{\frac{a}{b}} = \frac{\sqrt{a}}{\sqrt{b}} \quad \text{and} \quad \frac{\sqrt{a}}{\sqrt{b}} = \sqrt{\frac{a}{b}}.$$

The radical of a quotient is the quotient of the radicals, and the quotient of two radicals is the radical of the quotient.

EXAMPLE 4 Using the Quotient Rule to Simplify Radicals

Simplify each radical.

(a) $\sqrt{\frac{25}{9}} = \frac{\sqrt{25}}{\sqrt{9}} = \frac{5}{3}$ Quotient rule

(b) $\frac{\sqrt{288}}{\sqrt{2}} = \sqrt{\frac{288}{2}} = \sqrt{144} = 12$ Quotient rule

(c) $\sqrt{\frac{3}{4}} = \frac{\sqrt{3}}{\sqrt{4}} = \frac{\sqrt{3}}{2}$ Quotient rule ■

EXAMPLE 5 Using the Quotient Rule

Simplify $\frac{27\sqrt{15}}{9\sqrt{3}}$.

Use the quotient rule as follows.

$$\frac{27\sqrt{15}}{9\sqrt{3}} = \frac{27}{9} \cdot \frac{\sqrt{15}}{\sqrt{3}} = \frac{27}{9} \cdot \sqrt{\frac{15}{3}} = 3\sqrt{5}$$ ■

WORK PROBLEM 4 AT THE SIDE.

Some problems require both the product and the quotient rules, as Example 6 shows.

EXAMPLE 6 Using Both the Product and Quotient Rules

Simplify $\sqrt{\frac{3}{5}} \cdot \sqrt{\frac{4}{5}}$.

Use the product and quotient rules.

$$\sqrt{\frac{3}{5}} \cdot \sqrt{\frac{4}{5}} = \frac{\sqrt{3}}{\sqrt{5}} \cdot \frac{\sqrt{4}}{\sqrt{5}}$$ Quotient rule

$$= \frac{\sqrt{3} \cdot \sqrt{4}}{\sqrt{5} \cdot \sqrt{5}}$$ Multiply fractions.

$$= \frac{\sqrt{3} \cdot 2}{\sqrt{25}}$$ Product rule; $\sqrt{4} = 2$

$$= \frac{2\sqrt{3}}{5}$$ $\sqrt{25} = 5$ ■

WORK PROBLEM 5 AT THE SIDE.

4. Use the quotient rule to simplify each radical.

(a) $\sqrt{\frac{81}{16}}$

(b) $\sqrt{\frac{100}{9}}$

(c) $\sqrt{\frac{10}{49}}$

(d) $\frac{8\sqrt{50}}{4\sqrt{5}}$

5. Multiply and then simplify each product.

(a) $\sqrt{\frac{5}{6}} \cdot \sqrt{120}$

(b) $\sqrt{\frac{3}{8}} \cdot \sqrt{\frac{7}{2}}$

ANSWERS

4. (a) $\frac{9}{4}$ (b) $\frac{10}{3}$ (c) $\frac{\sqrt{10}}{7}$ (d) $2\sqrt{10}$

5. (a) 10 (b) $\frac{\sqrt{21}}{4}$

6. Simplify each radical. Assume all variables represent positive real numbers.

(a) $\sqrt{36y^6}$

(b) $\sqrt{100p^8}$

(c) $\sqrt{a^5}$

(d) $\sqrt{\frac{7}{p^2}}$

Finally, the properties of this section are also valid when variables appear under the radical sign, as long as all the variables represent only nonnegative real numbers. For example, $\sqrt{5^2} = 5$, but $\sqrt{(-5)^2} = \sqrt{25} \neq -5$.

For a real number a,

$$\sqrt{a^2} = a$$

only if a is nonnegative.

EXAMPLE 7 Simplifying Radicals Involving Variables

Simplify each radical. Assume all variables represent positive real numbers.

(a) $\sqrt{25m^4} = \sqrt{25} \cdot \sqrt{m^4}$ Product rule
$= 5m^2$

(b) $\sqrt{64p^{10}} = 8p^5$ Product rule

(c) $\sqrt{r^9} = \sqrt{r^8 \cdot r}$
$= \sqrt{r^8} \cdot \sqrt{r} = r^4\sqrt{r}$ Product rule

(d) $\sqrt{\frac{5}{x^2}} = \frac{\sqrt{5}}{\sqrt{x^2}} = \frac{\sqrt{5}}{x}$ Quotient rule ■

WORK PROBLEM 6 AT THE SIDE.

ANSWERS

6. (a) $6y^3$ (b) $10p^4$ (c) $a^2\sqrt{a}$ (d) $\frac{\sqrt{7}}{p}$

name date hour

9.2 EXERCISES

Use the product rule to simplify each radical or radical expression. See Examples 1–3.

1. $\sqrt{8} \cdot \sqrt{2}$

2. $\sqrt{27} \cdot \sqrt{3}$

3. $\sqrt{6} \cdot \sqrt{6}$

4. $\sqrt{11} \cdot \sqrt{11}$

5. $\sqrt{21} \cdot \sqrt{21}$

6. $\sqrt{17} \cdot \sqrt{17}$

7. $\sqrt{3} \cdot \sqrt{7}$

8. $\sqrt{2} \cdot \sqrt{5}$

9. $\sqrt{27}$

10. $\sqrt{45}$

11. $\sqrt{28}$

12. $\sqrt{40}$

13. $\sqrt{18}$

14. $\sqrt{75}$

15. $\sqrt{48}$

16. $\sqrt{80}$

17. $\sqrt{125}$

18. $\sqrt{150}$

19. $\sqrt{700}$

20. $\sqrt{1100}$

21. $10\sqrt{27}$

22. $4\sqrt{8}$

23. $2\sqrt{20}$

24. $5\sqrt{80}$

25. $\sqrt{27} \cdot \sqrt{48}$

26. $\sqrt{75} \cdot \sqrt{27}$

27. $\sqrt{50} \cdot \sqrt{72}$

28. $\sqrt{98} \cdot \sqrt{8}$

29. $\sqrt{7} \cdot \sqrt{21}$

30. $\sqrt{12} \cdot \sqrt{48}$

31. $\sqrt{15} \cdot \sqrt{45}$

32. $\sqrt{20} \cdot \sqrt{45}$

33. $\sqrt{80} \cdot \sqrt{15}$

34. $\sqrt{60} \cdot \sqrt{12}$

35. $\sqrt{50} \cdot \sqrt{20}$

36. $\sqrt{72} \cdot \sqrt{12}$

Use the quotient rule and the product rule, as necessary, to simplify each radical or radical expression. See Examples 4–6.

37. $\sqrt{\frac{100}{9}}$

38. $\sqrt{\frac{225}{16}}$

39. $\sqrt{\frac{36}{49}}$

40. $\sqrt{\frac{256}{9}}$

name date hour

41. $\sqrt{\frac{5}{16}}$

42. $\sqrt{\frac{11}{25}}$

43. $\sqrt{\frac{30}{49}}$

44. $\sqrt{\frac{10}{121}}$

45. $\sqrt{\frac{1}{5}} \cdot \sqrt{\frac{4}{5}}$

46. $\sqrt{\frac{2}{3}} \cdot \sqrt{\frac{2}{27}}$

47. $\sqrt{\frac{2}{5}} \cdot \sqrt{\frac{8}{125}}$

48. $\sqrt{\frac{3}{8}} \cdot \sqrt{\frac{3}{2}}$

49. $\frac{\sqrt{75}}{\sqrt{3}}$

50. $\frac{\sqrt{200}}{\sqrt{2}}$

51. $\frac{\sqrt{48}}{\sqrt{3}}$

52. $\frac{\sqrt{72}}{\sqrt{8}}$

53. $\frac{15\sqrt{10}}{5\sqrt{2}}$

54. $\frac{18\sqrt{20}}{2\sqrt{10}}$

55. $\frac{25\sqrt{50}}{5\sqrt{5}}$

56. $\frac{26\sqrt{10}}{13\sqrt{5}}$

Simplify each radical or radical expression. Assume that all variables represent positive real numbers. See Example 7.

57. $\sqrt{y} \cdot \sqrt{y}$

58. $\sqrt{m} \cdot \sqrt{m}$

59. $\sqrt{x} \cdot \sqrt{z}$

60. $\sqrt{p} \cdot \sqrt{q}$

61. $\sqrt{x^2}$

62. $\sqrt{y^2}$

63. $\sqrt{x^4}$

64. $\sqrt{y^4}$

65. $\sqrt{x^2y^4}$

66. $\sqrt{x^4y^8}$

67. $\sqrt{x^3}$

68. $\sqrt{y^3}$

69. $\sqrt{\frac{16}{x^2}}$

70. $\sqrt{\frac{100}{m^4}}$

71. $\sqrt{\frac{11}{r^4}}$

72. $\sqrt{\frac{23}{y^6}}$

SKILL SHARPENERS

Rewrite each fraction with the given denominator. See Section 6.3.

73. $\frac{9}{13} = \frac{}{39}$

74. $\frac{5}{11} = \frac{}{66}$

75. $\frac{2}{3} = \frac{}{15}$

76. $\frac{3}{14} = \frac{}{56}$

77. $\frac{2}{x} = \frac{}{3x}$

78. $\frac{5}{2y} = \frac{}{2y^2}$

79. $\frac{5}{\sqrt{3}} = \frac{}{2\sqrt{3}}$

80. $\frac{8}{\sqrt{5}} = \frac{}{3\sqrt{5}}$

9.3 ADDITION AND SUBTRACTION OF RADICALS

OBJECTIVES

1. Add and subtract radical expressions.
2. Simplify radical expressions before adding or subtracting.
3. Simplify radical expressions with multiplication.

1 We add or subtract radical expressions by using the distributive property. For example,

$$8\sqrt{3} + 6\sqrt{3} = (8 + 6)\sqrt{3} = 14\sqrt{3}.$$

Also,

$$2\sqrt{11} - 7\sqrt{11} = -5\sqrt{11}.$$

Only **like radicals,** those that are multiples of the same root of the same number, can be combined in this way.

EXAMPLE 1 Adding and Subtracting Like Radicals

Add or subtract, as indicated.

(a) $3\sqrt{6} + 5\sqrt{6} = (3 + 5)\sqrt{6} = 8\sqrt{6}$ Distributive property

(b) $5\sqrt{10} - 7\sqrt{10} = (5 - 7)\sqrt{10} = -2\sqrt{10}$

(c) $\sqrt{7} + 2\sqrt{7} = 1\sqrt{7} + 2\sqrt{7} = 3\sqrt{7}$

(d) $\sqrt{5} + \sqrt{5} = 1\sqrt{5} + 1\sqrt{5} = (1 + 1)\sqrt{5} = 2\sqrt{5}$

(e) $\sqrt{3} + \sqrt{7}$ Cannot be further simplified. ■

WORK PROBLEM 1 AT THE SIDE.

1. Add or subtract, as indicated.

(a) $8\sqrt{5} + 2\sqrt{5}$

(b) $-4\sqrt{3} + 9\sqrt{3}$

(c) $12\sqrt{11} - 3\sqrt{11}$

(d) $\sqrt{15} + \sqrt{15}$

(e) $2\sqrt{7} + 2\sqrt{10}$

2 Sometimes each radical expression in a sum or difference must be simplified first. Doing this might cause like radicals to appear, which then can be added or subtracted.

EXAMPLE 2 Adding and Subtracting Radicals that Require Simplification

Add or subtract as indicated.

(a)
$$\begin{aligned} 3\sqrt{2} + \sqrt{8} &= 3\sqrt{2} + \sqrt{4 \cdot 2} && \text{Simplify } \sqrt{8}. \\ &= 3\sqrt{2} + \sqrt{4} \cdot \sqrt{2} && \text{Product rule} \\ &= 3\sqrt{2} + 2\sqrt{2} && \sqrt{4} = 2 \\ &= 5\sqrt{2} && \text{Add like radicals.} \end{aligned}$$

(b)
$$\begin{aligned} \sqrt{18} - \sqrt{27} &= \sqrt{9 \cdot 2} - \sqrt{9 \cdot 3} && \text{Simplify } \sqrt{18} \text{ and } \sqrt{27}. \\ &= \sqrt{9} \cdot \sqrt{2} - \sqrt{9} \cdot \sqrt{3} && \text{Product rule} \\ &= 3\sqrt{2} - 3\sqrt{3} && \sqrt{9} = 3 \end{aligned}$$

Since $\sqrt{2}$ and $\sqrt{3}$ are unlike radicals, this difference cannot be simplified further.

(c)
$$\begin{aligned} 2\sqrt{12} + 3\sqrt{75} &= 2(\sqrt{4} \cdot \sqrt{3}) + 3(\sqrt{25} \cdot \sqrt{3}) && \text{Product rule} \\ &= 2(2\sqrt{3}) + 3(5\sqrt{3}) && \sqrt{4} = 2 \text{ and } \sqrt{25} = 5 \\ &= 4\sqrt{3} + 15\sqrt{3} && \text{Multiply.} \\ &= 19\sqrt{3} && \text{Add like radicals.} \end{aligned}$$

ANSWERS

1. (a) $10\sqrt{5}$ (b) $5\sqrt{3}$ (c) $9\sqrt{11}$ (d) $2\sqrt{15}$ (e) cannot be simplified further

2. Add or subtract as indicated.

(a) $\sqrt{8} + 4\sqrt{2}$

(b) $\sqrt{27} + \sqrt{12}$

(c) $5\sqrt{200} - 6\sqrt{18}$

(d) $3\sqrt{45} + 14\sqrt{20}$

3. Multiply and combine terms. Assume all variables represent nonnegative real numbers.

(a) $\sqrt{7} \cdot \sqrt{21} + 2\sqrt{27}$

(b) $\sqrt{3} \cdot \sqrt{48} + 5\sqrt{3}$

(c) $\sqrt{3r} \cdot \sqrt{6} + \sqrt{8r}$

(d) $6\sqrt{5m} - 9\sqrt{m} \cdot \sqrt{125}$

ANSWERS
2. (a) $6\sqrt{2}$ (b) $5\sqrt{3}$ (c) $32\sqrt{2}$ (d) $37\sqrt{5}$
3. (a) $13\sqrt{3}$ (b) $12 + 5\sqrt{3}$ (c) $5\sqrt{2r}$ (d) $-39\sqrt{5m}$

(d) $2\sqrt{125} + 3\sqrt{80} = 2(\sqrt{25 \cdot 5}) + 3(\sqrt{16 \cdot 5})$ Product rule

$= 2(5\sqrt{5}) + 3(4\sqrt{5})$ $\sqrt{25} = 5$ and $\sqrt{16} = 4$

$= 10\sqrt{5} + 12\sqrt{5}$ Multiply.

$= 22\sqrt{5}$ Add like radicals. ■

WORK PROBLEM 2 AT THE SIDE.

3 Some radical expressions require both multiplication and addition (or subtraction). The order of operations presented earlier still applies.

EXAMPLE 3 Multiplying and Combining Terms in Radical Expressions

Multiply and combine terms. Assume all variables represent nonnegative real numbers.

(a) $\sqrt{5} \cdot \sqrt{15} + 4\sqrt{3} = \sqrt{5 \cdot 15} + 4\sqrt{3}$ Product rule

$= \sqrt{75} + 4\sqrt{3}$ Multiply.

$= \sqrt{25 \cdot 3} + 4\sqrt{3}$ 25 is a perfect square.

$= \sqrt{25} \cdot \sqrt{3} + 4\sqrt{3}$ Product rule

$= 5\sqrt{3} + 4\sqrt{3}$ $\sqrt{25} = 5$

$= 9\sqrt{3}$ Add like radicals.

(b) $\sqrt{2} \cdot \sqrt{6k} + \sqrt{27k} = \sqrt{12k} + \sqrt{27k}$ Product rule

$= \sqrt{4 \cdot 3k} + \sqrt{9 \cdot 3k}$ Factor.

$= \sqrt{4} \cdot \sqrt{3k} + \sqrt{9} \cdot \sqrt{3k}$ Product rule

$= 2\sqrt{3k} + 3\sqrt{3k}$ $\sqrt{4} = 2$ and $\sqrt{9} = 3$

$= 5\sqrt{3k}$ Add like radicals.

(c) $3\sqrt{7x} - 8\sqrt{x} \cdot \sqrt{28} = 3\sqrt{7x} - 8\sqrt{x} \cdot \sqrt{4} \cdot \sqrt{7}$ Product rule

$= 3\sqrt{7x} - 8\sqrt{x} \cdot 2 \cdot \sqrt{7}$ $\sqrt{4} = 2$

$= 3\sqrt{7x} - 16\sqrt{7x}$ Multiply.

$= -13\sqrt{7x}$ Subtract like radicals. ■

WORK PROBLEM 3 AT THE SIDE.

CAUTION A sum or difference of radicals can be simplified only if the radicals are **like radicals.**

For example,

$$\sqrt{5} + 3\sqrt{5} = 4\sqrt{5},$$

but $\sqrt{5} + 5\sqrt{3}$ cannot be simplified further.

name date hour

9.3 EXERCISES

Simplify and add or subtract wherever possible. See Examples 1–2.

1. $2\sqrt{3} + 5\sqrt{3}$

2. $6\sqrt{5} + 8\sqrt{5}$

3. $4\sqrt{7} - 9\sqrt{7}$

4. $6\sqrt{2} - 8\sqrt{2}$

5. $\sqrt{6} + \sqrt{6}$

6. $\sqrt{11} + \sqrt{11}$

7. $\sqrt{17} + 2\sqrt{17}$

8. $3\sqrt{19} + \sqrt{19}$

9. $5\sqrt{7} - \sqrt{7}$

10. $12\sqrt{14} - \sqrt{14}$

11. $\sqrt{45} + 2\sqrt{20}$

12. $\sqrt{24} + 5\sqrt{54}$

13. $-\sqrt{12} + \sqrt{75}$

14. $2\sqrt{27} - \sqrt{300}$

15. $5\sqrt{72} - 2\sqrt{50}$

16. $6\sqrt{18} - 4\sqrt{32}$

17. $-5\sqrt{32} + \sqrt{98}$

18. $4\sqrt{75} + 3\sqrt{12}$

19. $5\sqrt{7} - 2\sqrt{28} + 6\sqrt{63}$

20. $3\sqrt{11} + 5\sqrt{44} - 3\sqrt{99}$

21. $6\sqrt{5} + 3\sqrt{20} - 8\sqrt{45}$

22. $7\sqrt{3} + 2\sqrt{12} - 5\sqrt{27}$

23. $2\sqrt{8} - 5\sqrt{32} + 2\sqrt{48}$

24. $5\sqrt{72} - 3\sqrt{48} - 4\sqrt{128}$

25. $4\sqrt{50} + 3\sqrt{12} + 5\sqrt{45}$

26. $6\sqrt{18} + 2\sqrt{48} - 6\sqrt{28}$

27. $\frac{1}{4}\sqrt{288} - \frac{1}{6}\sqrt{72}$

28. $\frac{2}{3}\sqrt{27} - \frac{3}{4}\sqrt{48}$

29. $\frac{3}{5}\sqrt{75} - \frac{2}{3}\sqrt{45}$

30. $\frac{5}{8}\sqrt{128} - \frac{3}{4}\sqrt{160}$

Perform the indicated operations. Assume that all variables represent nonnegative real numbers. See Example 3.

31. $\sqrt{3} \cdot \sqrt{7} + 2\sqrt{21}$

32. $\sqrt{13} \cdot \sqrt{2} + 3\sqrt{26}$

33. $\sqrt{6} \cdot \sqrt{2} + 3\sqrt{3}$

34. $4\sqrt{15} \cdot \sqrt{3} - 2\sqrt{5}$

35. $\sqrt{9x} + \sqrt{49x} - \sqrt{16x}$

36. $\sqrt{4a} - \sqrt{16a} + \sqrt{9a}$

37. $\sqrt{4a} + 6\sqrt{a} + \sqrt{25a}$

38. $\sqrt{6x^2} + x\sqrt{54}$

39. $\sqrt{75x^2} + x\sqrt{300}$

40. $\sqrt{20y^2} - 3y\sqrt{5}$

41. $3\sqrt{8x^2} - 4x\sqrt{2}$

42. $\sqrt{2b^2} + 3b\sqrt{18}$

43. $5\sqrt{75p^2} - 4\sqrt{27p^2}$

44. $-3\sqrt{32k} + 6\sqrt{8k}$

45. $2\sqrt{125x^2z} + 8x\sqrt{80z}$

46. $4p\sqrt{14m} - 6\sqrt{28mp^2}$

SKILL SHARPENERS

Find each product. See Section 4.6.

47. $(m + 3)(m + 5)$

48. $(p - 2)(p + 7)$

49. $(2k - 3)(3k - 4)$

50. $(4z + 2)(5z - 8)$

51. $(3x - 1)^2$

52. $(2y + 5)^2$

53. $(a + 3)(a - 3)$

54. $(2x + 3)(2x - 3)$

55. $(4k + 5)(4k - 5)$

9.4 RATIONALIZING THE DENOMINATOR

1 Decimal approximations for radicals were found in the first section of this chapter. For more complicated radical expressions, it is easier to find these decimals if the denominators do not contain any radicals. For example, the radical in the denominator of

$$\frac{\sqrt{3}}{\sqrt{2}}$$

can be eliminated by multiplying the numerator and the denominator by $\sqrt{2}$.

$$\frac{\sqrt{3}}{\sqrt{2}} = \frac{\sqrt{3} \cdot \sqrt{2}}{\sqrt{2} \cdot \sqrt{2}} = \frac{\sqrt{6}}{2} \qquad \sqrt{2} \cdot \sqrt{2} = \sqrt{4} = 2$$

This process of changing the denominator from a radical (irrational number) to a rational number is called **rationalizing the denominator.** The value of the radical expression is not changed; only the form is changed, because the expression has been multiplied by 1 in the form of $\sqrt{2}/\sqrt{2}$.

EXAMPLE 1 Rationalizing Denominators

Rationalize each denominator.

(a) $\frac{9}{\sqrt{6}}$

Multiply both numerator and denominator by $\sqrt{6}$.

$$\frac{9}{\sqrt{6}} = \frac{9 \cdot \sqrt{6}}{\sqrt{6} \cdot \sqrt{6}}$$

$$= \frac{9\sqrt{6}}{6} \qquad \sqrt{6} \cdot \sqrt{6} = 6$$

$$= \frac{3\sqrt{6}}{2} \qquad \text{Reduce.}$$

(b) $\frac{12}{\sqrt{8}}$

The denominator could be rationalized here by multiplying by $\sqrt{8}$. However, the result can be found more directly by multiplying numerator and denominator by $\sqrt{2}$. This is because $\sqrt{8} \cdot \sqrt{2} = \sqrt{16} = 4$, a rational number.

$$\frac{12}{\sqrt{8}} = \frac{12 \cdot \sqrt{2}}{\sqrt{8} \cdot \sqrt{2}} \qquad \text{Multiply by } \sqrt{2} \text{ in numerator and denominator.}$$

$$= \frac{12\sqrt{2}}{\sqrt{16}} \qquad \text{Product rule}$$

$$= \frac{12\sqrt{2}}{4} \qquad \sqrt{16} = 4$$

$$= 3\sqrt{2} \qquad \text{Reduce.}$$ ■

WORK PROBLEM 1 AT THE SIDE.

OBJECTIVES

1 Rationalize denominators with square roots.

2 Write radicals in simplified form.

1. Rationalize each denominator.

(a) $\frac{3}{\sqrt{5}}$

(b) $\frac{-6}{\sqrt{11}}$

(c) $\frac{\sqrt{7}}{\sqrt{2}}$

(d) $\frac{20}{\sqrt{18}}$

(e) $\frac{-12}{\sqrt{32}}$

ANSWERS

1. (a) $\frac{3\sqrt{5}}{5}$ (b) $\frac{-6\sqrt{11}}{11}$ (c) $\frac{\sqrt{14}}{2}$ (d) $\frac{10\sqrt{2}}{3}$ (e) $\frac{-3\sqrt{2}}{2}$

2. Simplify by rationalizing each denominator.

(a) $\sqrt{\frac{1}{7}}$

(b) $\sqrt{\frac{3}{2}}$

(c) $\sqrt{\frac{5}{18}}$

(d) $\sqrt{\frac{16}{11}}$

ANSWERS
2. (a) $\frac{\sqrt{7}}{7}$ (b) $\frac{\sqrt{6}}{2}$ (c) $\frac{\sqrt{10}}{6}$
(d) $\frac{4\sqrt{11}}{11}$

2 A radical is considered to be in simplified form if the following three conditions are met.

SIMPLIFIED FORM OF A RADICAL

1. The radicand contains no factor (except 1) that is a perfect square.
2. The radicand has no fractions.
3. No denominator contains a radical.

In the following examples, radicals are simplified according to these conditions.

EXAMPLE 2 Simplifying a Radical

Simplify $\sqrt{\frac{27}{5}}$.

First use the quotient rule for radicals.

$$\sqrt{\frac{27}{5}} = \frac{\sqrt{27}}{\sqrt{5}}$$

Now multiply both numerator and denominator by $\sqrt{5}$.

$$\frac{\sqrt{27}}{\sqrt{5}} = \frac{\sqrt{27} \cdot \sqrt{5}}{\sqrt{5} \cdot \sqrt{5}}$$

$$= \frac{\sqrt{9 \cdot 3} \cdot \sqrt{5}}{5} \quad \text{Product rule; } \sqrt{5} \cdot \sqrt{5} = 5$$

$$= \frac{\sqrt{9} \cdot \sqrt{3} \cdot \sqrt{5}}{5} \quad \text{Product rule}$$

$$= \frac{3 \cdot \sqrt{3} \cdot \sqrt{5}}{5} \quad \sqrt{9} = 3$$

$$= \frac{3\sqrt{15}}{5} \quad \text{Product rule} \ \blacksquare$$

WORK PROBLEM 2 AT THE SIDE.

EXAMPLE 3 Simplifying a Radical Product

Simplify $\sqrt{\frac{5}{8}} \cdot \sqrt{\frac{1}{6}}$.

Use both the quotient rule and the product rule.

$$\sqrt{\frac{5}{8}} \cdot \sqrt{\frac{1}{6}} = \sqrt{\frac{5}{8} \cdot \frac{1}{6}} \quad \text{Product rule}$$

$$= \sqrt{\frac{5}{48}} \quad \text{Multiply fractions.}$$

$$= \frac{\sqrt{5}}{\sqrt{48}} \quad \text{Quotient rule}$$

Now rationalize the denominator by multiplying both the numerator and the denominator by $\sqrt{3}$ (since $\sqrt{48} \cdot \sqrt{3} = \sqrt{48 \cdot 3} = \sqrt{144} = 12$). One way to find the smallest radical to multiply by is to first factor the denominator: $\sqrt{48} = \sqrt{3 \cdot 16} = \sqrt{3} \cdot 4$. Since $\sqrt{3} \cdot \sqrt{3} = 3$, multiplying by $\sqrt{3}$ will produce a rational denominator, as required.

$$\frac{\sqrt{5}}{\sqrt{48}} = \frac{\sqrt{5} \cdot \sqrt{3}}{\sqrt{48} \cdot \sqrt{3}}$$

$$= \frac{\sqrt{15}}{\sqrt{144}} \qquad \text{Product rule}$$

$$= \frac{\sqrt{15}}{12} \qquad \sqrt{144} = 12 \quad \blacksquare$$

WORK PROBLEM 3 AT THE SIDE.

EXAMPLE 4 Simplifying a Radical Expression

Simplify $\frac{\sqrt{4x}}{\sqrt{y}}$. Assume that x and y are positive real numbers.

Multiply numerator and denominator by $\sqrt{y}$.

$$\frac{\sqrt{4x}}{\sqrt{y}} = \frac{\sqrt{4x} \cdot \sqrt{y}}{\sqrt{y} \cdot \sqrt{y}} = \frac{\sqrt{4xy}}{y} = \frac{2\sqrt{xy}}{y} \quad \blacksquare$$

WORK PROBLEM 4 AT THE SIDE.

EXAMPLE 5 Simplifying a Radical

Simplify the expression $\sqrt{\frac{2x^2y}{3}}$. Assume that x and y are positive real numbers.

First use the quotient rule.

$$\sqrt{\frac{2x^2y}{3}} = \frac{\sqrt{2x^2y}}{\sqrt{3}} \qquad \text{Quotient rule}$$

Next, multiply both the numerator and denominator by $\sqrt{3}$ to rationalize the denominator.

$$\frac{\sqrt{2x^2y}}{\sqrt{3}} = \frac{\sqrt{2x^2y} \cdot \sqrt{3}}{\sqrt{3} \cdot \sqrt{3}}$$

$$= \frac{\sqrt{6x^2y}}{3} \qquad \text{Product rule}$$

$$= \frac{\sqrt{x^2}\sqrt{6y}}{3} \qquad \text{Product rule}$$

$$= \frac{x\sqrt{6y}}{3} \qquad \sqrt{x^2} = x, \text{ since } x > 0 \quad \blacksquare$$

WORK PROBLEM 5 AT THE SIDE.

3. Simplify.

(a) $\sqrt{\frac{1}{2}} \cdot \sqrt{\frac{5}{6}}$

(b) $\sqrt{\frac{1}{10}} \cdot \sqrt{20}$

(c) $\sqrt{\frac{5}{8}} \cdot \sqrt{\frac{24}{10}}$

4. Simplify $\frac{\sqrt{5p}}{\sqrt{q}}$. Assume that p and q are positive real numbers.

5. Simplify $\sqrt{\frac{5r^2t^2}{7}}$. Assume that r and t represent positive real numbers.

ANSWERS

3. (a) $\frac{\sqrt{15}}{6}$ (b) $\sqrt{2}$ (c) $\frac{\sqrt{6}}{2}$

4. $\frac{\sqrt{5pq}}{q}$

5. $\frac{rt\sqrt{35}}{7}$

Historical Reflections

Historical Reflections

"The Man Who Loved Numbers": Srinivasa Ramanujan (1887–1920)

The Indian mathematician Srinivasa Ramanujan developed many ideas in the branch of mathematics known as number theory. His friend and collaborator on occasion was G. H. Hardy, also a number theorist and professor at Cambridge University in England.

Ramanujan introduced himself to Hardy in 1913 by a letter in which he stated without proof a number of complicated formulas. Hardy at first thought it was a crank letter, but he soon realized that the formulas had to have been the work of a genius. Hardy arranged for Ramanujan to receive a modest stipend, enough to live fairly well at Cambridge.

A story has been told about Ramanujan that illustrates his genius. Hardy once mentioned to Ramanujan that he had just taken a taxicab with a rather dull number: 1729. Ramanujan countered by saying that this number isn't dull at all: it is the smallest natural number that can be expressed as the sum of two cubes in two different ways:

$$1^3 + 12^3 = 1729 \quad \text{and} \quad 9^3 + 10^3 = 1729.$$

Ramanujan was the subject of a 1988 installment of the *Nova* series aired on the Public Broadcasting System. It was aptly titled "The Man Who Loved Numbers."

name date hour

9.4 EXERCISES

Perform the indicated operations, and write all answers in simplest form. Rationalize all denominators. See Examples 1–3.

1. $\dfrac{6}{\sqrt{5}}$

2. $\dfrac{4}{\sqrt{2}}$

3. $\dfrac{5}{\sqrt{5}}$

4. $\dfrac{15}{\sqrt{15}}$

5. $\dfrac{3}{\sqrt{7}}$

6. $\dfrac{12}{\sqrt{10}}$

7. $\dfrac{8\sqrt{3}}{\sqrt{5}}$

8. $\dfrac{9\sqrt{6}}{\sqrt{5}}$

9. $\dfrac{12\sqrt{10}}{8\sqrt{3}}$

10. $\dfrac{9\sqrt{15}}{6\sqrt{2}}$

11. $\dfrac{8}{\sqrt{27}}$

12. $\dfrac{12}{\sqrt{18}}$

13. $\dfrac{3}{\sqrt{50}}$

14. $\dfrac{5}{\sqrt{75}}$

15. $\dfrac{12}{\sqrt{72}}$

16. $\dfrac{21}{\sqrt{45}}$

17. $\dfrac{9}{\sqrt{32}}$

18. $\dfrac{50}{\sqrt{125}}$

19. $\dfrac{\sqrt{8}}{\sqrt{2}}$

20. $\dfrac{\sqrt{27}}{\sqrt{3}}$

21. $\dfrac{\sqrt{10}}{\sqrt{5}}$

22. $\dfrac{\sqrt{6}}{\sqrt{3}}$

23. $\dfrac{\sqrt{40}}{\sqrt{3}}$

24. $\dfrac{\sqrt{5}}{\sqrt{8}}$

25. $\sqrt{\dfrac{1}{2}}$

26. $\sqrt{\dfrac{1}{8}}$

27. $\sqrt{\dfrac{10}{7}}$

28. $\sqrt{\dfrac{2}{3}}$

29. $\sqrt{\frac{9}{5}}$ **30.** $\sqrt{\frac{16}{7}}$ **31.** $\sqrt{\frac{7}{5}} \cdot \sqrt{\frac{5}{3}}$ **32.** $\sqrt{\frac{1}{3}} \cdot \sqrt{\frac{2}{5}}$

33. $\sqrt{\frac{3}{4}} \cdot \sqrt{\frac{1}{5}}$ **34.** $\sqrt{\frac{1}{10}} \cdot \sqrt{\frac{10}{3}}$ **35.** $\sqrt{\frac{21}{7}} \cdot \sqrt{\frac{21}{8}}$ **36.** $\sqrt{\frac{1}{11}} \cdot \sqrt{\frac{33}{16}}$

37. $\sqrt{\frac{2}{5}} \cdot \sqrt{\frac{3}{10}}$ **38.** $\sqrt{\frac{9}{8}} \cdot \sqrt{\frac{7}{16}}$ **39.** $\sqrt{\frac{5}{8}} \cdot \sqrt{\frac{5}{6}}$

Perform the indicated operations, and write all answers in simplest form. Rationalize all denominators. Assume that all variables represent positive real numbers. See Examples 4 and 5.

40. $\sqrt{\frac{5}{x}}$ **41.** $\sqrt{\frac{6}{p}}$ **42.** $\sqrt{\frac{4r^3}{s}}$ **43.** $\sqrt{\frac{6p^3}{3m}}$ **44.** $\sqrt{\frac{a^3b}{6}}$

45. $\sqrt{\frac{x^2}{4y}}$ **46.** $\sqrt{\frac{m^2n}{2}}$ **47.** $\sqrt{\frac{9a^2r}{5}}$ **48.** $\sqrt{\frac{2x^2z^4}{3y}}$

SKILL SHARPENERS

Combine like terms. See Section 2.7.

49. $2x + 3x - x$ **50.** $-5a + 2a + 7a$ **51.** $m + 2 - 3m - 5$

52. $-2p + 4p + 7 - 9 - 3p$ **53.** $4y + 3z + 5z - y$ **54.** $5w + 2t - w + 3t$

9.5 SIMPLIFYING RADICAL EXPRESSIONS

It can be difficult to decide on the "simplest" form of a radical. The conditions for which a square root radical is in simplest form were listed in the previous section. Below is a set of guidelines that should be followed when you are simplifying radical expressions.

OBJECTIVES

1. Simplify radical expressions with sums.
2. Simplify radical expressions with products.
3. Simplify radical expressions with quotients.
4. Write radical expressions with quotients in lowest terms.

SIMPLIFYING RADICAL EXPRESSIONS

1. If a radical represents a rational number, then that rational number should be used in place of the radical.

   For example, $\sqrt{49}$ is simplified by writing 7; $\sqrt{64}$ by writing 8; $\sqrt{\frac{169}{9}}$ by writing $\frac{13}{3}$.

2. If a radical expression contains products of radicals, the product rule for radicals, $\sqrt{x} \cdot \sqrt{y} = \sqrt{xy}$, should be used to get a single radical.

   For example, $\sqrt{3} \cdot \sqrt{2}$ is simplified to $\sqrt{6}$; $\sqrt{5} \cdot \sqrt{x}$ to $\sqrt{5x}$.

3. If a radicand has a factor that is a perfect square, the radical should be expressed as the product of the positive square root of the perfect square and the remaining radical factor.

   For example, $\sqrt{20}$ is simplified to $\sqrt{20} = 2\sqrt{5}$; $\sqrt{75}$ to $5\sqrt{3}$.

4. If a radical expression contains sums or differences of radicals, the distributive property should be used to combine like radicals.

   For example, $3\sqrt{2} + 4\sqrt{2}$ is combined as $7\sqrt{2}$, but $3\sqrt{2} + 4\sqrt{3}$ cannot be further combined.

5. Any radicals in a denominator should be changed to rational numbers.

   For example, $\frac{5}{\sqrt{3}}$ is rationalized as $\frac{5}{\sqrt{3}} = \frac{5\sqrt{3}}{\sqrt{3} \cdot \sqrt{3}} = \frac{5\sqrt{3}}{3}$.

1 The first example involves sums of rational expressions.

EXAMPLE 1 Adding Radical Expressions

Perform the indicated operations.

(a) $\sqrt{16} + \sqrt{9}$

Here $\sqrt{16} + \sqrt{9} = 4 + 3 = 7$.

(b) $5\sqrt{2} + 2\sqrt{18}$

First simplify $\sqrt{18}$.

$$
\begin{aligned}
5\sqrt{2} + 2\sqrt{18} &= 5\sqrt{2} + 2(\sqrt{9} \cdot \sqrt{2}) && \text{9 is a perfect square.} \\
&= 5\sqrt{2} + 2(3\sqrt{2}) && \sqrt{9} = 3 \\
&= 5\sqrt{2} + 6\sqrt{2} && \text{Multiply.} \\
&= 11\sqrt{2} && \text{Add like radicals.} \ \blacksquare
\end{aligned}
$$

WORK PROBLEM 1 AT THE SIDE.

1. Perform the indicated operations.

(a) $\sqrt{36} + \sqrt{25}$

(b) $3\sqrt{3} + 2\sqrt{27}$

(c) $4\sqrt{8} - 2\sqrt{32}$

(d) $2\sqrt{12} - 5\sqrt{48}$

ANSWERS
1. (a) 11 (b) $9\sqrt{3}$ (c) 0 (d) $-16\sqrt{3}$

2. Find each product. Simplify the answers.

(a) $\sqrt{7}(\sqrt{2} + \sqrt{5})$

(b) $\sqrt{2}(\sqrt{8} + \sqrt{20})$

(c) $(\sqrt{2} + 5\sqrt{3})(\sqrt{3} - 2\sqrt{2})$

(d) $(2\sqrt{7} + \sqrt{10}) \cdot (3\sqrt{7} - 2\sqrt{10})$

(e) $(\sqrt{2} - \sqrt{5})(\sqrt{10} + \sqrt{2})$

ANSWERS
2. (a) $\sqrt{14} + \sqrt{35}$ (b) $4 + 2\sqrt{10}$
(c) $11 - 9\sqrt{6}$ (d) $22 - \sqrt{70}$
(e) $2\sqrt{5} + 2 - 5\sqrt{2} - \sqrt{10}$

2 The next examples show how to simplify radical expressions with products.

EXAMPLE 2 Multiplying Radical Expressions
Multiply $\sqrt{5}(\sqrt{8} - \sqrt{32})$ and simplify the product.

Start by simplifying $\sqrt{8}$ and $\sqrt{32}$.

$$\sqrt{8} = 2\sqrt{2} \quad \text{and} \quad \sqrt{32} = 4\sqrt{2}$$

Now simplify inside the parentheses.

$$\begin{aligned}
\sqrt{5}(\sqrt{8} - \sqrt{32}) &= \sqrt{5}(2\sqrt{2} - 4\sqrt{2}) \\
&= \sqrt{5}(-2\sqrt{2}) && \text{Subtract like radicals.} \\
&= -2\sqrt{5 \cdot 2} && \text{Product rule} \\
&= -2\sqrt{10} && \text{Multiply.}
\end{aligned}$$

■

EXAMPLE 3 Multiplying Radical Expressions
Find each product and simplify the answers.

(a) $(\sqrt{3} + 2\sqrt{5})(\sqrt{3} - 4\sqrt{5})$

The products of these sums of radicals can be found in the same way that we found the product of binomials in Chapter 4. The pattern of multiplication is the same, using the FOIL method.

$$\begin{aligned}
&(\sqrt{3} + 2\sqrt{5})(\sqrt{3} - 4\sqrt{5}) \\
&= \underbrace{\sqrt{3} \cdot \sqrt{3}}_{\text{First}} + \underbrace{\sqrt{3}(-4\sqrt{5})}_{\text{Outside}} + \underbrace{2\sqrt{5}(\sqrt{3})}_{\text{Inside}} + \underbrace{2\sqrt{5}(-4\sqrt{5})}_{\text{Last}} \\
&= 3 - 4\sqrt{15} + 2\sqrt{15} - 8 \cdot 5 && \text{Product rule} \\
&= 3 - 2\sqrt{15} - 40 && \text{Add like radicals.} \\
&= -37 - 2\sqrt{15} && \text{Combine terms.}
\end{aligned}$$

(b) $(\sqrt{3} + \sqrt{21})(\sqrt{3} - \sqrt{7})$

$$\begin{aligned}
&= \sqrt{3}(\sqrt{3}) + \sqrt{3}(-\sqrt{7}) + \sqrt{21}(\sqrt{3}) && \text{FOIL} \\
&\quad + \sqrt{21}(-\sqrt{7}) \\
&= 3 - \sqrt{21} + \sqrt{63} - \sqrt{147} && \text{Product rule} \\
&= 3 - \sqrt{21} + \sqrt{9} \cdot \sqrt{7} - \sqrt{49} \cdot \sqrt{3} && \text{9 and 49 are perfect squares.} \\
&= 3 - \sqrt{21} + 3\sqrt{7} - 7\sqrt{3} && \sqrt{9} = 3 \text{ and } \sqrt{49} = 7
\end{aligned}$$

Since there are no like radicals, no terms may be combined. ■

WORK PROBLEM 2 AT THE SIDE.

Since radicals represent real numbers, the special products of binomials discussed in Chapters 4 and 5 can be used to find products of radicals. Example 4 uses the rule for the product that gives a difference of two squares,

$$(a + b)(a - b) = a^2 - b^2.$$

EXAMPLE 4 Using a Special Product with Radicals
Find each product.

(a) $(4 + \sqrt{3})(4 - \sqrt{3})$

Follow the pattern given above. Let $a = 4$ and $b = \sqrt{3}$.

$$\begin{aligned}(4 + \sqrt{3})(4 - \sqrt{3}) &= (4)^2 - (\sqrt{3})^2 \\ &= 16 - 3 \qquad 4^2 = 16 \text{ and } (\sqrt{3})^2 = 3 \\ &= 13\end{aligned}$$

(b) $$\begin{aligned}(\sqrt{12} - \sqrt{6})(\sqrt{12} + \sqrt{6}) &= (\sqrt{12})^2 - (\sqrt{6})^2 \\ &= 12 - 6 \qquad (\sqrt{12})^2 = 12 \text{ and } (\sqrt{6})^2 = 6 \\ &= 6\end{aligned}$$

■

WORK PROBLEM 3 AT THE SIDE.

Both products in Example 4 resulted in rational numbers. The pairs of expressions in those products, such as $4 + \sqrt{3}$ and $4 - \sqrt{3}$, and $\sqrt{12} - \sqrt{6}$ and $\sqrt{12} + \sqrt{6}$, are called **conjugates** of each other.

3 Products of radicals similar to those in Example 4 can be used to rationalize the denominators in more complicated quotients, such as

$$\frac{2}{4 - \sqrt{3}}.$$

By Example 4(a), if this denominator, $4 - \sqrt{3}$, is multiplied by $4 + \sqrt{3}$, then the product $(4 - \sqrt{3})(4 + \sqrt{3})$ is the rational number 13. Multiplying numerator and denominator of the quotient by $4 + \sqrt{3}$ gives

$$\frac{2}{4 - \sqrt{3}} = \frac{2(4 + \sqrt{3})}{(4 - \sqrt{3})(4 + \sqrt{3})} = \frac{2(4 + \sqrt{3})}{13}.$$

The denominator now has been rationalized; it contains no radical signs.

USING CONJUGATES TO SIMPLIFY A RADICAL EXPRESSION

To simplify a radical expression with two terms in the denominator, where at least one of those terms is a radical, multiply both the numerator and the denominator by the conjugate of the denominator.

EXAMPLE 5 Using Conjugates to Rationalize a Denominator
Rationalize the denominator in the quotient

$$\frac{5}{3 + \sqrt{5}}.$$

3. Find each product. Simplify the answers.

(a) $(3 + \sqrt{5})(3 - \sqrt{5})$

(b) $(\sqrt{3} - 2)(\sqrt{3} + 2)$

(c) $(\sqrt{5} + \sqrt{3})(\sqrt{5} - \sqrt{3})$

(d) $(\sqrt{11} + \sqrt{14})(\sqrt{11} - \sqrt{14})$

ANSWERS
3. (a) 4 (b) −1 (c) 2 (d) −3

4. Rationalize each denominator.

(a) $\dfrac{5}{4 + \sqrt{2}}$

(b) $\dfrac{3}{2 - \sqrt{5}}$

(c) $\dfrac{1}{6 + \sqrt{3}}$

5. Write each quotient in lowest terms.

(a) $\dfrac{4 + 8\sqrt{2}}{2}$

(b) $\dfrac{5\sqrt{3} - 15}{10}$

(c) $\dfrac{8\sqrt{5} + 12}{16}$

ANSWERS

4. (a) $\dfrac{5(4 - \sqrt{2})}{14}$ (b) $-3(2 + \sqrt{5})$

(c) $\dfrac{6 - \sqrt{3}}{33}$

5. (a) $2 + 4\sqrt{2}$ (b) $\dfrac{\sqrt{3} - 3}{2}$

(c) $\dfrac{2\sqrt{5} + 3}{4}$

The radical in the denominator can be eliminated by multiplying both numerator and denominator by $3 - \sqrt{5}$.

$$\frac{5}{3 + \sqrt{5}} = \frac{5(3 - \sqrt{5})}{(3 + \sqrt{5})(3 - \sqrt{5})}$$

$$= \frac{5(3 - \sqrt{5})}{3^2 - (\sqrt{5})^2} \qquad (a + b)(a - b) = a^2 - b^2$$

$$= \frac{5(3 - \sqrt{5})}{9 - 5} \qquad 3^2 = 9 \text{ and } (\sqrt{5})^2 = 5$$

$$= \frac{5(3 - \sqrt{5})}{4}$$

■

EXAMPLE 6 Using Conjugates to Rationalize a Denominator

Simplify $\dfrac{6 + \sqrt{2}}{\sqrt{2} - 5}$.

Multiply numerator and denominator by $\sqrt{2} + 5$.

$$\frac{6 + \sqrt{2}}{\sqrt{2} - 5} = \frac{(6 + \sqrt{2})(\sqrt{2} + 5)}{(\sqrt{2} - 5)(\sqrt{2} + 5)}$$

$$= \frac{6\sqrt{2} + 30 + 2 + 5\sqrt{2}}{2 - 25} \qquad \text{FOIL}$$

$$= \frac{11\sqrt{2} + 32}{-23} \qquad \text{Combine terms.}$$

$$= \frac{-11\sqrt{2} - 32}{23} \qquad \frac{a}{-b} = \frac{-a}{b}$$

■

WORK PROBLEM 4 AT THE SIDE.

4 The final example shows how to write certain quotients in lowest terms.

EXAMPLE 7 Writing a Radical Quotient in Lowest Terms

Write $\dfrac{3\sqrt{3} + 9}{12}$ in lowest terms.

Factor the numerator, and then divide numerator and denominator by any common factors.

$$\frac{3\sqrt{3} + 9}{12} = \frac{3(\sqrt{3} + 3)}{12} = \frac{\sqrt{3} + 3}{4}$$

■

CAUTION A common error is to try to reduce an expression like the one in Example 7 to lowest terms before factoring. For example,

$$\frac{4 + 8\sqrt{5}}{4} \neq 1 + 8\sqrt{5}.$$

The correct simplification is $1 + 2\sqrt{5}$. Do you see why?

WORK PROBLEM 5 AT THE SIDE.

name date hour

9.5 EXERCISES

Simplify each expression. Use the five guidelines given in the text. See Examples 1–4.

1. $3\sqrt{5} + 8\sqrt{45}$

2. $6\sqrt{2} + 4\sqrt{18}$

3. $9\sqrt{50} - 4\sqrt{72}$

4. $3\sqrt{80} - 5\sqrt{45}$

5. $\sqrt{2}(\sqrt{8} - \sqrt{32})$

6. $\sqrt{3}(\sqrt{27} - \sqrt{3})$

7. $\sqrt{5}(\sqrt{3} + \sqrt{7})$

8. $\sqrt{7}(\sqrt{10} - \sqrt{3})$

9. $2\sqrt{5}(\sqrt{2} + \sqrt{5})$

10. $3\sqrt{7}(2\sqrt{7} - 4\sqrt{5})$

11. $-\sqrt{14} \cdot \sqrt{2} - \sqrt{28}$

12. $\sqrt{6} \cdot \sqrt{3} - 2\sqrt{50}$

13. $(2\sqrt{6} + 3)(3\sqrt{6} - 5)$

14. $(4\sqrt{5} - 2)(2\sqrt{5} + 3)$

15. $(5\sqrt{7} - 2\sqrt{3})(3\sqrt{7} + 3\sqrt{3})$

16. $(2\sqrt{10} + 5\sqrt{2})(3\sqrt{10} - 4\sqrt{2})$

17. $(3\sqrt{2} + 4)(3\sqrt{2} + 4)$

18. $(4\sqrt{5} - 1)(4\sqrt{5} - 1)$

19. $(2\sqrt{7} - 3)^2$

20. $(3\sqrt{5} + 5)^2$

21. $(3 - \sqrt{2})(3 + \sqrt{2})$

22. $(7 - \sqrt{5})(7 + \sqrt{5})$

23. $(2 + \sqrt{8})(2 - \sqrt{8})$

24. $(3 + \sqrt{11})(3 - \sqrt{11})$

25. $(\sqrt{6} - \sqrt{5})(\sqrt{6} + \sqrt{5})$

26. $(\sqrt{11} + \sqrt{10})(\sqrt{11} - \sqrt{10})$

27. $(\sqrt{2} + \sqrt{3})(\sqrt{6} - \sqrt{2})$

28. $(\sqrt{3} + \sqrt{5})(\sqrt{15} - \sqrt{5})$

name date hour

29. $(\sqrt{8} - \sqrt{2})(\sqrt{2} + \sqrt{4})$

30. $(\sqrt{6} - \sqrt{3})(\sqrt{3} + \sqrt{18})$

31. $(\sqrt{5} + \sqrt{30})(\sqrt{6} + \sqrt{3})$

32. $(\sqrt{10} - \sqrt{20})(\sqrt{2} - \sqrt{5})$

Rationalize the denominators. See Examples 5 and 6.

33. $\dfrac{1}{3 + \sqrt{2}}$

34. $\dfrac{1}{4 - \sqrt{3}}$

35. $\dfrac{5}{2 + \sqrt{5}}$

36. $\dfrac{6}{3 + \sqrt{7}}$

37. $\dfrac{7}{2 - \sqrt{11}}$

38. $\dfrac{38}{5 - \sqrt{6}}$

39. $\dfrac{\sqrt{2}}{1 + \sqrt{2}}$

40. $\dfrac{\sqrt{7}}{2 - \sqrt{7}}$

41. $\dfrac{\sqrt{5}}{1 - \sqrt{5}}$

42. $\dfrac{\sqrt{3}}{2 + \sqrt{3}}$

43. $\dfrac{\sqrt{12}}{\sqrt{3} + 1}$

44. $\dfrac{\sqrt{18}}{\sqrt{2} - 1}$

45. $\dfrac{2\sqrt{3}}{\sqrt{3} + 5}$

46. $\dfrac{\sqrt{12}}{2 - \sqrt{10}}$

47. $\dfrac{\sqrt{2} + 3}{\sqrt{3} - 1}$

48. $\dfrac{\sqrt{5} + 2}{2 - \sqrt{3}}$

Write each quotient in lowest terms. See Example 7.

49. $\dfrac{5\sqrt{7} - 10}{5}$

50. $\dfrac{6\sqrt{5} - 9}{3}$

51. $\dfrac{2\sqrt{3} + 10}{8}$

52. $\dfrac{4\sqrt{6} + 6}{10}$

53. $\dfrac{12 - 2\sqrt{10}}{4}$

54. $\dfrac{9 - 6\sqrt{2}}{12}$

SKILL SHARPENERS

Solve each equation. See Section 5.5.

55. $y^2 - 4y + 3 = 0$

56. $x^2 - x - 20 = 0$

57. $k - 1 = (k - 1)^2$

9.6 EQUATIONS WITH RADICALS

The addition and multiplication properties of equality are not enough to solve an equation with radicals such as

$$\sqrt{x+1} = 3.$$

1 Solving equations with square roots requires a new property, the *squaring property*.

SQUARING PROPERTY OF EQUALITY

If each side of a given equation is squared, all solutions of the original equation are *among* the solutions of the squared equation.

CAUTION Be very careful with the squaring property: Using this property can give a new equation with *more* solutions than the original equation. For example, starting with the equation $y = 4$ and squaring each side gives

$$y^2 = 4^2, \quad \text{or} \quad y^2 = 16.$$

This last equation, $y^2 = 16$, has *two* solutions, 4 or -4, while the original equation, $y = 4$, has only *one* solution, 4. Because of this possibility, checking is more than just a guard against algebraic errors when solving an equation with radicals. It is an essential part of the solution process. *All potential solutions from the squared equation must be checked in the original equation.*

EXAMPLE 1 Using the Squaring Property of Equality

Solve the equation $\sqrt{p+1} = 3$.

Use the squaring property of equality to square each side of the equation.

$$(\sqrt{p+1})^2 = 3^2$$

$$p + 1 = 9 \qquad (\sqrt{p+1})^2 = p + 1$$

$$p = 8 \qquad \text{Subtract 1.}$$

Now check this answer in the original equation.

$$\sqrt{p+1} = 3$$

$$\sqrt{8+1} = 3 \qquad \text{Let } p = 8.$$

$$\sqrt{9} = 3$$

$$3 = 3 \qquad \text{True}$$

Since this statement is true, 8 is the solution of $\sqrt{p+1} = 3$. In this case the squared equation had just one solution, which also satisfied the original equation. ■

WORK PROBLEM 1 AT THE SIDE.

EXAMPLE 2 Using the Squaring Property of Equality

Solve $3\sqrt{x} = \sqrt{x+8}$.

Squaring each side gives

$$(3\sqrt{x})^2 = (\sqrt{x+8})^2$$

$$3^2(\sqrt{x})^2 = (\sqrt{x+8})^2 \qquad (ab)^2 = a^2b^2$$

$$9x = x + 8 \qquad (\sqrt{x})^2 = x;\ (\sqrt{x+8})^2 = x + 8$$

$$8x = 8 \qquad \text{Subtract } x.$$

$$x = 1. \qquad \text{Divide by 8.}$$

OBJECTIVES

1. Solve equations with radicals.
2. Identify equations with no solutions.
3. Solve equations by squaring a binomial.

1. Solve each equation.

(a) $\sqrt{k} = 3$

(b) $\sqrt{m-2} = 4$

(c) $\sqrt{9-y} = 4$

ANSWERS

1. (a) 9 (b) 18 (c) -7

Check this potential solution.

$$3\sqrt{\mathbf{x}} = \sqrt{\mathbf{x} + 8}$$
$$3\sqrt{\mathbf{1}} = \sqrt{\mathbf{1} + 8} \quad \text{Let } x = 1.$$
$$3(1) = \sqrt{9} \quad \sqrt{1} = 1$$
$$3 = 3 \quad \text{True}$$

The solution of $3\sqrt{x} = \sqrt{x + 8}$ is the number 1. ■

WORK PROBLEM 2 AT THE SIDE.

2 Not all equations with radicals have a solution, as shown by the equations in Examples 3 and 4.

EXAMPLE 3 Using the Squaring Property of Equality

Solve the equation $\sqrt{y} = -3$.

Square each side.

$$(\sqrt{y})^2 = (-3)^2$$
$$y = 9$$

Check this proposed answer in the original equation.

$$\sqrt{y} = -3$$
$$\sqrt{\mathbf{9}} = -3 \quad \text{Let } y = 9.$$
$$3 = -3 \quad \text{False}$$

Since the statement $3 = -3$ is false, the number 9 is not a solution of the given equation and is said to be *extraneous*. In fact, $\sqrt{y} = -3$ has no solution. Since $\sqrt{y}$ represents the *nonnegative* square root of y, we might have seen immediately that there is no solution. ■

EXAMPLE 4 Using the Squaring Property of Equality

Solve $a = \sqrt{a^2 + 5a + 10}$.

Square each side.

$$(a)^2 = (\sqrt{a^2 + 5a + 10})^2$$
$$a^2 = a^2 + 5a + 10 \quad (\sqrt{a^2 + 5a + 10})^2 = a^2 + 5a + 10$$
$$0 = 5a + 10 \quad \text{Subtract } a^2.$$
$$-10 = 5a \quad \text{Subtract 10.}$$
$$a = -2 \quad \text{Divide by 5.}$$

Check this potential solution in the original equation.

$$a = \sqrt{a^2 + 5a + 10}$$
$$\mathbf{-2} = \sqrt{(\mathbf{-2})^2 + 5(\mathbf{-2}) + 10} \quad \text{Let } a = -2.$$
$$-2 = \sqrt{4 - 10 + 10} \quad \text{Multiply.}$$
$$-2 = 2 \quad \text{False}$$

Since $a = -2$ leads to a false result, the equation has no solution. ■

WORK PROBLEM 3 AT THE SIDE.

2. Solve each equation.

(a) $\sqrt{2k + 1} = 5$

(b) $\sqrt{3x + 9} = 2\sqrt{x}$

(c) $5\sqrt{a} = \sqrt{20a + 5}$

3. Solve each equation that has a solution. (*Hint:* In (c) subtract 4 from each side.)

(a) $\sqrt{m} = -5$

(b) $\sqrt{x} = 2$

(c) $\sqrt{y} + 4 = 0$

(d) $m = \sqrt{m^2 - 4m - 16}$

ANSWERS
2. (a) 12 (b) 9 (c) 1
3. (a) no solution (b) 4
(c) no solution (d) no solution

3 The next examples use the following facts from Section 4.6.

$$(a + b)^2 = a^2 + 2ab + b^2$$

and

$$(a - b)^2 = a^2 - 2ab + b^2.$$

By these patterns, for example,

$$(y - 3)^2 = y^2 - 2(y)(3) + (3)^2$$
$$= y^2 - 6y + 9.$$

WORK PROBLEM 4 AT THE SIDE.

EXAMPLE 5 Using the Squaring Property of Equality

Solve the equation $\sqrt{2y - 3} = y - 3$.

Square each side, using the result above on the right side of the equation.

$$(\sqrt{2y - 3})^2 = (y - 3)^2$$
$$2y - 3 = y^2 - 6y + 9$$

This equation is quadratic because of the y^2 term. As shown in Section 5.5, solving this equation requires that one side be equal to 0. Subtract $2y$ and add 3, getting

$$0 = y^2 - 8y + 12.$$

Solve this equation by factoring.

$$0 = (y - 6)(y - 2)$$

Set each factor equal to 0.

$$y - 6 = 0 \quad \text{or} \quad y - 2 = 0$$
$$y = 6 \quad \text{or} \quad y = 2$$

Check both of these potential solutions in the original equation.

If $y = 6$,		If $y = 2$,	
$\sqrt{2y - 3} = y - 3$		$\sqrt{2y - 3} = y - 3$	
$\sqrt{2(6) - 3} = 6 - 3$	Let $y = 6$.	$\sqrt{2(2) - 3} = 2 - 3$	Let $y = 2$.
$\sqrt{12 - 3} = 3$	Multiply.	$\sqrt{4 - 3} = -1$	Multiply.
$\sqrt{9} = 3$		$\sqrt{1} = -1$	
$3 = 3.$	True	$1 = -1.$	False

Only 6 is a valid solution of the equation. ■

WORK PROBLEM 5 AT THE SIDE.

Sometimes it is necessary to write an equation in a different form before squaring each side. The next example shows why.

EXAMPLE 6 Using the Squaring Property of Equality

Solve the equation $\sqrt{x} + 1 = 2x$.

Squaring each side gives

$$(\sqrt{x} + 1)^2 = (2x)^2$$
$$x + 2\sqrt{x} + 1 = 4x^2,$$

an equation that is more complicated, and still contains a radical. It would be better instead to rewrite the original equation so that the radical is alone

4. Square each expression.

(a) $m - 5$

(b) $2k - 5$

(c) $3m - 2p$

(d) $4z + 9y$

5. Solve each equation.

(a) $\sqrt{6w + 6} = w + 1$

(b) $x + 3 = \sqrt{12x + 1}$

(c) $2u - 1 = \sqrt{10u + 9}$

ANSWERS

4. (a) $m^2 - 10m + 25$
(b) $4k^2 - 20k + 25$
(c) $9m^2 - 12mp + 4p^2$
(d) $16z^2 + 72zy + 81y^2$

5. (a) 5, −1 (b) 4, 2 (c) 4

on one side of the equals sign. Get the radical alone by subtracting 1 from each side to get

$$\sqrt{x} = 2x - 1.$$

Now square each side.

$$(\sqrt{x})^2 = (2x - 1)^2$$
$$x = 4x^2 - 4x + 1$$

Subtract x from each side.

$$0 = 4x^2 - 5x + 1$$

This quadratic equation can be solved by factoring.

$$0 = (4x - 1)(x - 1)$$
$$4x - 1 = 0 \quad \text{or} \quad x - 1 = 0$$
$$x = \frac{1}{4} \quad \text{or} \quad x = 1$$

Both of these potential solutions must be checked in the original equation.

If $x = \frac{1}{4}$,

$$\sqrt{x} + 1 = 2x$$
$$\sqrt{\frac{1}{4}} + 1 = 2\left(\frac{1}{4}\right) \quad \text{Let } x = \frac{1}{4}.$$
$$\frac{1}{2} + 1 = \frac{1}{2}. \quad \text{False}$$

If $x = 1$,

$$\sqrt{x} + 1 = 2x$$
$$\sqrt{1} + 1 = 2(1) \quad \text{Let } x = 1.$$
$$2 = 2. \quad \text{True}$$

The only solution to the original equation is 1. ■

CAUTION Errors often occur when each side of an equation is squared. For example, in Example 6 after the equation is rewritten as

$$\sqrt{x} = 2x - 1,$$

it would be *incorrect* to write the next step as

$$x = 4x^2 + 1.$$

Don't forget that the binomial $2x - 1$ must be squared to get $4x^2 - 4x + 1$.

WORK PROBLEM 6 AT THE SIDE.

The steps to use in solving an equation with radicals are summarized below.

SOLVING AN EQUATION WITH RADICALS

Step 1 Arrange the terms so that a radical is alone on one side of the equation.

Step 2 Square each side.

Step 3 Combine like terms.

Step 4 Solve the equation for potential solutions.

Step 5 Check all solutions from Step 4 in the original equation.

6. Solve each equation.

(a) $\sqrt{x} - 3 = x - 15$

(b) $\sqrt{z + 5} + 2 = z + 5$

ANSWERS
6. (a) 16 (b) -1

name date hour

9.6 EXERCISES

Find all solutions for each equation. See Examples 1–4.

1. $\sqrt{x} = 2$

2. $\sqrt{m} = 5$

3. $\sqrt{y + 3} = 2$

4. $\sqrt{z + 1} = 5$

5. $\sqrt{t - 3} = 2$

6. $\sqrt{r + 5} = 4$

7. $\sqrt{n + 8} = 1$

8. $\sqrt{k + 10} = 2$

9. $\sqrt{m + 5} = 0$

10. $\sqrt{y - 4} = 0$

11. $\sqrt{z + 5} = -2$

12. $\sqrt{t - 3} = -2$

13. $\sqrt{k} - 2 = 5$

14. $\sqrt{p} - 3 = 7$

15. $\sqrt{y} + 4 = 2$

16. $\sqrt{m} + 6 = 5$

17. $\sqrt{5t - 9} = 2\sqrt{t}$

18. $\sqrt{3n + 4} = 2\sqrt{n}$

19. $3\sqrt{r} = \sqrt{8r + 16}$

20. $2\sqrt{r} = \sqrt{3r + 9}$

21. $\sqrt{5y - 5} = \sqrt{4y + 1}$

22. $\sqrt{2x + 2} = \sqrt{3x - 5}$

23. $\sqrt{x + 2} = \sqrt{2x - 5}$

24. $\sqrt{3m + 3} = \sqrt{5m - 1}$

25. $p = \sqrt{p^2 - 3p - 12}$

26. $k = \sqrt{k^2 - 2k + 10}$

27. $2r = \sqrt{4r^2 + 5r - 30}$

28. $3t = \sqrt{9t^2 - 6t + 12}$

name date hour

Find all solutions for each equation. See Examples 5 and 6. Remember that $(a + b)^2 = a^2 + 2ab + b^2$ *and* $(\sqrt{a})^2 = a$.

29. $\sqrt{2x + 1} = x - 7$

30. $\sqrt{5x + 1} = x + 1$

31. $\sqrt{3x + 10} = 2x - 5$

32. $\sqrt{4x + 13} = 2x - 1$

33. $\sqrt{x + 1} - 1 = x$

34. $\sqrt{3x + 3} + 5 = x$

35. $\sqrt{4x + 5} - 2 = 2x - 7$

36. $\sqrt{6x + 7} - 1 = x + 1$

37. $3\sqrt{x + 13} = x + 9$

38. $2\sqrt{x + 7} = x - 1$

39. $\sqrt{4x} - x + 3 = 0$

40. $\sqrt{2x} - x + 4 = 0$

41. $\sqrt{3x} - 4 = x - 10$

42. $\sqrt{x} + 9 = x + 3$

Solve each problem.

43. Police sometimes use the following procedure to estimate the speed at which a car was traveling at the time of an accident. A police officer drives the car involved in the accident under conditions similar to those during which the accident took place and then skids to a stop. If the car is driven at 30 miles per hour, then the speed at the time of the accident is given by

$$s = 30\sqrt{\frac{a}{p}}$$

where a is the length of the skid marks left at the time of the accident and p is the length of the skid marks in the police test. Find s for the following values of a and p.

(a) $a = 900$ feet; $p = 100$ feet

(b) $a = 400$ feet; $p = 25$ feet

(c) $a = 80$ feet; $p = 20$ feet

(d) $a = 120$ feet; $p = 30$ feet

44. A formula for calculating the distance, d, one can see from an airplane to the horizon on a clear day is

$$d = 1.22\sqrt{x}$$

where x is the altitude of the plane in feet and d is given in miles. How far can one see to the horizon in a plane flying at the following altitudes? Round to the nearest tenth.

(a) 20,000 feet **(b)** 30,000 feet

SKILL SHARPENERS

Find all square roots of each number. Simplify where possible. See Section 9.1.

45. 25 **46.** 49 **47.** 14 **48.** 29

49. 18 **50.** 48 **51.** 80 **52.** 75

CHAPTER 9 SUMMARY

KEY TERMS

9.1	**square root**	The square roots of a^2 are a and $-a$ (a is nonnegative).
	perfect square	A number with a rational square root is called a perfect square.
	radicand	The number or expression under a radical sign is called the radicand.
	radical	A radical sign with a radicand is called a radical.
	radical expression	An algebraic expression containing a radical is called a radical expression.
9.3	**like radicals**	Like radicals are multiples of the same radical.
9.4	**rationalizing the denominator**	The process of changing the denominator of a fraction from a radical (irrational number) to a rational number is called rationalizing the denominator.
	simplified form	A radical is in simplified form if the radicand contains no factor (except 1) that is a perfect square, the radicand contains no fractions, and no denominator contains a radical.
9.5	**conjugates**	The conjugate of $a + b$ is $a - b$.

NEW SYMBOLS

$\sqrt{}$ radical sign

$\approx$ is approximately equal to

QUICK REVIEW

Section	Concepts	Examples
9.1 Finding Square Roots	If a is a positive real number, $\sqrt{a}$ is the positive square root of a; $-\sqrt{a}$ is the negative square root of a; $\sqrt{0} = 0$. If a is a negative real number, $\sqrt{a}$ is not a real number. If a is a positive rational number, $\sqrt{a}$ is rational if a is a perfect square. $\sqrt{a}$ is irrational if a is not a perfect square.	$\sqrt{49} = 7$ $-\sqrt{81} = -9$ $\sqrt{-25}$ is not a real number. $\sqrt{\frac{4}{9}}$, $\sqrt{16}$ are rational. $\sqrt{\frac{2}{3}}$, $\sqrt{21}$ are irrational.
9.2 Products and Quotients of Radicals	**Product Rule for Radicals** For nonnegative real numbers a and b, $\sqrt{a} \cdot \sqrt{b} = \sqrt{ab}$ and $\sqrt{ab} = \sqrt{a} \cdot \sqrt{b}$. **Quotient Rule for Radicals** If a and b are nonnegative real numbers and b is not 0, $\frac{\sqrt{a}}{\sqrt{b}} = \sqrt{\frac{a}{b}}$ and $\sqrt{\frac{a}{b}} = \frac{\sqrt{a}}{\sqrt{b}}$.	$\sqrt{5} \cdot \sqrt{7} = \sqrt{35}$ $\sqrt{8} \cdot \sqrt{2} = \sqrt{16} = 4$ $\sqrt{48} = \sqrt{16} \cdot \sqrt{3} = 4\sqrt{3}$ $\sqrt{\frac{25}{64}} = \frac{\sqrt{25}}{\sqrt{64}} = \frac{5}{8}$ $\frac{\sqrt{8}}{\sqrt{2}} = \sqrt{\frac{8}{2}} = \sqrt{4} = 2$

Section	Concepts	Examples
9.3 Addition and Subtraction of Radicals	Add and subtract like radicals by using the distributive property. Only like radicals can be combined in this way.	$2\sqrt{5} + 4\sqrt{5} = (2 + 4)\sqrt{5}$ $= 6\sqrt{5}$ $\sqrt{8} + \sqrt{32} = 2\sqrt{2} + 4\sqrt{2}$ $= 6\sqrt{2}$
9.4 Rationalizing the Denominator	The denominator of a radical may be rationalized by multiplying both the numerator and denominator by the same number.	$\frac{2}{\sqrt{3}} = \frac{2 \cdot \sqrt{3}}{\sqrt{3} \cdot \sqrt{3}} = \frac{2\sqrt{3}}{3}$ $\sqrt{\frac{5}{11}} = \frac{\sqrt{5} \cdot \sqrt{11}}{\sqrt{11} \cdot \sqrt{11}} = \frac{\sqrt{55}}{11}$
9.5 Simplifying Radical Expressions	When appropriate, use the rules for adding and multiplying polynomials to simplify radical expressions. If a radical expression contains two terms in the denominator and at least one of those terms is a radical, multiply both the numerator and the denominator by the conjugate of the denominator.	$\sqrt{6}(\sqrt{5} - \sqrt{7}) = \sqrt{30} - \sqrt{42}$ $(\sqrt{5} - \sqrt{3})(\sqrt{5} + \sqrt{3}) = 5 - 3 = 2$ $\frac{6}{\sqrt{7} - \sqrt{2}} = \frac{6}{\sqrt{7} - \sqrt{2}} \cdot \frac{\sqrt{7} + \sqrt{2}}{\sqrt{7} + \sqrt{2}}$ $= \frac{6(\sqrt{7} + \sqrt{2})}{7 - 2}$ Multiply fractions. $= \frac{6(\sqrt{7} + \sqrt{2})}{5}$ Simplify.
9.6 Equations with Radicals	**Solving an Equation with Radicals** *Step 1* Arrange the terms so that a radical is alone on one side of the equation. *Step 2* Square each side. (By the squaring property of equality, all solutions of the original equation are *among* the solutions of the squared equation.) *Step 3* Combine like terms. *Step 4* Solve the equation for potential solutions. *Step 5* Check all potential solutions from Step 4 in the original equation.	Solve: $\sqrt{2x - 3} + x = 3$ $\sqrt{2x - 3} = 3 - x$ Isolate radical. $(\sqrt{2x - 3})^2 = (3 - x)^2$ Square. $2x - 3 = 9 - 6x + x^2$ $0 = x^2 - 8x + 12$ Get one side = 0. $0 = (x - 2)(x - 6)$ Factor. $x - 2 = 0$ or $x - 6 = 0$ Set each factor = 0. $x = 2$ $x = 6$ Solve. Verify that 2 is the only solution (6 is extraneous).

name date hour

CHAPTER 9 REVIEW EXERCISES

[9.1] *Find all square roots of each number.*

1. 49 **2.** 81 **3.** 196 **4.** 121

5. 225 **6.** 729 **7.** 2401 **8.** 7569

Find each square root that is a real number.

9. $\sqrt{16}$ **10.** $-\sqrt{36}$ **11.** $-\sqrt{400}$ **12.** $\sqrt{2704}$

13. $\sqrt{-8100}$ **14.** $-\sqrt{4225}$ **15.** $\sqrt{\frac{49}{36}}$ **16.** $\sqrt{\frac{100}{81}}$

Write rational, irrational, *or* not a real number *for each number. If a number is rational, give its exact value. If a number is irrational, give a decimal approximation for the number. Round approximations to the nearest thousandth.*

17. $\sqrt{15}$ **18.** $\sqrt{64}$ **19.** $-\sqrt{169}$ **20.** $\sqrt{-81}$

[9.2] *Use the product rule to simplify each expression.*

21. $\sqrt{2} \cdot \sqrt{5}$ **22.** $\sqrt{12} \cdot \sqrt{3}$ **23.** $\sqrt{5} \cdot \sqrt{15}$

24. $\sqrt{8} \cdot \sqrt{8}$ **25.** $\sqrt{27}$ **26.** $\sqrt{48}$

27. $\sqrt{160}$ **28.** $\sqrt{320}$ **29.** $\sqrt{98} \cdot \sqrt{50}$

30. $\sqrt{12} \cdot \sqrt{27}$ **31.** $\sqrt{32} \cdot \sqrt{48}$ **32.** $\sqrt{50} \cdot \sqrt{125}$

Use the product rule and the quotient rule, as necessary, to simplify each expression.

33. $\sqrt{\frac{9}{4}}$

34. $\sqrt{\frac{121}{400}}$

35. $\sqrt{\frac{3}{25}}$

36. $\sqrt{\frac{10}{169}}$

37. $\sqrt{\frac{1}{6}} \cdot \sqrt{\frac{5}{6}}$

38. $\sqrt{\frac{2}{5}} \cdot \sqrt{\frac{2}{45}}$

39. $\frac{3\sqrt{10}}{\sqrt{2}}$

40. $\frac{24\sqrt{12}}{6\sqrt{3}}$

41. $\frac{12\sqrt{75}}{4\sqrt{3}}$

Simplify each expression. Assume that all variables represent positive real numbers.

42. $\sqrt{p} \cdot \sqrt{p}$

43. $\sqrt{k} \cdot \sqrt{m}$

44. $\sqrt{y^2}$

45. $\sqrt{r^{18}}$

46. $\sqrt{x^{10}y^{16}}$

47. $\sqrt{x^9}$

48. $\sqrt{\frac{36}{p^2}}$

49. $\sqrt{\frac{13}{k^4}}$

50. $\sqrt{\frac{100}{y^{10}}}$

[9.3] *Simplify and combine terms where possible.*

51. $\sqrt{7} + \sqrt{7}$

52. $3\sqrt{2} + 5\sqrt{2}$

53. $3\sqrt{75} + 2\sqrt{27}$

name date hour

54. $4\sqrt{12} + \sqrt{48}$

55. $4\sqrt{24} - 3\sqrt{54} + \sqrt{6}$

56. $2\sqrt{7} - 4\sqrt{28} + 3\sqrt{63}$

57. $\frac{2}{5}\sqrt{75} - \frac{3}{4}\sqrt{160}$

58. $\frac{1}{3}\sqrt{18} + \frac{1}{4}\sqrt{32}$

59. $\sqrt{15} \cdot \sqrt{2} + 5\sqrt{30}$

Simplify each expression. Assume that all variables represent nonnegative real numbers.

60. $\sqrt{4x} + \sqrt{36x} + \sqrt{9x}$

61. $\sqrt{16p} + 3\sqrt{p} - \sqrt{49p}$

62. $\sqrt{20m^2} - m\sqrt{45}$

63. $3k\sqrt{8k^2n} + 5k^2\sqrt{2n}$

[9.4] *Perform the indicated operations, and write all answers in simplest form. Rationalize all denominators. Assume that all variables represent positive real numbers.*

64. $\frac{10}{\sqrt{3}}$

65. $\frac{5}{\sqrt{2}}$

66. $\frac{3\sqrt{2}}{\sqrt{5}}$

67. $\frac{8}{\sqrt{8}}$

68. $\frac{12}{\sqrt{24}}$

69. $\frac{\sqrt{2}}{\sqrt{15}}$

70. $\sqrt{\frac{3}{5}}$

71. $\sqrt{\frac{5}{14}} \cdot \sqrt{28}$

72. $\sqrt{\frac{2}{3}} \cdot \sqrt{\frac{1}{5}}$ **73.** $\sqrt{\frac{1}{6}} \cdot \sqrt{\frac{18}{5}}$ **74.** $\sqrt{\frac{t^2}{9x}}$ **75.** $\sqrt{\frac{x^4y}{2}}$

[9.5] *Simplify each expression.*

76. $2\sqrt{6} + 4\sqrt{54}$ **77.** $-\sqrt{3}(\sqrt{5} - \sqrt{27})$ **78.** $3\sqrt{2}(\sqrt{3} + 2\sqrt{2})$

79. $(2\sqrt{3} - 4)(5\sqrt{3} + 2)$ **80.** $(5\sqrt{7} + 2)^2$ **81.** $(\sqrt{5} - \sqrt{7})(\sqrt{5} + \sqrt{7})$

82. $(2\sqrt{3} + 5)(2\sqrt{3} - 5)$ **83.** $(\sqrt{7} + 2\sqrt{6})(\sqrt{12} - \sqrt{2})$

Rationalize the denominators.

84. $\frac{1}{2 + \sqrt{5}}$ **85.** $\frac{2}{\sqrt{2} - 3}$ **86.** $\frac{\sqrt{8}}{\sqrt{2} + 6}$

87. $\frac{\sqrt{3}}{1 + \sqrt{3}}$ **88.** $\frac{\sqrt{5} - 1}{\sqrt{2} + 3}$ **89.** $\frac{2 + \sqrt{6}}{\sqrt{3} - 1}$

name date hour

Write each quotient in lowest terms.

90. $\dfrac{5 + 10\sqrt{6}}{5}$

91. $\dfrac{3 - 9\sqrt{7}}{12}$

92. $\dfrac{6 + 8\sqrt{3}}{2}$

[9.6] *Find all solutions for each equation.*

93. $\sqrt{m} = 5$

94. $\sqrt{p} = -2$

95. $\sqrt{k + 1} = 10$

96. $\sqrt{y - 8} = 0$

97. $\sqrt{r} - 3 = 8$

98. $\sqrt{x} + 2 = 1$

99. $\sqrt{5m + 4} = 3\sqrt{m}$

100. $\sqrt{2p + 3} = \sqrt{5p - 3}$

101. $\sqrt{4y + 1} = y - 1$

102. $\sqrt{-2k - 4} = k + 2$

103. $\sqrt{2 - x} + 3 = x + 7$

104. $\sqrt{x} - x + 2 = 0$

MIXED REVIEW EXERCISES

Simplify each expression if possible. Assume all variables represent positive real numbers.

105. $\sqrt{3} \cdot \sqrt{27}$

106. $2\sqrt{27} + 3\sqrt{75} - \sqrt{300}$

107. $\sqrt{\dfrac{121}{t^2}}$

108. $\dfrac{1}{5 + \sqrt{2}}$

109. $\sqrt{\dfrac{1}{3}} \cdot \sqrt{\dfrac{24}{5}}$

110. $\sqrt{50y^2}$

111. $\sqrt{7} + \sqrt{21}$

112. $-\sqrt{5}(\sqrt{2} - \sqrt{75})$

113. $\sqrt{\dfrac{16r^3}{3s}}$

114. $\dfrac{2 + 6\sqrt{13}}{2}$

115. $-\sqrt{162} + \sqrt{8}$

116. $(\sqrt{5} - \sqrt{2})^2$

117. $(6\sqrt{7} + 2)(4\sqrt{7} - 1)$

118. $-\sqrt{121}$

119. $\sqrt{98}$

120. $\dfrac{9}{\sqrt{18}}$

121. $\sqrt{\dfrac{3}{8}} \cdot \sqrt{\dfrac{3}{2}}$

122. $\sqrt{13} \cdot \sqrt{3} + 2\sqrt{39}$

name date hour

CHAPTER 9 TEST

Find each square root that is a real number. Use the square root table or a calculator to give a decimal approximation if necessary. Round to the nearest thousandth.

1. $\sqrt{100}$

2. $\sqrt{190}$

3. $-\sqrt{\frac{25}{49}}$

4. $\sqrt{-36}$

Simplify where possible.

5. $\sqrt{\frac{64}{169}}$

6. $\sqrt{50}$

7. $\frac{20\sqrt{18}}{5\sqrt{3}}$

8. $2\sqrt{3} + 4\sqrt{3}$

9. $\sqrt{20} - \sqrt{45}$

10. $3\sqrt{6} + \sqrt{14}$

1. __________

2. __________

3. __________

4. __________

5. __________

6. __________

7. __________

8. __________

9. __________

10. __________

11. ______________________

12. ______________________

13. ______________________

14. ______________________

15. ______________________

16. ______________________

17. ______________________

18. ______________________

19. ______________________

20. ______________________

11. $3\sqrt{28} + \sqrt{63}$

12. $(6 - \sqrt{5})(6 + \sqrt{5})$

13. $(2 - \sqrt{7})(3\sqrt{2} + 1)$

14. $(\sqrt{5} + \sqrt{6})^2$

Rationalize each denominator.

15. $\frac{1}{\sqrt{3}}$

16. $\sqrt{\frac{3}{5}}$

17. $\frac{-3}{4 - \sqrt{3}}$

Solve each equation.

18. $\sqrt{m} + 5 = 0$

19. $\sqrt{k + 2} = 5$

20. $\sqrt{y + 3} = y + 1$

name date hour

CUMULATIVE REVIEW EXERCISES: CHAPTERS 7–9

Complete the given ordered pairs for each equation.

1. $4x + 7y = 28$ (0,), (, 0), (3,), (, 5)

2. $6 + 2x = y$ (0,), (, 0), (−2,), (, 3)

Graph each linear equation.

3. $x - y = 4$

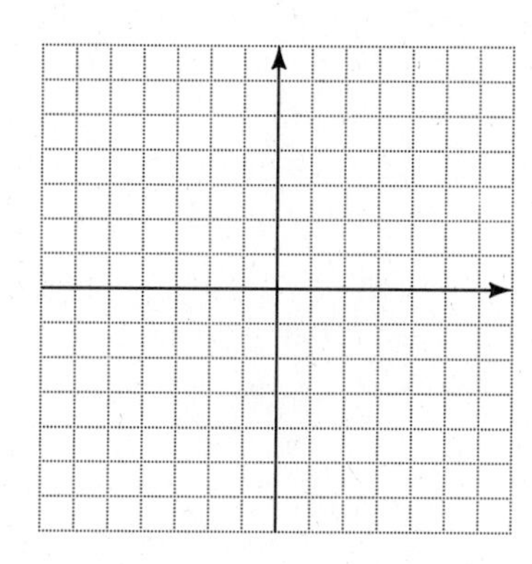

4. $3x + y = 6$

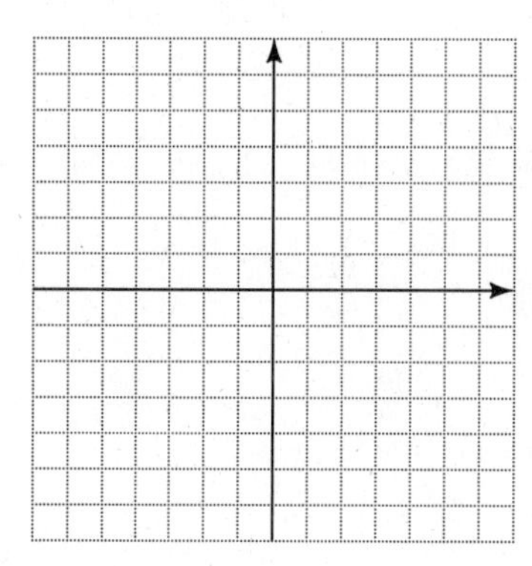

5. $x - 4 = 0$

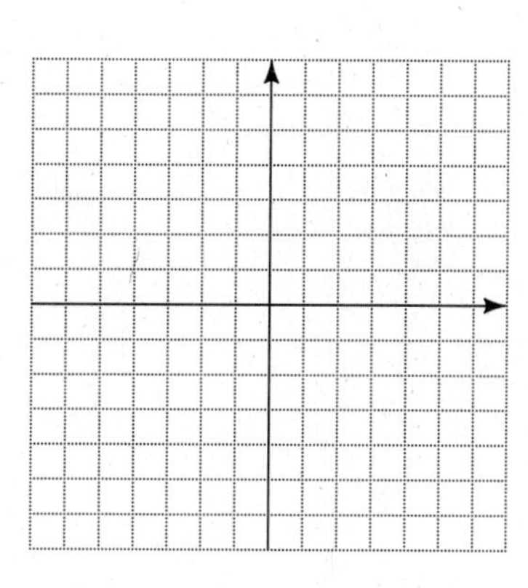

6. $y + 5 = 0$

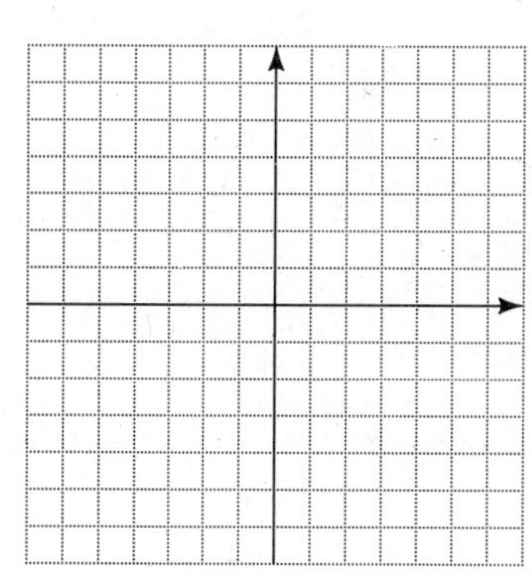

Find the slope of each of the following lines.

7. Through $(-5, 6)$ and $(1, -2)$

8. Through $(4, -1)$ and $(3, 5)$

9. $y = 4x - 3$

10. $x + 2y = 9$

Decide whether the lines in each of the following pairs are parallel, perpendicular, *or* neither.

11. $4x + y = 8$
$8x + 2y = 3$

12. $3x - 2y = 4$
$2x + 3y = 9$

13. $3x + 2y = 8$
$6x - 4y = 1$

Write an equation for each line. Write answers in the form $Ax + By = C$.

14. $m = -2; b = 3$

15. $m = \frac{4}{3}; b = 1$

16. Through $(2, -5)$; $m = 3$

17. Through $(1, -4)$; $m = \frac{2}{3}$

Graph each linear inequality.

18. $x - 3y \geq 6$

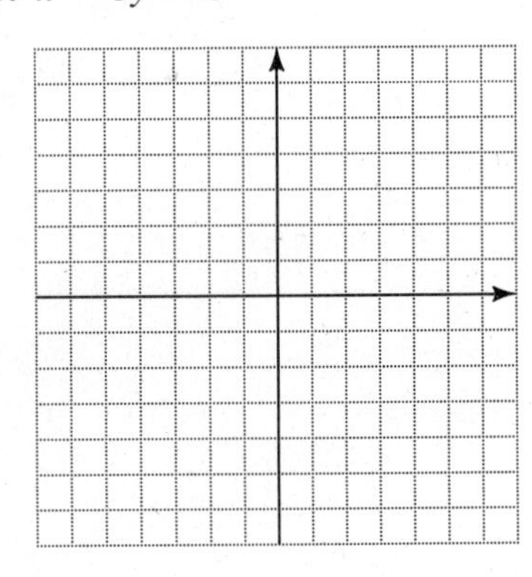

19. $2x + 3y < 12$

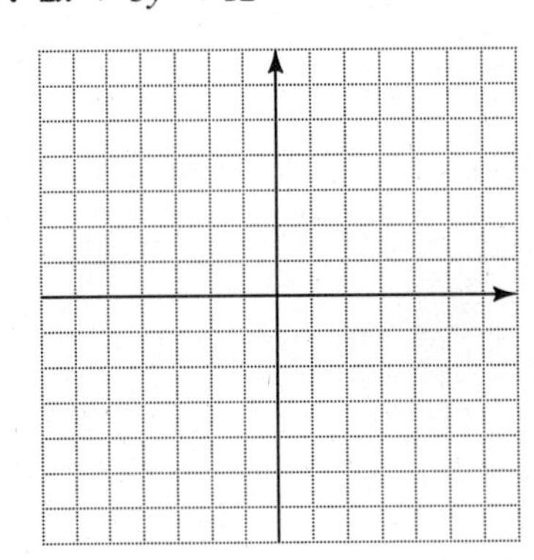

20. $3x + 2y \geq 0$

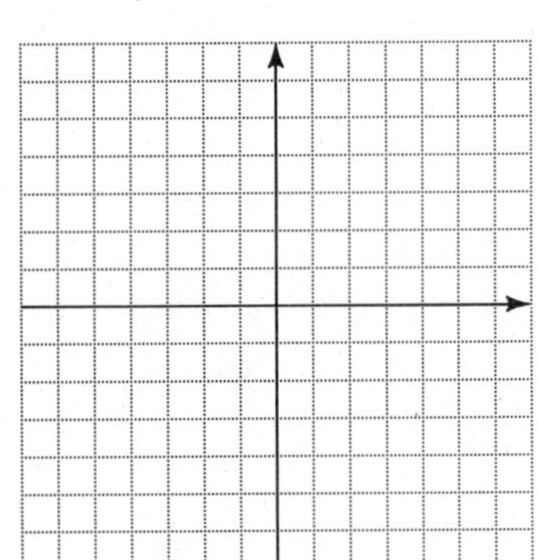

21. Is $(-6, 2)$ a solution for the following system?

$3x + 2y = -14$
$4x + y = -26$

Solve each system by any method.

22. $2x - y = -8$
$x + 2y = 11$

23. $4x + 5y = -8$
$3x + 4y = -7$

24. $3x + 5y = 1$
$x = y + 3$

25. $3x + 4y = 2$
$6x + 8y = 1$

Solve each word problem by any method. Use two variables.

26. Two numbers have a sum of 24 and a difference of 6. Find the numbers.

27. A rectangle has a perimeter of 40 meters. The length is 2 meters more than the width. Find the width of the rectangle.

Let x represent the width.

Length = ______

$P = 2L + 2W$

x

28. The cashier at a small motel has 17 bills, all of which are fives and tens. The value of the money is $125. How many of each type of bill are there?

Type of bill	*Number of bills*	*Value of bills*
Fives	x	
Tens	y	
Totals		

Graph each system of inequalities.

29. $x - y \le 2$
$2x + 3y \ge 6$

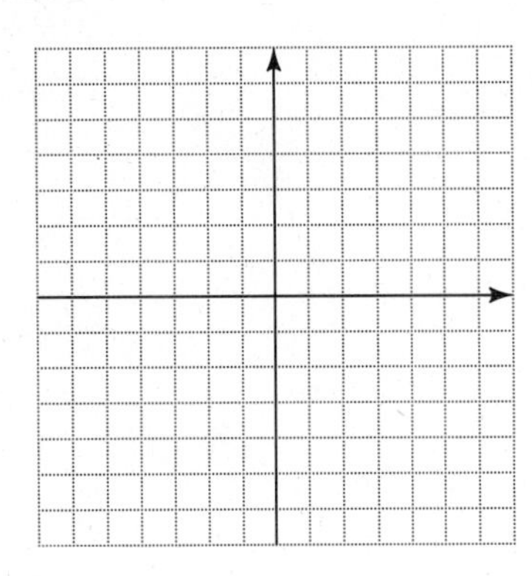

30. $y \le 2x$
$x \ge -1$

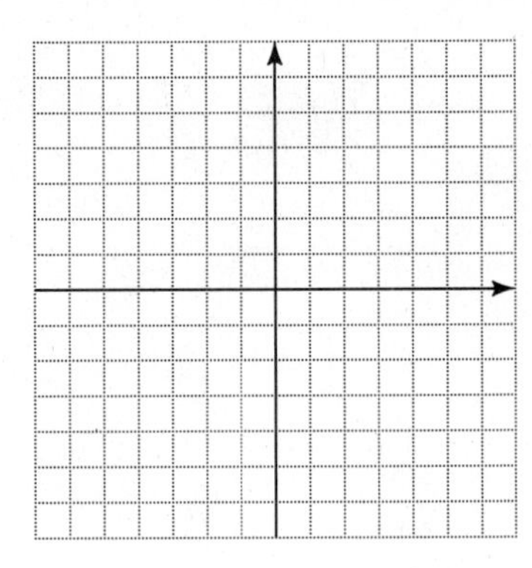

Simplify each expression if possible. Assume all variables represent positive real numbers.

31. $\dfrac{\sqrt{56}}{\sqrt{7}}$

32. $\sqrt{27} - 2\sqrt{12} + 6\sqrt{75}$

33. $\sqrt{\dfrac{5}{12}}$

34. $\dfrac{2}{\sqrt{3} + \sqrt{5}}$

35. $\sqrt{200x^2y^5}$

36. $\sqrt{17} + \sqrt{26}$

37. $\dfrac{5 + \sqrt{75}}{10}$

38. $(3\sqrt{2} + 1)(4\sqrt{2} - 3)$

39. $\sqrt{\dfrac{2}{3}} \cdot \sqrt{\dfrac{6}{7}}$

40. $\sqrt{11} \cdot \sqrt{2} + 3\sqrt{22}$

41. $\sqrt{-100}$

42. $(\sqrt{2} + \sqrt{3})^2$

Solve each equation.

43. $y = \sqrt{3y + 4}$

44. $\sqrt{x} + 2 = x - 10$

QUADRATIC EQUATIONS

10.1 SOLVING QUADRATIC EQUATIONS BY THE SQUARE ROOT METHOD

OBJECTIVES

1. Solve equations of the form x^2 = a number.
2. Solve equations of the form $(ax + b)^2$ = a number.

Recall that a *quadratic equation* is an equation that can be written in the form $ax^2 + bx + c = 0$ for real numbers a, b, and c, with $a \neq 0$. In Chapter 5 these equations were solved by factoring. As mentioned there, not all quadratic equations can easily be solved by factoring (for instance, $x^2 - x + 1 = 0$ cannot). Other ways to solve quadratic equations are shown in this chapter. A quadratic equation such as

$$(x - 3)^2 = 16,$$

with the square of a binomial equal to a number, can be solved with square roots.

1 The *square root property of equations* justifies taking square roots of both sides of an equation.

SQUARE ROOT PROPERTY OF EQUATIONS

If b is a positive number and if $a^2 = b$, then

$$\boldsymbol{a = \sqrt{b}} \quad \text{or} \quad \boldsymbol{a = -\sqrt{b}}.$$

NOTE When we solve an equation, we want to find *all* values of the variable that satisfy the equation. Therefore, we want both the positive and the negative square roots of b.

EXAMPLE 1 Solving a Quadratic Equation by the Square Root Property

Solve each equation. Write radicals in simplified form.

(a) $x^2 = 16$

By the square root property, if $x^2 = 16$, then

$$x = \sqrt{16} = 4 \quad \text{or} \quad x = -\sqrt{16} = -4.$$

An abbreviation for $x = 4$ or $x = -4$ is written $x = \pm 4$ (read "plus or minus"). Check each solution by substituting back in the original equation.

(b) $z^2 = 5$

The solutions are $z = \sqrt{5}$ or $z = -\sqrt{5}$, which may be written $\pm\sqrt{5}$.

(c) $m^2 = 8$

Use the square root property to get $m = \sqrt{8}$ or $m = -\sqrt{8}$. Simplify $\sqrt{8}$ as $\sqrt{8} = 2\sqrt{2}$, so $m = 2\sqrt{2}$ or $m = -2\sqrt{2}$.

(d) $y^2 = -4$

Since -4 is a negative number and since the square of a real number cannot be negative, there is no real number solution for this equation. (The square root property cannot be used because of the requirement that b must be positive.) ■

WORK PROBLEM 1 AT THE SIDE.

2 The equation $(x - 3)^2 = 16$ also can be solved with the square root property of equations. If $(x - 3)^2 = 16$, then

$$x - 3 = 4 \quad \text{or} \quad x - 3 = -4 \qquad \text{Square root property}$$
$$x = 7 \quad \text{or} \quad x = -1. \qquad \text{Add 3.}$$

Check both answers in the original equation.

$$(7 - 3)^2 = 4^2 = 16 \quad \text{and} \quad (-1 - 3)^2 = (-4)^2 = 16$$

Both 7 and -1 are solutions.

EXAMPLE 2 Solving a Quadratic Equation by the Square Root Property

Solve $(x - 1)^2 = 6$.

$$x - 1 = \sqrt{6} \quad \text{or} \quad x - 1 = -\sqrt{6} \qquad \text{Square root property}$$
$$x = 1 + \sqrt{6} \quad \text{or} \quad x = 1 - \sqrt{6}. \qquad \text{Add 1.}$$

Check: $(1 + \sqrt{6} - 1)^2 = (\sqrt{6})^2 = 6$;
$(1 - \sqrt{6} - 1)^2 = (-\sqrt{6})^2 = 6$.

The solutions are $1 + \sqrt{6}$ and $1 - \sqrt{6}$. ■

WORK PROBLEM 2 AT THE SIDE.

EXAMPLE 3 Solving a Quadratic Equation by the Square Root Property

Solve the equation $(3r - 2)^2 = 27$.

$$3r - 2 = \sqrt{27} \quad \text{or} \quad 3r - 2 = -\sqrt{27} \qquad \text{Square root property}$$

Now simplify the radical: $\sqrt{27} = \sqrt{9 \cdot 3} = \sqrt{9} \cdot \sqrt{3} = 3\sqrt{3}$, so

$$3r - 2 = 3\sqrt{3} \quad \text{or} \quad 3r - 2 = -3\sqrt{3}$$
$$3r = 2 + 3\sqrt{3} \quad \text{or} \quad 3r = 2 - 3\sqrt{3} \qquad \text{Add 2.}$$
$$r = \frac{2 + 3\sqrt{3}}{3} \quad \text{or} \quad r = \frac{2 - 3\sqrt{3}}{3} \qquad \text{Divide by 3.}$$

The solutions are $\frac{2 + 3\sqrt{3}}{3}$ and $\frac{2 - 3\sqrt{3}}{3}$. ■

WORK PROBLEM 3 AT THE SIDE.

EXAMPLE 4 Recognizing a Quadratic Equation with No Solution

Solve $(x + 3)^2 = -9$.

The square root of -9 is not a real number. There is no real number solution for this equation. ■

WORK PROBLEM 4 AT THE SIDE.

1. Solve each equation. Write radicals in simplified form.

(a) $k^2 = 49$

(b) $b^2 = 11$

(c) $c^2 = 12$

(d) $x^2 = -9$

2. Solve each equation.

(a) $(m + 2)^2 = 36$

(b) $(p - 4)^2 = 3$

3. Solve each equation.

(a) $(2x - 5)^2 = 18$

(b) $(5m + 1)^2 = 7$

4. Solve each equation.

(a) $(x + 8)^2 = -25$

(b) $(7z - 1)^2 = -1$

ANSWERS

1. (a) $7, -7$ (b) $\sqrt{11}, -\sqrt{11}$ (c) $2\sqrt{3}, -2\sqrt{3}$ (d) no real number solution
2. (a) $4, -8$ (b) $4 + \sqrt{3}, 4 - \sqrt{3}$
3. (a) $\frac{5 + 3\sqrt{2}}{2}, \frac{5 - 3\sqrt{2}}{2}$ (b) $\frac{-1 + \sqrt{7}}{5}, \frac{-1 - \sqrt{7}}{5}$
4. (a) no real number solution (b) no real number solution

name date hour

10.1 EXERCISES

Solve each equation by using the square root property. Express all radicals in simplest form. See Example 1.

1. $x^2 = 25$

2. $y^2 = 100$

3. $x^2 = 64$

4. $z^2 = 81$

5. $m^2 = 13$

6. $x^2 = 7$

7. $p^2 = 2$

8. $q^2 = 6$

9. $x^2 = 24$

10. $t^2 = 27$

11. $m^2 = \frac{169}{9}$

12. $r^2 = \frac{144}{25}$

13. $k^2 = \frac{9}{16}$

14. $q^2 = \frac{225}{16}$

15. $r^2 = 1.21$

16. $k^2 = 1.96$

17. $k^2 = 2.56$

18. $z^2 = 9.61$

19. $r^2 = 77.44$

20. $y^2 = 43.56$

Solve each equation by using the square root property. Express all radicals in simplest form. See Examples 2–4.

21. $(x - 2)^2 = 16$

22. $(r + 4)^2 = 25$

23. $(a + 4)^2 = 10$

24. $(r - 3)^2 = 15$

25. $(x - 1)^2 = 32$

26. $(y + 5)^2 = 28$

27. $(2m - 1)^2 = 9$

28. $(3y - 7)^2 = 4$

29. $(3z + 5)^2 = 9$

30. $(2y - 7)^2 = 49$

31. $(2a - 5)^2 = 30$

32. $(2y + 3)^2 = 45$

33. $(3p - 1)^2 = 18$

34. $(5r - 6)^2 = 75$

35. $(2k - 5)^2 = 98$

36. $(4x - 1)^2 = 48$

37. $(3m + 4)^2 = 8$

38. $(2p - 5)^2 = 180$

39. The amount A that P dollars invested at a rate of interest r will amount to in two years is

$$A = P(1 + r)^2.$$

At what interest rate will \$1 grow to \$1.21 in two years?

40. This exercise uses the formula for the distance traveled by a falling object,

$$d = 16t^2$$

where d is the distance the object falls in t seconds. One expert at marksmanship can hold a silver dollar above his head, drop it, draw his gun, and shoot the coin as it passes waist level. If the coin falls about 4 feet, estimate the time that elapses between the dropping of the coin and the shot.

SKILL SHARPENERS

Simplify all radicals and combine terms. See Sections 9.3 and 9.4.

41. $\frac{3}{2} + \sqrt{\frac{27}{4}}$

42. $-\frac{1}{4} + \sqrt{\frac{5}{16}}$

43. $6 + \sqrt{\frac{2}{3}}$

44. $5 - \sqrt{\frac{7}{2}}$

Factor each perfect square trinomial. See Section 5.4.

45. $x^2 + 8x + 16$

46. $p^2 - 5p + \frac{25}{4}$

47. $z^2 + 3z + \frac{9}{4}$

1 The properties studied so far are not enough to solve the equation

$$x^2 + 6x + 7 = 0.$$

If the equation $x^2 + 6x + 7 = 0$ could be rewritten in a form like $(x + 3)^2 = 2$, it could be solved with the square root property discussed in the previous section. The next example shows how to rewrite the equation so it can be solved by that method.

WORK PROBLEM 1 AT THE SIDE.

EXAMPLE 1 Rewriting an Equation to Use the Square Root Property

Solve $x^2 + 6x + 7 = 0$.

Start by subtracting 7 from each side of the equation.

$$x^2 + 6x = -7$$

The quantity on the left-hand side of $x^2 + 6x = -7$ must be made into a perfect square trinomial. The expression $x^2 + 6x + 9$ is a perfect square, since

$$x^2 + 6x + 9 = (x + 3)^2.$$

Therefore, if 9 is added to each side, the equation will have a perfect square trinomial on the left-hand side, as needed.

$$x^2 + 6x + 9 = -7 + 9 \quad \text{Add 9.}$$

$$(x + 3)^2 = 2 \quad \text{Factor.}$$

Now use the square root property to complete the solution.

$$x + 3 = \sqrt{2} \quad \text{or} \quad x + 3 = -\sqrt{2}$$

$$x = -3 + \sqrt{2} \quad \text{or} \quad x = -3 - \sqrt{2}$$

The solutions of the original equation are $-3 + \sqrt{2}$ and $-3 - \sqrt{2}$. Check by substituting $-3 + \sqrt{2}$ and $-3 - \sqrt{2}$ for x in the original equation. ■

The process of changing the form of the equation in Example 1 from

$$x^2 + 6x + 7 = 0 \quad \text{to} \quad (x + 3)^2 = 2$$

is called **completing the square.** When completing the square, only the *form* of the equation is changed. To see this, simplify $(x + 3)^2 = 2$; the result will be $x^2 + 6x + 7 = 0$.

EXAMPLE 2 Solving a Quadratic Equation by Completing the Square

Solve $m^2 - 8m = 5$.

A suitable number must be added to each side to make the left side a perfect square. This number can be found as follows: recall from Chapter 4 that

$$(m + a)^2 = m^2 + 2am + a^2.$$

We want to find the value of a^2, the number to be added to each side. Here, the middle term of the trinomial $m^2 - 8m + a^2$ is $-8m$, so

$$-8m = 2am$$

$$-8 = 2a \quad \text{Divide each side by } m.$$

$$-4 = a. \quad \text{Divide each side by 2.}$$

OBJECTIVES

1. Solve quadratic equations by completing the square when the coefficient of the squared term is 1.
2. Solve quadratic equations by completing the square when the coefficient of the squared term is not 1.
3. Simplify an equation before solving.

1. As a review, factor each of these perfect square trinomials.

(a) $x^2 + 6x + 9$

(b) $q^2 - 20q + 100$

(c) $9m^2 - 30m + 25$

(d) $100r^2 - 60r + 9$

ANSWERS
1. (a) $(x + 3)^2$ (b) $(q - 10)^2$ (c) $(3m - 5)^2$ (d) $(10r - 3)^2$

2. Solve $a^2 + 3a = 1$ by completing the square.

Then $a^2 = (-4)^2 = 16$, and 16 should be added to each side of the given equation.

$$m^2 - 8m = 5$$

$$m^2 - 8m + 16 = 5 + 16 \qquad (1) \qquad \text{Add 16.}$$

The trinomial $m^2 - 8m + 16$ is a perfect square trinomial. Factor this trinomial to get

$$m^2 - 8m + 16 = (m - 4)^2.$$

Equation (1) becomes

$$(m - 4)^2 = 21.$$

Now use the square root property.

$$m - 4 = \sqrt{21} \quad \text{or} \quad m - 4 = -\sqrt{21}$$

$$m = 4 + \sqrt{21} \quad \text{or} \quad m = 4 - \sqrt{21}$$

The solutions are

$$4 + \sqrt{21} \quad \text{and} \quad 4 - \sqrt{21}. \ \blacksquare$$

Let us summarize what we did to find $a = -4$ above. The coefficient of m in the middle term was -8.

1. We multiplied -8 by $\frac{1}{2}$ (took half of -8).
2. We squared -4 to get 16.
3. We added 16 to each side of the given equation.

WORK PROBLEM 2 AT THE SIDE.

2 The process of completing the square discussed above requires the coefficient of the squared term to be 1. For an equation of the form $ax^2 + bx + c = 0$, to get 1 as a coefficient of x^2 first divide each side of the equation by a. The next examples illustrate this approach.

EXAMPLE 3 Solving a Quadratic Equation by Completing the Square

Solve $4y^2 + 16y = 9$.

Before completing the square, the coefficient of y^2 must be 1. Here the coefficient of y^2 is 4. Make the coefficient 1 by dividing each side of the equation by 4.

$$y^2 + 4y = \frac{9}{4} \qquad \text{Divide by 4.}$$

Next, complete the square by taking half the coefficient of y, or $(\frac{1}{2})(4) = 2$, and squaring the result: $2^2 = 4$. Add 4 to each side of the equation, and perform the addition on the right-hand side.

$$y^2 + 4y + 4 = \frac{9}{4} + 4 \qquad \text{Add 4.}$$

$$y^2 + 4y + 4 = \frac{25}{4}$$

$$(y + 2)^2 = \frac{25}{4} \qquad \text{Factor.}$$

ANSWERS

2. $\frac{-3 + \sqrt{13}}{2}, \frac{-3 - \sqrt{13}}{2}$

Use the square root property of equations and solve for y.

$$y + 2 = \frac{5}{2} \quad \text{or} \quad y + 2 = -\frac{5}{2} \qquad \text{Square root property}$$

$$y = -2 + \frac{5}{2} \quad \text{or} \quad y = -2 - \frac{5}{2} \qquad \text{Subtract 2.}$$

$$y = \frac{1}{2} \quad \text{or} \quad y = -\frac{9}{2}$$

The two solutions are $\frac{1}{2}$ and $-\frac{9}{2}$. ■

The steps in solving a quadratic equation by completing the square are summarized below.

Use *completing the square* to solve the quadratic equation $ax^2 + bx + c = 0$ as follows.

Step 1 If the coefficient of the squared term is 1, proceed to Step 2. If the coefficient of the squared term is not 1 but some other nonzero number, divide each side of the equation by this coefficient. This gives an equation that has 1 as coefficient of x^2.

Step 2 Make sure that all terms with variables are on one side of the equals sign and that all constants are on the other side.

Step 3 Take half the coefficient of x and square the result. Add the square to each side of the equation. By factoring, the side containing the variables can now be written as a perfect square.

Step 4 Apply the square root property of equations.

WORK PROBLEM 3 AT THE SIDE.

EXAMPLE 4 Solving a Quadratic Equation by Completing the Square

Solve $2x^2 - 7x = 9$.

Divide each side of the equation by 2 to get a coefficient of 1 for the x^2 term.

$$x^2 - \frac{7}{2}x = \frac{9}{2} \qquad \text{Divide by 2.}$$

Now take half the coefficient of x and square it. Half of $-\frac{7}{2}$ is $-\frac{7}{4}$, and $(-\frac{7}{4})^2 = \frac{49}{16}$. Add $\frac{49}{16}$ to each side of the equation, and write the left side as a perfect square.

$$x^2 - \frac{7}{2}x + \frac{49}{16} = \frac{9}{2} + \frac{49}{16} \qquad \text{Add } \frac{49}{16}.$$

$$\left(x - \frac{7}{4}\right)^2 = \frac{121}{16} \qquad \text{Factor.}$$

3. Solve by completing the square.

(a) $9m^2 + 18m + 5 = 0$

(b) $4k^2 - 24k + 11 = 0$

ANSWERS

3. (a) $-\frac{1}{3}, -\frac{5}{3}$ (b) $\frac{11}{2}, \frac{1}{2}$

Use the square root property.

$$x - \frac{7}{4} = \sqrt{\frac{121}{16}} \quad \text{or} \quad x - \frac{7}{4} = -\sqrt{\frac{121}{16}}$$

Since $\sqrt{\frac{121}{16}} = \frac{11}{4}$,

$$x - \frac{7}{4} = \frac{11}{4} \quad \text{or} \quad x - \frac{7}{4} = -\frac{11}{4}$$

$$x = \frac{18}{4} \quad \text{or} \quad x = -\frac{4}{4} \qquad \text{Add } \frac{7}{4}.$$

$$x = \frac{9}{2} \quad \text{or} \quad x = -1.$$

The solutions are $\frac{9}{2}$ and -1. ■

WORK PROBLEM 4 AT THE SIDE.

EXAMPLE 5 Solving a Quadratic Equation by Completing the Square

Solve $4p^2 + 8p + 5 = 0$.

First divide each side by 4 to get a coefficient of 1 for p^2.

$$p^2 + 2p + \frac{5}{4} = 0 \qquad \text{Divide each side by 4.}$$

Subtract $\frac{5}{4}$ from each side, which gives

$$p^2 + 2p = -\frac{5}{4}. \qquad \text{Subtract } \frac{5}{4} \text{ from each side.}$$

The coefficient of p is 2. Take half of 2, square the result, and add this square to each side. The left-hand side can then be written as a perfect square.

$$p^2 + 2p + 1 = -\frac{5}{4} + 1 \qquad \text{Add 1 on each side.}$$

$$(p + 1)^2 = -\frac{1}{4} \qquad \text{Factor.}$$

The square root of $-\frac{1}{4}$ is not a real number, so the square root property does not apply. This equation has no real number solutions. ■

WORK PROBLEM 5 AT THE SIDE.

3 The next example shows how to simplify an equation before solving it.

EXAMPLE 6 Solving a Quadratic Equation by Completing the Square

Solve $(m - 2)(m + 1) = 5$.

Before we can use the method of completing the square, the equation must be in the form $ax^2 + bx + c = 0$. Start by multiplying on the left.

$$(m - 2)(m + 1) = 5$$

$$m^2 - m - 2 = 5 \qquad \text{Use FOIL.}$$

$$m^2 - m = 7 \qquad \text{Add 2.}$$

4. Solve by completing the square.

(a) $3x^2 + 5x - 2 = 0$

(b) $2p^2 - 3p = 1$

5. Solve by completing the square.

(a) $a^2 + 3a + 10 = 0$

(b) $5v^2 + 3v + 1 = 0$

ANSWERS

4. (a) $-2, \frac{1}{3}$ (b) $\frac{3 + \sqrt{17}}{4}, \frac{3 - \sqrt{17}}{4}$

5. (a) no real number solution
(b) no real number solution

Now complete the square. Half of -1 is $-\frac{1}{2}$, and $(-\frac{1}{2})^2 = \frac{1}{4}$. Add $\frac{1}{4}$ to each side.

$$m^2 - m + \frac{1}{4} = 7 + \frac{1}{4} \qquad \text{Add } \frac{1}{4}.$$

Factor on the left, and add on the right.

$$\left(m - \frac{1}{2}\right)^2 = \frac{29}{4}$$

Complete the solution.

$$m - \frac{1}{2} = \sqrt{\frac{29}{4}} \quad \text{or} \quad m - \frac{1}{2} = -\sqrt{\frac{29}{4}} \qquad \text{Square root property}$$

$$m - \frac{1}{2} = \frac{\sqrt{29}}{2} \quad \text{or} \quad m - \frac{1}{2} = -\frac{\sqrt{29}}{2}$$

$$m = \frac{1}{2} + \frac{\sqrt{29}}{2} \quad \text{or} \quad m = \frac{1}{2} - \frac{\sqrt{29}}{2} \qquad \text{Add } \frac{1}{2}.$$

$$m = \frac{1 + \sqrt{29}}{2} \quad \text{or} \quad m = \frac{1 - \sqrt{29}}{2}$$

Check each of these solutions by substituting it in the original equation. ■

WORK PROBLEM 6 AT THE SIDE.

6. Solve each equation.

(a) $r^2 + 1 = 3r$

(b) $3p^2 + 7p = p^2 + p + 3$

ANSWERS

6. (a) $\frac{3 + \sqrt{5}}{2}, \frac{3 - \sqrt{5}}{2}$

(b) $\frac{-3 + \sqrt{15}}{2}, \frac{-3 - \sqrt{15}}{2}$

Historical Reflections

WOMEN IN MATHEMATICS: Dr. Grace M. Hopper (1905–)

Dr. Grace Murray Hopper has been called the world's second computer programmer (the first having been Ada Augusta, who preceded her by some 100 years). Trained as a mathematics teacher, Dr. Hopper did original programming crucial to the Mark I project. The Mark I, an electromechanical machine built by Howard Aiken in 1944, was the first machine to successfully implement many of the old ideas of Charles Babbage.

Dr. Hopper's work laid the foundation for the COBOL language, and she did extensive work in developing programs known as compilers. While working with the Mark I in 1945, she and other researchers encountered a circuit malfunction that turned out to be caused by an "ill-positioned" moth. This was the origin of the term "bug" for anything that causes a problem in a computer or a program.

When first recognized for her computer expertise in 1943, Grace Hopper was a lieutenant in the U.S. Navy. In 1984, President Ronald Reagan promoted her from captain to commodore.

Art: Official U.S. Navy Photograph

name date hour

10.2 EXERCISES

Find the number that should be added to each expression to make it a perfect square. See Example 2.

1. $x^2 + 2x$

2. $y^2 - 4y$

3. $x^2 + 18x$

4. $m^2 - 6m$

5. $x^2 + 3x$

6. $r^2 + 7r$

Solve each equation by completing the square. See Examples 1–6.

7. $x^2 + 4x = -3$

8. $a^2 + 2a = 5$

9. $m^2 + 4m = -1$

10. $z^2 + 6z = -8$

11. $q^2 - 8q = -16$

12. $x^2 - 6x + 1 = 0$

13. $b^2 - 2b - 2 = 0$

14. $c^2 + 3c = 2$

15. $k^2 + 5k - 3 = 0$

16. $2m^2 + 4m = -7$

17. $4y^2 + 4y - 3 = 0$

18. $-x^2 + 6x = 4$

19. $2m^2 - 4m - 5 = 0$

20. $-x^2 + 4 = 2x$

21. $m^2 - 4m + 8 = 6m$

22. $2z^2 = 8z + 5 - 4z^2$

23. $4p - 3 = p^2 + 2p$

24. $(r - 2)(r + 2) = 5$

Work each problem.

25. A rancher has determined that the number of cattle in his herd has increased over a two-year period at a rate r given by the equation $5r^2 + 10r = 1$. Find r. Do both answers make sense?

26. A rule for estimating the number of board feet of lumber that can be cut from a log of a certain length depends on the diameter of the log. The diameter d required to get 9 board feet is found from the equation

$$\left(\frac{d - 4}{4}\right)^2 = 9.$$

Solve this equation for d. Are both answers reasonable?

SKILL SHARPENERS

Write each quotient in lowest terms. Simplify the radicals if necessary. See Section 9.5.

27. $\frac{2 + 2\sqrt{3}}{2}$

28. $\frac{3 + 6\sqrt{5}}{3}$

29. $\frac{4 + 2\sqrt{7}}{8}$

30. $\frac{5 + 5\sqrt{2}}{10}$

31. $\frac{8 + 6\sqrt{3}}{4}$

32. $\frac{4 + \sqrt{28}}{6}$

33. $\frac{6 + \sqrt{45}}{12}$

34. $\frac{8 + \sqrt{32}}{8}$

10.3 SOLVING QUADRATIC EQUATIONS BY THE QUADRATIC FORMULA

Completing the square can be used to solve any quadratic equation, but the method often is not very handy. This section introduces a general formula, the *quadratic formula,* that gives the solution for any quadratic equation.

We can get the quadratic formula by starting with the **standard form** of a quadratic equation,

$$ax^2 + bx + c = 0, \quad a \neq 0.$$

(The restriction $a \neq 0$ is important in order to make sure that the equation is quadratic. If $a = 0$, then the equation becomes $0x^2 + bx + c = 0$, or $bx + c = 0$, which is a linear, not a quadratic equation.)

1 The first step in solving a quadratic equation with the quadratic formula is to identify the letters a, b, and c in the standard form of the quadratic equation.

EXAMPLE 1 Determining Values of *a, b,* and *c* in a Quadratic Equation

Match the coefficients of each of the following quadratic equations with the letters a, b, and c of the standard quadratic equation

$$ax^2 + bx + c = 0.$$

(a) $2x^2 + 3x - 5 = 0$

In this example $a = 2$, $b = 3$, and $c = -5$.

(b) $-x^2 + 2 = 6x$

First rewrite the equation with 0 on one side to match the standard form $ax^2 + bx + c = 0$.

$$-x^2 + 2 = 6x$$

$$-x^2 - 6x + 2 = 0 \qquad \text{Subtract } 6x.$$

Now identify $a = -1$, $b = -6$, and $c = 2$.

(c) $2(x + 3)(x - 1) = 5$

Multiply on the left.

$$2(x + 3)(x - 1) = 2(x^2 + 2x - 3) \qquad \text{FOIL}$$

$$= 2x^2 + 4x - 6 \qquad \text{Distributive property}$$

The equation becomes $2x^2 + 4x - 6 = 5$, or

$$2x^2 + 4x - 11 = 0$$

in standard form, so $a = 2$, $b = 4$, and $c = -11$. ■

WORK PROBLEM 1 AT THE SIDE.

The quadratic formula is developed by solving the equation $ax^2 + bx + c = 0$ by completing the square. First, get the coefficient 1 for the x^2 term by dividing each side by a.

$$x^2 + \frac{b}{a}x + \frac{c}{a} = 0 \qquad \text{Divide by } a.$$

Next, subtract $\frac{c}{a}$ from each side.

$$x^2 + \frac{b}{a}x = -\frac{c}{a} \qquad \text{Subtract } \frac{c}{a}.$$

OBJECTIVES

1. Identify the letters *a, b,* and *c* in a quadratic equation.
2. Use the quadratic formula to solve quadratic equations.
3. Solve quadratic equations with only one solution.
4. Solve quadratic equations with fractions.

1. Match the coefficients of each of the following quadratic equations with the letters a, b, and c of the standard quadratic equation $ax^2 + bx + c = 0$.

(a) $5x^2 + 2x - 1 = 0$

(b) $3m^2 = m - 2$

(c) $p(p + 5) = 4$

(d) $3(m - 2)(m + 1) = 7$

ANSWERS

1. (a) $a = 5$, $b = 2$, $c = -1$
(b) $a = 3$, $b = -1$, $c = 2$
(c) $a = 1$, $b = 5$, $c = -4$
(d) $a = 3$, $b = -3$, $c = -13$

2. Complete these steps to simplify the right side of the equation.

(a) $-\frac{c}{a} + \frac{b^2}{4a^2} = \frac{b^2}{4a^2} - ?$

(b) $\frac{b^2}{4a^2} - \frac{c}{a} = \frac{b^2}{4a^2} - \frac{?}{4a^2}$

(c) $\frac{b^2}{4a^2} - \frac{4ac}{4a^2} = \frac{?}{4a^2}$

Complete the square on the left. Take half the coefficient of x, or $\frac{1}{2} \cdot \frac{b}{a} = \frac{b}{2a}$. Square $\frac{b}{2a}$ to get $\frac{b^2}{4a^2}$. Add $\frac{b^2}{4a^2}$ to each side of the equation.

$$x^2 + \frac{b}{a}x + \frac{b^2}{4a^2} = -\frac{c}{a} + \frac{b^2}{4a^2} \qquad \text{Add } \frac{b^2}{4a^2}.$$

Rewrite the left-hand side as a perfect square.

$$\left(x + \frac{b}{2a}\right)^2 = -\frac{c}{a} + \frac{b^2}{4a^2} \qquad \text{Factor.}$$

Now simplify the right-hand side of the equation.

WORK PROBLEM 2 AT THE SIDE.

From the result of Problem 2 at the side,

$$\left(x + \frac{b}{2a}\right)^2 = \frac{b^2 - 4ac}{4a^2}.$$

Apply the square root property. For convenience, we use the ± symbol giving both solutions in one expression.

$$x + \frac{b}{2a} = \pm\sqrt{\frac{b^2 - 4ac}{4a^2}}$$

Simplify the radical.

$$\sqrt{\frac{b^2 - 4ac}{4a^2}} = \frac{\sqrt{b^2 - 4ac}}{\sqrt{4a^2}} = \frac{\sqrt{b^2 - 4ac}}{2a}$$

Write the solutions as follows.

$$x + \frac{b}{2a} = \pm\frac{\sqrt{b^2 - 4ac}}{2a}$$

$$x = -\frac{b}{2a} \pm \frac{\sqrt{b^2 - 4ac}}{2a}$$

$$x = \frac{-b \pm \sqrt{b^2 - 4ac}}{2a}$$

The solutions of the standard quadratic equation $ax^2 + bx + c = 0$ (with $a \neq 0$) are

$$\frac{-b + \sqrt{b^2 - 4ac}}{2a} \quad \text{and} \quad \frac{-b - \sqrt{b^2 - 4ac}}{2a}.$$

The result is called the **quadratic formula,** a key formula that should be memorized.

THE QUADRATIC FORMULA

The solutions of the quadratic equation $ax^2 + bx + c = 0$ are given by the **quadratic formula,**

$$\boldsymbol{x = \frac{-b \pm \sqrt{b^2 - 4ac}}{2a}}, \quad a \neq 0.$$

CAUTION Notice that the fraction bar is under $-b$ as well as the radical. In using this formula, be sure to find the value of $-b \pm \sqrt{b^2 - 4ac}$ first, then divide that result by the value of $2a$.

ANSWERS

2. (a) $\frac{c}{a}$ (b) $4ac$ (c) $b^2 - 4ac$

2 The following examples show how to use the quadratic formula.

EXAMPLE 2 Solving a Quadratic Equation by Using the Quadratic Formula

Use the quadratic formula to solve $2x^2 - 7x - 9 = 0$.

Match the coefficients of the variables with those of the standard quadratic equation

$$ax^2 + bx + c = 0.$$

Here, $a = 2$, $b = -7$, and $c = -9$. Substitute these numbers into the quadratic formula and simplify the result.

$$x = \frac{-b \pm \sqrt{b^2 - 4ac}}{2a}$$

$$x = \frac{-(-7) \pm \sqrt{(-7)^2 - 4(2)(-9)}}{2(2)} \quad \text{Let } a = 2,\ b = -7,\ c = -9.$$

$$x = \frac{7 \pm \sqrt{49 + 72}}{4} = \frac{7 \pm \sqrt{121}}{4}$$

$$x = \frac{7 \pm 11}{4} \quad \sqrt{121} = 11$$

Write the two separate solutions by first using the plus sign, and then using the minus sign:

$$x = \frac{7 + 11}{4} = \frac{18}{4} = \frac{9}{2} \quad \text{or} \quad x = \frac{7 - 11}{4} = \frac{-4}{4} = -1.$$

The solutions of $2x^2 - 7x - 9 = 0$ are $\frac{9}{2}$ and -1. Check by substituting back in the original equation. ■

WORK PROBLEM 3 AT THE SIDE.

3. Solve by using the quadratic formula.

(a) $2x^2 + 3x - 5 = 0$

(b) $6p^2 + p = 1$

EXAMPLE 3 Solving a Quadratic Equation by Using the Quadratic Formula

Solve $x^2 = 2x + 1$.

Find a, b, and c by rewriting the equation in standard form (with 0 on one side). Add $-2x - 1$ to each side of the equation to get

$$x^2 - 2x - 1 = 0.$$

Then $a = 1$, $b = -2$, and $c = -1$. The solution is found by substituting these values into the quadratic formula.

$$x = \frac{-b \pm \sqrt{b^2 - 4ac}}{2a}$$

$$x = \frac{-(-2) \pm \sqrt{(-2)^2 - 4(1)(-1)}}{2(1)} \quad \text{Let } a = 1,\ b = -2,\ c = -1.$$

$$x = \frac{2 \pm \sqrt{4 + 4}}{2} = \frac{2 \pm \sqrt{8}}{2}$$

Since $\sqrt{8} = \sqrt{4 \cdot 2} = \sqrt{4} \cdot \sqrt{2} = 2\sqrt{2}$,

$$x = \frac{2 \pm 2\sqrt{2}}{2}.$$

ANSWERS

3. (a) $1, -\frac{5}{2}$ (b) $-\frac{1}{2}, \frac{1}{3}$

4. Solve by using the quadratic formula.

(a) $-y^2 = 8y + 1$

(b) $4m^2 - 12m + 2 = 0$

5. Solve each equation.

(a) $9y^2 - 12y + 4 = 0$

(b) $16m^2 - 24m + 9 = 0$

6. Solve by using the quadratic formula.

(a) $m^2 = \frac{2}{3}m + \frac{4}{9}$

(b) $x^2 - \frac{4}{3}x + \frac{2}{3} = 0$

ANSWERS

4. (a) $-4 + \sqrt{15}, -4 - \sqrt{15}$

(b) $\frac{3 + \sqrt{7}}{2}, \frac{3 - \sqrt{7}}{2}$

5. (a) $\frac{2}{3}$ (b) $\frac{3}{4}$

6. (a) $\frac{1 + \sqrt{5}}{3}, \frac{1 - \sqrt{5}}{3}$

(b) no real number solution

Write these solutions in lowest terms by factoring $2 \pm 2\sqrt{2}$ as $2(1 \pm \sqrt{2})$ to get

$$x = \frac{2(1 \pm \sqrt{2})}{2} = 1 \pm \sqrt{2}.$$

The two solutions of the original equation are

$$1 + \sqrt{2} \quad \text{and} \quad 1 - \sqrt{2}. \blacksquare$$

WORK PROBLEM 4 AT THE SIDE.

3 When the quantity under the radical, $b^2 - 4ac$, equals 0, the equation has just one rational number solution. In this case, the trinomial $ax^2 + bx + c$ is a perfect square.

EXAMPLE 4 Solving a Quadratic Equation by Using the Quadratic Formula

Solve $4x^2 + 25 = 20x$.

Write the equation as

$$4x^2 - 20x + 25 = 0. \qquad \text{Subtract } 20x.$$

Here, $a = 4$, $b = -20$, and $c = 25$. By the quadratic formula,

$$x = \frac{-(-20) \pm \sqrt{(-20)^2 - 400}}{8} = \frac{20 \pm 0}{8} = \frac{5}{2}.$$

Since there is just one solution, the trinomial $4x^2 - 20x + 25$ is a perfect square. ■

WORK PROBLEM 5 AT THE SIDE.

4 The final example shows how to solve quadratic equations with fractions.

EXAMPLE 5 Solving a Quadratic Equation by the Quadratic Formula

Solve the equation

$$\frac{1}{10}t^2 = \frac{2}{5}t - \frac{1}{2}.$$

Eliminate the denominators by multiplying each side of the equation by the common denominator, 10.

$$10\left(\frac{1}{10}t^2\right) = 10\left(\frac{2}{5}t - \frac{1}{2}\right)$$

$$t^2 = 4t - 5$$

$$t^2 - 4t + 5 = 0. \qquad \text{Add } -4t \text{ and } 5.$$

From this form, identify $a = 1$, $b = -4$, and $c = 5$. Use the quadratic formula to complete the solution.

$$t = \frac{4 \pm \sqrt{(-4)^2 - 4(1)(5)}}{2(1)} = \frac{4 \pm \sqrt{16 - 20}}{2} = \frac{4 \pm \sqrt{-4}}{2}$$

The radical $\sqrt{-4}$ is not a real number, so the equation has no real number solution. ■

WORK PROBLEM 6 AT THE SIDE.

name date hour

10.3 EXERCISES

For each equation, identify the letters a, b, and c of the standard quadratic equation, $ax^2 + bx + c = 0$. Do not solve. See Example 1.

1. $3x^2 + 4x - 8 = 0$ $a =$ ____ $b =$ ____ $c =$ ____

2. $9x^2 + 2x - 3 = 0$ $a =$ ____ $b =$ ____ $c =$ ____

3. $2x^2 = 3x - 2$ $a =$ ____ $b =$ ____ $c =$ ____

4. $9x^2 - 2 = 4x$ $a =$ ____ $b =$ ____ $c =$ ____

5. $3x^2 - 8x = 0$ $a =$ ____ $b =$ ____ $c =$ ____

6. $5x^2 = 2x$ $a =$ ____ $b =$ ____ $c =$ ____

7. $9(x - 1)(x + 2) = 8$ $a =$ ____ $b =$ ____ $c =$ ____

8. $(3x - 1)(2x + 5) = x(x - 1)$ $a =$ ____ $b =$ ____ $c =$ ____

Use the quadratic formula to solve each equation. Write all radicals in simplified form. Write answers in lowest terms. See Examples 2–4.

9. $p^2 + 2p - 2 = 0$

10. $6k^2 + 6k + 1 = 0$

11. $y^2 + 4y + 4 = 0$

12. $3r^2 - 5r + 1 = 0$

13. $z^2 = 13 - 12z$

14. $x^2 = 8x + 9$

15. $2w^2 + 12w + 5 = 0$

16. $k^2 = 20k - 19$

17. $5x^2 + 4x - 1 = 0$

18. $5n^2 + n - 1 = 0$

19. $2z^2 = 3z + 5$

20. $7r - 2r^2 + 30 = 0$

21. $9r^2 + 6r + 1 = 0$

22. $5m^2 + 5m = 0$

23. $4y^2 - 8y = 0$

24. $6p^2 = 10p$

25. $3r^2 = 16r$

26. $m^2 - 20 = 0$

27. $k^2 - 5 = 0$

28. $9r^2 - 16 = 0$

29. $4y^2 - 25 = 0$

name date hour

30. $2x^2 + 2x + 4 = 4 - 2x$

31. $3x^2 - 4x + 3 = 8x - 1$

32. $2x^2 + x + 7 = 0$

33. $x^2 + x + 1 = 0$

34. $2x^2 = 3x - 2$

35. $x^2 = 5x - 20$

Use the quadratic formula to solve each equation. See Example 5.

36. $\frac{1}{2}x^2 = 1 - \frac{1}{6}x$

37. $\frac{3}{2}r^2 - r = \frac{4}{3}$

38. $\frac{2}{3}m^2 - \frac{4}{9}m - \frac{1}{3} = 0$

39. $\frac{3}{5}x - \frac{2}{5}x^2 = -1$

40. $\frac{r^2}{2} = r + \frac{1}{2}$

41. $\frac{m^2}{4} + \frac{3m}{2} + 1 = 0$

42. $\frac{3k^2}{8} - k = -\frac{17}{24}$

43. $2 = \frac{4}{x} - \frac{3}{x^2}$

44. $y^2 + \frac{8}{3}y = -\frac{7}{3}$

Work each problem.

45. A certain projectile is located $d = 2t^2 - 5t + 2$ feet from its starting point after t seconds. How many seconds will it take the projectile to travel 14 feet? Are both answers reasonable?

46. The time t in seconds for a car to skid 48 feet is given (approximately) by

$$48 = 64t - 16t^2.$$

Solve this equation for t. Are both answers reasonable?

SKILL SHARPENERS

Complete the ordered pairs using the given equation. Then graph the line. See Sections 7.1–7.3.

47. $2x - 3y = 6$

(0,   ) (, 0)
(, 2) (−3,   )

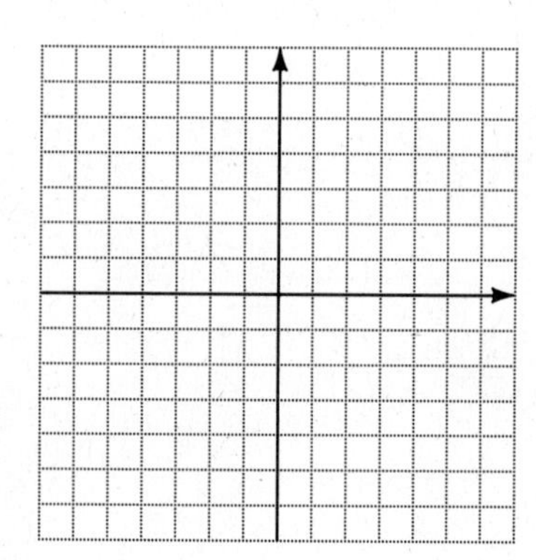

48. $4x - y = 8$

(0,   ) (, 0)
(4,   ) (, −4)

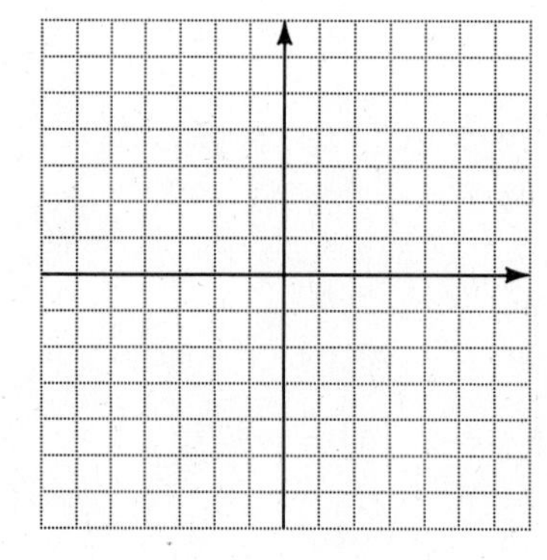

49. $3x + 5y = 15$

(0,   ) (, 0)
(−5,   ) (, −3)

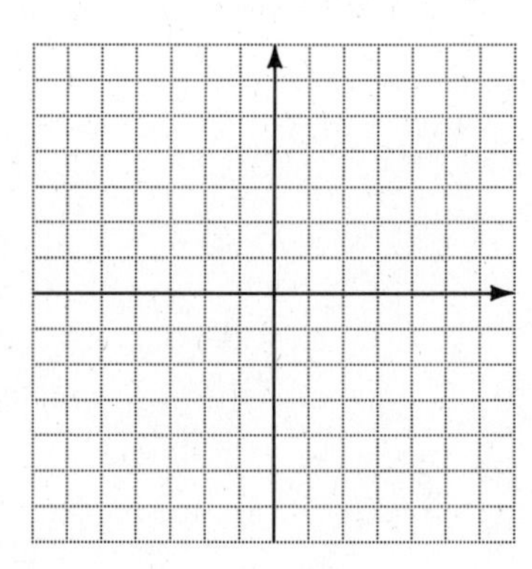

name date hour

SUMMARY EXERCISES ON QUADRATIC EQUATIONS

Four methods have now been introduced for solving quadratic equations written in the form $ax^2 + bx + c = 0$. The chart below shows some advantages and some disadvantages of each method.

Method	Advantages	Disadvantages
1. Factoring	Usually the fastest method.	Not all equations can be solved by factoring. Some factorable polynomials are hard to factor.
2. Completing the square	Can always be used (also, the procedure is useful in other areas of mathmetics).	Requires more steps than other methods.
3. Quadratic formula	Can always be used.	More difficult than factoring because of the $\sqrt{b^2 - 4ac}$ expression.
4. Square root method	Simplest method for solving equations of the form $(ax + b)^2 =$ a number.	Few equations are given in this form.

Solve each quadratic equation by the best method.

1. $y^2 + 3y + 1 = 0$

2. $r^2 = 25$

3. $k^2 = \frac{49}{81}$

4. $p^2 = \frac{100}{81}$

5. $x^2 + 3x + 2 = 0$

6. $m^2 - 4m + 3 = 0$

7. $p^2 + 3p - 2 = 0$

8. $z^2 - 9z + 20 = 0$

9. $(2p - 1)^2 = 10$

10. $(3r - 2)^2 = 9$

11. $(5m + 1)^2 = 36$

12. $(y + 6)^2 = 121$

13. $(7m - 1)^2 = 32$

14. $(3k - 7)^2 = 24$

15. $(5r - 7)^2 = -1$

16. $2x^2 - x = 1$

17. $2m^2 = 3m + 2$

18. $2p^2 + 1 = p$

19. $3r^2 + 8r = 3$

20. $8m^2 = 2m + 15$

21. $3p^2 + 5p + 1 = 0$

22. $m^2 - 2m - 1 = 0$

23. $y^2 + 6y + 4 = 0$

24. $5k^2 + 8 = 22k$

25. $z^2 + 12z + 25 = 0$

26. $3z^2 = 4z + 1$

27. $2m^2 = 2m + 1$

28. $4x^2 = 5x - 1$

29. $4x^2 + 5x = 1$

30. $9m^2 + 12m = 7$

31. $15t^2 + 58t + 48 = 0$

32. $4r^2 + 14r + 11 = 0$

33. $p^2 + 5p + 5 = 0$

34. $9k^2 = 48k + 64$

35. $\frac{1}{5}z^2 = z + \frac{14}{5}$

36. $3m^2 - 4m = 4$

37. $\frac{1}{2}r^2 = r + \frac{15}{2}$

38. $5k^2 + 17k = 12$

39. $m^2 = \frac{5}{12}m + \frac{1}{6}$

40. $z^2 - z + 3 = 0$

41. $k^2 + \frac{4}{15}k = \frac{4}{15}$

42. $4m^2 - 11m + 10 = 0$

10.4 GRAPHING QUADRATIC EQUATIONS IN TWO VARIABLES

OBJECTIVES

1 Graph quadratic equations.

2 Find the vertex of a parabola.

1 In Chapter 7 we saw that the graph of a linear equation in two variables is a straight line that represents all the solutions of the equation. Quadratic equations in two variables, of the form $y = ax^2 + bx + c$, are graphed in this section. Perhaps the simplest quadratic equation is $y = x^2$ (or $y = 1x^2 + 0x + 0$). The graph of this equation cannot be a straight line since only linear equations of the form $Ax + By = C$ have graphs that are straight lines. However, $y = x^2$ can be graphed in much the same way that straight lines were graphed, by finding ordered pairs that satisfy the equation $y = x^2$.

EXAMPLE 1 Graphing a Quadratic Equation

Graph $y = x^2$.

Select several values for x; then find the corresponding y-values. For example, selecting $x = 2$ gives

$$y = 2^2 = 4,$$

and so the point (2, 4) is on the graph of $y = x^2$. (Recall that in an ordered pair such as (2, 4), the x-value comes first and the y-value second.)

WORK PROBLEM 1 AT THE SIDE.

1. Complete the following table of values for $y = x^2$.

x	y
3	
2	4
1	
0	
−1	
−2	
−3	

If the points from Problem 1 at the side are plotted on a coordinate system and a smooth curve drawn through them, the graph is as shown in Figure 1. The table of values completed in Problem 1 is shown with the graph. ■

x	y
3	9
2	4
1	1
0	0
−1	1
−2	4
−3	9

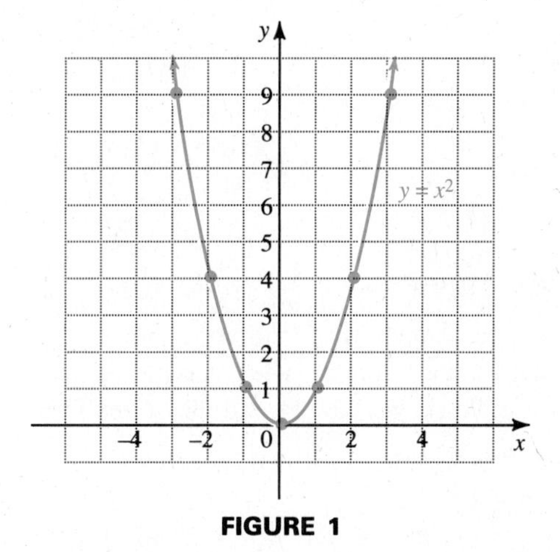

FIGURE 1

The curve in Figure 1 is called a **parabola.** The point (0, 0), the lowest point on this graph, is called the **vertex** of the parabola. The vertical line through the vertex (the y-axis here) is called the **axis** of the parabola.

Every equation of the form

$$y = ax^2 + bx + c,$$

with $a \neq 0$, has a graph that is a parabola. Because of its many useful properties, the parabola occurs frequently in real-life applications. For example, if an object is thrown into the air, the path that the object follows is a parabola (ignoring wind resistance). The cross sections of radar, spotlight, and telescope reflectors also form parabolas.

ANSWERS

1.

x	3	1	0	−1	−2	−3
y	9	1	0	1	4	9

2. Complete each ordered pair for $y = x^2 - 3$.

$(-2, \quad)$

$(-1, \quad)$

$(1, \quad)$

$(2, \quad)$

3. Graph $y = x^2 + 2$. Identify the vertex.

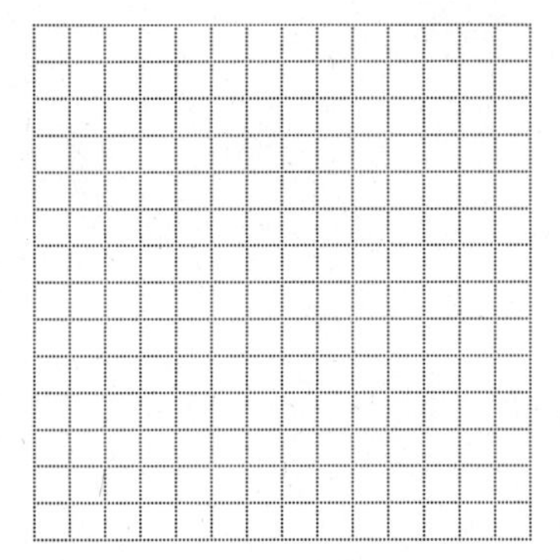

ANSWERS

2. $(-2, 1)$, $(-1, -2)$, $(1, -2)$, $(2, 1)$

3.

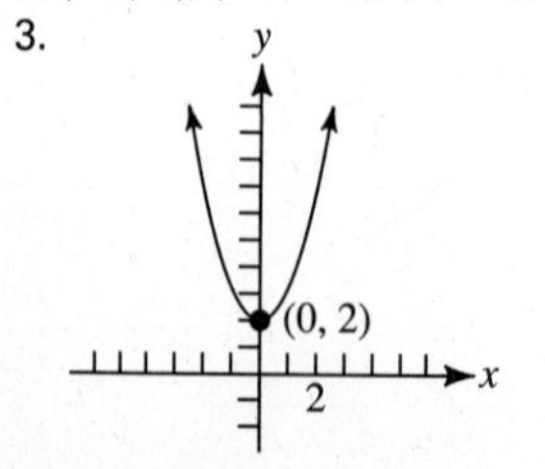

EXAMPLE 2 Graphing a Parabola

Graph $y = x^2 - 3$.

Find several ordered pairs. Begin with the intercepts. If $x = 0$,

$$y = x^2 - 3 = \mathbf{0}^2 - 3 = -3,$$

giving the ordered pair $(0, -3)$. If $y = 0$,

$$y = x^2 - 3$$
$$\mathbf{0} = x^2 - 3$$
$$x^2 = 3$$
$$x = \sqrt{3} \quad \text{or} \quad -\sqrt{3},$$

giving the two ordered pairs $(-\sqrt{3}, 0)$ and $(\sqrt{3}, 0)$. Now choose additional x-values near the x-values of these three points.

WORK PROBLEM 2 AT THE SIDE.

The ordered pairs from above and from Problem 2 at the side are listed in the table of values shown with Figure 2. Plot all these points and connect them with a smooth curve as shown in Figure 2. The vertex of this parabola is $(0, -3)$. ■

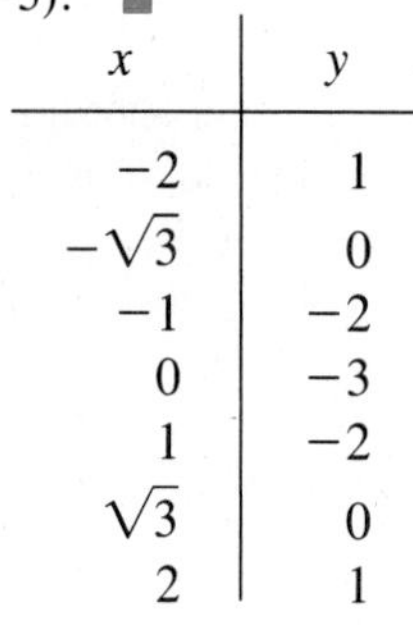

x	y
-2	1
$-\sqrt{3}$	0
-1	-2
0	-3
1	-2
$\sqrt{3}$	0
2	1

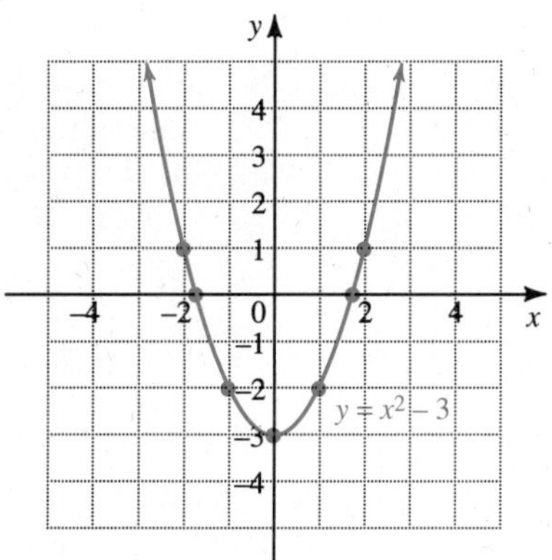

FIGURE 2

WORK PROBLEM 3 AT THE SIDE.

2 As the graphs we found above suggest, the vertex is the most important point to locate when you are graphing a quadratic equation. The next example shows how to find the vertex in a more general case.

EXAMPLE 3 Finding the Vertex to Graph a Parabola

Graph $y = x^2 - 2x - 3$.

We want to find the vertex of the graph. Note in Figure 2 that the vertex is exactly halfway between the x-intercepts. If a parabola has two x-intercepts this is always the case. Therefore, let us begin by finding the x-intercepts. Let $y = 0$ in the equation and solve for x.

$$0 = x^2 - 2x - 3$$
$$0 = (x + 1)(x - 3) \qquad \text{Factor.}$$
$$x + 1 = 0 \quad \text{or} \quad x - 3 = 0 \qquad \text{Set each factor equal to 0.}$$
$$x = -1 \quad \text{or} \quad x = 3$$

There are two x-intercepts, $(-1, 0)$ and $(3, 0)$. Now find any y-intercepts. Substitute $x = 0$ in the equation.

$$y = \mathbf{0}^2 - 2(\mathbf{0}) - 3 = -3$$

There is one y-intercept, $(0, -3)$.

As mentioned above, the x-value of the vertex is halfway between the x-values of the two x-intercepts. Thus, it is $\frac{1}{2}$ their sum.

$$x = \frac{1}{2}(-1 + 3) = 1$$

Find the corresponding y-value by substituting 1 for x in the equation.

$$y = \mathbf{1}^2 - 2(\mathbf{1}) - 3 = -4$$

The vertex is $(1, -4)$. The axis is the line $x = 1$. Plot the three intercepts and the vertex. Find additional ordered pairs as needed. For example, if $x = 2$,

$$y = \mathbf{2}^2 - 2(\mathbf{2}) - 3 = -3,$$

leading to the ordered pair $(2, -3)$. A table of values with the ordered pairs we have found is shown with the graph in Figure 3. ■

x	y
−2	5
−1	0
0	−3
1	−4
2	−3
3	0
4	5

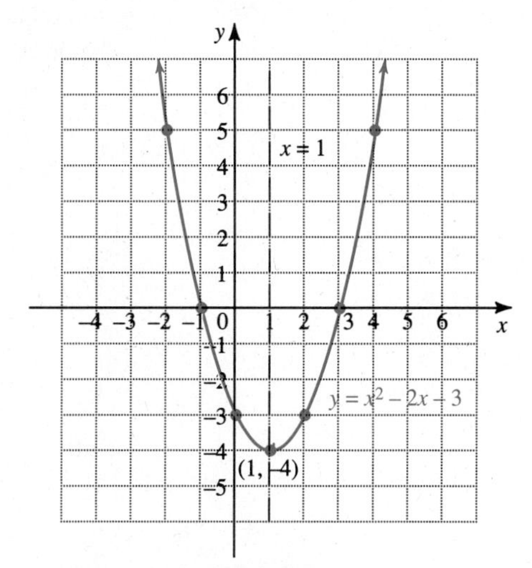

FIGURE 3

WORK PROBLEM 4 AT THE SIDE.

We can generalize from Example 3. The x-intercepts for the equation $0 = ax^2 + bx + c$, by the quadratic formula, have

$$x = \frac{-b + \sqrt{b^2 - 4ac}}{2a} \quad \text{and} \quad x = \frac{-b - \sqrt{b^2 - 4ac}}{2a}.$$

Thus, the x-value of the vertex is

$$x = \frac{1}{2}\left(\frac{-b + \sqrt{b^2 - 4ac}}{2a} + \frac{-b - \sqrt{b^2 - 4ac}}{2a}\right)$$

$$x = \frac{1}{2}\left(\frac{-b + \sqrt{b^2 - 4ac} - b - \sqrt{b^2 - 4ac}}{2a}\right)$$

$$x = \frac{1}{2}\left(\frac{-2b}{2a}\right) = -\frac{b}{2a}.$$

For the equation in Example 3, $y = x^2 - 2x - 3$, $a = 1$, and $b = -2$. Thus,

$$x = -\frac{b}{2a} = -\frac{-2}{2(1)} = 1,$$

which is the same x-value for the vertex we found in Example 3. (The x-value of the vertex is $x = -\frac{b}{2a}$ even if the graph has no x-intercepts.) A procedure for graphing quadratic equations is given on the next page.

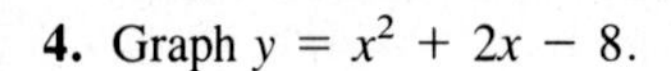

4. Graph $y = x^2 + 2x - 8$.

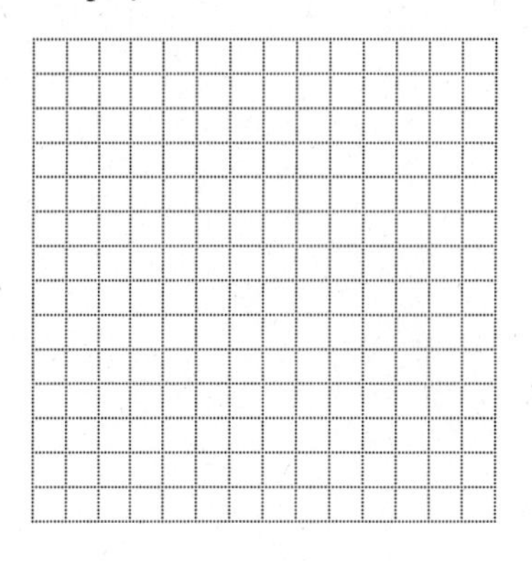

ANSWERS

4.

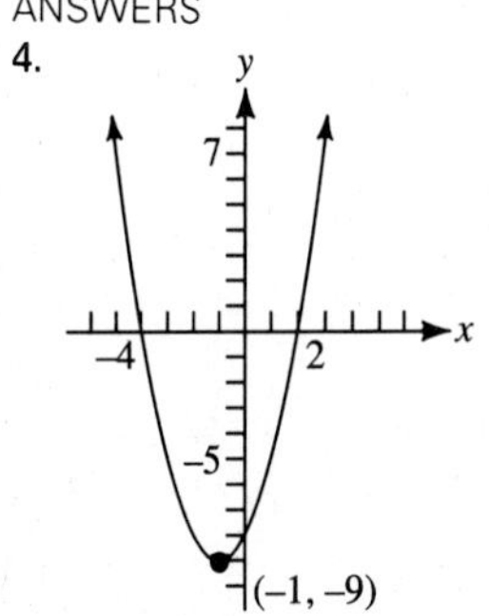

5. Complete the following ordered pairs for $y = x^2 - 4x + 1$.

(5,) (1,) (4,)

(3,) (−1,)

6. Graph the parabola

$$y = x^2 - 2x + 4.$$

Identify the vertex.

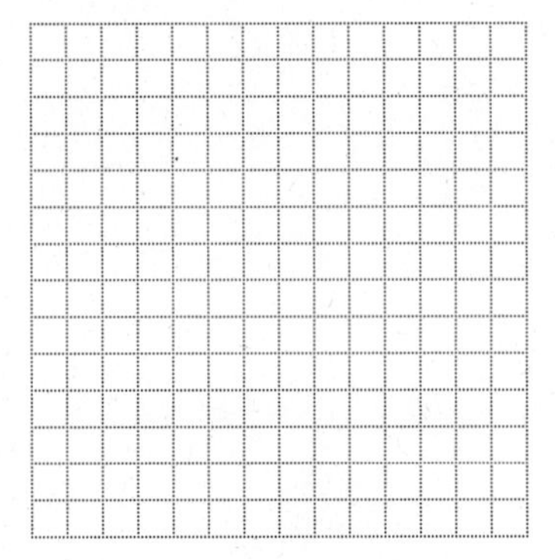

ANSWERS

5. (5, 6), (4, 1), (3, −2), (1, −2), (−1, 6)

6.

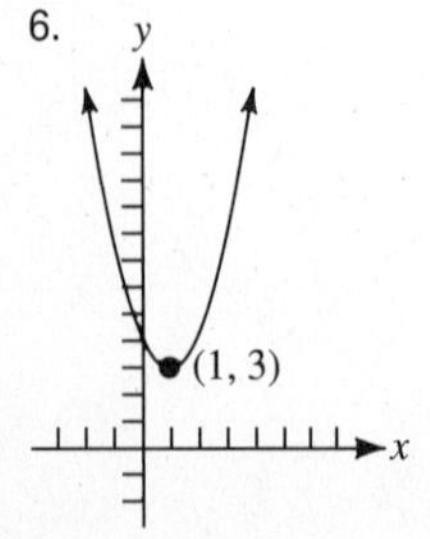

GRAPHING THE PARABOLA $y = ax^2 + bx + c$

Step 1 Find the y-intercept.

Step 2 Find the x-intercepts.

Step 3 Find the vertex. Let $x = -\frac{b}{2a}$ and find the corresponding y-value by substituting for x in the equation.

Step 4 Plot the intercepts and the vertex.

Step 5 Find and plot additional ordered pairs near the vertex and intercepts as needed.

EXAMPLE 4 Graphing a Parabola

Graph $y = x^2 - 4x + 1$.

Find the intercepts. Let $x = 0$ in $y = x^2 - 4x + 1$ to get the y-intercept (0, 1). Let $y = 0$ to get the x-intercepts. If $y = 0$, the equation is $0 = x^2 - 4x + 1$, which cannot be factored. Use the quadratic formula to solve for x.

$$x = \frac{4 \pm \sqrt{16 - 4}}{2} \qquad \text{Let } a = 1,\ b = -4,\ c = 1.$$

$$x = \frac{4 \pm \sqrt{12}}{2}$$

$$x = \frac{4 \pm 2\sqrt{3}}{2} \qquad \sqrt{12} = 2\sqrt{3}$$

$$x = \frac{2(2 \pm \sqrt{3})}{2} = 2 \pm \sqrt{3}$$

Use a calculator or Table 2 to find that the x-intercepts are (3.7, 0) and (.3, 0) to the nearest tenth. The x-value of the vertex is

$$x = -\frac{b}{2a} = -\frac{-4}{2(1)} = 2.$$

The y-value of the vertex is

$$y = 2^2 - 4(2) + 1 = -3,$$

so the vertex is $(2, -3)$. The axis is the line $x = 2$.

WORK PROBLEM 5 AT THE SIDE.

Plot the intercepts, vertex, and the points found in Problem 5. Connect these points with a smooth curve. The graph is shown in Figure 4. ■

x	y
−1	6
0	1
1	−2
2	−3
3	−2
4	1
5	6

FIGURE 4

WORK PROBLEM 6 AT THE SIDE.

name date hour

10.4 EXERCISES

Sketch the graph of each equation. Identify each vertex. See Examples 1–4.

1. $y = 2x^2$

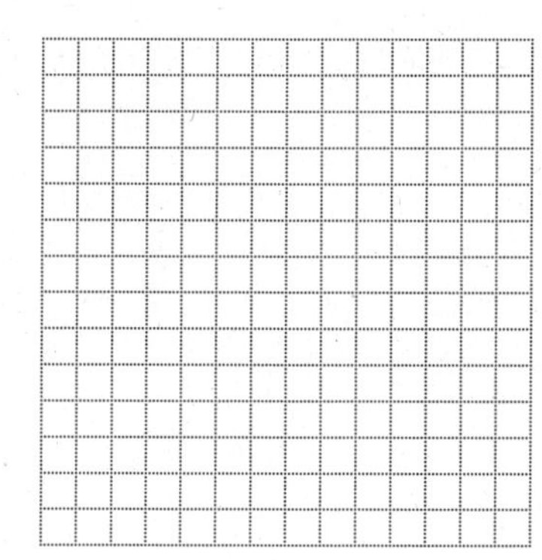

2. $y = -2x^2$

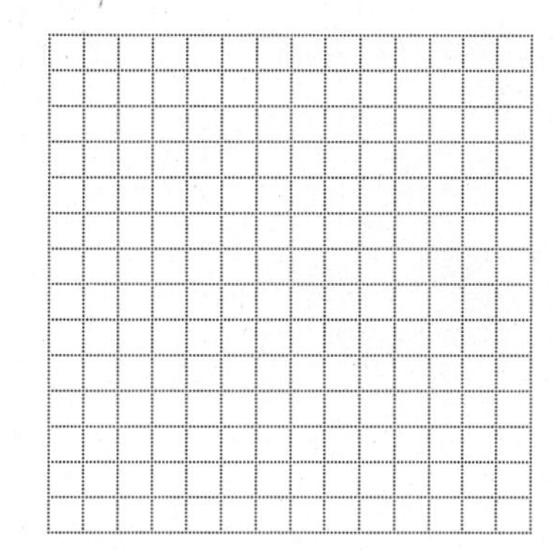

3. $y = x^2 + 2x + 1$

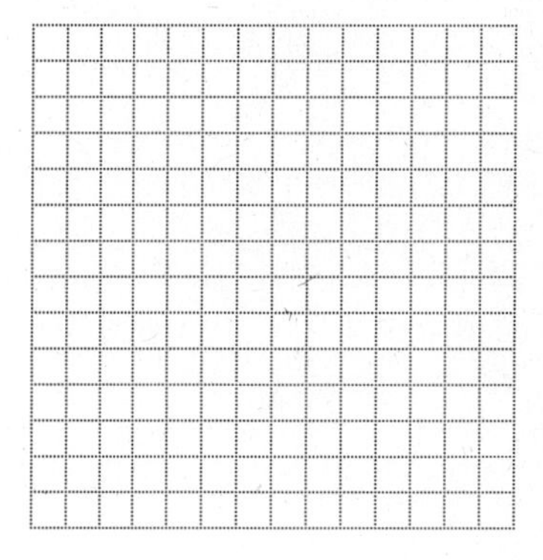

4. $y = x^2 - 4x + 4$

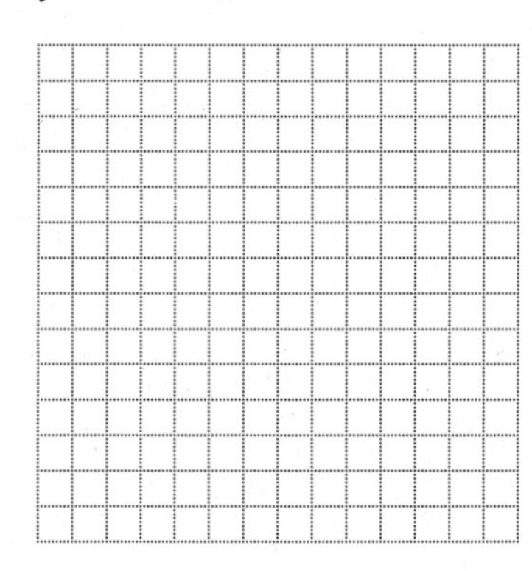

5. $y = -x^2 - 2x - 1$

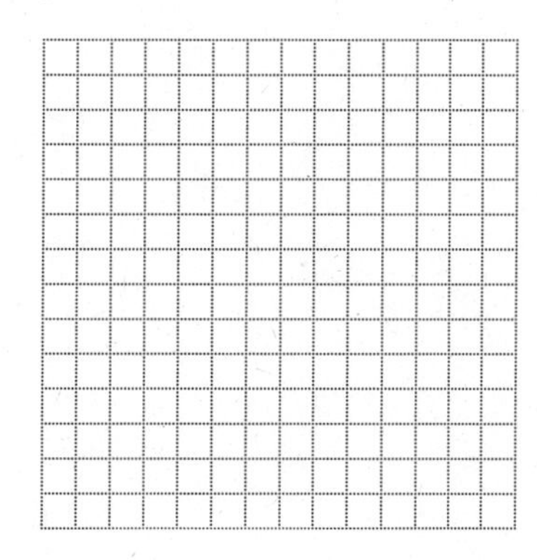

6. $y = -x^2 + 4x - 4$

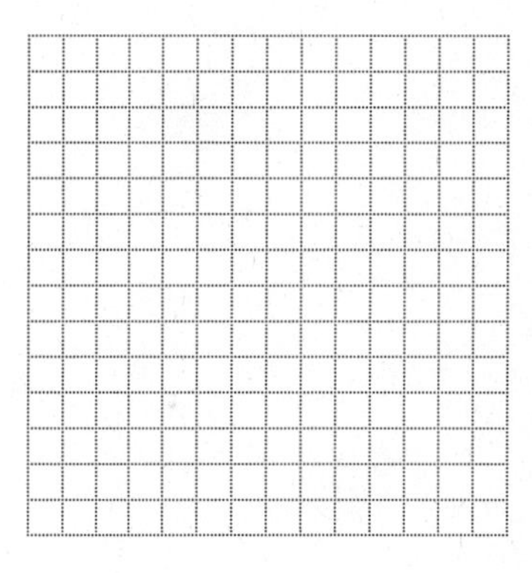

7. $y = x^2 + 1$

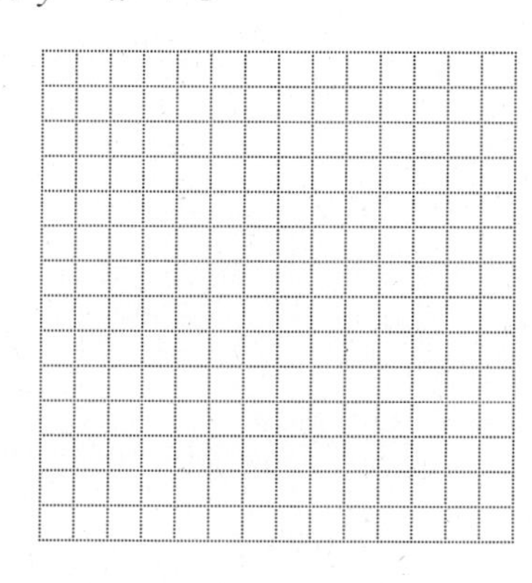

8. $y = -x^2 - 2$

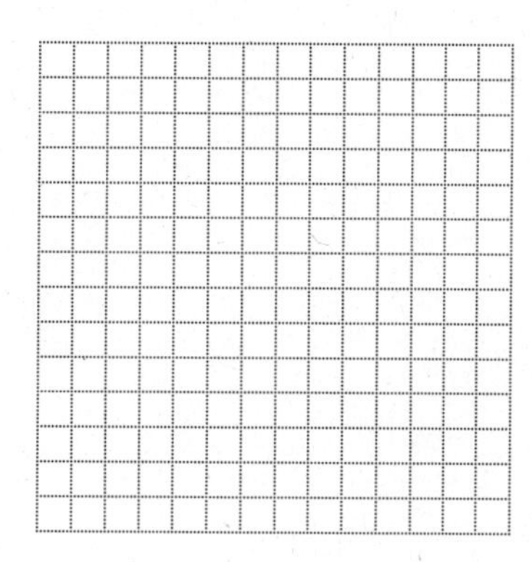

9. $y = 2 - x^2$

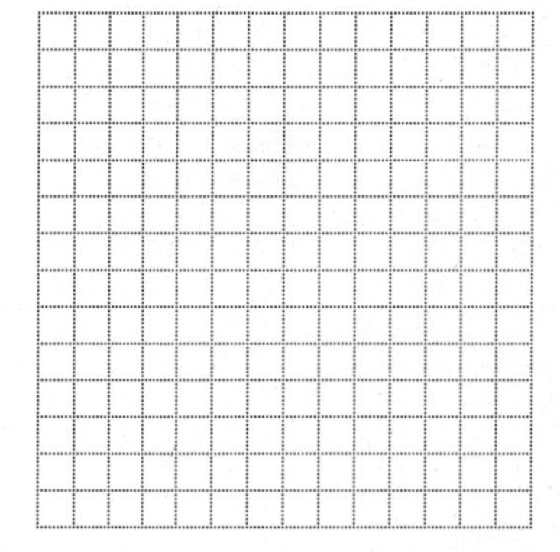

10. $y = x^2 - x - 2$

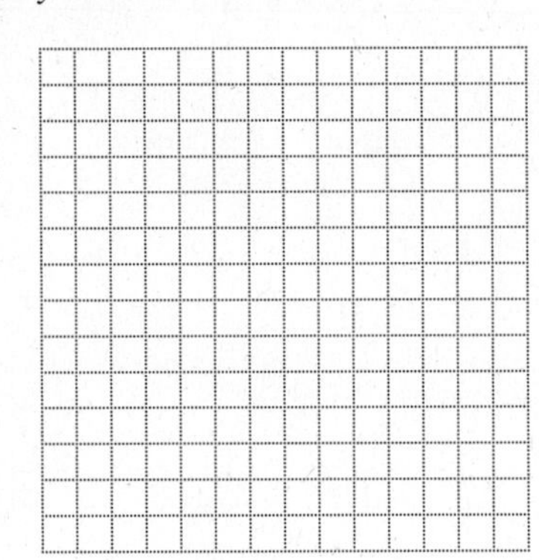

11. $y = x^2 + 6x + 8$

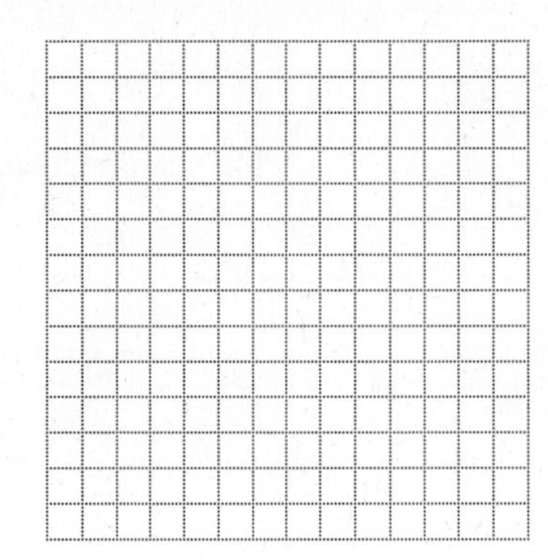

12. $y = x^2 - 3x - 4$

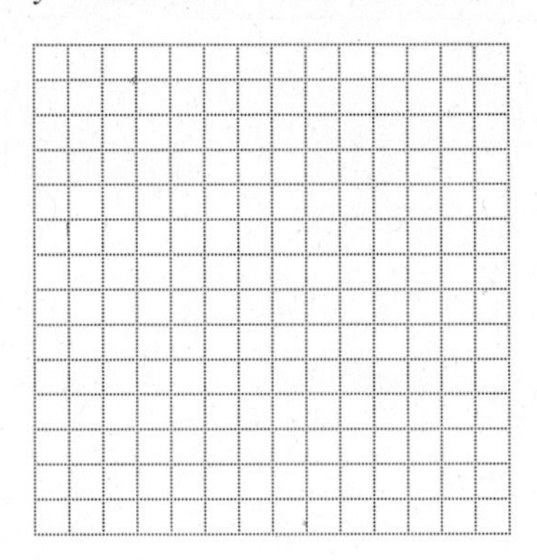

13. $y = x^2 + 2x + 3$

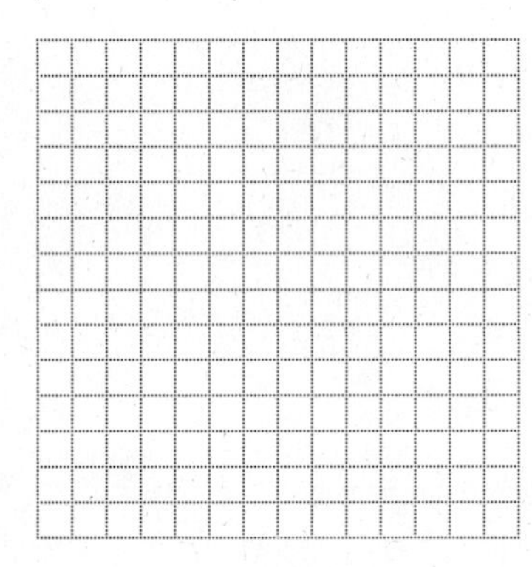

14. $y = x^2 - 4x + 3$

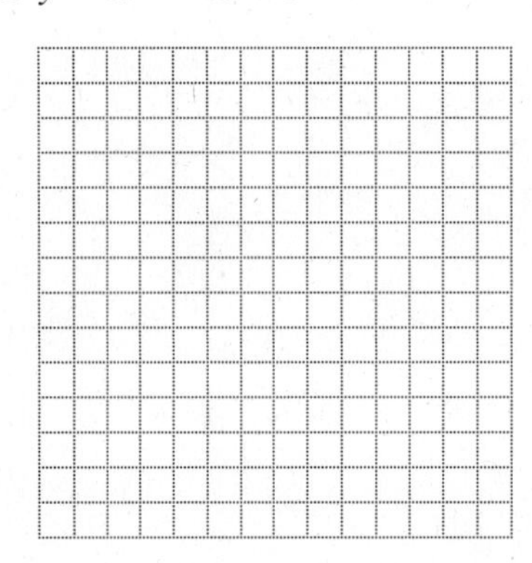

15. $y = x^2 - 8x + 15$

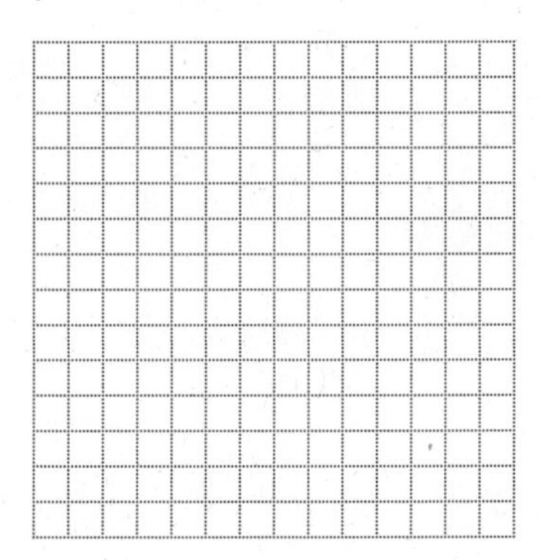

16. $y = x^2 + 6x + 12$

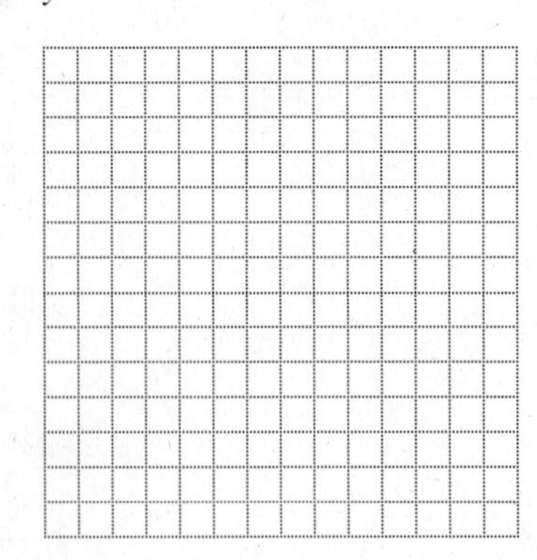

17. $y = -x^2 - 4x - 3$

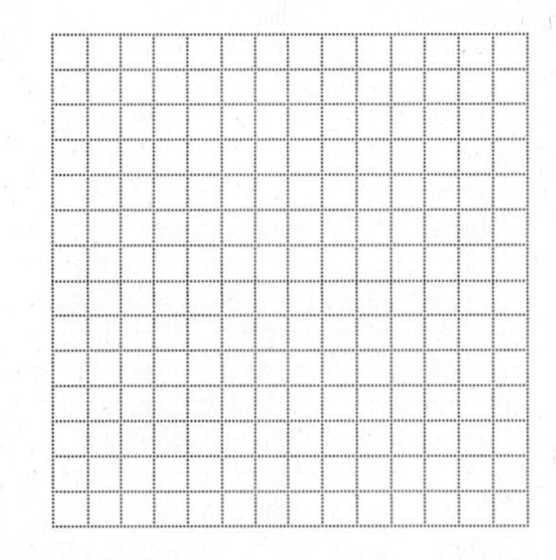

18. $y = -x^2 + 6x - 5$

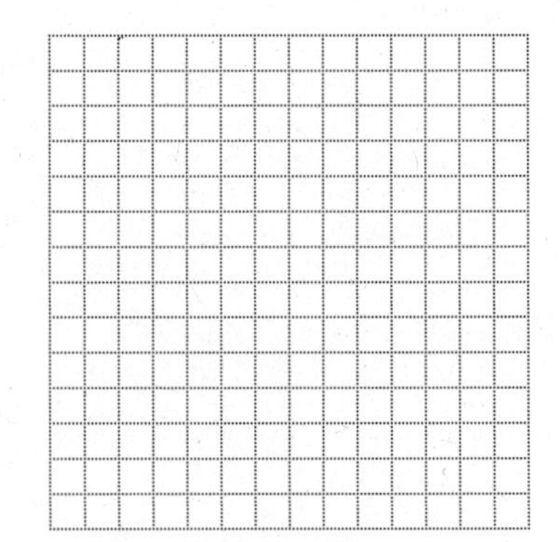

CHAPTER 10 SUMMARY

KEY TERMS

10.3	**standard form**	A quadratic equation written as $ax^2 + bx + c = 0$ $(a \neq 0)$ is in standard form.
10.4	**parabola**	The graph of a quadratic equation is called a parabola.
	vertex	The vertex of a parabola is the highest or lowest point on the graph.
	axis	The axis of a parabola is a vertical line through the vertex.

NEW SYMBOLS

$\pm$ plus or minus

QUICK REVIEW

Section	Concepts	Examples
10.1 Solving Quadratic Equations by the Square Root Method	**Square Root Property of Equations** If b is positive, and if $a^2 = b$, then $a = \sqrt{b}$ or $a = -\sqrt{b}$.	Solve $(2x + 1)^2 = 5$. $2x + 1 = \pm\sqrt{5}$ $2x = -1 \pm \sqrt{5}$ $x = \frac{-1 \pm \sqrt{5}}{2}$
10.2 Solving Quadratic Equations by Completing the Square	**Completing the Square**	Solve $2x^2 + 4x - 1 = 0$.
	1. If the coefficient of the squared term is 1, go to Step 2. If it is not 1, divide each side of the equation by this coefficient.	**1.** $x^2 + 2x - \frac{1}{2} = 0$
	2. Make sure that all variable terms are on one side of the equation and all constant terms are on the other.	**2.** $x^2 + 2x = \frac{1}{2}$
	3. Take half the coefficient of x, square it, and add the square to each side of the equation.	**3.** $x^2 + 2x + 1 = \frac{1}{2} + 1$
	4. Factor the variable side.	**4.** $(x + 1)^2 = \frac{3}{2}$
	5. Use the square root property to solve the equation.	**5.** $x + 1 = \pm\sqrt{\frac{3}{2}} = \pm\frac{\sqrt{6}}{2}$ $x = -1 \pm \frac{\sqrt{6}}{2}$ $x = \frac{-2 \pm \sqrt{6}}{2}$

Section	Concepts	Examples
10.3 Solving Quadratic Equations by the Quadratic Formula	**Quadratic Formula:** The solutions of $ax^2 + bx + c = 0$, $(a \neq 0)$, are $x = \frac{-b \pm \sqrt{b^2 - 4ac}}{2a}$.	Solve $3x^2 - 4x - 2 = 0$. $x = \frac{-(-4) \pm \sqrt{(-4)^2 - 4(3)(-2)}}{2(3)}$ $x = \frac{4 \pm \sqrt{16 + 24}}{6}$ $x = \frac{4 \pm \sqrt{40}}{6} = \frac{4 \pm 2\sqrt{10}}{6}$ $x = \frac{2(2 \pm \sqrt{10})}{6} = \frac{2 \pm \sqrt{10}}{3}$
10.4 Graphing Quadratic Equations in Two Variables	**Graphing $y = ax^2 + bx + c$** **1.** Find the y-intercept.	Graph $y = 2x^2 - 5x - 3$. **1.** $y = 2(0)^2 - 5(0) - 3 = -3$ The y-intercept is $(0, -3)$.
	2. Find the x-intercepts.	**2.** $0 = 2x^2 - 5x - 3$ $0 = (2x + 1)(x - 3)$ $2x + 1 = 0$ or $x - 3 = 0$ $2x = -1$ or $x = 3$ $x = -\frac{1}{2}$ or $x = 3$ The x-intercepts are $(-\frac{1}{2}, 0)$ and $(3, 0)$.
	3. Find the vertex: $x = -\frac{b}{2a}$; find y by substituting for x in the equation.	**3.** For the vertex: $x = -\frac{b}{2a} = -\frac{-5}{2(2)} = \frac{5}{4}$ $y = 2\left(\frac{5}{4}\right)^2 - 5\left(\frac{5}{4}\right) - 3$ $y = 2\left(\frac{25}{16}\right) - \frac{25}{4} - 3$ $y = \frac{25}{8} - \frac{50}{8} - \frac{24}{8} = -\frac{49}{8}$ $y = -6\frac{1}{8}$. The vertex is $(\frac{5}{4}, -\frac{49}{8})$.
	4. Plot the intercepts and the vertex. **5.** Find and plot additional ordered pairs near the vertex and intercepts as needed.	**4.** and **5.** $(-\frac{1}{2}, 0)$, $(3, 0)$, $(0, -3)$, $(2, -5)$, $(\frac{5}{4}, -\frac{49}{8})$, $(1, -6)$

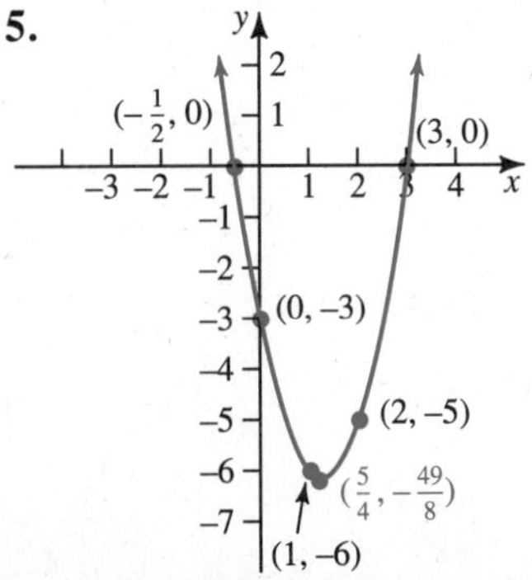

name date hour

CHAPTER 10 REVIEW EXERCISES

[10.1] *Solve each equation by using the square root property. Express all radicals in simplest form.*

1. $y^2 = 49$

2. $x^2 = 15$

3. $m^2 = 48$

4. $(k + 2)^2 = 9$

5. $(r - 3)^2 = 7$

6. $(2p + 1)^2 = 11$

7. $(3k + 2)^2 = 12$

8. $\left(x + \frac{1}{2}\right)^2 = \frac{3}{4}$

[10.2] *Solve each equation by completing the square.*

9. $m^2 + 6m + 5 = 0$

10. $p^2 + 4p = 7$

11. $-x^2 + 5 = 2x$

12. $2y^2 + 8y = 3$

13. $5k^2 - 3k - 2 = 0$

14. $(4a + 1)(a - 1) = -3$

[10.3] *Use the quadratic formula to solve each equation.*

15. $x^2 - 2x - 4 = 0$

16. $-m^2 + 3m + 5 = 0$

17. $3k^2 + 2k + 3 = 0$

18. $5p^2 = p + 1$

19. $2p^2 - 3 = 4p$

20. $-4a^2 + 7 = 2a$

21. $\frac{c^2}{4} = 2 - \frac{3}{4}c$

22. $\frac{3}{2}r^2 = \frac{1}{2} - r$

[10.4] *Sketch the graph of each equation. Identify each vertex.*

23. $y = 3x^2$

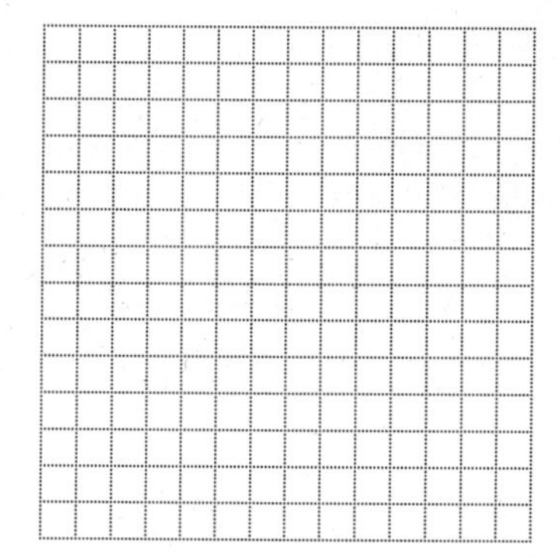

24. $y = x^2 - 2x + 1$

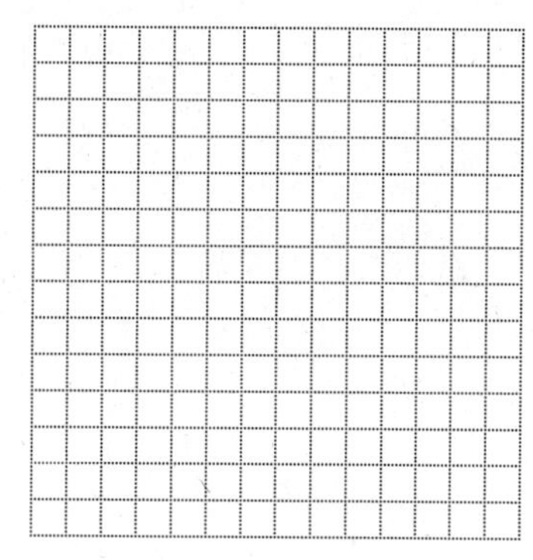

25. $y = -x^2 + 3$

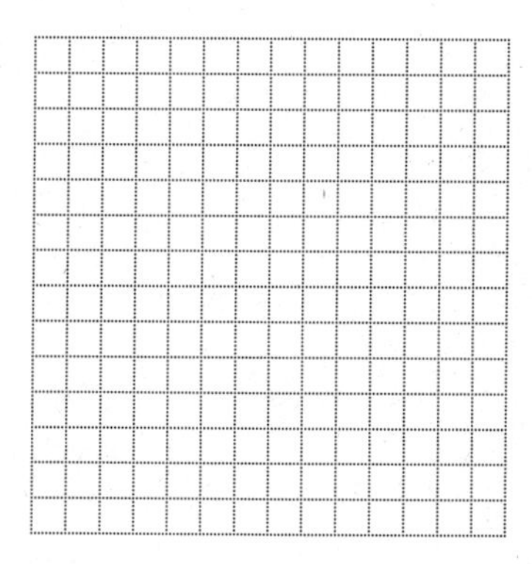

26. $y = x^2 - 1$

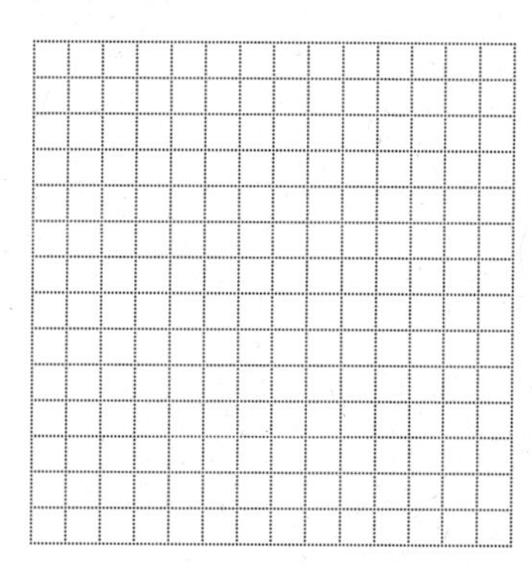

27. $y = -x^2 + 2x + 3$

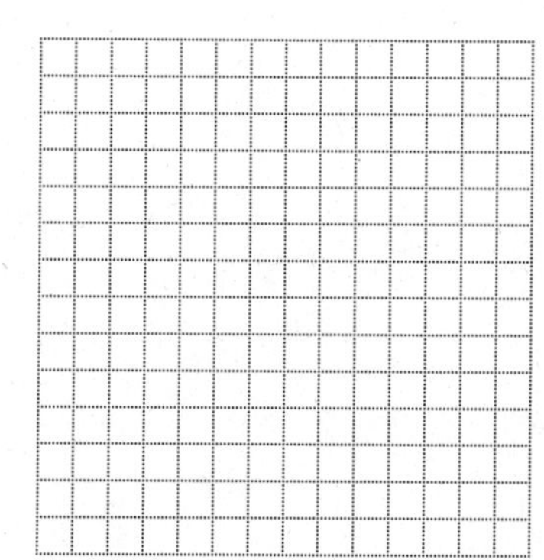

28. $y = x^2 + 4x + 1$

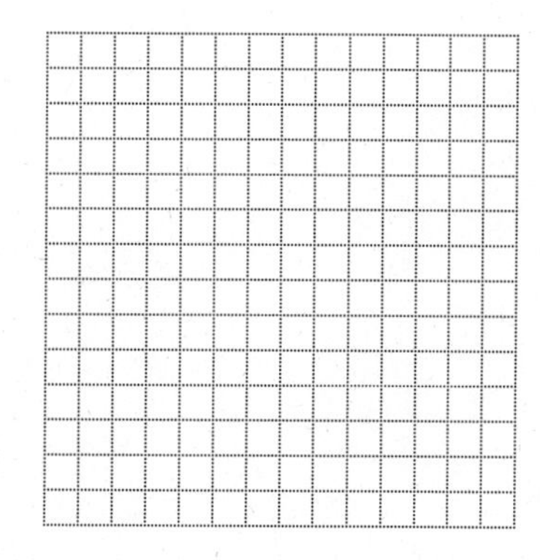

MIXED REVIEW EXERCISES
Solve by any method.

29. $(2t - 1)(t + 5) = 0$

30. $(2p + 1)^2 = 4$

31. $(k + 2)(k - 1) = 3$

32. $6t^2 + 7t - 3 = 0$

33. $2x^2 + 3x + 2 = x^2 - 2x$

34. $x^2 + 2x + 1 = 3$

35. $m^2 - 4m + 5 = 0$

36. $k^2 - 9k + 10 = 0$

37. $(3x + 5)(3x + 5) = 7$

38. $\frac{1}{2}r^2 = \frac{7}{2} - r$

39. $x^2 + 4x - 1 = 0$

40. $7x^2 - 8 = 5x^2 + 16$

name date hour

CHAPTER 10 TEST

Solve by completing the square.

1. $x^2 = 3x - 2$

2. $2x^2 - 5x = 0$

Solve by using the square root property.

3. $x^2 = 5$

4. $(k - 3)^2 = 2$

5. $(3r - 2)^2 = 72$

Solve by the quadratic formula.

6. $m^2 = 3m + 10$

7. $2k^2 + 5k = 3$

8. $3z^2 + 2 = 7z$

9. $4n^2 + 8n + 5 = 0$

10. $y^2 - \frac{5}{3}y + \frac{1}{3} = 0$

Solve by the best method.

11. $y^2 - 2y = 1$

12. $(2x - 1)^2 = 18$

1. ______
2. ______
3. ______
4. ______
5. ______
6. ______
7. ______
8. ______
9. ______
10. ______
11. ______
12. ______

13. ______________________

13. $(x - 5)(3x + 2) = 0$

14. ______________________

14. $(x - 5)(3x - 2) = 4$

15. ______________________

15. $(x - 5)^2 = 8$

16. ______________________

16. $x^2 = 6x - 2$

17. ______________________

17. $p^2 + 16 = 8p$

18. ______________________

18. $2x^2 = 3x + 5$

Sketch the graph of each equation. Identify each vertex.

19.

19. $y = -x^2 + 4$

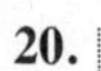

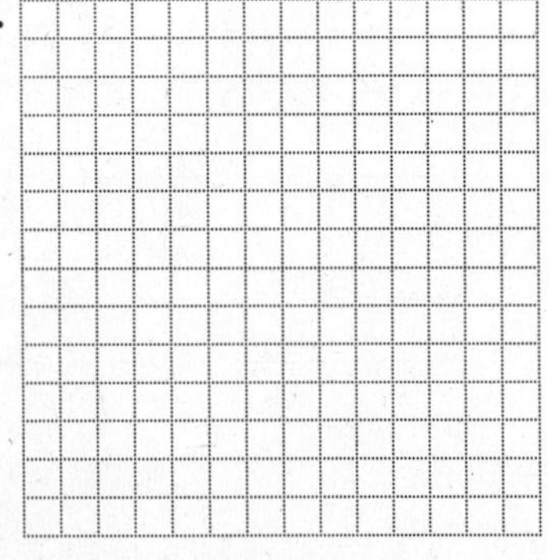

20. $y = x^2 - 6x + 10$

name date hour

FINAL EXAMINATION

Perform each operation, wherever possible.

1. $\dfrac{-4 \cdot 3^2 + 6}{2 - 4 \cdot 1}$

2. $-9 - (-8)(2) + 6 - (3 + 5)$

1. ______
2. ______

Perform the indicated operations.

3. $|-3| - |-5|$

4. $-4r + 14 - 18$

5. $13k - 4k + k - 14k - 2k$

6. $5(4m - 2) - (m + 9)$

3. ______
4. ______
5. ______
6. ______

Solve each equation.

7. $6x - 5 = 7$

8. $3k - 9k - 8k + 6 - 3 = 31$

9. $2(m - 1) - 6(3 - m) + 5 = 1$

10. A salad dressing requires 3 times as much oil as vinegar. If the dressing contains only oil, vinegar, and seasonings, how much vinegar is needed to make 8 ounces of dressing?

11. Solve the formula $I = prt$ for t.

12. Solve the formula $A = \frac{1}{2}bh$ for b.

7. ______
8. ______
9. ______
10. ______
11. ______
12. ______

Solve each inequality. Graph the solution.

13. $-8m < 16$

14. $-9p + 2(8 - p) - 6 \geq 3p - 14 - 4$

13. ——→
14. ——→

15. ____________

16. ____________

17. ____________

18. ____________

19. ____________

20. ____________

21. ____________

22. ____________

23. ____________

24. ____________

25. ____________

26. ____________

27. ____________

28. ____________

29. ____________

30. ____________

31. ____________

32. ____________

Simplify each of the following. Leave answers in exponential form, with positive exponents.

15. $\dfrac{6^{12} \cdot 6^8}{6^9 \cdot 6^{10}}$ **16.** $\dfrac{(5^2)^3(5^3)^4}{5^{12}5^8}$ **17.** 7^{-2}

Perform the indicated operations.

18. $(5x^5 - 9x^4 + 8x^2) - (-3x^5 + 8x^4 - 9x^2)$

19. $(2x - 5)(x^3 + 3x^2 - 2x - 4)$

20. $(6r + 1)^2$ **21.** $\dfrac{3x^3 + 10x^2 - 7x + 4}{x + 4}$

Factor each of the following as completely as possible.

22. $9x - x^2$ **23.** $x^2 - 6x - 16$

24. $2a^2 - 5a - 3$ **25.** $25m^2 - 20m + 4$

Solve each of the following equations by factoring.

26. $x^2 + 3x = 10$ **27.** $3x^2 = x + 4$

28. The length of a rectangle is 2.5 times its width. The area is 1,000 square meters. Find the length.

Perform the following operations. Write all answers in lowest terms.

29. $\dfrac{2}{a - 3} \div \dfrac{5}{2a - 6}$ **30.** $\dfrac{1}{k} - \dfrac{2}{k - 1}$

31. $\dfrac{2}{a^2 - 4} + \dfrac{3}{a^2 - 4a + 4}$ **32.** $\dfrac{2 + \frac{1}{x}}{3 - \frac{1}{x}}$

name date hour

FINAL EXAMINATION

33. Solve $\frac{3k - 1}{4} = \frac{k + 1}{5}$.

33. ____________________

Graph each of the following.

34. $2x + 3y = 6$

34.

35. $5x - 3y = 15$

35.

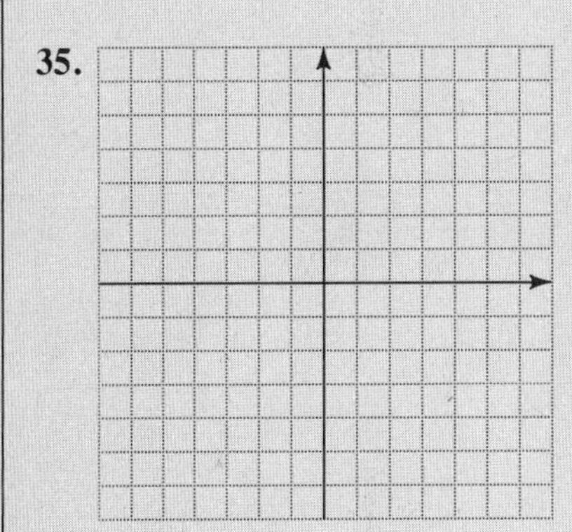

36. $x = -4$

36.

37. $2x - 5y < 10$

37.

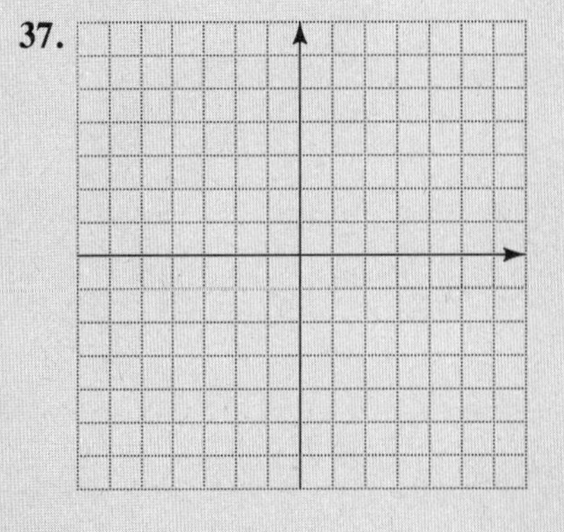

38. ______

39. ______

40. ______

41. ______

42. ______

43. ______

44. ______

45. ______

46. ______

47. ______

48. ______

49. ______

50.

38. Find the slope of the line through $(-1, 4)$ and $(5, 2)$.

39. Write an equation of a line with slope 2 and y-intercept $(0, 3)$.

Solve each system of equations.

40. $2x + y = -4$
$-3x + 2y = 13$

41. $3x - 5y = 8$
$-6x + 10y = 16$

42. The sum of two numbers is 82. One number is 5 less than twice the other. Find the two numbers.

Simplify each of the following as much as possible.

43. $\sqrt{100}$

44. $-\sqrt{841}$

45. $\dfrac{5\sqrt{3}}{\sqrt{2}}$

46. $3\sqrt{5} - 2\sqrt{20}$

47. $-2\sqrt{10} + 3\sqrt{40}$

Solve each quadratic equation.

48. $m^2 - 2m - 1 = 0$

49. $2a^2 - 2a = 1$

50. Graph the parabola $y = x^2 - 4$. Identify the vertex.

APPENDICES

APPENDIX A: SETS

1 A **set** is a collection of things. The objects in a set are called the **elements** of the set. A set is represented by listing its elements between **set braces,** { }. The order in which the elements of a set are listed is unimportant.

EXAMPLE 1 Listing the Elements of a Set
Represent the following sets by listing the elements.

(a) The set of states in the United States that border on the Pacific Ocean = {California, Oregon, Washington, Hawaii, Alaska}.

(b) The set of all counting numbers less than 6 = {1, 2, 3, 4, 5}. ■

WORK PROBLEM 1 AT THE SIDE.

2 Capital letters are used to name sets. To state that 5 is an element of

$$S = \{1, 2, 3, 4, 5\},$$

write $5 \in S$. The statement $6 \notin S$ means that 6 is not an element of S.

WORK PROBLEM 2 AT THE SIDE.

A set with no elements is called the **empty set,** or the **null set.** The symbols $\emptyset$ or { } are used for the empty set. If we let A be the set of all cats that fly, then A is the empty set.

$$A = \emptyset \quad \text{or} \quad A = \{\ \}$$

CAUTION *Do not make the common error* of writing the empty set as $\{\emptyset\}$.

In any discussion of sets, there is some set that includes all the elements under consideration. This set is called the **universal set** for that situation. For example, if the discussion is about presidents of the United States, then the set of all presidents of the United States is the universal set. The universal set is denoted U.

3 In Example 1, there are five elements in the set in part (a), and five in part (b). If the number of elements in a set is either 0 or a counting number, then the set is **finite.** On the other hand, the set of natural numbers, for example, is an **infinite set,** because there is no final number. We can list the elements of the set of natural numbers as

$$N = \{1, 2, 3, 4, . . .\}$$

where the three dots indicate that the set continues indefinitely. Not all infinite sets can be listed in this way. For example, there is no way to list the elements in the set of all real numbers between 1 and 2.

OBJECTIVES

1. List the elements of a set.
2. Learn the vocabulary and symbols used to discuss sets.
3. Decide whether a set is finite or infinite.
4. Decide whether a given set is a subset of another set.
5. Find the complement of a set.
6. Find the union and the intersection of two sets.

1. Represent the following sets by listing the elements.

(a) The set of whole numbers between 2.5 and 4.8

(b) The set of all the days of the week

2. Decide whether each statement is true or false for the set $T = \{m, n, p, q\}$.

(a) $m \in T$

(b) $n \in T$

(c) $k \notin T$

(d) $h \notin T$

ANSWERS
1. (a) {3, 4} (b) {Sunday, Monday, Tuesday, Wednesday, Thursday, Friday, Saturday}
2. (a) true (b) true (c) true (d) true

EXAMPLE 2 Distinguishing Between Finite and Infinite Sets
List the elements of each set, if possible. Decide whether each set is finite or infinite.

(a) The set of all integers

One way to list the elements is $\{\ldots, -2, -1, 0, 1, 2, \ldots\}$. The set is infinite.

(b) The set of all natural numbers between 0 and 5

$\{1, 2, 3, 4\}$ The set is finite.

(c) The set of all irrational numbers

This is an infinite set whose elements cannot be listed. ■

WORK PROBLEM 3 AT THE SIDE.

Two sets are **equal** if they have exactly the same elements. Thus, the set of natural numbers and the set of positive integers are equal sets. Also, the sets

$$\{1, 2, 4, 7\} \quad \text{and} \quad \{4, 2, 7, 1\}$$

are equal. The order of the elements does not make a difference.

4 If all elements of a set A are also elements of some new set B, then we say A is a **subset** of B, written $A \subset B$. We use the symbol $A \not\subset B$ to mean that A is not a subset of B.

EXAMPLE 3 Using Subset Notation
Let $A = \{1, 2, 3, 4\}$, $B = \{1, 4\}$, and $C = \{1\}$. Then $B \subset A$, $C \subset A$, and $C \subset B$, but $A \not\subset B$, $A \not\subset C$, and $B \not\subset C$ ■

WORK PROBLEM 4 AT THE SIDE.

The set $M = \{a, b\}$ has four subsets: $\{a, b\}$, $\{a\}$, $\{b\}$, and $\emptyset$. The empty set is defined to be a subset of any set. How many subsets does $N = \{a, b, c\}$ have? There is one subset with 3 elements: $\{a, b, c\}$. There are three subsets with 2 elements:

$$\{a, b\}, \quad \{a, c\}, \quad \text{and} \quad \{b, c\}.$$

There are three subsets with 1 element:

$$\{a\}, \quad \{b\}, \quad \text{and} \quad \{c\}.$$

There is one subset with 0 elements: $\emptyset$. Thus, set N has eight subsets.

The following generalization can be made.

A set with n elements has 2^n subsets.

WORK PROBLEM 5 AT THE SIDE.

To illustrate the relationships between sets, **Venn diagrams** are often used. A rectangle represents the universal set, U. The sets under discussion are represented by regions within the rectangle. The Venn diagram in Figure 1 shows that $B \subset A$.

3. List the elements of each set, if possible. Decide whether each set is finite or infinite.

(a) The set of integers between -2 and 2

(b) The set of all even integers

(c) The set of all real numbers between 0 and 1

4. Let $P = \{5, 10, 15\}$, $Q = \{5\}$, $R = \{10, 15\}$, and $S = \{15, 10, 5\}$. Use $=$, $\subset$, or $\not\subset$ to make each statement true.

(a) $Q \quad P$

(b) $R \quad P$

(c) $S \quad P$

(d) $P \quad Q$

5. Find all subsets of $\{2, 4, 10\}$.

ANSWERS
3. (a) $\{-1, 0, 1\}$; finite
(b) $\{\ldots, -2, 0, 2, 4, \ldots\}$; infinite
(c) cannot be listed; infinite
4. (a) $\subset$ (b) $\subset$ (c) $=$ (d) $\not\subset$
5. $\{2, 4, 10\}$, $\{2, 4\}$, $\{2, 10\}$, $\{4, 10\}$, $\{2\}$, $\{4\}$, $\{10\}$, $\emptyset$

5 For every set A, there is a set A', the **complement** of A, that contains all the elements of U that are not in A. The shaded region in the Venn diagram in Figure 2 represents A'.

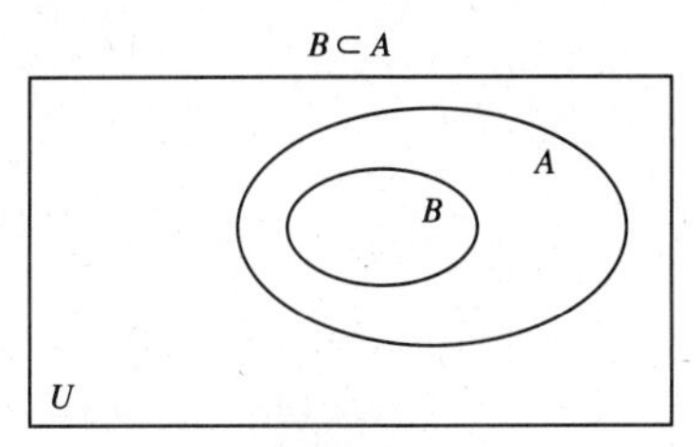

FIGURE 1

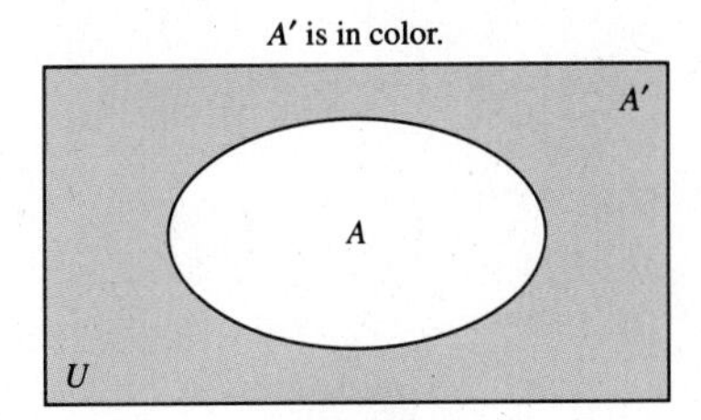

FIGURE 2

EXAMPLE 4 Determining the Complement of a Set

Given $U = \{a, b, c, d, e, f, g\}$, $A = \{a, b, c\}$, $B = \{a, d, f, g\}$, and $C = \{d, e\}$, find A', B', and C'.

(a) $A' = \{d, e, f, g\}$

(b) $B' = \{b, c, e\}$

(c) $C' = \{a, b, c, f, g\}$ ■

WORK PROBLEM 6 AT THE SIDE.

6. Given $U = \{1, 2, 3, 4, 5, 6\}$, $A = \{2, 4, 6\}$, $B = \{1, 2, 3, 4\}$, and $C = \{1, 2, 5, 6\}$, find A', B', and C'.

6 The **union** of two sets A and B, written $A \cup B$, is the set of all elements of A together with all elements of B. Thus, for the sets in Example 4,

$$A \cup B = \{a, b, c, d, f, g\}$$

and

$$A \cup C = \{a, b, c, d, e\}.$$

In Figure 3 the shaded region is the union of sets A and B.

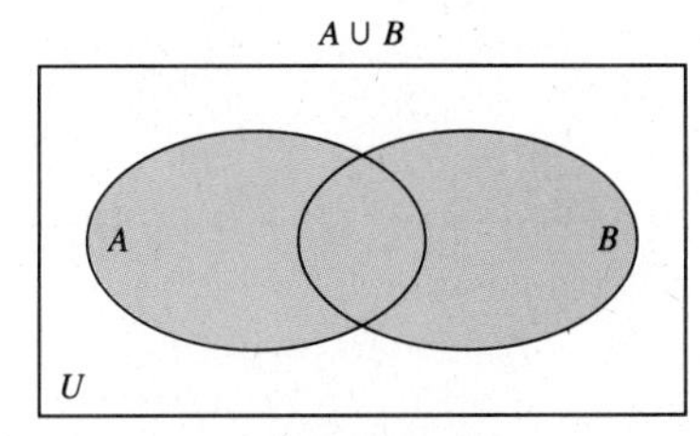

FIGURE 3

EXAMPLE 5 Finding the Union of Two Sets

If $M = \{2, 5, 7\}$ and $N = \{1, 2, 3, 4, 5\}$, then

$$M \cup N = \{1, 2, 3, 4, 5, 7\}. \quad ■$$

WORK PROBLEM 7 AT THE SIDE.

7. Given $A = \{2, 4, 6\}$ and $B = \{1, 2, 3, 4\}$, find $A \cup B$.

ANSWERS

6. $A' = \{1, 3, 5\}$, $B' = \{5, 6\}$, $C' = \{3, 4\}$

7. $\{1, 2, 3, 4, 6\}$

8. Let $S = \{a, b, c, d, e, f\}$, $T = \{a, c, k\}$, and $W = \{d, g, h\}$. Find the following.

(a) $S \cap T$

(b) $S \cap W$

(c) $T \cap W$

9. Use the sets in Example 7 to find the following.

(a) $A \cap C$

(b) $A \cup C$

(c) C'

ANSWERS
8. (a) {a, c} (b) {d} (c) $\emptyset$
9. (a) {2} (b) $\{2, 5, 7, 10, 14, 20\} = U$
(c) {10, 14, 20}

The **intersection** of two sets A and B, written $A \cap B$, is the set of all elements that belong to both A and B. For example, if

$$A = \{\text{Jose, Ellen, Marge, Kevin}\}$$

and

$$B = \{\text{Jose, Patrick, Ellen, Sue}\},$$

then

$$A \cap B = \{\text{Jose, Ellen}\}.$$

The shaded region in Figure 4 represents the intersection of the two sets A and B.

EXAMPLE 6 Finding the Intersection of Two Sets
Suppose that $P = \{3, 9, 27\}$, $Q = \{2, 3, 10, 18, 27, 28\}$, and $R = \{2, 10, 28\}$.

(a) $P \cap Q = \{3, 27\}$

(b) $Q \cap R = \{2, 10, 28\} = R$

(c) $P \cap R = \emptyset$ ■

Sets like P and R in Example 6 that have no elements in common are called **disjoint sets.** The Venn diagram in Figure 5 shows a pair of disjoint sets.

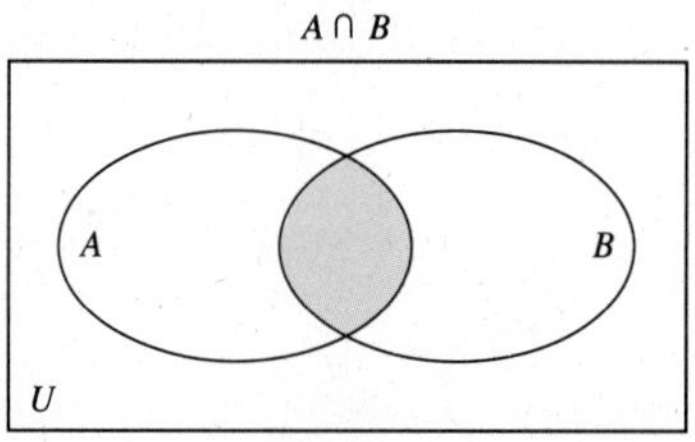

$A \cap B$

A B U

FIGURE 4

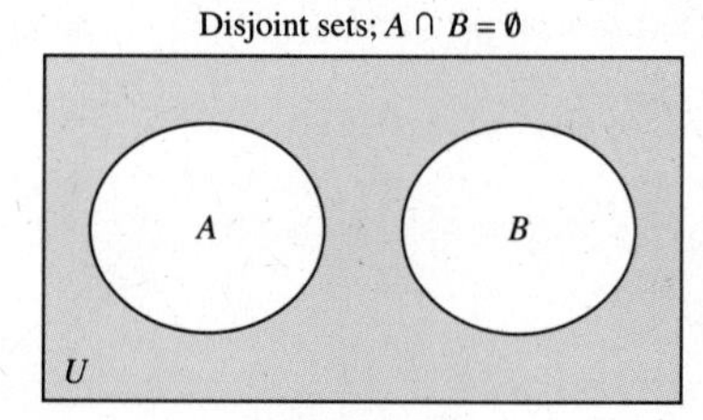

FIGURE 5

WORK PROBLEM 8 AT THE SIDE.

EXAMPLE 7 Using Set Operations
Let $U = \{2, 5, 7, 10, 14, 20\}$, $A = \{2, 10, 14, 20\}$, $B = \{5, 7\}$, and $C = \{2, 5, 7\}$. Find each of the following.

(a) $A \cup B = \{2, 5, 7, 10, 14, 20\} = U$

(b) $A \cap B = \emptyset$

(c) $B \cup C = \{2, 5, 7\} = C$

(d) $B \cap C = \{5, 7\} = B$

(e) $A' = \{5, 7\}$ ■

WORK PROBLEM 9 AT THE SIDE.

name date hour

APPENDIX A EXERCISES

List the elements of each of the following sets. See Examples 1 and 2.

1. The set of all natural numbers less than 8

2. The set of all integers between 4 and 10

3. The set of seasons

4. The set of months of the year

5. The set of women presidents of the United States

6. The set of all living humans who are more than 200 years old

7. The set of letters of the alphabet between K and M

8. The set of letters of the alphabet between D and H

9. The set of positive even numbers

10. The set of multiples of 5

11. Which of the sets described in Exercises 1–10 are infinite sets?

12. Which of the sets described in Exercises 1–10 are finite sets?

Tell whether each of the following is true or false.

13. $5 \in \{1, 2, 5, 8\}$

14. $6 \in \{1, 2, 3, 4, 5\}$

15. $2 \in \{1, 3, 5, 7, 9\}$

16. $1 \in \{6, 2, 5, 1\}$

17. $7 \not\in \{2, 4, 6, 8\}$

18. $7 \not\in \{1, 3, 5, 7\}$

19. $\{2, 4, 9, 12, 13\} = \{13, 12, 9, 4, 2\}$

20. $\{7, 11, 4\} = \{7, 11, 4, 0\}$

Let $A = \{1, 3, 4, 5, 7, 8\}$
$B = \{2, 4, 6, 8\}$
$C = \{1, 3, 5, 7\}$
$D = \{1, 2, 3\}$
$E = \{3, 7\}$
$U = \{1, 2, 3, 4, 5, 6, 7, 8, 9, 10\}$.

Tell whether each of the following is true or false. See Examples 3, 5, 6, and 7.

21. $A \subset U$

22. $D \subset A$

23. $\emptyset \subset A$

24. $\{1, 2\} \subset D$

25. $C \subset A$ **26.** $A \subset C$ **27.** $D \subset B$ **28.** $E \subset C$

29. $D \not\subset E$

30. $E \not\subset A$

31. There are exactly 4 subsets of E.

32. There are exactly 8 subsets of D.

33. There are exactly 12 subsets of C.

34. There are exactly 16 subsets of B.

35. $\{4, 6, 8, 12\} \cap \{6, 8, 14, 17\} = \{6, 8\}$

36. $\{2, 5, 9\} \cap \{1, 2, 3, 4, 5\} = \{2, 5\}$

37. $\{3, 1, 0\} \cap \{0, 2, 4\} = \{0\}$

38. $\{4, 2, 1\} \cap \{1, 2, 3, 4\} = \{1, 2, 3\}$

39. $\{3, 9, 12\} \cap \emptyset = \{3, 9, 12\}$

40. $\{3, 9, 12\} \cup \emptyset = \emptyset$

41. $\{4, 9, 11, 7, 3\} \cup \{1, 2, 3, 4, 5\}$
$= \{1, 2, 3, 4, 5, 7, 9, 11\}$

42. $\{1, 2, 3\} \cup \{1, 2, 3\} = \{1, 2, 3\}$

43. $\{3, 5, 7, 9\} \cup \{4, 6, 8\} = \emptyset$

44. $\{5, 10, 15, 20\} \cup \{5, 15, 30\} = \{5, 15\}$

Let $U = \{a, b, c, d, e, f, g, h\}$
$A = \{a, b, c, d, e, f\}$
$B = \{a, c, e\}$
$C = \{a, f\}$
$D = \{d\}$.

List the elements in the following sets. See Examples 4–7.

45. A' **46.** B' **47.** C'

48. D' **49.** $A \cap B$ **50.** $B \cap A$

51. $A \cap D$ **52.** $B \cap D$ **53.** $B \cap C$

54. $A \cup B$ **55.** $B \cup D$ **56.** $B \cup C$

57. $C \cup B$ **58.** $C \cup D$ **59.** $A \cap \emptyset$ **60.** $B \cup \emptyset$

61. Name every pair of disjoint sets among A–D above.

APPENDIX B: FORMULAS

FORMULAS FROM GEOMETRY

Square

Perimeter: $P = 4s$
Area: $A = s^2$

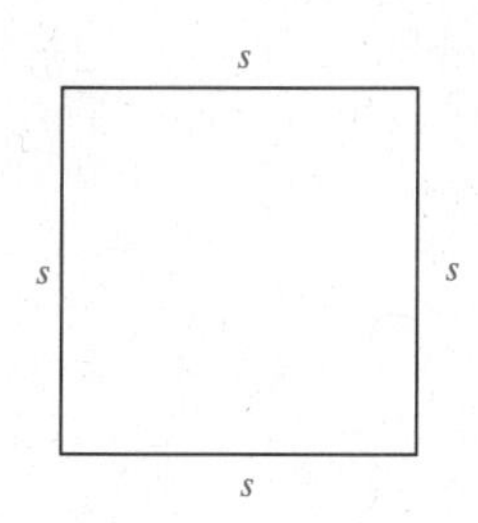

Rectangle

Perimeter: $P = 2L + 2W$
Area: $A = LW$

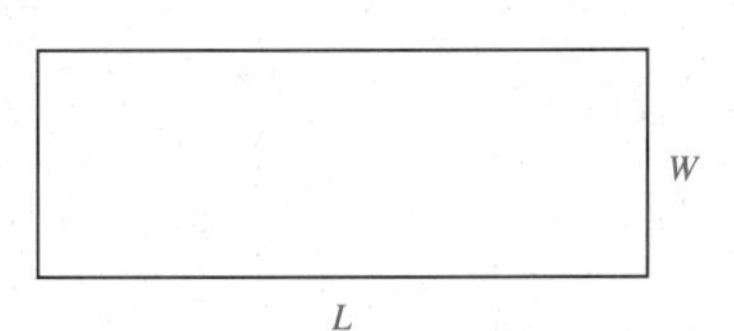

Triangle

Perimeter: $P = a + b + c$
Area: $A = \frac{1}{2}bh$

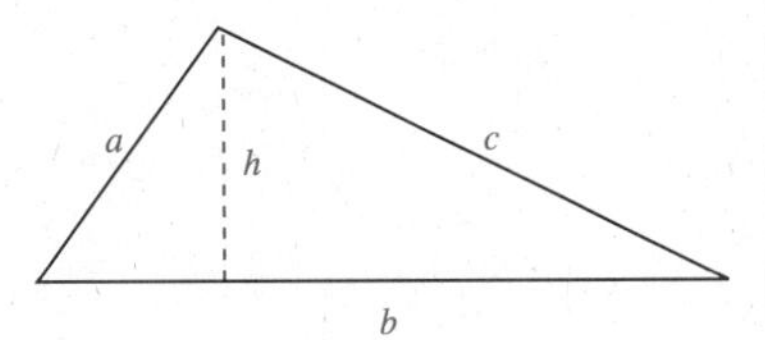

Isosceles Triangle

Two sides equal

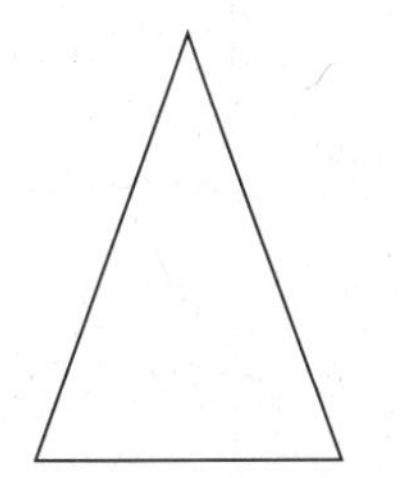

Equilateral Triangle

All sides equal

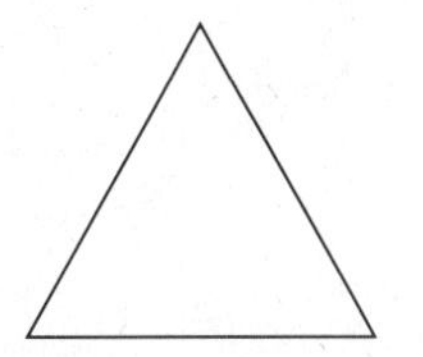

Right Triangle

One 90° (right) angle

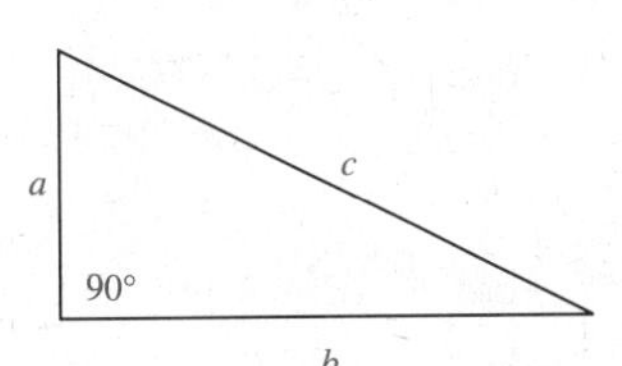

Pythagorean Formula (for right triangles)

$c^2 = a^2 + b^2$

Sum of the Angles

$A + B + C = 180°$

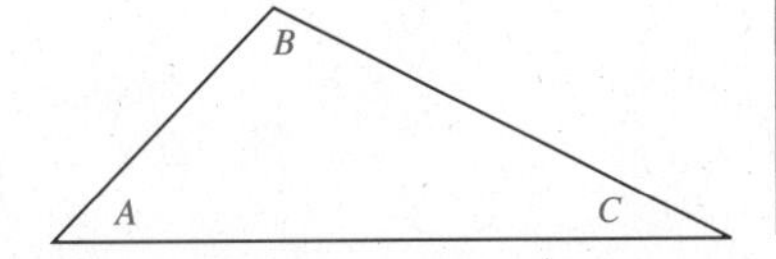

Circle

Diameter: $d = 2r$
Circumference: $C = 2\pi r = \pi d$
Area: $A = \pi r^2$

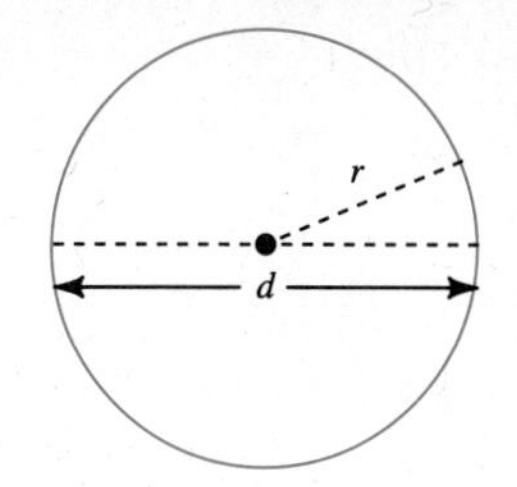

Parallelogram

Area: $A = bh$

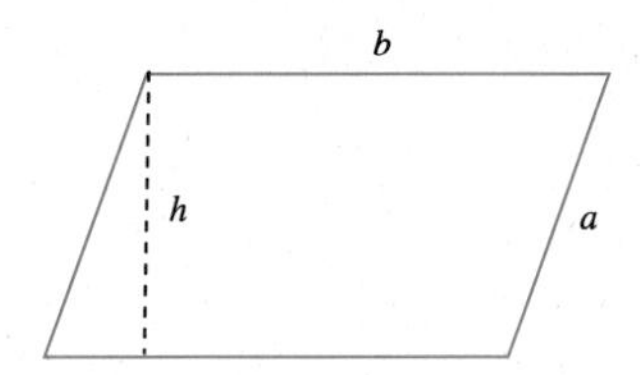

Trapezoid

Area: $A = \frac{1}{2}(B + b)h$

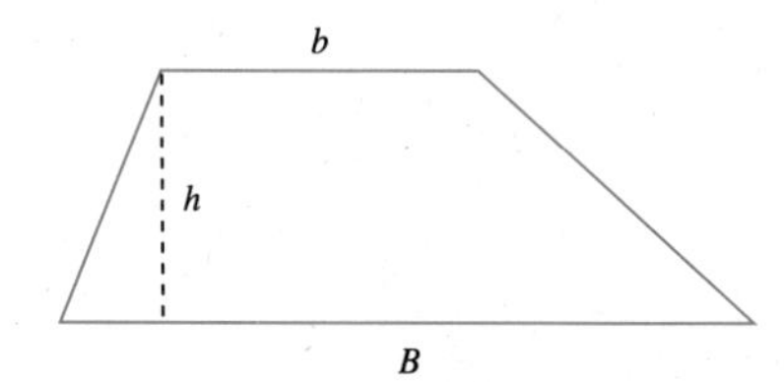

Cone

Volume: $V = \frac{1}{3}\pi r^2 h$
Surface area: $S = \pi r\sqrt{r^2 + h^2}$

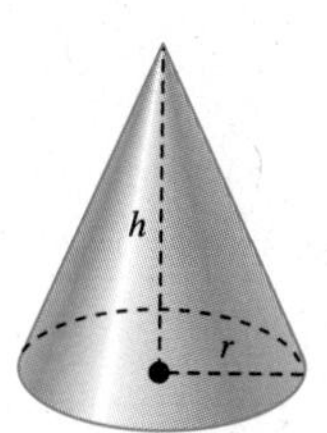

Sphere

Volume: $V = \frac{4}{3}\pi r^3$
Surface area: $S = 4\pi r^2$

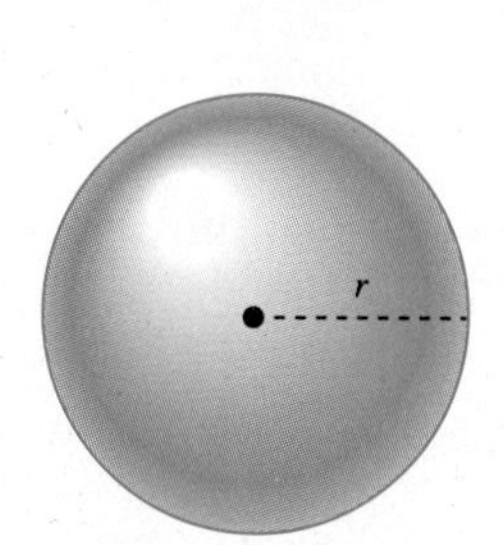

Rectangular Solid

Volume: $V = LWH$
Surface area: $A = 2HW + 2LW + 2LH$

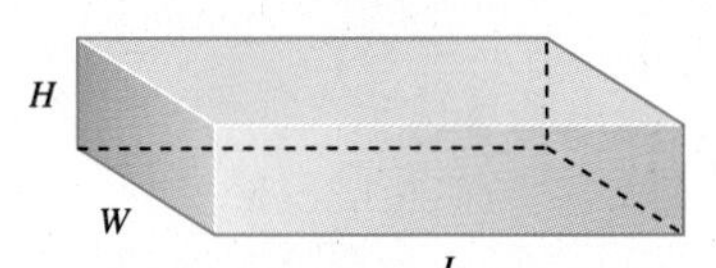

Right Circular Cylinder

Volume: $V = \pi r^2 h$
Surface area: $S = 2\pi rh + 2\pi r^2$

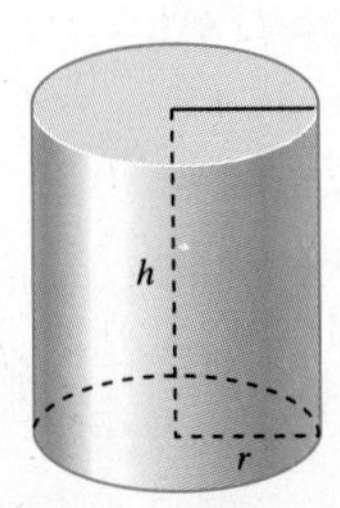

Right Pyramid

Volume: $V = \frac{1}{3}Bh$

B = area of the base

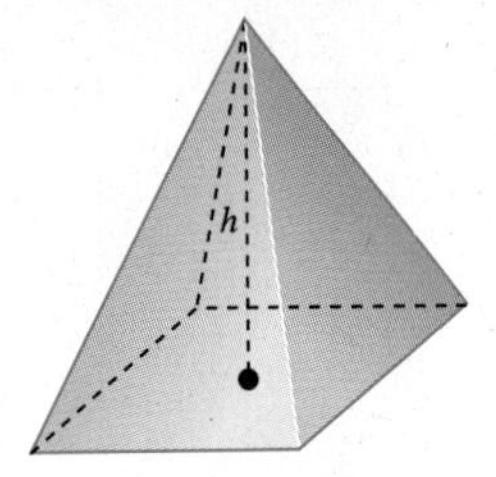

OTHER FORMULAS

Distance: $d = rt$; r = rate or speed, t = time

Percent: $p = br$; p = percentage
b = base
r = rate

Temperature: $F = \frac{9}{5}C + 32$

$C = \frac{5}{9}(F - 32)$

Simple Interest: $I = prt$; p = principal or amount invested
r = rate or percent
t = time in years

APPENDIX C: TABLES

TABLE 1 SELECTED POWERS OF NUMBERS

n	n^2	n^3	n^4	n^5	n^6
2	4	8	16	32	64
3	9	27	81	243	729
4	16	64	256	1024	4096
5	25	125	625	3125	
6	36	216	1296		
7	49	343	2401		
8	64	512	4096		
9	81	729	6561		
10	100	1000	10,000		

TABLE 2 POWERS AND ROOTS

n	n^2	n^3	$\sqrt{n}$	$\sqrt[3]{n}$	$\sqrt{10n}$
1	1	1	1.000	1.000	3.162
2	4	8	1.414	1.260	4.472
3	9	27	1.732	1.442	5.477
4	16	64	2.000	1.587	6.325
5	25	125	2.236	1.710	7.071
6	36	216	2.449	1.817	7.746
7	49	343	2.646	1.913	8.367
8	64	512	2.828	2.000	8.944
9	81	729	3.000	2.080	9.487
10	100	1,000	3.162	2.154	10.000
11	121	1,331	3.317	2.224	10.488
12	144	1,728	3.464	2.289	10.954
13	169	2,197	3.606	2.351	11.402
14	196	2,744	3.742	2.410	11.832
15	225	3,375	3.873	2.466	12.247
16	256	4,096	4.000	2.520	12.649
17	289	4,913	4.123	2.571	13.038
18	324	5,832	4.243	2.621	13.416
19	361	6,859	4.359	2.688	13.784
20	400	8,000	4.472	2.714	14.142
21	441	9,261	4.583	2.759	14.491
22	484	10,648	4.690	2.802	14.832
23	529	12,167	4.796	2.844	15.166
24	576	13,824	4.899	2.884	15.492
25	625	15,625	5.000	2.924	15.811
26	676	17,576	5.099	2.962	16.125
27	729	19,683	5.196	3.000	16.432
28	784	21,952	5.292	3.037	16.733
29	841	24,389	5.385	3.072	17.029
30	900	27,000	5.477	3.107	17.321
31	961	29,791	5.568	3.141	17.607
32	1024	32,768	5.657	3.175	17.889
33	1089	35,937	5.745	3.208	18.166
34	1156	39,304	5.831	3.240	18.439
35	1225	42,875	5.916	3.271	18.708
36	1296	46,656	6.000	3.302	18.974
37	1369	50,653	6.083	3.332	19.235
38	1444	54,872	6.164	3.362	19.494
39	1521	59,319	6.245	3.391	19.748
40	1600	64,000	6.325	3.420	20.000
41	1681	68,921	6.403	3.448	20.248
42	1764	74,088	6.481	3.476	20.494
43	1849	79,507	6.557	3.503	20.736
44	1936	85,184	6.633	3.530	20.976
45	2025	91,125	6.708	3.557	21.213
46	2116	97,336	6.782	3.583	21.448
47	2209	103,823	6.856	3.609	21.679
48	2304	110,592	6.928	3.634	21.909
49	2401	117,649	7.000	3.659	22.136
50	2500	125,000	7.071	3.684	22.361
51	2601	132,651	7.141	3.708	22.583
52	2704	140,608	7.211	3.733	22.804
53	2809	148,877	7.280	3.756	23.022
54	2916	157,464	7.348	3.780	23.238
55	3025	166,375	7.416	3.803	23.452
56	3136	175,616	7.483	3.826	23.664
57	3249	185,193	7.550	3.849	23.875
58	3364	195,112	7.616	3.871	24.083
59	3481	205,379	7.681	3.893	24.290
60	3600	216,000	7.746	3.915	24.495
61	3721	226,981	7.810	3.936	24.698
62	3844	238,328	7.874	3.958	24.900
63	3969	250,047	7.937	3.979	25.100
64	4096	262,144	8.000	4.000	25.298
65	4225	274,625	8.062	4.021	25.495
66	4356	287,496	8.124	4.041	25.690
67	4489	300,763	8.185	4.062	25.884
68	4624	314,432	8.246	4.082	26.077
69	4761	328,509	8.307	4.102	26.268
70	4900	343,000	8.367	4.121	26.458
71	5041	357,911	8.426	4.141	26.646
72	5184	373,248	8.485	4.160	26.833
73	5329	389,017	8.544	4.179	27.019
74	5476	405,224	8.602	4.198	27.203
75	5625	421,875	8.660	4.217	27.386
76	5776	438,976	8.718	4.236	27.568
77	5929	456,533	8.775	4.254	27.749
78	6084	474,552	8.832	4.273	27.928
79	6241	493,039	8.888	4.291	28.107
80	6400	512,000	8.944	4.309	28.284
81	6561	531,441	9.000	4.327	28.460
82	6724	551,368	9.055	4.344	28.636
83	6889	571,787	9.110	4.362	28.810
84	7056	592,704	9.165	4.380	28.983
85	7225	614,125	9.220	4.397	29.155
86	7396	636,056	9.274	4.414	29.326
87	7569	658,503	9.327	4.431	29.496
88	7744	681,472	9.381	4.448	29.665
89	7921	704,969	9.434	4.465	29.833
90	8100	729,000	9.487	4.481	30.000
91	8281	753,571	9.539	4.498	30.166
92	8464	778,688	9.592	4.514	30.332
93	8649	804,357	9.644	4.531	30.496
94	8836	830,584	9.695	4.547	30.659
95	9025	857,375	9.747	4.563	30.822
96	9216	884,736	9.798	4.579	30.984
97	9409	912,673	9.849	4.595	31.145
98	9604	941,192	9.899	4.610	31.305
99	9801	970,299	9.950	4.626	31.464
100	10,000	1,000,000	10.000	4.642	31.623

ANSWERS TO SELECTED EXERCISES

The solutions to selected odd-numbered exercises are given in the section beginning on page A-25.

In this section we provide the answers that we think most students will obtain when they work the exercises using the methods explained in the text. If your answer does not look exactly like the one given here, it is not necessarily wrong. In many cases there are equivalent forms of the answer that are correct. For example, if the answer section shows $\frac{3}{4}$ and your answer is .75, you have obtained the right answer but written it in a different (yet equivalent) form. Unless the directions specify otherwise, .75 is just as valid an answer as $\frac{3}{4}$.

In general, if your answer does not agree with the one given in the text, see whether it can be transformed into the other form. If it can, then it is the correct answer. If you still have doubts, talk with your instructor.

Diagnostic Pretest (page xvii)

1. $\frac{2}{3}$ **2.** $\frac{61}{24}$ or $2\frac{13}{24}$ **3.** $\frac{9}{10}$ **4.** 27 **5.** 7
6. 5 **7.** −14 **8.** −11 **9.** 25 **10.** 8
11. 11 **12.** 15 **13.** 3 **14.** $13\frac{1}{2}$ meters
15. $z > 5$ **16.** $y \geq -9$
17. 625; base is 5, exponent is 4 **18.** $-8x^{15}$
19. $\frac{y^{15}}{3^6 \cdot 4^3}$ **20.** $5x^3 + 6x^2 - 10$
21. $6p^2 + 8p - 30$ **22.** $4x^2 - 8x + 5$
23. $2 \cdot 5^2$ **24.** $2^6 \cdot 5$ **25.** $(m + 7)(m + 2)$
26. $(4a + 5b)(3a - 4b)$ **27.** $(11z - 2)^2$ **28.** 2, 3
29. 9, 11, or −1, 1 **30.** $\frac{x}{x + 1}$
31. $\frac{x^2}{(x + 1)(x - 1)}$ **32.** −2

33.

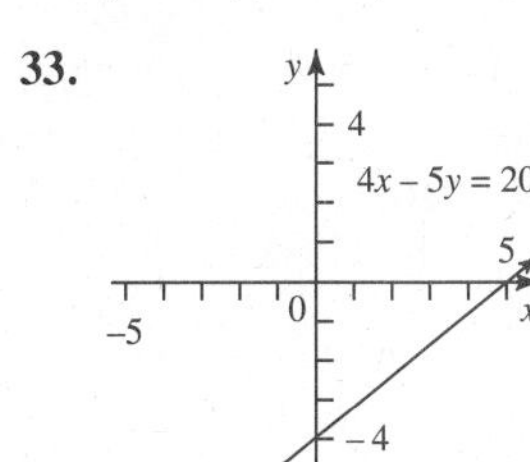

34.

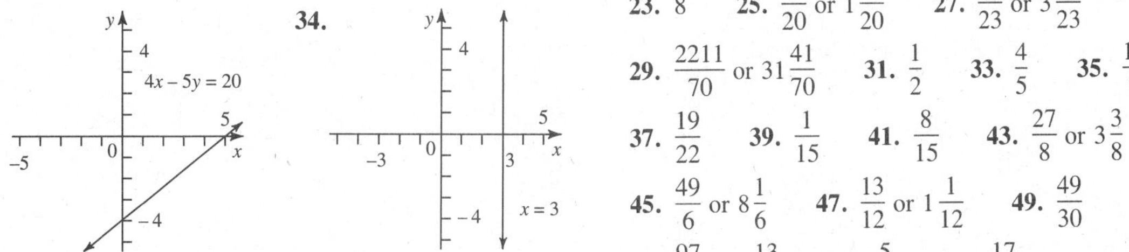

35.

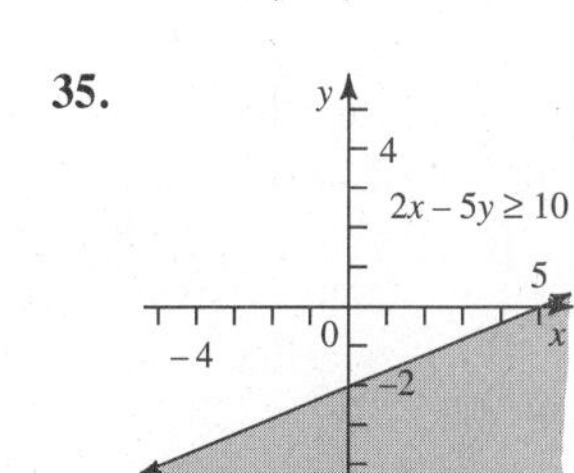

36. $-\frac{5}{6}$ **37.** (3, −7)
38. same line
39. −32
40. 6
41. $6\sqrt{2}$ **42.** $2\sqrt[3]{5}$
43. $19\sqrt{3}$ **44.** $15\sqrt{6}$
45. $2\sqrt{7}$
46. $\frac{-2(3 + \sqrt{2})}{7}$
47. 1 **48.** $1 + \sqrt{2}$, $1 - \sqrt{2}$

49. $\frac{-5 + \sqrt{41}}{2}, \frac{-5 - \sqrt{41}}{2}$

50.

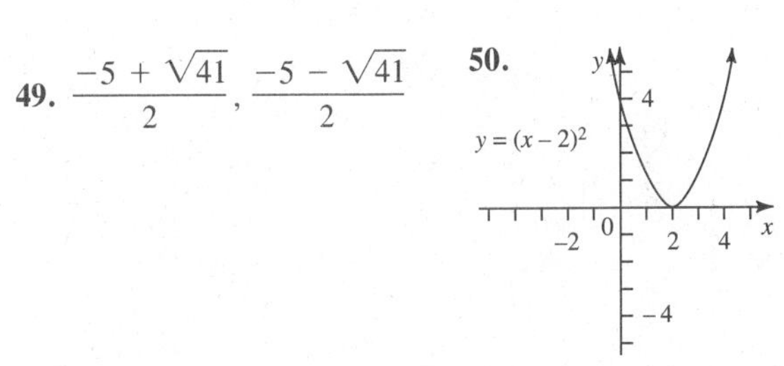

CHAPTER 1

Section 1.1 (page 9)

1. $\frac{1}{2}$ **3.** $\frac{5}{6}$ **5.** $\frac{8}{9}$ **7.** $\frac{2}{3}$ **9.** $\frac{2}{3}$ **11.** $\frac{5}{6}$
13. $\frac{9}{20}$ **15.** $\frac{3}{25}$ **17.** $\frac{6}{5}$ **19.** $\frac{3}{10}$ **21.** $\frac{1}{9}$
23. 8 **25.** $\frac{21}{20}$ or $1\frac{1}{20}$ **27.** $\frac{78}{23}$ or $3\frac{9}{23}$
29. $\frac{2211}{70}$ or $31\frac{41}{70}$ **31.** $\frac{1}{2}$ **33.** $\frac{4}{5}$ **35.** $\frac{10}{9}$
37. $\frac{19}{22}$ **39.** $\frac{1}{15}$ **41.** $\frac{8}{15}$ **43.** $\frac{27}{8}$ or $3\frac{3}{8}$
45. $\frac{49}{6}$ or $8\frac{1}{6}$ **47.** $\frac{13}{12}$ or $1\frac{1}{12}$ **49.** $\frac{49}{30}$
51. $\frac{97}{28}$ or $3\frac{13}{28}$ **53.** $\frac{5}{24}$ **55.** $\frac{17}{24}$ of the debt
57. $618\frac{3}{4}$ feet **59.** $8\frac{23}{24}$ hours **61.** 8 cakes

Section 1.2 (page 19)

1. 80 + 6 **3.** 600 + 90 + 4
5. 5000 + 200 + 30 + 7 **7.** 30 + 6 + .8 + .01
9. .5 + .06 + .007 **11.** $\frac{4}{5}$ **13.** $\frac{18}{25}$ **15.** $\frac{43}{200}$

17. $\frac{5}{8}$ **19.** 887.859 **21.** 19.57 **23.** 270.58
25. 31.48 **27.** 14.218 **29.** 43.01
31. 80.0904 **33.** 1655.9 **35.** 8.21 **37.** 6.14
39. 16.76 **41.** .375 **43.** .438 **45.** .938
47. .833 **49.** .733 **51.** .632 **53.** 75%
55. 140% **57.** .3% **59.** 98.3% **61.** .38
63. 1.74 **65.** .11 **67.** .001 **69.** 42
71. 300 **73.** 65% **75.** 120% **77.** 2%
79. 3.84 **81.** 7.50 **83.** 76.60% **85.** \$2400
87. 35% **89.** 15% **91.** \$559 **93.** 6%
95. 11,844 square feet **97.** \$161.34

Section 1.3 (page 25)

1. $<$ **3.** $<$ **5.** $>$ **7.** $<$ **9.** $<$ **11.** $>$
13. $\leq$ **15.** $\geq$ **17.** $\leq$ **19.** $\leq$
21. $<$ and $\leq$ **23.** $>$ and $\geq$ **25.** $\leq$ and $\geq$
27. $>$ and $\geq$ **29.** $>$ and $\geq$ **31.** $<$ and $\leq$
33. $<$ and $\leq$ **35.** $<$ and $\leq$ **37.** $7 = 5 + 2$
39. $3 < \frac{50}{5}$ **41.** $12 \neq 5$ **43.** $0 \geq 0$ **45.** true
47. true **49.** true **51.** true **53.** true
55. false **57.** false **59.** false **61.** false
63. $14 > 6$ **65.** $3 \leq 15$ **67.** $8 < 9$
69. $6 \geq 0$ **71.** $15 \leq 18$ **73.** $.439 \leq .481$

Section 1.4 (page 31)

1. 36 **3.** 64 **5.** 289 **7.** 125 **9.** 1296
11. 32 **13.** 729 **15.** $\frac{1}{4}$ **17.** $\frac{8}{125}$ **19.** $\frac{64}{125}$
21. .475 **23.** 3.112 **25.** 16 **27.** $\frac{49}{30}$
29. 11 **31.** 87.34 **33.** 65 **35.** 2 **37.** 3
39. $\frac{1}{2}$ **41.** true **43.** true **45.** true
47. false **49.** false **51.** false **53.** false
55. $10 - (7 - 3) = 6$
57. no parentheses needed, or $(3 \cdot 5) + 7 = 22$
59. $3 \cdot (5 - 4) = 3$ **61.** $3 \cdot (5 - 2) \cdot 4 = 36$
63. no parentheses needed, or $(360 \div 18) \div 4 = 5$
65. $(6 + 5) \cdot 3^2 = 99$ **67.** $(8 - 2^2) \cdot 2 = 8$
69. no parentheses needed, or $\frac{1}{2} + \left(\frac{5}{3} \cdot \frac{9}{7}\right) - \frac{3}{2} = \frac{8}{7}$
71. $(5 \cdot 52) - 4$ or $4(52) + (52 - 4)$
73. $5 + (6 - 1) + (3 - 2)$

Section 1.5 (page 37)

1. **(a)** 12 **(b)** 24 **3.** **(a)** 15 **(b)** 75 **5.** **(a)** $\frac{7}{3}$
(b) $\frac{31}{3}$ **7.** **(a)** $\frac{4}{3}$ **(b)** $\frac{16}{3}$ **9.** **(a)** $\frac{2}{3}$ **(b)** $\frac{4}{3}$
11. **(a)** 30 **(b)** 690 **13.** **(a)** 19.377 **(b)** 96.885
15. **(a)** .8195 **(b)** 16.9115 **17.** **(a)** 43 **(b)** 28
19. **(a)** 24 **(b)** 33 **21.** **(a)** 6 **(b)** $\frac{9}{5}$
23. **(a)** $\frac{19}{6}$ **(b)** $\frac{8}{3}$ **25.** **(a)** $\frac{185}{3}$ **(b)** $\frac{565}{6}$
27. **(a)** 2 **(b)** $\frac{17}{7}$ **29.** **(a)** $\frac{5}{6}$ **(b)** $\frac{16}{27}$
31. **(a)** 28 **(b)** 55 **33.** **(a)** $\frac{10}{3}$ **(b)** $\frac{13}{3}$
35. **(a)** $\frac{8}{7}$ **(b)** $\frac{28}{17}$ **37.** **(a)** 14.736 **(b)** 8.841
39. $8x$ **41.** $5 \cdot x$ or $5x$ **43.** $x + 4$ **45.** $x - 9$
47. $\frac{2}{3}x + 6$ **49.** yes **51.** no **53.** yes
55. yes **57.** yes **59.** yes **61.** no
63. $x + 8 = 12$; 4 **65.** $16 - \frac{3}{4}x = 10$; 8
67. $2x + 5 = 13$; 4 **69.** $3x = 2x + 2$; 2
71. $\frac{20}{5x} = 2$; 2

Chapter 1 Review Exercises (page 45)

1. $\frac{1}{2}$ **2.** $\frac{1}{3}$ **3.** $\frac{5}{9}$ **4.** $\frac{3}{5}$ **5.** $\frac{1}{3}$ **6.** $\frac{5}{6}$
7. $\frac{6}{7}$ **8.** $\frac{4}{5}$ **9.** $\frac{5}{24}$ **10.** $\frac{7}{50}$ **11.** $\frac{4}{15}$ **12.** $\frac{3}{2}$
13. $\frac{3}{4}$ **14.** $\frac{1}{5}$ **15.** $\frac{7}{3}$ **16.** $\frac{8}{7}$ **17.** $\frac{11}{24}$
18. $\frac{9}{20}$ **19.** $\frac{161}{12}$ or $13\frac{5}{12}$ **20.** $472\frac{1}{6}$
21. $\frac{7}{12}$ of the room **22.** $31\frac{1}{4}$ cubic yards
23. $138\frac{1}{16}$ square feet **24.** $42\frac{16}{63}$ square meters
25. $\frac{2}{5}$ **26.** $\frac{27}{50}$ **27.** $\frac{17}{20}$ **28.** $\frac{101}{200}$ **29.** $\frac{469}{200}$
30. $\frac{21879}{5000}$ **31.** .875 **32.** .25 **33.** .688
34. .167 **35.** .853 **36.** .652 **37.** 287.954
38. 31.3 **39.** 1.675 **40.** 1001.698 **41.** 5.22
42. 7.88 **43.** 96% **44.** 142% **45.** 50%
46. 513.6% **47.** 3600 **48.** 218.24
49. 15.22% **50.** \$1495.20 **51.** \$2385
52. $<, \leq$ **53.** $\leq, \geq$ **54.** $>, \geq$ **55.** $>, \geq$
56. $>, \geq$ **57.** $>, \geq$ **58.** $<, \leq$ **59.** $<, \leq$
60. $9 \geq 4$ **61.** $3 \leq 16$ **62.** $10 > 9$
63. $25 \geq 23$ **64.** 64 **65.** 25 **66.** 32
67. $\frac{25}{64}$ **68.** $\frac{196}{9}$ or $21\frac{7}{9}$ **69.** $\frac{343}{27}$ or $12\frac{19}{27}$
70. false **71.** true **72.** true **73.** true
74. true **75.** true **76.** 6 **77.** 17 **78.** $\frac{1}{4}$
79. 9 **80.** 60 **81.** 18 **82.** $12x$ **83.** $x + 9$
84. $4 - x$ **85.** $3x - 6$ **86.** $x + 12$
87. $x - 3$ **88.** $x(x + 1)$ **89.** $\frac{x + 4}{8}$
90. $(x + 28) - \frac{3}{10}x$ **91.** $\frac{10}{3}x - (x + 16)$ **92.** 68
93. $\frac{13}{18}$ **94.** $\frac{4}{15}$ **95.** 40 awards **96.** \$520

97. 216 **98.** 7.9454 **99.** $\frac{5}{14}$ **100.** $103\frac{25}{36}$

101. 41 **102.** 219.78 **103.** $\frac{8}{9}$ **104.** $\frac{2}{3}$

105. $\frac{1}{16}$ **106.** 6% **107.** $6\frac{11}{12}$ feet

108. 223 **109.** $\frac{5}{14}$ **110.** $\frac{6}{5}$ **111.** 124%

112. 3 **113.** $\frac{1}{2}$

Chapter 1 Test (page 49)

[1.1] **1.** $\frac{3}{8}$ **2.** $\frac{7}{11}$ **3.** $\frac{61}{40}$ **4.** $\frac{203}{120}$

5. $\frac{111}{8}$ or $13\frac{7}{8}$ **6.** $\frac{2}{3}$ **7.** $\frac{18}{19}$ **8.** $58\frac{4}{9}$ acres

[1.2] **9.** $\frac{5}{8}$ **10.** .5625 **11.** 21.77 **12.** 24.7

13. 15.8256 **14.** 11.56 **15.** 19% **16.** .762
17. 13.6 **18.** 14.2% **19.** $9.60 **[1.4]** **20.** 8
21. 2 **22.** 23 **[1.5]** **23.** 29 **24.** 30

25. 2 **26.** $4x$ **27.** $2x + 11$ **28.** $\frac{9}{x - 8}$

29. yes **30.** no

CHAPTER 2

Section 2.1 (page 57)

1. −12 **3.** 4 **5.** −6 **7.** 18 **9.** −10
11. −7 **13.** −6 **15.** −7 **17.** false
19. true **21.** true **23.** true **25.** **(a)** 0, 3, 7
(b) −9, 0, 3, 7 **(c)** $-9, -1\frac{1}{4}, -\frac{3}{5}, 0, 3, 5.9, 7$
(d) $-\sqrt{7}, \sqrt{5}$ **(e)** All are real numbers. **27.** true

29. true **31.**
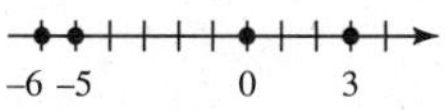

33.
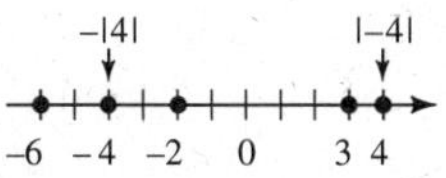

35.
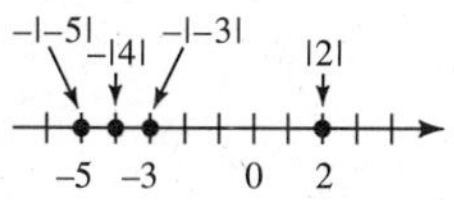

37.
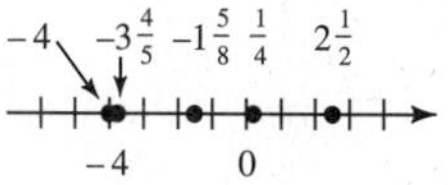

39. $\frac{29}{21}$ **41.** $\frac{13}{28}$

43. $\frac{11}{30}$

Section 2.2 (page 63)

1. 2 **3.** −2 **5.** −8 **7.** −11 **9.** −12
11. 4.18 **13.** 12 **15.** 5 **17.** 2 **19.** −9

21. 13 **23.** −11 **25.** $\frac{1}{2}$ **27.** $-\frac{19}{24}$ **29.** $-\frac{3}{4}$

31. −.5 **33.** −7.7 **35.** −8 **37.** 0
39. −20 **41.** true **43.** false **45.** false
47. true **49.** false **51.** true **53.** true
55. false **57.** true **59.** $-5 + 12 + 4$; 11
61. $[-19 + (-3)] + 14$; −8
63. $[-4 + (-10)] + (-3)$; −17
65. $[8 + (-23)] + 4$; −11 **67.** $6 **69.** −$4
71. 87° **73.** gain of 13 yards **75.** $286.60
77. −2 **79.** −3 **81.** −2 **83.** −2 **85.** 2
87. 30 **89.** 18 **91.** 2

Section 2.3 (page 71)

1. −3 **3.** −4 **5.** −8 **7.** −14 **9.** 9

11. 17 **13.** −4 **15.** 4 **17.** 1 **19.** $\frac{3}{4}$

21. $-\frac{11}{8}$ **23.** $\frac{15}{8}$ **25.** 11.6 **27.** −9.9

29. −1.8 **31.** 10.1 **33.** 10 **35.** −5

37. 11 **39.** −10 **41.** −18 **43.** $\frac{37}{12}$

45. −16 **47.** −12 **49.** −6.9 **51.** −42.04
53. $4 - (-12)$; 16 **55.** $-5 - 8$; −13
57. $[9 + (-4)] - 5$; 0 **59.** $[8 - (-7)] - 12$; 3
61. −$17 **63.** $34 **65.** −$17,000
67. down by 12.756 (−12.756) **69.** 8 **71.** 42
73. 36

Section 2.4 (page 79)

1. 12 **3.** −12 **5.** 5 **7.** 44 **9.** 120

11. −48 **13.** −30 **15.** 0 **17.** $-\frac{8}{9}$

19. $-\frac{1}{2}$ **21.** 7.77 **23.** 1, −1, 2, −2, 3, −3, 4,
−4, 6, −6, 9, −9, 12, −12, 18, −18, 36, −36
25. 1, −1, 5, −5, 25, −25 **27.** 1, −1, 2, −2, 4,
−4, 5, −5, 8, −8, 10, −10, 20, −20, 40, −40
29. 1, −1, 17, −17 **31.** 1, −1, 29, −29
33. −15 **35.** 17 **37.** −15 **39.** 27 **41.** 15
43. 28 **45.** 27 **47.** −1 **49.** 26 **51.** −43

53. −128 **55.** $\frac{34}{9}$ **57.** −64 **59.** −74

61. 28 **63.** $-9 + 4(-7)$; −37
65. $-4 - 2(-8 \cdot 2)$; 28 **67.** $-2(3) - 3$; −9

69. $-3[4 + (-7)]$; 9 **71.** $\frac{3}{10}[-2 - (-22)]$; 6

73. −2 **75.** 0 **77.** −3 **79.** −2 **81.** 14
83. 6 **85.** 2

Section 2.5 (page 89)

1. $\frac{1}{9}$ **3.** $-\frac{1}{4}$ **5.** $\frac{3}{2}$ **7.** $-\frac{10}{9}$

9. no multiplicative inverse **11.** 5 **13.** −2

15. −6 **17.** 15 **19.** 0 **21.** $\frac{2}{3}$ **23.** 2.1

25. −4 **27.** 5 **29.** −4 **31.** −6 **33.** 8

35. 4 37. 3 39. 8 41. −1 43. $\frac{52}{75}$

45. 1 47. −1 49. −2 51. 4 53. −2

55. $\frac{-48}{-8}$; 6 57. $\frac{-20}{-12+2}$; 2

59. $\frac{-18+(-6)}{3(-1)}$; 8 61. $-25(9-4)$; −125

63. $4x = -44$; −11 65. $\frac{x}{4} = -3$; −12

67. $x - 7 = 1$; 8 69. $x + 6 = -5$; −11

71. $\frac{x}{4} = -2$; −8 73. 0 75. 1 77. −3

Section 2.6 (page 99)

1. commutative property 3. associative property
5. associative property 7. commutative property
9. commutative property 11. inverse property
13. identity property 15. inverse property
17. identity property 19. distributive property
21. $k + 9$ 23. m 25. $3r + 3m$ 27. 1
29. 0 31. $(-5) + 5$; 0 33. $-3r - 6$ 35. 9
37. $k + [5 + (-6)]$; $k - 1$ 39. $4z + (2r + 3k)$
41. $5m + 10$ 43. $-4r - 8$ 45. $-8k + 16$
47. $-\frac{2}{3}a - 6$ 49. $\frac{3}{4}r + 2$ 51. $-16 + 2k$
53. $10r + 12m$ 55. $-12x + 16y$
57. $5(8 + 9)$; 85 59. $7.12(2.3 + 7.7)$; 71.2
61. $9(p + q)$ 63. $5(7z + 8w)$ 65. $-3k - 5$
67. $-4y + 8$ 69. $4 - p$ 71. $1 + 15r$
73. not commutative 75. commutative 77. no
79. 17 81. 15

Section 2.7 (page 107)

1. 15 3. −22 5. 35 7. −9 9. 1
11. −1 13. like 15. unlike 17. like
19. like 21. like 23. $27m$ 25. $7k + 15$
27. $m - 1$ 29. $14 - 7m$ 31. $-\frac{5}{4}x - \frac{28}{3}$
33. $-3.1r + 4.5$ 35. $5p^2 - 14p^3$
37. $-2.8q^2 - 6.7q - 21.3$ 39. $30t + 66$
41. $-3n - 15$ 43. $9t - 10$ 45. $4r + 15$
47. $12k - 5$ 49. $-2k - 3$ 51. $-16y + 14.4$
53. $[x + (-12)] + 4x$; $5x - 12$
55. $(12 + 8x) - 4x$; $4x + 12$
57. $[9(3x + 5)] - [(-8) + 6x]$; $21x + 53$ 59. −6
61. 4 63. 7 65. −6

Chapter 2 Review Exercises (page 113)

1. −9 2. −5 3. −8 4. $-\frac{7}{8}$ 5. $\left|-\frac{3}{2}\right|$
6. $|-7|$ 7. $-|-9|$ 8. $-|-7|$ 9. true
10. false 11. false 12. true 13. true
14. false 15. false 16. true

17. 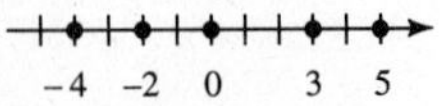

18. 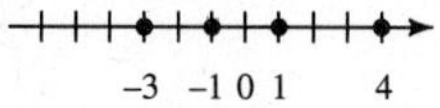

19. 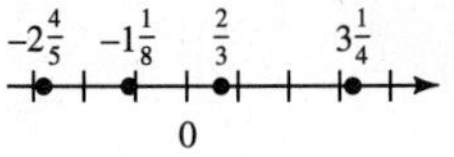

20. 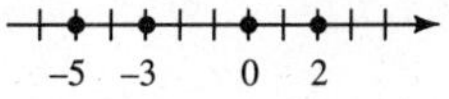

21. −6 22. −3 23. −15 24. $\frac{23}{40}$ 25. $\frac{13}{36}$
26. −14.2 27. −1.33 28. −1 29. −13
30. −16 31. −18 32. −15 33. −12
34. −\$5 35. −8° 36. 3° 37. \$1450
38. −2 39. 9 40. 16 41. 18 42. −17
43. −39 44. $\frac{17}{12}$ 45. $\frac{1}{2}$ 46. 2.6
47. −25.1 48. 0 49. −13 50. −14
51. −27 52. 33 53. −85 54. −108
55. $\frac{8}{7}$ 56. $\frac{2}{5}$ 57. −28.25 58. 26.32
59. −16 60. −4 61. −20 62. −60
63. −12 64. −13 65. −27 66. 34
67. −2 68. −378 69. −73 70. 5
71. −40 72. $-\frac{4}{3}$ 73. −11.2 74. 6
75. 10 76. −1 77. −3 78. 1 79. −3
80. $\frac{13+(-5)}{5(8)}$; $\frac{1}{5}$ 81. $\frac{-7(4)}{-2-2}$; 7
82. identity property 83. inverse property
84. commutative property 85. associative property
86. inverse property 87. identity property
88. distributive property 89. $16p^2$
90. $16p^2 + 2p$ 91. $-4k + 12$ 92. $-2m + 29$
93. 15 94. $\frac{36}{49}$ 95. −28 96. −64 97. 2
98. $\frac{1}{24}$ 99. $\frac{7}{2}$ 100. 4 101. 64 102. $1\frac{1}{2}$
103. 16 104. $-\frac{15}{28}$ 105. −1 106. 14
107. 6 108. 14 109. 12,500 − 1700; \$10,800
110. $\frac{x}{x+5}$

Chapter 2 Test (page 117)

[2.1] 1. 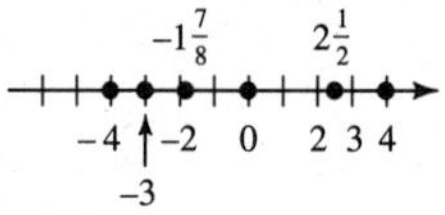

2. 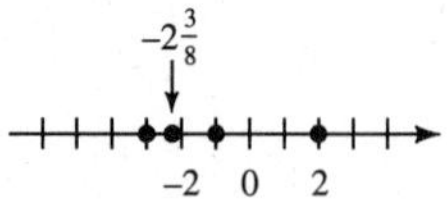

3. −.742 4. $-|-8|$

[2.2–2.5] 5. $11 - 2(17)$; −23 6. $\frac{9}{6-8}$; $-\frac{9}{2}$
7. −7 8. $\frac{61}{20}$ or $3\frac{1}{20}$ 9. 0 10. −1 11. 58
12. 3 13. 4 14. undefined 15. −9
16. −4 17. −2 [2.4] 18. −116 19. 6
[2.3] 20. −\$13 [2.6] 21. b 22. a 23. d
24. e 25. c [2.7] 26. $7m - 1$

CHAPTER 3

Section 3.1 (page 123)

1. 10 **3.** −2 **5.** 10 **7.** −8 **9.** 4
11. −5 **13.** −11 **15.** −6 **17.** $\frac{1}{2}$ **19.** −5
21. no solution **23.** all real numbers **25.** 7
27. 8 **29.** −10 **31.** no solution **33.** 6
35. −2 **37.** 17 **39.** 18 **41.** 26
43. all real numbers **45.** no solution **47.** m
49. y **51.** x **53.** r

Section 3.2 (page 129)

1. 5 **3.** 25 **5.** −8 **7.** −7 **9.** $-\frac{8}{3}$
11. −6 **13.** 0 **15.** −6 **17.** 4 **19.** 4
21. 8 **23.** 20 **25.** −3 **27.** 4 **29.** 7
31. 3 **33.** 49 **35.** 9 **37.** $\frac{8}{3}$ **39.** −80
41. $\frac{15}{2}$ **43.** $\frac{49}{2}$ **45.** $-\frac{7}{2}$ **47.** $-\frac{27}{28}$ **49.** 3
51. −6.1 **53.** $18q + 63$ **55.** $-20p + 10$
57. $-2 + 11r$ **59.** $-3 - 12a$

Section 3.3 (page 137)

1. 2 **3.** 24 **5.** 9 **7.** $-\frac{5}{3}$ **9.** −1
11. $\frac{9}{11}$ **13.** −12.5 **15.** no solution
17. all real numbers **19.** $-\frac{10}{3}$ **21.** −3 **23.** 8
25. −2 **27.** 6 **29.** $-\frac{5}{2}$ **31.** 0 **33.** 0
35. $\frac{10}{7}$ **37.** −4 **39.** 3 **41.** all real numbers
43. no solution **45.** $x + (-1)$ **47.** $x + (-18)$
49. $x - 6$ **51.** $x - 16$ **53.** $2x$ **55.** $\frac{3}{5}x$
57. $\frac{-9}{x}$ **59.** $\frac{7}{x}$ **61.** $\frac{1}{x} - x$ **63.** $8(x - 8)$
65. $7x = -84$; −12 **67.** $x + 13 = 9$; −4
69. $\frac{1}{2}x = 4$; 8

Section 3.4 (page 145)

1. 9 **3.** 1 **5.** 133 votes
7. Brian: 26; Katherine: 54 **9.** 9 men, 11 women
11. Jean: 30 miles; Larry: 60 miles
13. 4, 9, and 27 centimeters
15. Boggs: 551 times; Evans: 541 times; Burks: 558 times
17. *A* and *B*: 52°; *C*: 76° **19.** 18 prescriptions
21. cashews: $4\frac{1}{2}$ ounces; peanuts: $22\frac{1}{2}$ ounces
23. 68 and 69 **25.** 10 and 12 **27.** $450
29. 6 pints; 36 quarts **31.** 42 **33.** 640 **35.** 60

Section 3.5 (page 153)

1. $P = 16$ **3.** $A = 108$ **5.** $r = 20$ **7.** $h = 10$
9. $W = 40$ **11.** $B = 12$ **13.** $r = 4$
15. $A = 706.5$ **17.** $t = 5$ **19.** $H = 2$
21. $h = 4$ **23.** $72 = 4s$; 18 centimeters
25. $40 = 24 + 2W$; 8 inches
27. $A = \frac{1}{2}(8 + 12)4$; 40 square centimeters
29. $60 = \frac{1}{2}(24)h$; 5 centimeters
31. $502.4 = (3.14)(4)^2h$; 10 centimeters
33. $339.12 = 2(3.14)(3)h + 2(3.14)(3)^2$; 15 inches
35. $L = \frac{A}{W}$ **37.** $t = \frac{d}{r}$ **39.** $H = \frac{V}{LW}$
41. $t = \frac{I}{pr}$ **43.** $a^2 = c^2 - b^2$ **45.** $b = \frac{2A}{h}$
47. $r^2 = \frac{V}{\pi h}$ **49.** $W = \frac{P - 2L}{2}$
51. $t = \frac{P - A}{-Ar}$ or $\frac{A - P}{Ar}$ **53.** $v = \frac{d - gt^2}{t}$
55. $F = \frac{9}{5}C + 32$ or $\frac{9C + 160}{5}$ **57.** 6 **59.** $\frac{22}{9}$
61. −28

Section 3.6 (page 161)

1. $\frac{8}{5}$ **3.** $\frac{2}{1}$ **5.** $\frac{12}{5}$ **7.** $\frac{5}{36}$ **9.** $\frac{1}{12}$ **11.** $\frac{11}{6}$
13. $\frac{9}{2}$ **15.** $\frac{1}{4}$ **17.** true **19.** true **21.** false
23. false **25.** true **27.** true **29.** true
31. false **33.** 49 **35.** 27 **37.** 26 **39.** $\frac{33}{8}$
41. $\frac{17}{6}$ **43.** $\frac{35}{12}$ **45.** $-\frac{3}{2}$ **47.** $\frac{25}{19}$ **49.** $\frac{87}{4}$
51. $11.25 **53.** $2\frac{1}{2}$ Hershey bars **55.** 9 pounds
57. 9 inches **59.** $67.50 **61.** 420 miles
63. $144 **65.** $<$ **67.** $>$ **69.** $<$ **71.** $>$

Section 3.7 (page 173)

1.
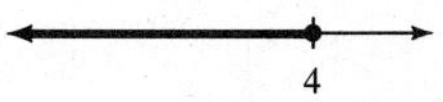

3.
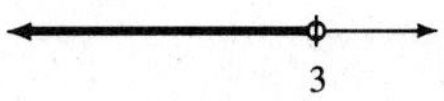

5.
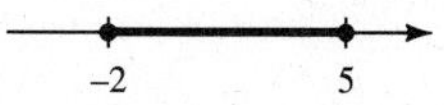

7.
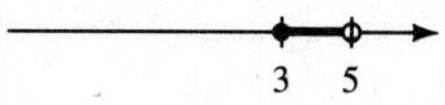

9. $a < 2$ **11.** $z \geq 1$ **13.** $k \geq 5$ **15.** $x < 9$
17. $k \geq -6$ **19.** $y > 9$
21. $n \leq -11$

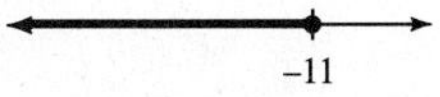

23. $k \geq 44$

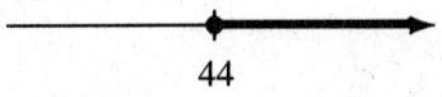

25. $k \geq -5$

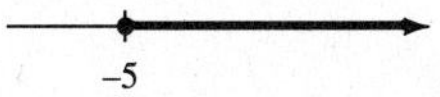

27. $q > -2$

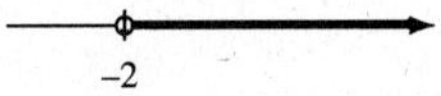

29. $z < 1$

1

31. $w \geq 3$

3

33. $k \leq 0$

0

35. $x < -11$

−11

37. $6 \leq p \leq 13$

6 13

39. $5 < y < 12$

5 12

41. $-1 \leq x \leq 6$

−1 6

43. $-26 \leq z \leq 6$

−26 6

45. all numbers greater than or equal to -16
47. at least 83 **49.** at least \$275
51. 11 feet and 22 feet **53.** 32 **55.** 625
57. $\frac{8}{27}$ **59.** 3 **61.** 5 **63.** 13 **65.** 3
67. 6

Chapter 3 Review Exercises (page 181)

1. 6 **2.** -12 **3.** 7 **4.** $\frac{2}{3}$ **5.** 11 **6.** 17
7. 5 **8.** -4 **9.** 5 **10.** -12 **11.** $\frac{64}{5}$
12. -6 **13.** $\frac{13}{4}$ **14.** all real numbers
15. no solution **16.** 6 **17.** 7 **18.** 40 units
19. 61, 96 **20.** 12 miles **21.** 50 golf balls
22. $h = 11$ **23.** $A = 84$ **24.** $r = 2$
25. $V = 4.19$ (rounded) **26.** $W = \frac{A}{L}$ **27.** $r = \frac{I}{pt}$
28. $m = \frac{x^2 - a}{2}$ **29.** $h = \frac{2A}{b + B}$ **30.** 5 meters
31. 360 square centimeters **32.** $\frac{5}{3}$ **33.** $\frac{6}{7}$
34. $\frac{3}{4}$ **35.** $\frac{1}{18}$ **36.** true **37.** false **38.** $\frac{7}{2}$
39. $\frac{36}{5}$ **40.** 7 **41.** $-\frac{2}{7}$ **42.** 8 quarts
43. 63 hours **44.** 75 yards **45.** 375 kilometers

46.

−2

47.

−5 6

48. $y \geq -3$

−3

49. $y > 8$

8

50. $k \geq 53$

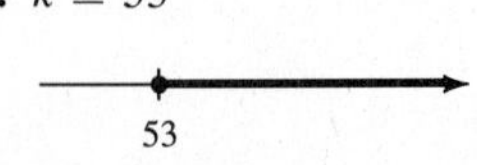

51. $z \leq -37$

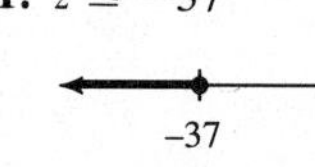

52. $-8 \leq x \leq -2$

−8 −2

53. $7 < y < 9$

7 9

54. $k \geq -3$

−3

55. $y > -2$

−2

56. $z < 4$

4

57. $m \leq -6$

−6

58. $p < -5$

−5

59. $y \leq \frac{27}{10}$

$\frac{27}{10}$

60. $-2 \leq m \leq \frac{3}{2}$

−2 $\frac{3}{2}$

61. $\frac{4}{3} < m \leq 5$

$\frac{4}{3}$ 5

62. -3 **63.** $t = \frac{7x}{2}$ **64.** $z < 3$ **65.** 9
66. $m \leq -6$ **67.** $a = P - b - c$ **68.** 3
69. $-5 \leq z \leq 4$ **70.** 6 **71.** 12
72. Bob: 600 votes; Buddy: 1200 votes
73. Gwen: 28 miles; John: 84 miles **74.** 12 pounds
75. 125 miles **76.** 35 **77.** 37 **78.** 15
79. $\frac{2}{3}$ cup **80.** 70 feet **81.** 44 meters
82. 26 inches **83.** 26 inches
84. $71\frac{3}{5}$° F or 71.6° F **85.** 50° C
86. 250 kilometers **87.** \$13.98 **88.** \$14.63
89. $34\frac{7}{8}$ yards **90.** \$630 **91.** at least 92

Chapter 3 Test (page 189)

[3.1] **1.** 4 [3.2] **2.** 5 [3.3] **3.** -5
4. 0 **5.** 10 **6.** $\frac{8}{3}$ **7.** no solution **8.** -3
[3.4] **9.** \$100 **10.** \$12 [3.5] **11.** 9
12. $p = \frac{I}{rt}$ **13.** $h = \frac{2A}{b + B}$ **14.** 4 meters
[3.6] **15.** true **16.** 1 **17.** $-\frac{22}{3}$ **18.** \$17.60
[3.7] **19.** $x \leq 4$

4

20. $m > 7$

7

21. $r \leq 4$

4

22. $-2 < k \leq \frac{14}{3}$

−2 $\frac{14}{3}$

23. 450 miles

Cumulative Review Exercises (page 191)

1. $\frac{3}{8}$ **2.** $\frac{3}{4}$ **3.** $\frac{31}{20}$ **4.** $\frac{551}{40}$ or $13\frac{31}{40}$ **5.** 6
6. $\frac{6}{5}$ **7.** $20\frac{23}{24}$ pounds **8.** 35 yards **9.** $4\frac{1}{6}$ cups
10. 34.03 **11.** 27.31 **12.** 30.51 **13.** 56.3
14. 21.6 **15.** 174 **16.** 37.5% **17.** 25%
18. \$1187.65 **19.** \$1261.88 **20.** \$3750
21. true **22.** true **23.** 7 **24.** 1 **25.** 13
26. −40 **27.** −12 **28.** undefined **29.** −6
30. 28 **31.** 1 **32.** undefined **33.** $\frac{73}{18}$
34. −64 **35.** −11 **36.** −8.23 **37.** −3
38. 24 **39.** −162 **40.** −8 **41.** −134
42. $-\frac{29}{6}$ **43.** 3 **44.** −2
45. $\frac{1}{2}x + 3 = 2$; −2 **46.** $\frac{x+6}{x^2} = 1$; −2, 3
47. $-x + 7 = 4$; 3 **48.** $4x^2 = 0$; 0
49. commutative property **50.** identity property
51. distributive property **52.** commutative property
53. inverse property **54.** inverse property
55. identity property **56.** $7p - 14$ **57.** $2k - 11$
58. 7 **59.** −4 **60.** −1 **61.** $-\frac{3}{5}$ **62.** 2
63. 6 **64.** −13 **65.** 26 **66.** $x = 3y - 2$
67. $x = \frac{10 + a}{2}$
68. $z \leq 2$ **69.** $r \leq 1$

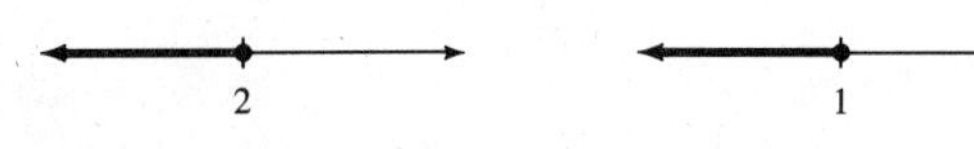

70. $-7 < x < 5$ **71.** $-2 \leq x \leq 1$

−7 5 −2 1

72. 12 **73.** 8 **74.** \$49.50 **75.** \$98.45
76. 30 centimeters **77.** 16 inches

CHAPTER 4

Section 4.1 (page 201)

1. base: 5; exp: 12 **3.** base: $3m$; exp: 4
5. base: 125; exp: 3 **7.** base: $-2x$; exp: 2
9. base: m; exp: 2 **11.** base: r; exp: 5 **13.** 3^5
15. $(-2)^5$ **17.** p^5 **19.** $\frac{1}{(-2)^3}$ **21.** $\frac{1}{2^5}$
23. $(-2z)^3$ **25.** 90 **27.** 80 **29.** 36 **31.** 12
33. 4^5 **35.** 3^{11} **37.** 4^{18} **39.** $(-3)^5$
41. y^{14} **43.** r^{17} **45.** $-63r^9$ **47.** $10p^{13}$
49. $13m^3$; $36m^6$ **51.** $-p$; $-132p^2$ **53.** $15r$; $105r^3$
55. $-2a^2$; $-15a^4$ **57.** 6^6 **59.** 9^6 **61.** 5^8
63. -4^6 **65.** 5^3m^3 **67.** $(-2)^4p^4q^4$
69. $\frac{(-3)^2x^{10}}{4^2}$ **71.** $\frac{5^3a^6b^3}{c^{12}}$ **73.** $\frac{4^8}{3^5}$ **75.** $\frac{3^5}{4^5x^5}$
77. 3^7m^7 **79.** 8^9z^9 **81.** $2^3m^7n^5$ **83.** $5^7a^{11}b^{12}$
85. (a) −3 (b) −28 **87.** (a) −8 (b) 32

Section 4.2 (page 209)

1. 2 **3.** 2 **5.** $\frac{1}{27}$ **7.** $-\frac{1}{12}$ **9.** 32 **11.** 2
13. 4^5 **15.** $\frac{1}{8^6}$ **17.** $\frac{1}{6^6}$ **19.** x^{15} **21.** 2^5
23. $2k^2$ **25.** $\frac{1}{4^9}$ **27.** 5^9 **29.** 2^6 **31.** 3^5x^5
33. $\frac{5^2}{m^4}$ **35.** $\frac{1}{x^9}$ **37.** $\frac{a^2}{3^3}$ **39.** $4^2k^4m^8$
41. $\frac{a^8}{4^2b^2}$ **43.** 2^2yz^6 **45.** $\frac{9z^6x^5}{5^2}$ **47.** 64,270
49. 1230 **51.** 3.4 **53.** .237

Section 4.3 (page 213)

1. 6.835×10^9 **3.** 2.5×10^4 **5.** 1.01×10^{-2}
7. 9.82×10^{-7} **9.** 6.9×10^{-3} **11.** 350
13. 324,000,000 **15.** .0000576 **17.** 1500
19. 1.2 **21.** .00045 **23.** .00000003
25. 50,000 **27.** .00019 **29.** .0000000000000012
31. 4.037×10^9; 3×10^{-4} **33.** 8.65×10^5
35. 400,000,000,000,000 **37.** 1,000,000,000
39. $20y$ **41.** $9x$ **43.** $4x + 3$

Section 4.4 (page 219)

1. $8m^5$ **3.** $-r^5$ **5.** cannot be simplified **7.** x^5
9. $-p^7$ **11.** $6y^2$ **13.** 0 **15.** $9y^4 + 7y^2$
17. $14z^5 - 9z^3 + 8z^2$ **19.** cannot be simplified
21. always **23.** never **25.** never
27. already simplified; degree 4; binomial
29. already simplified; degree 9; trinomial
31. already simplified; degree 8; trinomial
33. x^5; degree 5; monomial
35. already simplified; degree 0; monomial
37. $5m^2 + 3m$ **39.** $4x^4 - 4x^2$
41. $-n^5 - 12n^3 - 2$ **43.** $12m^3 + m^2 + 12m - 14$
45. $15m^2 - 3m + 4$ **47.** $8b^2 + 2b + 7$
49. $-r^2 - 2r$ **51.** $5m^2 - 14m$
53. $-6s^2 + 5s + 1$ **55.** $2s^2 + 4s$
57. $4x^3 + 2x^2 + 5x$ **59.** $-11y^4 + 8y^2 + 3y$
61. $a^4 - a^2 + 1$ **63.** $5m^2 + 8m - 10$
65. $(-9x + 2) + (4 + x^2) > 8$
67. $(5 + x^2) + (3 - 2x) \neq 5$ **69.** $2p^2$ **71.** $10x^3$
73. $56m^5$ **75.** $30p^9$

Section 4.5 (page 227)

1. $-32x^7$ **3.** $15y^{11}$ **5.** $30a^9$ **7.** $6m^2 + 4m$
9. $-6p^4 + 12p^3$ **11.** $-16z^2 - 24z^3 - 24z^4$
13. $6y + 4y^2 + 10y^5$ **15.** $32z^5 + 20xz^4 - 12x^2z^3$
17. $x^2 - 25$ **19.** $3r^2 - 13r + 4$
21. $6p^2 - p - 5$ **23.** $16m^2 + 24m + 9$
25. $6b^2 + 46b - 16$ **27.** $16b^2 + 2ab - 3a^2$
29. $-8h^2 + 6hk - k^2$ **31.** $12x^3 + 26x^2 + 10x + 1$

33. $20m^4 - m^3 - 8m^2 - 17m - 15$
35. $6x^6 - 3x^5 - 4x^4 + 4x^3 - 5x^2 + 8x - 3$
37. $5x^4 - 13x^3 + 20x^2 + 7x + 5$
39. $x^2 + 14x + 49$ 41. $4p^2 - 20p + 25$
43. $25k^2 + 80k + 64$ 45. $m^3 - 15m^2 + 75m - 125$
47. $k^4 + 4k^3 + 6k^2 + 4k + 1$
49. $27r^3 - 54r^2s + 36rs^2 - 8s^3$ 51. 3, 2
53. −7, 3 55. 9, −6

Section 4.6 (page 233)

1. $r^2 + 2r - 3$ 3. $x^2 - 10x + 21$
5. $6x^2 + x - 2$ 7. $6z^2 - 13z - 15$
9. $2a^2 + 9a + 4$ 11. $8r^2 + 2r - 3$
13. $6a^2 + 8a - 8$ 15. $20 + 9x - 20x^2$
17. $-12 + 5r + 2r^2$ 19. $15 + a - 2a^2$
21. $p^2 + 4pq + 3q^2$ 23. $10y^2 - 3yz - z^2$
25. $8y^2 + 31yz - 45z^2$ 27. $-8r^2 + 2rs + 45s^2$
29. $x^2 + 16x + 64$ 31. $9m^2 - 6mn + n^2$
33. $36a^2 - 18ab + \frac{9}{4}b^2$ 35. $64a^2 - 48ab + 9b^2$
37. $p^2 - 4$ 39. $4b^2 - 25$ 41. $36a^2 - p^2$
43. $4m^2 - \frac{25}{9}$ 45. $49y^4 - 100z^2$
47. $(x + 6)^2 < 3$ 49. $[(2x + 4) \cdot 6x] - 5 = 8$
51. $4p$ 53. $-5p$ 55. $\frac{2}{5k^2}$ 57. $-\frac{9p^5}{16q^2}$

Section 4.7 (page 237)

1. $2x$ 3. $2a^2$ 5. $\frac{9k^3}{m}$ 7. $30m^3 - 10m$
9. $5m^4 - 8m + 4m^2$ 11. $4m^4 - 2m^2 + 2m$
13. $m^4 - 2m + 4$ 15. $m - 1 + \frac{5}{2m}$
17. $4x^2 - x + 1$ 19. $9x - 3x^2 + 6x^3$
21. $\frac{12}{x} + 8 + x$ 23. $\frac{1}{3}x + 2 - \frac{1}{3x}$
25. $4k^3 - 6k^2 - k + \frac{7}{2} - \frac{3}{2k}$
27. $-10p^3 + 5p^2 - 3p + \frac{3}{p}$ 29. $4y^3 - 2 + \frac{3}{y}$
31. $\frac{12}{x} - \frac{6}{x^2} + \frac{14}{x^3} - \frac{10}{x^4}$
33. $12x^5 + 9x^4 - 12x^3 + 6x^2$
35. $-20k^4 + 12k^3 - 8k^2$ 37. $6m^{10} - 12m^8 + 3m^7$
39. $2x + 1$

Section 4.8 (page 243)

1. $x + 2$ 3. $2y - 5$ 5. $p - 4 + \frac{4}{p + 6}$
7. $r - 5$ 9. $6m - 1$ 11. $a - 7 + \frac{37}{2a + 3}$
13. $3m + 5 + \frac{13}{5m + 3}$ 15. $3t^2 - 5t + 6$
17. $9r^3 - 2r + 6$ 19. $2r^2 + 6r + 12$
21. $5z - 1 + \frac{4}{z^2 + 2}$ 23. $r^2 - 1 + \frac{4}{r^2 - 1}$
25. $3r^2 - 5r + 4 + \frac{4}{2r^2 - 3}$ 27. $x^3 + 3x^2 - x + 5$
29. $x^2 + 1$ 31. 1, 2, 3, 6, 9, 18
33. 1, 2, 3, 4, 6, 9, 12, 18, 36
35. 1, 2, 5, 7, 10, 14, 35, 70 37. 1, 41

Chapter 4 Review Exercises (page 247)

1. 80 2. 2 3. $\frac{1}{16}$ 4. $\frac{1}{6}$ 5. $\frac{64}{25}$ 6. $-\frac{8}{9}$
7. 5^{11} 8. $\frac{1}{9^2}$ 9. $(-4)^8$ 10. 15^5 11. $\frac{1}{5^{11}}$
12. 6^2 13. x^2 14. $\frac{1}{p^{12}}$ 15. r^4 16. 2^8
17. $\frac{1}{9^6}$ 18. 5^8 19. $\frac{1}{8^{12}}$ 20. $\frac{1}{m^2}$ 21. y^7
22. r^{13} 23. $(-5)^2m^6$ 24. $\frac{y^{12}}{2^3}$ 25. $\frac{1}{a^3b^5}$
26. $2 \cdot 6^2 \cdot r^5$ 27. $\frac{2^3n^{10}}{3m^{13}}$ 28. 6.4×10^4
29. 1.58×10^7 30. 2.6954×10^{10}
31. 4.251×10^{-4} 32. 9.76×10^{-5}
33. 7.84×10^{-1} 34. 12,000 35. 689,000,000
36. .0004253 37. .877 38. 90,000,000
39. 800 40. .0003 41. 4,000,000 42. .025
43. .01 44. $22m^2$; degree 2
45. $p^3 - p^2 + 4p + 2$; degree 3 46. $4r^4$; degree 4
47. already simplified; degree 5
48. $-8x^5 + 9x^3 - 7x$; degree 5
49. $-5z^3 - 6z^2 + 8z + 7$; degree 3 50. $a^3 + 4a^2$
51. $2r^3 - 3r^2 + 3r$ 52. $y^2 - 10y + 9$
53. $-13k^4 - 15k^2 + 18k - 6$ 54. $5m^3 - 6m^2 - 3$
55. $-y^2 - 4y + 4$ 56. $10p^2 - 3p - 5$
57. $7r^4 - 4r^3 + 1$ 58. $10x^2 - 55x$
59. $-6p^5 + 15p^4$ 60. $m^2 - 7m - 18$
61. $6k^2 - 9k - 6$ 62. $6r^3 + 8r^2 - 17r + 6$
63. $5p^3 + 8p^2 - 7p - 6$ 64. $r^3 + 6r^2 + 12r + 8$
65. $6k^2 - 7k - 3$ 66. $2a^2 + 5ab - 3b^2$
67. $12k^2 - 32kq - 35q^2$ 68. $a^2 + 8a + 16$
69. $9p^2 - 12p + 4$ 70. $4r^2 + 20rs + 25s^2$
71. $64z^2 - 48zy + 9y^2$ 72. $36m^2 - 25$
73. $4z^2 - 49$ 74. $25a^2 - 36b^2$ 75. $81y^2 - 64z^2$
76. $-\frac{5y^2}{3}$ 77. $2x^2y$ 78. $y^3 - 2y + 3$
79. $p - 3 + \frac{5}{2p}$ 80. $-2m^2n + mn^2 + \frac{6n^3}{5}$
81. $2r + 7$ 82. $4m + 3 + \frac{5}{3m - 5}$
83. $2a^2 + 3a - 1 + \frac{6}{5a - 3}$
84. $k^3 + k^2 + 4k - 2 + \frac{-6}{2k + 1}$
85. $m^2 + 2m - \frac{1}{2} + \frac{-6m + \frac{9}{2}}{2m^2 - 3}$ 86. $\frac{3^5}{p^3}$ 87. $\frac{1}{7^2}$

88. $49 - 28k + 4k^2$ **89.** $y^2 + 5y + 1$
90. $\frac{6^3r^6s^3}{5^3}$ **91.** $-8m^7 + 10m^6 - 6m^5$ **92.** 2^5
93. $\frac{5}{2} - \frac{4}{5xy} + \frac{3x}{2y^2}$ **94.** $\frac{r^2}{6}$ **95.** $-11 + p$
96. $\frac{3}{4}$ **97.** $a^3 - 2a^2 - 7a + 2$
98. $8y^3 - 9y^2 + 5$ **99.** $10r^2 + 21r - 10$
100. $144a^2 - 1$ **101.** $-22y^4 + 4y^3 + 18y^2$

Chapter 4 Test (page 251)

[4.2] **1.** $\frac{1}{81}$ **2.** -1 **3.** $\frac{1}{5}$ **4.** 8^5 **5.** p^{24}
6. $\frac{1}{a^{10}b^4}$ [4.3] **7.** 2.45×10^8 **8.** 3.79×10^{-4}
9. .0048 **10.** 400
[4.4] **11.** $-x^2 + 6x$; degree 2; binomial
12. $2m^4 + 11m^3 - 8m^2$; degree 4; trinomial
13. already simplified; degree 3; none of these
14. $x^5 - x^2 - 2x + 12$ **15.** $3y^2 - 2y - 2$
[4.5, 4.6] **16.** $6m^5 + 12m^4 - 18m^3 + 42m^2$
17. $r^2 - 3r - 10$ **18.** $6t^2 - tw - 12w^2$
19. $25r^2 - 30rs + 9s^2$ **20.** $36p^2 - 64q^2$
21. $2x^3 + x^2 - 16x + 15$ [4.7] **22.** $3y^2 - 5y + 2$
23. $2r^2 + 5r - 3 + \frac{8}{5r}$
[4.8] **24.** $3y - 6 + \frac{7}{4y + 3}$ **25.** $3x^2 + 4x + 2$

CHAPTER 5

Section 5.1 (page 259)

1. 1, 3, 9, 27
3. 1, 2, 3, 4, 5, 6, 10, 12, 15, 20, 30, 60
5. 1, 2, 4, 5, 10, 20, 25, 50, 100 **7.** 1, 29
9. $2^3 \cdot 3 \cdot 5$ **11.** $2^2 \cdot 3^2 \cdot 5$ **13.** $5^2 \cdot 11$
15. $5^2 \cdot 19$ **17.** 12 **19.** $10p^2$ **21.** $6m^2n$
23. 2 **25.** x **27.** $3m^2$ **29.** $2z^4$ **31.** $2mn^4$
33. $7x^3y^2$ **35.** $12(x + 2)$ **37.** $9a(a - 2)$
39. $5y^5(13y^4 - 7)$ **41.** no common factor (except 1)
43. $19y^2p^2(y + 2p)$ **45.** $13y^3(y^3 + 2y^2 - 3)$
47. $9qp^3(5q^3p^2 - 4p^3 + 9q)$
49. $5z^3a^3(25z^2 - 12za^2 + 17a)$ **51.** $(m + 2)(m + 5)$
53. $(y - 6)(y + 4)$ **55.** $(4k + 3q)(2k + 3q)$
57. $(3r + 2y)(6r - x)$ **59.** $(4m - p^2)(4m^2 - p)$
61. $(p - 4)(2q^2 + 1)$ **63.** $y^2 + 4y - 21$
65. $q^2 - 12q + 35$ **67.** $a^2 - 49$

Section 5.2 (page 265)

1. $x + 3$ **3.** $r + 8$ **5.** $t - 12$ **7.** $x - 8$
9. $m + 6$ **11.** $p - 1$ **13.** $x - 5y$
15. $(y + 1)(y + 8)$ **17.** $(b + 3)(b + 5)$
19. $(m + 5)(m - 4)$ **21.** $(n - 2)(n + 6)$
23. $(r - 6)(r + 5)$ **25.** prime
27. $(a + 9)(a - 11)$ **29.** $(x - 4)(x - 5)$
31. $(z - 7)(z - 7)$ or $(z - 7)^2$ **33.** $(x + 3a)(x + a)$
35. $(y - 6b)(y + 5b)$ **37.** $(x + 6y)(x - 5y)$
39. $(r - s)(r - s)$ or $(r - s)^2$ **41.** $(p - 5q)(p + 2q)$
43. $3m(m + 3)(m + 1)$ **45.** $6(a - 10)(a + 2)$
47. $3j(j - 6)(j - 4)$ **49.** $3x^2(x - 6)(x + 5)$
51. $a^3(a + 4b)(a - b)$ **53.** $yz(y + 3z)(y - 2z)$
55. The factor $3x + 12$ can be factored further as $3(x + 4)$.
57. $a^2 + 15a + 54$ **59.** $6a^2 + 7a + 2$
61. $8m^2 + 14m - 15$ **63.** $18r^2 - 2$

Section 5.3 (page 271)

1. $x - 1$ **3.** $b - 3$ **5.** $4y - 3$ **7.** $5x + 4$
9. $m + 10$ **11.** $3a - 4b$ **13.** $k + 3m$
15. $2x^2 - 5x - 3$; $x - 3$
17. $6m^2 + 7m - 20$; $2m + 5$ **19.** $(2x + 1)(x + 3)$
21. $(a + 1)(3a + 7)$ **23.** $(4r - 3)(r + 1)$
25. $(5m + 2)(3m - 1)$ **27.** $(2m - 3)(4m + 1)$
29. $(a - 2)(5a + 3)$ **31.** $(3r - 5)(r + 2)$
33. $(4y + 1)(y + 17)$ **35.** $(19x + 2)(2x + 1)$
37. $(2x + 3)(5x - 2)$ **39.** $(2w + 5)(3w + 2)$
41. $(2q + 3)(3q + 7)$ **43.** $(5m - 4)(2m - 3)$
45. $(2k + 3)(4k - 5)$ **47.** $3(2x - 3)(4x - 1)$
49. $q(5m + 2)(8m - 3)$ **51.** $2m(m + 5)(m - 4)$
53. $2a^2(4a - 1)(3a + 2)$ **55.** $4z^3(8z + 3)(z - 1)$
57. $(4p - 3q)(3p + 4q)$ **59.** $(5a + 3b)(5a + 2b)$
61. $(3a - 5b)(2a + b)$ **63.** $m^4n(2m + n)(3m + 2n)$
65. $9r^2 - 1$ **67.** $a^2 - 36b^2$ **69.** $4z^2 - 12z + 9$

Section 5.4 (page 279)

1. $(x + 4)(x - 4)$ **3.** $(p + 2)(p - 2)$
5. $(3m + 1)(3m - 1)$ **7.** $\left(5m + \frac{4}{7}\right)\left(5m - \frac{4}{7}\right)$
9. $4(3t + 2)(3t - 2)$ **11.** $(5a + 4r)(5a - 4r)$
13. prime **15.** $(p^2 + 6)(p^2 - 6)$
17. $(a^2 + 1)(a + 1)(a - 1)$
19. $(m^2 + 9)(m + 3)(m - 3)$
21. $(4k^2 + 1)(2k + 1)(2k - 1)$ **23.** $(p + 1)^2$
25. $(y - 4)^2$ **27.** $(m - 10)^2$ **29.** $\left(r + \frac{1}{3}\right)^2$
31. not a perfect square **33.** $(4a - 5b)^2$
35. $25(2m + 1)^2$ **37.** $(7x + 2y)^2$ **39.** $(2c + 3d)^2$
41. $(5h - 2y)^2$ **43.** 10 **45.** 9 **47.** 2
49. $\frac{2}{3}$ **51.** $-\frac{9}{7}$ **53.** $-\frac{5}{8}$

Summary Exercises on Factoring (page 281)

1. $8m^3(4m^6 + 2m^2 + 3)$ **3.** $7k(2k + 5)(k - 2)$
5. $(z + 5)(6z + 1)$ **7.** $(7z + 4y)(7z - 4y)$
9. $4(4x + 5)$ **11.** $(5y - 6z)(2y + z)$
13. $(m + 5)(m - 3)$ **15.** $8z(4z - 1)(z + 2)$
17. $(z - 6)^2$ **19.** $(y + 2k)(y - 6k)$
21. $6(y - 2)(y + 1)$ **23.** $(p - 6)(p - 11)$
25. prime **27.** $(z - 5a)(z + 2a)$ **29.** $(2k - 3)^2$
31. $(4r + 3m)^2$ **33.** $3k(k - 5)(k + 1)$
35. $4(2k - 3)^2$ **37.** $6y^4(3y + 4)(2y - 5)$
39. $5z(z - 7)(z - 2)$ **41.** $6(3m + 2z)(3m - 2z)$

43. $2(3a - 1)(a + 2)$ **45.** $(m + 9)(m - 9)$
47. $5m^2(5m - 3n)(5m - 13n)$ **49.** $(m - 2)^2$
51. $6k^2p(4k^2 + 10kp + 25p^2)$
53. $(4p + 3q)(3p - 2q)$ **55.** $4(4p + 5m)(4p - 5m)$
57. $(10a + 9y)(10a - 9y)$ **59.** $(a + 4)^2$

Section 5.5 (page 289)

1. $2, -4$ **3.** $-7, \frac{8}{3}$ **5.** $-\frac{1}{5}, \frac{1}{2}$ **7.** $-2, -3$
9. $-7, 4$ **11.** $-8, 3$ **13.** $3, -1$ **15.** $-1, -2$
17. -4 **19.** $\frac{1}{3}, -2$ **21.** $\frac{5}{2}, -2$ **23.** $-\frac{4}{3}, \frac{1}{2}$
25. $\frac{1}{3}, -\frac{5}{2}$ **27.** $-4, \frac{5}{2}$ **29.** $-6, 6$ **31.** $-\frac{3}{2}, \frac{3}{2}$
33. $\frac{1}{2}, -4$ **35.** $-\frac{2}{5}, 2$ **37.** $\frac{2}{3}, 6$ **39.** 1
41. 2 **43.** $\frac{4}{3}, -\frac{2}{3}, \frac{1}{2}$ **45.** $1, -\frac{3}{2}, \frac{4}{3}$
47. $0, -\frac{3}{4}, \frac{3}{4}$ **49.** 5

Section 5.6 (page 295)

1. length: 11 centimeters; width: 6 centimeters
3. length: 18 inches; width: 9 inches
5. length: 5 feet; width: 2 feet
7. width: 4 inches; length: 8 inches
9. height: 5 centimeters; base: 10 centimeters
11. height: 6 meters; area of base: 16 square meters
13. 19 feet and 18 feet
15. length: 6 feet; width: 5 feet
17. 4 and 5 or -1 and 0
19. $-1, 1, 3$ or 11, 13, 15
21. -6 and -5 or 5 and 6 **23.** 12 centimeters
25. 8 feet **27.** 256 feet **29.** 10 seconds
31. $y^2 - 4xy + 4x^2$ **33.** false **35.** true
37. false

Chapter 5 Review Exercises (page 301)

1. $2^2 \cdot 3$ **2.** $2^4 \cdot 3$ **3.** $2^2 \cdot 3^2 \cdot 5$
4. $6(p + 2)$ **5.** $20(2r^2 + 1)$ **6.** $25m^3(m - 2)$
7. $6(1 - 3r^5 + 2r^3)$ **8.** $50y^3(2y^3 - 1 + 6y)$
9. $(2p + 3)(3p + 2)$ **10.** $(4y + 3)(y + 2)$
11. $(m + 2)(m + 3)$ **12.** $(y + 8)(y + 5)$
13. $(r - 9)(r + 3)$ **14.** $(p + 6)(p - 5)$
15. $(z - 11)(z + 4)$ **16.** $(r - 12s)(r + 8s)$
17. $(p + 12q)(p - 10q)$ **18.** $4p(p + 2)(p - 5)$
19. $3z^2(z - 2)(z - 8)$ **20.** $(m - 6n)(m + 3n)$
21. $(y - 5z)(y - 3z)$ **22.** $p^5(p - 2q)(p + q)$
23. $3r^3(r + 3s)(r - 5s)$ **24.** $5p^4(p - 7)(p - 2)$
25. $6m^7(m - 9)(m - 5)$ **26.** $(2k - 1)(k - 2)$
27. $(3z - 1)(z + 4)$ **28.** $(3r + 2)(2r - 3)$
29. $(5p + 1)(2p - 1)$ **30.** $(8y - 7)(y + 3)$
31. $4k^3(2k - 1)(3k - 1)$ **32.** $(7m - 2n)(m + 3n)$
33. $rs(2r + s)(5r + 6s)$ **34.** $z(2z - 1)(z + 5)$
35. $(m + 5)(m - 5)$ **36.** $(5p + 11)(5p - 11)$
37. $(7y + 5z)(7y - 5z)$ **38.** $36(2p + q)(2p - q)$
39. prime **40.** $(z - 4)^2$ **41.** $(3r - 7)^2$
42. $(4m + 5n)^2$ **43.** $6x(3x - 2)^2$ **44.** $\frac{1}{3}, -2$
45. $\frac{2}{5}, -4$ **46.** $-\frac{5}{2}, \frac{7}{3}$ **47.** $-1, -3$ **48.** $4, 1$
49. $5, 3$ **50.** $5, 6$ **51.** $5, -\frac{4}{3}$ **52.** $\frac{7}{10}, -\frac{7}{10}$
53. $6, -1$ **54.** 6 **55.** $\frac{1}{3}, -1, -2$
56. $-\frac{3}{2}, 3, 1$ **57.** $-3, 3$ **58.** $-9, 9$
59. width: 4 meters; length: 10 meters
60. width: 2 meters; length: 6 meters
61. width: 8 centimeters; length: 10 centimeters
62. length: 6 meters; height: 5 meters
63. 4 and 6 or -2 and 0
64. 6 and 8 or -8 and -6
65. width: 3 meters; length: 7 meters
66. width: 4 meters; length: 9 meters **67.** 1
68. 2 or 3 **69.** -4 and -2 or 8 and 10
70. 16 and 18 or -6 and -4 **71.** 34 miles
72. 8 feet **73.** 112 feet **74.** 192 feet
75. 256 feet **76.** after 8 seconds **77.** 2 inches
78. 2 centimeters **79.** $(z - 10x)(z - x)$
80. $(3k + 5)(k + 2)$ **81.** $(3m + 4p)(5m - 4p)$
82. $(y^2 + 25)(y + 5)(y - 5)$
83. $3m(m - 5)(2m + 3)$
84. $8abc(3b^2c - 7ac^2 + 9ab)$ **85.** prime
86. $6xyz(2xz^2 + 2y - 5x^2yz^3)$
87. $2a^3(a - 6)(a + 2)$ **88.** $(2r + 3q)(6r - 5q)$
89. $(10a + 3)(10a - 3)$ **90.** $(k + 9)(k - 7)$
91. $2, -5$ **92.** $\frac{2}{5}$ **93.** $\frac{3}{2}, -\frac{1}{4}$ **94.** 25 miles
95. length: 6 meters; width: 4 meters
96. 5, 6, 7 or $-5, -4, -3$
97. 15 meters; 36 meters; 39 meters **98.** -1 or 3
99. base: 6 meters; height: 4 meters

Chapter 5 Test (page 307)

[5.1] **1.** $8m(2m - 3)$ **2.** $6y(x + 2y)$
3. $14p(2q + 1 + 4p)$ **4.** $(m + 3n)(3 + k)$
5. prime **[5.2]** **6.** $(x + 5)(x + 6)$
7. $(p + 7)(p - 1)$ **[5.3]** **8.** $(2y + 3)(y - 5)$
9. $(2m - 1)(2m + 3)$ **10.** $(3x - 2)(x + 5)$
11. $(2z + 1)(5z + 1)$ **12.** $(5a + 1)(2a - 5)$
13. $(4r - 5)(3r + 1)$ **14.** $(m + 7)(m - 2)$
15. $(a + 5b)(a - 2b)$ **16.** $(3r - 2s)(2r + s)$
[5.4] **17.** $(x + 5)(x - 5)$ **18.** $(5m + 7)(5m - 7)$
19. $(2p + 3)^2$ **20.** $(5z - 1)^2$ **21.** $4p(p + 2)^2$
[5.3] **22.** $5m^2(2m + 1)(m + 5)$
[5.5] **23.** $-1, -2$ **24.** $\frac{1}{3}, -2$ **25.** $\frac{5}{2}, -4$
26. $2, -2$ **27.** $0, 4, -4$
[5.6] **28.** length: 5 inches; width: 3 inches
29. -1 and 0
30. 5 meters, 12 meters, and 13 meters

CHAPTER 6

Section 6.1 (page 313)

1. 0 **3.** 4 **5.** -5 **7.** 3, 5 **9.** $-1, \frac{3}{2}$

11. never undefined **13. (a)** 1 **(b)** $\frac{14}{9}$

15. (a) 2 **(b)** $-\frac{14}{3}$ **17. (a)** $\frac{256}{15}$ **(b)** undefined

19. (a) -5 **(b)** $\frac{5}{23}$ **21. (a)** -5 **(b)** $\frac{15}{2}$

23. (a) undefined **(b)** $\frac{1}{10}$ **25.** $2k$ **27.** $\frac{b}{4a}$

29. $3z + 2$ **31.** $3p^2 - 2$ **33.** $\frac{x-1}{x+1}$ **35.** $\frac{m}{2}$

37. $2(2r + s)$ or $4r + 2s$ **39.** $\frac{m-2}{m+3}$ **41.** $\frac{x+4}{x+1}$

43. -1 **45.** $-(x + 1)$ or $-x - 1$ **47.** -1

49. $\frac{15}{32}$ **51.** $\frac{32}{9}$ **53.** 4

Section 6.2 (page 319)

1. $\frac{3m}{4}$ **3.** $\frac{3}{32}$ **5.** $2a^4$ **7.** $\frac{1}{4}$ **9.** $\frac{1}{6}$ **11.** $\frac{2}{r^5}$

13. $\frac{6}{a+b}$ **15.** 2 **17.** $\frac{2}{9}$ **19.** $\frac{3}{10}$ **21.** $\frac{2r}{3}$

23. $\frac{6(y-4)}{z+3}$ **25.** $-\frac{7}{8}$ **27.** -1 **29.** $\frac{y+3}{y+4}$

31. $\frac{m+6}{m+3}$ **33.** $\frac{p+5}{p-2}$ **35.** $\frac{m-3}{2m-3}$

37. $\frac{r+6s}{r+s}$ **39.** $\frac{m-8}{8}$ **41.** $2 \cdot 5^2$

43. $3 \cdot 5 \cdot 7$ **45.** $5a^2$

Section 6.3 (page 325)

1. 60 **3.** 120 **5.** 1800 **7.** x^5 **9.** $30p$

11. $150m$ **13.** $180y^4$ **15.** $300r^5$

17. $32r^2(r - 2)$ **19.** $18(r - 2)$ **21.** $12p(p + 5)$

23. $p - q$ or $q - p$ **25.** $a(a + 6)(a - 3)$

27. $(k + 7)(k - 5)(k + 8)$

29. $(2y - 1)(y + 4)(y - 3)$ **31.** $\frac{42}{66}$ **33.** $\frac{54}{6r}$

35. $\frac{-88}{8m}$ **37.** $\frac{24y^2}{70y^3}$ **39.** $\frac{60m^2k^3}{32k^4}$ **41.** $\frac{57z}{6z-18}$

43. $\frac{-4a}{18a-36}$ **45.** $\frac{6(k+1)}{k(k-4)(k+1)}$ **47.** $\frac{7}{3}$

49. $\frac{9}{10}$ **51.** $\frac{2}{3}$ **53.** $\frac{19}{72}$

Section 6.4 (page 333)

1. $\frac{7}{p}$ **3.** $-\frac{3}{k}$ **5.** b **7.** y **9.** $z - 5$

11. $\frac{p-4}{6}$ **13.** $\frac{9-5r}{10}$ **15.** $\frac{8a-10}{5a}$ **17.** $\frac{11}{2x}$

19. $\frac{7}{24}$ **21.** $\frac{13r+15}{9}$ **23.** $\frac{5x-11}{2x}$

25. $\frac{3p+5}{p^2}$ **27.** $\frac{15-12k}{4k^2}$ **29.** $\frac{9x+4p}{p^2x}$

31. $\frac{4(m+10)}{(m-5)(m+5)}$ **33.** $\frac{4-a}{4(a-b)}$ **35.** $\frac{-3}{2(y-2)}$

37. $\frac{4y-3}{(y-2)(y+2)(y+3)}$ **39.** $\frac{2(10z+1)}{7z(z+2)}$

41. $\frac{7k-9}{k(k-1)(k+3)}$ **43.** $\frac{-5q^2-13q+7}{(3q-2)(q+4)(2q-3)}$

45. $\frac{5}{12}$ **47.** $\frac{10}{7}$

Section 6.5 (page 339)

1. $\frac{31}{50}$ **3.** $\frac{5}{18}$ **5.** $\frac{mp}{40}$ **7.** $\frac{x(x+1)}{4(x-3)}$ **9.** $\frac{2}{y}$

11. $\frac{8}{x}$ **13.** $\frac{y^2-6}{y^2+2}$ **15.** $\frac{r^2+1}{(1-r)(1+r)}$

17. $\frac{2m^2+p}{mp(4m+1)}$ **19.** $\frac{m(m+3)}{2(m-3)}$ **21.** $\frac{a(a-1)}{2}$

23. $\frac{x}{2-x}$ **25.** $\frac{4(r-2)(r+2)}{(5r+12)(r-1)}$ **27.** $\frac{13}{5}$

29. -4 **31.** $-\frac{13}{2}$ **33.** $\frac{1}{6}$

Section 6.6 (page 345)

1. $\frac{2}{5}$ **3.** 24 **5.** 2 **7.** -7 **9.** 1 **11.** 2

13. 5 **15.** -8 **17.** 8 **19.** 5 **21.** 13

23. 12 **25.** 5 **27.** -2 **29.** -3 **31.** 1

33. $-\frac{1}{3}, 3$ **35.** no solution **37.** 6 **39.** $\frac{4}{3}, 3$

41. $-\frac{6}{7}, 3$ **43.** $R = \frac{kE}{I}$

45. $D = \frac{k-Fd}{-F}$ or $D = \frac{Fd-k}{F}$ **47.** $r = \frac{E-IR}{I}$

49. $b = \frac{2A-hB}{h}$ or $b = \frac{2A}{h} - B$ **51.** $r = \frac{800}{p}$

53. $t = \frac{780}{z}$ **55.** $\frac{1}{h}$

Summary Exercises on Rational Expressions (page 349)

1. $\frac{8}{m}$ **3.** $\frac{3}{4x^2(x+3)}$ **5.** $\frac{r+4}{r+1}$ **7.** -43

9. $\frac{17}{3(y-1)}$ **11.** $\frac{1}{5}, 2$ **13.** $-\frac{19}{18z}$

15. $\frac{3m+5}{(m+2)(m+3)(m+1)}$

Section 6.7 (page 355)

1. 9 **3.** 6 and 9 **5.** $\frac{2}{3}$ or -3

7. mother: \$240; daughter: \$96 **9.** \$32,000

11. 2 miles per hour **13.** 150 miles each way

15. $\frac{37}{2}$ or $18\frac{1}{2}$ miles per hour **17.** $\frac{6}{5}$ or $1\frac{1}{5}$ hours

19. $\frac{84}{19}$ or $4\frac{8}{19}$ hours **21.** $\frac{28}{5}$ or $5\frac{3}{5}$ days

23. $\frac{100}{11}$ or $9\frac{1}{11}$ minutes **25.** 3 hours **27.** $\frac{63}{2}$

29. 50 **31.** $\frac{5}{2}$ **33.** 69.08 centimeters

35. 100 amperes **37.** (a) 1 (b) -23

39. (a) 7 (b) -17 **41.** (a) $-\frac{1}{3}$ (b) $-\frac{13}{3}$

Chapter 6 Review Exercises (page 365)

1. 0 **2.** 5 **3.** $-2, 4$ **4.** $-3, \frac{1}{2}$

5. (a) $\frac{9}{5}$ (b) 1 **6.** (a) $\frac{18}{5}$ (b) $\frac{2}{3}$

7. (a) $\frac{24}{7}$ (b) 8 **8.** (a) undefined (b) $\frac{3}{2}$ **9.** $\frac{2z}{3y^2}$

10. $\frac{3x-4}{2}$ **11.** -1 **12.** $\frac{2k+5y}{4k-y}$ **13.** $36y$

14. $\frac{16}{9}$ **15.** $\frac{1}{z^5}$ **16.** $\frac{5}{4}$ **17.** $\frac{7}{2}$

18. $\frac{k-2}{2(3k-5)}$ **19.** $\frac{a-2}{a-1}$ **20.** $\frac{z-1}{z-3}$

21. $\frac{(2p-1)(2p+3)}{(p+3)^2}$ **22.** $\frac{8r+1}{r-7}$ **23.** 90

24. $48y^3$ **25.** $y(y+2)(y+4)$

26. $(z+3)(z-2)(z+1)$ **27.** $\frac{20}{45}$ **28.** $\frac{60}{5m}$

29. $\frac{9m}{24m^3}$ **30.** $\frac{96}{8y-32}$ **31.** $\frac{-10k}{15k+75}$

32. $\frac{12y(y-4)}{(y-2)(y+1)(y-4)}$ or $\frac{12y^2-48y}{(y-2)(y+1)(y-4)}$

33. $\frac{13}{m}$ **34.** $\frac{1}{r}$ **35.** $\frac{12+p}{3p}$ **36.** $\frac{35-2k}{5k}$

37. $\frac{6+9m}{4}$ **38.** $\frac{16-3r}{2r^2}$ **39.** $\frac{14p+q}{(2p-q)(2p+q)}$

40. $\frac{16-3p}{5(p+2)(p-2)}$ **41.** $\frac{5m-12}{m(m+5)(m-2)}$

42. $\frac{8p-30}{p(p-2)(p-3)}$ or $\frac{2(4p-15)}{p(p-2)(p-3)}$ **43.** $\frac{1}{q^3r^3}$

44. $\frac{3(m-2)}{m+2}$ **45.** $\frac{8(5k-1)}{4k+3}$ **46.** $\frac{(n-m)^2}{mn}$

47. $\frac{zx+1}{zx-1}$ **48.** $\frac{1-a-b}{1+a+b}$ **49.** $\frac{10}{3}$ **50.** -4

51. -6 **52.** no solution **53.** 3 **54.** $V = \frac{kT}{P}$

55. $y = \frac{3-xz}{2x}$ **56.** $m = \frac{4+p^2q}{3p^2}$ **57.** 1

58. $\frac{6}{5}$ or $\frac{1}{2}$ **59.** small car, \$120; large car, \$180

60. 3 miles per hour **61.** 6 days **62.** 2 days

63. $\frac{36}{5}$ **64.** 12 **65.** $\frac{(p+q)(2+3p-3q)}{(p-q)(5p+5q-4)}$

66. $\frac{-y-5}{(y+1)(y-1)}$ **67.** $\frac{40}{3p^2}$ **68.** $\frac{1}{9}$ **69.** 4

70. $\frac{z+7}{(z+1)(z-1)^2}$ **71.** $d = \frac{k+FD}{F}$ or $d = \frac{k}{F} + D$

72. $-2, 3$ **73.** $\frac{40}{13}$ or $3\frac{1}{13}$ hours

74. 385 kilograms **75.** 150 kilometers per hour

76. $\frac{9}{14}$ billion dollars

Chapter 6 Test (page 371)

[6.1] **1.** 1, 3 **2.** $\frac{5s(s-1)}{2}$ **3.** $\frac{4p-3q}{p+q}$

[6.2] **4.** x **5.** $\frac{40}{27}$ **6.** $\frac{3m-2}{3m+2}$ **7.** $\frac{a+3}{a+4}$

[6.3] **8.** $\frac{77r}{49r^2}$ **9.** $\frac{15}{24m-48}$ [6.4] **10.** $-\frac{1}{x}$

11. $-\frac{13}{6(a+1)}$ **12.** $\frac{5k+2}{(k+2)(2k-1)(k+1)}$

13. $\frac{4m+20}{(2m-5)(m+2)(2m+1)}$ or $\frac{4(m+5)}{(2m-5)(m+2)(2m+1)}$

[6.5] **14.** $\frac{3(1-2k)}{k(1+3k)}$ **15.** $\frac{-2p-7}{2p+9}$

[6.6] **16.** 6 **17.** 1 [6.7] **18.** $\frac{1}{4}$ or $\frac{1}{2}$

19. $\frac{20}{9}$ or $2\frac{2}{9}$ hours **20.** $\frac{56}{3}$

Cumulative Review Exercises (page 373)

1. -27 **2.** $\frac{1}{2}$ **3.** $\frac{16}{9}$ **4.** $\frac{81}{64}$ **5.** $\frac{1}{49}$ **6.** 1

7. $\frac{1}{4^9}$ **8.** 256 **9.** $\frac{1}{2^4x^7}$ **10.** $\frac{1}{p^2}$ **11.** $\frac{1}{m^6}$

12. $5x^3 + 2x^2 - 5x + 7$ **13.** $-4k^2 + 2k + 3$

14. $72x^6y^7$ **15.** $6m^7 - 15m^5 + 3m^3$

16. $2y^2 - 7y - 4$ **17.** $12z^2 - 2wz - 4w^2$

18. $3y^3 + 8y^2 + 12y - 5$ **19.** $9p^2 + 12p + 4$

20. $16a^2 - 8ab + b^2$ **21.** $25k^2 - 16$

22. $4p^2 - 9q^2$ **23.** $a^2 - 3a + 5$

24. $4x^3 + 6x^2 - 3x + 10$ **25.** $6p^2 + 7p + 1 + \frac{6}{2p-2}$

26. $3z^2 - z + 4 + \frac{16}{5z-2}$ **27.** $8x(x-3)$

28. $9q^5(r + 2qp)$ **29.** prime

30. $(m+2)(m+7)$ **31.** $(r+5)(r-8)$

32. $(p+2q)(p-9q)$ **33.** $(2a-1)(a+4)$

34. $(2m+3)(5m+2)$ **35.** $(5x+3y)(3x-2y)$

36. $(4t+3v)(2t+v)$ **37.** $(3x+1)^2$

38. $(2p-3)^2$ **39.** $(4t+7z)^2$

40. $(5r+9t)(5r-9t)$ **41.** $(10p-7q)(10p+7q)$

42. $(10y-13p^2)(10y+13p^2)$

43. $m(2a - 1)(3a + 2)$ **44.** $2pq(3p + 1)(p + 1)$ **45.** $-2, \frac{3}{2}$ **46.** $-3, 2$ **47.** $-3, 5$ **48.** $-7, 11$ **49.** $-\frac{2}{3}, \frac{1}{2}$ **50.** $0, 8$ **51.** $-2, 2$ **52.** $5, -\frac{1}{2}, \frac{5}{3}$ **53.** $0, -4, 4$ **54.** -2 or -1 **55.** 6 meters **56.** 4 centimeters **57.** 3 centimeters **58.** $\frac{m}{4}$ **59.** $\frac{1}{4p^2}$ **60.** $\frac{y+2}{6}$ **61.** $\frac{5}{r+1}$ **62.** $\frac{1}{xy^6}$ **63.** $\frac{s^9}{r^{11}}$ **64.** $\frac{7}{m}$ **65.** $\frac{4}{q}$ **66.** $\frac{4z-3}{3z}$ **67.** $\frac{3r+28}{7r}$ **68.** $\frac{p-7}{p-5}$ **69.** $\frac{9}{2(m+3)}$ **70.** $\frac{7}{15(q-4)}$ **71.** $\frac{3a-3}{a(a-2)(a+1)}$ or $\frac{3(a-1)}{a(a-2)(a+1)}$ **72.** $\frac{-k-5}{k(k+1)(k-1)}$ **73.** $\frac{1}{3(m-1)}$ **74.** $\frac{7(2z+1)}{24}$ **75.** $\frac{x+3}{x+1}$ **76.** $\frac{5}{x+1}$ **77.** $\frac{z^4}{3y^5}$ **78.** $\frac{195}{29}$ **79.** $\frac{21}{2}$ **80.** $-2, 1$ **81.** $\frac{2}{5}, 3$ **82.** $\frac{6}{5}$ or $1\frac{1}{5}$ hours **83.** $\frac{18}{5}$ or $3\frac{3}{5}$ hours **84.** 3 days **85.** 4 hours

CHAPTER 7

Section 7.1 (page 381)

1. yes **3.** yes **5.** no **7.** yes **9.** yes **11.** no **13.** (2, 11) **15.** (8, 29) **17.** (−3, −4) **19.** (3, 14) **21.** (1, 8) **23.** (2, 0) **25.** (−4, 24) **27.** (4, −8) **29.** (2, 1), (0, −5), (−3, −14) **31.** (4, −22), (0, −2), (−4, 18)

33.

m	1	0	−2
n	7	4	−2

35.

t	0	3	−3
w	4	0	8

37.

x	9	0	18
y	0	−4	4

39.

x	2	2	2
y	3	8	0

41.

x	4	0	−4
y	−8	−8	−8

43.

x	9	2	0
y	−4	−4	−4

45. **47.**

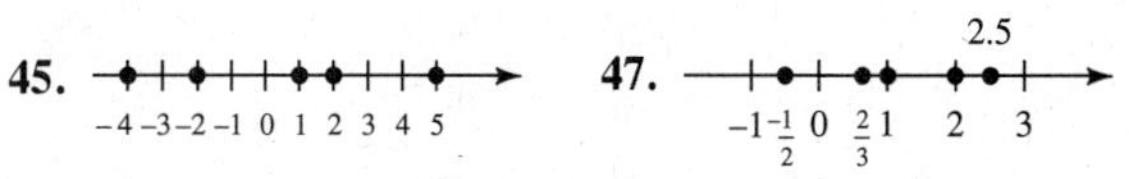

Section 7.2 (page 387)

1. (2, 5) **3.** (−5, 5) **5.** (7, 3)

7.–17.

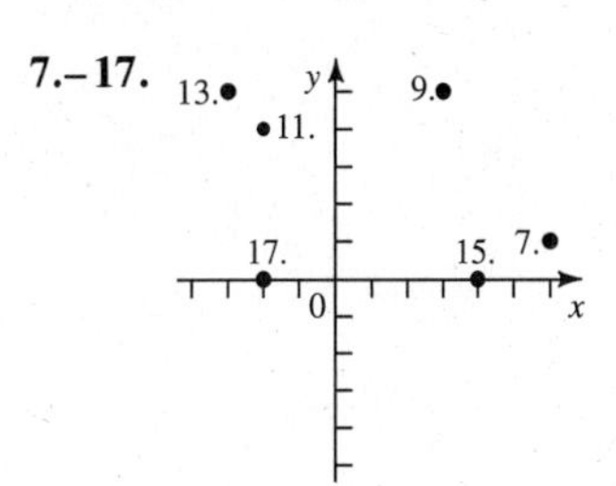

19. I **21.** II **23.** III **25.** II **27.** IV

29.

x	y
0	3
6	0
4	1
−1	$\frac{7}{2}$

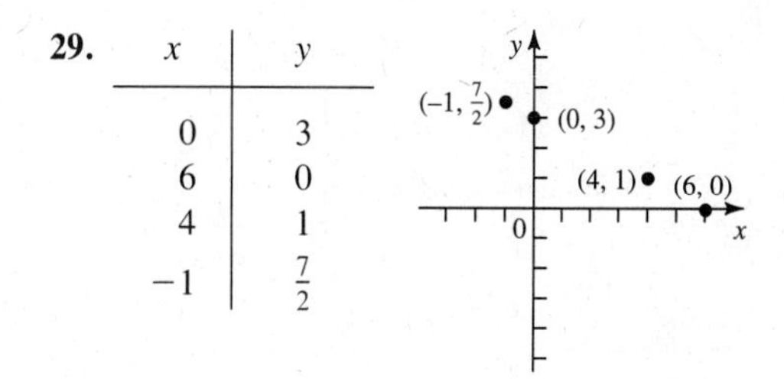

31.

x	y
0	−4
3	0
6	4
$1\frac{1}{2}$	−2

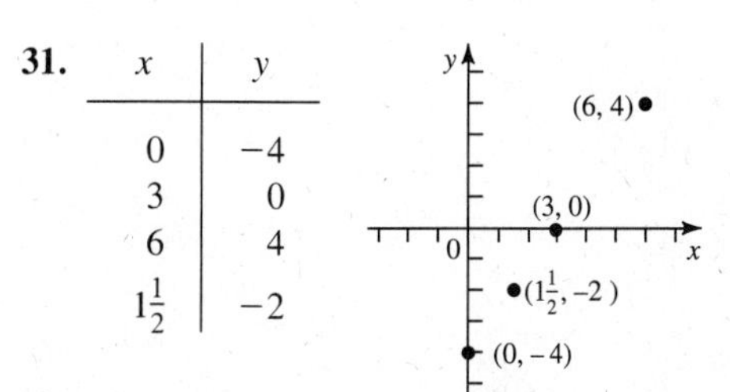

33.

x	y
5	−2
0	−2
−3	−2
−2	−2

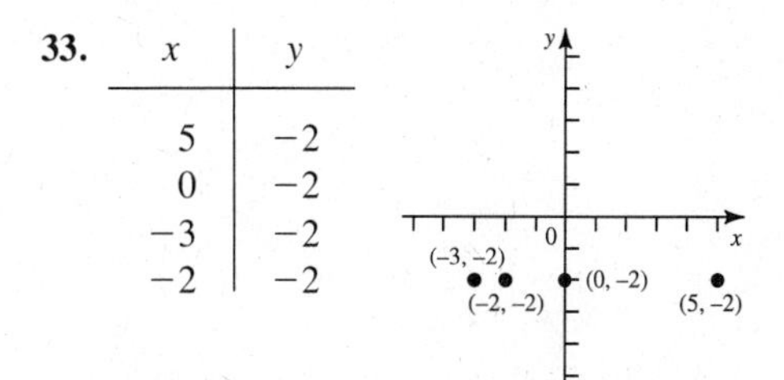

35. **(a)**

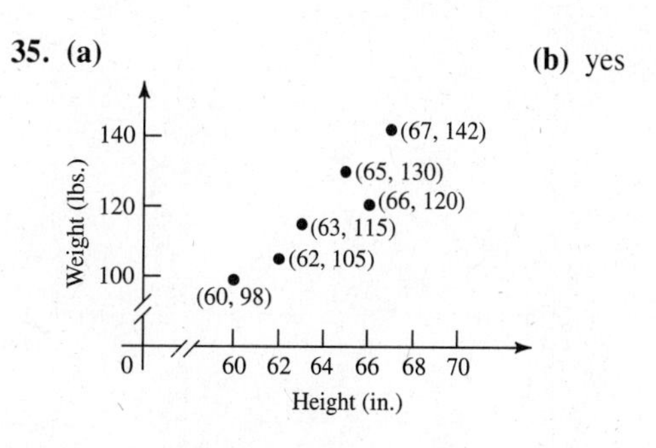

(b) yes

37. 8 **39.** −10 **41.** −2

Section 7.3 (page 395)

1. (0, 5), (5, 0), (2, 3)

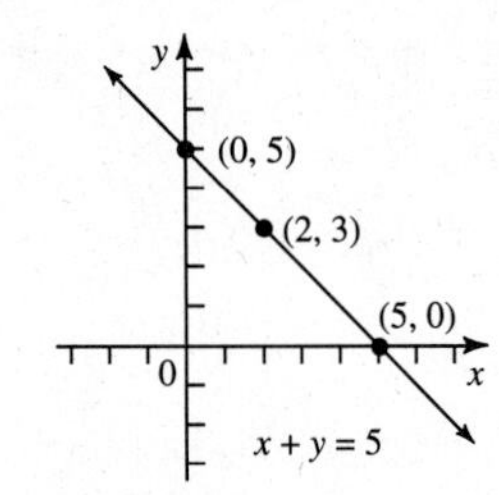

3. (0, 4), (−4, 0), (−2, 2)

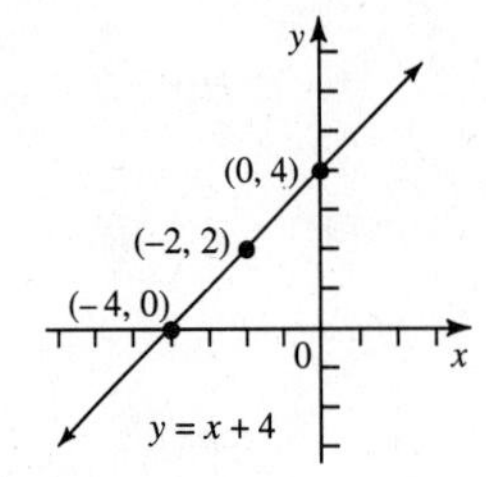

5. (0, −6), (2, 0), (3, 3)

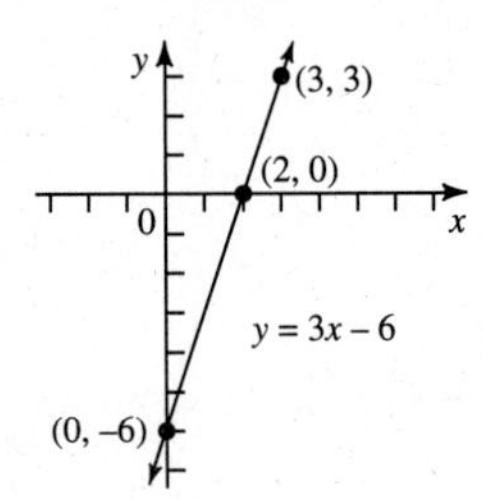

7. (3, 0), (0, 2)

9. (3, 0), $\left(0, -\frac{9}{5}\right)$ **11.** (0, 0), (0, 0)

13.

15.

17.

19.

21.

23.

25.

27.

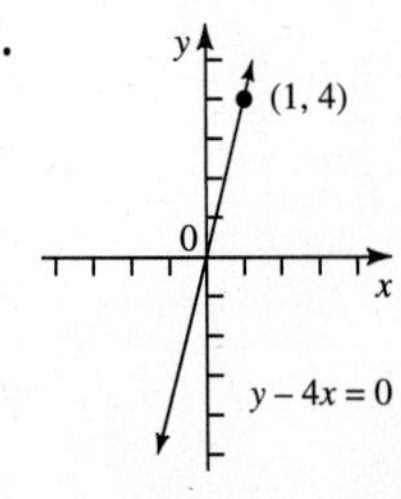

29.

31.

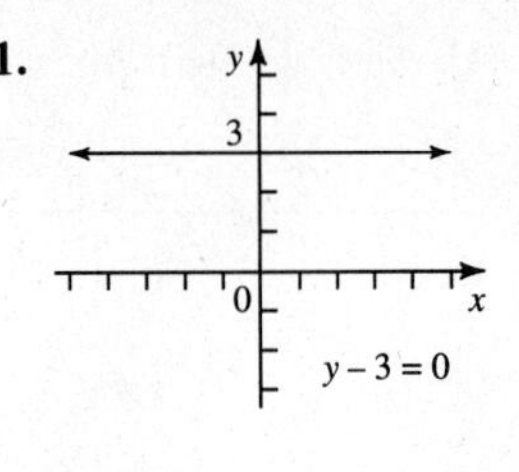

33.

35. (a)

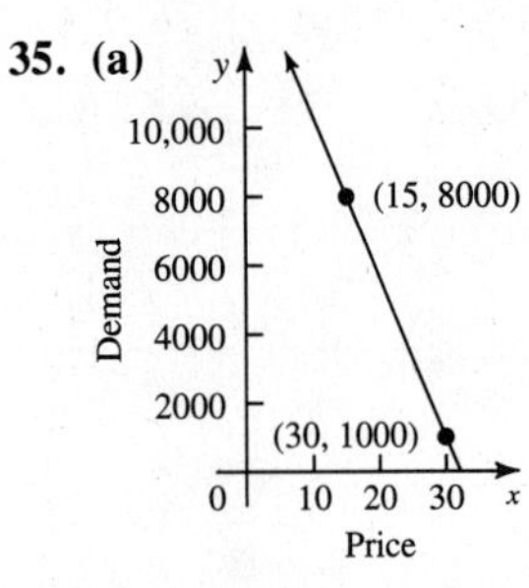

(b) 10,000
(c) about $23

37. $\frac{1}{2}$ **39.** −1 **41.** undefined

Section 7.4 (page 405)

1. $-\frac{1}{2}$ **3.** $\frac{2}{7}$ **5.** 0 **7.** $\frac{7}{6}$ **9.** $-\frac{5}{8}$ **11.** 0

13. $-\frac{1}{2}$ **15.** undefined **17.** 5 **19.** 1

21. 9 **23.** $\frac{3}{2}$ **25.** $\frac{2}{3}$ **27.** 0

29. −5; $\frac{1}{5}$; perpendicular **31.** −1; 1; perpendicular

33. $\frac{2}{5}$; $\frac{2}{5}$; parallel **35.** $\frac{3}{5}$; $-\frac{5}{3}$; perpendicular

37. $\frac{1}{4}$; $-\frac{1}{2}$; neither **39.** $\frac{5}{3}$; $\frac{3}{5}$; neither **41.** $\frac{1}{6}$

43. $\frac{6}{7}$ **45.** $y = 2x - 11$ **47.** $y = -x - 5$

49. $y = -3x - 1$

Section 7.5 (page 413)

1. $y = 3x + 5$ **3.** $y = -x - 6$ **5.** $y = \frac{5}{3}x + \frac{1}{2}$

7. $y = 8x$

9.

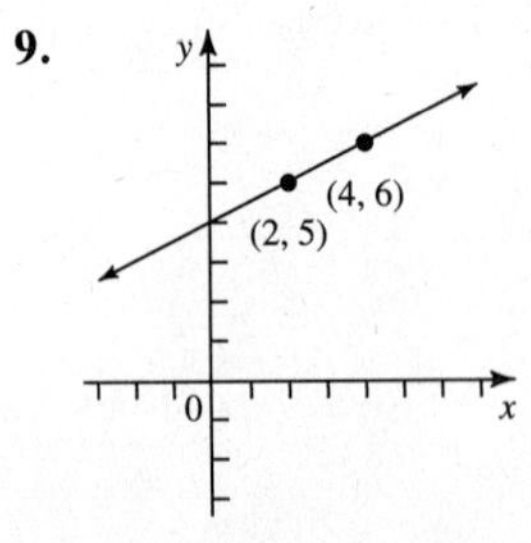

11.

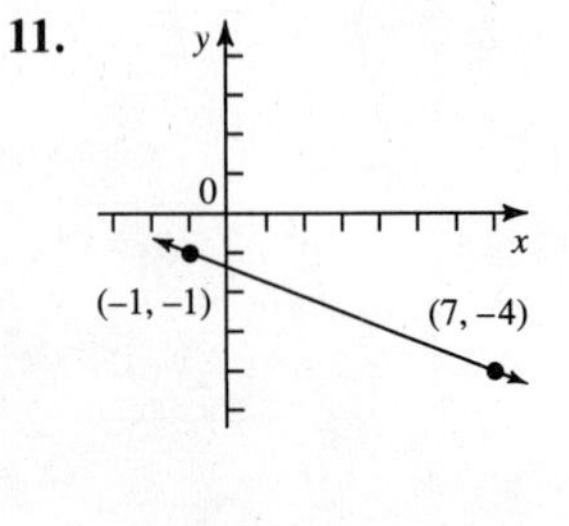

13.

15. 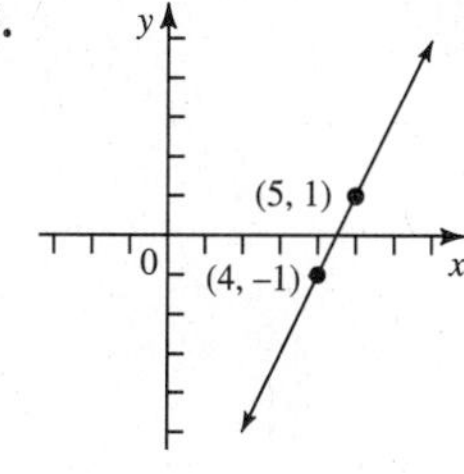

17. 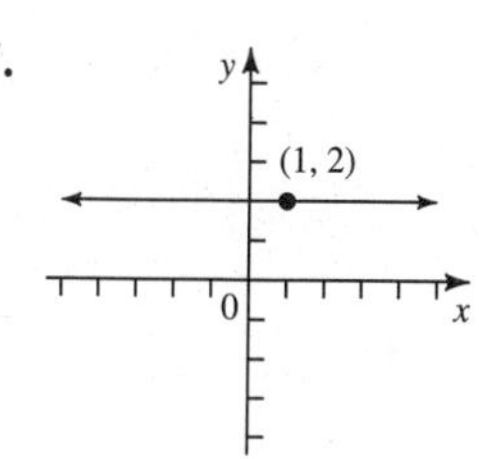

19. 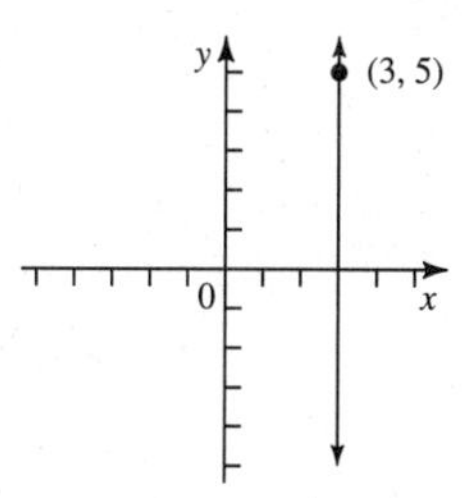

21. $2x - y = 7$ **23.** $2x + y = -4$
25. $2x - 3y = -9$ **27.** $3x + 4y = -17$
29. $8x + 11y = 48$ **31.** $3x - 5y = -11$
33. $3x + 5y = -11$ **35.** $2x + 3y = 6$
37. $2x + y = -1$ **39.** $x - 3y = -4$
41. $88x - 72y = -133$ **43.** $y = 12x + 100$
45. **(a)** 700 **(b)** 1600
47. **(a)** (1, 18); (4, 27) **(b)** $3x - y = -15$

49.

51. 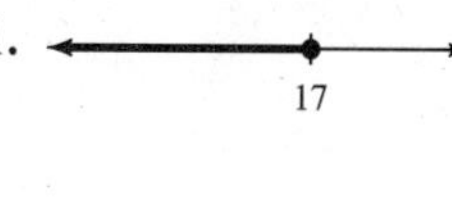

53.

Section 7.6 (page 421)

1. 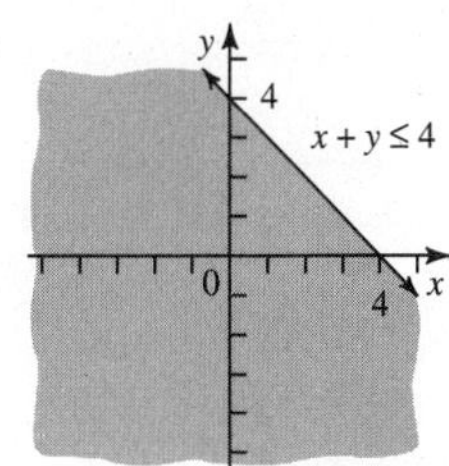

3. 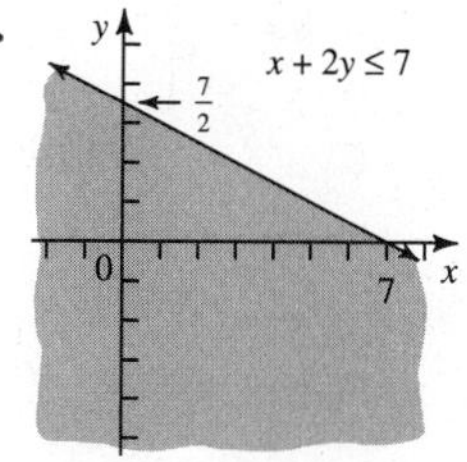

5. 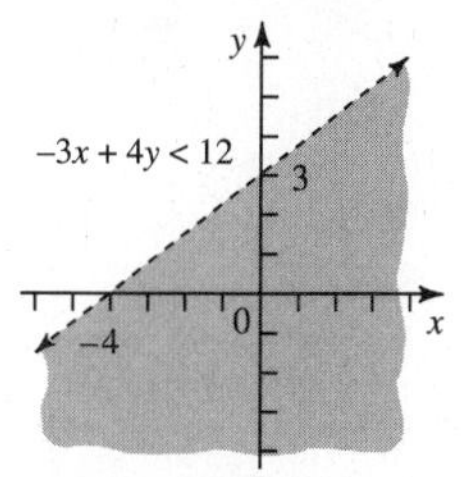

7. 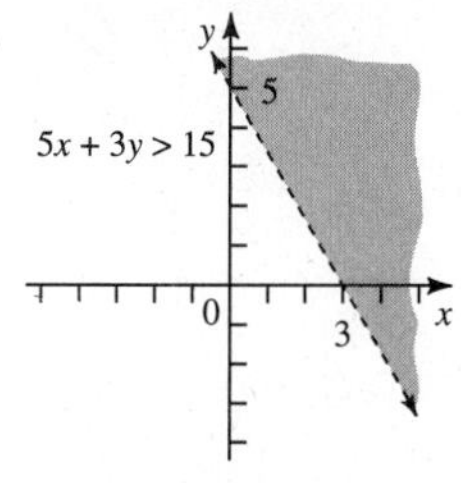

9. 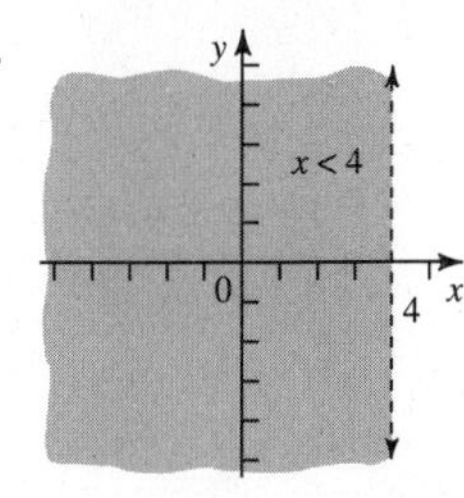

11. 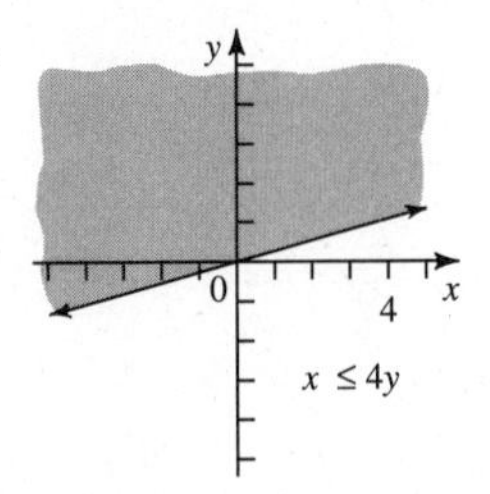

13. 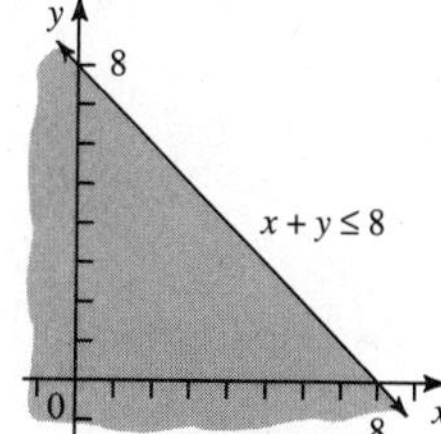

15. 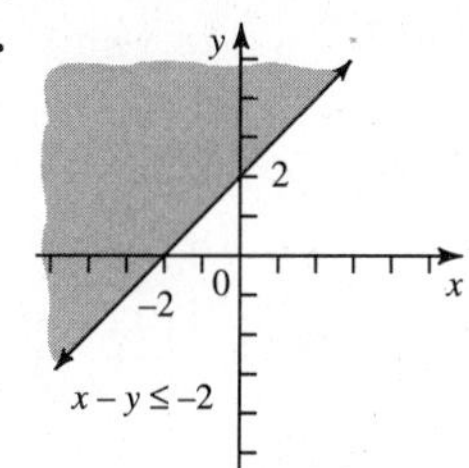

17. 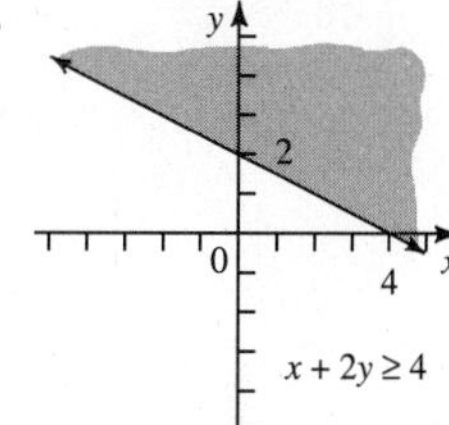

19. 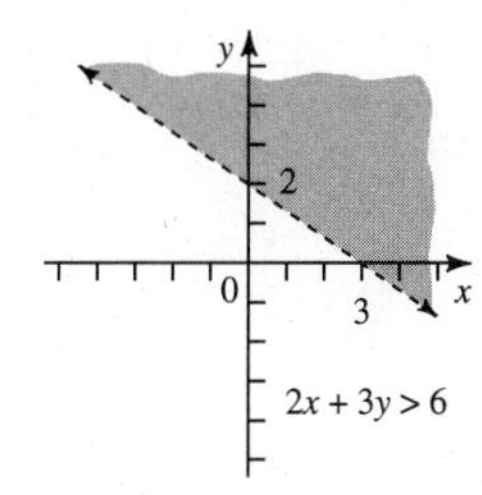

21. 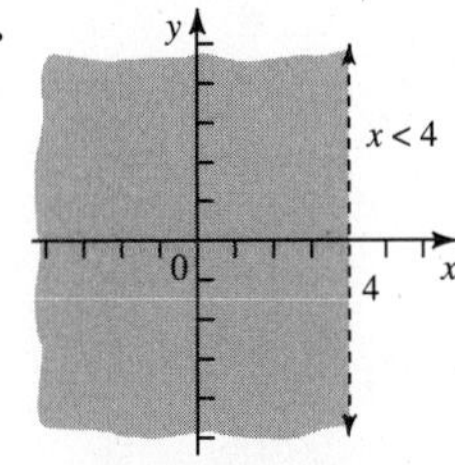

23. 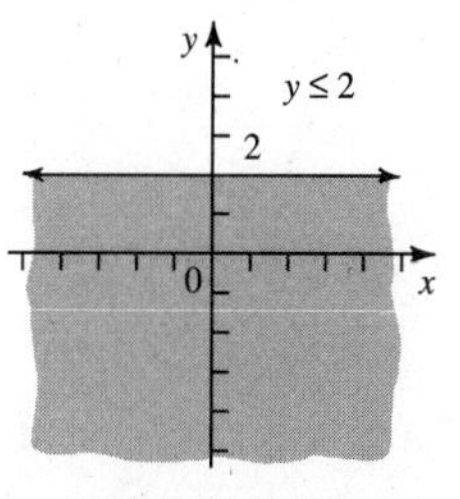

25. 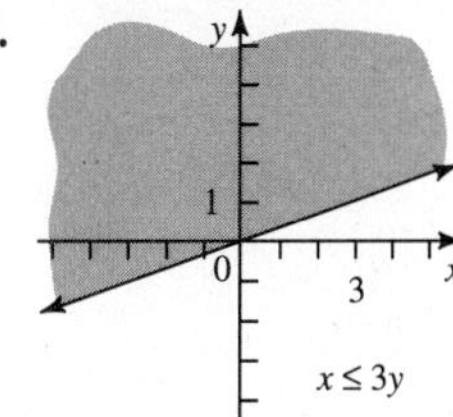

27. 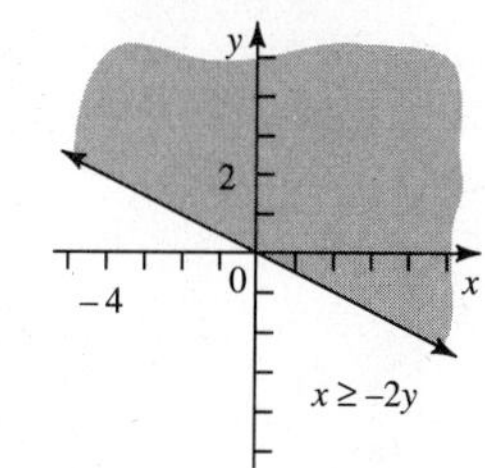

29. **(a)** 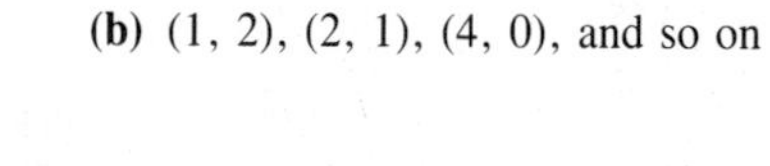
(b) (1, 2), (2, 1), (4, 0), and so on

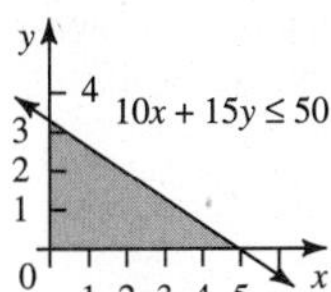

31. $17x$ **33.** $-q$

Chapter 7 Review Exercises (page 429)

1. $(-1, -5)$, $(0, -2)$, $\left(\frac{7}{3}, 5\right)$
2. $(0, 3)$, $\left(\frac{9}{4}, 0\right)$, $\left(-2, \frac{17}{3}\right)$
3. (0, 0), (8, 4), $(-6, -3)$
4. $(-4, -3)$, $(-4, 0)$, $(-4, 5)$ **5.** yes **6.** no
7. yes **8.** yes

9.–14.

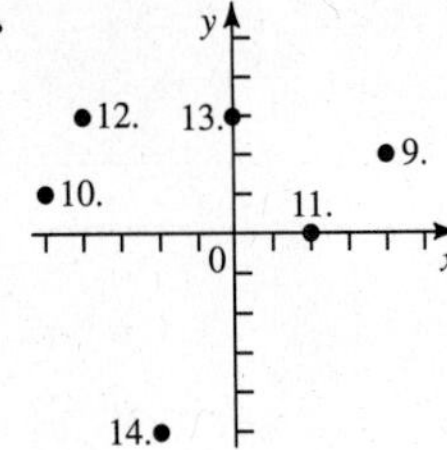

15. II **16.** III **17.** none **18.** none

19. $\left(\frac{5}{2}, 0\right)$, $(0, -5)$ **20.** $\left(\frac{7}{2}, 0\right)$, $(0, 7)$

21. $\left(\frac{8}{3}, 0\right)$, $(0, -4)$

22.

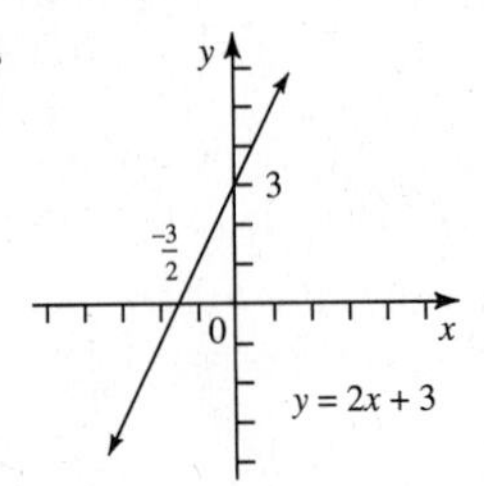

23.

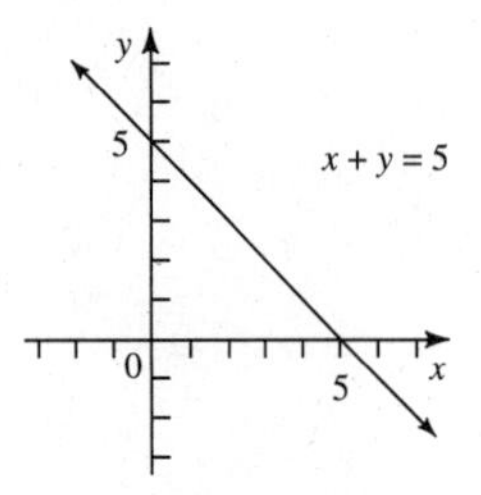

24.

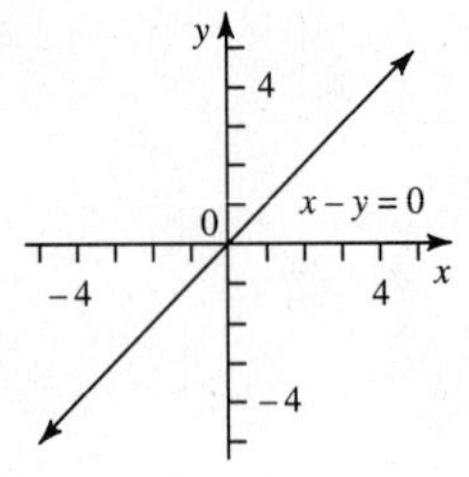

25. $\frac{2}{3}$ **26.** 2 **27.** 0 **28.** undefined **29.** 3

30. $\frac{2}{3}$ **31.** 0 **32.** undefined **33.** $-\frac{6}{7}$

34. parallel **35.** perpendicular **36.** perpendicular
37. neither **38.** $3x - y = 2$ **39.** $4x + 4y = 3$
40. $2x - 3y = -15$ **41.** $x - y = 7$
42. $2x - 3y = -14$ **43.** $3x + 4y = -1$

44. $x + 4y = 6$ **45.** $y = 6$ **46.** $x = \frac{1}{2}$

47.

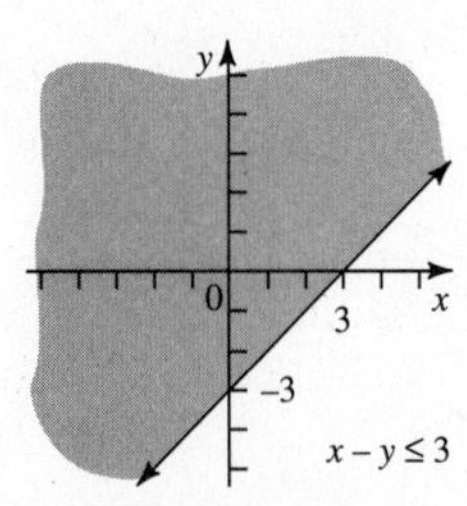

48.

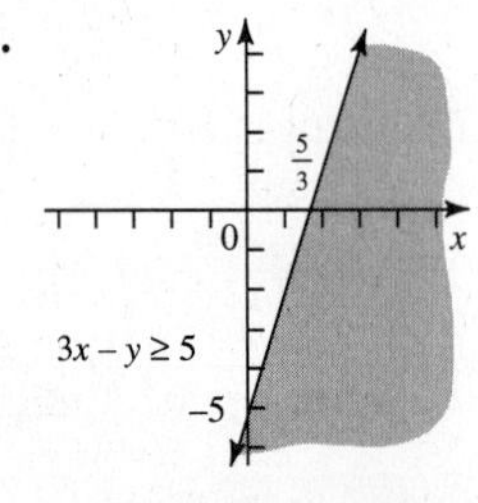

49.

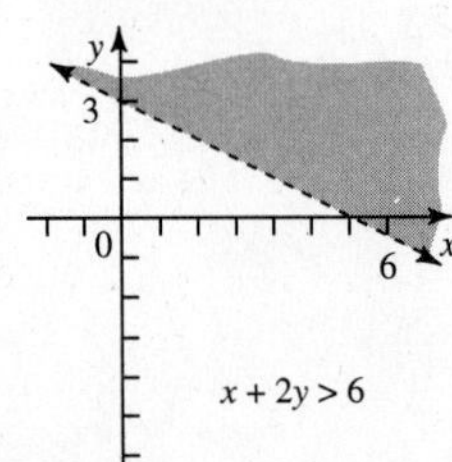

50.

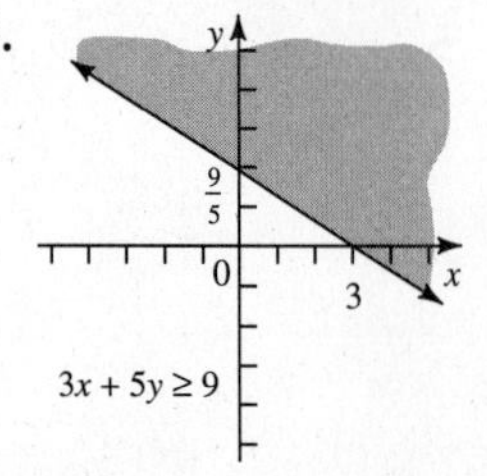

51.

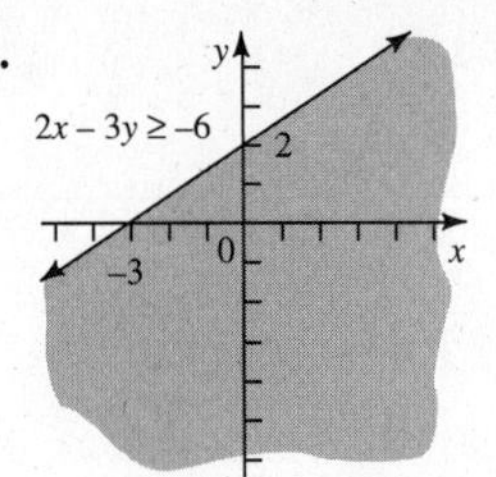

52.

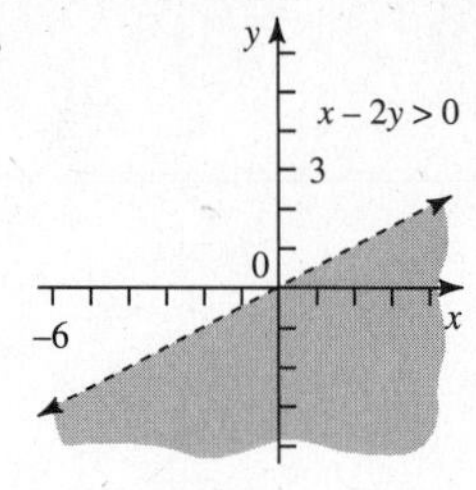

53. $m = \frac{11}{3}$; x-intercept: $\left(\frac{4}{11}, 0\right)$; y-intercept: $\left(0, -\frac{4}{3}\right)$

54. $m = \frac{1}{3}$; x-intercept: $(7, 0)$; y-intercept: $\left(0, -\frac{7}{3}\right)$

55. $m = -\frac{4}{9}$; x-intercept: $\left(-\frac{9}{4}, 0\right)$; y-intercept: $(0, -1)$

56. $m = -\frac{8}{3}$; x-intercept: $\left(\frac{3}{4}, 0\right)$; y-intercept: $(0, 2)$

57. $x = 5$ **58.** $x + 4y = -5$ **59.** $3x + y = 30$
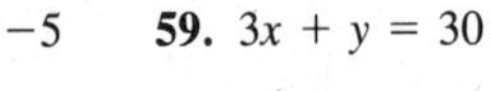
60. $4x + 7y = -23$

61.

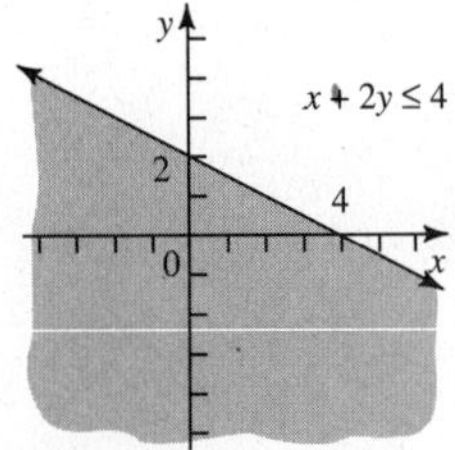

62.

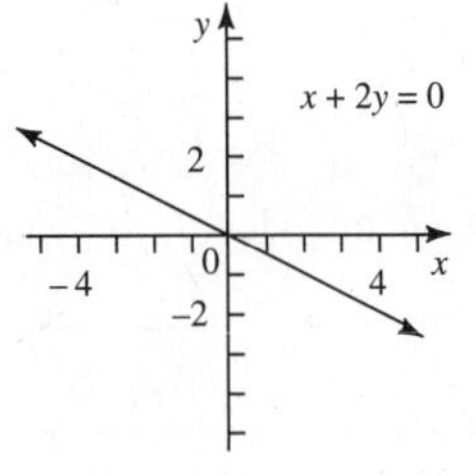

63.

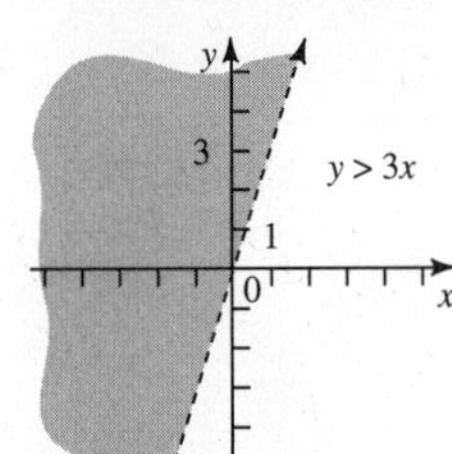

64.

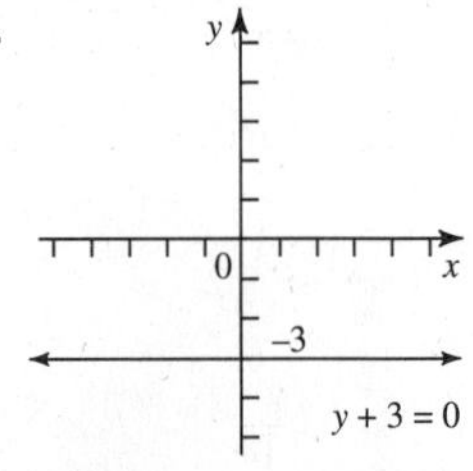

65.

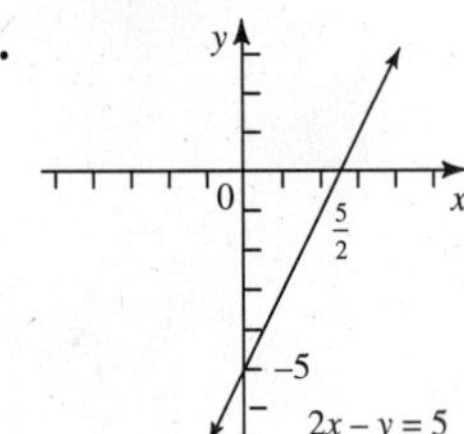

66.

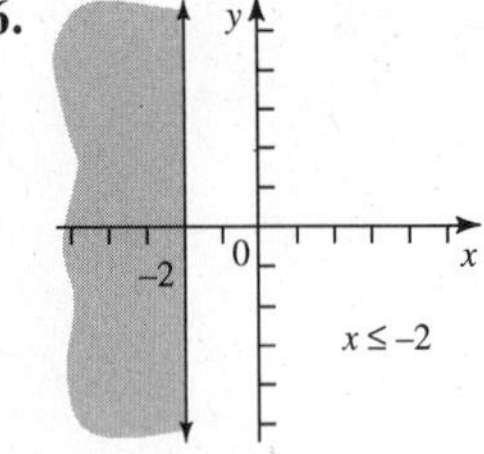

Chapter 7 Test (page 433)

[7.1] **1.** $(0, -6)$, $(-2, -16)$, $(4, 14)$
2. $(0, -6)$, $(10, 0)$, $(5, -3)$

3. $(0, 0)$, $(6, 2)$, $\left(8, \frac{8}{3}\right)$, $(-12, -4)$

4. $(5, 2)$, $(4, 2)$, $(0, 2)$, $(-3, 2)$

6. (3, 0), (0, 6)

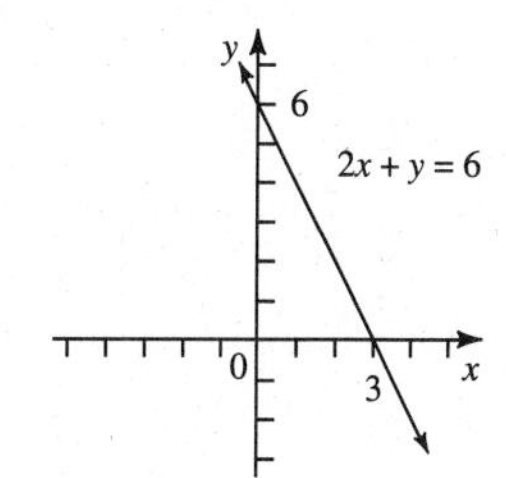

7. (6, 0), $\left(0, -\frac{9}{2}\right)$

8. (0, 0), (0, 0)

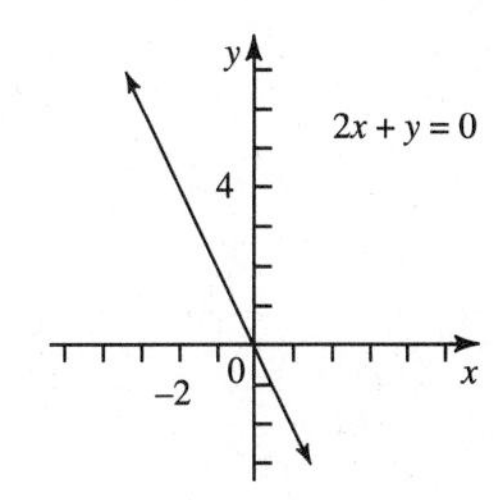

9. (−5, 0); no y-intercept

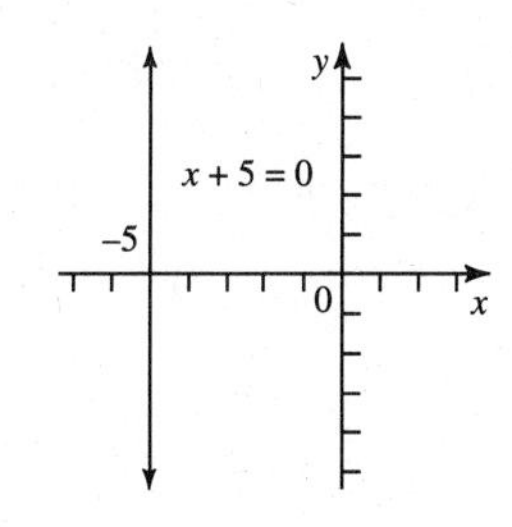

10. no x-intercept; (0, 2)

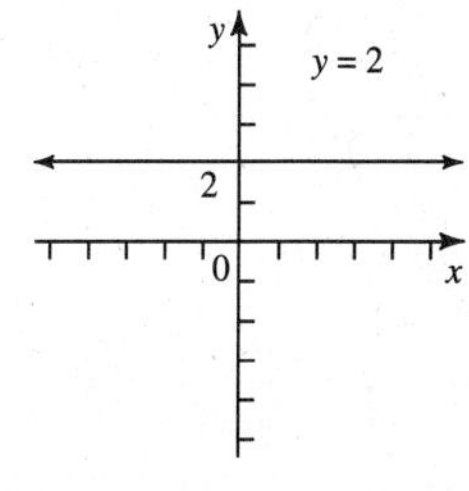

[7.4] 11. $-\frac{3}{7}$ 12. $-\frac{3}{4}$ [7.5] 13. $4x + y = 1$
14. $3x - y = 1$ 15. $3x + 4y = 8$
16. $9x - y = -12$

[7.6] 17.

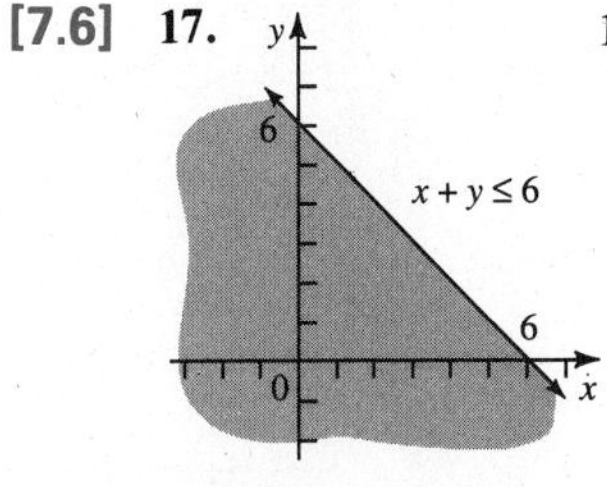

18.

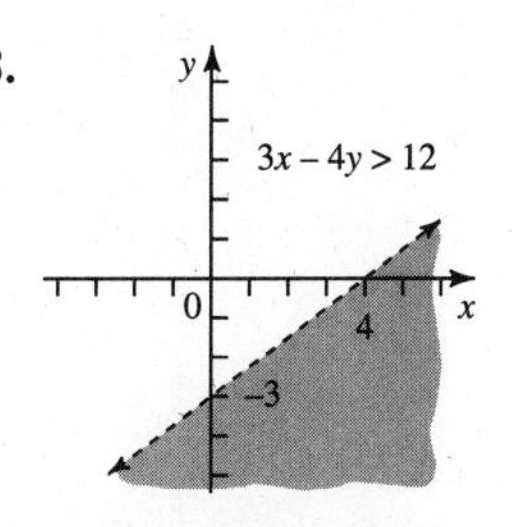

19.

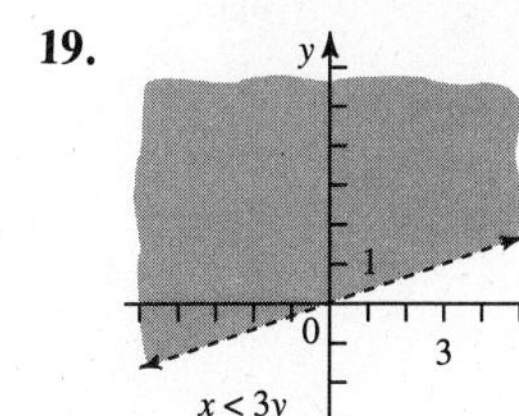

20.

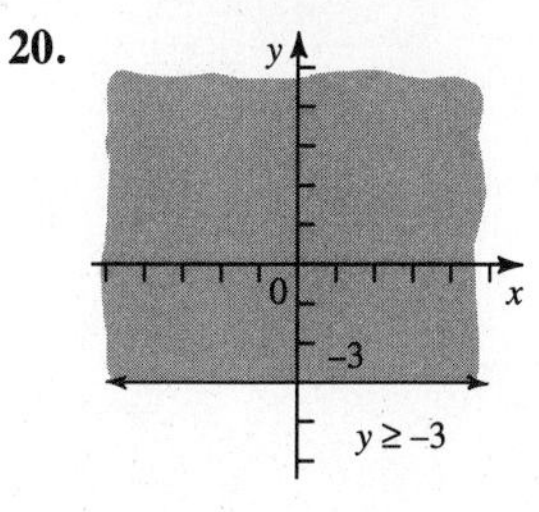

CHAPTER 8

Section 8.1 (page 441)

1. yes 3. no 5. no 7. yes 9. no
11. yes 13. (4, 2) 15. (0, 4) 17. (4, −1)
19. (1, 3) 21. (0, 2) 23. (4, 0) 25. (−4, 1)
27. (−4, 3) 29. (−4, 2) 31. (0, −2)
33. (1, 2) 35. no solution 37. no solution
39. same line 41. $10p + 7q$ 43. $8r$ 45. $2a$
47. 0

Section 8.2 (page 449)

1. (1, −2) 3. (2, 0) 5. (6, 2) 7. (−2, 6)
9. $\left(\frac{1}{2}, 2\right)$ 11. $\left(\frac{3}{2}, -2\right)$ 13. (2, −3)
15. (4, 4) 17. (2, −5) 19. (3, 0) 21. (4, 9)
23. (−4, 0) 25. (4, −3) 27. (−9, −11)
29. (6, 3) 31. (6, −5) 33. (−6, 0)
35. $\left(\frac{1}{2}, 3\right)$ 37. $\left(\frac{3}{8}, \frac{5}{6}\right)$ 39. (11, 15)
41. $\left(\frac{22}{9}, \frac{8}{9}\right)$ 43. (2, −3) 45. no solution
47. all real numbers 49. all real numbers

Section 8.3 (page 455)

1. no solution 3. no solution 5. same line
7. no solution 9. no solution 11. same line
13. (0, 0) 15. (3, 2) 17. no solution
19. same line 21. 2 23. 3 25. 1 27. −7
29. $-\frac{31}{26}$

Section 8.4 (page 461)

1. (2, 4) 3. (6, 4) 5. (8, −1) 7. (2, −4)
9. (5, 1) 11. no solution 13. (4, 1)
15. (2, −3) 17. same line 19. (5, −3)
21. (2, 8) 23. (4, −6) 25. (3, 2) 27. (7, 0)
29. same line 31. (6, 5) 33. (0, 3)
35. (18, −12) 37. (3, 2) 39. 21 41. 8 feet

Section 8.5 (page 471)

1. 46 and 23 3. smaller: 40°; larger: 140°
5. 12 centimeters; 12 centimeters; 18 centimeters
7. 53 ones; 28 fives
9. 83 student tickets; 110 non-student tickets
11. \$2500 at 8%; \$5000 at 10%
13. 21 large; 18 small
15. 40 liters of 40% solution; 20 liters of 70% solution
17. 30 pounds of 60% copper; 10 pounds of 80% copper
19. 60 pounds at \$1.20 per pound; 30 pounds at \$1.80 per pound

7. 53 ones; 28 fives
9. 83 student tickets; 110 non-student tickets
11. \$2500 at 8%; \$5000 at 10%
13. 21 large; 18 small
15. 40 liters of 40% solution; 20 liters of 70% solution
17. 30 pounds of 60% copper; 10 pounds of 80% copper
19. 60 pounds at \$1.20 per pound; 30 pounds at \$1.80 per pound
21. 50 bags at \$70 per bag; 30 bags at \$90 per bag
23. boat: 5 miles per hour; current: 3 miles per hour
25. plane: 80 miles per hour; wind: 20 miles per hour
27. faster car: 65 miles per hour; slower car: 35 miles per hour
29. Anderson: 80 miles per hour; Bentley: 40 miles per hour

31.
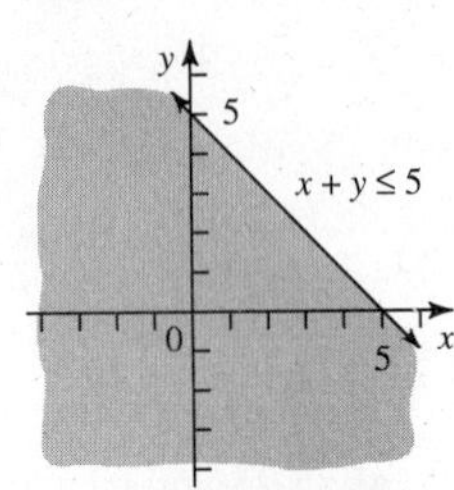

33.
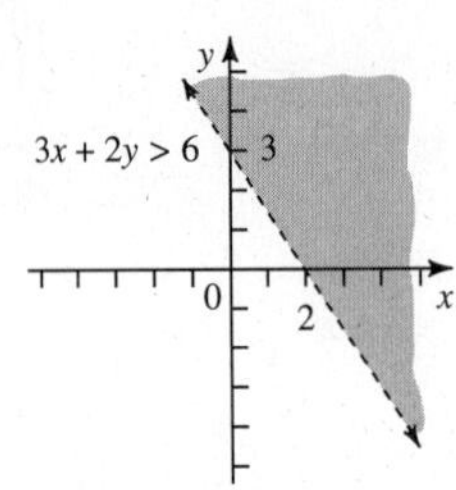

35.
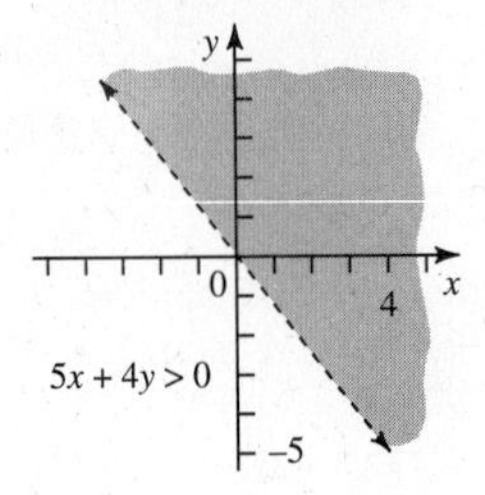

Section 8.6 (page 479)

1.
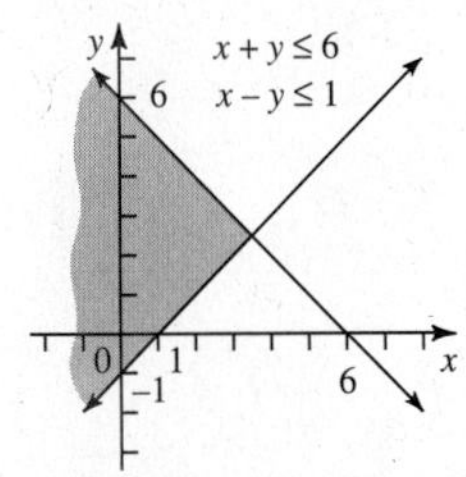

3.
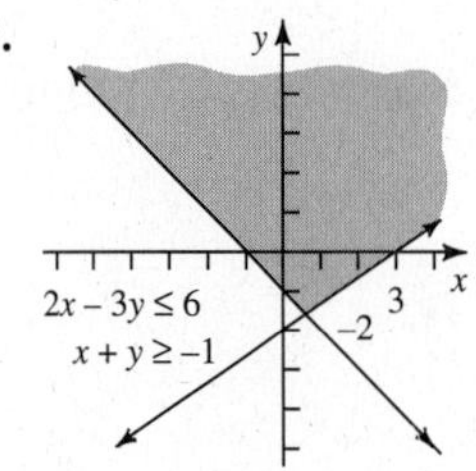

5.
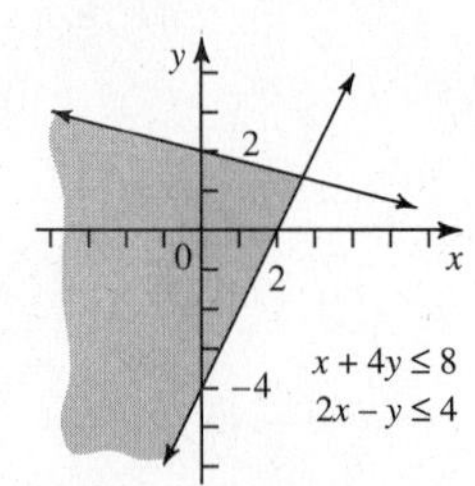

7.
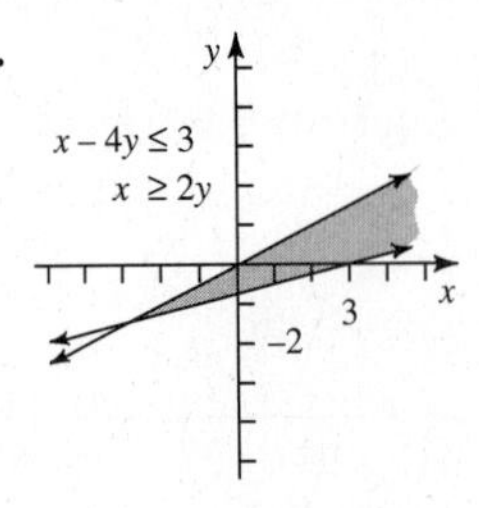

9.
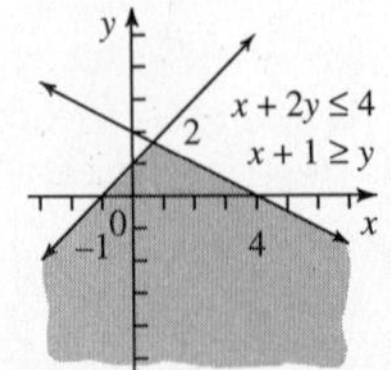

11.
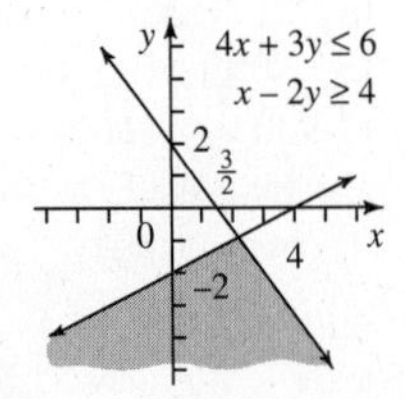

13.
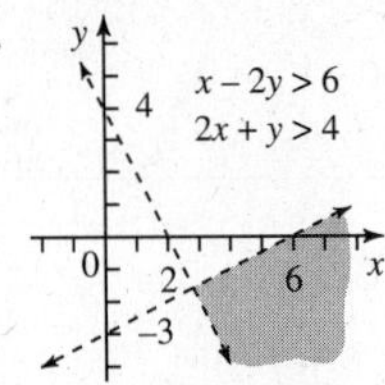

15.
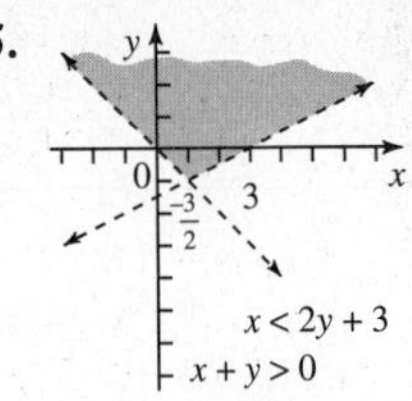

17.
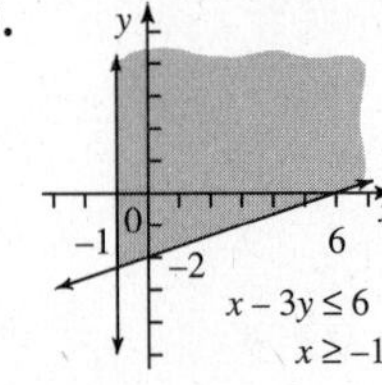

19. 25 21. 121 23. 400

Chapter 8 Review Exercises (page 485)

1. no 2. yes 3. no 4. yes 5. (2, 1)
6. (−2, 2) 7. (−1, 1) 8. same line
9. (−1, −2) 10. no solution 11. (2, 4)
12. (5, 2) 13. (3, −2) 14. (1, −1)
15. (4, −1) 16. (3, 2) 17. same line
18. no solution 19. (4, 2) 20. (−5, −3)
21. (5, 3) 22. (−1, −1) 23. (4, −1)
24. (4, −3) 25. (9, 2) 26. $\left(\frac{10}{7}, -\frac{9}{7}\right)$
27. (−6, 1) 28. (2, 1) 29. 28, 14 30. 5, 12
31. length: 12 meters; width: 8 meters
32. 5 twenties; 15 tens
33. \$7000 at 12%; \$11,000 at 16%
34. 12.5 pounds of \$1.30 candy; 37.5 pounds of \$.90 candy
35. 40 liters of 40% solution; 20 liters of 70% solution
36. plane: 250 miles per hour; wind: 20 miles per hour

37. $x + y \le 3$, $2x \ge y$

38.
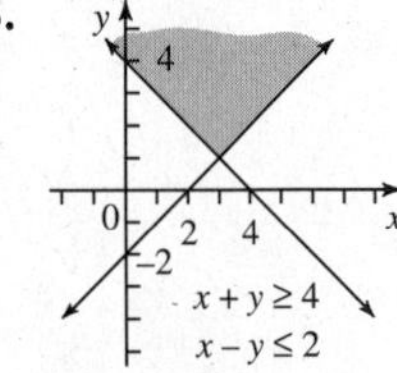

39.
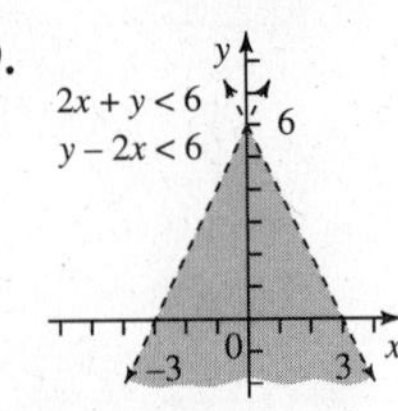

40.
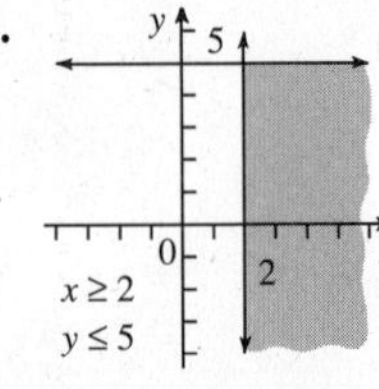

41.
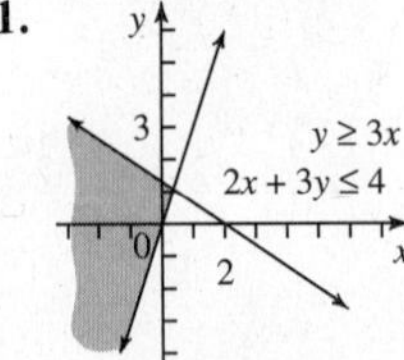

42.
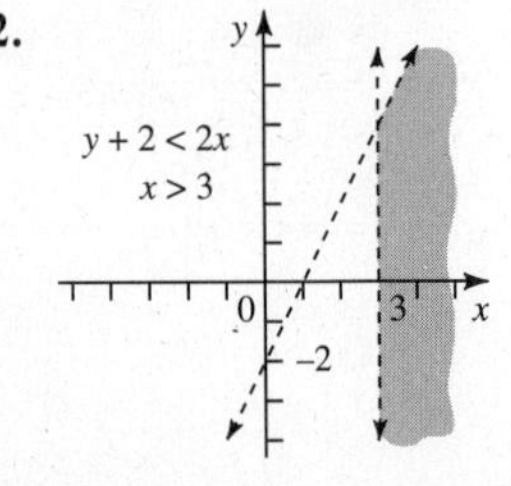

43. (4, 8) **44.** same line
45. 17 inches; 17 inches; 24 inches **46.** (0, 2)
47. (−2, 0)

48.

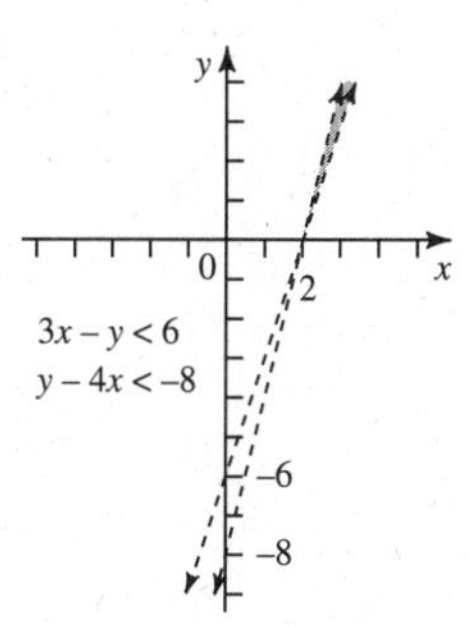

49.

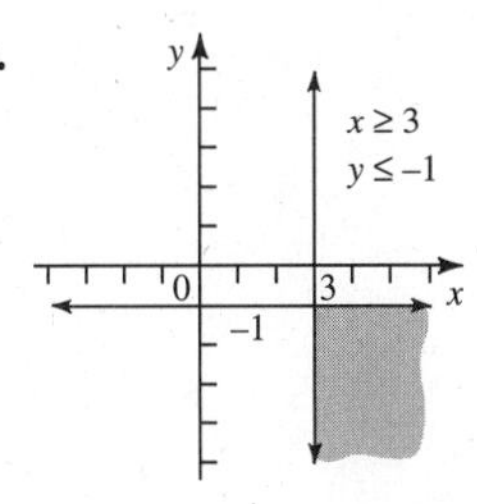

50. no solution **51.** same line **52.** −7 and −5

Chapter 8 Test (page 491)

[8.1] **1.** (4, −3) **2.** (6, −5) **3.** (4, 1)
[8.2–8.3] **4.** (6, 5) **5.** (8, −2) **6.** (2, −5)
7. no solution **8.** (2, 2) **9.** $\left(-\frac{11}{14}, -\frac{8}{7}\right)$
[8.4] **10.** (3, −5) **11.** (−6, 8)
[8.2–8.4] **12.** same line **13.** (−6, 4)
[8.5] **14.** 12 and 27
15. 2 cassettes at \$2.50; 4 cassettes at \$3.75
16. 75 liters of 40% solution; 25 liters of 60% solution
17. 45 miles per hour; 60 miles per hour
[8.6] **18.** **19.**

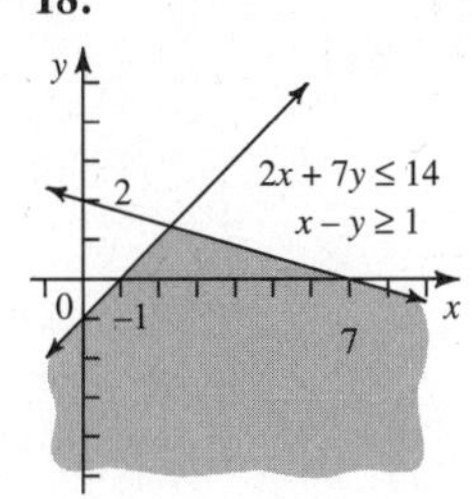

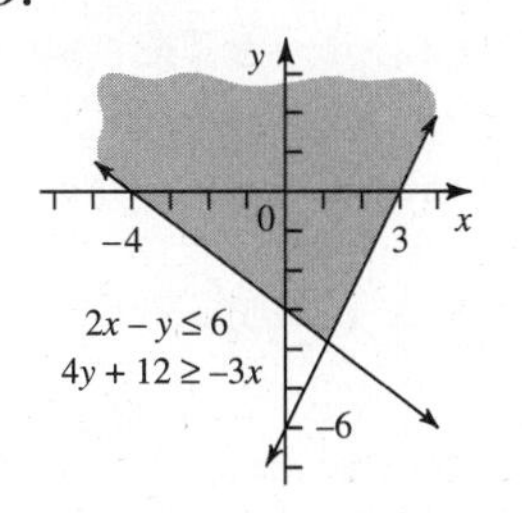

20.

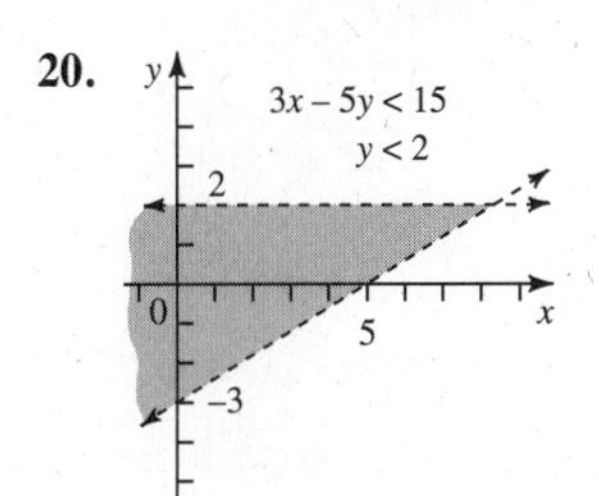

CHAPTER 9

Section 9.1 (page 501)

1. 3, −3 **3.** 11, −11 **5.** $\frac{20}{9}, -\frac{20}{9}$
7. 25, −25 **9.** 16 **11.** 57 **13.** $4x^2 + 3$
15. 5 **17.** 8 **19.** −13 **21.** 30 **23.** −41
25. $\frac{6}{7}$ **27.** $-\frac{10}{13}$ **29.** not a real number
31. rational; 4 **33.** irrational; 3.873
35. irrational; 6.856 **37.** rational; −11
39. not a real number **41.** rational; 20
43. irrational; −14.142 **45.** not a real number
47. $c = 10$ **49.** $b = 15$
51. $c = \sqrt{164} \approx 12.806$ **53.** $a = \sqrt{95} \approx 9.747$
55. 7 inches **57.** 60 feet **59.** 9 feet
61. 8.485 inches **63.** $2^3 \cdot 3$ **65.** $2^2 \cdot 3 \cdot 5^2$
67. $2^3 \cdot 3^2$ **69.** $2 \cdot 3 \cdot 5^2$

Section 9.2 (page 509)

1. 4 **3.** 6 **5.** 21 **7.** $\sqrt{21}$ **9.** $3\sqrt{3}$
11. $2\sqrt{7}$ **13.** $3\sqrt{2}$ **15.** $4\sqrt{3}$ **17.** $5\sqrt{5}$
19. $10\sqrt{7}$ **21.** $30\sqrt{3}$ **23.** $4\sqrt{5}$ **25.** 36
27. 60 **29.** $7\sqrt{3}$ **31.** $15\sqrt{3}$ **33.** $20\sqrt{3}$
35. $10\sqrt{10}$ **37.** $\frac{10}{3}$ **39.** $\frac{6}{7}$ **41.** $\frac{\sqrt{5}}{4}$
43. $\frac{\sqrt{30}}{7}$ **45.** $\frac{2}{5}$ **47.** $\frac{4}{25}$ **49.** 5 **51.** 4
53. $3\sqrt{5}$ **55.** $5\sqrt{10}$ **57.** y **59.** $\sqrt{xz}$
61. x **63.** x^2 **65.** xy^2 **67.** $x\sqrt{x}$ **69.** $\frac{4}{x}$
71. $\frac{\sqrt{11}}{r^2}$ **73.** $\frac{27}{39}$ **75.** $\frac{10}{15}$ **77.** $\frac{6}{3x}$
79. $\frac{10}{2\sqrt{3}}$

Section 9.3 (page 515)

1. $7\sqrt{3}$ **3.** $-5\sqrt{7}$ **5.** $2\sqrt{6}$ **7.** $3\sqrt{17}$
9. $4\sqrt{7}$ **11.** $7\sqrt{5}$ **13.** $3\sqrt{3}$ **15.** $20\sqrt{2}$
17. $-13\sqrt{2}$ **19.** $19\sqrt{7}$ **21.** $-12\sqrt{5}$
23. $-16\sqrt{2} + 8\sqrt{3}$ **25.** $20\sqrt{2} + 6\sqrt{3} + 15\sqrt{5}$
27. $2\sqrt{2}$ **29.** $3\sqrt{3} - 2\sqrt{5}$ **31.** $3\sqrt{21}$
33. $5\sqrt{3}$ **35.** $6\sqrt{x}$ **37.** $13\sqrt{a}$ **39.** $15x\sqrt{3}$
41. $2x\sqrt{2}$ **43.** $13p\sqrt{3}$ **45.** $42x\sqrt{5z}$
47. $m^2 + 8m + 15$ **49.** $6k^2 - 17k + 12$
51. $9x^2 - 6x + 1$ **53.** $a^2 - 9$ **55.** $16k^2 - 25$

Section 9.4 (page 521)

1. $\frac{6\sqrt{5}}{5}$ **3.** $\sqrt{5}$ **5.** $\frac{3\sqrt{7}}{7}$ **7.** $\frac{8\sqrt{15}}{5}$
9. $\frac{\sqrt{30}}{2}$ **11.** $\frac{8\sqrt{3}}{9}$ **13.** $\frac{3\sqrt{2}}{10}$ **15.** $\sqrt{2}$

17. $\frac{9\sqrt{2}}{8}$ 19. 2 21. $\sqrt{2}$ 23. $\frac{2\sqrt{30}}{3}$
25. $\frac{\sqrt{2}}{2}$ 27. $\frac{\sqrt{70}}{7}$ 29. $\frac{3\sqrt{5}}{5}$ 31. $\frac{\sqrt{21}}{3}$
33. $\frac{\sqrt{15}}{10}$ 35. $\frac{3\sqrt{14}}{4}$ 37. $\frac{\sqrt{3}}{5}$ 39. $\frac{5\sqrt{3}}{12}$
41. $\frac{\sqrt{6p}}{p}$ 43. $\frac{p\sqrt{2pm}}{m}$ 45. $\frac{x\sqrt{y}}{2y}$ 47. $\frac{3a\sqrt{5r}}{5}$
49. $4x$ 51. $-2m - 3$ 53. $3y + 8z$

Section 9.5 (page 527)

1. $27\sqrt{5}$ 3. $21\sqrt{2}$ 5. -4 7. $\sqrt{15} + \sqrt{35}$
9. $2\sqrt{10} + 10$ 11. $-4\sqrt{7}$ 13. $21 - \sqrt{6}$
15. $87 + 9\sqrt{21}$ 17. $34 + 24\sqrt{2}$
19. $37 - 12\sqrt{7}$ 21. 7 23. -4 25. 1
27. $2\sqrt{3} - 2 + 3\sqrt{2} - \sqrt{6}$ 29. $2 + 2\sqrt{2}$
31. $\sqrt{30} + \sqrt{15} + 6\sqrt{5} + 3\sqrt{10}$ 33. $\frac{3 - \sqrt{2}}{7}$
35. $-10 + 5\sqrt{5}$ 37. $-2 - \sqrt{11}$
39. $-\sqrt{2} + 2$ 41. $\frac{-\sqrt{5} - 5}{4}$ 43. $3 - \sqrt{3}$
45. $\frac{-3 + 5\sqrt{3}}{11}$ 47. $\frac{\sqrt{6} + \sqrt{2} + 3\sqrt{3} + 3}{2}$
49. $\sqrt{7} - 2$ 51. $\frac{\sqrt{3} + 5}{4}$ 53. $\frac{6 - \sqrt{10}}{2}$
55. 1, 3 57. 1, 2

Section 9.6 (page 535)

1. 4 3. 1 5. 7 7. -7 9. -5
11. no solution 13. 49 15. no solution
17. 9 19. 16 21. 6 23. 7
25. no solution 27. 6 29. 12 31. 5
33. 0, -1 35. 5 37. 3 39. 9 41. 12
43. (a) 90 miles per hour (b) 120 miles per hour
(c) 60 miles per hour (d) 60 miles per hour
45. 5, -5 47. $\sqrt{14}$, $-\sqrt{14}$
49. $3\sqrt{2}$, $-3\sqrt{2}$ 51. $4\sqrt{5}$, $-4\sqrt{5}$

Chapter 9 Review Exercises (page 541)

1. 7, -7 2. 9, -9 3. 14, -14 4. 11, -11
5. 15, -15 6. 27, -27 7. 49, -49
8. 87, -87 9. 4 10. -6 11. -20
12. 52 13. not a real number 14. -65 15. $\frac{7}{6}$
16. $\frac{10}{9}$ 17. irrational; 3.873 18. rational; 8
19. rational; -13 20. not a real number
21. $\sqrt{10}$ 22. 6 23. $5\sqrt{3}$ 24. 8
25. $3\sqrt{3}$ 26. $4\sqrt{3}$ 27. $4\sqrt{10}$ 28. $8\sqrt{5}$
29. 70 30. 18 31. $16\sqrt{6}$ 32. $25\sqrt{10}$
33. $\frac{3}{2}$ 34. $\frac{11}{20}$ 35. $\frac{\sqrt{3}}{5}$ 36. $\frac{\sqrt{10}}{13}$ 37. $\frac{\sqrt{5}}{6}$
38. $\frac{2}{15}$ 39. $3\sqrt{5}$ 40. 8 41. 15 42. p
43. $\sqrt{km}$ 44. y 45. r^9 46. x^5y^8
47. $x^4\sqrt{x}$ 48. $\frac{6}{p}$ 49. $\frac{\sqrt{13}}{k^2}$ 50. $\frac{10}{y^5}$
51. $2\sqrt{7}$ 52. $8\sqrt{2}$ 53. $21\sqrt{3}$ 54. $12\sqrt{3}$
55. 0 56. $3\sqrt{7}$ 57. $2\sqrt{3} - 3\sqrt{10}$
58. $2\sqrt{2}$ 59. $6\sqrt{30}$ 60. $11\sqrt{x}$ 61. 0
62. $-m\sqrt{5}$ 63. $11k^2\sqrt{2n}$ 64. $\frac{10\sqrt{3}}{3}$
65. $\frac{5\sqrt{2}}{2}$ 66. $\frac{3\sqrt{10}}{5}$ 67. $2\sqrt{2}$ 68. $\sqrt{6}$
69. $\frac{\sqrt{30}}{15}$ 70. $\frac{\sqrt{15}}{5}$ 71. $\sqrt{10}$ 72. $\frac{\sqrt{30}}{15}$
73. $\frac{\sqrt{15}}{5}$ 74. $\frac{t\sqrt{x}}{3x}$ 75. $\frac{x^2\sqrt{2y}}{2}$ 76. $14\sqrt{6}$
77. $-\sqrt{15} + 9$ 78. $3\sqrt{6} + 12$ 79. $22 - 16\sqrt{3}$
80. $179 + 20\sqrt{7}$ 81. -2 82. -13
83. $2\sqrt{21} + 12\sqrt{2} - \sqrt{14} - 4\sqrt{3}$ 84. $-2 + \sqrt{5}$
85. $\frac{-2\sqrt{2} - 6}{7}$ 86. $\frac{-2 + 6\sqrt{2}}{17}$ 87. $\frac{-\sqrt{3} + 3}{2}$
88. $\frac{-\sqrt{10} + 3\sqrt{5} + \sqrt{2} - 3}{7}$
89. $\frac{2\sqrt{3} + 3\sqrt{2} + 2 + \sqrt{6}}{2}$ 90. $1 + 2\sqrt{6}$
91. $\frac{1 - 3\sqrt{7}}{4}$ 92. $3 + 4\sqrt{3}$ 93. 25
94. no solution 95. 99 96. 8 97. 121
98. no solution 99. 1 100. 2 101. 6
102. -2 103. -2 104. 4 105. 9
106. $11\sqrt{3}$ 107. $\frac{11}{t}$ 108. $\frac{5 - \sqrt{2}}{23}$
109. $\frac{2\sqrt{10}}{5}$ 110. $5y\sqrt{2}$
111. cannot be simplified 112. $-\sqrt{10} + 5\sqrt{15}$
113. $\frac{4r\sqrt{3rs}}{3s}$ 114. $1 + 3\sqrt{13}$ 115. $-7\sqrt{2}$
116. $7 - 2\sqrt{10}$ 117. $166 + 2\sqrt{7}$ 118. -11
119. $7\sqrt{2}$ 120. $\frac{3\sqrt{2}}{2}$ 121. $\frac{3}{4}$ 122. $3\sqrt{39}$

Chapter 9 Test (page 547)

[9.1] 1. 10 2. 13.784 3. $-\frac{5}{7}$
4. not a real number [9.2] 5. $\frac{8}{13}$ 6. $5\sqrt{2}$
7. $4\sqrt{6}$ [9.3] 8. $6\sqrt{3}$ 9. $-\sqrt{5}$
10. cannot be simplified 11. $9\sqrt{7}$
[9.5] 12. 31 13. $6\sqrt{2} + 2 - 3\sqrt{14} - \sqrt{7}$
14. $11 + 2\sqrt{30}$ [9.4] 15. $\frac{\sqrt{3}}{3}$ 16. $\frac{\sqrt{15}}{5}$
[9.5] 17. $\frac{-12 - 3\sqrt{3}}{13}$ [9.6] 18. no solution
19. 23 20. 1

Cumulative Review Exercises (page 549)

1. (0, 4), (7, 0), $\left(3, \frac{16}{7}\right)$, $\left(-\frac{7}{4}, 5\right)$
2. (0, 6), (−3, 0), (−2, 2), $\left(-\frac{3}{2}, 3\right)$

3.

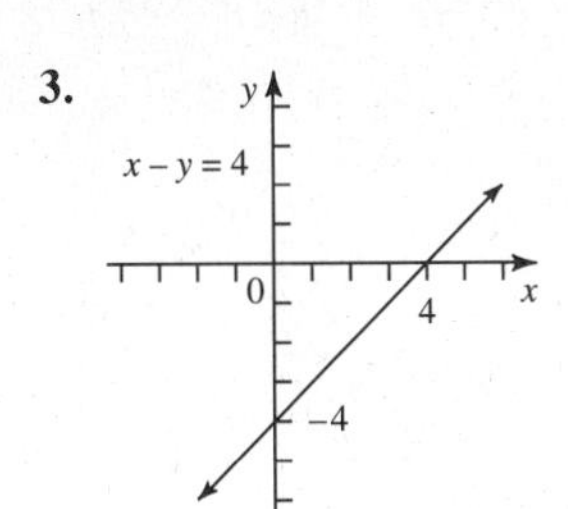

4.

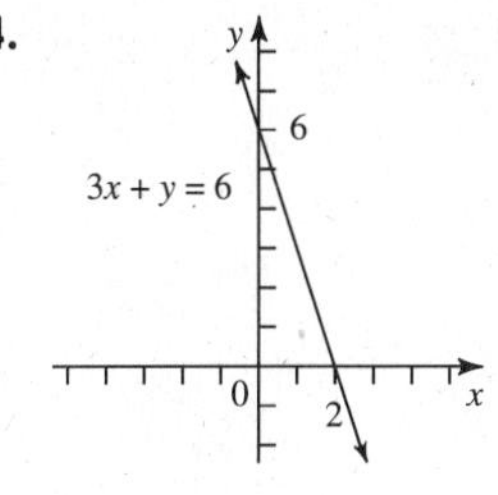

5.

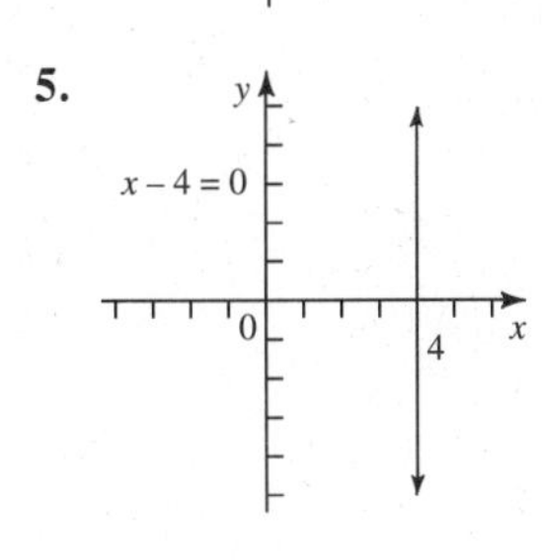

6.

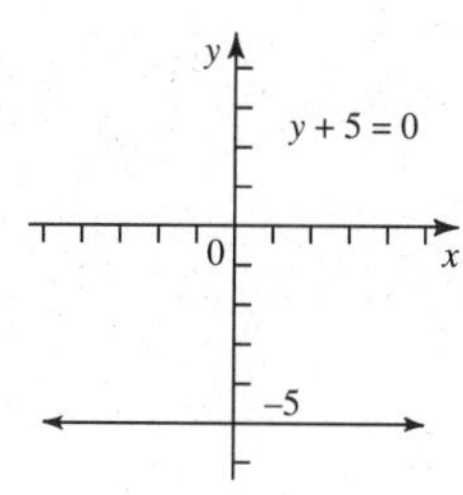

7. $-\frac{4}{3}$ **8.** -6 **9.** 4 **10.** $-\frac{1}{2}$ **11.** parallel
12. perpendicular **13.** neither **14.** $2x + y = 3$
15. $4x - 3y = -3$ **16.** $3x - y = 11$
17. $2x - 3y = 14$

18.

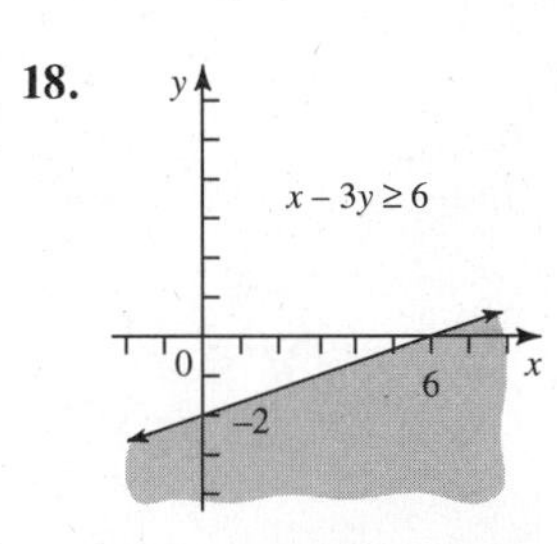

19.

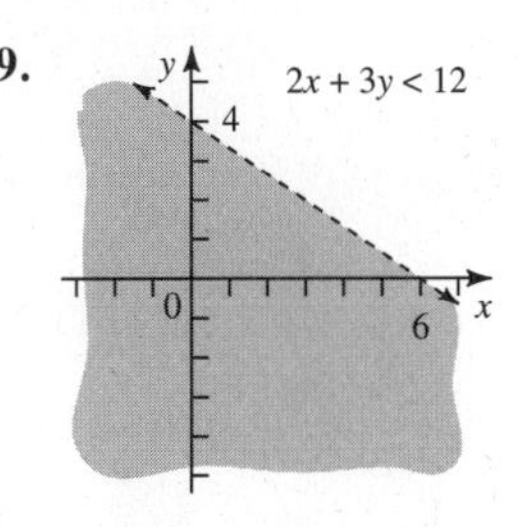

20.

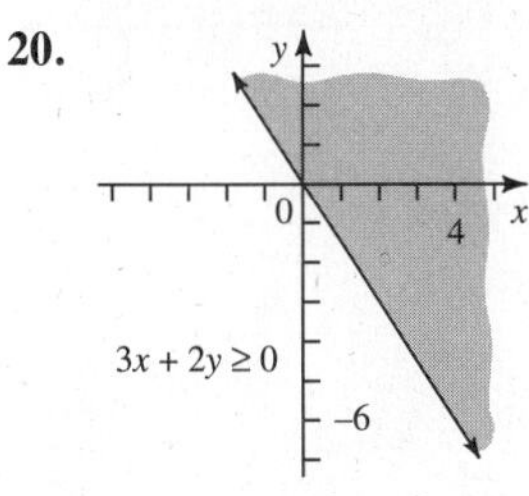

21. no **22.** $(-1, 6)$ **23.** $(3, -4)$ **24.** $(2, -1)$
25. no solution **26.** 15 and 9 **27.** 9 meters
28. 9 fives and 8 tens

29.

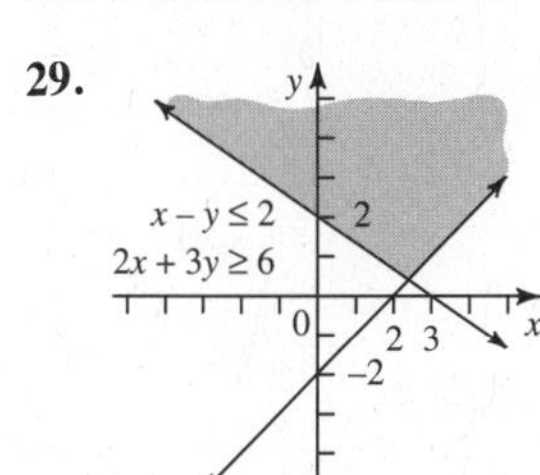

30.

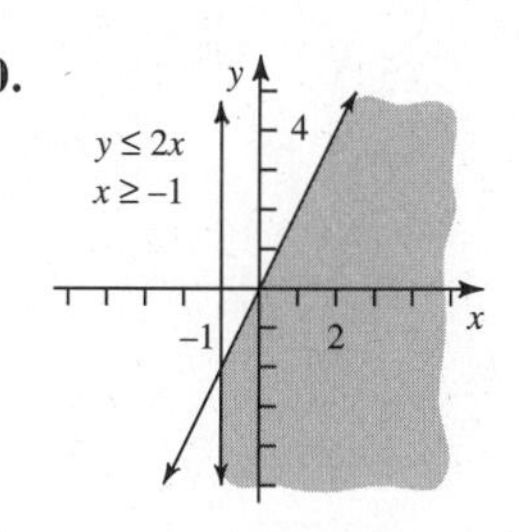

31. $2\sqrt{2}$ **32.** $29\sqrt{3}$ **33.** $\frac{\sqrt{15}}{6}$
34. $-\sqrt{3} + \sqrt{5}$ **35.** $10xy^2\sqrt{2y}$
36. cannot be simplified **37.** $\frac{1 + \sqrt{3}}{2}$

38. $21 - 5\sqrt{2}$ **39.** $\frac{2\sqrt{7}}{7}$ **40.** $4\sqrt{22}$
41. not a real number **42.** $5 + 2\sqrt{6}$
43. 4 **44.** 16

CHAPTER 10

Section 10.1 (page 555)

1. $5, -5$ **3.** $8, -8$ **5.** $\sqrt{13}, -\sqrt{13}$
7. $\sqrt{2}, -\sqrt{2}$ **9.** $2\sqrt{6}, -2\sqrt{6}$ **11.** $\frac{13}{3}, -\frac{13}{3}$
13. $\frac{3}{4}, -\frac{3}{4}$ **15.** $1.1, -1.1$ **17.** $1.6, -1.6$
19. $8.8, -8.8$ **21.** $6, -2$
23. $-4 - \sqrt{10}, -4 + \sqrt{10}$
25. $1 + 4\sqrt{2}, 1 - 4\sqrt{2}$ **27.** $2, -1$
29. $-\frac{8}{3}, -\frac{2}{3}$ **31.** $\frac{5 + \sqrt{30}}{2}, \frac{5 - \sqrt{30}}{2}$
33. $\frac{1 + 3\sqrt{2}}{3}, \frac{1 - 3\sqrt{2}}{3}$ **35.** $\frac{5 + 7\sqrt{2}}{2}, \frac{5 - 7\sqrt{2}}{2}$
37. $\frac{-4 + 2\sqrt{2}}{3}, \frac{-4 - 2\sqrt{2}}{3}$ **39.** 10%
41. $\frac{3 + 3\sqrt{3}}{2}$ **43.** $\frac{18 + \sqrt{6}}{3}$ **45.** $(x + 4)^2$
47. $\left(z + \frac{3}{2}\right)^2$

Section 10.2 (page 563)

1. 1 **3.** 81 **5.** $\frac{9}{4}$ **7.** $-3, -1$
9. $-2 + \sqrt{3}, -2 - \sqrt{3}$ **11.** 4
13. $1 + \sqrt{3}, 1 - \sqrt{3}$
15. $\frac{-5 + \sqrt{37}}{2}, \frac{-5 - \sqrt{37}}{2}$ **17.** $-\frac{3}{2}, \frac{1}{2}$
19. $\frac{2 + \sqrt{14}}{2}, \frac{2 - \sqrt{14}}{2}$ **21.** $5 + \sqrt{17}, 5 - \sqrt{17}$
23. no real number solution
25. $\frac{-5 - \sqrt{30}}{5}, \frac{-5 + \sqrt{30}}{5}$; no, $\frac{-5 - \sqrt{30}}{5}$ cannot be a value of r, since it is negative. "Increased" indicates $r > 0$.
27. $1 + \sqrt{3}$ **29.** $\frac{2 + \sqrt{7}}{4}$ **31.** $\frac{4 + 3\sqrt{3}}{2}$
33. $\frac{2 + \sqrt{5}}{4}$

Section 10.3 (page 569)

1. $a = 3, b = 4, c = -8$ **3.** $a = 2, b = -3, c = 2$
5. $a = 3, b = -8, c = 0$
7. $a = 9, b = 9, c = -26$
9. $-1 + \sqrt{3}, -1 - \sqrt{3}$ **11.** -2 **13.** $-13, 1$
15. $\frac{-6 + \sqrt{26}}{2}, \frac{-6 - \sqrt{26}}{2}$ **17.** $-1, \frac{1}{5}$

19. $-1, \frac{5}{2}$ 21. $-\frac{1}{3}$ 23. 0, 2 25. $0, \frac{16}{3}$

27. $\sqrt{5}, -\sqrt{5}$ 29. $\frac{5}{2}, -\frac{5}{2}$

31. $\frac{6 + 2\sqrt{6}}{3}, \frac{6 - 2\sqrt{6}}{3}$

33. no real number solution

35. no real number solution 37. $-\frac{2}{3}, \frac{4}{3}$

39. $-1, \frac{5}{2}$ 41. $-3 + \sqrt{5}, -3 - \sqrt{5}$

43. no real number solution

45. 4 seconds and $-\frac{3}{2}$ seconds; only 4 seconds is reasonable, since t must be positive.

47. (0, −2), (3, 0), (6, 2), (−3, −4)

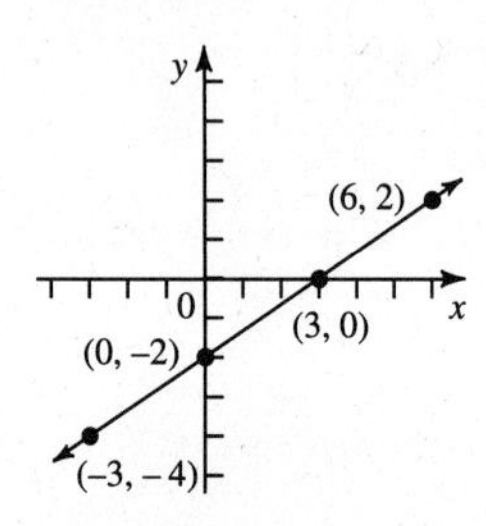

49. (0, 3), (5, 0), (−5, 6), (10, −3)

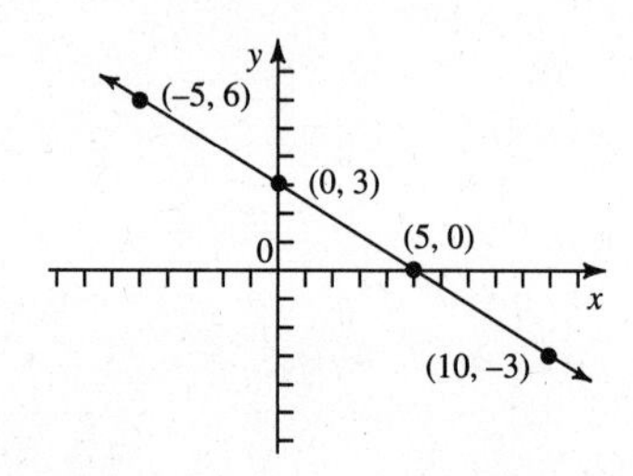

Summary Exercises on Quadratic Equations (page 573)

1. $\frac{-3 + \sqrt{5}}{2}, \frac{-3 - \sqrt{5}}{2}$ 3. $-\frac{7}{9}, \frac{7}{9}$ 5. −2, −1

7. $\frac{-3 + \sqrt{17}}{2}, \frac{-3 - \sqrt{17}}{2}$

9. $\frac{1 + \sqrt{10}}{2}, \frac{1 - \sqrt{10}}{2}$ 11. $-\frac{7}{5}, 1$

13. $\frac{1 + 4\sqrt{2}}{7}, \frac{1 - 4\sqrt{2}}{7}$

15. no real number solution 17. $-\frac{1}{2}, 2$

19. $-3, \frac{1}{3}$ 21. $\frac{-5 + \sqrt{13}}{6}, \frac{-5 - \sqrt{13}}{6}$

23. $-3 + \sqrt{5}, -3 - \sqrt{5}$

25. $-6 + \sqrt{11}, -6 - \sqrt{11}$

27. $\frac{1 + \sqrt{3}}{2}, \frac{1 - \sqrt{3}}{2}$

29. $\frac{-5 + \sqrt{41}}{8}, \frac{-5 - \sqrt{41}}{8}$ 31. $-\frac{8}{3}, -\frac{6}{5}$

33. $\frac{-5 + \sqrt{5}}{2}, \frac{-5 - \sqrt{5}}{2}$ 35. −2, 7

37. −3, 5 39. $-\frac{1}{4}, \frac{2}{3}$ 41. $-\frac{2}{3}, \frac{2}{5}$

Section 10.4 (page 579)

1. (0, 0)

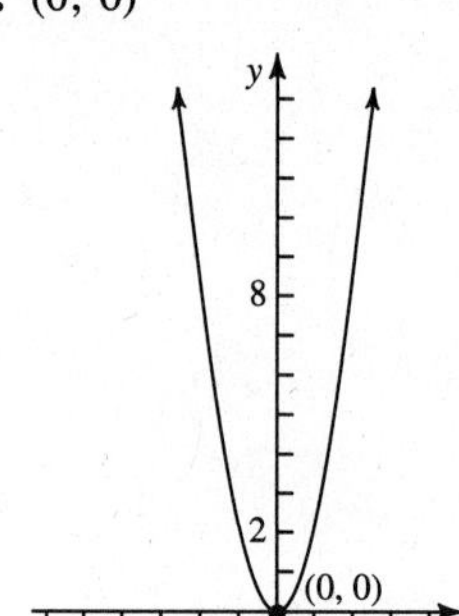

3. (−1, 0)

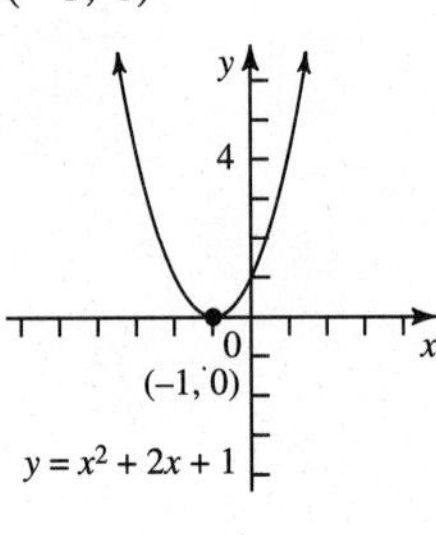

5. (−1, 0)

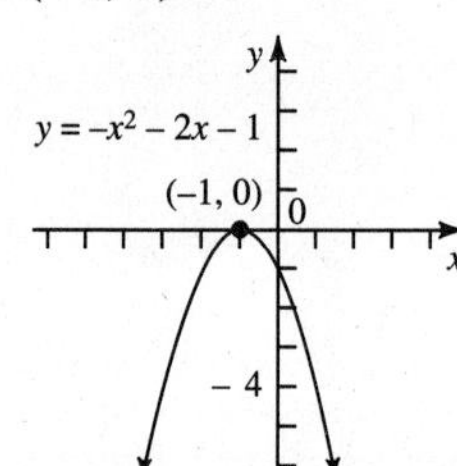

7. (0, 1)

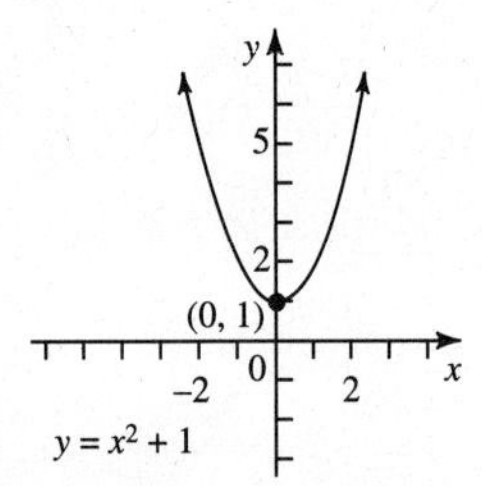

9. (0, 2)

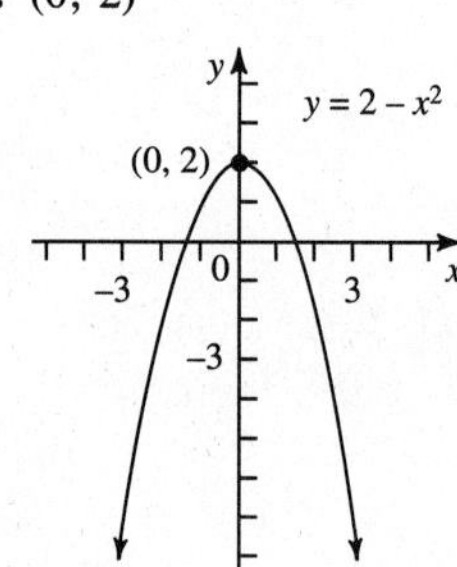

11. (−3, −1)

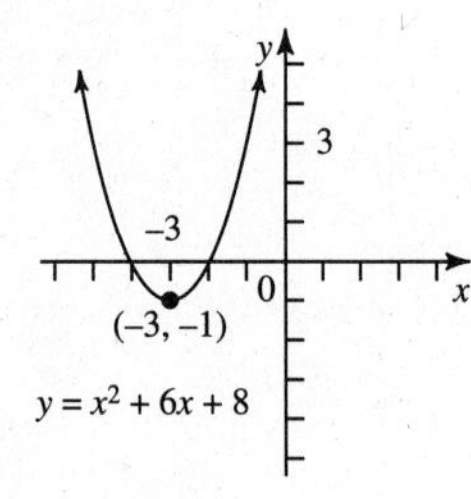

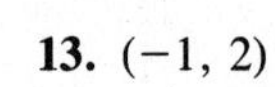

13. (−1, 2)

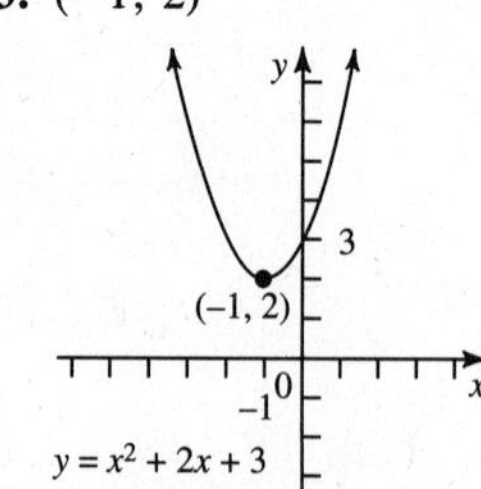

15. (4, −1)

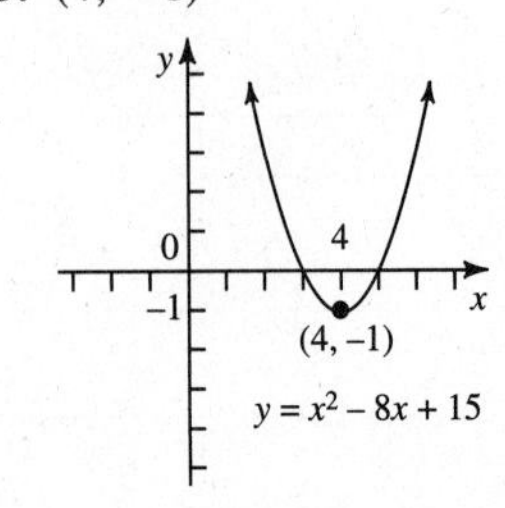

17. $(-2, 1)$

27. $(1, 4)$

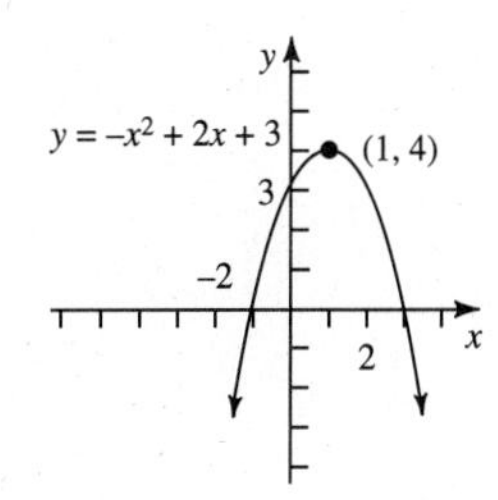

28. $(-2, -3)$

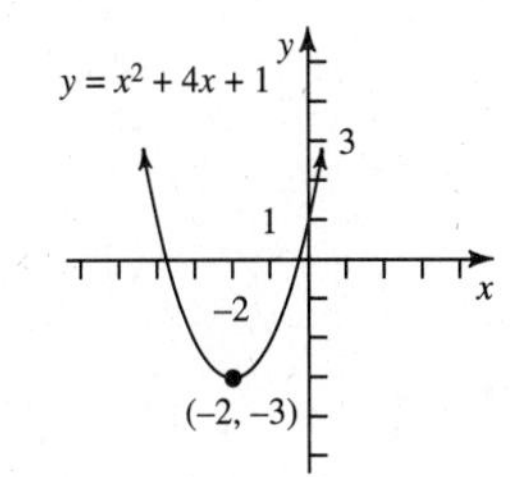

Chapter 10 Review Exercises (page 583)

1. $-7, 7$ **2.** $-\sqrt{15}, \sqrt{15}$ **3.** $-4\sqrt{3}, 4\sqrt{3}$
4. $-5, 1$ **5.** $3 + \sqrt{7}, 3 - \sqrt{7}$
6. $\dfrac{-1 + \sqrt{11}}{2}, \dfrac{-1 - \sqrt{11}}{2}$
7. $\dfrac{-2 + 2\sqrt{3}}{3}, \dfrac{-2 - 2\sqrt{3}}{3}$
8. $\dfrac{-1 + \sqrt{3}}{2}, \dfrac{-1 - \sqrt{3}}{2}$ **9.** $-5, -1$
10. $-2 + \sqrt{11}, -2 - \sqrt{11}$
11. $-1 + \sqrt{6}, -1 - \sqrt{6}$
12. $\dfrac{-4 + \sqrt{22}}{2}, \dfrac{-4 - \sqrt{22}}{2}$ **13.** $-\dfrac{2}{5}, 1$
14. no real number solution **15.** $1 + \sqrt{5}, 1 - \sqrt{5}$
16. $\dfrac{3 + \sqrt{29}}{2}, \dfrac{3 - \sqrt{29}}{2}$
17. no real number solution
18. $\dfrac{1 + \sqrt{21}}{10}, \dfrac{1 - \sqrt{21}}{10}$ **19.** $\dfrac{2 + \sqrt{10}}{2}, \dfrac{2 - \sqrt{10}}{2}$
20. $\dfrac{-1 + \sqrt{29}}{4}, \dfrac{-1 - \sqrt{29}}{4}$
21. $\dfrac{-3 + \sqrt{41}}{2}, \dfrac{-3 - \sqrt{41}}{2}$ **22.** $-1, \dfrac{1}{3}$
23. $(0, 0)$

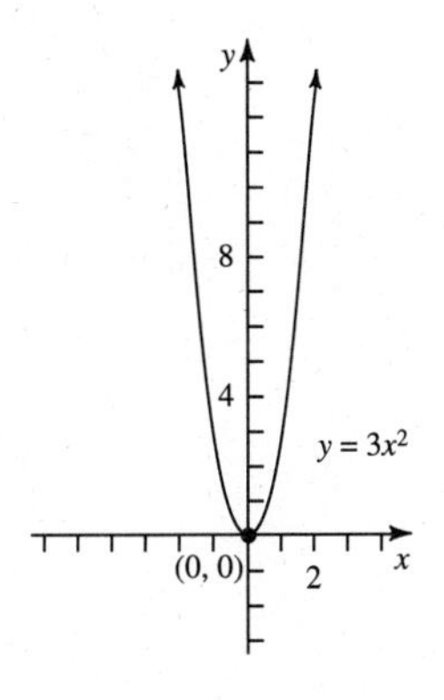

24. $(1, 0)$

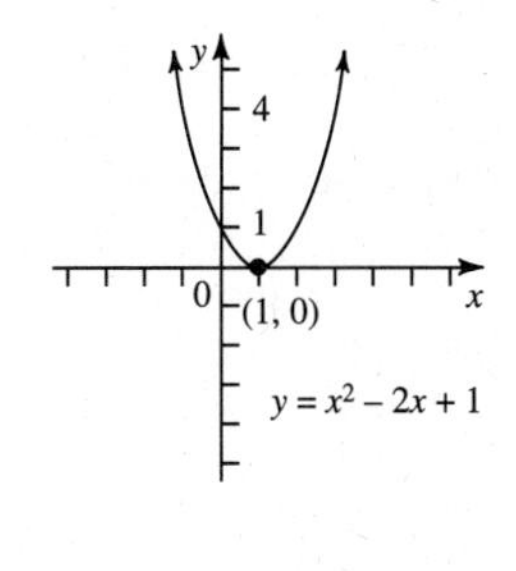

25. $(0, 3)$

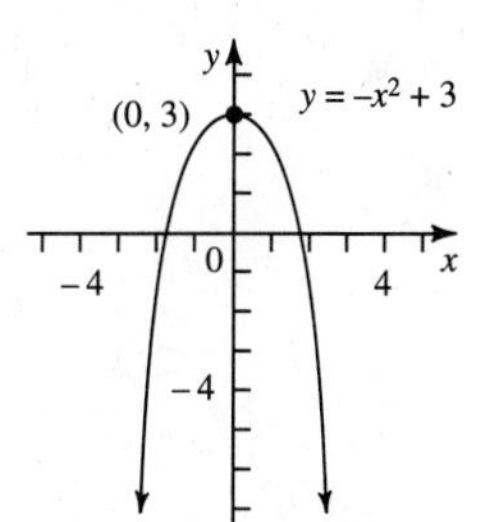

26. $(0, -1)$

29. $-5, \dfrac{1}{2}$ **30.** $-\dfrac{3}{2}, \dfrac{1}{2}$
31. $\dfrac{-1 - \sqrt{21}}{2}, \dfrac{-1 + \sqrt{21}}{2}$ **32.** $-\dfrac{3}{2}, \dfrac{1}{3}$
33. $\dfrac{-5 + \sqrt{17}}{2}, \dfrac{-5 - \sqrt{17}}{2}$
34. $-1 + \sqrt{3}, -1 - \sqrt{3}$
35. no real number solution
36. $\dfrac{9 + \sqrt{41}}{2}, \dfrac{9 - \sqrt{41}}{2}$
37. $\dfrac{-5 + \sqrt{7}}{3}, \dfrac{-5 - \sqrt{7}}{3}$
38. $-1 + 2\sqrt{2}, -1 - 2\sqrt{2}$
39. $-2 + \sqrt{5}, -2 - \sqrt{5}$ **40.** $-2\sqrt{3}, 2\sqrt{3}$

Chapter 10 Test (page 587)

[10.2] **1.** $1, 2$ **2.** $0, \dfrac{5}{2}$ **[10.1]** **3.** $-\sqrt{5}, \sqrt{5}$
4. $3 + \sqrt{2}, 3 - \sqrt{2}$ **5.** $\dfrac{2 + 6\sqrt{2}}{3}, \dfrac{2 - 6\sqrt{2}}{3}$
[10.3] **6.** $-2, 5$ **7.** $-3, \dfrac{1}{2}$ **8.** $\dfrac{1}{3}, 2$
9. no real number solution **10.** $\dfrac{5 + \sqrt{13}}{6}, \dfrac{5 - \sqrt{13}}{6}$
[10.1–10.3] **11.** $1 + \sqrt{2}, 1 - \sqrt{2}$
12. $\dfrac{1 + 3\sqrt{2}}{2}, \dfrac{1 - 3\sqrt{2}}{2}$ **13.** $-\dfrac{2}{3}, 5$
14. $\dfrac{17 + \sqrt{217}}{6}, \dfrac{17 - \sqrt{217}}{6}$
15. $5 + 2\sqrt{2}, 5 - 2\sqrt{2}$ **16.** $3 + \sqrt{7}, 3 - \sqrt{7}$
17. 4 **18.** $-1, \dfrac{5}{2}$
[10.4] **19.** $(0, 4)$ **20.** $(3, 1)$

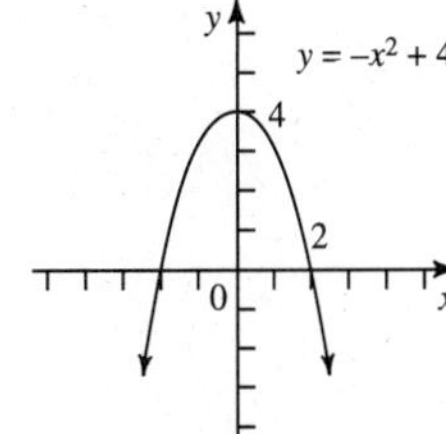

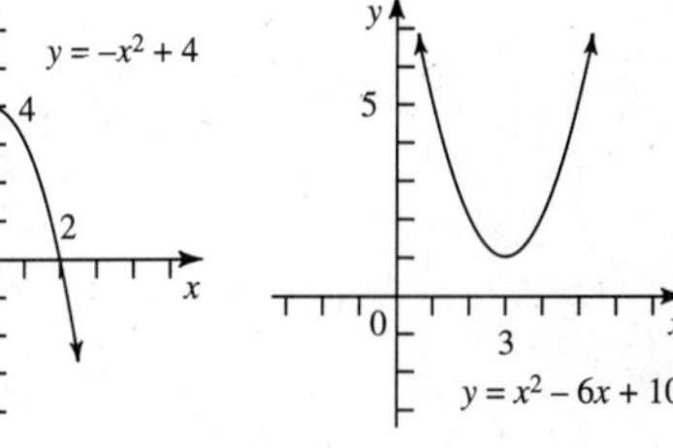

Final Examination (page 589)

[1.4] **1.** 15 **[2.1–2.4]** **2.** 5 **[2.3]** **3.** -2
[2.7] **4.** $-4r - 4$ **5.** $-6k$ **6.** $19m - 19$
[3.1, 3.2] **7.** 2 **8.** -2 **[3.3]** **9.** 2
[3.4] **10.** 2 ounces **[3.5]** **11.** $t = \frac{I}{pr}$
12. $b = \frac{2A}{h}$
[3.7] **13.** $m > -2$ **14.** $p \leq 2$

-2 2

[4.1–4.2] **15.** 6 **16.** $\frac{1}{5^2}$ **17.** $\frac{1}{7^2}$
[4.4] **18.** $8x^5 - 17x^4 + 17x^2$
[4.5] **19.** $2x^4 + x^3 - 19x^2 + 2x + 20$
[4.6] **20.** $36r^2 + 12r + 1$
[4.8] **21.** $3x^2 - 2x + 1$ **[5.1]** **22.** $x(9 - x)$
[5.2] **23.** $(x - 8)(x + 2)$
[5.3] **24.** $(2a + 1)(a - 3)$ **[5.4]** **25.** $(5m - 2)^2$
[5.5] **26.** $-5, 2$ **27.** $-1, \frac{4}{3}$
[5.6] **28.** 50 meters **[6.2]** **29.** $\frac{4}{5}$
[6.4] **30.** $\frac{-k - 1}{k(k - 1)}$ **31.** $\frac{5a + 2}{(a - 2)^2(a + 2)}$
[6.5] **32.** $\frac{2x + 1}{3x - 1}$ **[6.6]** **33.** $\frac{9}{11}$

[7.2–7.3] **34.**

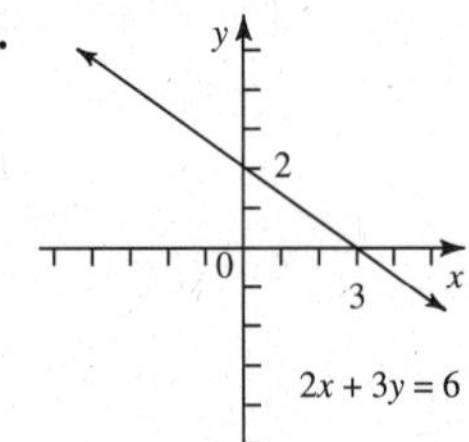

35.

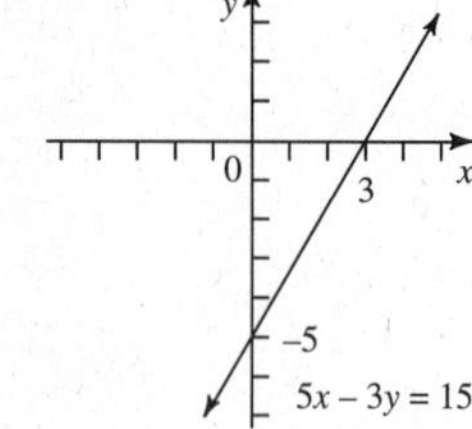

36.

y
–4
0
x
x = –4

[7.6] **37.**

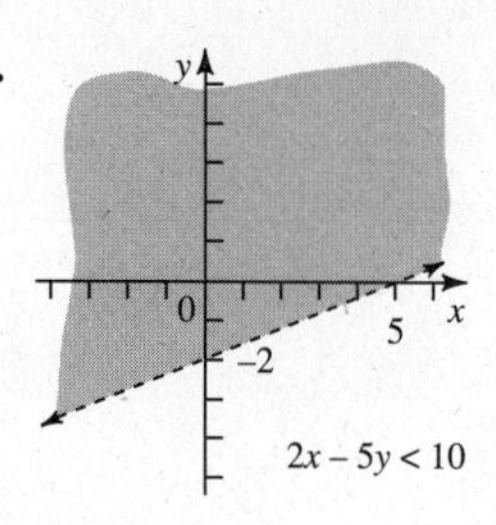

[7.4] **38.** $-\frac{1}{3}$ **[7.5]** **39.** $y = 2x + 3$
[8.2–8.4] **40.** $(-3, 2)$ **[8.3]** **41.** no solution
[8.5] **42.** 29 and 53 **[9.1]** **43.** 10 **44.** -29
[9.4] **45.** $\frac{5\sqrt{6}}{2}$ **[9.3]** **46.** $-\sqrt{5}$
47. $4\sqrt{10}$
[10.2–10.3] **48.** $1 + \sqrt{2}, 1 - \sqrt{2}$
49. $\frac{1 + \sqrt{3}}{2}, \frac{1 - \sqrt{3}}{2}$
[10.4] **50.** $(0, -4)$

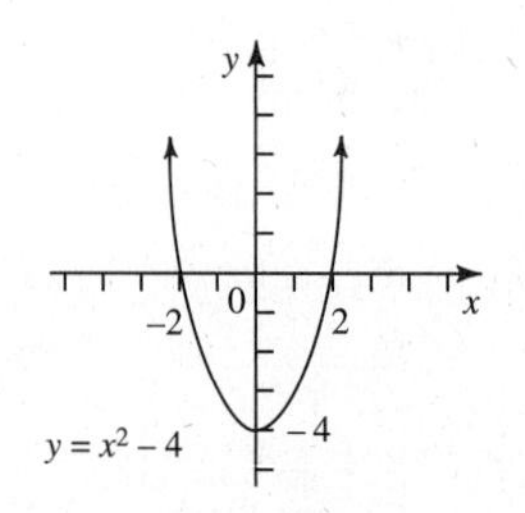

Appendix Exercises (page 597)

1. {1, 2, 3, 4, 5, 6, 7}
3. {winter, spring, summer, fall} **5.** ∅ **7.** {L}
9. {2, 4, 6, 8, . . .}
11. the sets in Exercises 9 and 10, since each contains an unlimited number of elements **13.** true **15.** false
17. true **19.** true **21.** true **23.** true
25. true **27.** false **29.** true **31.** true
33. false **35.** true **37.** true **39.** false
41. true **43.** false **45.** {g, h}
47. {b, c, d, e, g, h} **49.** {a, c, e} **51.** {d}
53. {a} **55.** {a, c, d, e} **57.** {a, c, e, f} **59.** ∅
61. *B* and *D*; *C* and *D*

SOLUTIONS TO SELECTED EXERCISES

SOLUTIONS

For the answers to all odd-numbered section exercises, all chapter review exercises, and all chapter tests, see the section beginning on page A-1.

If you would like to see more solutions, you may order the *Student's Solutions Manual for Introductory Algebra* from your college bookstore. It contains solutions to the odd-numbered exercises that do not appear in this section, as well as solutions to all chapter review exercises and chapter tests.

As you are looking at these solutions, remember that many algebraic exercises and problems can be solved in a variety of ways. In this section we provide only one method of solving selected exercises from each section; space does not permit showing other methods that may be equally correct.

If you work the exercise differently but obtain the same answer, then *as long as your steps are mathematically valid,* your work is correct. Mathematical thinking is a creative process, and solving problems in more than one way is an example of this creativity.

CHAPTER 1

Section 1.1 (page 9)

1. $\frac{7}{14} = \frac{7 \cdot 1}{7 \cdot 2} = \frac{1}{2}$

5. $\frac{16}{18} = \frac{8 \cdot 2}{9 \cdot 2} = \frac{8}{9}$

9. $\frac{72}{108} = \frac{2 \cdot 6 \cdot 6}{3 \cdot 6 \cdot 6} = \frac{2}{3}$

13. $\frac{3}{4} \cdot \frac{3}{5} = \frac{3 \cdot 3}{4 \cdot 5} = \frac{9}{20}$

17. $\frac{9}{4} \cdot \frac{8}{15} = \frac{9 \cdot 8}{4 \cdot 15} = \frac{72}{60} = \frac{6}{5}$

21. $\frac{5}{12} \div \frac{15}{4} = \frac{5}{12} \cdot \frac{4}{15} = \frac{1}{9}$

25. $3\frac{15}{16} \div \frac{15}{4} = \frac{63}{16} \div \frac{15}{4} = \frac{63}{16} \cdot \frac{4}{15} = \frac{21}{20}$ or $1\frac{1}{20}$

29. $6\frac{7}{10} \cdot 4\frac{5}{7} = \frac{67}{10} \cdot \frac{33}{7} = \frac{2211}{70}$ or $31\frac{41}{70}$

33. $\frac{1}{10} + \frac{7}{10} = \frac{1 + 7}{10} = \frac{8}{10} = \frac{4}{5}$

37. $\frac{8}{11} + \frac{3}{22} = \frac{8 \cdot 2}{11 \cdot 2} + \frac{3}{22} = \frac{16}{22} + \frac{3}{22}$
$= \frac{16 + 3}{22} = \frac{19}{22}$

41. $\frac{5}{6} - \frac{3}{10} = \frac{5 \cdot 5}{6 \cdot 5} - \frac{3 \cdot 3}{10 \cdot 3} = \frac{25}{30} - \frac{9}{30} = \frac{25 - 9}{30}$
$= \frac{16}{30} = \frac{8}{15}$

45. $4\frac{1}{2} + 3\frac{2}{3} = \frac{9}{2} + \frac{11}{3} = \frac{9 \cdot 3}{2 \cdot 3} + \frac{11 \cdot 2}{3 \cdot 2} = \frac{27}{6} + \frac{22}{6}$
$= \frac{27 + 22}{6}$
$= \frac{49}{6}$ or $8\frac{1}{6}$

49. $\frac{2}{5} + \frac{1}{3} + \frac{9}{10} = \frac{2 \cdot 6}{5 \cdot 6} + \frac{1 \cdot 10}{3 \cdot 10} + \frac{9 \cdot 3}{10 \cdot 3}$
$= \frac{12}{30} + \frac{10}{30} + \frac{27}{30} = \frac{12 + 10 + 27}{30} = \frac{49}{30}$

53. $\frac{3}{4} + \frac{1}{8} - \frac{2}{3} = \frac{3 \cdot 6}{4 \cdot 6} + \frac{1 \cdot 3}{8 \cdot 3} - \frac{2 \cdot 8}{3 \cdot 8}$
$= \frac{18}{24} + \frac{3}{24} - \frac{16}{24} = \frac{18 + 3 - 16}{24} = \frac{5}{24}$

57. The total distance around the property is the sum of the measures of the 5 sides.
$196 + 76\frac{5}{8} + 100\frac{7}{8} + 146\frac{1}{2} + 98\frac{3}{4}$
$= 196 + \frac{613}{8} + \frac{807}{8} + \frac{293}{2} + \frac{395}{4}$
$= \frac{196 \cdot 8}{8} + \frac{613}{8} + \frac{807}{8} + \frac{293 \cdot 4}{2 \cdot 4} + \frac{395 \cdot 2}{4 \cdot 2}$
$= \frac{1568}{8} + \frac{613}{8} + \frac{807}{8} + \frac{1172}{8} + \frac{790}{8}$
$= \frac{4950}{8} = \frac{2475}{4}$ or $618\frac{3}{4}$
The distance around is $618\frac{3}{4}$ feet.

61. Divide the total amount of sugar on hand by the amount needed for one cake.
$15\frac{1}{2} \div 1\frac{3}{4} = \frac{31}{2} \div \frac{7}{4} = \frac{31}{2} \cdot \frac{4}{7} = \frac{31 \cdot 2 \cdot 2}{2 \cdot 7}$
$= \frac{31 \cdot 2}{7} = \frac{62}{7}$ or $8\frac{6}{7}$
Eight cakes can be made.

Section 1.2 (page 19)

1. $80 + 6$

5. $5000 + 200 + 30 + 7$

9. $.5 + .06 + .007$

13. $.72 = \frac{72}{100} = \frac{18}{25}$

17. $.625 = \frac{625}{1000} = \frac{5}{8}$

21.
$$\begin{array}{r} 27.96 \\ -\ 8.39 \\ \hline 19.57 \end{array}$$

25.
$$\begin{array}{r} 9.71 \\ 4.80 \\ 3.60 \\ 5.20 \\ +\ 8.17 \\ \hline 31.48 \end{array}$$

29. $18.7 \times 2.3 = 43.01$

33.
$$\begin{array}{r} 571 \\ \times\ 2.9 \\ \hline 1655.9 \end{array}$$

37. $\frac{14.9202}{2.43}$ becomes $\frac{1492.02}{243}$

or
$$\begin{array}{r} 6.14 \\ 243\overline{)1492.02} \\ \underline{1458} \\ 340 \\ \underline{243} \\ 972 \\ \underline{972} \\ 0 \end{array}$$

41.
$$\begin{array}{r} .375 \\ 8\overline{)3.000} \\ \underline{2\ 4} \\ 60 \\ \underline{56} \\ 40 \\ \underline{40} \\ 0 \end{array}$$
← Attach zeros.

45.
$$\begin{array}{r} .9375 \\ 16\overline{)15.0000} \\ \underline{14\ 4} \\ 60 \\ \underline{48} \\ 120 \\ \underline{112} \\ 80 \\ \underline{80} \\ 0 \end{array}$$
← Attach zeros.

Round answer to .938.

49.
$$\begin{array}{r} .7333 \\ 15\overline{)11.0000} \\ \underline{10\ 5} \\ 50 \\ \underline{45} \\ 50 \\ \underline{45} \\ 50 \\ \underline{45} \\ 5 \end{array}$$
← Attach zeros.

Round answer to .733.

53. $.75 = 75(.01) = 75(1\%) = 75\%$

57. $.003 = .3(.01) = .3(1\%) = .3\%$

61. $38\% = 38(1\%) = 38(.01) = .38$

65. $11\% = 11(1\%) = 11(.01) = .11$

69. 12% of $350 = .12(350) = 42$

73. Divide the percentage by the amount:
$$\frac{1300}{2000} = .65 = 65\%.$$

77. Divide the percentage by the amount:
$$\frac{64}{3200} = .02 = 2\%.$$

81. 12.741% of 58.902 is $.12741 \times 58.902$
$= 7.50$ (rounded)

85. The family spends 90% of its income, so it saves 10%. Find 10% of \$2000.
$$10\% \times \$2000 = .10 \times \$2000 = \$200$$
The family saves \$200 per month, or, in a year,
$$12 \times \$200 = \$2400.$$

89. The discount was
$$\frac{\$37.50}{\$250} = .15 \text{ or } 15\%.$$

93. Divide:
$$\frac{\$4620}{\$77{,}000} = .06 \text{ or } 6\%.$$

97. Change 7.65% to .0765 and multiply.
$$\begin{array}{r} 2109 \\ \underline{.0765} \\ 1\ 0545 \\ 12\ 654 \\ \underline{147\ 63\ \ } \\ 161.3385 \end{array}$$
The tax is \$161.34 (rounded).

Section 1.3 (page 25)

1. Since 6 is smaller than 9, insert $<$.

5. 25 is greater than 12, so insert $>$.

9. $\frac{3}{4}$ is less than 1, so insert $<$.

13. 12 is less than 17, so insert $\leq$.

17. 8 is less than 28, so insert $\leq$.

21. 6 is less than 9, so both $<$ and $\leq$ may be used.

25. 5 equals 5, so $\leq$ and $\geq$ may be used.

29. 16 is greater than 10, so $>$ and $\geq$ may be used.

33. Write .61 as .610 to see that .609 is less than .610. Both $<$ and $\leq$ may be used.

37. $7 = 5 + 2$

41. Use $\neq$ to get $12 \neq 5$.

45. Since $8 + 2 = 10$, the statement is true.

49. 0 is less than 15, so the statement is true.

53. 25 is greater than 19, so the statement is true.

57. Since $6 = 5 + 1$, the statement is false.

61. Since 8 is greater than 0, the statement is false.

65. Point the symbol toward the smaller number:
$3 \leq 15$.

69. Point the symbol toward the smaller number:
$6 \geq 0$.

73. Point the symbol toward the smaller number:
$.439 \leq .481$.

Section 1.4 (page 31)

1. $6^2 = 6 \cdot 6 = 36$

5. $17^2 = 17 \cdot 17 = 289$

9. $6^4 = 6 \cdot 6 \cdot 6 \cdot 6 = 1296$

13. $3^6 = 3 \cdot 3 \cdot 3 \cdot 3 \cdot 3 \cdot 3 = 729$

17. $\left(\frac{2}{5}\right)^3 = \frac{2}{5} \cdot \frac{2}{5} \cdot \frac{2}{5} = \frac{8}{125}$

21. $(.83)^4 = (.83) \cdot (.83) \cdot (.83) \cdot (.83)$
$= .475$ (rounded)

25. $9 \cdot 3 - 11 = 27 - 11$ Multiply first.
$= 16$ Subtract.

29. $13 \cdot 2 - 15 \cdot 1 = 26 - 15$ Multiply first.
$= 11$ Subtract.

33. $5[2^3 + (2 + 3)] = 5[8 + (2 + 3)]$ Apply the exponent.
$= 5[8 + 5]$ Work in innermost parentheses.
$= 5[13]$ Work in brackets.
$= 65$ Multiply.

37. $\frac{2(5+1) - 3(1+1)}{5(8-6) - 2^3} = \frac{2(5+1) - 3(1+1)}{5(8-6) - 8}$ Apply the exponent.
$= \frac{2(6) - 3(2)}{5(2) - 8}$ Work in parentheses.
$= \frac{12 - 6}{10 - 8}$ Multiply.
$= \frac{6}{2}$ Subtract.
$= 3$ Divide.

41. On the left,
$2 \cdot 20 - 8 \cdot 5 = 40 - 40 = 0,$
and $0 \geq 0$ is true.

45. $3[5(2) - 3] > 20$
$3[10 - 3] > 20$
$3[7] > 20$
$21 > 20$ True

49. $\frac{9(7-1) - 8 \cdot 2}{4(6-1)}$
$= \frac{9(6) - 8 \cdot 2}{4(5)}$ Subtract inside parentheses on top and bottom.
$= \frac{54 - 16}{20}$ Multiply on top and bottom.
$= \frac{38}{20}$ Subtract on top.
$= 1\frac{9}{10}$ Divide.
The inequality is now
$1\frac{9}{10} > 2,$
which is false.

53. $21.92 \leq 7.43^2 - 5.77^2$
$21.92 \leq 55.2049 - 33.2929$
$21.92 \leq 21.912$ False

57. The statement $3 \cdot 5 + 7 = 22$ is true as it stands, since
$3 \cdot 5 + 7 = 15 + 7 = 22.$
Use parentheses as follows (if desired):
$(3 \cdot 5) + 7 = 22.$

61. The statement $3 \cdot 5 - 2 \cdot 4 = 36$ will be true if parentheses are inserted around $5 - 2$.
$3 \cdot (5 - 2) \cdot 4 = 3(3) \cdot 4$ Subtract inside parentheses.
$= 9 \cdot 4$ Multiply 3 by 3 first.
$= 36$ Multiply.
36 is what we wanted.

65. The statement $6 + 5 \cdot 3^2 = 99$ will be true if parentheses are inserted around $6 + 5$.
$(6 + 5) \cdot 3^2 = 11 \cdot 3^2$ Add inside parentheses.
$= 11 \cdot 3 \cdot 3$ Use the exponent.
$= 33 \cdot 3$ Multiply 11 by 3.
$= 99$ Multiply.
99 is what we wanted.

69. No parentheses are needed, or
$\frac{1}{2} + \left(\frac{5}{3} \cdot \frac{9}{7}\right) - \frac{3}{2} = \frac{8}{7}.$

73. Getting on is +; getting off is −. This gives
$5 + (6 - 1) + (3 - 2).$

Section 1.5 (page 37)

1. **(a)** Let $x = 3$; then $x + 9 = 3 + 9 = 12$.
(b) Let $x = 15$; then $x + 9 = 15 + 9 = 24$.

5. **(a)** Let $x = 3$; then
$\frac{2}{3}x + \frac{1}{3} = \frac{2}{3}(3) + \frac{1}{3} = \frac{6}{3} + \frac{1}{3} = \frac{7}{3}.$
(b) Let $x = 15$; then
$\frac{2}{3}x + \frac{1}{3} = \frac{2}{3}(15) + \frac{1}{3} = \frac{30}{3} + \frac{1}{3} = \frac{31}{3}.$

9. **(a)** $\frac{3x - 5}{2x} = \frac{3 \cdot 3 - 5}{2 \cdot 3}$ Replace x with 3.
$= \frac{9 - 5}{6}$ Multiply on the top and bottom where indicated.
$= \frac{4}{6}$ Subtract on top.
$= \frac{2}{3}$ Lowest terms

(b) $\frac{3x - 5}{2x} = \frac{3 \cdot 15 - 5}{2 \cdot 15}$ Let $x = 15$.
$= \frac{45 - 5}{30}$ Multiply.
$= \frac{40}{30}$ Subtract.
$= \frac{4}{3}$ Lowest terms

13. **(a)** $6.459x = 6.459 \cdot 3$ Let $x = 3$.
$= 19.377$ Multiply.
(b) $6.459x = 6.459 \cdot 15$ Let $x = 15$.
$= 96.885$ Multiply.

17. **(a)** $8x + 3y + 5 = 8 \cdot 4 + 3 \cdot 2 + 5$
Let $x = 4$ and $y = 2$.
$= 32 + 6 + 5$ Multiply first.
$= 43$ Add.
(b) $8x + 3y + 5 = 8 \cdot 1 + 3 \cdot 5 + 5$
Let $x = 1$ and $y = 5$
$= 8 + 15 + 5$ Multiply.
$= 28$ Add.

21. (a) $x + \frac{4}{y} = 4 + \frac{4}{2}$ Let $x = 4$ and $y = 2$.

$= 4 + 2$ Divide 4 by 2 first.

$= 6$ Add.

(b) $x + \frac{4}{y} = 1 + \frac{4}{5}$ Let $x = 1$ and $y = 5$.

$= \frac{9}{5}$ Add.

25. (a) Let $x = 4$ and $y = 2$; then

$$5\left(\frac{4}{3}x + \frac{7}{2}y\right) = 5\left(\frac{4}{3} \cdot 4 + \frac{7}{2} \cdot 2\right)$$
$$= 5\left(\frac{16}{3} + 7\right)$$
$$= 5\left(\frac{16}{3} + \frac{21}{3}\right)$$
$$= 5\left(\frac{37}{3}\right)$$
$$= \frac{185}{3}$$

(b) Let $x = 1$ and $y = 5$; then

$$5\left(\frac{4}{3} \cdot 1 + \frac{7}{2} \cdot 5\right) = 5\left(\frac{4}{3} + \frac{35}{2}\right)$$
$$= 5\left(\frac{8}{6} + \frac{105}{6}\right)$$
$$= 5\left(\frac{113}{6}\right) = \frac{565}{6}.$$

29. (a) $\frac{2x + 4y - 6}{5y + 2} = \frac{2 \cdot 4 + 4 \cdot 2 - 6}{5 \cdot 2 + 2}$ Let $x = 4$ and $y = 2$.

$= \frac{8 + 8 - 6}{10 + 2}$ Multiply where indicated on the top and bottom.

$= \frac{16 - 6}{12}$ Add on the top, then the bottom.

$= \frac{10}{12}$ Subtract.

$= \frac{5}{6}$ Lowest terms

(b) $\frac{2x + 4y - 6}{5y + 2} = \frac{2 \cdot 1 + 4 \cdot 5 - 6}{5 \cdot 5 + 2}$ Let $x = 1$ and $y = 5$.

$= \frac{2 + 20 - 6}{25 + 2}$ Multiply.

$= \frac{16}{27}$ Add and subtract.

33. (a) $\frac{x^2 + y^2}{x + y} = \frac{4^2 + 2^2}{4 + 2}$ Let $x = 4$ and $y = 2$.

$= \frac{16 + 4}{4 + 2}$ Use exponents.

$= \frac{20}{6}$ Add.

$= \frac{10}{3}$ Lowest terms

(b) $\frac{x^2 + y^2}{x + y} = \frac{1^2 + 5^2}{1 + 5}$ Let $x = 1$ and $y = 5$.

$= \frac{1 + 25}{1 + 5}$ Use exponents.

$= \frac{26}{6}$ Add.

$= \frac{13}{3}$ Lowest terms.

37. (a) $.841x^2 + .32y^2 = .841(4)^2 + .32(2)^2$ Let $x = 4$ and $y = 2$.

$= .841(16) + .32(4)$ Use exponents.

$= 13.456 + 1.28$ Multiply.

$= 14.736$ Add.

(b) $.841x^2 + .32y^2 = .841(1)^2 + .32(5)^2$ Let $x = 1$ and $y = 5$.

$= .841(1) + .32(25)$ Use exponents.

$= .841 + 8$ Multiply.

$= 8.841$ Add.

41. Five times a number

$5 \cdot x$ or $5x$

45. Nine subtracted from a number

$x - 9$

49. $p - 5 = 12$

$17 - 5 = 12$ Let $p = 17$.

$12 = 12$ True

17 is a solution.

53. $2y + 3(y - 2) = 14$

$2 \cdot 4 + 3(4 - 2) = 14$ Let $y = 4$.

$2 \cdot 4 + 3 \cdot 2 = 14$ Work in parentheses.

$8 + 6 = 14$ Multiply.

$14 = 14$ True

4 is a solution.

57. $3r^2 - 2 = 46$

$3 \cdot 4^2 - 2 = 46$ Let $r = 4$.

$3 \cdot 16 - 2 = 46$ Use exponent.

$48 - 2 = 46$ Multiply.

$46 = 46$ True

4 is a solution.

61. $9.54x + 3.811 = .4273x + 16.57718$

$9.54(1.4) + 3.811 = .4273(1.4) + 16.57718$ Let $x = 1.4$.

$13.356 + 3.811 = .59822 + 16.57718$ Multiply.

$17.167 = 17.1754$ Add.

Results are different, so 1.4 is not a solution.

65. Sixteen minus three-fourths of a number is ten.

$16 - \frac{3}{4}x = 10$

Since $16 - \frac{3}{4}x = 16 - \frac{3}{4}(8) = 16 - 6 = 10$, and since 8 is in the list, 8 is the solution.

69. "Three times" indicates multiplication by 3; "more than" indicates addition; "twice" indicates multiplication by 2. So "three times a number is equal to two more than twice the number" translates to $3x = 2x + 2$.

Since $3 \cdot 2 = 2 \cdot 2 + 2$

or $6 = 4 + 2$

$= 6$

and since 2 is in the list, 2 is the solution.

CHAPTER 2

Section 2.1 (page 57)

1. The additive inverse of 12 is -12.
5. Since $|6| = 6$, the additive inverse of $|6|$ is -6.
9. Since -10 is to the left of -3 on the number line, -10 is smaller.
13. Since -6 is to the left of 5 on the number line, -6 is smaller.
17. 7 is greater than -4, since 7 is to the right of -4 on the number line. The statement $7 < -4$ is false.
21. -7 is greater than -10, since -7 is to the right of -10 on the number line. The statement $-7 \geq -10$ is true.
25. **(a)** The set of whole numbers is $\{0, 1, 2, 3, 4, \ldots\}$. The elements from the given set that belong to the set of whole numbers are 0, 3, and 7.
 (b) The set of integers is $\{\ldots, -3, -2, -1, 0, 1, 2, 3, \ldots\}$. The elements from the given set that belong to the set of integers are -9, 0, 3, and 7.
 (c) A rational number is a number that can be written as the quotient of two integers. The elements from the given set that belong to the set of rational numbers are $-9, -1\frac{1}{4}, -\frac{3}{5}, 0, 3,$ 5.9, and 7.
 (d) An irrational number is a non-rational number that can be represented by a point on the number line. The elements from the given set that belong to the set of irrational numbers are $-\sqrt{7}$ and $\sqrt{5}$.
 (e) A real number is a number that can be represented by a point on the number line. All numbers in the given set are real numbers.
29. By the definitions of rational number and of irrational number, no number can be both rational and irrational. The statement is true.
33. Graph the numbers $-2, -6, |-4|, 3, -|4|$ using the facts that $|-4| = 4$, and $-|4| = -4$. See the graph in the answer section.
37. Graph $\frac{1}{4}, 2\frac{1}{2}, -3\frac{4}{5}, -4, -1\frac{5}{8}$. See the graph in the answer section.
41. The least common denominator for $\frac{3}{4}$ and $\frac{2}{7}$ is 28.
 $$\frac{3}{4} = \frac{21}{28} \text{ and } \frac{2}{7} = \frac{8}{28}.$$
 Therefore,
 $$\frac{3}{4} - \frac{2}{7} = \frac{21}{28} - \frac{8}{28} = \frac{13}{28}.$$

Section 2.2 (page 63)

1. Find the difference in the absolute values of both numbers:
 $5 - 3 = 2.$
 Since the larger number in absolute value is 5, the answer is positive. Therefore, the answer is 2.
5. Since numbers with the same sign are being added, add the absolute values of both numbers:
 $6 + 2 = 8.$
 Since both numbers are negative, the sign of the answer is negative:
 $-6 + (-2) = -8.$
9. Since you are adding numbers with the same sign, simply add the absolute values of both numbers:
 $3 + 9 = 12.$
 Since both numbers are negative, the sign of the answer is negative:
 $-3 + (-9) = -12.$
13. Do the work inside brackets first:
 $[13 + (-5)] = 8.$
 Now, $4 + [13 + (-5)] = 4 + 8 = 12.$
17. $-2 + [5 + (-1)] = -2 + 4 = 2$
21. $[9 + (-2)] + 6 = 7 + 6 = 13$
25. $-\frac{1}{6} + \frac{2}{3} = -\frac{1}{6} + \frac{4}{6}$ Write each fraction with a common denominator.
 $= \frac{3}{6}$ Add the numerators: $-1 + 4 = 3$.
 $= \frac{1}{2}$ Lowest terms
29. $2\frac{1}{2} + \left(-3\frac{1}{4}\right) = \frac{5}{2} + \left(\frac{-13}{4}\right)$ Change each fraction to an improper fraction.
 $= \frac{10}{4} + \left(\frac{-13}{4}\right)$ Write each fraction with a common denominator.
 $= -\frac{3}{4}$ Add the numerators: $10 + (-13) = -3$.
33. Do the work inside the brackets first:
 $[3.2 + (-4.8)] = -1.6.$
 So $-6.1 + [3.2 + (-4.8)] = -6.1 + (-1.6)$
 $= -7.7.$
37. Do the work inside the brackets first:
 $[-4 + (-3)] = -7$ and $[8 + (-1)] = 7.$
 So $[-4 + (-3)] + [8 + (-1)] = -7 + 7 = 0.$
41. $-4 + 0 = -4$
 $-4 = -4$ Add on left side.
 True
45. $-9 + 5 + 6 = -2$
 $-4 + 6 = 2$ Add on left side.
 $2 = -2$ Add again on left.
 False
49. $\left|-\frac{8}{13} + \frac{3}{4}\right| = \frac{8}{13} + \frac{3}{4}$
 $\left|-\frac{32}{52} + \frac{39}{52}\right| = \frac{32}{52} + \frac{39}{52}$ Write fractions with a common denominator.
 $\left|\frac{7}{52}\right| = \frac{32}{52} + \frac{39}{52}$ Add inside absolute value symbol.
 $\frac{7}{52} = \frac{32}{52} + \frac{39}{52}$ Remove absolute value symbol.
 $\frac{7}{52} = \frac{71}{52}$ or $1\frac{19}{52}$ Add on right side.
 False

SOLUTIONS

53. $[4 + (-6)] + 6 = 4 + (-6 + 6)$

$-2 + 6 = 4 + 0$ Work within brackets and parentheses.

$4 = 4$ Add on each side.

True

57. $-5 + (-|5|) = -10$

$-5 + (-5) = -10$

$-10 = -10$ Add on left side.

True

61. First, translate "the sum of -19 and -3" as $-19 + (-3)$. Since 14 is added to this, we obtain $[-19 + (-3)] + 14$. Simplify in the brackets first to get $-22 + 14$. Add to get -8.

65. First, translate "the sum of 8 and -23" as $8 + (-23)$. "4 more than" this sum indicates that 4 is added to $8 + (-23)$. We obtain $[8 + (-23)] + 4$. Simplify in the brackets first to get $-15 + 4$. Add to get -11.

69. To obtain the balance, add -18 (which represents the check) to 14.

$14 + (-18) = -4$

His new balance is $-\$4$.

73. Think of gains as positive numbers and losses as negative numbers.

$12 + (-3) + 4 = 13$

His gain was 13 yards.

77. Try all numbers in the list. The only number that works is -2, which is the solution.

81. Try all numbers in the list. The only number that works is -2, which is the solution.

85. Try all numbers in the list. The only number that works is 2, which is the solution.

89. $12 + 3(8 - 6) = 12 + 3(2)$ Work within parentheses.

$= 12 + 6$ Multiply.

$= 18$ Add.

Section 2.3 (page 71)

1. $3 - 6 = 3 + (-6) = -3$

5. $-6 - 2 = -6 + (-2) = -8$

9. $6 - (-3) = 6 + (3) = 9$

13. $-6 - (-2) = -6 + (2) = -4$

17. $-2 - (5 - 8) = -2 - (-3) = -2 + (3) = 1$

21. $-\frac{3}{4} - \frac{5}{8} = -\frac{3}{4} + \left(-\frac{5}{8}\right)$

$= -\frac{6}{8} + \left(-\frac{5}{8}\right)$

$= -\frac{11}{8}$

25. $3.4 - (-8.2) = 3.4 + (8.2) = 11.6$

29. $-4.2 - (7.4 - 9.8) = -4.2 - [7.4 + (-9.8)]$

$= -4.2 - (-2.4)$

$= -4.2 + 2.4$

$= -1.8$

33. $(4 - 6) + 12 = (-2) + 12 = 10$

37. $6 - (-8 + 3) = 6 - (-5) = 6 + (5) = 11$

41. $(-5 - 6) - (9 - 2)$

$= [-5 + (-6)] - [9 + (-2)]$

$= -11 - (7)$

$= -11 + (-7)$

$= -18$

45. $-9 - [(3 - 2) - (-4 - 2)]$

$= -9 - [1 - (-6)]$

$= -9 - (7) = -16$

49. $-9.2 + [(-4.9 - 4.1) + 11.3]$

$= -9.2 + (-9 + 11.3)$

$= -9.2 + 2.3$

$= -6.9$

53. *Difference* indicates subtraction. The phrase translates as $4 - (-12)$. It simplifies to $4 - (-12) = 4 + 12 = 16$.

57. We must add 9 and -4, and subtract 5 from the sum. The phrase translates as $[9 + (-4)] - 5$. It simplifies to $5 - 5 = 0$.

61. Jack's initial status is -19. Think of paying as adding and borrowing as subtracting.

$-19 + 12 - 10 = -7 - 10 = -17$

His present financial status is $-\$17$.

65. We must subtract 3500 from $-13,500$.

$-13,500 - 3500 = -17,000$

The profit in 1989 was $-\$17,000$. (He actually lost money.)

69. $3(4) - 2(2) = 12 - 4 = 8$

73. $(4 + 2)[2(4) - 2] = 6(8 - 2) = 6 \cdot 6 = 36$

Section 2.4 (page 79)

1. $(-3)(-4) = 3 \cdot 4 = 12$

5. $(-1)(-5) = 1 \cdot 5 = 5$

9. $(-10)(-12) = 10 \cdot 12 = 120$

13. $(-6)(5) = -(6 \cdot 5) = -30$

17. $\left(-\frac{7}{3}\right)\left(\frac{8}{21}\right) = \frac{-7 \cdot 8}{3 \cdot 21} = \frac{-56}{63} = -\frac{8}{9}$

21. $(-3.7)(-2.1) = 7.77$

25. The factors of 25 are $-25, -5, -1, 1, 5,$ and 25.

29. The factors of 17 are $-17, -1, 1,$ and 17.

33. $3 - 2 \cdot 9 = 3 - 18$

$= 3 + (-18)$

$= -15$

37. $5(12 - 15)$

$= 5(-3)$ Do subtraction inside parentheses.

$= -15$ Multiply.

41. $(6 - 11)(3 - 6)$

$= (-5)(-3)$ Subtract inside parentheses first.

$= 15$ Multiply.

45. $(-5 - 2)(-3) + 6$

$= (-7)(-3) + 6$ Subtract inside parentheses first.

$= 21 + 6$ Multiply.

$= 27$ Add.

49. $(-9 - 1)(-2) - (-6)$

$= (-10)(-2) - (-6)$ Subtract inside parentheses.

$= 20 - (-6)$ Multiply.

$= 20 + 6$ Change subtraction to addition.

$= 26$ Add.

53. Let $x = -2$, $y = 3$, and $a = -4$ to get

$(5x - 2y)(-2a)$

$= [5(-2) - 2(3)][-2(-4)]$

$= [-10 - 6][8]$

$= (-16)(8)$

$= -128.$

57. Let $x = -2$, $y = 3$, and $a = -4$ to get
$$(6 - x)(5 + y)(3 + a) = [6 - (-2)][5 + 3][3 + (-4)] = 8(8)(-1) = -64.$$

61. Let $x = -2$ and $y = 3$ to get
$$4y^2 - 2x^2 = 4(3)^2 - 2(-2)^2 = 4(9) - 2(4) = 36 - 8 = 28.$$

65. $-4 - 2(-8 \cdot 2) = -4 - 2(-16) = -4 + 32 = 28$

69. $-3[4 + (-7)] = -3(-3) = 9$

In Exercises 73 and 77, mentally replace the variable with values from the domain to find the value that makes the equation a true statement.

73. $3k = -6$
$3(-2) = -6$ Let $k = -2$.
$-6 = -6$ True
The solution is -2.

77. $-9r = 27$
$-9(-3) = 27$ Let $r = -3$.
$27 = 27$ True
The solution is -3.

81. $\dfrac{3(1) + 5(5)}{2} = \dfrac{3 + 25}{2} = \dfrac{28}{2} = 14$

85. $\dfrac{1^2 + 5^2}{3(5) - 2} = \dfrac{1 + 25}{15 - 2} = \dfrac{26}{13} = 2$

Section 2.5 (page 89)

1. Since $9 \cdot \dfrac{1}{9} = 1$, the multiplicative inverse is $\dfrac{1}{9}$.

5. Since $\dfrac{2}{3} \cdot \dfrac{3}{2} = 1$, the multiplicative inverse is $\dfrac{3}{2}$.

9. 0 has no multiplicative inverse, since there is no number that when multiplied by 0 gives a product of 1.

13. $\dfrac{-10}{5} = -10 \cdot \dfrac{1}{5} = -2$

17. $\dfrac{-150}{-10} = -150 \cdot \left(-\dfrac{1}{10}\right) = 15$

21. $-\dfrac{1}{2} \div \left(-\dfrac{3}{4}\right) = -\dfrac{1}{2} \cdot \left(-\dfrac{4}{3}\right)$ Multiply by the reciprocal of $-\dfrac{3}{4}$.
$= \dfrac{2}{3}$ Multiply.

25. $\dfrac{12}{2 - 5} = \dfrac{12}{-3} = 12 \cdot \left(-\dfrac{1}{3}\right) = -4$

29. $\dfrac{-40}{8 - (-2)} = \dfrac{-40}{8 + 2} = \dfrac{-40}{10} = -4$

33. $\dfrac{-3.8 - (-2.2)}{-.2} = \dfrac{-3.8 + 2.2}{-.2} = \dfrac{-1.6}{-.2} = 8$

37. $\dfrac{-15(2)}{-7 - 3} = \dfrac{-30}{-10} = 3$

41. $\dfrac{-9(-2) - (-4)(-2)}{-2(3) - 2(2)} = \dfrac{18 - 8}{-6 - 4} = \dfrac{10}{-10} = -1$

45. $\dfrac{10^2 - 5^2}{8^2 + 3^2 + 2} = \dfrac{100 - 25}{64 + 9 + 2} = \dfrac{75}{75} = 1$

In Exercises 49 and 53, mentally replace the variable with values from the domain to find the value that makes the equation a true statement.

49. $\dfrac{x}{2} = -1$
$\dfrac{-2}{2} = -1$ Let $x = -2$.
$-2 \cdot \left(\dfrac{1}{2}\right) = -1$
$-1 = -1$ True
The solution is -2.

53. $\dfrac{y}{-1} = 2$
$\dfrac{-2}{-1} = 2$ Let $y = -2$.
$-2 \cdot (-1) = 2$
$2 = 2$ True
The solution is -2.

57. The phrase translates as $\dfrac{-20}{-12 + 2}$. This simplifies to $\dfrac{-20}{-10} = 2$.

61. The phrase translates as $-25(9 - 4)$. This simplifies to $-25(5) = -125$.

65. The sentence translates as $\dfrac{x}{4} = -3$. -12 is the only integer between -12 and 12, inclusive, that is a solution, since $\dfrac{-12}{4} = -3$.

69. The sentence translates as $x + 6 = -5$. -11 is the only integer between -12 and 12, inclusive, that is a solution, since $-11 + 6 = -5$.

73. $3 + (-3) = 0$

77. $(5 + 6) + (-14) = 11 + (-14) = -3$

Section 2.6 (page 99)

1. The numbers are being added in a different order, indicating the commutative property.

5. The numbers are being multiplied using different groupings, indicating the associative property.

9. The numbers are being added in a different order, indicating the commutative property.

13. The sum of a number and 0 is the number, indicating the identity property.

17. $\dfrac{2}{3}$ is multiplied by 1 in the form of $\dfrac{6}{6}$, indicating the identity property.

21. $9 + k = k + 9$ Add in opposite order.

25. $3(r + m) = 3 \cdot r + 3 \cdot m$ Multiply both terms in parentheses by 3.
$= 3r + 3m$

29. The sum of a number and its inverse is 0.

33. Multiply -3 times r and times 2, to get $-3r - 6$.

37. $(k + 5) + (-6) = k + [5 + (-6)] = k - 1$
Keep the numbers in the same order, but use a different grouping.

41. $5(m + 2) = 5 \cdot m + 5 \cdot 2 = 5m + 10$

45. $-8(k - 2) = -8k - (-8)(2) = -8k + 16$

49. $\left(r + \frac{8}{3}\right)\frac{3}{4} = r \cdot \frac{3}{4} + \frac{8}{3} \cdot \frac{3}{4}$
$= \frac{3}{4}r + 2$

53. $2(5r + 6m) = 2(5r) + 2(6m) = 10r + 12m$

57. $5 \cdot 8 + 5 \cdot 9 = 5(8 + 9) = 5(17) = 85$

61. $9p + 9q = 9(p + q)$

65. $-(3k + 5) = -1(3k + 5) = -1(3k) + (-1)5$
$= -3k - 5$

69. $-(-4 + p) = -1(-4) + (-1)(p) = 4 - p$

73. It makes a difference which activity you do first, so they are not commutative.

77. $25 - (6 - 2) = 25 - 4 = 21$
and $(25 - 6) - 2 = 19 - 2 = 17$,
so subtraction is not associative.

81. $14 - [(-6) - (4 - 9)] = 14 - [(-6) - (-5)]$
$= 14 - (-6 + 5) = 14 - (-1) = 14 + 1 = 15$

Section 2.7 (page 107)

1. $\underbrace{15}y$
└Numerical coefficient

5. $\underbrace{35}a^4b^2$
└Numerical coefficient

9. $y^2 = \underbrace{1} \cdot y^2$
└Numerical coefficient

13. $6m$ and $-14m$ are like terms since the variables are the same.

17. $25y$, $-14y$, and $8y$ are like terms since the variables are the same.

21. p, $-5p$, and $12p$ are like terms since the variables are the same.

25. $2k + 9 + 5k + 6$
$= (2 + 5)k + (9 + 6) = 7k + 15$

29. $16 - 5m - 4m - 2 + 2m$
$= -5m - 4m + 2m + 16 - 2$
$= (-5 - 4 + 2)m + 14$
$= (-7)m + 14 = -7m + 14$ or $14 - 7m$

33. $-4.3r + 3.9 - r + .6 + 2.2r$
$= (-4.3r - r + 2.2r) + (3.9 + .6)$
$= (-4.3 - 1 + 2.2)r + 4.5$
$= -3.1r + 4.5$

37. $-7.9q^2 + 2.8q - 14.4 + 5.1q^2 - 9.5q - 6.9$
$= (-7.9q^2 + 5.1q^2) + (2.8q - 9.5q) + (-14.4 - 6.9)$
$= (-7.9 + 5.1)q^2 + (2.8 - 9.5)q + (-21.3)$
$= -2.8q^2 - 6.7q - 21.3$

41. $-3(n + 5) = -3n + (-3)5 = -3n - 15$

45. $-3(2r - 3) + 2(5r + 3)$
$= -3(2r) + (-3)(-3) + 2(5r) + 2(3)$
$= -6r + 9 + 10r + 6$
$= -6r + 10r + 9 + 6$
$= (-6 + 10)r + (9 + 6)$
$= 4r + 15$

49. $-2(-3k + 2) - (5k - 6) - 3k - 5$
$= 6k - 4 - 5k + 6 - 3k - 5$
$= (6 - 5 - 3)k + (-4 + 6 - 5)$
$= -2k - 3$

53. The phrase translates as $[x + (-12)] + 4x$. It simplifies as follows.
$[x + (-12)] + 4x = x + 4x + (-12) = 5x - 12$

57. The phrase translates as $[9(3x + 5)] - [(-8) + 6x]$. It simplifies as follows.
$[9(3x + 5)] - [(-8) + 6x]$
$= 27x + 45 + 8 - 6x$
$= 21x + 53$

61. The additive inverse of -4 is $-(-4) = 4$.

65. In order to get a result of x, we must add -6 to $x + 6$, since $(x + 6) + (-6) = x + [6 + (-6)] = x + 0 = x$. The answer is -6.

CHAPTER 3

Section 3.1 (page 123)

Each of these answers should be checked by substituting in the original equation.

1. $x - 3 = 7$
$x - 3 + 3 = 7 + 3$ Add 3 to both sides.
$x = 10$

5. $3r = 2r + 10$
$3r - 2r = 2r + 10 - 2r$ Subtract $2r$.
$r = 10$

9. $2p + 6 = 10 + p$
$2p + 6 - p = 10 + p - p$ Subtract p.
$p + 6 = 10$ Simplify.
$p + 6 - 6 = 10 - 6$ Subtract 6.
$p = 4$

13. $x - 5 = 2x + 6$
$x - 5 - x = 2x + 6 - x$ Subtract x.
$-5 = x + 6$
$-5 - 6 = x + 6 - 6$ Subtract 6.
$-11 = x$

17. $2p = p + \frac{1}{2}$
$2p - p = p + \frac{1}{2} - p$ Subtract p.
$p = \frac{1}{2}$

21. $6x = 6x + 5$
$6x - 6x = 6x + 5 - 6x$ Subtract $6x$.
$0 = 5$ False.
There is no solution.

25. $4x + 3 + 2x - 5x = 2 + 8$
$x + 3 = 10$ Simplify each side.
$x + 3 - 3 = 10 - 3$ Subtract 3 on both sides.
$x = 7$

29. $11z + 2 + 4z - 3z = 5z - 8 + 6z$
$12z + 2 = 11z - 8$ Simplify.
$-11z + 12z + 2 = -11z + 11z - 8$
Subtract $11z$.
$z + 2 = -8$ Simplify.
$-2 + z + 2 = -2 + (-8)$ Subtract 2.
$z = -10$

33. $(5y + 6) - (3 + 4y) = 9$
$5y + 6 - 3 - 4y = 9$ Distributive property
$y + 3 = 9$ Simplify.
$-3 + y + 3 = -3 + 9$ Subtract 3.
$y = 6$

37. $-6(2a + 1) + 1(13a - 7) = 4$
$-6(2a) - 6(1) + 1(13a) + 1(-7) = 4$
Distributive property
$-12a - 6 + 13a - 7 = 4$
$a - 13 = 4$
Simplify left side.
$a - 13 + 13 = 4 + 13$
Add 13 to both sides.
$a = 17$

41. $-2(8p + 7) - 3(4 - 7p) = 2(3 + 2p) - 6$
$-2(8p) - 2(7) - 3(4) - 3(-7p) = 2(3) + 2(2p) - 6$
Distributive property
$-16p - 14 - 12 + 21p = 6 + 4p - 6$
$5p - 26 = 4p$ Simplify each side.
$5p - 26 - 4p = 4p - 4p$
Subtract $4p$ on each side.
$p - 26 = 0$
$p - 26 + 26 = 0 + 26$
Add 26 to each side.
$p = 26$

45. $2(3x + 4) - 3(7 + x) = 3(x - 4) + 7$
$6x + 8 - 21 - 3x = 3x - 12 + 7$
Distributive property
$3x - 13 = 3x - 5$ Combine terms.
$-3x + 3x - 13 = -3x + 3x - 5$
Subtract $3x$ from each side.
$-13 = -5$ False

There is no solution.

49. $\frac{1}{7}(7y) = \left(\frac{1}{7} \cdot 7\right)y$ Associative property
$= 1y$ Inverse property
$= y$ Identity property

53. $-\frac{5}{8}\left(-\frac{8}{5}r\right) = \left(-\frac{5}{8} \cdot -\frac{8}{5}\right)r$ Associative property
$= 1r$ Inverse property
$= r$ Identity property

Section 3.2 (page 129)

Each of these answers should be checked by substituting in the original equation.

1. $5x = 25$
$\frac{5x}{5} = \frac{25}{5}$ Divide both sides by 5.
$1x = 5$ Simplify.
$x = 5$

5. $3a = -24$
$\frac{3a}{3} = \frac{-24}{3}$ Divide by 3.
$1a = -8$ Simplify.
$a = -8$

9. $-6x = 16$
$\frac{-6x}{-6} = \frac{16}{-6}$ Divide both sides by -6.
$x = -\frac{8}{3}$

13. $5r = 0$
$\frac{5r}{5} = \frac{0}{5}$ Divide both sides by 5.
$1r = 0$ 0 times any number is 0.
$r = 0$

17. $-n = -4$
$(-1) \cdot (-n) = (-1) \cdot (-4)$ Multiply by -1.
$n = 4$

21. $5m + 6m - 2m = 72$
$9m = 72$ Simplify the left side.
$\frac{9m}{9} = \frac{72}{9}$ Divide each side by 9.
$m = 8$

25. $3r - 5r = 6$
$-2r = 6$ Simplify the left side.
$\frac{-2r}{-2} = \frac{6}{-2}$ Divide each side by -2.
$r = -3$

29. $-7y + 8y - 9y = -56$
$-8y = -56$ Simplify.
$y = 7$ Divide each side by -8.

33. $\frac{x}{7} = 7$
$\frac{1}{7}x = 7$ Definition of division
$7 \cdot \frac{1}{7}x = 7 \cdot 7$ Multiply each side by 7.
$1x = 49$ Simplify.
$x = 49$

37. $\frac{15}{2}z = 20$
$\frac{2}{15} \cdot \frac{15}{2}z = \frac{2}{15} \cdot 20$ Multiply each side by $\frac{2}{15}$.
$1z = \frac{8}{3}$
$z = \frac{8}{3}$

41. $\frac{2}{3}k = 5$
$\frac{3}{2}\left(\frac{2}{3}k\right) = \frac{3}{2} \cdot 5$ Multiply each side by $\frac{3}{2}$.
$1k = \frac{15}{2}$
$k = \frac{15}{2}$

SOLUTIONS

45. $-\frac{4}{7}r = 2$

$-\frac{7}{4}\left(-\frac{4}{7}r\right) = -\frac{7}{4}(2)$ Multiply each side by $-\frac{7}{4}$.

$1r = -\frac{7}{2}$

$r = -\frac{7}{2}$

49. $1.7p = 5.1$

$\frac{1.7p}{1.7} = \frac{5.1}{1.7}$ Divide each side by 1.7.

$1p = 3$

$p = 3$

53. $9(2q + 7) = 9 \cdot 2q + 9 \cdot 7$ Distributive property

$= 18q + 63$

57. $-(2 - 5r) + 6r = -2 + 5r + 6r$ Distributive property

$= -2 + 11r$

Section 3.3 (page 137)

Each of these answers should be checked by substituting in the original equation.

1. $4h + 8 = 16$

$4h + 8 - 8 = 16 - 8$ Subtract 8 on each side.

$4h = 8$

$\frac{4h}{4} = \frac{8}{4}$ Divide each side by 4.

$h = 2$

5. $12p + 18 = 14p$

$-12p + 12p + 18 = -12p + 14p$ Subtract $12p$.

$18 = 2p$

$\frac{18}{2} = \frac{2p}{2}$ Divide by 2.

$9 = p$

9. $2(2r - 1) = -3(r + 3)$

$4r - 2 = -3r - 9$ Distributive property

$7r - 2 = -9$ Add $3r$ to each side.

$7r = -7$ Add 2 to each side.

$\frac{7r}{7} = \frac{-7}{7}$ Divide each side by 7.

$r = -1$ Simplify.

13. $3(5 + 1.4x) = 3x$

$15 + 4.2x = 3x$ Distributive property

$15 + 4.2x - 3x = 3x - 3x$ Subtract $3x$ from each side.

$15 + 1.2x = 0$

$-15 + 15 + 1.2x = -15 + 0$ Subtract 15 from each side.

$1.2x = -15$

$\frac{1.2x}{1.2} = \frac{-15}{1.2}$ Divide each side by 1.2.

$x = -\frac{15}{1.2}$

$x = -12.5$

17. $9(2x - 4) = 18(x - 2)$

$18x - 36 = 18x - 36$ Distributive property

$-18x + 18x - 36 = -18x + 18x - 36$ Subtract $18x$ on each side.

$-36 = -36$ True

The solution includes all real numbers.

21. $-5k - 8 = 2(k + 6) + 1$

$-5k - 8 = 2k + 12 + 1$ Distributive property

$-5k - 8 = 2k + 13$ Simplify.

$5k - 5k - 8 = 5k + 2k + 13$ Add $5k$.

$-8 = 7k + 13$ Simplify

$-13 - 8 = -13 + 7k + 13$ Subtract 13.

$-21 = 7k$ Simplify.

$\frac{-21}{7} = \frac{7k}{7}$ Divide by 7.

$-3 = k$

25. $5(4t + 3) = 6(3t + 2) - 1$

$20t + 15 = 18t + 12 - 1$ Distributive property

$20t + 15 = 18t + 11$ Simplify.

$-18t + 20t + 15 = -18t + 18t + 11$ Subtract $18t$.

$2t + 15 = 11$

$-15 + 2t + 15 = -15 + 11$ Subtract 15.

$2t = -4$ Simplify.

$\frac{2t}{2} = \frac{-4}{2}$ Divide by 2.

$t = -2$

29. $-2(3s + 9) - 6 = -3(4s + 11) - 6$

$-6s - 18 - 6 = -12s - 33 - 6$ Distributive property

$-6s - 24 = -12s - 39$ Simplify.

$12s - 6s - 24 = 12s - 12s - 39$ Add $12s$.

$6s - 24 = -39$ Simplify.

$24 + 6s - 24 = 24 - 39$ Add 24.

$6s = -15$ Simplify.

$\frac{6s}{6} = \frac{-15}{6}$ Divide by 6.

$s = -\frac{5}{2}$

33. $-(4m + 2) - (-3m - 5) = 3$

$-4m - 2 + 3m + 5 = 3$

$-m + 3 = 3$ Simplify.

$-3 - m + 3 = -3 + 3$ Subtract 3.

$-m = 0$

$(-1)(-m) = (-1)0$ Multiply by -1.

$m = 0$

37. $3(4x + 2) - 2(5x - 1) = 0$

$12x + 6 - 10x + 2 = 0$ Simplify.

$2x + 8 = 0$

$2x + 8 - 8 = 0 - 8$ Subtract 8.

$2x = -8$

$\frac{2x}{2} = \frac{-8}{2}$ Divide by 2.

$x = -4$

41. $4(x + 8) = 2(2x + 5) + 22$

$4x + 32 = 4x + 10 + 22$ Distributive property

$4x + 32 = 4x + 32$ Combine terms.

$-4x + 4x + 32 = -4x + 4x + 32$ Subtract $4x$ on each side.

$32 = 32$ True

The solution includes all real numbers.

45. "-1 added to a number" becomes
$x + (-1)$.

49. "A number decreased by 6" becomes
$x - 6$.

53. "Double a number" becomes
$2x$.

57. "The quotient of -9 and a number" becomes
$\frac{-9}{x}$.

61. "A number subtracted from its reciprocal" becomes
$\frac{1}{x} - x$.

65. Seven times a number is -84.

$$7x = -84$$
$$\frac{7x}{7} = \frac{-84}{7} \quad \text{Divide by 7 on each side.}$$
$$x = -12$$

69. A number multiplied by $\frac{1}{2}$ is 4.

$$\frac{1}{2}x = 4$$
$$2\left(\frac{1}{2}x\right) = 2(4) \quad \text{Multiply each side by 2.}$$
$$x = 8$$

Section 3.4 (page 145)

1. Let x be the number.

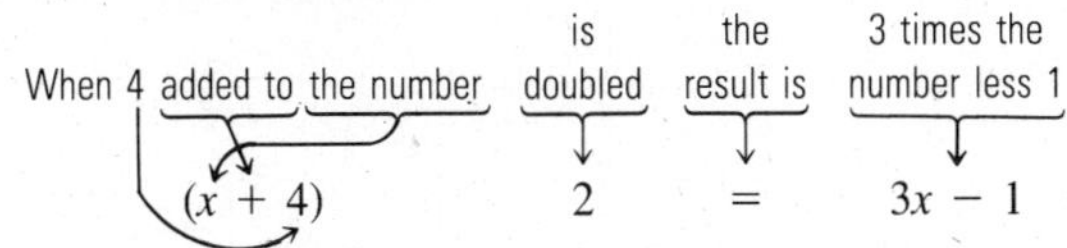

Solve the equation:

$$2x + 8 = 3x - 1$$
$$-2x + 2x + 8 = -2x + 3x - 1$$
$$8 = x - 1$$
$$8 + 1 = x - 1 + 1$$
$$9 = x$$

The number is 9.

5. Let x be the number of votes for Hartman.
Then $x + 80$ is the number of votes for Sprague.

Number for Hartman → x; plus → $+$; number for Sprague → $x + 80$; is → $=$; total votes cast → 346

Solve the equation:

$$x + x + 80 = 346$$
$$2x + 80 = 346$$
$$2x + 80 - 80 = 346 - 80$$
$$2x = 266$$
$$\frac{2x}{2} = \frac{266}{2}$$
$$x = 133$$

Hartman received 133 votes.

9. Let x be the number of women.
Then $x - 2$ is the number of men.

Number of men → $x - 2$; plus → $+$; number of women → x; gives → $=$; total customers. → 20

Solve the equation:

$$x - 2 + x = 20$$
$$2x - 2 = 20$$
$$2x - 2 + 2 = 20 + 2$$
$$2x = 22$$
$$\frac{2x}{2} = \frac{22}{2}$$
$$x = 11 \quad \text{and} \quad x - 2 = 11 - 2 = 9$$

There were 11 women and 9 men.

13. Let x be the length of the middle-sized piece.
Then $3x$ is the length of the longest piece,
and $3x - 23$ is the length of the shortest piece.

Length of shortest piece → $3x - 23$; plus → $+$; length of middle-sized piece → x; plus → $+$; length of longest piece → $3x$; equals → $=$; total length. → 40

Solve the equation:

$$3x - 23 + x + 3x = 40$$
$$7x - 23 = 40$$
$$7x - 23 + 23 = 40 + 23$$
$$7x = 63$$
$$\frac{7x}{7} = \frac{63}{7}$$
$$x = 9$$

$3x = 3 \cdot 9 = 27$, $3x - 23 = 3 \cdot 9 - 23 = 4$.
The lengths are 4, 9, and 27 centimeters.

17. Let x be the measure of angles A and B.
Then $x + 24$ is the measure of angle C.

Measure of A → x; plus → $+$; measure of B → x; plus → $+$; measure of C → $x + 24$; is → $=$; 180°. → 180

Solve the equation:

$$x + x + x + 24 = 180$$
$$3x + 24 = 180$$
$$3x + 24 - 24 = 180 - 24$$
$$3x = 156$$
$$\frac{3x}{3} = \frac{156}{3}$$

$x = 52$, $x + 24 = 52 + 24 = 76$.
The angle measures are 52°, 52°, and 76°.

21. Let x be the number of ounces of cashews.
Then $5x$ is the number of ounces of peanuts.

Ounces of cashews → x; plus → $+$; ounces of peanuts → $5x$; equals → $=$; total ounces. → 27

Solve the equation:

$$x + 5x = 27$$
$$6x = 27$$
$$x = \frac{27}{6} \text{ or } 4\frac{1}{2}, \quad 5x = 22\frac{1}{2}.$$

There are $4\frac{1}{2}$ ounces of cashews and $22\frac{1}{2}$ ounces of peanuts in the mixture.

SOLUTIONS

25. Let x be the first even integer.
Then $x + 2$ is the next even integer.

The smaller integer	added to	three times the larger	equals	46.
↓	↓	↓	↓	↓
x	$+$	$3(x + 2)$	$=$	46

Solve the equation:

$$x + 3(x + 2) = 46$$
$$x + 3x + 6 = 46$$
$$4x + 6 = 46$$
$$4x = 40$$
$$x = 10, \quad x + 2 = 12.$$

The integers are 10 and 12.

29. Let x be the number of pint cartons.
Then $6x$ is the number of quart cartons.

Number of quarts in pint cartons	plus	number of quarts in quart cartons	equals	39.
↓	↓	↓	↓	↓
$\frac{x}{2}$	$+$	$6x$	$=$	39

Solve the equation:

$$\frac{x}{2} + 6x = 39$$
$$\frac{1}{2}x + \frac{12}{2}x = 39$$
$$\frac{13}{2}x = 39$$
$$x = 6, \quad 6x = 36.$$

There are 6 pint cartons and 36 quart cartons.

33. $prt = (4000)(.08)(2)$ Let $p = 4000$, $r = .08$, $t = 2$.
$= 640$

Section 3.5 (page 153)

1. $P = 2L + 2W$
$P = 2(5) + 2(3)$ Substitute $L = 5$, $W = 3$.
$P = 10 + 6$
$P = 16$

5. $d = rt$
$100 = (r)(5)$ Substitute $d = 100$, $t = 5$.
$100 = 5r$ Commutative property
$\frac{100}{5} = \frac{5r}{5}$ Divide both sides by 5.
$20 = r$

9. $P = 2L + 2W$
$180 = 2(50) + 2W$ Substitute $P = 180$, $L = 50$.
$180 = 100 + 2W$
$180 - 100 = 100 + 2W - 100$ Subtract 100 from both sides.
$80 = 2W$
$\frac{80}{2} = \frac{2W}{2}$ Divide both sides by 2.
$40 = W$

13. $C = 2\pi r$
$25.12 = (2)(3.14)r$ Substitute $C = 25.12$, $\pi = 3.14$.
$25.12 = 6.28r$ Simplify.
$\frac{25.12}{6.28} = \frac{6.28r}{6.28}$ Divide both sides by 6.28.
$4 = r$

17. $I = prt$
$60 = (150)(.08)t$ Substitute $I = 60$, $p = 150$, $r = .08$.
$60 = 12t$ Simplify.
$5 = t$ Solve.

21. $A = \frac{1}{2}(b + B)h$
$70 = \frac{1}{2}(15 + 20)h$ Substitute $b = 15$, $B = 20$, $A = 70$.
$70 = \frac{1}{2}(35)h$
$\left(\frac{2}{35}\right)(70) = \left(\frac{2}{35}\right)\left(\frac{35}{2}h\right)$ Multiply both sides by $\frac{2}{35}$.
$4 = h$

25. To find the width, substitute $P = 40$ and $L = 12$ in the formula $P = 2L + 2W$.
$P = 2L + 2W$
$40 = 2(12) + 2W$
$40 = 24 + 2W$
$16 = 2W$
$8 = W$
The width is 8 inches.

29. To find the height, substitute $A = 60$ and $b = 24$ in the formula $A = \frac{1}{2}bh$.
$A = \frac{1}{2}bh$
$60 = \frac{1}{2}(24)h$
$60 = 12h$
$5 = h$
The height is 5 centimeters.

33. To find the height, substitute $S = 339.12$, $\pi = 3.14$, and $r = 3$ in the formula $S = 2\pi r^2 + 2\pi rh$.
$S = 2\pi r^2 + 2\pi rh$
$S = 2\pi r(r + h)$ Distributive property
$339.12 = 2(3.14)(3)(3 + h)$
Substitute $S = 339.12$, $\pi = 3.14$, $r = 3$.
$339.12 = 18.84(3 + h)$
$\frac{339.12}{18.84} = \frac{18.84(3 + h)}{18.84}$
$18 = 3 + h$
$18 - 3 = 3 + h - 3$
$15 = h$
The height is 15 inches.

37. $d = rt$

$\frac{d}{r} = \frac{rt}{r}$ Divide each side by r.

$\frac{d}{r} = t$ or $t = \frac{d}{r}$

41. $I = prt$

$\frac{I}{pr} = \frac{prt}{pr}$ Divide each side by pr.

$\frac{I}{pr} = t$ or $t = \frac{I}{pr}$

45. $A = \frac{1}{2}bh$

$2A = 2\left(\frac{1}{2}bh\right)$ Multiply each side by 2.

$2A = bh$

$\frac{2A}{h} = \frac{bh}{h}$ Divide each side by h.

$\frac{2A}{h} = b$ or $b = \frac{2A}{h}$

49. $P = 2L + 2W$

$P - 2L = 2W$ Subtract $2L$.

$\frac{P - 2L}{2} = W$ Divide by 2.

53. $d = gt^2 + vt$

$d - gt^2 = vt$ Subtract gt^2 on each side.

$\frac{d - gt^2}{t} = v$ Divide each side by t.

57. $9y = 54$

$\frac{9y}{9} = \frac{54}{9}$ Divide each side by 9.

$y = 6$

61. $-\frac{4}{7}r = 16$

$\left(-\frac{7}{4}\right)\left(-\frac{4}{7}\right)r = \left(-\frac{7}{4}\right)(16)$ Multiply each side by $-\frac{7}{4}$.

$r = -28$

Section 3.6 (page 161)

1. $\frac{80}{50} = \frac{8}{5}$

5. Change yards to feet. Since 1 yard = 3 feet, 12 yards = 12(3) feet = 36 feet.

$\frac{36}{15} = \frac{12}{5}$

9. Change hours to minutes. Since 1 hour = 60 minutes, 3 hours = 3(60) minutes = 180 minutes.

$\frac{15}{180} = \frac{1}{12}$

13. Change days to hours. Since 1 day = 24 hours, 3 days = 3(24) hours = 72 hours.

$\frac{72}{16} = \frac{9}{2}$

17. $\frac{9}{10} \times \frac{18}{20}$ $10 \cdot 18 = 180$, $9 \cdot 20 = 180$ Same, so proportion is true

21. $\frac{12}{7} = \frac{36}{20}$

$12 \cdot 20 = 240$ and $7 \cdot 36 = 252$

$240 \neq 252$, so proportion is false.

25. $\frac{19}{30} = \frac{57}{90}$

$19 \cdot 90 = 1710$ and $30 \cdot 57 = 1710$

Since $1710 = 1710$, the proportion is true.

29. $\frac{6.3}{4.5} = \frac{3.5}{2.5}$

$(6.3)(2.5) = 15.75$ and $(4.5)(3.5) = 15.75$

Since $15.75 = 15.75$, the proportion is true.

33. $\frac{z}{56} = \frac{7}{8}$

$8z = 7(56)$

$8z = 392$

$z = \frac{392}{8} = 49$

37. $\frac{25}{100} = \frac{8}{m + 6}$

$25(m + 6) = 8(100)$

$25m + 150 = 800$

$25m = 650$

$m = \frac{650}{25} = 26$

41. $\frac{3}{4} = \frac{3n - 1}{10}$

$3(10) = 4(3n - 1)$

$30 = 12n - 4$

$34 = 12n$

$n = \frac{34}{12} = \frac{17}{6}$

45. $\frac{r + 1}{r} = \frac{1}{3}$

$3(r + 1) = 1(r)$

$3r + 3 = r$

$3r = r - 3$

$2r = -3$

$r = -\frac{3}{2}$

49. $\frac{r + 8}{r - 9} = \frac{7}{3}$

$3(r + 8) = 7(r - 9)$

$3r + 24 = 7r - 63$

$-4r = -87$

$r = \frac{-87}{-4} = \frac{87}{4}$

53. Let x be the number of bars.

Then $\frac{x \text{ bars}}{500 \text{ calories}} = \frac{1 \text{ bar}}{200 \text{ calories}}$.

Solve the equation:

$\frac{x}{500} = \frac{1}{200}$

$(x)(200) = (500)(1)$

$200x = 500$

$x = \frac{5}{2}$.

You need to eat $\frac{5}{2}$ or $2\frac{1}{2}$ Hershey bars.

SOLUTIONS

57. Let x be the number of inches.

Then $\frac{x \text{ inches}}{24 \text{ miles}} = \frac{3 \text{ inches}}{8 \text{ miles}}$.

Solve the equation:

$$\frac{x}{24} = \frac{3}{8}$$
$$(x)(8) = (24)(3)$$
$$8x = 72$$
$$x = 9.$$

It takes 9 inches.

61. Let x be the distance in miles. Then

$\frac{x \text{ miles}}{15 \text{ inches}} = \frac{308 \text{ miles}}{11 \text{ inches}}$.

Solve the equation:

$$\frac{x}{15} = \frac{308}{11}$$
$$(x)(11) = (15)(308)$$
$$11x = 4620$$
$$x = 420.$$

The two cities are 420 miles apart.

65. Since -10 is less than 7, use $<$.

69. Since -6 is less than -4, use $<$.

Section 3.7 (page 173)

All the graphs for this section can be found in the answer section.

1. Use a dot at 4, and draw an arrow to the left.

5. Use a dot at -2 and one at 5; draw a line segment between them.

9.
$$a + 6 < 8$$
$$-6 + a + 6 < -6 + 8 \quad \text{Subtract 6.}$$
$$a < 2$$

13.
$$-3 + k \geq 2$$
$$3 - 3 + k \geq 3 + 2 \quad \text{Add 3.}$$
$$k \geq 5$$

17.
$$-2k \leq 12$$
$$\frac{-2k}{-2} \geq \frac{12}{-2} \quad \text{Divide by } -2\text{; reverse the inequality symbol.}$$
$$k \geq -6$$

21.
$$3n + 5 \leq 2n - 6$$
$$-2n + 3n + 5 \leq -2n + 2n - 6$$
Subtract $2n$ on each side.
$$n + 5 \leq -6$$
$$n + 5 - 5 \leq -6 - 5$$
Subtract 5 on each side.
$$n \leq -11$$

25.
$$4k + 1 \geq 2k - 9$$
$$4k + 1 - 2k \geq 2k - 9 - 2k \quad \text{Subtract } 2k.$$
$$2k + 1 \geq -9$$
$$2k + 1 - 1 \geq -9 - 1 \quad \text{Subtract 1.}$$
$$2k \geq -10$$
$$\frac{2k}{2} \geq \frac{-10}{2} \quad \text{Divide by 2.}$$
$$k \geq -5$$

29.
$$3 - 2z + 2 > 4 - z$$
$$5 - 2z > 4 - z$$
$$5 - 2z + z > 4 - z + z \quad \text{Add } z.$$
$$5 - z > 4$$
$$5 - z - 5 > 4 - 5 \quad \text{Subtract 5.}$$
$$-z > -1$$
$$(-1)(-z) < (-1)(-1) \quad \text{Multiply by } -1 \text{ and reverse symbol.}$$
$$z < 1$$

33.
$$-k + 4 + 7k \leq -1 + 3k + 5$$
$$6k + 4 \leq 4 + 3k$$
$$6k + 4 - 3k \leq 4 + 3k - 3k \quad \text{Subtract } 3k.$$
$$3k + 4 \leq 4$$
$$3k + 4 - 4 \leq 4 - 4 \quad \text{Subtract 4.}$$
$$3k \leq 0$$
$$\frac{3k}{3} \leq \frac{0}{3}$$
$$k \leq 0$$

37.
$$8 \leq p + 2 \leq 15$$
$$8 - 2 \leq p + 2 - 2 \leq 15 - 2 \quad \text{Subtract 2 from each part of the inequality.}$$
$$6 \leq p \leq 13$$

41.
$$-5 \leq 2x - 3 \leq 9$$
$$-2 \leq 2x \leq 12 \quad \text{Add 3.}$$
$$-1 \leq x \leq 6 \quad \text{Divide by 2.}$$

45. Let x be the unknown number.

Half a number	added to	5	is greater than or equal to	-3.
↓	↓	↓	↓	↓
$\frac{1}{2}x$	$+$	5	$\geq$	-3

Solve: $\frac{1}{2}x + 5 \geq -3$

$$\frac{1}{2}x \geq -8$$
$$x \geq -16.$$

The solution includes all numbers greater than or equal to -16.

49. Let x be the amount she must earn in October. The average salary for the four months is equal to the total salary for the four months divided by 4.

Average salary	is at least	250.
↓	↓	↓
$\frac{200 + 300 + 225 + x}{4}$	$\geq$	250

Solve: $\frac{725 + x}{4} \geq 250$

$$4\left(\frac{725 + x}{4}\right) \geq 4(250)$$
$$725 + x \geq 1000$$
$$x \geq 275.$$

She must earn at least $275 in October.

53. $2 \cdot 2 \cdot 2 \cdot 2 \cdot 2 = 32$

57. $\frac{2}{3} \cdot \frac{2}{3} \cdot \frac{2}{3} = \frac{8}{27}$

61. $3 - (-2) = 3 + 2 = 5$

65. $-2 - [-4 - (3 - 2)] = -2 - [-4 - 1]$
$= -2 - [-5]$
$= -2 + 5$
$= 3$

CHAPTER 4

Section 4.1 (page 201)

1. 5^{12} ← Exponent; 5 ← Base

5. The negative sign in front of 125 is not part of the base, so the base is 125 and the exponent is 3.

9. Only m is under the exponent, so m is the base and 2 the exponent. The number 3 is a coefficient here.

13. 3 is used as a factor 5 times, so 3 is the base and 5 is the exponent: 3^5.

17. p is used as a factor 5 times, so p is the base and 5 the exponent: p^5.

21. 2 is used as a factor 5 times, giving $\frac{1}{2^5}$.

25. $3^2 + 3^4 = 3 \cdot 3 + 3 \cdot 3 \cdot 3 \cdot 3$
$= 9 + 81$
$= 90$

29. $2^2 + 2^5 = 2 \cdot 2 + 2 \cdot 2 \cdot 2 \cdot 2 \cdot 2$
$= 4 + 32$
$= 36$

33. $4^2 \cdot 4^3 = 4^{2+3}$ Add exponents.
$= 4^5$

37. $4^3 \cdot 4^5 \cdot 4^{10} = 4^{3+5+10}$ Add exponents.
$= 4^{18}$

41. $y^3 \cdot y^4 \cdot y^7 = y^{3+4+7}$ Add exponents.
$= y^{14}$

45. $(-9r^3)(7r^6) = (-9 \cdot 7)(r^3 \cdot r^6)$ Associative property
$= -63r^9$

49. $4m^3 + 9m^3 = (4 + 9)m^3 = 13m^3$
$(4m^3)(9m^3) = (4 \cdot 9)(m^3 \cdot m^3) = 36m^6$

53. $7r + 3r + 5r = (7 + 3 + 5)r$
$= 15r$
$(7r)(3r)(5r) = (7 \cdot 3 \cdot 5)r^{1+1+1}$
$= 105r^3$

57. $(6^3)^2 = 6^{3 \cdot 2}$ Use the power rule.
$= 6^6$

61. $(-5^2)^4 = [(-1)(5)^2]^4$ $\quad -a = (-1)(a)$
$= (-1)^4(5^2)^4$ Use the power rule.
$= 1 \cdot 5^8$ $\quad (-1)^4 = 1$
$= 5^8$

65. $(5m)^3 = (5^1m^1)^3$
$= 5^{1 \cdot 3}m^{1 \cdot 3}$ Use the power rule.
$= 5^3m^3$ Multiply the exponents.

69. $\left(\frac{-3x^5}{4}\right)^2 = \frac{(-3)^2(x^{10})}{4^2}$

73. $\left(\frac{4}{3}\right)^5 \cdot (4)^3 = \frac{4^5}{3^5} \cdot \frac{4^3}{1} = \frac{4^{5+3}}{3^5 \cdot 1} = \frac{4^8}{3^5}$

77. $(3m)^2(3m)^5 = (3^2m^2)(3^5m^5)$
$= (3^2 \cdot 3^5)(m^2m^5) = 3^7m^7$

81. $(2m^2n)^3(mn^2) = [2^3(m^2)^3n^3](mn^2)$
$= (2^3m^6n^3)(mn^2)$
$= 2^3m^{6+1}n^{3+2}$ Add exponents.
$= 2^3m^7n^5$

85. (a) $-(2)^2 + 4(2) - 7 = -4 + 8 - 7 = -3$
(b) $-(-3)^2 + 4(-3) - 7 = -9 + (-12) - 7 = -28$

Section 4.2 (page 209)

1. $4^0 + 5^0 = 1 + 1 = 2$

In Exercises 5 and 9, use the definition of negative exponents.

5. $3^{-3} = \frac{1}{3^3}$
$= \frac{1}{27}$

9. $\left(\frac{1}{2}\right)^{-5} = \left(\frac{2}{1}\right)^5 = \frac{2^5}{1} = \frac{2 \cdot 2 \cdot 2 \cdot 2 \cdot 2}{1} = 32$

13. $\frac{4^7}{4^2} = 4^{7-2}$ Subtract exponents.
$= 4^5$

17. $\frac{6^{-4}}{6^2} = 6^{-4-2}$ Subtract exponents.
$= 6^{-6}$
$= \frac{1}{6^6}$

21. $\frac{1}{2^{-5}} = 2^5$

25. $\frac{4^3 \cdot 4^{-5}}{4^7} = \frac{4^{3+(-5)}}{4^7}$
$= \frac{4^{-2}}{4^7} = 4^{-2-7} = 4^{-9} = \frac{1}{4^9}$

29. $\frac{64^6}{32^6} = \left(\frac{64}{32}\right)^6 = 2^6$

33. $(5m^{-2})^2 = 5^2(m^{-2})^2$
$= 5^2m^{-4}$
$= \frac{5^2}{m^4}$

37. $\frac{3^{-1}a^{-2}}{3^2\,a^{-4}} = \frac{a^4}{3^13^2\,a^2}$
$= \frac{a^4}{3^3a^2}$
$= \frac{a^2}{3^3}$

41. $\frac{(4a^2b^3)^{-2}}{(a^3b)^{-4}} = \frac{4^{-2}a^{-4}b^{-6}}{a^{-12}b^{-4}}$
$= \frac{a^{12}b^4}{4^2a^4b^6}$
$= \frac{a^8}{4^2b^2}$

45. $\frac{(9^{-1}z^{-2}x)^{-1}}{(5z^{-2}x^{-3})^2} = \frac{9^1z^2x^{-1}}{5^2z^{-4}x^{-6}}$
$= \frac{9x^6z^2z^4}{5^2x}$
$= \frac{9x^6z^6}{5^2x}$
$= \frac{9x^5z^6}{5^2}$

49. $1000(1.23) = 1230$ (To multiply by 1000, move the decimal point 3 places to the right. Use a 0 placeholder.)

53. $237 \div 1000 = .237$ (To divide by 1000, move the decimal point 3 places to the left.)

Section 4.3 (page 213)

1. In 6,835,000,000, count from the right of the first nonzero digit (the 6), all the way to the decimal point (9 places). Using 9 as the exponent of 10, we have

6.835×10^9.

5. Place a caret after the first nonzero digit.

$.01_{\wedge}01$ 2 places

Moving the decimal will make the number larger, so the exponent on 10 is negative.

$.0101 = 1.01 \times 10^{-2}$

9. Place a caret after the first nonzero digit.

$.006_{\wedge}9$ 3 places

Moving the decimal will make the number larger, so the exponent on 10 is negative.

$.0069 = 6.9 \times 10^{-3}$

13. $3.24 \times 10^8 = 324{,}000{,}000$ Move decimal point 8 places to the right.

17. $(5 \times 10^4) \times (3 \times 10^{-2})$

$= (5 \times 3) \times (10^4 \times 10^{-2})$

$= 15 \times 10^{4+(-2)}$

$= 15 \times 10^2$

$= 1500$

21. $(3 \times 10^{-5}) \times (3 \times 10^2) \times (5 \times 10^{-2})$

$= (3 \times 3 \times 5) \times (10^{-5} \times 10^2 \times 10^{-2})$

$= 45 \times 10^{-5+2+(-2)}$

$= 45 \times 10^{-5}$

$= .00045$

25. $\dfrac{5 \times 10^{-1}}{1 \times 10^{-5}} = \dfrac{5}{1} \times \dfrac{10^{-1}}{10^{-5}}$

$= 5 \times 10^{-1-(-5)}$

$= 5 \times 10^{-1+5}$

$= 5 \times 10^4$

$= 50{,}000$

29. $\dfrac{8.7 \times 10^{-2} \times 1.2 \times 10^{-6}}{3 \times 10^{-4} \times 2.9 \times 10^{11}}$

$= \dfrac{8.7 \times 1.2}{3 \times 2.9} \times \dfrac{10^{-2} \times 10^{-6}}{10^{-4} \times 10^{11}}$

$= \dfrac{8.7}{2.9} \times \dfrac{1.2}{3} \times \dfrac{10^{-2+(-6)}}{10^{-4+11}}$

$= 1.2 \times \dfrac{10^{-8}}{10^7}$

$= 1.2 \times 10^{-8-7}$

$= 1.2 \times 10^{-15}$

$= .0000000000000012$

33. $8_{\wedge}65{,}000$ Count from left to right 5 places.

$865{,}000 = 8.65 \times 10^5$

37. $1 \times 10^9 = 1{,}000{,}000{,}000$

Move decimal point 9 places to the right.

41. $8x - 9x + 10x = (8 - 9 + 10)x = 9x$

Section 4.4 (page 219)

1. $3m^5 + 5m^5 = (3 + 5)m^5 = 8m^5$

5. $2m^5 - 5m^2$

Since the exponents are different, the terms are unlike terms, so the expression cannot be simplified.

9. $-4p^7 + 8p^7 - 5p^7 = (-4 + 8 - 5)p^7$

$= -1p^7$

$= -p^7$

13. $-5p^5 + 8p^5 - 2p^5 - 1p^5 = (-5 + 8 - 2 - 1)p^5$

$= 0 \cdot p^5$

$= 0$ Zero times any number is zero.

17. $4z^5 - 9z^3 + 8z^2 + 10z^5$

$= (4z^5 + 10z^5) - 9z^3 + 8z^2$

$= 14z^5 - 9z^3 + 8z^2$

21. A binomial is a special kind of polynomial, so the statement is always true.

25. A binomial must have just two terms and a trinomial has three, so the statement is never true.

29. The polynomial is already simplified; the degree is the highest exponent, 9; there are three terms, so it is a trinomial.

33. $\dfrac{3}{5}x^5 + \dfrac{2}{5}x^5 = \dfrac{5}{5}x^5 = 1x^5 = x^5$

The degree is 5; it is a monomial (one term).

37.

$$\begin{array}{l} 3m^2 + 5m \\ \underline{2m^2 - 2m} \\ 5m^2 + 3m \end{array}$$

Add column by column.

41.

$$\begin{array}{l} 2n^5 - 5n^3 + 6 \\ \underline{3n^5 + 7n^3 + 8} \end{array}$$

Change all signs in the second row, then add.

$$\begin{array}{l} 2n^5 - 5n^3 + 6 \\ \underline{-3n^5 - 7n^3 - 8} \\ -n^5 - 12n^3 - 2 \end{array}$$

45.

$$\begin{array}{l} 12m^2 - 8m + 6 \\ \underline{3m^2 + 5m - 2} \\ 15m^2 - 3m + 4 \end{array}$$

Add column by column.

49. $(2r^2 + 3r) - (3r^2 + 5r)$

$= (2r^2 + 3r) + (-3r^2 - 5r)$

$= 2r^2 + 3r - 3r^2 - 5r$

$= 2r^2 - 3r^2 + 3r - 5r$

$= -r^2 - 2r$

53. $8 - (6s^2 - 5s + 7)$

$= 8 + (-6s^2 + 5s - 7)$

$= 8 - 6s^2 + 5s - 7$

$= -6s^2 + 5s + 8 - 7$

$= -6s^2 + 5s + 1$

57. $(16x^3 - x^2 + 3x) + (-12x^3 + 3x^2 + 2x)$

$= 16x^3 - x^2 + 3x - 12x^3 + 3x^2 + 2x$

$= 16x^3 - 12x^3 - x^2 + 3x^2 + 3x + 2x$

$= 4x^3 + 2x^2 + 5x$

61. $(9a^4 - 3a^2 + 2) + (4a^4 - 4a^2 + 2) + (-12a^4 + 6a^2 - 3)$

$= 9a^4 - 3a^2 + 2 + 4a^4 - 4a^2 + 2 - 12a^4 + 6a^2 - 3$

$= 9a^4 + 4a^4 - 12a^4 - 3a^2 - 4a^2 + 6a^2 + 2 + 2 - 3$

$= a^4 - a^2 + 1$

65. Add the two binomials. The result is larger than 8, so use the symbol $>$.

$$(-9x + 2) + (4 + x^2) > 8$$

69. $p(2p) = 2p^{1+1} = 2p^2$

73. $7m^3(8m^2) = (7 \cdot 8)(m^3m^2) = 56m^{3+2} = 56m^5$

Section 4.5 (page 227)

1. $(-4x^5)(8x^2) = (-4)(8)(x^5)(x^2) = -32x^7$

5. $(15a^4)(2a^5) = (15 \cdot 2)a^{4+5} = 30a^9$

9. $3p(-2p^3 + 4p^2) = (3p)(-2p^3) + (3p)(4p^2)$
$= -6p^4 + 12p^3$

13. $2y(3 + 2y + 5y^4)$
$= (2y)(3) + (2y)(2y) + (2y)(5y^4)$
$= 6y + 4y^2 + 10y^5$

17.
$$\begin{array}{r} x + 5 \\ x - 5 \\ \hline -5x - 25 \\ x^2 + 5x \quad\;\; \\ \hline x^2 \qquad - 25 \end{array}$$

21.
$$\begin{array}{r} 6p + 5 \\ p - 1 \\ \hline -6p - 5 \\ 6p^2 + 5p \quad\;\; \\ \hline 6p^2 - p - 5 \end{array}$$

25.
$$\begin{array}{r} b + 8 \\ 6b - 2 \\ \hline -2b - 16 \\ 6b^2 + 48b \quad\;\;\; \\ \hline 6b^2 + 46b - 16 \end{array}$$

29.
$$\begin{array}{r} -4h + k \\ 2h - k \\ \hline 4hk - k^2 \\ -8h^2 + 2hk \quad\;\;\; \\ \hline -8h^2 + 6hk - k^2 \end{array}$$

33.
$$\begin{array}{r} 5m^3 - 4m^2 + m - 5 \\ 4m + 3 \\ \hline 15m^3 - 12m^2 + 3m - 15 \\ 20m^4 - 16m^3 + 4m^2 - 20m \quad\;\;\; \\ \hline 20m^4 - m^3 - 8m^2 - 17m - 15 \end{array}$$

37.
$$\begin{array}{r} 5x^2 + 2x + 1 \\ x^2 - 3x + 5 \\ \hline 25x^2 + 10x + 5 \\ -15x^3 - 6x^2 - 3x \quad\;\; \\ 5x^4 + 2x^3 + x^2 \quad\;\;\;\;\;\;\;\;\;\; \\ \hline 5x^4 - 13x^3 + 20x^2 + 7x + 5 \end{array}$$

41. $(2p - 5)^2 = (2p - 5)(2p - 5)$
$$\begin{array}{r} 2p - 5 \\ 2p - 5 \\ \hline -10p + 25 \\ 4p^2 - 10p \quad\;\;\; \\ \hline 4p^2 - 20p + 25 \end{array}$$

45. $(m - 5)^3 = (m - 5)(m - 5)(m - 5)$

First multiply $(m - 5)(m - 5)$.
$$\begin{array}{r} m - 5 \\ m - 5 \\ \hline -5m + 25 \\ m^2 - 5m \quad\;\;\; \\ \hline m^2 - 10m + 25 \end{array}$$

Now multiply this last result by $(m - 5)$.
$$\begin{array}{r} m^2 - 10m + 25 \\ m - 5 \\ \hline -5m^2 + 50m - 125 \\ m^3 - 10m^2 + 25m \quad\;\;\;\;\; \\ \hline m^3 - 15m^2 + 75m - 125 \end{array}$$

49. $(3r - 2s)^3 = (3r - 2s)(3r - 2s)(3r - 2s)$
$$\begin{array}{r} 3r - 2s \\ 3r - 2s \\ \hline -6rs + 4s^2 \\ 9r^2 - 6rs \quad\;\;\; \\ \hline 9r^2 - 12rs + 4s^2 \end{array}$$

Now multiply this last result by $3r - 2s$.
$$\begin{array}{r} 9r^2 - 12rs + 4s^2 \\ 3r - 2s \\ \hline -18r^2s + 24rs^2 - 8s^3 \\ 27r^3 - 36r^2s + 12rs^2 \quad\;\;\;\;\; \\ \hline 27r^3 - 54r^2s + 36rs^2 - 8s^3 \end{array}$$

53. The two numbers are -7 and 3, since $-7(3) = -21$ and $-7 + 3 = -4$.

Section 4.6 (page 233)

1. $(r - 1)(r + 3)$

F O I L

$= (r)(r) + (r)(3) + (-1)(r) + (-1)(3)$
$= r^2 + 3r - 1r - 3$
$= r^2 + 2r - 3$ Simplify.

5. $(2x - 1)(3x + 2) = 2x(3x) + (2x)(2) - 1(3x) - 1(2)$

F O I L

$= 6x^2 + 4x - 3x - 2$
$= 6x^2 + x - 2$

9. $(a + 4)(2a + 1)$
$= (a)(2a) + (a)(1) + (4)(2a) + (4)(1)$
$= 2a^2 + 1a + 8a + 4$
$= 2a^2 + 9a + 4$

13. $(2a + 4)(3a - 2)$
$= (2a)(3a) + (2a)(-2) + (4)(3a) + (4)(-2)$
$= 6a^2 - 4a + 12a - 8$
$= 6a^2 + 8a - 8$

17. $(-3 + 2r)(4 + r)$
$= (-3)(4) + (-3)(r) + (2r)(4) + (2r)(r)$
$= -12 - 3r + 8r + 2r^2$
$= -12 + 5r + 2r^2$

21. $(p + 3q)(p + q)$
$= (p)(p) + (p)(q) + (3q)(p) + (3q)(q)$
$= p^2 + 1pq + 3pq + 3q^2$
$= p^2 + 4pq + 3q^2$

25. $(8y - 9z)(y + 5z)$
$= (8y)(y) + (8y)(5z) + (-9z)(y) + (-9z)(5z)$
$= 8y^2 + 40yz - 9yz - 45z^2$
$= 8y^2 + 31yz - 45z^2$

29. $(x + 8)^2 = x^2 + 2(x)(8) + 8^2$
$= x^2 + 16x + 64$

33. $\left(6a - \frac{3}{2}b\right)^2$
$= (6a)^2 - 2(6a)\left(\frac{3}{2}b\right) + \left(\frac{3}{2}b\right)^2$
$= 36a^2 - 18ab + \frac{9}{4}b^2$

37. $(p + 2)(p - 2) = (p)^2 - (2)^2$
$= p^2 - 4$

41. $(6a - p)(6a + p) = (6a)^2 - (p)^2$
$= 36a^2 - p^2$

45. $(7y^2 + 10z)(7y^2 - 10z) = (7y^2)^2 - (10z)^2$
$= 49y^4 - 100z^2$

49. Twice a number plus 4 | multiplied by 6 times the number | less 5 | gives | 8

$$[(2x + 4) \cdot 6x] - 5 = 8$$

53. $\dfrac{-10p^4}{2p^3} = \left(\dfrac{-10}{2}\right) \cdot \left(\dfrac{p^4}{p^3}\right) = -5p$

57. $\dfrac{-18p^7q^{10}}{32p^2q^{12}} = \left(\dfrac{-18}{32}\right)\left(\dfrac{p^7}{p^2}\right)\left(\dfrac{q^{10}}{q^{12}}\right) = \dfrac{-9p^5}{16q^2}$

Section 4.7 (page 237)

1. $\dfrac{4x^2}{2x} = \dfrac{4}{2} \cdot \dfrac{x^2}{x^1} = 2x^{2-1} = 2x^1 = 2x$

5. $\dfrac{27k^4m^5}{3km^6} = \dfrac{27}{3} \cdot \dfrac{k^4}{k^1} \cdot \dfrac{m^5}{m^6}$
$= 9 \cdot k^{4-1} \cdot m^{5-6} = 9k^3m^{-1} = \dfrac{9k^3}{m}$

9. $\dfrac{10m^5 - 16m^2 + 8m^3}{2m} = \dfrac{10m^5}{2m} - \dfrac{16m^2}{2m} + \dfrac{8m^3}{2m}$
$= 5m^4 - 8m + 4m^2$

13. $\dfrac{2m^5 - 4m^2 + 8m}{2m} = \dfrac{2m^5}{2m} - \dfrac{4m^2}{2m} + \dfrac{8m}{2m}$
$= m^4 - 2m + 4$

17. $\dfrac{12x^4 - 3x^3 + 3x^2}{3x^2} = \dfrac{12x^4}{3x^2} - \dfrac{3x^3}{3x^2} + \dfrac{3x^2}{3x^2}$
$= 4x^2 - x + 1$

21. $\dfrac{36x + 24x^2 + 3x^3}{3x^2} = \dfrac{36x}{3x^2} + \dfrac{24x^2}{3x^2} + \dfrac{3x^3}{3x^2}$
$= \dfrac{12}{x} + 8 + x$

25. $\dfrac{8k^4 - 12k^3 - 2k^2 + 7k - 3}{2k}$
$= \dfrac{8k^4}{2k} - \dfrac{12k^3}{2k} - \dfrac{2k^2}{2k} + \dfrac{7k}{2k} - \dfrac{3}{2k}$
$= 4k^3 - 6k^2 - k + \dfrac{7}{2} - \dfrac{3}{2k}$

29. $(16y^5 - 8y^2 + 12y) \div (4y^2)$
$= \dfrac{16y^5 - 8y^2 + 12y}{4y^2}$
$= \dfrac{16y^5}{4y^2} - \dfrac{8y^2}{4y^2} + \dfrac{12y}{4y^2}$
$= 4y^3 - 2 + \dfrac{3}{y}$

33. Multiplying the quotient by the divisor yields the original polynomial.
$3x^2(4x^3 + 3x^2 - 4x + 2)$
$= 12x^5 + 9x^4 - 12x^3 + 6x^2$

37. $3m^5(2m^5 - 4m^3 + m^2)$
$= 3m^5(2m^5) + 3m^5(-4m^3) + 3m^5(m^2)$
$= 6m^{10} - 12m^8 + 3m^7$

Section 4.8 (page 243)

1.
$$\begin{array}{r} x + 2 \\ x - 3\overline{)x^2 - x - 6} \\ x^2 - 3x \\ \hline 2x - 6 \\ 2x - 6 \\ \hline 0 \end{array}$$

Subtract by changing signs in the second row and then adding.

There is no remainder. The quotient is $x + 2$.

5.
$$\begin{array}{r} p - 4 \\ p + 6\overline{)p^2 + 2p - 20} \\ p^2 + 6p \\ \hline -4p - 20 \\ -4p - 24 \\ \hline 4 \end{array}$$

Change signs and add.

The remainder is 4. The quotient is $p - 4 + \dfrac{4}{p + 6}$.

9.
$$\begin{array}{r} 6m - 1 \\ 2m - 3\overline{)12m^2 - 20m + 3} \\ 12m^2 - 18m \\ \hline -2m + 3 \\ -2m + 3 \\ \hline 0 \end{array}$$

Change signs and add.

There is no remainder. The quotient is $6m - 1$.

13.
$$\begin{array}{r} 3m + 5 \\ 5m + 3\overline{)15m^2 + 34m + 28} \\ 15m^2 + 9m \\ \hline 25m + 28 \\ 25m + 15 \\ \hline 13 \end{array}$$

Change signs and add.

The remainder is 13. The quotient is
$3m + 5 + \dfrac{13}{5m + 3}$.

17.
$$\begin{array}{r} 9r^3 \qquad -2r + 6 \\ 3r - 4\overline{)27r^4 - 36r^3 - 6r^2 + 26r - 24} \\ 27r^4 - 36r^3 \\ \hline -6r^2 + 26r \\ -6r^2 + 8r \\ \hline 18r - 24 \\ 18r - 24 \\ \hline 0 \end{array}$$

There is no remainder. The quotient is $9r^3 - 2r + 6$.

21.
$$\begin{array}{r} 5z - 1 \\ z^2 + 0z + 2\overline{)5z^3 - z^2 + 10z + 2} \\ 5z^3 + 0z^2 + 10z \\ \hline -z^2 + 0z + 2 \\ -z^2 + 0z - 2 \\ \hline 4 \end{array}$$

The remainder is 4. The quotient is $5z - 1 + \dfrac{4}{z^2 + 2}$.

25.
$$\begin{array}{r} 3r^2 - 5r + 4 \\ 2r^2 + 0r - 3\overline{)6r^4 - 10r^3 - r^2 + 15r - 8} \\ 6r^4 + 0r^3 - 9r^2 \\ \hline -10r^3 + 8r^2 + 15r \\ -10r^3 + 0r^2 + 15r \\ \hline 8r^2 + 0r - 8 \\ 8r^2 + 0r - 12 \\ \hline 4 \end{array}$$

The remainder is 4. The quotient is

$3r^2 - 5r + 4 + \dfrac{4}{2r^2 - 3}$.

29. $$\begin{array}{r} x^2 \qquad\quad + 1 \\ x^2 + 0x - 1\overline{)x^4 + 0x^3 + 0x^2 + 0x - 1} \\ \underline{x^4 + 0x^3 - x^2} \qquad\qquad \\ x^2 + 0x - 1 \\ \underline{x^2 + 0x - 1} \\ 0 \end{array}$$

There is no remainder. The quotient is $x^2 + 1$.

33. The positive integers that are factors of 36 are 1, 2, 3, 4, 6, 9, 12, 18, and 36. Each divides 36 with no remainder.

37. The positive integers that are factors of 41 are 1 and 41. Each divides 41 with no remainder.

CHAPTER 5

Section 5.1 (page 259)

1. Write all the different multiplication problems having 2 positive factors and 27 as an answer:
$1 \cdot 27$
$3 \cdot 9$
List all the numbers used in increasing order: 1, 3, 9, 27.
These are the positive integer factors of 27.

5. The multiplication problems with 100 as an answer are
$1 \cdot 100 = 100$
$2 \cdot 50 = 100$
$4 \cdot 25 = 100$
$5 \cdot 20 = 100$
$10 \cdot 10 = 100$
The positive factors of 100 are 1, 2, 4, 5, 10, 20, 25, 50, and 100.

9. A quick and simple way of finding prime factors of a number is to use "short division." Find the prime factors of 120 by first dividing 120 by the smallest possible prime, 2.
$2\underline{|120}$
$\quad 60$
Now use 2 again, and divide it into 60.
$2\underline{|120}$
$2\underline{|60}$
$\quad 30$
We continue the process until we obtain a prime as an answer.
$2\underline{|120}$
$2\underline{|60}$
$2\underline{|30}$
$3\underline{|15}$
$\quad 5$
The prime factored form of 120 is
$2 \cdot 2 \cdot 2 \cdot 3 \cdot 5$ or $2^3 \cdot 3 \cdot 5$.

13. Use the "short division method" explained in the solution to problem 9.
$5\underline{|275}$
$5\underline{|55}$
$\quad 11 \qquad 275 = 5^2 \cdot 11$

17. In factored form,
$12y = 2 \cdot 2 \cdot 3 \cdot y$ and $24 = 2 \cdot 2 \cdot 2 \cdot 3$.
To get the greatest common factor, take each prime the least number of times it appears in factored form. The least number of times the 2 appears is twice, and the 3 appears one time. The y is not in 24 so it will not be in the greatest common factor. Thus,
greatest common factor $= 2 \cdot 2 \cdot 3 = 12$.

21. In factored form,
$18m^2n^2 = 2 \cdot 3 \cdot 3 \cdot m^2 \cdot n^2$
$36m^4n^5 = 2 \cdot 2 \cdot 3 \cdot 3 \cdot m^4 \cdot n^5$
$12m^3n = 2 \cdot 2 \cdot 3 \cdot m^3 \cdot n.$
The greatest common numerical factor is $2 \cdot 3 = 6$. We use the lowest exponent of the common variable factors to get m^2n. Therefore, the greatest common factor is $6m^2n$.

25. Since $3x^2 = 3x \cdot x$, place x in the blank.

29. Since $-8z^9 = -4z^5 \cdot (2z^4)$, place $2z^4$ in the blank.

33. Since $14x^4y^3 = 2xy(7x^3y^2)$, place $7x^3y^2$ in the blank.

37. The greatest common factor for $9a^2$ and $18a$ is $9a$.
$9a^2 - 18a = 9a \cdot a - 9a \cdot 2 = 9a(a - 2)$

41. The greatest common factor for $11z^2$ and 100 is 1. Therefore, we cannot factor a greatest common factor from $11z^2 - 100$.

45. The greatest common factor of $13y^6 + 26y^5 - 39y^3$ is $13y^3$. Thus,
$13y^6 + 26y^5 - 39y^3$
$= 13y^3(y^3) + 13y^3(2y^2) - 13y^3(3)$
$= 13y^3(y^3 + 2y^2 - 3).$

49. The greatest common factor is $5z^3a^3$. Thus,
$125z^5a^3 - 60z^4a^5 + 85z^3a^4$
$= 5z^3a^3(25z^2) - 5z^3a^3(12za^2) + 5z^3a^3(17a)$
$= 5z^3a^3(25z^2 - 12za^2 + 17a).$

53. $y^2 - 6y + 4y - 24$
The first two terms have a common factor of y, and the last two terms have a common factor of 4. Thus,
$y^2 - 6y + 4y - 24$
$= y(y - 6) + 4(y - 6).$
Now, $y - 6$ is a common factor.
$= (y - 6)(y + 4)$

57. $18r^2 + 12ry - 3xr - 2xy$
The first two terms have a common factor of $6r$, and the last two terms have a common factor of $(-x)$. Thus,
$18r^2 + 12ry - 3xr - 2xy$
$= 6r(3r + 2y) - x(3r + 2y)$
$= (3r + 2y)(6r - x).$

61. $2pq^2 - 8q^2 + p - 4$
The first two terms have a common factor of $2q^2$, and the last two terms have a common factor of 1. Thus,
$2pq^2 - 8q^2 + p - 4$
$= 2q^2(p - 4) + 1(p - 4)$
$= (p - 4)(2q^2 + 1)$

65. $(q - 5)(q - 7) = q^2 - 7q - 5q + 35$ Use FOIL.
$= q^2 - 12q + 35$

Section 5.2 (page 265)

1. To get the last term of the trinomial, we use $7 \cdot 3 = 21$ and notice that $7 + 3 = 10$, which matches the coefficient of the middle term, $10x$. So
$$x^2 + 10x + 21 = (x + 7)(x + 3).$$

5. To get the last term of the trinomial, we use $-2 \cdot (-12) = 24$ and notice that $-2 + (-12) = -14$, which matches the coefficient of the middle term, $-14t$. So
$$t^2 - 14t + 24 = (t - 2)(t - 12).$$

9. To get the last term of the trinomial, we use $-4 \cdot 6 = -24$ and notice that $-4 + 6 = 2$, which matches the coefficient of the middle term, $2m$. So
$$m^2 + 2m - 24 = (m - 4)(m + 6).$$

13. Since $(-2y) \cdot (-5y) = 10y^2$ and since $-2y + (-5y) = -7y$, the missing expression is $x - 5y$, and
$$x^2 - 7xy + 10y^2 = (x - 2y)(x - 5y).$$

17. Write all pairs of factors whose product is 15.

$1 \cdot 15 = 15 \qquad -1(-15) = 15$

$3 \cdot 5 = 15 \qquad -3(-5) = 15$

The pair 3 and 5 has a sum of 8, so
$$b^2 + 8b + 15 = (b + 3)(b + 5).$$

21. The factors of -12 are as follows.

$-3 \cdot 4 = -12 \qquad 3(-4) = -12$

$-2 \cdot 6 = -12 \qquad 2(-6) = -12$

$-1 \cdot 12 = -12 \qquad 1(-12) = -12$

Since $-2 + 6 = 4$, and 4 is the coefficient of n,
$$n^2 + 4n - 12 = (n - 2)(n + 6).$$

25. List the pairs of factors of 12.

$3 \cdot 4 = 12 \qquad -3(-4) = 12$

$2 \cdot 6 = 12 \qquad -2(-6) = 12$

$1 \cdot 12 = 12 \qquad -1(-12) = 12$

None of these pairs of numbers has a sum of 11, so $h^2 + 11h + 12$ is prime.

29. The only pair of numbers having a product of 20 and a sum of -9 is -4 and -5, so
$$x^2 - 9x + 20 = (x - 4)(x - 5).$$

33. $x^2 + 4ax + 3a^2$

Product is $3a^2$.	*Sum*
$(3a)(a)$	$3a + a = 4a$

$$x^2 + 4ax + 3a^2 = (x + 3a)(x + a)$$

37. $x^2 + xy - 30y^2$

Product is $-30y^2$.	*Sum*
$30y(-y)$	$30y - y = 29y$
$15y(-2y)$	$15y - 2y = 13y$
$10y(-3y)$	$10y - 3y = 7y$
$6y(-5y)$	$6y - 5y = y$

$$x^2 + xy - 30y^2 = (x + 6y)(x - 5y)$$

41. $p^2 - 3pq - 10q^2$

Product is $-10q^2$.	*Sum*
$(-10q)q$	$-10q + q = -9q$
$(-5q)(2q)$	$-5q + 2q = -3q$

$$p^2 - 3pq - 10q^2 = (p - 5q)(p + 2q)$$

45. $6a^2 - 48a - 120$

The greatest common factor is 6.

So $\quad 6a^2 - 48a - 120 = 6(a^2 - 8a - 20).$

Now factor $a^2 - 8a - 20$.

Product is -20.	*Sum*
$(-20)1$	$-20 + 1 = -19$
$(-10)2$	$-10 + 2 = -8$

$$a^2 - 8a - 20 = (a - 10)(a + 2)$$

Thus,
$$6a^2 - 48a - 120 = 6(a - 10)(a + 2).$$

49. $3x^4 - 3x^3 - 90x^2$

The greatest common factor is $3x^2$.

So
$$3x^4 - 3x^3 - 90x^2 = 3x^2(x^2 - x - 30).$$

Now factor $x^2 - x - 30$.

Product is -30.	*Sum*
$(-30)1$	$-30 + 1 = -29$
$(-15)2$	$-15 + 2 = -13$
$(-10)3$	$-10 + 3 = -7$
$(-6)5$	$-6 + 5 = -1$

$$x^2 - x - 30 = (x - 6)(x + 5)$$

Thus,
$$3x^4 - 3x^3 - 90x^2 = 3x^2(x - 6)(x + 5).$$

53. $y^3z + y^2z^2 - 6yz^3$

The greatest common factor is yz.

So
$$y^3z + y^2z^2 - 6yz^3 = yz(y^2 + yz - 6z^2).$$

Now factor $y^2 + yz - 6z^2$.

Product is $-6z^2$.	*Sum*
$6z(-z)$	$6z - z = 5z$
$3z(-2z)$	$3z - 2z = z$

$$y^2 + yz - 6z^2 = (y + 3z)(y - 2z)$$

Thus,
$$y^3z + y^2z^2 - 6yz^3 = yz(y + 3z)(y - 2z).$$

57. The product of $(a + 9)$ and $(a + 6)$ is $a^2 + 15a + 54$. Thus, $a^2 + 15a + 54$ may be factored as $(a + 9)(a + 6)$.

61. $(4m - 3)(2m + 5) = 8m^2 + 20m - 6m - 15$ Use FOIL.

$\qquad = 8m^2 + 14m - 15$

Section 5.3 (page 271)

1. Since $2x \cdot x = 2x^2$ and $1 \cdot -1 = -1$, let the missing factor be $x - 1$. We can see that in
$$(2x + 1)(x - 1)$$
the inner product is $(1)(x) = x$, and the outer product is $(2x)(-1) = -2x$. Since $x + (-2x) = -x$, the middle term of the original trinomial,
$$2x^2 - x - 1 = (2x + 1)(x - 1).$$

5. Since $4y^2 = y \cdot 4y$ and $-15 = 5(-3)$, try $4y - 3$ as the missing factor:
$$(y + 5)(4y - 3).$$
Multiply to see that this product is correct.

9. Since $2m^2 = m \cdot 2m$ and $-10 = -1 \cdot 10$, check to see that the product is $(2m - 1)(m + 10)$.

13. Since $4k^2 = k \cdot 4k$ and $3m^2 = m \cdot 3m$, the product is

$$(4k + m)(k + 3m).$$

17. First, factor out a common factor of m^4.

$$6m^6 + 7m^5 - 20m^4 = m^4(6m^2 + 7m - 20)$$

Now factor within the parentheses to get

$$6m^6 + 7m^5 - 20m^4 = m^4(3m - 4)(2m + 5).$$

In Exercises 21–61, we show only one method.

21. $3a^2 + 10a + 7$

$3 \cdot 7 = 21$

We need two numbers whose product is 21 and whose sum is 10. The numbers are 3 and 7. Write $10a$ as $3a + 7a$.

$$3a^2 + 10a + 7 = 3a^2 + 3a + 7a + 7$$

Factor by grouping.

$$3a^2 + 3a + 7a + 7 = 3a(a + 1) + 7(a + 1) = (a + 1)(3a + 7)$$

25. $15m^2 + m - 2$

$15 \cdot (-2) = -30$

We need two numbers whose product is -30 and whose sum is 1. The numbers are 6 and -5. Write m as $1m$, or $6m - 5m$.

$$15m^2 + m - 2 = 15m^2 + 6m - 5m - 2$$
$$= 3m(5m + 2) - 1(5m + 2) \quad \text{Factor by grouping.}$$
$$= (5m + 2)(3m - 1)$$

29. $5a^2 - 7a - 6$

$5 \cdot (-6) = -30$

Two numbers whose product is -30 and whose sum is -7 are -10 and 3. Therefore,

$$5a^2 - 7a - 6 = 5a^2 - 10a + 3a - 6 = 5a(a - 2) + 3(a - 2) = (a - 2)(5a + 3).$$

33. Here we factor by the alternate method. Write the first term $4y^2 = 4y \cdot y$ and the last term $17 = 1 \cdot 17$. We have $(4y + 1)(y + 17)$. The outer product $= (4y)(17) = 68y$, and the inner product $(1)(y) = y$. The sum of the inner and outer products is $68y + 1y = 69y$, which matches the middle term. So

$$4y^2 + 69y + 17 = (4y + 1)(y + 17).$$

37. Write the first term $10x^2 = 2x \cdot 5x$ and the last term $-6 = 3 \cdot -2$. We have

$$(2x + 3)(5x - 2).$$

The inner product $= (3)(5x) = 15x$, and the outer product $= (2x)(-2) = -4x$. Since $15x + (-4x) = 11x$,

$$10x^2 + 11x - 6 = (2x + 3)(5x - 2).$$

41. $6q^2 + 23q + 21$

$6 \cdot 21 = 126$

Two numbers whose product is 126 and whose sum is 23 are 9 and 14. So

$$6q^2 + 23q + 21 = 6q^2 + 9q + 14q + 21 = 3q(2q + 3) + 7(2q + 3) = (2q + 3)(3q + 7).$$

45. $8k^2 + 2k - 15$

$8 \cdot (-15) = -120$

Two numbers whose product is -120 and whose sum is 2 are 12 and -10, so

$$8k^2 + 2k - 15 = 8k^2 + 12k - 10k - 15 = 4k(2k + 3) - 5(2k + 3) = (2k + 3)(4k - 5).$$

49. $40m^2q + mq - 6q$

$$= q(40m^2 + m - 6) \quad \text{Factor out the common factor.}$$

Look for integers whose product is $40(-6)$ or -240 and whose sum is 1. The integers are 16 and -15.

$$= q(40m^2 + 16m - 15m - 6)$$
$$= q[8m(5m + 2) - 3(5m + 2)]$$
$$= q(5m + 2)(8m - 3)$$

53. $24a^4 + 10a^3 - 4a^2$

$$= 2a^2(12a^2 + 5a - 2) \quad \text{Factor out the greatest common factor.}$$

Look for integers whose product is $12(-2)$ or -24 and whose sum is 5. The integers are 8 and -3.

$$= 2a^2(12a^2 + 8a - 3a - 2)$$
$$= 2a^2[4a(3a + 2) - 1(3a + 2)]$$
$$= 2a^2(3a + 2)(4a - 1)$$

57. $12p^2 + 7pq - 12q^2$

Look for integers whose product is $12(-12)$ or -144 and whose sum is 7. The integers are 16 and -9.

$$12p^2 + 7pq - 12q^2 = 12p^2 + 16pq - 9pq - 12q^2 = 4p(3p + 4q) - 3q(3p + 4q) = (3p + 4q)(4p - 3q)$$

61. $6a^2 - 7ab - 5b^2$

Possible factors of $6a^2$ are $6a \cdot a$, $3a \cdot 2a$.

Possible factors of $-5b^2$ are $(-5b) \cdot b$, $5b(-b)$.

$(6a - 5b)(a + b) = 6a^2 + ab - 5b^2$ Incorrect

$(3a - 5b)(2a + b) = 6a^2 - 7ab - 5b^2$ Correct

65. $(3r - 1)(3r + 1) = (3r)^2 - (1)^2 = 9r^2 - 1$

69. $(2z - 3)^2 = (2z)^2 - 2(2z)(3) + 3^2 = 4z^2 - 12z + 9$

Section 5.4 (page 279)

1. Since $x^2 - 16 = (x)^2 - (4)^2$, the expression $x^2 - 16$ is a difference of two squares. Therefore,

$$x^2 - 16 = (x + 4)(x - 4).$$

5. $9m^2 - 1 = (3m)^2 - (1)^2 = (3m + 1)(3m - 1)$

9. The greatest common factor is 4, so

$$36t^2 - 16 = 4(9t^2 - 4).$$

Factor $9t^2 - 4$.

$$9t^2 - 4 = (3t)^2 - (2)^2 = (3t + 2)(3t - 2)$$

Putting the above results together, we have

$$36t^2 - 16 = 4(3t + 2)(3t - 2).$$

13. $81x^2 + 16$ is the *sum* of two squares, which is prime.

17. $a^4 - 1 = (a^2)^2 - (1)^2 = (a^2 + 1)(a^2 - 1)$

Factor $a^2 - 1$.

$$a^2 - 1 = (a)^2 - (1)^2 = (a + 1)(a - 1)$$

Putting the above results together, we have

$$a^4 - 1 = (a^2 + 1)(a + 1)(a - 1).$$

21. $16k^4 - 1 = (4k^2)^2 - 1^2 = (4k^2 + 1)(4k^2 - 1)$

Now factor $4k^2 - 1$:

$$4k^2 - 1 = (2k)^2 - 1^2 = (2k + 1)(2k - 1).$$

$4k^2 + 1$ is prime. Putting all this together gives

$$16k^4 - 1 = (4k^2 + 1)(2k + 1)(2k - 1).$$

25. Since $y^2 = (y)^2$ and $16 = (4)^2$, try $(y - 4)^2$. Since $2(-4)(y) = -8y$, which is the middle term,
$$y^2 - 8y + 16 = (y - 4)^2.$$

29. Since $r^2 = (r)^2$ and $\frac{1}{9} = \left(\frac{1}{3}\right)^2$, try $\left(r + \frac{1}{3}\right)^2$. Since $2(r)\left(\frac{1}{3}\right) = \frac{2}{3}r$, which is the middle term,
$$r^2 + \frac{2}{3}r + \frac{1}{9} = \left(r + \frac{1}{3}\right)^2.$$

33. Since $16a^2 = (4a)^2$ and $25b^2 = (5b)^2$, try $(4a - 5b)^2$. Since $2(4a)(-5b) = -40ab$, the middle term,
$$16a^2 - 40ab + 25b^2 = (4a - 5b)^2.$$

37. Since $49x^2 = (7x)^2$ and $4y^2 = (2y)^2$, try $(7x + 2y)^2$. Since $2(7x)(2y) = 28xy$, the middle term,
$$49x^2 + 28xy + 4y^2 = (7x + 2y)^2.$$

41. Since $25h^2 = (5h)^2$ and $4y^2 = (2y)^2$, try $(5h - 2y)^2$. Since $2(5h)(-2y) = -20hy$, the middle term,
$$25h^2 - 20hy + 4y^2 = (5h - 2y)^2.$$

45. $ay^2 - 12y + 4 = (3y - 2)^2$
$ay^2 - 12y + 4 = 9y^2 - 2(3y)(2) + 2^2$
$ay^2 - 12y + 4 = 9y^2 - 12y + 4$
Compare the two sides of the equation to see that $a = 9$.

49. $3k - 2 = 0$
$3k = 2$ Add 2 to each side.
$k = \frac{2}{3}$ Divide each side by 3.

53. $8y + 5 = 0$
$8y = -5$ Subtract 5 on each side.
$y = -\frac{5}{8}$ Divide each side by 8.

Summary Exercises on Factoring (page 281)

1. The greatest common factor is $8m^3$.
$$32m^9 + 16m^5 + 24m^3 = 8m^3(4m^6 + 2m^2 + 3)$$

5. $6z^2 + 31z + 5$
$6 \cdot 5 = 30$
Find a pair of numbers whose product is 30 and whose sum is 31. The numbers are 30 and 1. Write $31z$ as $30z + 1z$:
$$6z^2 + 31z + 5 = 6z^2 + 30z + 1z + 5.$$
Factor by grouping.
$$6z^2 + 30z + 1z + 5 = 6z(z + 5) + 1(z + 5) = (z + 5)(6z + 1)$$
Then $6z^2 + 31z + 5 = (z + 5)(6z + 1)$.

9. $16x + 20$
The greatest common factor is 4.
$$16x + 20 = 4(4x + 5)$$

13. Look for a pair of numbers whose product is -15 and whose sum is 2. The numbers are 5 and -3, so
$$m^2 + 2m - 15 = (m + 5)(m - 3).$$

17. Since $z^2 = (z)^2$ and $36 = 6^2$, try $(z - 6)^2$. Since $2(z)(-6) = -12z$, the middle term,
$$z^2 - 12z + 36 = (z - 6)^2.$$

21. Factoring out the greatest common factor, 6, gives
$$6y^2 - 6y - 12 = 6(y^2 - y - 2).$$
Now factor $y^2 - y - 2$. Look for a pair of numbers whose product is -2 and whose sum is -1. The numbers are -2 and 1, so
$$y^2 - y - 2 = (y - 2)(y + 1).$$
Putting everything together gives
$$6y^2 - 6y - 12 = 6(y - 2)(y + 1).$$

25. The polynomial $k^2 + 9$ cannot be factored, since it is the sum of two squares. Thus, $k^2 + 9$ is prime.

29. Since $4k^2 = (2k)^2$ and $9 = 3^2$, try $(2k - 3)^2$. Since $2(2k)(-3) = -12k$, the middle term,
$$4k^2 - 12k + 9 = (2k - 3)^2.$$

33. Factoring out the greatest common factor of $3k$ gives
$$3k^3 - 12k^2 - 15k = 3k(k^2 - 4k - 5).$$
To factor $k^2 - 4k - 5$, look for a pair of numbers whose product is -5 and whose sum is -4. The numbers are -5 and 1, so
$$k^2 - 4k - 5 = (k - 5)(k + 1).$$
Putting everything together gives
$$3k^3 - 12k^2 - 15k = 3k(k - 5)(k + 1).$$

37. Factoring out the greatest common factor of $6y^4$ gives
$$36y^6 - 42y^5 - 120y^4 = 6y^4(6y^2 - 7y - 20).$$
To factor $6y^2 - 7y - 20$, use trial and error. The correct factors are $(3y + 4)(2y - 5)$, so
$$36y^6 - 42y^5 - 120y^4 = 6y^4(3y + 4)(2y - 5).$$

41. Factor out the greatest common factor of 6 to get
$$54m^2 - 24z^2 = 6(9m^2 - 4z^2).$$
Since $9m^2 = (3m)^2$ and $4z^2 = (2z)^2$, use the difference of squares pattern to factor $9m^2 - 4z^2$.
$$9m^2 - 4z^2 = (3m)^2 - (2z)^2 = (3m + 2z)(3m - 2z)$$
Now put this all together to get
$$54m^2 - 24z^2 = 6(3m + 2z)(3m - 2z).$$

45. $m^2 - 81 = m^2 - 9^2 = (m + 9)(m - 9)$

49. Since $m^2 = (m)^2$ and $4 = (2)^2$, try $(m - 2)^2$. Since $2(m)(-2) = -4m$, the middle term,
$$m^2 - 4m + 4 = (m - 2)^2.$$

53. $12p^2 + pq - 6q^2$
$12(-6) = -72$
Look for two numbers whose product is -72 and whose sum is 1. The numbers are 9 and -8. Write pq as $9pq - 8pq$. Then
$$12p^2 + pq - 6q^2 = 12p^2 + 9pq - 8pq - 6q^2$$
$$= 3p(4p + 3q) - 2q(4p + 3q)$$ Factor by grouping.
$$= (4p + 3q)(3p - 2q).$$

57. $100a^2 - 81y^2 = (10a)^2 - (9y)^2 = (10a + 9y)(10a - 9y)$

Section 5.5 (page 289)

1. $(x - 2)(x + 4) = 0$
Set each factor equal to 0.
$x - 2 = 0$ or $x + 4 = 0$
Solve each equation.
$x = 2$ or $x = -4$
The solutions are $x = 2$ or $x = -4$.

SOLUTIONS

5. $(5p + 1)(2p - 1) = 0$
Set each factor equal to 0 and solve each equation.
$5p + 1 = 0$ or $2p - 1 = 0$
$5p = -1$ $\quad 2p = 1$
$p = -\frac{1}{5}$ or $p = \frac{1}{2}$
Solutions: $-\frac{1}{5}, \frac{1}{2}$.

9. $m^2 + 3m - 28 = 0$
Factor: $(m + 7)(m - 4) = 0$. Set each factor equal to 0 and solve.
$m + 7 = 0$ or $m - 4 = 0$
$m = -7$ or $m = 4$
Solutions: $-7, 4$.

13. $x^2 = 3 + 2x$
Get 0 alone on one side.
$x^2 - 2x - 3 = 0$
Factor the left side.
$(x - 3)(x + 1) = 0$
Set each factor equal to 0 and solve.
$x - 3 = 0$ or $x + 1 = 0$
$x = 3$ or $x = -1$
Solutions: $3, -1$.

17. $m^2 + 8m + 16 = 0$
Factor the left side.
$(m + 4)(m + 4) = 0$
Set $m + 4$ equal to 0 and solve.
$m + 4 = 0$
$m = -4$
Solution: -4.

21. $2k^2 - k - 10 = 0$
Factor: $(2k - 5)(k + 2) = 0$. Set each factor equal to 0 and solve.
$2k - 5 = 0$ or $k + 2 = 0$
$2k = 5$
$k = \frac{5}{2}$ or $k = -2$
Solutions: $\frac{5}{2}, -2$.

25. $6a^2 = 5 - 13a$
Get one side equal to 0.
$6a^2 + 13a - 5 = 0$
Factor: $(3a - 1)(2a + 5) = 0$. Set factors equal to 0 and solve.
$3a - 1 = 0$ or $2a + 5 = 0$
$a = \frac{1}{3}$ or $a = -\frac{5}{2}$
Solutions: $\frac{1}{3}, -\frac{5}{2}$.

29. $m^2 - 36 = 0$
Factor: $(m + 6)(m - 6) = 0$. Set each factor equal to 0 and solve.
$m + 6 = 0$ or $m - 6 = 0$
$m = -6$ or $m = 6$
Solutions: $-6, 6$.

33. $z(2z + 7) = 4$
$2z^2 + 7z = 4$ Simplify.
Rewrite with all terms on the left side.
$2z^2 + 7z - 4 = 0$
Factor the left side.
$(2z - 1)(z + 4) = 0$
Set each factor equal to zero and solve.
$2z - 1 = 0$ or $z + 4 = 0$
$2z = 1$ $\quad z = -4$
$z = \frac{1}{2}$
The solutions are $z = \frac{1}{2}$ and $z = -4$.

37. $3(m^2 + 4) = 20m$
$3m^2 + 12 = 20m$ Simplify.
Rewrite with all terms on the left side.
$3m^2 - 20m + 12 = 0$
Factor the left side.
$(3m - 2)(m - 6) = 0$
Set each factor equal to zero and solve.
$3m - 2 = 0$ or $m - 6 = 0$
$3m = 2$ $\quad m = 6$
$m = \frac{2}{3}$
The solutions are $m = \frac{2}{3}$ and $m = 6$.

41. $y^2 = 4(y - 1)$
$y^2 = 4y - 4$ Simplify.
Rewrite with all terms on the left side.
$y^2 - 4y + 4 = 0$
Factor the left side.
$(y - 2)^2 = 0$
Set $y - 2 = 0$ and solve.
$y = 2$ is the only solution.

45. $(x - 1)(6x^2 + x - 12) = 0$
Factor $6x^2 + x - 12$ as $(2x + 3)(3x - 4)$.
Then rewrite the original equation as
$(x - 1)(2x + 3)(3x - 4) = 0$.
Set each factor equal to zero and solve.
$x - 1 = 0$ $\quad 2x + 3 = 0$ $\quad 3x - 4 = 0$
$x = 1$ $\quad 2x = -3$ $\quad 3x = 4$
$x = -\frac{3}{2}$ $\quad x = \frac{4}{3}$
The solutions are $x = 1$, $x = -\frac{3}{2}$, and $x = \frac{4}{3}$.

49. Let x be the number of cats.
Then $x - 3$ is the number of dogs.

Number of cats	plus	number of dogs	equals	total number of cats and dogs.
x	$+$	$x - 3$	$=$	7

Solve: $x + x - 3 = 7$
$2x - 3 = 7$
$2x = 10$
$x = 5$.
She has 5 cats.

Section 5.6 (page 295)

1. Let x represent the width. Then $x + 5$ represents the length. The area of a rectangle equals length times width. So, with an area of 66 square centimeters,

Area	equals	length	times	width,
↓	↓	↓	↓	↓
66	=	$(x + 5)$	$\cdot$	x.

Solve.

$66 = x^2 + 5x$ Simplify.
$0 = x^2 + 5x - 66$ Get 0 alone on one side.
$0 = (x + 11)(x - 6)$ Factor.

Set each factor equal to zero and solve.

$x + 11 = 0$ or $x - 6 = 0$
$x = -11$ or $x = 6$

We reject $x = -11$ because a negative width is nonsense. So the width is 6 centimeters and the length is $6 + 5 = 11$ centimeters.

5. Let x represent the width.
Then $x + 3$ represents the length. Use the formula for the area of a rectangle.

$A = LW$
$A = (x + 3)x$ Substitute $L = x + 3$, $W = x$.

Use the formula for the perimeter of a rectangle.

$P = 2L + 2W$
$P = 2(x + 3) + 2x$ Substitute.

The area	is numerically equal to	the perimeter	less 4.
↓	↓	↓	↓
$(x + 3)x$	=	$2(x + 3) + 2x$	$- 4$

Solve:

$(x + 3)x = 2(x + 3) + 2x - 4$
$x^2 + 3x = 2x + 6 + 2x - 4$
$x^2 + 3x = 4x + 2$ Simplify.
$x^2 + 3x - 4x - 2 = 0$
$x^2 - x - 2 = 0$ Simplify.
$(x - 2)(x + 1) = 0$ Factor.
$x - 2 = 0$ or $x + 1 = 0$
$x = 2$ or $x = -1$. Solve.

Reject $x = -1$ since width cannot be negative. The dimensions of the rectangle are 2 feet by $2 + 3 = 5$ feet.

9. Let x represent the height.
Then $2x$ represents the base.
Use the formula

$$A = \frac{1}{2}bh.$$

Substitute $A = 25$, $b = 2x$, $h = x$.

$$25 = \frac{1}{2}(2x)x$$

$25 = x^2$ Simplify.
$x^2 - 25 = 0$
$(x + 5)(x - 5) = 0$ Factor.
$x + 5 = 0$ or $x - 5 = 0$
$x = -5$ or $x = 5$ Solve.

Reject $x = -5$ since height cannot be negative. The height is 5 centimeters and the base is $2 \cdot 5 = 10$ centimeters.

13. Let x represent the length of a side of the second square.
Then $x - 1$ represents the length of a side of the first square.
Use the formula

$A = s^2$.

Area of first square $= (x - 1)^2$
Area of second square $= x^2$

The difference of the areas	is	37.
↓	↓	↓
$x^2 - (x - 1)^2$	=	37

Solve:

$x^2 - (x - 1)^2 = 37$
$x^2 - x^2 + 2x - 1 = 37$
$2x - 1 = 37$ Simplify.
$2x = 38$
$x = 19$.

The length of a side of the first square is $19 - 1$ or 18 feet and the length of a side of the second square is 19 feet.

17. Let x represent one integer.
Then $x + 1$ represents the next consecutive integer.

The product of the integers	is	2 more than	twice their sum
↓	↓	↓	↓
$x(x + 1)$	=	$2 +$	$2(x + x + 1)$

Solve:

$x(x + 1) = 2 + 2(x + x + 1)$
$x^2 + x = 2 + 2(2x + 1)$
$x^2 + x = 2 + 4x + 2$
$x^2 + x = 4 + 4x$
$x^2 + x - 4x - 4 = 0$
$x^2 - 3x - 4 = 0$
$(x - 4)(x + 1) = 0$
$x - 4 = 0$ or $x + 1 = 0$
$x = 4$ or $x = -1$.

If one integer is 4, the other integer is $4 + 1$ or 5. If one integer is -1, the other integer is $-1 + 1$ or 0.

21. Let x represent one integer.
Then $x + 1$ represents the next consecutive integer.

The square of the sum of the integers	reduced by	three times their product	is	31.
↓	↓	↓	↓	↓
$(x + x + 1)^2$	$-$	$3(x)(x + 1)$	=	31

Solve:

$(x + x + 1)^2 - 3(x)(x + 1) = 31$
$(2x + 1)^2 - 3x(x + 1) = 31$
$4x^2 + 4x + 1 - 3x^2 - 3x = 31$
$x^2 + x + 1 - 31 = 0$
$x^2 + x - 30 = 0$
$(x + 6)(x - 5) = 0$
$x + 6 = 0$ or $x - 5 = 0$
$x = -6$ or $x = 5$.

If the first integer is -6, the other integer is $-6 + 1$ or -5. If the first integer is 5, the other integer is $5 + 1$ or 6.

25. Let x represent the length of the shorter leg. Then $2x + 1$ represents the length of the hypotenuse, and $2x - 1$ represents the length of the longer leg. In a right triangle, the square of the hypotenuse equals the sum of the squares of the legs, so

$$(2x + 1)^2 = x^2 + (2x - 1)^2$$
$$4x^2 + 4x + 1 = x^2 + 4x^2 - 4x + 1$$
$$4x = x^2 - 4x$$

Subtract $4x^2$ and 1 on both sides.

$$0 = x^2 - 8x$$

Get 0 on one side.

$$0 = x(x - 8) \quad \text{Factor.}$$
$$x = 0 \quad \text{or} \quad x - 8 = 0$$
$$x = 0 \quad \text{or} \quad x = 8.$$

Reject 0 because the side of a triangle cannot measure 0. The shorter leg has a length of 8 feet.

29. Let d, the distance, equal 1600 feet in the formula. Also, $g = 32$.

$$d = \frac{1}{2}gt^2$$
$$1600 = \frac{1}{2}(32)t^2 \quad \text{Let } d = 1600 \text{ and } g = 32.$$
$$3200 = 32t^2 \quad \text{Multiply by 2.}$$
$$100 = t^2 \quad \text{Divide by 32.}$$
$$0 = t^2 - 100 \quad \text{Get 0 on one side.}$$
$$0 = (t + 10)(t - 10) \quad \text{Factor.}$$
$$t = -10 \quad \text{or} \quad t = 10$$

Reject -10; the time is 10 seconds.

33. $x^2 - x + 4 \le 0$

Replace x with -2. Use parentheses to keep the signs correct.

$$(-2)^2 - (-2) + 4 \le 0$$
$$4 + 2 + 4 \le 0 \quad \text{Simplify.}$$
$$10 \le 0 \quad \text{False}$$

37. $x^2 + 3x + 7 \le 0$

Replace x with -2. Use parentheses to keep the signs correct.

$$(-2)^2 + 3(-2) + 7 \le 0$$
$$4 - 6 + 7 \le 0 \quad \text{Simplify.}$$
$$5 \le 0 \quad \text{False}$$

CHAPTER 6

Section 6.1 (page 313)

1. The denominator $4x$ will be zero when $x = 0$, so $\frac{3}{4x}$ is undefined for $x = 0$.

5. Set the denominator $x + 5$ equal to 0. Then $x + 5 = 0$, and $x = -5$, so $\frac{x^2}{x + 5}$ is undefined for $x = -5$.

9. To find the numbers that make the denominator 0, we must solve

$$2r^2 - r - 3 = 0.$$
$$(2r - 3)(r + 1) = 0 \quad \text{Factor.}$$
$$2r - 3 = 0 \quad \text{or} \quad r + 1 = 0 \quad \text{Set each factor equal to zero and solve.}$$
$$2r = 3 \qquad r = -1$$
$$r = \frac{3}{2}$$

The expression $\frac{8r + 2}{2r^2 - r - 3}$ is undefined for $r = \frac{3}{2}$ or $r = -1$.

13. (a) Let $x = 2$.

$$\frac{4x - 2}{3x} = \frac{4 \cdot 2 - 2}{3 \cdot 2} = \frac{8 - 2}{6} = \frac{6}{6} = 1$$

(b) Let $x = -3$.

$$\frac{4x - 2}{3x} = \frac{4(-3) - 2}{3(-3)} = \frac{-12 - 2}{-9} = \frac{-14}{-9} = \frac{14}{9}$$

17. (a)

$$\frac{(-8x)^2}{3x + 9} = \frac{(-8 \cdot 2)^2}{3(2) + 9} \quad \text{Let } x = 2.$$
$$= \frac{(-16)^2}{6 + 9}$$
$$= \frac{256}{15}$$

(b)

$$\frac{(-8x)^2}{3x + 9} = \frac{(-8 \cdot -3)^2}{3(-3) + 9} \quad \text{Substitute } -3 \text{ for } x.$$
$$= \frac{(24)^2}{-9 + 9} \quad \text{Simplify.}$$
$$= \frac{(24)^2}{0}$$

The expression is undefined because of the 0 in the denominator.

21. (a)

$$\frac{5x^2}{6 - 3x - x^2} = \frac{5 \cdot 2^2}{6 - 3 \cdot 2 - 2^2} \quad \text{Let } x = 2.$$
$$= \frac{5 \cdot 4}{6 - 6 - 4}$$
$$= \frac{20}{-4}$$
$$= -5$$

(b)

$$\frac{5x^2}{6 - 3x - x^2} = \frac{5 \cdot (-3)^2}{6 - 3 \cdot (-3) - (-3)^2} \quad \text{Let } x = -3.$$
$$= \frac{5 \cdot 9}{6 + 9 - 9}$$
$$= \frac{45}{6}$$
$$= \frac{15}{2}$$

25.

$$\frac{12k^2}{6k} = \frac{2 \cdot 2 \cdot 3 \cdot k \cdot k}{2 \cdot 3 \cdot k} \quad \text{Factor.}$$
$$= \frac{(2 \cdot 3 \cdot k)2 \cdot k}{(2 \cdot 3 \cdot k)}$$
$$= 2k \quad \text{Lowest terms}$$

29.

$$\frac{9z + 6}{3} = \frac{3(3z + 2)}{3} \quad \text{Factor.}$$
$$= 3z + 2 \quad \text{Use fundamental property.}$$

33.

$$\frac{x^2 - 1}{(x + 1)^2}$$
$$= \frac{(x + 1)(x - 1)}{(x + 1)(x + 1)} \quad \text{Factor.}$$
$$= \frac{x - 1}{x + 1} \quad \text{Use fundamental property.}$$

37.

$$\frac{16r^2 - 4s^2}{4r - 2s}$$
$$= \frac{4(2r + s)(2r - s)}{2(2r - s)} \quad \text{Factor.}$$
$$= 2(2r + s) \quad \text{Use fundamental property.}$$

or $4r + 2s$

SOLUTIONS

41. $\dfrac{x^2 + 3x - 4}{x^2 - 1} = \dfrac{(x + 4)(x - 1)}{(x + 1)(x - 1)}$ Factor.

$= \dfrac{x + 4}{x + 1}$ Use fundamental property.

45. $\dfrac{x^2 - 1}{1 - x} = \dfrac{(x + 1)(x - 1)}{-1(x - 1)}$ Factor.

$= \dfrac{x + 1}{-1}$ Use fundamental property.

$= -(x + 1)$

or $-x - 1$

49. $\dfrac{3}{4} \cdot \dfrac{5}{8} = \dfrac{3 \cdot 5}{4 \cdot 8} = \dfrac{15}{32}$

53. $\dfrac{6}{5} \div \dfrac{3}{10} = \dfrac{6}{5} \cdot \dfrac{10}{3} = \dfrac{60}{15} = 4$

Section 6.2 (page 319)

1. $\dfrac{9m^2}{16} \cdot \dfrac{4}{3m} = \dfrac{9m^2 \cdot 4}{16 \cdot 3m}$

$= \dfrac{36m^2}{48m}$ Multiply.

$= \dfrac{3m}{4}$ Divide numerator and denominator by $12m$.

5. $\dfrac{8a^4}{12a^3} \cdot \dfrac{9a^5}{3a^2} = \dfrac{8 \cdot 9a^{4+5}}{12 \cdot 3a^{3+2}}$

$= \dfrac{72a^9}{36a^5}$

$= \dfrac{72}{36} \cdot \dfrac{a^9}{a^5}$

$= 2a^4$

9. $\dfrac{3m^2}{(4m)^3} \div \dfrac{9m^3}{32m^4} = \dfrac{3m^2}{64m^3} \cdot \dfrac{32m^4}{9m^3}$ Invert and multiply.

$= \dfrac{96m^6}{576m^6}$

$= \dfrac{1}{6}$ Divide numerator and denominator by $96m^6$.

13. $\dfrac{a + b}{2} \cdot \dfrac{12}{(a + b)^2} = \dfrac{12(a + b)}{2(a + b)^2}$ Multiply.

$= \dfrac{6}{a + b}$ Lowest terms

17. $\dfrac{2k + 8}{6} \div \dfrac{3k + 12}{2}$

$= \dfrac{2(k + 4)}{6} \div \dfrac{3(k + 4)}{2}$ Factor.

$= \dfrac{2(k + 4)}{6} \cdot \dfrac{2}{3(k + 4)}$ Invert and multiply.

$= \dfrac{2 \cdot 2(k + 4)}{6 \cdot 3(k + 4)}$

$= \dfrac{2}{3 \cdot 3} = \dfrac{2}{9}$

21. $\dfrac{3r + 12}{8} \cdot \dfrac{16r}{9r + 36}$

$= \dfrac{3(r + 4)}{8} \cdot \dfrac{16r}{9(r + 4)}$ Factor.

$= \dfrac{3 \cdot 16r \cdot (r + 4)}{8 \cdot 9 \cdot (r + 4)}$ Multiply.

$= \dfrac{2r}{3}$

25. $\dfrac{2 - y}{8} \cdot \dfrac{7}{y - 2}$

$= \dfrac{-1(y - 2)}{8} \cdot \dfrac{7}{y - 2}$ Factor out -1.

$= \dfrac{-7(y - 2)}{8(y - 2)}$ Multiply.

$= -\dfrac{7}{8}$ Use fundamental property.

29. $\dfrac{y^2 + y - 2}{y^2 + 3y - 4} \div \dfrac{y + 2}{y + 3}$

$= \dfrac{(y + 2)(y - 1)}{(y + 4)(y - 1)} \cdot \dfrac{y + 3}{y + 2}$

$= \dfrac{y + 3}{y + 4}$

33. $\dfrac{p^2 - 4p + 3}{p^2 - 3p + 2} \div \dfrac{p - 3}{p + 5}$

$= \dfrac{p^2 - 4p + 3}{p^2 - 3p + 2} \cdot \dfrac{p + 5}{p - 3}$

$= \dfrac{(p - 3)(p - 1) \cdot (p + 5)}{(p - 2)(p - 1) \cdot (p - 3)}$

$= \dfrac{p + 5}{p - 2}$

37. $\dfrac{r^2 + rs - 12s^2}{r^2 - rs - 20s^2} \div \dfrac{r^2 - 2rs - 3s^2}{r^2 + rs - 30s^2}$

$= \dfrac{(r - 3s)(r + 4s)}{(r - 5s)(r + 4s)} \div \dfrac{(r - 3s)(r + s)}{(r + 6s)(r - 5s)}$

$= \dfrac{(r - 3s)(r + 4s)}{(r - 5s)(r + 4s)} \cdot \dfrac{(r + 6s)(r - 5s)}{(r - 3s)(r + s)}$

$= \dfrac{r + 6s}{r + s}$

41. $50 = 2 \cdot 25 = 2 \cdot 5^2$

45. $45a^3 = 3^2 \cdot 5 \cdot a^3$

$85a^2 = 5 \cdot 17 \cdot a^2$

$105a^4 = 3 \cdot 5 \cdot 7 \cdot a^4$

To find the greatest common factor, multiply the factors that are common to each factorization. These are 5 and a^2. The greatest common factor is $5a^2$.

Section 6.3 (page 325)

1. Factor each denominator.

$12 = 2 \cdot 2 \cdot 3; \quad 10 = 2 \cdot 5$

Take each factor the greatest number of times it appears:

least common denominator $= 2 \cdot 2 \cdot 3 \cdot 5 = 60$.

5. Factor each denominator: $100 = 2^2 \cdot 5^2$, $120 = 2^3 \cdot 3 \cdot 5$, $180 = 2^2 \cdot 3^2 \cdot 5$. Take each factor the greatest number of times it appears:

least common denominator $= 2^3 \cdot 3^2 \cdot 5^2 = 1800$.

9. The least common denominator for 5 and 6 is 30. The only variable is p, so the least common denominator is $30p$.

13. Factor each denominator.

$15y^2 = 3 \cdot 5 \cdot y^2$

$36y^4 = 2 \cdot 2 \cdot 3 \cdot 3 \cdot y^4$

Take each factor the greatest number of times it appears, and use the highest exponent on y:

least common denominator $= 2 \cdot 2 \cdot 3 \cdot 3 \cdot 5 \cdot y^4$

$= 180y^4$.

17. Factor each denominator.

$$32r^2 = 2^5 \cdot r^2$$
$$16r - 32 = 16(r - 2) = 2^4(r - 2)$$
$$\text{least common denominator} = 2^5r^2(r - 2) = 32r^2(r - 2)$$

21. $12p + 60 = 12(p + 5) = 2 \cdot 2 \cdot 3 \cdot (p + 5);$

$p^2 + 5p = p(p + 5)$

least common denominator $= 2 \cdot 2 \cdot 3 \cdot p \cdot (p + 5) = 12p(p + 5)$

25. $a^2 + 6a = a(a + 6);$

$a^2 + 3a - 18 = (a + 6)(a - 3)$

least common denominator $= a(a + 6)(a - 3)$

29. $2y^2 + 7y - 4 = (2y - 1)(y + 4)$

$2y^2 - 7y + 3 = (2y - 1)(y - 3)$

least common denominator $= (2y - 1)(y + 4)(y - 3)$

33. $\frac{9}{r} = \frac{\quad}{6r}$

To get a denominator of $6r$, multiply numerator and denominator by 6.

$$\frac{6 \cdot 9}{6 \cdot r} = \frac{54}{6r}$$

37. $\frac{12}{35y} = \frac{\quad}{70y^3}$

To get a denominator of $70y^3$, multiply numerator and denominator by $2y^2$.

$$\frac{2y^2 \cdot 12}{2y^2 \cdot 35y} = \frac{24y^2}{70y^3}$$

41. $\frac{19z}{2z - 6} = \frac{\quad}{6z - 18}$

$\frac{19z}{2(z - 3)} = \frac{\quad}{6(z - 3)}$ Factor denominators.

$\frac{3 \cdot 19z}{3 \cdot 2(z - 3)} = \frac{57z}{6z - 18}$ Multiply by 3 in both numerator and denominator.

45. $\frac{6}{k^2 - 4k} = \frac{\quad}{k(k - 4)(k + 1)}$

$\frac{6}{k(k - 4)} = \frac{\quad}{k(k - 4)(k + 1)}$ Factor first denominator.

$\frac{6 \cdot (k + 1)}{k(k - 4) \cdot (k + 1)} = \frac{6(k + 1)}{k(k - 4)(k + 1)}$

Multiply by $k + 1$ in both numerator and denominator.

49. $\frac{1}{2} + \frac{2}{5} = \frac{5}{10} + \frac{4}{10} = \frac{9}{10}$

53. $\frac{5}{24} + \frac{1}{18} = \frac{15}{72} + \frac{4}{72} = \frac{19}{72}$

Section 6.4 (page 333)

1. $\frac{2}{p} + \frac{5}{p} = \frac{2 + 5}{p} = \frac{7}{p}$

5. $\frac{a + b}{2} - \frac{a - b}{2} = \frac{a + b - (a - b)}{2} = \frac{a + b - a + b}{2} = \frac{2b}{2} = b$

9. $\frac{z^2 - 10z}{z - 5} + \frac{25}{z - 5}$

$= \frac{z^2 - 10z + 25}{z - 5}$ Add.

$= \frac{(z - 5)(z - 5)}{(z - 5)}$ Factor.

$= z - 5$ Use fundamental property.

13. Rewrite the fractions with the least common denominator, 10.

$$\frac{9}{10} - \frac{r}{2} = \frac{9}{10} - \frac{r \cdot 5}{2 \cdot 5} = \frac{9}{10} - \frac{5r}{10} = \frac{9 - 5r}{10}$$

17. Rewrite the fractions with the least common denominator, $2x$.

$$\frac{3}{x} + \frac{5}{2x} = \frac{3 \cdot 2}{x \cdot 2} + \frac{5}{2x} = \frac{6}{2x} + \frac{5}{2x} = \frac{11}{2x}$$

21. Rewrite the fractions with the least common denominator, 9.

$$\frac{5r + 4}{3} - \frac{2r - 3}{9} = \frac{(5r + 4) \cdot 3}{3 \cdot 3} - \frac{2r - 3}{9} = \frac{15r + 12}{9} - \frac{2r - 3}{9} = \frac{15r + 12 - (2r - 3)}{9} = \frac{15r + 12 - 2r + 3}{9} = \frac{13r + 15}{9}$$

25. Rewrite the fractions with the least common denominator, p^2.

$$\frac{3}{p} + \frac{5}{p^2} = \frac{3 \cdot p}{p \cdot p} + \frac{5}{p^2} = \frac{3p}{p^2} + \frac{5}{p^2} = \frac{3p + 5}{p^2}$$

29. Rewrite each fraction with the least common denominator, p^2x.

$$\frac{9}{p^2} + \frac{4}{px} = \frac{9 \cdot x}{p^2 \cdot x} + \frac{4 \cdot p}{px \cdot p} = \frac{9x}{p^2x} + \frac{4p}{p^2x} = \frac{9x + 4p}{p^2x}$$

SOLUTIONS

33. Rewrite the fractions with the least common denominator, $4(a - b)$.

$$\frac{1}{a-b} - \frac{a}{4a-4b} = \frac{4 \cdot 1}{4(a-b)} - \frac{a}{4(a-b)}$$
$$= \frac{4-a}{4(a-b)}$$

37. Factor the denominators.

$$\frac{1}{y^2-4} + \frac{3}{y^2+5y+6}$$
$$= \frac{1}{(y-2)(y+2)} + \frac{3}{(y+2)(y+3)}$$

The least common denominator is $(y-2)(y+2)(y+3)$.

$$= \frac{1 \cdot (y+3)}{(y-2)(y+2) \cdot (y+3)} + \frac{3 \cdot (y-2)}{(y-2) \cdot (y+2)(y+3)}$$
$$= \frac{y+3+3y-6}{(y-2)(y+2)(y+3)} \quad \text{Add fractions.}$$
$$= \frac{4y-3}{(y-2)(y+2)(y+3)}$$

41. Factor the denominators.

$$\frac{6}{k^2+3k} - \frac{1}{k^2-k} + \frac{2}{k^2+2k-3}$$
$$= \frac{6}{k(k+3)} - \frac{1}{k(k-1)} + \frac{2}{(k+3)(k-1)}$$

The least common denominator is $k(k-1)(k+3)$.

$$= \frac{6(k-1)}{k(k-1)(k+3)} - \frac{1(k+3)}{k(k-1)(k+3)} + \frac{2 \cdot k}{k(k-1)(k+3)}$$
$$= \frac{6k-6-k-3+2k}{k(k-1)(k+3)} \quad \text{Add fractions.}$$
$$= \frac{7k-9}{k(k-1)(k+3)}$$

45. $$\frac{\frac{5}{8}-\frac{3}{8}}{\frac{2}{5}+\frac{1}{5}} = \frac{\frac{2}{8}}{\frac{3}{5}}$$

Rewrite as a division problem.

$$\frac{2}{8} \div \frac{3}{5} = \frac{1}{4} \div \frac{3}{5} = \frac{1}{4} \cdot \frac{5}{3} = \frac{1 \cdot 5}{4 \cdot 3} = \frac{5}{12}$$

Section 6.5 (page 339)

Only one method is shown for each exercise in this section.

1. $$\frac{\frac{5}{8}+\frac{2}{3}}{\frac{7}{3}-\frac{1}{4}} = \frac{24\left(\frac{5}{8}+\frac{2}{3}\right)}{24\left(\frac{7}{3}-\frac{1}{4}\right)} \quad \text{Multiply by LCD = 24 in numerator and denominator.}$$
$$= \frac{24\left(\frac{5}{8}\right)+24\left(\frac{2}{3}\right)}{24\left(\frac{7}{3}\right)-24\left(\frac{1}{4}\right)} \quad \text{Distributive property}$$
$$= \frac{15+16}{56-6} = \frac{31}{50}$$

5. $$\frac{\frac{m^3p^4}{5m}}{\frac{8mp^5}{p^2}} = \frac{m^3p^4}{5m} \cdot \frac{p^2}{8mp^5} \quad \text{Use Method 1.}$$
$$= \frac{m^3p^4 \cdot p^2}{5m \cdot 8mp^5}$$
$$= \frac{mp}{40}$$

9. $$\frac{\frac{3}{y}+1}{\frac{3+y}{2}} = \frac{2y\left(\frac{3}{y}+1\right)}{2y\left(\frac{3+y}{2}\right)} \quad \text{Multiply by LCD = } 2y \text{ in numerator and denominator.}$$
$$= \frac{2y\left(\frac{3}{y}\right)+2y(1)}{2y\left(\frac{3+y}{2}\right)}$$
$$= \frac{6+2y}{y(3+y)}$$
$$= \frac{2(3+y)}{y(3+y)} \quad \text{Factor.}$$
$$= \frac{2}{y} \quad \text{Lowest terms}$$

13. Multiply numerator and denominator by the least common denominator, y.

$$\frac{y-\frac{6}{y}}{y+\frac{2}{y}} = \frac{\left(y-\frac{6}{y}\right) \cdot y}{\left(y+\frac{2}{y}\right) \cdot y}$$
$$= \frac{y^2-6}{y^2+2}$$

17. The least common denominator is m^3p^2.

$$\frac{\frac{1}{m^3p}+\frac{2}{mp^2}}{\frac{4}{mp}+\frac{1}{m^2p}} = \frac{\left(\frac{1}{m^3p}+\frac{2}{mp^2}\right) \cdot m^3p^2}{\left(\frac{4}{mp}+\frac{1}{m^2p}\right) \cdot m^3p^2}$$
$$= \frac{p+2m^2}{4m^2p+mp}$$
$$= \frac{2m^2+p}{mp(4m+1)} \quad \text{Factor.}$$

21. Divide the numerator by the denominator.

$$\frac{a}{a+1} \div \frac{2}{a^2-1}$$
$$= \frac{a}{a+1} \cdot \frac{a^2-1}{2}$$
$$= \frac{a(a^2-1)}{2(a+1)}$$
$$= \frac{a(a-1)(a+1)}{2(a+1)} \quad \text{Factor.}$$
$$= \frac{a(a-1)}{2} \quad \text{Use fundamental property.}$$

25. The least common denominator is $(r - 1)(r + 2)(r + 3)$.

$$\frac{\frac{5}{r+3} - \frac{1}{r-1}}{\frac{2}{r+2} + \frac{3}{r+3}} = \frac{\left(\frac{5}{r+3} - \frac{1}{r-1}\right) \cdot (r-1)(r+2)(r+3)}{\left(\frac{2}{r+2} + \frac{3}{r+3}\right) \cdot (r-1)(r+2)(r+3)}$$

$$= \frac{5(r-1)(r+2) - 1(r+2)(r+3)}{2(r-1)(r+3) + 3(r-1)(r+2)}$$

$$= \frac{5r^2 + 5r - 10 - r^2 - 5r - 6}{2r^2 + 4r - 6 + 3r^2 + 3r - 6}$$

$$= \frac{4r^2 - 16}{5r^2 + 7r - 12} = \frac{4(r-2)(r+2)}{(5r+12)(r-1)} \quad \text{Factor.}$$

29.
$$9z + 2 = 7z - 6$$
$$9z + 2 - 2 - 7z = 7z - 6 - 2 - 7z \quad \text{Subtract 2 and } 7z.$$
$$2z = -8 \quad \text{Combine terms.}$$
$$\frac{2z}{2} = \frac{-8}{2} \quad \text{Divide by 2.}$$
$$z = -4$$

33.
$$-(8 - a) + 5a = -7$$
$$-8 + a + 5a = -7 \quad \text{Distributive property}$$
$$-8 + 6a = -7 \quad \text{Combine terms.}$$
$$-8 + 6a + 8 = -7 + 8 \quad \text{Add 8.}$$
$$6a = 1 \quad \text{Combine terms.}$$
$$\frac{6a}{6} = \frac{1}{6} \quad \text{Divide by 6.}$$
$$a = \frac{1}{6}$$

Section 6.6 (page 345)

1.
$$\frac{6}{x} - \frac{4}{x} = 5$$
$$x\left(\frac{6}{x} - \frac{4}{x}\right) = x \cdot 5 \quad \text{Multiply both sides by } x.$$
$$x\left(\frac{6}{x}\right) - x\left(\frac{4}{x}\right) = 5x \quad \text{Distributive property}$$
$$6 - 4 = 5x$$
$$2 = 5x$$
$$\frac{2}{5} = x \quad \text{Divide by 5.}$$

Check $\frac{2}{5}$ in the original equation.

$$\frac{6}{x} - \frac{4}{x} = 5$$
$$\frac{6}{\frac{2}{5}} - \frac{4}{\frac{2}{5}} = 5 \quad \text{Let } x = \frac{2}{5}.$$
$$6 \cdot \frac{5}{2} - 4 \cdot \frac{5}{2} = 5$$
$$15 - 10 = 5$$
$$5 = 5 \quad \text{True}$$

The solution is $\frac{2}{5}$.

5.
$$\frac{9}{m} = 5 - \frac{1}{m}$$
$$m\left(\frac{9}{m}\right) = m\left(5 - \frac{1}{m}\right) \quad \text{Multiply by } m.$$
$$9 = 5m - 1 \quad \text{Distributive property}$$
$$10 = 5m \quad \text{Add 1.}$$
$$2 = m \quad \text{Divide by 5.}$$

Check 2 in the original equation.

9.
$$\frac{x+1}{2} = \frac{x+2}{3}$$
$$6\left(\frac{x+1}{2}\right) = 6\left(\frac{x+2}{3}\right)$$

Multiply each term by the common denominator, 6.

$$3(x + 1) = 2(x + 2)$$
$$3x + 3 = 2x + 4$$
$$x + 3 = 4 \quad \text{Subtract } 2x.$$
$$x = 1 \quad \text{Subtract 3.}$$

Check this solution in the original equation.

13.
$$\frac{2p + 8}{9} = \frac{10p + 4}{27}$$
$$27\left(\frac{2p+8}{9}\right) = 27\left(\frac{10p+4}{27}\right)$$
$$3(2p + 8) = 10p + 4$$
$$6p + 24 = 10p + 4$$
$$24 = 4p + 4$$
$$20 = 4p$$
$$5 = p$$

Check this solution in the original equation.

17.
$$\frac{a-4}{4} = \frac{a}{16} + \frac{1}{2}$$
$$16\left(\frac{a-4}{4}\right) = 16\left(\frac{a}{16} + \frac{1}{2}\right) \quad \text{Multiply both sides by 16.}$$
$$4(a - 4) = 16\left(\frac{a}{16}\right) + 16\left(\frac{1}{2}\right) \quad \text{Distributive property}$$
$$4a - 16 = a + 8$$
$$3a - 16 = 8 \quad \text{Subtract } a.$$
$$3a = 24 \quad \text{Add 16.}$$
$$a = 8 \quad \text{Divide by 3.}$$

Check this solution in the original equation.

21. Multiply each side of the equation by the least common denominator, 35.

$$\frac{m+2}{5} - \frac{m-6}{7} = 2$$
$$35\left(\frac{m+2}{5} - \frac{m-6}{7}\right) = 35(2)$$
$$7(m + 2) - 5(m - 6) = 70$$
$$7m + 14 - 5m + 30 = 70$$
$$2m + 44 = 70$$
$$2m = 26$$
$$m = 13$$

Check this solution in the original equation.

SOLUTIONS

25. Multiply each side of the equation by the least common denominator, 10.

$$\frac{8k}{5} - \frac{3k-4}{2} = \frac{5}{2}$$
$$10\left(\frac{8k}{5} - \frac{3k-4}{2}\right) = 10\left(\frac{5}{2}\right)$$
$$2(8k) - 5(3k-4) = 5 \cdot 5$$
$$16k - 15k + 20 = 25$$
$$k + 20 = 25$$
$$k = 5$$

Check this solution in the original equation.

29. Factor each denominator; then multiply each side by the least common denominator, $24(p + 5)$.

$$\frac{1}{3p+15} + \frac{5}{4p+20} = \frac{19}{24}$$
$$\frac{1}{3(p+5)} + \frac{5}{4(p+5)} = \frac{19}{24}$$
$$24(p+5)\left(\frac{1}{3(p+5)} + \frac{5}{4(p+5)}\right) = 24(p+5)\left(\frac{19}{24}\right)$$
$$8 \cdot 1 + 6 \cdot 5 = (p+5)(19)$$
$$8 + 30 = 19p + 95$$
$$38 = 19p + 95$$
$$-57 = 19p$$
$$-3 = p$$

Check this solution in the original equation.

33. Multiply each side of the equation by the least common denominator, x.

$$\frac{8x+3}{x} = 3x$$
$$x\left(\frac{8x+3}{x}\right) = x(3x)$$
$$8x + 3 = 3x^2$$
$$-3x^2 + 8x + 3 = 0$$
$$3x^2 - 8x - 3 = 0 \quad \text{Multiply by } -1.$$
$$(3x+1)(x-3) = 0 \quad \text{Factor.}$$
$$3x + 1 = 0 \quad \text{or} \quad x - 3 = 0$$
$$3x = -1 \qquad x = 3$$
$$x = -\frac{1}{3}$$

The solutions are $-\frac{1}{3}$ and 3, since both these numbers satisfy the original equation.

37. Factor the denominators.

$$\frac{x+4}{x^2-3x+2} - \frac{5}{x^2-4x+3} = \frac{x-4}{x^2-5x+6}$$
$$\frac{x+4}{(x-2)(x-1)} - \frac{5}{(x-3)(x-1)} = \frac{x-4}{(x-3)(x-2)}$$

Multiply each side of the equation by the least common denominator, $(x-2)(x-1)(x-3)$.

$$(x-2)(x-1)(x-3)\left(\frac{x+4}{(x-1)(x-2)} - \frac{5}{(x-1)(x-3)}\right)$$
$$= (x-2)(x-1)(x-3)\frac{x-4}{(x-2)(x-3)}$$
$$(x-3)(x+4) - 5(x-2) = (x-1)(x-4)$$
$$x^2 + x - 12 - 5x + 10 = x^2 - 5x + 4$$
$$-4x - 2 = -5x + 4$$
$$x - 2 = 4$$
$$x = 6$$

Check this solution in the original equation.

41. Multiply each side of the equation by the least common denominator, $(z-2)(z+2)$.

$$\frac{5}{z-2} + \frac{10}{z+2} = 7$$
$$(z-2)(z+2) \cdot \frac{5}{z-2} + (z-2)(z+2) \cdot \frac{10}{z+2}$$
$$= (z-2)(z+2) \cdot 7$$
$$5(z+2) + 10(z-2) = 7(z^2-4)$$
$$5z + 10 + 10z - 20 = 7z^2 - 28$$
$$15z - 10 = 7z^2 - 28$$
$$0 = 7z^2 - 15z - 18$$
$$0 = (7z+6)(z-3)$$
$$7z + 6 = 0 \quad \text{or} \quad z - 3 = 0$$
$$7z = -6 \qquad z = 3$$
$$z = -\frac{6}{7}$$

Check these solutions in the original equation.

45.

$$F = \frac{k}{d-D}$$
$$F(d-D) = \frac{k}{d-D}(d-D) \quad \text{Multiply by } d - D.$$
$$Fd - FD = k \quad \text{Distributive property}$$
$$-FD = k - Fd \quad \text{Subtract } Fd.$$
$$D = \frac{k-Fd}{-F} \quad \text{or} \quad \frac{Fd-k}{F} \quad \text{Divide by } -F.$$

49. Multiply both sides of the equation by the least common denominator, $B + b$.

$$h = \frac{2A}{B+b}$$
$$(B+b) \cdot h = (B+b) \cdot \frac{2A}{B+b}$$
$$hB + hb = 2A$$

Get the term containing b alone on one side.

$$hb = 2A - hB$$
$$b = \frac{2A - hB}{h}$$
$$\text{or} \quad b = \frac{2A}{h} - B$$

53. Use the formula for distance, rate, and time: $d = rt$. Let $d = 780$ and $r = z$. Solve for t.

$$d = rt$$
$$780 = zt$$
$$\frac{780}{z} = t \quad \text{Divide by } z.$$
$$\text{or} \quad t = \frac{780}{z}$$

Summary Exercises on Rational Expressions (page 349)

1. $\frac{6}{m} + \frac{2}{m} = \frac{6+2}{m} = \frac{8}{m}$

5.
$$\frac{2r^2-3r-9}{2r^2-r-6} \cdot \frac{r^2+2r-8}{r^2-2r-3}$$
$$= \frac{(2r+3)(r-3)}{(2r+3)(r-2)} \cdot \frac{(r+4)(r-2)}{(r-3)(r+1)} \quad \text{Factor.}$$
$$= \frac{r+4}{r+1} \quad \text{Multiply; lowest terms.}$$

9. $\frac{5}{y-1} + \frac{2}{3y-3} = \frac{5}{y-1} + \frac{2}{3(y-1)}$ Factor.

$= \frac{3 \cdot 5}{3(y-1)} + \frac{2}{3(y-1)}$

$= \frac{15+2}{3(y-1)} = \frac{17}{3(y-1)}$

13. $\frac{4}{9z} - \frac{3}{2z} = \frac{2 \cdot 4}{2 \cdot 9z} - \frac{9 \cdot 3}{9 \cdot 2z}$

$= \frac{8}{18z} - \frac{27}{18z}$

$= \frac{8-27}{18z}$

$= \frac{-19}{18z} = -\frac{19}{18z}$

Section 6.7 (page 355)

1. Let x represent the number. Then

One-half	of	a number	is	3 more than	one-sixth	of	the same number.
↓	↓	↓	↓	↓	↓	↓	↓
$\frac{1}{2}$	$\cdot$	x	$=$	$3+$	$\frac{1}{6}$	$\cdot$	x

Solve:

$6\left(\frac{1}{2}x\right) = 6(3) + 6\left(\frac{1}{6}x\right)$ Multiply each term by the common denominator, 6.

$3x = 18 + x$

$2x = 18$

$x = 9.$

The number is 9.

5. Let x represent the number. Then the reciprocal is $\frac{1}{x}$, and twice the reciprocal subtracted from the number is

$x - 2\left(\frac{1}{x}\right) = x - \frac{2}{x}.$

Since this result is $-\frac{7}{3}$, solve

$x - \frac{2}{x} = -\frac{7}{3}.$

$3x\left(x - \frac{2}{x}\right) = 3x\left(-\frac{7}{3}\right)$ Multiply by $3x$.

$3x^2 - 6 = -7x$

$3x^2 + 7x - 6 = 0$ Add $7x$.

$(3x-2)(x+3) = 0$ Factor.

$3x - 2 = 0$ or $x + 3 = 0$

$3x = 2$ $\quad x = -3$

$x = \frac{2}{3}$

Both $\frac{2}{3}$ and -3 are solutions.

9. Let x represent the salary of an experienced journeyman. Then $\frac{3}{4}$ of x, or $\frac{3}{4}x$, represents the salary of an apprentice, which gives the following equation.

Journeyman's salary	plus	apprentice's salary	equals	total salary paid.
↓	↓	↓	↓	↓
x	$+$	$\frac{3}{4}x$	$=$	56,000

Solve:

$4(x) + 4\left(\frac{3}{4}x\right) = 4(56{,}000)$ Multiply by 4.

$4x + 3x = 224{,}000$

$7x = 224{,}000$

$x = 32{,}000.$

The journeyman's salary is \$32,000.

13. Let x represent the distance each way (the distance to the destination is the same as the distance from the destination). We fill in the chart as follows, realizing that the time column is filled in by using the formula $t = \frac{d}{r}$.

	d	r	t
To destination	x	60	$\frac{x}{60}$
From destination	x	50	$\frac{x}{50}$

Using the numbers in the time column in the chart above, we have the following equation.

"From" time	is	$\frac{1}{2}$ hour more than	"To" time.
↓	↓	↓	↓
$\frac{x}{50}$	$=$	$\frac{1}{2}+$	$\frac{x}{60}$

Solve:

$300\left(\frac{x}{50}\right) = 300\left(\frac{1}{2}\right) + 300\left(\frac{x}{60}\right)$ Multiply by 300.

$6x = 150 + 5x$

$x = 150.$

She traveled 150 miles each way.

17. Let x represent the number of hours it takes for Paul and Marco to tune the Toyota together. Since Paul can tune the car in 2 hours, he can complete $\frac{1}{2}$ of the tuneup in 1 hour. Also, since Marco can tune the car in 3 hours, he can complete $\frac{1}{3}$ of the job in 1 hour. The amount of work Paul can do in 1 hour plus the amount of work Marco can do in 1 hour must equal the amount of work they can do together in 1 hour, or $\frac{1}{x}$ of the job. So $\frac{1}{2} + \frac{1}{3} = \frac{1}{x}$.

Solve:

$6x\left(\frac{1}{2}\right) + 6x\left(\frac{1}{3}\right) = 6x\left(\frac{1}{x}\right)$

$3x + 2x = 6$

$5x = 6$

$x = \frac{6}{5}.$

Working together, it takes Paul and Marco

$\frac{6}{5} = 1\frac{1}{5}$ hours

to tune the Toyota.

SOLUTIONS

21. Let x represent the number of days it takes Sue to complete the job. Then she completes $\frac{1}{x}$ of the job in 1 day. Since Dennis can do the job in 4 days, then he does $\frac{1}{4}$ of the job in 1 day. Finally, if it takes $2\frac{1}{3} = \frac{7}{3}$ days to complete the job working together, then they can complete $\frac{1}{\frac{7}{3}} = \frac{3}{7}$ of the job in 1 day.

So we have
$$\frac{1}{x} + \frac{1}{4} = \frac{3}{7}.$$
Solve:
$$28x\left(\frac{1}{x}\right) + 28x\left(\frac{1}{4}\right) = 28x\left(\frac{3}{7}\right) \quad \text{Multiply by } 28x.$$
$$28 + 7x = 12x$$
$$28 = 5x$$
$$\frac{28}{5} = x.$$
It takes Sue $\frac{28}{5} = 5\frac{3}{5}$ days to complete the job.

25. Let x represent the amount of time required for an experienced employee. Then $\frac{1}{x}$ of the job is done in 1 hour. The new employee then requires $2x$ hours, and so completes $\frac{1}{2x}$ of the job in 1 hour working alone. They can complete $\frac{1}{2}$ of the job in 1 hour working together. So we have
$$\frac{1}{x} + \frac{1}{2x} = \frac{1}{2}.$$
Solve:
$$2x\left(\frac{1}{x}\right) + 2x\left(\frac{1}{2x}\right) = 2x\left(\frac{1}{2}\right) \quad \text{Multiply by } 2x.$$
$$2 + 1 = x$$
$$3 = x.$$
It will take the experienced employee 3 hours working alone.

29. Since m varies directly as p, there is a number k such that
$$m = kp.$$
Find k from the given fact that $m = 20$ when $p = 2$.
$$20 = k \cdot 2 \quad \text{Let } m = 20, p = 2.$$
$$k = 10$$
Let $k = 10$ in $m = kp$ to get
$$m = 10p.$$
If $p = 5$, then
$$m = 10 \cdot 5 \quad \text{Let } p = 5.$$
$$m = 50$$

33. If the circumference varies directly as the radius, and if c represents the circumference and r the radius, then there is a number k such that
$$c = kr.$$
Find k by letting $r = 7$ and $c = 43.96$.
$$43.96 = k \cdot 7$$
$$6.28 = k$$
Let $k = 6.28$ to get
$$c = 6.28r.$$
If $r = 11$, then
$$c = 6.28(11) = 69.08 \text{ centimeters.}$$

37. $y = 4x - 7$

(a) Let $x = 2$.
$$y = 4(2) - 7 = 8 - 7 = 1$$
(b) Let $x = -4$.
$$y = 4(-4) - 7 = -16 - 7 = -23$$

41. $2x - 3y = 5$

(a) Let $x = 2$.
$$2(2) - 3y = 5$$
$$4 - 3y = 5$$
$$-3y = 1$$
$$y = -\frac{1}{3}$$
(b) Let $x = -4$.
$$2(-4) - 3y = 5$$
$$-8 - 3y = 5$$
$$-3y = 13$$
$$y = -\frac{13}{3}$$

CHAPTER 7

Section 7.1 (page 381)

1. The equation is $x + y = 9$, and we want to know if (2, 7) is a solution. Replace x with 2 and y with 7 to get $2 + 7 = 9$. This is true, so (2, 7) is a solution.

5. To see whether (1, 2) is a solution of $4x - 3y = 6$, replace x with 1 and y with 2 to get $4(1) - 3(2) = 6$, a false statement. The ordered pair is not a solution.

9. To see whether $(-6, 8)$ is a solution of $x = -6$, replace x with -6 (there is no replacement for y), to get $-6 = -6$, a true statement. The ordered pair is a solution.

13. Let $x = 2$; then $y = 3x + 5 = 3(2) + 5 = 6 + 5 = 11$. The ordered pair is (2, 11).

17. Let $x = -3$; then $y = 3(-3) + 5 = -9 + 5 = -4$. The ordered pair is $(-3, -4)$.

21. Let $y = 8$; then $8 = 3x + 5$. Add -5 to both sides to get $3 = 3x$. Divide both sides by 3 to get $x = 1$. The ordered pair is (1, 8).

25. Let $y = 24$; then $24 = -4x + 8$. Add -8 to both sides to get $16 = -4x$. Divide both sides by -4 to get $x = -4$. The ordered pair is $(-4, 24)$.

29. Let $x = 2$; then $y = 3(2) - 5 = 6 - 5 = 1$. Let $x = 0$; then $y = 3(0) - 5 = 0 - 5 = -5$. Let $x = -3$; then $y = 3(-3) - 5 = -9 - 5 = -14$. The ordered pairs are (2, 1), $(0, -5)$, and $(-3, -14)$.

33. Let $m = 1$; then $-3(1) + n = 4$ or $-3 + n = 4$. Add 3 to each side to get $n = 7$. Let $m = 0$; then $-3(0) + n = 4$ or $0 + n = 4$ or $n = 4$. Let $m = -2$; then $-3(-2) + n = 4$ or $6 + n = 4$. Subtract 6 from each side to get $n = -2$. The table of values is shown below.

m	1	0	−2
n	7	4	−2

37. Let $y = 0$; then $4x - 9(0) = 36$ or $4x - 0 = 36$ or $4x = 36$. Divide both sides by 4 to get $x = 9$. Let $x = 0$; then $4(0) - 9y = 36$ or $0 - 9y = 36$ or $-9y = 36$. Divide both sides by -9 to get $y = -4$. Let $y = 4$; then $4x - 9(4) = 36$ or $4x - 36 = 36$. Add 36 to both sides to get $4x = 72$. Divide both sides by 4 to get $x = 18$. The table of values is shown below.

x	9	0	18
y	0	-4	4

41. Since the equation is $y = -8$, any value of x will give a y-value of -8. The table of values is shown below.

x	4	0	-4
y	-8	-8	-8

45. See the graph in the answer section.

Section 7.2 (page 387)

1. (2, 5)
5. (7, 3)
9.–17. See the graph in the answer section.
21. The x-coordinate is negative and the y-coordinate is positive, so the point is in Quadrant II.
25. The x-coordinate is negative and the y-coordinate is positive, so the point is in Quadrant II.

29.

x	0	6	4	-1
y	3	0	1	$\frac{7}{2}$

See the graph in the answer section.

33.

x	5	0	-3	-2
y	-2	-2	-2	-2

See the graph in the answer section.

37. $2k - 7 = 9$

$2k = 16$ Add 7 to each side.

$k = 8$ Divide each side by 2.

41. $-3 + 8z = -19$

$8z = -16$ Add 3 to each side.

$z = -2$ Divide each side by 8.

Section 7.3 (page 395)

All the graphs for these exercises may be found in the answer section.

1. (0, 5), (5, 0), (2, 3)
5. (0, −6), (2, 0), (3, 3)
9. To find the x-intercept, let $y = 0$. Then $3x - 5(0) = 9$; $3x - 0 = 9$; $3x = 9$. Divide both sides by 3 to get $x = 3$, giving (3, 0) as the x-intercept. To find the y-intercept, let $x = 0$. Then $3(0) - 5y = 9$; $0 - 5y = 9$; $-5y = 9$. Divide on both sides by -5 to get $y = -\frac{9}{5}$, giving $\left(0, -\frac{9}{5}\right)$ as the y-intercept.
13. To graph $x - y = 2$, find the intercepts. If $x = 0$, then $0 - y = 2$ or $y = -2$, which gives the point $(0, -2)$. For $y = 0$, then $x - 0 = 2$ or $x = 2$, which gives (2, 0). A third point can be found as a check. Let $x = 4$; then $4 - y = 2$, or $-y = -2$, or $y = 2$, which gives the point (4, 2). Plot these three points and draw a line through them.
17. To graph $x + 2y = 6$, find the intercepts. If $x = 0$, then $0 + 2y = 6$, $2y = 6$, or $y = 3$. If $y = 0$, then $x + 2(0) = 6$, $x + 0 = 6$, or $x = 6$. The two points are (0, 3) and (6, 0). Find one more point as a check. Let $x = 3$; then $3 + 2y = 6$, $2y = 3$, or $y = \frac{3}{2}$, giving the point $\left(3, \frac{3}{2}\right)$. Plot these three points and draw a line through them.
21. To graph $3x = 6 - 2y$, find the intercepts. Let $x = 0$; then $3(0) = 6 - 2y$, or $0 = 6 - 2y$ or $6 = 2y$ or $y = 3$. Let $y = 0$; then $3x = 6 - 2(0)$, $3x = 6 - 0$, $3x = 6$, $x = 2$. The points are (0, 3) and (2, 0). Find one more point as a check. Let $x = 4$; then $3(4) = 6 - 2y$, $12 = 6 - 2y$, $6 = -2y$, or $y = -3$, giving $(4, -3)$. Plot these three points and draw a line through them.
25. In the equation $y = -3x$, let $x = 0$. Then $y = 0$. This gives the ordered pair (0, 0). If we let $y = 0$ we get the same pair, so let $x = 1$ to get $y = -3$. This gives the ordered pair $(1, -3)$. The graph is the line passing through (0, 0) and $(1, -3)$.
29. In the equation $3x - 4y = 0$, let $x = 0$. Then $0 - 4y = 0$ and $y = 0$. This gives the ordered pair (0, 0). If we let $y = 0$ we get the same pair, so let $x = 4$ to get $12 - 4y = 0$ and $y = 3$. This gives the ordered pair (4, 3). The graph is the line passing through (0, 0) and (4, 3).
33. Since the x-value of an ordered pair doesn't change the equation $y = 0$, we can graph the equation by using any ordered pairs with $y = 0$, such as $(-1, 0)$, (0, 0), and (1, 0). The graph, which is the line passing through these ordered pairs, is the x-axis.

37. $$\frac{-2 - (-4)}{3 - (-1)} = \frac{-2 + 4}{3 + 1} = \frac{2}{4} = \frac{1}{2}$$

41. $$\frac{12 - (-4)}{3 - 3} = \frac{12 + 4}{0}$$ Undefined

Section 7.4 (page 405)

1. Locate two points on the graph. The graph goes through $(-5, 2)$ and $(5, -3)$. Use the slope formula:

$$m = \frac{-3 - 2}{5 - (-5)} = \frac{-5}{10} = -\frac{1}{2}.$$

5. Locate two points on the graph. The graph goes through $(-4, 3)$ and $(2, 3)$. Use the slope formula:

$$m = \frac{3 - 3}{2 - (-4)} = \frac{0}{6} = 0.$$

9. $m = \dfrac{0 - 5}{8 - 0} = -\dfrac{5}{8}$

13. $m = \dfrac{-\frac{3}{10} - \frac{1}{2}}{\frac{7}{5} - \left(-\frac{1}{5}\right)} = \dfrac{\frac{-3}{10} - \frac{5}{10}}{\frac{7}{5} + \frac{1}{5}} = \dfrac{\frac{-8}{10}}{\frac{8}{5}}$

$$= \frac{-8}{10} \cdot \frac{5}{8} = -\frac{5}{10} = -\frac{1}{2}$$

17. Since y is alone on one side of the equation $y = 5x + 2$, the slope is given by the coefficient of x. Thus, Slope $= 5$.

21. Since y is alone on one side of the equation $y = 3 + 9x$, the slope is given by the coefficient of x. Thus, Slope $= 9$.

25. Get y alone on one side of the equation.

$$6x - 9y = 5$$
$$-9y = 5 - 6x \qquad \text{Subtract } 6x.$$
$$y = \frac{5}{-9} - \frac{6}{-9}x \qquad \text{Divide by } -9.$$
$$y = -\frac{5}{9} + \frac{6}{9}x$$

The coefficient of x is $\frac{6}{9} = \frac{2}{3}$, so the slope is $\frac{2}{3}$.

29. The slope of the first line is -5, the coefficient of x. The slope of the second line is $\frac{1}{5}$. The product of the two slopes is $(-5)\left(\frac{1}{5}\right) = -1$, so the lines are perpendicular.

33. Solve the equation $2x - 5y = 4$ for y.

$$-5y = 4 - 2x$$
$$y = \frac{2}{5}x - \frac{4}{5}$$

The slope is $\frac{2}{5}$.

Solve $4x - 10y = 1$ for y.

$$-10y = 1 - 4x$$
$$y = \frac{4}{10}x - \frac{1}{10}$$
$$y = \frac{2}{5}x - \frac{1}{10} \qquad \text{Simplify.}$$

The slope is $\frac{2}{5}$. The two lines have the same slope, so they are parallel.

37. Solve the equation $x - 4y = 2$ for y.

$$x - 4y = 2$$
$$-4y = 2 - x$$
$$y = \frac{2}{-4} - \left(-\frac{1}{4}x\right)$$

or

$$y = \frac{1}{4}x - \frac{1}{2}$$

The slope of the line is $\frac{1}{4}$.

Solve $2x + 4y = 1$ for y.

$$2x + 4y = 1$$
$$4y = 1 - 2x$$
$$y = \frac{1}{4} - \frac{2}{4}x \quad \text{or} \quad y = -\frac{1}{2}x + \frac{1}{4}$$

The slope of the line is $-\frac{1}{2}$. The lines are neither parallel nor perpendicular.

41. The change in y is 3. The change in x is 18.

$$m = \frac{3}{18} = \frac{1}{6}$$

45.
$$y + 1 = 2(x - 5)$$
$$y + 1 = 2x - 10 \qquad \text{Distributive property}$$
$$y = 2x - 11 \qquad \text{Subtract 1 on each side.}$$

49.
$$y - 2 = -3[x - (-1)]$$
$$y - 2 = -3(x + 1) \qquad \text{Definition of subtraction}$$
$$y - 2 = -3x - 3 \qquad \text{Distributive property}$$
$$y = -3x - 1 \qquad \text{Add 2 on each side.}$$

Section 7.5 (page 413)

1. Substitute $m = 3$ and $b = 5$ into the slope-intercept form.

$$y = mx + b$$
$$y = 3x + 5$$

5. Substitute $m = \frac{5}{3}$ and $b = \frac{1}{2}$ into the slope-intercept form.

$$y = mx + b$$
$$y = \frac{5}{3}x + \frac{1}{2}$$

9. Starting at $(2, 5)$, the slope $\frac{1}{2}$ says to go 1 up and 2 to the right to find a second point on the line. See the graph in the answer section.

13. Starting at $(-3, 0)$, the slope $-\frac{5}{4}$ says to go 4 to the right and 5 down to find a second point on the line. See the graph in the answer section.

17. The slope $m = 0$ means that the line through $(1, 2)$ is a horizontal line. See the graph in the answer section.

21. Use the values $m = 2$, $x_1 = 5$, and $y_1 = 3$ in the point-slope form.

$$y - y_1 = m(x - x_1)$$
$$y - 3 = 2(x - 5)$$
$$y - 3 = 2x - 10$$
$$-2x + y = -10 + 3$$

or

$$2x - y = 7$$

25. Use the values $m = \frac{2}{3}$, $x_1 = 3$, and $y_1 = 5$ in the point-slope form.

$$y - y_1 = m(x - x_1)$$
$$y - 5 = \frac{2}{3}(x - 3)$$
$$3(y - 5) = 2(x - 3)$$
$$3y - 15 = 2x - 6$$
$$-2x + 3y = -6 + 15 \quad \text{or} \quad 2x - 3y = -9$$

29. Use the values $m = -\frac{8}{11}$, $x_1 = 6$, and $y_1 = 0$ in the point-slope form.

$$y - y_1 = m(x - x_1)$$
$$y - 0 = \frac{-8}{11}(x - 6)$$
$$y = \frac{-8}{11}(x - 6)$$
$$11y = -8(x - 6) \quad \text{Multiply by 11.}$$
$$11y = -8x + 48$$
$$8x + 11y = 48$$

33. First find the slope of the line, using the definition of slope.

$$\text{Slope} = \frac{-4 - (-1)}{3 - (-2)} = \frac{-4 + 1}{3 + 2} = \frac{-3}{5}$$

Now use either point and $m = -\frac{3}{5}$ in the point-slope form. If we use $(3, -4)$ we get

$$y - y_1 = m(x - x_1)$$
$$y - (-4) = \frac{-3}{5}(x - 3)$$
$$y + 4 = \frac{-3}{5}(x - 3)$$
$$5(y + 4) = 5 \cdot \frac{-3}{5}(x - 3) \quad \text{Multiply by 5.}$$
$$5y + 20 = -3x + 9$$
$$5y = -3x - 11 \quad \text{or} \quad 3x + 5y = -11.$$

37. Use the definition to find the slope.

$$m = \frac{-5 - 7}{2 - (-4)} = \frac{-12}{6} = -2$$

Now use either point and $m = -2$ in the point-slope form. Using $(2, -5)$ we get

$$y - y_1 = m(x - x_1)$$
$$y - (-5) = -2(x - 2)$$
$$y + 5 = -2x + 4$$
$$y = -2x - 1 \quad \text{or} \quad 2x + y = -1.$$

41. Use the definition to find the slope.

$$m = \frac{\frac{5}{8} - 2}{-1 - \frac{1}{8}} = \frac{\frac{5}{8} - \frac{16}{8}}{-\frac{8}{8} - \frac{1}{8}} = \frac{\frac{-11}{8}}{\frac{-9}{8}} = \frac{11}{9}$$

Use one of the points, say $\left(\frac{1}{8}, 2\right)$, and $m = \frac{11}{9}$ in the point-slope form.

$$y - y_1 = m(x - x_1)$$
$$y - 2 = \frac{11}{9}\left(x - \frac{1}{8}\right)$$
$$72(y - 2) = 72 \cdot \frac{11}{9}\left(x - \frac{1}{8}\right) \quad \text{Multiply by 72.}$$
$$72y - 144 = 88\left(x - \frac{1}{8}\right)$$
$$72y - 144 = 88x - 11$$
$$72y = 88x + 133$$

or

$$88x - 72y = -133$$

45. From Exercise 43, the cost equation is $y = 12x + 100$.

(a) Let $x = 50$.

$$y = 12(50) + 100$$
$$y = 600 + 100 = 700$$

The total cost is 700.

(b) Let $x = 125$.

$$y = 12(125) + 100$$
$$y = 1500 + 100$$
$$y = 1600$$

The total cost is 1600.

49. The graph should have a solid dot at 10 and a line to the right of 10.

53. The graph should have a solid dot at $\frac{13}{3}$ and a line to the right of $\frac{13}{3}$.

Section 7.6 (page 421)

All the graphs for these exercises may be found in the answer section.

1. Use (0, 0) as a test point.

$$x + y \le 4 \quad \text{Original inequality}$$
$$0 + 0 \le 4 \quad \text{Let } x = 0, y = 0.$$
$$0 \le 4 \quad \text{True}$$

Since this last statement is true, shade the side of the graph containing (0, 0).

5. Use (0, 0) as a test point.

$$-3x + 4y < 12$$
$$-3(0) + 4(0) < 12$$
$$0 + 0 < 12$$
$$0 < 12 \quad \text{True}$$

Shade the side of the graph containing (0, 0).

9. Use (0, 0) as a test point.

$$x < 4 \quad \text{Original inequality}$$
$$0 < 4 \quad \text{Use } x = 0\text{; true.}$$

Since this statement is true, shade the side of the graph that contains (0, 0).

13. *Step 1* Graph $x + y = 8$.
If $x = 0$, then $y = 8$, giving (0, 8).
If $y = 0$, then $x = 8$, giving (8, 0).
Graph these points and the line through them. The line is solid because of the "≤" symbol.
Step 2 Use (0, 0) as a test point.
$x + y \leq 8$
$0 + 0 \leq 8$
$0 \leq 8$ True
Step 3 Since the statement is true, shade the side of the graph containing (0, 0).

17. *Step 1* Graph $x + 2y = 4$.
If $x = 0$, then $0 + 2y = 4$ or $y = 2$, giving (0, 2).
If $y = 0$, then $x + 2(0) = 4$ or $x = 4$, giving (4, 0).
Graph these points and the line through them. The line is solid because of the "≥" symbol.
Step 2 Use (0, 0) as a test point.
$x + 2y \geq 4$
$0 + 2(0) \geq 4$
$0 \geq 4$ False
Step 3 Since the statement is false, shade the side that does not contain (0, 0).

21. *Step 1* Graph the line $x = 4$. The graph is a vertical line through the x-axis at $x = 4$. The line is dotted because of the < sign.
Step 2 Use (0, 0) as a test point.
$x < 4$ Original inequality
$0 < 4$ Let $x = 0$; true.
Step 3 Since $0 < 4$, shade the side of the graph containing (0, 0).

25. *Step 1* Graph $x = 3y$. If $x = 0$, then $y = 0$, so the x-intercept and y-intercept are (0, 0). To find another point, choose a different value of x, say $x = 6$. If $x = 6$, then $y = 2$, giving (6, 2). Graph these points and the line through them. This line is solid because of the ≤ sign.
Step 2 Use (0, 1) as a test point, since (0, 0) is on the line.
$x \leq 3y$ Original inequality
$0 \leq 3(1)$ Let $x = 0$, $y = 1$.
$0 \leq 3$ True
Step 3 Since $0 \leq 3$, shade the side of the graph that contains (0, 1).

29. **(a)** *Step 1* Graph $10x + 15y = 50$. If $x = 0$, then $y = \frac{10}{3}$, so the y-intercept is $\left(0, \frac{10}{3}\right)$. If $y = 0$, then $x = 5$, so the x-intercept is (5, 0). Graph these points and the line through them. This line is solid because of the ≤ sign.
Step 2 Use (0, 0) as a test point.
$10x + 15y \leq 50$ Original inequality
$10(0) + 15(0) \leq 50$ Let $x = 0$, $y = 0$.
$0 \leq 50$ True
Step 3 Since $0 \leq 50$, shade the side of the graph that contains (0, 0). Because of the restrictions $x \geq 0$ and $y \geq 0$, shade only the portion of the graph in the first quadrant.

(b) Any point in the shaded region satisfies the inequality: for example, (1, 2), (2, 1), (4, 0), and so on.

33.
$$\begin{array}{r} -11p + 8q \\ 11p - 9q \\ \hline -q \end{array}$$

CHAPTER 8

Section 8.1 (page 441)

1. Replace x with 2 and y with -5. In the first equation we get $3(2) + (-5) = 1$ or $6 + (-5) = 1$, which is true. In the second equation we get $2(2) + 3(-5) = -11$ or $4 + (-15) = -11$, which is true. Since $(2, -5)$ makes both equations true, it is the solution of the system.

5. Replace x with 6 and y with -8. In the first equation we get $2(-8) = -6 - 10$ or $-16 = -16$, which is true. In the second equation we get $3(-8) = 2(6) + 30$, or $-24 = 42$, which is false. This ordered pair is not a solution.

9. Replace x with 9 and y with 1. In the first equation we get $2(9) = 23 - 5(1)$ or $18 = 18$, which is true. In the second equation we get $3(9) = 24 - 2(1)$, or $27 = 22$, which is false. Because the ordered pair does not satisfy both equations in the system, it is not the solution of the system.

13. Graph both lines on the same axes. For $x + y = 6$, use the intercepts, (0, 6) and (6, 0). For $x - y = 2$, use the intercepts, $(0, -2)$ and (2, 0). These lines intersect at (4, 2).

17. Graph both lines on the same axes. For $x + 2y = 2$, use the intercepts, (0, 1) and (2, 0). For $x - 2y = 6$, use the intercepts, $(0, -3)$ and (6, 0). These lines intersect at $(4, -1)$.

21. Graph both lines on the same axes. For $-2x + 3y = 6$, use the intercepts, (0, 2) and $(-3, 0)$. For $3x + y = 2$, use the intercepts, (0, 2) and $\left(\frac{2}{3}, 0\right)$. These lines intersect at (0, 2).

25. Graph both lines on the same axes. For $3x + 2y = -10$ use the points $(0, -5)$ and $(-2, -2)$. For $x - 2y = -6$ use the intercepts, $(-6, 0)$ and (0, 3). These lines intersect at $(-4, 1)$.

29. Graph both lines on the same axes. For $y = x + 6$ use the intercepts, (0, 6) and $(-6, 0)$. For $y = -x - 2$, use the intercepts, $(0, -2)$ and $(-2, 0)$. These lines intersect at $(-4, 2)$.

33. Graph both lines on the same axes. For $2x + 5y = 12$, use the points (6, 0) and (1, 2). For $x + y = 3$, use the intercepts, (0, 3) and (3, 0). These lines intersect at (1, 2).

37. Graph each line. The two lines are parallel, so there is no solution.

41. Add by columns.
$$\begin{array}{r} 6p + 2q \\ 4p + 5q \\ \hline 10p + 7q \end{array}$$

45. Add horizontally.

$(4a + 3b) + (-3b - 2a)$
$= [4a + (-2a)] + [3b + (-3b)]$
$= 2a + 0$
$= 2a$

Section 8.2 (page 449)

1. Add the equations, getting $2x = 2$ or $x = 1$. Replace x with 1 in either equation. If we use $x - y = 3$, we get $1 - y = 3$ or $-y = 2$ or $y = -2$. The solution is $(1, -2)$.

5. Add the equations, getting $3x = 18$ or $x = 6$. Replace x with 6 in either equation. If we use $2x + y = 14$, we get $2(6) + y = 14$ or $12 + y = 14$ or $y = 2$. The solution is $(6, 2)$.

9. Add the equations, getting $4y = 8$ or $y = 2$. Replace y with 2 in either equation. If we use $6x - y = 1$, we get $6x - 2 = 1$ or $6x = 3$ or $x = \frac{1}{2}$. The solution is $\left(\frac{1}{2}, 2\right)$.

13. Multiply each side of the first equation by 2 and add.

$$\begin{aligned} 4x - 2y &= 14 \\ 3x + 2y &= 0 \\ \hline 7x \quad &= 14 \\ x \quad &= 2 \end{aligned}$$

Let $x = 2$ in the second equation to get

$$\begin{aligned} 3x + 2y &= 0 \\ 3(2) + 2y &= 0 \\ 6 + 2y &= 0 \\ 2y &= -6 \\ y &= -3. \end{aligned}$$

The solution is $(2, -3)$.

17. Multiply each side of the first equation by -3 and then add.

$$\begin{aligned} -3x - 12y &= 54 \\ 3x + 5y &= -19 \\ \hline -7y &= 35 \\ y &= -5 \end{aligned}$$

Let $y = -5$ in the first equation to get

$$\begin{aligned} x + 4y &= -18 \\ x + 4(-5) &= -18 \\ x - 20 &= -18 \\ x &= 2. \end{aligned}$$

The solution is $(2, -5)$.

21. One way to proceed is to multiply the first equation by 2 to get $6x - 4y = -12$, then add.

$$\begin{aligned} 6x - 4y &= -12 \\ -5x + 4y &= 16 \\ \hline x \quad &= 4 \end{aligned}$$

Replace x with 4 in either equation to find that $y = 9$, giving $(4, 9)$.

25. One way to proceed is to multiply the first equation by -3 to get $-6x - 3y = -15$, then add.

$$\begin{aligned} -6x - 3y &= -15 \\ 5x + 3y &= 11 \\ \hline -x \quad &= -4 \\ x &= 4 \end{aligned}$$

Replace x with 4 in either equation to find that $y = -3$, giving $(4, -3)$.

29. One way to proceed is to multiply the top equation by 3 to get $9x + 15y = 99$, and the bottom equation by 5, to get $20x - 15y = 75$. Add these two equations to get $29x = 174$ or $x = 6$. Replace x with 6 in either of the original equations to find $y = 3$, giving the solution $(6, 3)$.

33. One way to proceed is to multiply the first equation by 7 to get $14x + 21y = -84$, and the second equation by 3 to get $15x - 21y = -90$. Add these equations to get $29x = -174$ or $x = -6$. Replace x with -6 in either of the original equations to find $y = 0$, giving the solution $(-6, 0)$.

37. Multiply the first equation by 3 to get $72x + 36y = 57$, and the second equation by 2 to get $32x - 36y = -18$. Add these equations to get $104x = 39$ or $x = \frac{3}{8}$. Replace x with $\frac{3}{8}$ in one of the original equations to find $y = \frac{5}{6}$, giving the solution $\left(\frac{3}{8}, \frac{5}{6}\right)$.

41. Begin by writing the first equation as $5x - 7y = 6$. One way to proceed is to multiply the first equation by -3 to get $-15x + 21y = -18$. Next, multiply the second equation by 5 to get $15x - 30y = 10$. Add these equations to get $-9y = -8$, or $y = \frac{8}{9}$. Replace y with $\frac{8}{9}$ in one of the original equations to find $x = \frac{22}{9}$, giving the solution $\left(\frac{22}{9}, \frac{8}{9}\right)$.

45.

$$\begin{aligned} x + 7 &= x + 8 \\ x + 7 - 7 &= x + 8 - 7 \\ x &= x + 1 \\ x - x &= x + 1 - x \\ 0 &= 1 \quad \text{False} \end{aligned}$$

The false statement indicates that the equation has no solution.

49.

$$\begin{aligned} 5x + x + 1 &= 2x + 4x + 1 \\ 6x + 1 &= 6x + 1 \\ 0 &= 0 \quad \text{Subtract } 6x \text{ and } 1. \end{aligned}$$

The true statement $0 = 0$ indicates that all real numbers are solutions.

Section 8.3 (page 455)

1. Multiply the top equation by -1, giving $-x - y = -4$. Then add the two equations, giving $0 = -6$. This false statement shows that the system has no solution.

5. Multiply the top equation by -2, giving $-2x - 6y = -10$. Add the two equations, giving $0 = 0$. This true statement shows that the two equations represent the same line.

9. Multiply the first equation by -1 to get $-5x = -y - 4$. Add this result to the second equation to get $0 = -8$, a false statement. This shows that the system has no solution.

13. Multiply the top equation by -2, getting $-4x + 6y = 0$; then add to get $11y = 0$ or $y = 0$. Replace y with 0 to see that $x = 0$, giving the solution $(0, 0)$.

SOLUTIONS

17. Multiply the first equation by -2 to get $-8x + 4y = -2$. Add this result to the second equation to get $0 = -1$. This false statement shows that the system has no solution.

21.
$$\begin{aligned} -2(y - 2) + 5y &= 10 \\ -2y + 4 + 5y &= 10 && \text{Distributive property} \\ 3y + 4 &= 10 && \text{Combine terms.} \\ 3y &= 6 && \text{Subtract 4 from each side.} \\ y &= 2 && \text{Divide each side by 3.} \end{aligned}$$

Check the solution in the original equation.

25.
$$\begin{aligned} 4x - 2\left(\frac{1 - 3x}{2}\right) &= 6 \\ 4x - (1 - 3x) &= 6 && \text{Multiply.} \\ 4x - 1 + 3x &= 6 && \text{Distributive property} \\ 7x - 1 &= 6 && \text{Combine terms.} \\ 7x &= 7 && \text{Add 1 to each side.} \\ x &= 1 && \text{Divide each side by 7.} \end{aligned}$$

Check the solution in the original equation.

29.
$$\begin{aligned} 4r + 5\left(\frac{2r + 7}{4}\right) &= 1 \\ 4\left[4r + 5\left(\frac{2r + 7}{4}\right)\right] &= 4 \cdot 1 && \text{Multiply each side by 4.} \\ 16r + 20\left(\frac{2r + 7}{4}\right) &= 4 && \text{Distributive property} \\ 16r + 5(2r + 7) &= 4 && \text{Divide 20 by 4.} \\ 16r + 10r + 35 &= 4 && \text{Distributive property} \\ 26r + 35 &= 4 && \text{Combine terms.} \\ 26r &= -31 && \text{Subtract 35 from each side.} \\ r &= -\frac{31}{26} && \text{Divide each side by 26.} \end{aligned}$$

Check the solution in the original equation.

Section 8.4 (page 461)

1. The second equation says that $y = 2x$. Replace y with $2x$ in the first equation to get $x + 2x = 6$ or $3x = 6$ or $x = 2$. Replace x with 2 in the equation $y = 2x$ to get $y = 2(2) = 4$. The solution is $(2, 4)$.

5. Solve the second equation for x to get $x = 2y + 10$. Replace x with $2y + 10$ in the first equation to get $2y + 10 + 5y = 3$, from which $7y + 10 = 3$, $7y = -7$, or $y = -1$. Replace y with -1 in $x = 2y + 10$ to get $x = 2(-1) + 10 = 8$. The solution is $(8, -1)$.

9. Solve either equation for either variable. If we solve the first equation for x, we get $x = 6 - y$. Replace x with $6 - y$ in the second equation to get $6 - y - y = 4$, from which $6 - 2y = 4$, $-2y = -2$, or $y = 1$. Replace y with 1 in $x = 6 - y$ to get $x = 6 - 1 = 5$. The solution is $(5, 1)$.

13. Solve the second equation for y to get $y = 2x - 7$. Replace y with $2x - 7$ in the first equation.
$$\begin{aligned} 2x + 3y &= 11 \\ 2x + 3(2x - 7) &= 11 \\ 2x + 6x - 21 &= 11 \\ 8x &= 32 \\ x &= 4 \end{aligned}$$

Find y from $y = 2x - 7$.
$$y = 2(4) - 7 = 8 - 7 = 1$$

The solution is $(4, 1)$.

17. To solve the system by the substitution method, solve the second equation for x.
$$\begin{aligned} 3x &= 4y + 2 \\ x &= \frac{4}{3}y + \frac{2}{3} \end{aligned}$$

Substitute for x in the first equation.
$$\begin{aligned} 6\left(\frac{4}{3}y + \frac{2}{3}\right) - 8y &= 4 \\ 6\left(\frac{4}{3}y\right) + 6\left(\frac{2}{3}\right) - 8y &= 4 \\ 8y + 4 - 8y &= 4 \\ 4 &= 4 && \text{True} \end{aligned}$$

This true statement shows that the equations of the system represent the same line.

In Exercises 21–37, only one method is shown.

21. Replace y in the first equation with $4x$ to get $x + 4(4x) = 34$ or $x + 16x = 34$, from which $17x = 34$ or $x = 2$. Replace x with 2 in the second equation to get $y = 8$. The solution is $(2, 8)$.

25. Combine terms in each equation to get $x - 6y = -9$ for the first equation and $2x + y = 8$ for the second equation. Solve the first equation for x: $x = 6y - 9$. Replace x with $6y - 9$ in the second equation to get $2(6y - 9) + y = 8$, from which $12y - 18 + y = 8$ or $13y = 26$, so $y = 2$. Replace y with 2 in $x = 6y - 9$ to get $x = 6(2) - 9 = 12 - 9 = 3$. The solution is $(3, 2)$.

29. Combine terms in each equation to get $-2x + y = 12$ and $2x - y = -12$. Add these two equations, getting $0 = 0$, which indicates that the equations represent the same line.

33. Multiply both sides of the first equation by 3 to clear fractions, getting $3x + y = 3y - 6$. Rearrange terms to get $3x - 2y = -6$. Multiply the second equation on each side by 4 to clear fractions, getting $x - 4y = 4x - 4y$. Combine terms to get $-3x = 0$ or $x = 0$. Replace x with 0 in $3x - 2y = -6$ to get $-2y = -6$ or $y = 3$. The solution is $(0, 3)$.

37. Multiply the first equation on each side by the common denominator, 12, to get $4x - 9y = -6$. Multiply the second equation on each side by the common denominator, 6, to get $4x + 3y = 18$. Multiply the first result by -1 and add to the second result, getting $12y = 24$ or $y = 2$. Replace y with 2 in the first result to get $4x - 9(2) = -6$ or $4x - 18 = -6$, from which $4x = 12$ or $x = 3$. The solution is $(3, 2)$.

41. Let w represent the width. Then $w + 7$ represents the length. Use the formula $P = 2L + 2W$, with $P = 46$.
$$\begin{aligned} 46 &= 2(w + 7) + 2w \\ 46 &= 2w + 14 + 2w \\ 46 &= 4w + 14 \\ 32 &= 4w \\ 8 &= w \end{aligned}$$

The width is 8 feet.

Section 8.5 (page 471)

1. Let x and y represent the two numbers. Then $x + y = 69$ and $x - y = 23$. Add the two equations to get $2x = 92$, from which $x = 46$. From the first equation, $46 + y = 69$, so $y = 23$. The numbers are 46 and 23.

5. Let x represent the length of each of the two equal sides and let y represent the length of the third side. Since the perimeter is 42 centimeters, one equation is

$$2x + y = 42.$$

Since the third side measures 6 centimeters more than one of the equal sides, another equation is

$$y = x + 6.$$

We must solve the system

$$2x + y = 42$$
$$y = x + 6.$$

Solve using the substitution method.

$2x + (x + 6) = 42$ Replace y with $x + 6$ in the first equation.

$3x + 6 = 42$ Combine terms.

$3x = 36$ Subtract 6.

$x = 12$ Divide by 3.

Each of the two equal sides measure 12 centimeters and the third side measures $12 + 6 = 18$ centimeters.

9. Let x represent the number of student tickets sold and let y represent the number of non-student tickets sold. Since there were 193 tickets sold in all, one equation is

$$x + y = 193.$$

The monetary value of the receipts for the student tickets is $.50x$ (dollars) and for the non-student tickets is $1.50y$ (dollars). Since $206.50 was collected, another equation is

$$.50x + 1.50y = 206.50.$$

We must solve the system

$$x + y = 193$$
$$.50x + 1.50y = 206.50.$$

Multiply the bottom equation by -2 to get the system

$$x + y = 193$$
$$-x - 3y = -413$$

$-2y = -220$ Add.

$y = 110.$ Divide by -2.

Replace y with 110 in the equation $x + y = 193$ to find that $x = 83$. 83 student tickets were sold and 110 non-student tickets were sold.

13. Let x represent the number of large canvases and let y represent the number of small canvases. Since he bought 39 canvases in all, one equation is

$$x + y = 39.$$

Each large canvas cost $7, so he spent $7x$ dollars on the large canvases. Similarly, he spent $4y$ dollars on the small canvases. Since he spent $219 altogether, another equation is

$$7x + 4y = 219.$$

Solve this system to get $x = 21$ and $y = 18$. He bought 21 large canvases and 18 small canvases.

17. Let x represent the number of pounds of 60% copper alloy and y represent the number of pounds of 80% copper alloy. Make a chart.

Pounds of alloy	*Percent copper (as a decimal)*	*Pounds of pure copper*
x	.60	$.60x$
y	.80	$.80y$
40	.65	$.65(40) = 26$

Write two equations: $x + y = 40$ and $.60x + .80y = 26$. Solve this system to get $x = 30$ and $y = 10$. Thus, 30 pounds of 60% copper alloy and 10 pounds of 80% copper alloy are needed.

21. Let x represent the number of bags of $70 fertilizer and y represent the number of bags of $90 fertilizer. Make a chart.

Bags of fertilizer	*Price per bag (in dollars)*	*Total price (in dollars)*
x	70	$70x$
y	90	$90y$
80	77.50	$80(77.50) = 6200$

Write two equations: $x + y = 80$ and $70x + 90y = 6200$. Solve this system to get $x = 50$ and $y = 30$. She should use 50 bags worth $70 per bag and 30 bags worth $90 per bag.

25. Let x represent the plane's speed in still air and y the wind speed. The rate of the plane with the wind is $x + y$, so one equation is $x + y = 100$. The rate of the plane against the wind is $x - y$, so another equation is $x - y = 60$. Solve the system by addition.

$$x + y = 100$$
$$x - y = 60$$
$$2x = 160$$
$$x = 80$$

Substitute 80 for x in the first equation.

$$80 + y = 100$$
$$y = 20$$

The speed of the plane in still air is 80 miles per hour, and the wind speed is 20 miles per hour.

29. Let A represent Anderson's speed and B Bentley's speed. Then

$$A + B = 120. \quad \textbf{(1)}$$

Draw a diagram.

Farmersville ← 400 mi → Exeter

|— 4A —→← 2B —|

Anderson		Bentley
12:00 P.M.	4:00 P.M.	2:00 P.M.
Anderson		*Bentley*
time: 4 hours		time: 2 hours
rate: A		rate: B
d: $4A$		d: $2B$

From the diagram, $4A + 2B = 400$. Divide by 2 to get

$$2A + B = 200. \quad \textbf{(2)}$$

SOLUTIONS

To solve the system by substitution, solve for A in equation (1).

$A = 120 - B$

Substitute $(120 - B)$ for A in equation (2) and solve for B.

$$2(120 - B) + B = 200$$
$$240 - 2B + B = 200$$
$$240 - B = 200$$
$$-B = -40$$
$$B = 40$$

Replace B by 40 in equation (1).

$$A + 40 = 120$$
$$A = 80$$

Anderson's speed was 80 miles per hour, and Bentley's was 40 miles per hour.

33. *Step 1* Graph $3x + 2y = 6$.
This graph is a straight line with intercepts (0, 3) and (2, 0). Graph this as a dashed line because of the > symbol.
Step 2 Use (0, 0) as a test point.

$3x + 2y > 6$ Original inequality
$3(0) + 2(0) > 6$
$0 > 6$ False

Step 3 Since the statement is false, shade the side that does not contain (0, 0). See the graph in the answer section.

Section 8.6 (page 479)

1. *Step 1* Graph the line $x + y = 6$.
If $x = 0$, then $y = 6$, giving the ordered pair (0, 6).
If $y = 0$, then $x = 6$, giving the ordered pair (6, 0).
Graph these two points and the solid line through them.
Step 2 Use (0, 0) as a test point.

$x + y \leq 6$ Original inequality
$0 + 0 \leq 6$ Substitute $x = 0$, $y = 0$.
$0 \leq 6$ True.

Step 3 Since the inequality is true for (0, 0), shade the side of the graph containing (0, 0).

Go through the same steps above for the second inequality.
Step 1 Graph the line $x - y = 1$.
If $x = 0$, then $-y = 1$ or $y = -1$, giving the ordered pair $(0, -1)$.
If $y = 0$, then $x = 1$, giving the ordered pair (1, 0).
Graph these two points and the solid line through them.
Step 2 Use (0, 0) as a test point.

$x - y \leq 1$ Original inequality
$0 - 0 \leq 1$ Substitute $x = 0$, $y = 0$.
$0 \leq 1$ True

Step 3 Since the inequality is true for (0, 0), shade the side of the graph containing (0, 0). The solution is where the two shaded regions overlap, or the darkest shaded region, and the lines that border this region. See the graph in the answer section.

5. Graph the line $x + 4y = 8$. Make it a solid line. Use (0, 0) as a test point.

$x + 4y \leq 8$
$0 + 4 \cdot 0 \leq 8$
$0 + 0 \leq 8$
$0 \leq 8$ True

Shade the area on the side of the line containing (0, 0).
Graph the line $2x - y = 4$. Make it a solid line. Use (0, 0) for a test point.

$2x - y \leq 4$
$2 \cdot 0 - 0 \leq 4$
$0 - 0 \leq 4$
$0 \leq 4$ True

Shade the area on the side of this line containing (0, 0).
The solution is the overlapped portion of the shaded areas. See the graph in the answer section.

9. Graph the line $x + 2y = 4$. Make the line solid. Use (0, 0) as a test point. The inequality $x + 2y \leq 4$ is true at (0, 0). Shade the area on the side of the line containing (0, 0).
Graph the line $x + 1 = y$. Make the line solid. Use (0, 0) as a test point. The inequality $x + 1 \geq y$ is true at (0, 0). Shade the area on the side of this line containing (0, 0).
The solution is the overlapped portion of the shaded areas. See the graph in the answer section.

13. Graph the line $x - 2y = 6$. Make the line dashed. Use (0, 0) for a test point. The inequality $x - 2y > 6$ is false at (0, 0). Shade the area on the side of the line that doesn't contain (0, 0).
Graph the line $2x + y = 4$. Make the line dashed. Use (0, 0) for a test point. The inequality $2x + y > 4$ is false at (0, 0). Shade the area on the side of this line that doesn't contain (0, 0).
The solution is the overlapped portion of the shaded areas. See the graph in the answer section.

17. Graph the line $x - 3y = 6$. Make it a solid line. Use (0, 0) for a test point. The inequality $x - 3y \leq 6$ is true at (0, 0). Shade the area on the side of the line containing (0, 0).
Graph the line $x = -1$. Make it a solid line. Use (0, 0) for a test point. The inequality $x \geq -1$ is true at (0, 0). Shade the area on the side of this line that contains (0, 0).
The solution is the overlapped portion of the shaded areas. See the graph in the answer section.

21. $11^2 = 11 \cdot 11 = 121$

CHAPTER 9

Section 9.1 (page 501)

1. Since $3^2 = 9$ and $(-3)^2 = 9$, the square roots of 9 are 3 and -3.

5. Since $\left(\frac{20}{9}\right)^2 = \frac{400}{81}$ and $\left(-\frac{20}{9}\right)^2 = \frac{400}{81}$, the square roots of $\frac{400}{81}$ are $\frac{20}{9}$ and $-\frac{20}{9}$.

9. Since 16 is positive, $(\sqrt{16})^2 = 16$.

13. Since $4x^2 + 3$ is positive for all x-values, $(\sqrt{4x^2 + 3})^2 = 4x^2 + 3$.

17. $\sqrt{64}$ is the nonnegative square root of 64, which is 8.

21. $\sqrt{900}$ is the nonnegative square root of 900, which is 30.

25. $\sqrt{\dfrac{36}{49}} = \dfrac{6}{7}$ since $\left(\dfrac{6}{7}\right)^2 = \dfrac{36}{49}$, and $\dfrac{6}{7}$ is positive.

29. $\sqrt{-9}$ is not a real number, since there is no real number whose square is -9.

33. There is no rational number whose square is 15, so $\sqrt{15}$ is irrational. From the table, or a calculator $\sqrt{15} \approx 3.873$.

37. $-\sqrt{121}$ is rational, since it equals the rational number -11.

41. $\sqrt{400}$ is rational, since it equals the rational number 20.

45. There is no real number whose square is -31, so $-\sqrt{-31}$ is not a real number.

49. Let $c = 17$ and $a = 8$ in $a^2 + b^2 = c^2$.

$$a^2 + b^2 = c^2$$
$$8^2 + b^2 = 17^2$$
$$64 + b^2 = 289$$
$$b^2 = 225 \quad \text{Subtract 64.}$$
$$b = 15$$

Use only the positive square root for the length of a side of a triangle.

53. Let $c = 12$ and $b = 7$.

$$a^2 + b^2 = c^2$$
$$a^2 + 7^2 = 12^2$$
$$a^2 + 49 = 144$$
$$a^2 = 95$$
$$a = \sqrt{95} \approx 9.747$$

57. As the figure in the text shows, a right triangle is formed with legs of length 48 and 36 feet. The wire is the hypotenuse and has length c, where $48^2 + 36^2 = c^2$ by the Pythagorean theorem.

$$c = \sqrt{48^2 + 36^2} = \sqrt{2304 + 1296} = \sqrt{3600}$$
$$= 60$$

The wire is 60 feet long.

61. Use the Pythagorean theorem. Let a be the measure of the other leg. Then

$$a^2 + 3^2 = 9^2 \quad \text{Pythagorean theorem}$$
$$a^2 + 9 = 81$$
$$a^2 = 72 \quad \text{Subtract 9 on each side.}$$
$$a = \sqrt{72} \approx 8.485 \quad \text{Use a calculator.}$$

The other leg measures 8.485 inches.

65. Factor 300 into prime factors.

$$\begin{array}{r|l} 2 & 300 \\ 2 & 150 \\ 3 & 75 \\ 5 & 25 \\ & 5 \end{array}$$

Therefore,
$$300 = 2 \cdot 2 \cdot 3 \cdot 5 \cdot 5$$
$$= 2^2 \cdot 3 \cdot 5^2.$$

69. Factor 150 into prime factors.

$$\begin{array}{r|l} 2 & 150 \\ 3 & 75 \\ 5 & 25 \\ & 5 \end{array}$$

Therefore,
$$150 = 2 \cdot 3 \cdot 5 \cdot 5$$
$$= 2 \cdot 3 \cdot 5^2.$$

Section 9.2 (page 509)

1. $\sqrt{8} \cdot \sqrt{2} = \sqrt{8 \cdot 2} = \sqrt{16} = 4$

5. $\sqrt{21} \cdot \sqrt{21} = \sqrt{21^2} = 21$

9. $\sqrt{27} = \sqrt{9 \cdot 3} = \sqrt{9} \cdot \sqrt{3} = 3\sqrt{3}$

13. $\sqrt{18} = \sqrt{9} \cdot \sqrt{2} = 3\sqrt{2}$

17. $\sqrt{125} = \sqrt{25} \cdot \sqrt{5} = 5\sqrt{5}$

21. $10\sqrt{27} = 10(3\sqrt{3}) = 30\sqrt{3}$

25. $\sqrt{27} \cdot \sqrt{48} = 3\sqrt{3} \cdot 4\sqrt{3}$
$= (3 \cdot 4)(\sqrt{3} \cdot \sqrt{3}) = 12(3) = 36$

Alternate solution:
$\sqrt{27} \cdot \sqrt{48} = \sqrt{27 \cdot 48} = \sqrt{1296} = 36$

29. $\sqrt{7} \cdot \sqrt{21} = \sqrt{7 \cdot 21} = \sqrt{147} = \sqrt{49 \cdot 3}$
$= \sqrt{49} \cdot \sqrt{3} = 7\sqrt{3}$

33. $\sqrt{80} \cdot \sqrt{15} = \sqrt{16} \cdot \sqrt{5} \cdot \sqrt{15}$
$= 4\sqrt{5} \cdot \sqrt{5} \cdot \sqrt{3} = 4 \cdot \sqrt{25} \cdot \sqrt{3}$
$= 4 \cdot 5 \cdot \sqrt{3} = 20\sqrt{3}$

37. $\sqrt{\dfrac{100}{9}} = \dfrac{\sqrt{100}}{\sqrt{9}} = \dfrac{10}{3}$

41. $\sqrt{\dfrac{5}{16}} = \dfrac{\sqrt{5}}{\sqrt{16}} = \dfrac{\sqrt{5}}{4}$

45. First multiply $\dfrac{1}{5}$ and $\dfrac{4}{5}$, getting $\dfrac{4}{25}$. The square root of $\dfrac{4}{25}$ is $\dfrac{2}{5}$.

49. Divide 3 into 75, getting 25. The square root of 25 is 5.

53. $\dfrac{15\sqrt{10}}{5\sqrt{2}} = \dfrac{15}{5} \cdot \dfrac{\sqrt{10}}{\sqrt{2}} = 3 \cdot \sqrt{\dfrac{10}{2}} = 3\sqrt{5}$

57. $\sqrt{y} \cdot \sqrt{y} = \sqrt{y^2} = y$

Absolute value bars are not necessary, since we assume $y > 0$.

61. $\sqrt{x^2} = x$

Absolute value bars are not necessary, since we assume $x > 0$.

65. $\sqrt{x^2y^4} = xy^2$

Absolute value bars are not necessary, since we assume $x > 0$.

69. $\sqrt{\dfrac{16}{x^2}} = \dfrac{\sqrt{16}}{\sqrt{x^2}} = \dfrac{4}{x}$

Absolute value bars are not necessary, since we assume $x > 0$.

73. $\dfrac{9}{13} = \dfrac{9 \cdot 3}{13 \cdot 3}$ Multiply numerator and denominator by 3, since $13 \cdot 3 = 39$.
$= \dfrac{27}{39}$

77. $\dfrac{2}{x} = \dfrac{2 \cdot 3}{x \cdot 3}$ Multiply numerator and denominator by 3, since $x \cdot 3 = 3x$.
$= \dfrac{6}{3x}$

Section 9.3 (page 515)

1. $2\sqrt{3} + 5\sqrt{3} = (2 + 5)\sqrt{3} = 7\sqrt{3}$

5. $\sqrt{6} + \sqrt{6} = 1 \cdot \sqrt{6} + 1 \cdot \sqrt{6} = 2\sqrt{6}$

9. $5\sqrt{7} - \sqrt{7} = 5\sqrt{7} - 1\sqrt{7} = (5 - 1)\sqrt{7} = 4\sqrt{7}$

13. $-\sqrt{12} + \sqrt{75} = -(\sqrt{4} \cdot \sqrt{3}) + \sqrt{25} \cdot \sqrt{3}$
$= -(2 \cdot \sqrt{3}) + 5 \cdot \sqrt{3} = (-2 + 5)\sqrt{3} = 3\sqrt{3}$

17. $-5\sqrt{32} + \sqrt{98} = -5 \cdot \sqrt{16} \cdot \sqrt{2} + \sqrt{49} \cdot \sqrt{2}$
$= -5 \cdot 4 \cdot \sqrt{2} + 7 \cdot \sqrt{2} = -20\sqrt{2} + 7\sqrt{2}$
$= -13\sqrt{2}$

21. $6\sqrt{5} + 3\sqrt{20} - 8\sqrt{45}$
$= 6\sqrt{5} + 3 \cdot \sqrt{4} \cdot \sqrt{5} - 8 \cdot \sqrt{9} \cdot \sqrt{5}$
$= 6\sqrt{5} + 3 \cdot 2\sqrt{5} - 8 \cdot 3 \cdot \sqrt{5}$
$= 6\sqrt{5} + 6\sqrt{5} - 24\sqrt{5} = -12\sqrt{5}$

25. $4\sqrt{50} + 3\sqrt{12} + 5\sqrt{45}$
$= 4\sqrt{25} \cdot \sqrt{2} + 3\sqrt{4} \cdot \sqrt{3} + 5\sqrt{9} \cdot \sqrt{5}$
$= 4 \cdot 5 \cdot \sqrt{2} + 3 \cdot 2 \cdot \sqrt{3} + 5 \cdot 3 \cdot \sqrt{5}$
$= 20\sqrt{2} + 6\sqrt{3} + 15\sqrt{5}$

29. $\frac{3}{5}\sqrt{75} - \frac{2}{3}\sqrt{45} = \frac{3}{5}\sqrt{25} \cdot \sqrt{3} - \frac{2}{3}\sqrt{9} \cdot \sqrt{5}$
$= \frac{3}{5} \cdot 5\sqrt{3} - \frac{2}{3} \cdot 3\sqrt{5}$
$= 3\sqrt{3} - 2\sqrt{5}$

33. $\sqrt{6} \cdot \sqrt{2} + 3\sqrt{3} = \sqrt{12} + 3\sqrt{3}$
$= \sqrt{4} \cdot \sqrt{3} + 3\sqrt{3}$
$= 2\sqrt{3} + 3\sqrt{3}$
$= (2 + 3)\sqrt{3} = 5\sqrt{3}$

37. $\sqrt{4a} + 6\sqrt{a} + \sqrt{25a} = 2\sqrt{a} + 6\sqrt{a} + 5\sqrt{a}$
$= (2 + 6 + 5)\sqrt{a}$
$= 13\sqrt{a}$

41. $3\sqrt{8x^2} - 4x\sqrt{2}$
$= 3\sqrt{(4x^2)2} - 4x\sqrt{2}$
$= 3(\sqrt{4x^2} \cdot \sqrt{2}) - 4x\sqrt{2}$
$= 3(2x\sqrt{2}) - 4x\sqrt{2}$
$= 6x\sqrt{2} - 4x\sqrt{2}$
$= 2x\sqrt{2}$

45. $2\sqrt{125x^2z} + 8x\sqrt{80z}$
$= 2\sqrt{(25x^2)5z} + 8x\sqrt{16(5z)}$
$= 2(5x\sqrt{5z}) + 8x(4\sqrt{5z})$
$= 10x\sqrt{5z} + 32x\sqrt{5z}$
$= 42x\sqrt{5z}$

49. $(2k - 3)(3k - 4) = 2k \cdot 3k + 2k(-4) + (-3)(3k) + (-3)(-4)$
$= 6k^2 - 8k - 9k + 12$
$= 6k^2 - 17k + 12$

53. $(a + 3)(a - 3) = a^2 - (3)^2$
$= a^2 - 9$

Section 9.4 (page 521)

1. Multiply numerator and denominator by $\sqrt{5}$, getting a final answer of $\frac{6\sqrt{5}}{5}$.

5. $\frac{3}{\sqrt{7}} = \frac{3}{\sqrt{7}} \cdot \frac{\sqrt{7}}{\sqrt{7}} = \frac{3\sqrt{7}}{7}$

9. Multiply numerator and denominator by $\sqrt{3}$ and then write in lowest terms, getting $\frac{\sqrt{30}}{2}$.

13. Here it is only necessary to multiply by $\sqrt{2}$ (and not $\sqrt{50}$), giving an answer of $\frac{3\sqrt{2}}{10}$.

17. $\frac{9}{\sqrt{32}} = \frac{9}{\sqrt{32}} \cdot \frac{\sqrt{2}}{\sqrt{2}}$
$= \frac{9\sqrt{2}}{\sqrt{64}} = \frac{9\sqrt{2}}{8}$

21. $\frac{\sqrt{10}}{\sqrt{5}} = \frac{\sqrt{10}}{\sqrt{5}} \cdot \frac{\sqrt{5}}{\sqrt{5}} = \frac{\sqrt{50}}{5} = \frac{\sqrt{25} \cdot \sqrt{2}}{5}$
$= \frac{5\sqrt{2}}{5} = \sqrt{2}$

25. Multiply numerator and denominator, inside the radical, by 2, giving $\sqrt{\frac{2}{4}}$, or $\frac{\sqrt{2}}{\sqrt{4}}$, or $\frac{\sqrt{2}}{2}$.

29. $\sqrt{\frac{9}{5}} = \frac{\sqrt{9}}{\sqrt{5}} = \frac{3}{\sqrt{5}} \cdot \frac{\sqrt{5}}{\sqrt{5}} = \frac{3\sqrt{5}}{5}$

33. $\sqrt{\frac{3}{4}} \cdot \sqrt{\frac{1}{5}} = \sqrt{\frac{3}{4} \cdot \frac{1}{5}} = \sqrt{\frac{3}{4 \cdot 5}} = \frac{\sqrt{3}}{\sqrt{4 \cdot 5}}$
$= \frac{\sqrt{3}}{2\sqrt{5}} \cdot \frac{\sqrt{5}}{\sqrt{5}} = \frac{\sqrt{15}}{2 \cdot 5} = \frac{\sqrt{15}}{10}$

37. $\sqrt{\frac{2}{5}} \cdot \sqrt{\frac{3}{10}} = \sqrt{\frac{2}{5} \cdot \frac{3}{10}} = \sqrt{\frac{6}{50}} = \frac{\sqrt{6}}{\sqrt{50}} \cdot \frac{\sqrt{2}}{\sqrt{2}}$
$= \frac{\sqrt{12}}{\sqrt{100}} = \frac{\sqrt{4} \cdot \sqrt{3}}{10} = \frac{2\sqrt{3}}{10} = \frac{\sqrt{3}}{5}$

41. $\sqrt{\frac{6}{p}} = \frac{\sqrt{6}}{\sqrt{p}} \cdot \frac{\sqrt{p}}{\sqrt{p}} = \frac{\sqrt{6p}}{\sqrt{p^2}} = \frac{\sqrt{6p}}{p}$

45. $\sqrt{\frac{x^2}{4y}} = \frac{\sqrt{x^2}}{\sqrt{4y}} = \frac{x}{2\sqrt{y}} \cdot \frac{\sqrt{y}}{\sqrt{y}} = \frac{x\sqrt{y}}{2y}$

49. $2x + 3x - x = 2x + 3x - 1x$
$= (2 + 3 - 1)x$
$= 4x$

53. $4y + 3z + 5z - y$
$= 4y - y + 3z + 5z$
$= 4y - 1y + 3z + 5z$
$= (4 - 1)y + (3 + 5)z$
$= 3y + 8z$

Section 9.5 (page 527)

1. $3\sqrt{5} + 8\sqrt{45} = 3\sqrt{5} + 8(\sqrt{9 \cdot 5})$
$= 3\sqrt{5} + 8(\sqrt{9} \cdot \sqrt{5}) = 3\sqrt{5} + 8(3\sqrt{5})$
$= 3\sqrt{5} + 24\sqrt{5} = 27\sqrt{5}$

5. $\sqrt{2}(\sqrt{8} - \sqrt{32}) = \sqrt{2 \cdot 8} - \sqrt{2 \cdot 32}$
$= \sqrt{16} - \sqrt{64} = 4 - 8 = -4$

9. $2\sqrt{5}(\sqrt{2} + \sqrt{5}) = 2\sqrt{10} + 2\sqrt{25}$
$= 2\sqrt{10} + 2 \cdot 5 = 2\sqrt{10} + 10$

13. Use FOIL.
$(2\sqrt{6} + 3)(3\sqrt{6} - 5)$
$= (2\sqrt{6})(3\sqrt{6}) - 2\sqrt{6}(5) + 3(3\sqrt{6}) - 3(5)$
$= (2 \cdot 3)(\sqrt{6} \cdot \sqrt{6}) - 10\sqrt{6} + 9\sqrt{6} - 15$
$= 6(6) - \sqrt{6} - 15 = 21 - \sqrt{6}$

17. Use FOIL.
$(3\sqrt{2} + 4)(3\sqrt{2} + 4)$
$= 3\sqrt{2} \cdot 3\sqrt{2} + 3\sqrt{2} \cdot 4 + 4 \cdot 3\sqrt{2} + 4 \cdot 4$
$= 9\sqrt{4} + 12\sqrt{2} + 12\sqrt{2} + 16$
$= 9 \cdot 2 + 24\sqrt{2} + 16$
$= 18 + 24\sqrt{2} + 16$
$= 34 + 24\sqrt{2}$

21. Use the pattern $(a - b)(a + b) = a^2 - b^2$.
$(3 - \sqrt{2})(3 + \sqrt{2})$
$= 3^2 - (\sqrt{2})^2 = 9 - 2 = 7$

25. Use the pattern $(a - b)(a + b) = a^2 - b^2$.
$$(\sqrt{6} - \sqrt{5})(\sqrt{6} + \sqrt{5}) = (\sqrt{6})^2 - (\sqrt{5})^2 = 6 - 5 = 1$$

29. Use FOIL.
$$(\sqrt{8} - \sqrt{2})(\sqrt{2} + \sqrt{4}) = \sqrt{8} \cdot \sqrt{2} + \sqrt{8} \cdot \sqrt{4} - \sqrt{2} \cdot \sqrt{2} - \sqrt{2} \cdot \sqrt{4}$$
$$= \sqrt{16} + \sqrt{32} - \sqrt{4} - \sqrt{8}$$
$$= 4 + 4\sqrt{2} - 2 - 2\sqrt{2} = 2 + 2\sqrt{2}$$

33. Multiply numerator and denominator by $3 - \sqrt{2}$.
$$\frac{1}{3 + \sqrt{2}} = \frac{1(3 - \sqrt{2})}{(3 + \sqrt{2})(3 - \sqrt{2})} = \frac{3 - \sqrt{2}}{3^2 - (\sqrt{2})^2} = \frac{3 - \sqrt{2}}{9 - 2} = \frac{3 - \sqrt{2}}{7}$$

37. Multiply numerator and denominator by $2 + \sqrt{11}$.
$$\frac{7}{2 - \sqrt{11}} = \frac{7(2 + \sqrt{11})}{(2 - \sqrt{11})(2 + \sqrt{11})} = \frac{7(2 + \sqrt{11})}{2^2 - (\sqrt{11})^2} = \frac{7(2 + \sqrt{11})}{4 - 11} = \frac{7(2 + \sqrt{11})}{-7} = -2 - \sqrt{11}$$

41. Multiply numerator and denominator by $1 + \sqrt{5}$.
$$\frac{\sqrt{5}}{1 - \sqrt{5}} = \frac{\sqrt{5}(1 + \sqrt{5})}{(1 - \sqrt{5})(1 + \sqrt{5})} = \frac{\sqrt{5} + 5}{1^2 - (\sqrt{5})^2} = \frac{\sqrt{5} + 5}{1 - 5} = \frac{\sqrt{5} + 5}{-4} \text{ or } \frac{-\sqrt{5} - 5}{4}$$

45. Multiply numerator and denominator by $\sqrt{3} - 5$.
$$\frac{2\sqrt{3}}{\sqrt{3} + 5} = \frac{2\sqrt{3}(\sqrt{3} - 5)}{(\sqrt{3} + 5)(\sqrt{3} - 5)} = \frac{2\sqrt{3}(\sqrt{3}) + 2\sqrt{3}(-5)}{(\sqrt{3})^2 - 5^2} = \frac{2 \cdot 3 - 2 \cdot 5\sqrt{3}}{3 - 25} = \frac{2(3 - 5\sqrt{3})}{-22} = \frac{3 - 5\sqrt{3}}{-11} \text{ or } \frac{-3 + 5\sqrt{3}}{11}$$

49. Factor the numerator as $5(\sqrt{7} - 2)$. Then divide numerator and denominator by the common factor, 5, to get $\sqrt{7} - 2$.

53. Factor the numerator as $2(6 - \sqrt{10})$. Then divide numerator and denominator by the common factor, 2, to get $\frac{6 - \sqrt{10}}{2}$.

57.
$k - 1 = (k - 1)^2$
$k - 1 = k^2 - 2k + 1$ — Multiply.
$0 = k^2 - 3k + 2$ — Add $-k$ and 1 to each side.
$0 = (k - 2)(k - 1)$ — Factor.
$k - 2 = 0$ or $k - 1 = 0$ — Set each factor equal to 0.
$k = 2$ or $k = 1$ — Solve each equation.

The solutions are 2 and 1.

Section 9.6 (page 535)

1. Square both sides of $\sqrt{x} = 2$ to get $x = 4$. Check the original equation; $\sqrt{4} = 2$ is true. The solution is 4.

5. Square both sides of $\sqrt{t - 3} = 2$ to get $t - 3 = 4$ or $t = 7$. Check this solution in the original equation: $\sqrt{7 - 3} = 2$ is true. The solution is 7.

9. Square both sides of $\sqrt{m + 5} = 0$ to get $m + 5 = 0$ or $m = -5$. Check this solution in the original equation: $\sqrt{-5 + 5} = 0$ is true. The solution is -5.

13. To get the radical alone on one side of the equation, add 2 to both sides to get $\sqrt{k} = 7$. Square both sides: $k = 49$. Check this solution in the original equation: $\sqrt{49} - 2 = 5$ or $7 - 2 = 5$, which is true. The solution is 49.

17. Square both sides to get $5t - 9 = 4t$ (be careful on the right side). Solve this equation and check to see that the answer is 9.

21. Square both sides to get $5y - 5 = 4y + 1$. Solve this equation to get $y = 6$. Check this solution in the original equation.

25. Square both sides to get $p^2 = p^2 - 3p - 12$. Subtract p^2 on both sides to get $0 = -3p - 12$. Solve this equation; you should get $p = -4$. Check this solution in the original equation: $-4 = \sqrt{16 + 12 - 12}$ or $-4 = 4$, which is false. There is no solution.

29. Square both sides of the equation. On the left you get $2x + 1$, and on the right you get $(x - 7)^2 = x^2 - 14x + 49$. The new equation is $2x + 1 = x^2 - 14x + 49$. Make the equation equal to 0 by adding $-2x$ and -1 to both sides. This gives $x^2 - 16x + 48 = 0$. Factor this equation as $(x - 12)(x - 4) = 0$. Solve each of these equations: $x - 12 = 0$ gives $x = 12$, and $x - 4 = 0$ gives $x = 4$. Now go back to the original equation: $x = 12$ gives a true statement, but $x = 4$ does not. The only solution is 12.

33. Add 1 to each side to get $\sqrt{x + 1} = x + 1$. Square both sides, getting $x + 1 = x^2 + 2x + 1$. (Remember that $(x + 1)^2 = x^2 + 2x + 1$.) Set one side equal to 0 by adding $-x$ and -1 to both sides: $0 = x^2 + x$. Factor on the right: $0 = x(x + 1)$, from which $x = 0$ or $x = -1$. Check each solution in the original equation to see that both 0 and -1 are solutions.

37. Square both sides, to get $9(x + 13) = x^2 + 18x + 81$. Multiply on the left, and make the left side of the equation 0 by adding $-9x$ and -117 to both sides: $0 = x^2 + 9x - 36$. Factor on the right to get $0 = (x + 12)(x - 3)$, from which $x = -12$ or $x = 3$. Check both solutions in the original equation. Since -12 leads to a false statement and 3 leads to a true statement, the only solution is 3.

41. Add 4 to both sides, getting $\sqrt{3x} = x - 6$. Square both sides, to get $3x = x^2 - 12x + 36$. Add $-3x$ to both sides to get $0 = x^2 - 15x + 36$. Factor on the right, which gives $0 = (x - 3)(x - 12)$, from which $x = 3$ or $x = 12$. Check in the original equation. Since 3 leads to a false statement and 12 to a true statement, the only solution is 12.

45. The square roots of 25 are 5 and -5, since $5^2 = 25$ and $(-5)^2 = 25$.

49. Since $(\sqrt{18})^2 = 18$ and $(-\sqrt{18})^2 = 18$, the square roots of 18 are $\sqrt{18}$ and $-\sqrt{18}$. Simplify $\sqrt{18}$ as $\sqrt{18} = \sqrt{9 \cdot 2} = 3\sqrt{2}$. The square roots of 18 are $3\sqrt{2}$ and $-3\sqrt{2}$.

CHAPTER 10

Section 10.1 (page 555)

1. Take the square root of each side to get 5 and -5.

5. Take the square root of each side to get $\sqrt{13}$ and $-\sqrt{13}$.

9. Simplify $\sqrt{24}$: $\sqrt{24} = \sqrt{4 \cdot 6} = \sqrt{4} \cdot \sqrt{6} = 2\sqrt{6}$. The two solutions are $2\sqrt{6}$ and $-2\sqrt{6}$.

13. By the square root property.

$$k = \sqrt{\frac{9}{16}} \quad \text{or} \quad k = -\sqrt{\frac{9}{16}}.$$

Since $\sqrt{\frac{9}{16}} = \frac{3}{4}$, the solutions are $\frac{3}{4}$ and $-\frac{3}{4}$.

17. By the square root property.

$$k = \sqrt{2.56} \quad \text{or} \quad k = -\sqrt{2.56}.$$

Since $\sqrt{2.56} = 1.6$, the solutions are 1.6 and -1.6.

21. By the square root property,

$$x - 2 = \sqrt{16} \quad \text{or} \quad x - 2 = -\sqrt{16}.$$

Since $\sqrt{16} = 4$,

$$x - 2 = 4 \quad \text{or} \quad x - 2 = -4.$$
$$x = 6 \quad \text{or} \quad x = -2.$$

The solutions are 6 and -2.

25. By the square root property,

$$x - 1 = \sqrt{32} \quad \text{or} \quad x - 1 = -\sqrt{32}.$$

Then

$$x = 1 + \sqrt{32} \quad \text{or} \quad x = 1 - \sqrt{32}.$$

Simplify $\sqrt{32}$ as $\sqrt{32} = \sqrt{16 \cdot 2} = \sqrt{16} \cdot \sqrt{2} = 4\sqrt{2}$, so the solutions are $1 + 4\sqrt{2}$ and $1 - 4\sqrt{2}$.

29. Use the square root property to get

$$3z + 5 = \sqrt{9} \quad \text{or} \quad 3z + 5 = -\sqrt{9}$$
$$3z + 5 = 3 \quad \text{or} \quad 3z + 5 = -3$$
$$3z = -2 \quad \text{or} \quad 3z = -8$$
$$z = -\frac{2}{3} \quad \text{or} \quad z = -\frac{8}{3}.$$

The solutions are $-\frac{2}{3}$ and $-\frac{8}{3}$.

33. Use the square root property to get $3p - 1 = \sqrt{18}$ or $3p - 1 = -\sqrt{18}$. We can simplify $\sqrt{18}$ as $3\sqrt{2}$. Add 1 to both sides of each equation to give $3p = 1 + 3\sqrt{2}$ and $3p = 1 - 3\sqrt{2}$. Divide both sides of each equation by 3 to end up with the solutions, $\frac{1 + 3\sqrt{2}}{3}$ and $\frac{1 - 3\sqrt{2}}{3}$.

37. Use the square root property to get $3m + 4 = \sqrt{8}$ or $3m + 4 = -\sqrt{8}$. Add -4 to both sides of each equation to get $3m = -4 + \sqrt{8}$ or $3m = -4 - \sqrt{8}$. Simplify $\sqrt{8}$ as $\sqrt{8} = \sqrt{4} \cdot \sqrt{2} = 2\sqrt{2}$, giving $3m = -4 + 2\sqrt{2}$ and $3m = -4 - 2\sqrt{2}$. Finally, divide both sides of each equation by 3 to get the solutions, $\frac{-4 + 2\sqrt{2}}{3}$ and $\frac{-4 - 2\sqrt{2}}{3}$.

41. $\frac{3}{2} + \sqrt{\frac{27}{4}} = \frac{3}{2} + \frac{\sqrt{27}}{\sqrt{4}} = \frac{3}{2} + \frac{3\sqrt{3}}{2} = \frac{3 + 3\sqrt{3}}{2}$

45. Since $x^2 = (x)^2$ and $16 = 4^2$, try $(x + 4)^2$. Since $2(x)(4) = 8x$, the middle term, $x^2 + 8x + 16 = (x + 4)^2$.

Section 10.2 (page 563)

1. Take half of 2, which is 1. Square this to get the answer: $1^2 = 1$.

5. Half of 3 is $\frac{3}{2}$ and $\left(\frac{3}{2}\right)^2 = \frac{9}{4}$, so $\frac{9}{4}$ should be added.

9. Take half the coefficient of m and square it.

$$\frac{1}{2}(4) = 2$$
$$2^2 = 4$$

Add 4 to each side.

$$m^2 + 4m = -1$$
$$m^2 + 4m + 4 = -1 + 4$$
$$(m + 2)^2 = 3 \quad \text{Factor.}$$
$$m + 2 = \sqrt{3} \quad \text{or} \quad m + 2 = -\sqrt{3}$$
$$m = -2 + \sqrt{3} \quad \text{or} \quad m = -2 - \sqrt{3}$$

The solutions are $-2 + \sqrt{3}$ and $-2 - \sqrt{3}$.

13. Rewrite the equation with all variable terms on one side.

$$b^2 - 2b = 2$$

Square half the coefficient of b.

$$\frac{1}{2}(-2) = -1$$
$$(-1)^2 = 1$$

Add 1 to each side of the equation.

$$b^2 - 2b + 1 = 2 + 1$$
$$(b - 1)^2 = 3 \quad \text{Factor.}$$
$$b - 1 = \sqrt{3} \quad \text{or} \quad b - 1 = -\sqrt{3}$$
$$b = 1 + \sqrt{3} \quad \text{or} \quad b = 1 - \sqrt{3}$$

The solutions are $1 + \sqrt{3}$ and $1 - \sqrt{3}$.

17. Divide the equation by 4, the coefficient of y^2.

$$y^2 + y - \frac{3}{4} = 0$$

Rewrite the equation with only variable terms on one side.

$$y^2 + y = \frac{3}{4} \qquad \textbf{(1)}$$

Square half the coefficient of y.

$$\frac{1}{2}(1) = \frac{1}{2}$$
$$\left(\frac{1}{2}\right)^2 = \frac{1}{4}$$

Add $\frac{1}{4}$ to both sides of equation (1).

$$y^2 + y + \frac{1}{4} = \frac{3}{4} + \frac{1}{4}$$

$$\left(y + \frac{1}{2}\right)^2 = 1$$

$$y + \frac{1}{2} = \sqrt{1} \quad \text{or} \quad y + \frac{1}{2} = -\sqrt{1}$$

$$y + \frac{1}{2} = 1 \quad \text{or} \quad y + \frac{1}{2} = -1$$

$$y = \frac{1}{2} \quad \text{or} \quad y = -\frac{3}{2}$$

The solutions are $\frac{1}{2}$ and $-\frac{3}{2}$.

21. Get the variable terms alone on one side.

$m^2 - 10m = -8$

Square half the coefficient of m.

$\frac{1}{2}(-10) = -5$

$(-5)^2 = 25$

Add 25 to each side.

$m^2 - 10m + 25 = -8 + 25$

$(m - 5)^2 = 17$ Factor.

$m - 5 = \sqrt{17}$ or $m - 5 = -\sqrt{17}$

$m = 5 + \sqrt{17}$ or $m = 5 - \sqrt{17}$

The solutions are $5 + \sqrt{17}$ and $5 - \sqrt{17}$.

25. Divide both sides of the equation by 5 to get

$$r^2 + 2r = \frac{1}{5}.$$

Half the coefficient of r is $\frac{1}{2}(2) = 1$. Since $1^2 = 1$, we add 1 to each side of the equation.

$$r^2 + 2r + 1 = \frac{1}{5} + 1$$

$(r + 1)^2 = \frac{6}{5}$ Factor; combine terms.

$r + 1 = \pm\sqrt{\frac{6}{5}}$ Square root property

$\sqrt{\frac{6}{5}}$ simplifies as $\frac{\sqrt{6}}{\sqrt{5}} \cdot \frac{\sqrt{5}}{\sqrt{5}} = \frac{\sqrt{30}}{5}$.

$$r + 1 = \frac{\sqrt{30}}{5} \quad \text{or} \quad r + 1 = -\frac{\sqrt{30}}{5}$$

$$r = -1 + \frac{\sqrt{30}}{5} \qquad r = -1 - \frac{\sqrt{30}}{5}$$

$$r = \frac{-5}{5} + \frac{\sqrt{30}}{5} \qquad r = \frac{-5}{5} - \frac{\sqrt{30}}{5}$$

$$r = \frac{-5 + \sqrt{30}}{5} \quad \text{or} \quad r = \frac{-5 - \sqrt{30}}{5}$$

Only the positive value $\frac{-5 + \sqrt{30}}{5}$ makes sense.

The value of r cannot be $\frac{-5 - \sqrt{30}}{5}$, which is negative. "Increased" indicates $r > 0$.

29. $\frac{4 + 2\sqrt{7}}{8} = \frac{2(2 + \sqrt{7})}{2 \cdot 4}$ Factor numerator and denominator.

$= \frac{2 + \sqrt{7}}{4}$ Use fundamental property.

33. $\frac{6 + \sqrt{45}}{12} = \frac{6 + \sqrt{9 \cdot 5}}{12}$

$= \frac{6 + 3\sqrt{5}}{12}$ Simplify the radical.

$= \frac{3(2 + \sqrt{5})}{3 \cdot 4}$ Factor.

$= \frac{2 + \sqrt{5}}{4}$ Use fundamental property.

Section 10.3 (page 569)

1. The coefficient of the x^2 term is 3, so that $a = 3$. The coefficient of the x term is 4, so that $b = 4$. The constant is -8, so that $c = -8$. (*Note:* One side of the equation was equal to 0 before we started.)

5. Since 0 is on one side, the value of a is 3, while $b = -8$. No constant term is given, so $c = 0$.

9. The equation $p^2 + 2p - 2 = 0$ matches the general form, so $a = 1$, $b = 2$, and $c = -2$. Substitute these values in the quadratic formula.

$$p = \frac{-b \pm \sqrt{b^2 - 4ac}}{2a}$$

$$p = \frac{-2 \pm \sqrt{2^2 - 4(1)(-2)}}{2(1)}$$

$$= \frac{-2 \pm \sqrt{4 + 8}}{2}$$

$$= \frac{-2 \pm \sqrt{12}}{2} = \frac{-2 \pm 2\sqrt{3}}{2}$$

$$= \frac{2(-1 \pm \sqrt{3})}{2} = -1 \pm \sqrt{3}$$

The solutions are $-1 + \sqrt{3}$ and $-1 - \sqrt{3}$.

13. Rewrite the equation to match the general form.

$z^2 + 12z - 13 = 0$

Substitute $a = 1$, $b = 12$, and $c = -13$ in the quadratic formula.

$$z = \frac{-b \pm \sqrt{b^2 - 4ac}}{2a}$$

$$z = \frac{-12 + \sqrt{12^2 - 4(1)(-13)}}{2(1)}$$

$$= \frac{-12 \pm \sqrt{144 + 52}}{2}$$

$$= \frac{-12 \pm \sqrt{196}}{2}$$

$$= \frac{-12 \pm 14}{2}$$

The solutions are

$$\frac{-12 + 14}{2} = \frac{2}{2} = 1$$

and $\frac{-12 - 14}{2} = \frac{-26}{2} = -13$.

SOLUTIONS

17. The equation $5x^2 + 4x - 1 = 0$ matches the general form, so $a = 5$, $b = 4$, and $c = -1$. Substitute these values in the quadratic formula.

$$x = \frac{-b \pm \sqrt{b^2 - 4ac}}{2a}$$

$$x = \frac{-4 \pm \sqrt{4^2 - 4(5)(-1)}}{2(5)}$$

$$= \frac{-4 \pm \sqrt{16 + 20}}{10}$$

$$= \frac{-4 \pm \sqrt{36}}{10}$$

$$= \frac{-4 \pm 6}{10}$$

The solutions are

$$\frac{-4 + 6}{10} = \frac{2}{10} = \frac{1}{5} \text{ and } \frac{-4 - 6}{10} = \frac{-10}{10} = -1.$$

21. $9r^2 + 6r + 1 = 0$ matches the general quadratic form, so $a = 9$, $b = 6$, and $c = 1$. Substitute these values in the quadratic formula.

$$r = \frac{-b \pm \sqrt{b^2 - 4ac}}{2a}$$

$$r = \frac{-6 \pm \sqrt{6^2 - 4(9)(1)}}{2(9)}$$

$$= \frac{-6 \pm \sqrt{36 - 36}}{18}$$

$$= \frac{-6 \pm 0}{18}$$

$$= -\frac{1}{3}$$

The solution is $-\frac{1}{3}$.

25. Write $3r^2 = 16r$ as $3r^2 - 16r + 0 = 0$. Then $a = 3$, $b = -16$, and $c = 0$. Substitute these values in the quadratic formula.

$$r = \frac{-b \pm \sqrt{b^2 - 4ac}}{2a}$$

$$r = \frac{-(-16) \pm \sqrt{(-16)^2 - 4(3)(0)}}{2(3)}$$

$$= \frac{16 \pm \sqrt{256 - 0}}{6}$$

$$= \frac{16 \pm \sqrt{256}}{6}$$

$$= \frac{16 \pm 16}{6}$$

The solutions are

$$\frac{16 + 16}{6} = \frac{32}{6} = \frac{16}{3} \text{ and } \frac{16 - 16}{6} = 0.$$

29. Write $4y^2 - 25 = 0$ as $4y^2 + 0 \cdot y - 25 = 0$. Then $a = 4$, $b = 0$, and $c = -25$. Substitute these values in the quadratic formula.

$$y = \frac{-b \pm \sqrt{b^2 - 4ac}}{2a}$$

$$= \frac{-0 \pm \sqrt{0^2 - 4(4)(-25)}}{2(4)}$$

$$= \frac{\pm \sqrt{400}}{8} = \frac{\pm 20}{8} = \pm\frac{5}{2}$$

The solutions are $\frac{5}{2}$ and $-\frac{5}{2}$.

33. Write $x^2 + x + 1 = 0$ as $1 \cdot x^2 + 1 \cdot x + 1 = 0$. Then $a = 1$, $b = 1$, and $c = 1$. Substitute these values in the quadratic formula.

$$x = \frac{-b \pm \sqrt{b^2 - 4ac}}{2a}$$

$$x = \frac{-1 \pm \sqrt{1^2 - 4(1)(1)}}{2(1)}$$

$$= \frac{-1 \pm \sqrt{1 - 4}}{2}$$

$$= \frac{-1 \pm \sqrt{-3}}{2}$$

Since $\sqrt{-3}$ is not a real number, there is no real number solution.

37. To clear fractions, multiply by 6.

$$\frac{3}{2}r^2 - r = \frac{4}{3}$$

$$6\left(\frac{3}{2}r^2 - r\right) = 6\left(\frac{4}{3}\right)$$

$$9r^2 - 6r = 8$$

Write the equation in general quadratic form.

$$9r^2 - 6r - 8 = 0$$

Use $a = 9$, $b = -6$, and $c = -8$ in the quadratic formula.

$$r = \frac{-b \pm \sqrt{b^2 - 4ac}}{2a}$$

$$r = \frac{-(-6) \pm \sqrt{(-6)^2 - 4(9)(-8)}}{2(9)}$$

$$= \frac{6 \pm \sqrt{36 + 288}}{18}$$

$$= \frac{6 \pm \sqrt{324}}{18}$$

$$= \frac{6 \pm 18}{18}$$

The solutions are

$$\frac{6 + 18}{18} = \frac{24}{18} = \frac{4}{3}$$

$$\text{and } \frac{6 - 18}{18} = \frac{-12}{18} = -\frac{2}{3}.$$

41. Multiply by 4 to clear fractions.

$$\frac{m^2}{4} + \frac{3m}{2} + 1 = 0$$

$$4\left(\frac{m^2}{4} + \frac{3m}{2} + 1\right) = 4 \cdot 0$$

$$m^2 + 6m + 4 = 0$$

Use $a = 1$, $b = 6$, and $c = 4$ in the quadratic formula.

$$m = \frac{-b \pm \sqrt{b^2 - 4ac}}{2a}$$

$$m = \frac{-6 \pm \sqrt{6^2 - 4(1)(4)}}{2(1)}$$

$$= \frac{-6 \pm \sqrt{36 - 16}}{2}$$

$$= \frac{-6 \pm \sqrt{20}}{2} = \frac{-6 \pm 2\sqrt{5}}{2}$$

$$= \frac{2(-3 \pm \sqrt{5})}{2} = -3 \pm \sqrt{5}$$

The solutions are

$$-3 + \sqrt{5} \text{ and } -3 - \sqrt{5}.$$

45. When the projectile has traveled 14 feet, the value of d is 14. Substitute 14 for d in the equation $d = 2t^2 - 5t + 2$.

$$14 = 2t^2 - 5t + 2$$
$$0 = 2t^2 - 5t - 12 \quad \text{Subtract 14.}$$

This is now in the general quadratic form, with $a = 2$, $b = -5$, and $c = -12$. Use the quadratic formula.

$$t = \frac{-b \pm \sqrt{b^2 - 4ac}}{2a}$$
$$t = \frac{-(-5) \pm \sqrt{(-5)^2 - 4(2)(-12)}}{2(2)}$$
$$= \frac{5 \pm \sqrt{25 + 96}}{4}$$
$$= \frac{5 \pm \sqrt{121}}{4}$$
$$= \frac{5 \pm 11}{4}$$
$$t = \frac{5 + 11}{4} = \frac{16}{4} = 4 \quad \text{or}$$
$$t = \frac{5 - 11}{4} = \frac{-6}{4} = -\frac{3}{2}$$

The only reasonable answer is 4, since t must be positive. The answer is 4 seconds.

49. When $x = 0$,

$$3(0) + 5y = 15$$
$$y = 3.$$

When $y = 0$,

$$3x + 5(0) = 15$$
$$x = 5.$$

When $x = -5$,

$$3(-5) + 5y = 15$$
$$y = 6.$$

When $y = -3$,

$$3x + 5(-3) = 15$$
$$x = 10.$$

The points are (0, 3), (5, 0), (−5, 6), and (10, −3). See the graph in the answer section.

Summary Exercises on Quadratic Equations (page 573)

1. This polynomial will not factor, so use the quadratic formula, with $a = 1$, $b = 3$, $c = 1$.

$$y = \frac{-b \pm \sqrt{b^2 - 4ac}}{2a}$$
$$= \frac{-3 \pm \sqrt{3^2 - 4(1)(1)}}{2 \cdot 1}$$
$$= \frac{-3 \pm \sqrt{9 - 4}}{2}$$
$$= \frac{-3 \pm \sqrt{5}}{2}$$

The solutions are $\frac{-3 + \sqrt{5}}{2}$ and $\frac{-3 - \sqrt{5}}{2}$.

Check each solution.

5. Factor to get

$$(x + 2)(x + 1) = 0$$
$$x + 2 = 0 \quad \text{or} \quad x + 1 = 0$$
$$x = -2 \quad \text{or} \quad x = -1.$$

Check each solution.

9. Use the square root property to get

$$2p - 1 = \sqrt{10} \quad \text{or} \quad 2p - 1 = -\sqrt{10}$$
$$2p = 1 + \sqrt{10} \quad \text{or} \quad 2p = 1 - \sqrt{10}$$
$$p = \frac{1 + \sqrt{10}}{2} \quad \text{or} \quad p = \frac{1 - \sqrt{10}}{2}.$$

13. Use the square root property.

$$(7m - 1)^2 = 32$$
$$7m - 1 = \sqrt{32} \quad \text{or} \quad 7m - 1 = -\sqrt{32}$$
$$7m - 1 = 4\sqrt{2} \quad \text{or} \quad 7m - 1 = -4\sqrt{2}$$
$$7m = 1 + 4\sqrt{2} \quad \text{or} \quad 7m = 1 - 4\sqrt{2}$$
$$m = \frac{1 + 4\sqrt{2}}{7} \quad \text{or} \quad m = \frac{1 - 4\sqrt{2}}{7}$$

17. Rewrite the equation in the form of the general quadratic equation.

$$2m^2 = 3m + 2$$
$$2m^2 - 3m - 2 = 0$$

The left side can be factored, so use the method of factoring.

$$(2m + 1)(m - 2) = 0$$
$$2m + 1 = 0 \quad \text{or} \quad m - 2 = 0$$
$$2m = -1 \quad \text{or} \quad m = 2$$
$$m = -\frac{1}{2}$$

The solutions are $-\frac{1}{2}$ and 2.

21. Let $a = 3$, $b = 5$, and $c = 1$. The quadratic formula gives

$$p = \frac{-5 \pm \sqrt{5^2 - 4(3)(1)}}{2(3)}$$
$$p = \frac{-5 \pm \sqrt{25 - 12}}{6} = \frac{-5 \pm \sqrt{13}}{6}.$$

The solutions are

$$\frac{-5 + \sqrt{13}}{6} \quad \text{and} \quad \frac{-5 - \sqrt{13}}{6}.$$

25. The equation $z^2 + 12z + 25 = 0$ is in general quadratic form, so $a = 1$, $b = 12$, and $c = 25$. Substitute these values in the quadratic formula.

$$z = \frac{-b \pm \sqrt{b^2 - 4ac}}{2a}$$
$$z = \frac{-12 \pm \sqrt{12^2 - 4(1)25}}{2(1)}$$
$$= \frac{-12 \pm \sqrt{144 - 100}}{2}$$
$$= \frac{-12 \pm \sqrt{44}}{2} = \frac{-12 \pm 2\sqrt{11}}{2}$$
$$= \frac{2(-6 \pm \sqrt{11})}{2} = -6 \pm \sqrt{11}$$

The solutions are

$$-6 + \sqrt{11} \quad \text{and} \quad -6 - \sqrt{11}.$$

SOLUTIONS

29. Rewrite $4x^2 + 5x = 1$ as $4x^2 + 5x - 1 = 0$. The left side does not factor, so use the quadratic formula with $a = 4$, $b = 5$, and $c = -1$.

$$x = \frac{-b \pm \sqrt{b^2 - 4ac}}{2a}$$

$$x = \frac{-5 \pm \sqrt{5^2 - 4(4)(-1)}}{2(4)}$$

$$= \frac{-5 \pm \sqrt{25 + 16}}{8}$$

$$= \frac{-5 \pm \sqrt{41}}{8}$$

The solutions are $\frac{-5 + \sqrt{41}}{8}$ and $\frac{-5 - \sqrt{41}}{8}$.

33. The left side of $p^2 + 5p + 5 = 0$ does not factor, so use the quadratic formula with $a = 1$, $b = 5$, and $c = 5$.

$$p = \frac{-b \pm \sqrt{b^2 - 4ac}}{2a}$$

$$p = \frac{-5 \pm \sqrt{5^2 - 4(1)(5)}}{2(1)}$$

$$= \frac{-5 \pm \sqrt{25 - 20}}{2} = \frac{-5 \pm \sqrt{5}}{2}$$

The solutions are $\frac{-5 + \sqrt{5}}{2}$ and $\frac{-5 - \sqrt{5}}{2}$.

37. Multiply by 2 to eliminate denominators.

$$2\left(\frac{1}{2}r^2\right) = 2\left(r + \frac{15}{2}\right)$$

$$r^2 = 2r + 15$$

Rewrite the result in the form of the general quadratic equation.

$$r^2 - 2r - 15 = 0$$

The left side can be factored, so use the method of factoring.

$$(r - 5)(r + 3) = 0$$

$$r - 5 = 0 \quad \text{or} \quad r + 3 = 0$$

$$r = 5 \quad \text{or} \quad r = -3$$

The solutions are 5 and -3.

41.

$$k^2 + \frac{4}{15}k = \frac{4}{15}$$

$$15k^2 + 4k = 4 \quad \text{Multiply both sides by 15.}$$

$$15k^2 + 4k - 4 = 0 \quad \text{Subtract 4.}$$

$$(3k + 2)(5k - 2) = 0 \quad \text{Factor.}$$

$$3k + 2 = 0 \quad \text{or} \quad 5k - 2 = 0$$

$$3k = -2 \qquad 5k = 2$$

$$k = -\frac{2}{3} \quad \text{or} \quad k = \frac{2}{5}$$

Section 10.4 (page 579)

All the graphs in this section may be found in the answer section.

1. The graph of $y = 2x^2$ is a parabola with vertex at (0, 0). Because of the 2, the graph is narrower than that of $y = x^2$ and opens upward. Some additional points on the graph are (1, 2), (−1, 2), (2, 8), and (−2, 8).

5. The graph of $y = -x^2 - 2x - 1$ is a parabola with vertex at (−1, 0). Because the coefficient of x^2 is −1 (negative), the graph opens downward and has the same shape as that of $y = x^2$. Some additional points on the graph are (0, −1), (−2, −1), (1, −4), and (−3, −4).

9. The graph of $y = 2 - x^2$ is a parabola with vertex at (0, 2). Since the coefficient of x^2 is −1, the graph opens downward and has the same shape as that of $y = x^2$. Additional points on the graph are (−2, −2), (−1, 1), (1, 1), and (2, −2).

13. The graph of $y = x^2 + 2x + 3$ is a parabola with vertex at (−1, 2). Because the coefficient of x^2 is 1 (positive), the graph opens upward and has the same shape as that of $y = x^2$. Some additional points on the graph are (0, 3), (−2, 3), (1, 6), and (−3, 6).

17. The graph of $y = -x^2 - 4x - 3$ is a parabola with vertex at (−2, 1). Because the coefficient of x^2 is −1 (negative), the graph opens downward and has the same shape as that of $y = x^2$. Some additional points on the graph are (−1, 0), (−3, 0), (0, −3) and (−4, −3).

Appendix Exercises (page 597)

1. The natural numbers less than 8 are the counting numbers 1, 2, 3, 4, 5, 6, and 7. Remember to use set braces in the answer; the answer is {1, 2, 3, 4, 5, 6, 7}.

5. There have been no women presidents of the United States, so this set is empty: ∅.

9. The set of positive even numbers is the infinite set {2, 4, 6, 8, . . .}.

13. Since 5 is an element of {1, 2, 5, 8}, the statement is true.

17. Since 7 is not an element of {2, 4, 6, 8}, the statement is true.

21. True, since every element of set A is in set U.

25. True, because every element in set C is also in set A.

29. True, since the numbers 1 and 2 belong to set D, but not to set E.

33. False; set C has 4 elements and therefore has $2^4 = 16$ subsets, not 12 subsets.

37. True. The intersection of the given sets is the set {0}.

41. True. Every element of the first set and every element of the second set are in the union.

45. A', the complement of A, contains the elements of U that are not in A: $A' =$ {g, h}.

49. The intersection of sets A and B contains only those elements that are in both A and B: a, c, and e. Therefore, $A \cap B =$ {a, c, e}.

53. The intersection of sets B and C contains the elements that are in both B and C, so a is the only element in $B \cap C$. Therefore, $B \cap C =$ {a}.

57. The union of C and B contains every element in C and every element in B, so the elements are a, c, e, and f. Therefore, $C \cup B =$ {a, c, e, f}.

61. Disjoint sets have no elements in common, so sets B and D are disjoint, and sets C and D are disjoint.

GLOSSARY

This glossary provides an alphabetized listing of all the entries found in the "Key Terms" section of each Chapter Summary. For reference or further study, the corresponding section number is given at the end of each entry.

A

absolute value The absolute value of a number is the distance between 0 and the number on the number line. **[2.1]**

additive inverse The additive inverse of a number a is the number that is the same distance from 0 on the number line as a, but on the opposite side of 0. This number is also known as the **opposite** of a. **[2.1]**

algebraic expression An algebraic expression is a collection of numbers, variables, symbols for operations, and symbols for grouping. **[1.5]**

axis The axis of a parabola is a vertical line through the vertex. **[10.4]**

B

base The base is the number that is a repeated factor when written with an exponent. **[1.1]**

binomial A binomial is a polynomial with two terms. **[4.4]**

C

complex fraction A rational expression with one or more fractions in the numerator, denominator, or both is a complex fraction. **[6.5]**

conjugates The conjugate of $a + b$ is $a - b$. **[9.5]**

consistent system A system of equations with a solution is a consistent system. **[8.1]**

coordinate system An x-axis and y-axis at right angles is a coordinate system. **[7.2]**

cross products The method of cross products provides a way of determining whether a proportion is true. **[3.6]**

D

decimal fraction (decimal) A decimal fraction is a number written with a decimal point. **[1.2]**

decimal places The digits after the decimal point are called decimal places. **[1.2]**

degree of a polynomial The degree of a polynomial in one variable is the highest exponent found in any term of the polynomial. **[4.4]**

degree of a term The degree of a term with one variable is the exponent on the variable. **[4.4]**

denominator The denominator of a fraction is the number below the fraction bar. **[1.1]**

dependent equations Equations of a system that have the same graph (because they are different forms of the same equation) are called dependent equations. **[8.1]**

difference The answer to a subtraction problem is called the difference. **[2.3]**

direct variation y varies directly as x if there is a constant k such that $y = kx$. **[6.7]**

dividend A dividend is a number that is divided by another number. **[1.2]**

divisor A divisor is the number by which another number is divided. **[1.2]**

domain The domain of an equation is the set of numbers from which the solution is chosen. **[1.5]**

E

equation An equation is a statement that says two expressions are equal. **[1.5]**

exponent An exponent is a number that indicates how many times a factor is repeated. **[1.1]**

exponential expression A number written with an exponent is an exponential expression. **[1.4]**

F

factor An expression A is a factor of an expression B if B can be divided by A with zero remainder. **[5.1]**

factored A number is factored if it is written as a product. **[1.1]**

factored form An expression is in factored form when it is written as a product. **[5.1]**

G

greatest common factor The greatest common factor is the largest quantity that is a factor of each of a group of quantities. **[5.1]**

I

identity element for addition When the identity element for addition, which is 0, is added to a number, the number is unchanged. **[2.6]**

identity element for multiplication When a number is multiplied by the identity element for multiplication, which is 1, the number is unchanged. **[2.6]**

inconsistent system An inconsistent system of equations is a system with no solutions. **[8.1]**

independent equations Equations of a system that have different graphs are called independent equations. **[8.1]**

inequality An inequality is a statement with algebraic expressions related by $<$, $\leq$, $>$, or $\geq$. **[3.7]**

integers The set of integers is $\{\ldots, -3, -2, -1, 0, 1, 2, 3, \ldots\}$. **[2.1]**

inverse variation y varies inversely as x if there is a constant k such that $y = \frac{k}{x}$. **[6.7]**

irrational numbers Irrational numbers are non-rational numbers represented by points on the number line. **[2.1]**

L

least common denominator The least common denominator is the smallest number that all denominators in a problem divide into without remainder. **[1.1]**

like radicals Like radicals are multiples of the same radical. **[9.3]**

like terms Terms with exactly the same variables (including the same exponents) are called like terms. **[2.7]**

linear equation A linear equation is an equation that can be written in the form $ax + by = c$, for real numbers a, b, and c, with $a \neq 0$. **[3.1]**

lowest terms A fraction or rational expression is written in lowest terms when the numerator and denominator have no common factor (except 1). **[1.1]**

M

monomial A monomial is a polynomial with one term. **[4.4]**

multiplicative inverse Pairs of numbers whose product is 1 are called multiplicative inverses of each other. Multiplicative inverses are also known as **reciprocals.** **[2.5]**

N

negative number A negative number is located to the *left* of 0 on the number line. **[2.1]**

numerator The numerator of a fraction is the number above the fraction bar. **[1.1]**

numerical coefficient The numerical factor in a term is its coefficient. **[2.7]**

O

ordered pair A pair of numbers written between parentheses in which the order is important is called an ordered pair. **[7.1]**

P

parabola The graph of a quadratic equation is called a parabola. **[10.4]**

parallel lines Two lines in a plane that never intersect are parallel. **[7.5]**

percent A percent indicates the number of one-hundredths in a quantity. **[1.2]**

percentage A percentage is part of a total amount. **[1.2]**

perfect square A number with a rational square root is called a perfect square. **[9.1]**

perfect square trinomial A perfect square trinomial is a trinomial that can be factored as the square of a binomial. **[5.4]**

perimeter The perimeter of a geometric figure is the distance around the figure. **[5.6]**

perpendicular lines Perpendicular lines intersect at a 90° angle. **[7.5]**

plot To plot an ordered pair is to find the corresponding point on a coordinate system. **[7.2]**

polynomial A polynomial is the sum of a finite number of terms. **[4.4]**

positive number A positive number is located to the *right* of 0 on the number line. **[2.1]**

prime number A natural number greater than 1 is prime if it has only itself and 1 as factors. **[5.1]**

prime polynomial A prime polynomial is a polynomial that cannot be factored. **[5.2]**

product The answer to a multiplication problem is called the product. **[2.4]**

proportion A proportion is a statement that two ratios are equal. **[3.6]**

Pythagorean formula If c is the length of the hypotenuse of a right triangle and a and b are the lengths of the legs, then $a^2 + b^2 = c^2$. **[9.1]**

Q

quadrants A coordinate system divides the plane into four regions called quadrants. **[7.2]**

quadratic equation A quadratic equation is an equation that can be written in the form $ax^2 + bx + c = 0$, with $a \neq 0$. **[5.5]**

quotient The answer to a division problem is called the quotient. **[2.5]**

R

radical A radical sign with a radicand is called a radical. **[9.1]**

radical expression An algebraic expression containing a radical is called a radical expression. **[9.1]**

radicand The number or expression inside a radical sign is called the radicand. **[9.1]**

ratio A ratio is a quotient of two quantities with the same units. **[3.6]**

rational expression The quotient of two polynomials with denominator not zero is called a rational expression. **[6.1]**

rational number A rational number can be written as the quotient of two integers, with denominator not 0. **[2.1]**

rationalizing the denominator The process of changing the denominator of a fraction from a radical (irrational number) to a rational number is called rationalizing the denominator. **[9.4]**

real numbers Real numbers include all numbers that can be represented by points on the number line. **[2.1]**

reciprocals Two numbers whose product is 1 are reciprocals. **[1.1]**

S

satisfy An ordered pair or number that makes an equation true is said to satisfy the equation. **[7.1]**

scientific notation A number written as $a \times 10^n$, where $1 \le |a| < 10$ and n is an integer, is in scientific notation. **[4.3]**

signed numbers Signed numbers are either positive or negative. **[2.1]**

simplified form A radical is in simplified form if the radicand contains no factor (except 1) that is a perfect square, the radicand contains no fractions, and no denominator contains a radical. **[9.4]**

slope The slope of a line is the ratio of the change in y compared to the change in x when moving along the line. **[7.4]**

solution A solution of an equation is any replacement for the variable that makes the equation true. **[1.5]**

solution of a system of linear equations The solution of a system of linear equations includes all the ordered pairs that make all the equations of the system true at the same time. **[8.1]**

solution of a system of linear inequalities The solution of a system of linear inequalities consists of all ordered pairs that make all inequalities of the system true at the same time. **[8.6]**

square root The square roots of a^2 are a and $-a$ (a is nonnegative). **[9.1]**

standard form A quadratic equation written as $ax^2 + bx + c = 0$ ($a \neq 0$) is in standard form. **[10.3]**

sum The answer to an addition problem is called the sum. **[2.2]**

system of linear equations A system of linear equations consists of two or more linear equations with the same variables. **[8.1]**

system of linear inequalities A system of linear inequalities contains two or more linear inequalities (and no other kinds of inequalities). **[8.6]**

T

table of values A table showing ordered pairs of numbers is called a table of values. **[7.1]**

term A term is a single number, or a product of a number and one or more variables raised to powers. **[2.7]**

trinomial A trinomial is a polynomial with three terms. **[4.4]**

V

variable A variable is a symbol, usually a letter, used to represent an unknown number. **[1.5]**

vertex The vertex of a parabola is the highest or lowest point on the graph. **[10.4]**

W

whole numbers The set of whole numbers is $\{0, 1, 2, 3, 4, 5, \ldots\}$. **[2.1]**

X

***x*-intercept** If a graph crosses the x-axis at k, then the x-intercept is $(k, 0)$. **[7.3]**

Y

***y*-intercept** If a graph crosses the y-axis at k, then the y-intercept is $(0, k)$. **[7.3]**

INDEX

R

S

T

U

V

W

X

Y

Z